LIBER AMICORUM JUDGE SHIGERU ODA

LIBER AMICORUM

JUDGE SHIGERU ODA

edited by

NISUKE ANDO
EDWARD MCWHINNEY
RÜDIGER WOLFRUM

BETSY BAKER RÖBEN
Assistant General Editor

Volume 2

KLUWER LAW INTERNATIONAL
THE HAGUE / LONDON / NEW YORK

A C.I.P. Catalogue record for this book is available from the Library of Congress.

ISBN 90-411-1797-0 (vol. 1)
ISBN 90-411-1798-9 (vol. 2)
ISBN 90-411-1790-3 (set)

Published by Kluwer Law International,
P.O. Box 85889, 2508 CN The Hague, The Netherlands.

Sold and distributed in North, Central and South America
by Kluwer Law International,
101 Philip Drive, Norwell, MA 02061, U.S.A.
kluwerlaw@wkap.com

In all other countries, sold and distributed
by Kluwer Law International,
P.O. Box 85889, 2508 CN The Hague, The Netherlands.

Printed on acid-free paper

All Rights Reserved
© 2002 Kluwer Law International
Kluwer Law International incorporates the imprint of
Martinus Nijhoff Publishers.

No part of the material protected by this copyright notice may be reproduced or
utilized in any form or by any means, electronic or mechanical,
including photocopying, recording or by any information storage and
retrieval system, without written permission from the copyright owner.

Printed in the Netherlands.

Table of Contents

VOLUME II

VIII. International Litigation: Strategies, Rules and Procedures

David Anderson
    Trust Funds in International Litigation   793

Sir Franklin Berman
    The Uses and Abuses of Advisory Opinions   809

Ian Brownlie
    Why Do States Take Disputes to the International Court?   829

Jean-Pierre Cot
    Appearing "for" or "on behalf of" a State: The Role of
    Private Counsel Before International Tribunals   835

Yoram Dinstein
    Deprivation of Property of Foreigners under International Law   849

Yuji Iwasawa
    Third Parties Before International Tribunals: The ICJ and the
    WTO   871

Sir Robert Jennings
    The Differences Between Conducting a Case in the ICJ and
    in an ad hoc Arbitration Tribunal - An Inside View   893

Akira Kotera
    On the Legal Character of Retaliation in the World Trade
    Organization System   911

Roy S. Lee
    Towards a More Proactive System of Dispute Settlement   923

François Rigaux
Les demandes reconventionelles devant la Cour internationale
de Justice 935

Henry G. Schermers
Exhaustion of Domestic Remedies 947

Christian Tomuschat
The 1928 General Act for the Pacific Settlement of
International Disputes Revisited 977

Santiago Torres Barnárdez
Bilateral, Plural and Multipartite Elements in International
Judicial Settlement 995

IX. Land and Maritime Boundaries, International Watercourses and Other Waters

Jonathan I. Charney
International Maritime Boundaries for the Continental Shelf:
The Relevance of Natural Prolongation 1011

Hungdah Chiu
The Problem of Delimiting the Maritime Zone Between
China (Mainland and Taiwan) and Japan 1031

Ryuichi Ida
The Role of Proportionality in Maritime Delimitation Revisited:
The Origin and the Meaning of the Principle from the Early
Decisions of the Court 1037

Stephen C. McCaffrey
International Watercourses in the Jurisprudence of the World
Court 1055

Maurice Mendelson
On the Quasi-Normative Effect of Maritime Boundary
Agreements 1069

Masahiro Miyoshi
Considerations of Equity in Maritime Boundary Cases Before
the International Court of Justice 1087

Kazuhiro Nakatani
    Oil and Gas in the Caspian Sea and International Law   1103

M. C. W. Pinto
    Maritime Boundary Issues and Their Resolution: An Overview   1115

Walter Rudolf
    Deutsche Seegrenzen   1143

Oscar Schachter
    Linking Equity and Law in Maritime Delimitation   1163

Ignaz Seidl-Hohenveldern
    The Austrian Continental Shelf   1169

Bruno Simma and Daniel-Erasmus Khan
    Peaceful Settlement of Boundary Disputes under the Auspices of the Organisation of African Unity and the United Nations: The Case of the Frontier Dispute Between Eritrea and Ethiopia   1179

Jon M. Van Dyke
    Judge Shigeru Oda and Maritime Boundary Delimitation   1197

X. The Law of the Sea

Milenko Kreća
    A Few Remarks on the Theoretical Basis of the New Law of the Sea   1207

Chiyuki Mizukami
    Changes in Japan's Policy on the Law of the Sea   1217

L. D. M. Nelson
    The Continental Shelf: Interplay of Law and Science   1235

Jordan J. Paust
    Waves Within and Outside the Law of the Sea: Traversing Gaps, Ambiguities and Priorities   1255

Natalino Ronzitti
    Coastal State Jurisdiction over Refugees and Migrants at Sea   1271

Sompong Sucharitkul
  The Intertemporal Character of International Law Regarding
  the Ocean   1287

Budislav Vukas
  The Definition of the Law of the Sea   1303

XI. The Environment and the Law of the Sea

Elisabeth Mann Borgese
  The Common Heritage of Mankind: From Non-living to
  Living Resources and Beyond   1313

Michael Bothe
  Das Seerecht als Motor des internationalen Umweltrechts
  Gedanken zu neueren Entwicklungen im Bereich des
  Tiefseebergbaus   1335

William T. Burke
  Evolution in the Fishery Provisions of UNCLOS   1355

Vladimir-Djuro Degan
  The Common Heritage of Mankind in the Present Law of the
  Sea   1363

L. Yves Fortier
  From Confrontation to Cooperation on the High Seas:
  Recent Developments in International Law Concerning
  the Conservation of Marine Resources   1377

Thomas A. Mensah
  Civil Liability and Compensation for Vessel-Source Pollution
  of the Marine Environment and the United Nations Convention
  on the Law of the Sea (1982)   1391

XII. Human Rights

Nisuke Ando
  The Follow-Up Procedure of the Human Rights Committee's
  Views   1437

Héctor Gros Espiell
    Les droits de l'homme et la Cour internationale de Justice :
    une vision latino-américaine   1449

Alexandre Kiss
    The Impact of Judgments and Advisory Opinions of the PCIJ-
    ICJ on Regional Courts of Human Rights   1469

C. G. Weeramantry
    Cultural and Ideological Pluralism in Public International Law   1491

XIII. Defence, the Use of Force and the Law of Armed Conflict

L. C. Vohrah, Kelly Dawn Askin and Daryl A. Mundis
    Contemporary Law Regulating Armed Conflict at Sea   1523

Shinya Murase
    The Relationship Between the UN Charter and General
    International Law Regarding Non-Use of Force: The Case of
    NATO's Air Campaign in the Kosovo Crisis of 1999   1543

Myron H. Nordquist
    A Review of the Debate in the United States Senate over the
    Deployment of Ground Troops to Kosovo   1555

Sreenivasa Rao Pemmaraju
    International Organizations and Use of Force   1575

Daniel Vignes
    Réflexions sur l'intégration européenne dans les domaines
    des relations extérieures et de la défense   1609

VIII

INTERNATIONAL LITIGATION: STRATEGIES, RULES AND PROCEDURES

Trust Funds in International Litigation

David Anderson

I. Introduction

The costs incurred by the parties in international litigation can be high, even though the expenses of the principal permanent courts and tribunals are met not from fees paid by parties to cases, but rather from a regular budget to which the Member States contribute. The normal rule before international courts and tribunals is that each party bears its own costs, irrespective of the outcome.[1] These two factors may combine to discourage a State with limited resources from seeking to settle a legal dispute with another State by submitting it to an international court or tribunal for its decision. These considerations may also inhibit a State from instituting proceedings in respect of an incident in which it has suffered loss or damage. In the latter example, if the State does decide to proceed, the cost factor may reduce or even render Pyrrhic an award of compensation. To take a further example, following recourse to a form of judicial settlement which has proved to be efficacious, such as a decision resolving a disputed boundary, the cost of implementing the decision on the ground by erecting border markers in some difficult terrain may be financially prohibitive for one or both of the States concerned.

It was against this sort of background that the Secretary-General of the United Nations[2] took the initiative in 1989 to create a Trust Fund for

---

[1] Article 64 of the Statute of the International Court of Justice reads: "Unless otherwise decided by the Court, each party shall bear its own costs". The same rule appears as Article 34 of the Statute of the International Tribunal for the Law of the Sea. Exceptions are very rare in interstate litigation before courts and tribunals of many types.

[2] At that time, the office was held by Señor Javier Perez de Cuellar, a distinguished lawyer-diplomat. His account of his tenure, entitled *Pilgrimage for Peace*, 1997, refers frequently to international law and to the International Court of Justice as a means of resolving disputes. His Legal

the International Court of Justice. On the occasion of the consideration of the Report of the Court, the Secretary-General announced the initiative to the General Assembly, referring to his responsibility to promote the settlement of disputes by the Court. Initiatives by the Secretary-General are rare. On this particular occasion, there was no proposal, debate or decision in the General Assembly. Instead, the Secretary-General acted as the Chief Administrative Officer of the Organisation, of which the Court is the principal judicial organ, and established by his own motion a permanent Trust Fund with its own terms of reference.[3]

The purpose of the Trust Fund was (and is still today) to provide financial assistance where this is needed in order to enable a State to have access to the Court for resolving a dispute. Following this initiative, during the 1990s, Judge Oda and his colleagues in The Hague have decided several cases in which one of the parties was in receipt of assistance from the Trust Fund.[4] Sir Robert Jennings has described the creation of the Fund as an important development. He added the following:

> "The fund is clearly an important step in enabling poorer countries to take advantage of the Court for the disposal of disputes within its contentious jurisdiction, where the parties are able to make a special agreement".[5]

Arrangements for providing financial or other assistance to needy litigants, similar to the Trust Fund, have been made in regard to several other international courts and tribunals, and for similar reasons. Thus, for example, the Permanent Court of Arbitration has established a

---

Counsel in 1989 was Dr. Carl-August Fleischhauer, a most distinguished international lawyer and now a member of the International Court of Justice.

[3] Annex to UN Doc. A/47/444 of 7 October 1992. See also *ILM* 28 (1989), 1589 et seq.

[4] Information about the working of the Trust Fund may be found in several sources, including P. Bekker, "International Legal Aid in Practice: The ICJ Trust Fund", *AJIL* 87 (1993), 659 et seq., citing earlier sources; R. S. Lee, letter dated 30 November 1998, as part of the preparatory work for The Hague Centennial Conference, information about which is available at http://www.minbuza.nl/English; D. Vignes, "Aide au Développement et assistance judiciaire pour le règlement des différends devant la Cour internationale de Justice", *AFDI* 35 (1989), 321 et seq.; and comments and contributions by H. Corell, A. Pellet, K. Highet and R. S. Lee, in C. Peck and R. S. Lee (eds.), *Increasing the Effectiveness of the ICJ*, 1997, at 8, 118, 134 and 360, respectively.

[5] R. Y. Jennings, "*The International Court of Justice after Fifty Years*", *AJIL* 89 (1995), 499 et seq.

Financial Assistance Fund for the Settlement of International Disputes. Assistance is available from the Fund to States named on the List of Aid Recipients compiled by the Development Assistance Committee of the Organisation for Economic Cooperation and Development.[6] In the case of the World Trade Organisation, the Secretariat has been authorised to provide technical and legal assistance to developing Member States, upon request, in regard to cases before the dispute settlement mechanisms of the Organisation. Assistance may take the form of providing the services of a legal expert from within the Secretariat.[7] The European Court of Human Rights has created a legal aid scheme. Applicants for legal aid must show financial need in relation to an actual or a proposed case. Decisions are made by Chambers of the Court. The International Criminal Tribunals for former Yugoslavia and for Rwanda have arrangements in place to assist accused persons who are indigent. The future International Criminal Court is to have the power to grant legal assistance where the interests of justice so require.

II. The Establishment of the Trust Fund for the International Tribunal for the Law of the Sea

In May 1999, in written comments submitted to the Netherlands Government in preparation for the Conference to mark the Centennial of the First International Peace Conference, the Government of the United Kingdom included a passage in positive terms concerning the operation and effects of the Secretary-General's Trust Fund for the Court. Three suggestions were made: consideration should be given, first, to broadening the base by encouraging donations from charitable foundations and, secondly, to "the idea of Counsel working on a reduced fee *(pro bono)* basis in certain cases". Thirdly, the following tentative suggestion was advanced:

> "Thought should also be given to whether similar funds should be considered for other standing dispute resolution bodies (such as the International Tribunal for the Law of the Sea)".[8]

---

[6] Except as otherwise indicated, the information in this paragraph has been taken from Ph. Sands (ed.), *Manual on International Courts and Tribunals*, 1999.

[7] E.-U. Petersmann (ed.), *International Trade Law and the GATT/WTO Dispute Settlement System*, 1997, 78.

[8] Para. 15 of the Comments on the Centennial Themes, May 1999, for the Centennial of the First International Peace Conference, is available at http://www.minbuza.nl/English (comment of the United Kingdom).

It is safe to assume that this last suggestion did not pass unnoticed. Later that year, the suggestion was considered briefly by the members of the Tribunal, who formed a favourable attitude towards the idea of creating a Trust Fund for the Tribunal, provided the Fund would be administered by an outside body such as the Secretary-General, rather than by the Registry.

In May 2000, the delegation of the United Kingdom to the Meeting of the States Parties to the United Nations Convention on the Law of the Sea ("the Convention") proposed that a Trust Fund for the Tribunal should be established and used "principally for proceedings on the merits of a case and where the jurisdiction of the Tribunal was not an issue". The delegation suggested that the Fund should be available "for expenses incurred in connection with any phase of the proceedings in cases submitted, or to be submitted, to the Tribunal, including its Chambers".[9] The expenses would include the preparation of pleadings, counsel's fees, travel expenses, etc., but not any award of compensation. The proposal was welcomed by the President of the Tribunal, Judge P. Chandrasekhara Rao, who noted that the availability of financial assistance would help "to overcome financial impediments to the peaceful settlement of disputes".[10] He suggested that the Fund should not be restricted to providing assistance only on the merits of a case. Other delegations also welcomed the proposal to create a trust fund supported by voluntary contributions and administered by the UN Secretary-General, with the advice of a Panel of Experts. With regard to the procedure, the general conclusion of the Meeting was that, having regard to the relationship[11] between the United Nations and the Tribunal, the General Assembly was the most appropriate organ to create the Fund by means of a resolution. After a full discussion, it was decided to recommend to the General Assembly that it should give consideration to the establishment of a Trust Fund.[12]

In October 2000, a draft resolution under the item entitled Oceans and Law of the Sea on the agenda of the General Assembly's fifty-fifth session was submitted by a large group of co-sponsors. The proposals included a request to the Secretary-General "to establish a voluntary trust fund to assist States in the settlement of disputes through the tribunal...".[13] Paragraph 10 of the resolution invited "States,

---

[9] UN Doc. SPLOS/60 of 22 June 2000, para. 41
[10] *Ibid.*, paras. 42 to 44.
[11] The Agreement on Cooperation and Relationship Between the United Nations and the International Tribunal for the Law of the Sea of 18 December 1997 defines the relationship. For details, see G. Eiriksson, *The International Tribunal for the Law of the Sea*, 2000, 20.
[12] UN Doc. SPLOS/60, paras. 45 to 47.
[13] UN Doc. A/55/L.10, para. 9.

intergovernmental organizations, non-governmental organizations, as well as natural and juridical persons, to make voluntary financial contributions to the fund". The proposal was accepted in GA Resolution 55/7.[14] The full terms of reference of the Trust Fund were annexed to the resolution.[15] Its adoption was immediately welcomed by the President of the Tribunal.[16]

III. Similarities Between the Funds for the Court and the Tribunal

A perusal of the Annex to Resolution 55/7 shows that the scheme has many points in common with the Trust Fund for the International Court of Justice. For instance, both Funds are administered in accordance with the Financial Regulations of the United Nations[17] by the Secretary-General, who has to report to the appropriate political organ on the status of each Fund. In the case of the Court, reports go to the General Assembly,[18] and in that of the Tribunal, to the Meeting of States Parties. In considering requests for support from either Fund, the Secretary-General is assisted by a Panel of Experts, which plays an advisory role. Decisions are made by the Secretary-General. It seems reasonable to expect some overlap in membership between the two Panels. Assistance is available on request from States which need it for the purposes of a case before either the Court or the Tribunal. Potential contributors to the two Funds are the same: they are listed as States, intergovernmental organisations, national institutions, non-governmental organisations, as well as natural and juridical persons. These lists are cast in terms which appear to be wide enough to embrace foundations, many of which have contributed generously in the past to projects designed to improve arrangements for the peaceful settlement of international disputes.[19] Contributions to both Funds are voluntary and usually in the form of cash. The moneys may be disbursed in order to defray the costs of preparing the application and the written pleadings, translations, counsel's fees, and the expenses of hearings, including travel expenses.

---

[14] The debates were recorded in UN Doc. A/55/PV 42 and 43, of 26 and 27 October 2000, respectively.

[15] For ease of reference, the Annex is reproduced at the end of this article.

[16] ITLOS/Press Release 39 of 14 November 2000.

[17] Financial Regulations and Rules of the United Nations, UN Doc. ST/SGB/188 (1985).

[18] For an example of such a Report, see UN Doc. A/47/444 of 7 October 1992.

[19] Indeed, an important landmark concerning the settlement of disputes during the Third United Nations Conference on the Law of the Sea was the informal meeting held in Montreux in 1975 under the sponsorship of the Rockefeller Foundation.

In addition, the Funds are available to defray the costs of executing a judgment, such as demarcating a boundary.

IV. Some Differences Between the Two Funds

At the same time, there are some important differences between the terms of reference of the two Funds. These differences concern the respective reasons for establishing the two Funds, their objectives and purposes, the requirements to be met by applicants for assistance, and the facility of offering to provide professional assistance on a reduced fee basis. There are several reasons for these differences.

1. The Respective Reasons for Establishing the Two Funds

The rationale for establishing the Court's Trust Fund was explained by the Secretary-General in paragraphs 1 to 5 of the terms of reference. Problems were identified with the inability of States to proceed to submit a dispute to the Court, or to implement a decision of the Court (e.g., the demarcation of a land boundary), because of a lack of expertise or funds.

In the case of the Tribunal, the reasons were basically the same as those for the Court's Trust Fund and require no further elaboration. In addition, however, there were some structural reasons having to do with Part XV of the Convention on the Law of the Sea, as follows. Paragraph 1 of the terms of reference, after setting out the four means of settlement listed in Article 287 of the Convention, recalls that States are free to choose one or more of those means. The next paragraph of the terms of reference notes that both the Court and the Permanent Court of Arbitration have Trust Funds which are available, in accordance with their terms, for those eligible States Parties which choose either the Court or arbitration under Article 287. Thus, needy litigants, whether appearing before the Court or the Permanent Court of Arbitration, may be provided with financial assistance. In order to enable indigent States to exercise properly their freedom of choice under Article 287, the Tribunal should be placed in a similar situation. This was the thinking behind paragraph 2 of the terms of reference, reading:

> "The burden of costs should not be a factor for States in making choices under Article 287, in deciding whether a dispute should be submitted to the Tribunal, or in deciding upon the response to an application made to the Tribunal by others".

In other words, the creation of a Trust Fund for the Tribunal means that assistance is potentially available whether an eligible State chooses the Court or opts for certain types of arbitration (where the Permanent Court of Arbitration is clearly one possibility for both the forum and the venue) or prefers the Tribunal. In the new circumstances resulting from the establishment of the Fund for the Tribunal, cost should no longer be a significant factor in the making of choices by States eligible for assistance.

2. Objects and Purposes of the Trust Funds

The object and purpose of the Fund for the International Court of Justice are to provide financial assistance to States for expenses incurred in connection with a dispute submitted to it by means of a special agreement, or for the purpose of executing a judgment resulting from such an agreement. It is apparent that the requirement of a special agreement limits the availability of the Fund quite considerably. For instance, the requirement means that the Fund is not available to the impecunious State which has been made a respondent in a case in which the Court's jurisdiction is founded not upon a special agreement under Article 36(1) of the Court's Statute but instead upon two matching declarations made under Article 36(2). Nor would the Fund appear to be available for a case where jurisdiction is based on the disputes article of a treaty, even when jurisdiction is not in issue.[20] No doubt, in establishing the Fund, the Secretary-General wished to avoid all possible controversy over jurisdictional issues. According to a member of the Court's Registry, Arthur Witteveen, "That approach was taken in order that the Fund not be used to finance cases in which the Respondent State contested the jurisdiction of the Court but at the same time, through its participation in the Fund, was contributing to the financing of that case".[21] The risk of such an unfortunate coincidence could, perhaps, have been averted by administrative means, and the restriction may have had unintended consequences. Shabtai Rosenne has written: "It is not clear why the purpose of the Fund is limited to cases submitted jointly by the parties, or why it cannot be used to contest the jurisdiction of the

---

[20] Although there may be some flexibility in certain circumstances. According to R. S. Lee, the Fund provided assistance in the *Senegal/Guinea-Bissau* case where there was no agreement but the Court's jurisdiction was no longer contested: see his letter cited in note 2 *supra*. It is unclear whether or not the Fund would be available to an otherwise eligible party to a case in which the Court's jurisdiction was based on two choices of the Court made under Article 287 of the LOS Convention.

[21] See A. Eyffinger, *The International Court of Justice 1946-1996*, 1996, 365.

Court whether by way of preliminary objection or in any other way. The Statute (Article 40) does not give preference to any particular method of conferring jurisdiction on the Court...".[22] Conceivably, a State could be placed in the position of being unable, for financial reasons, to conduct its legal affairs in the manner of its choice. That situation would appear to be undesirable in principle. More generally, it would be unfortunate were developing States to hesitate to make declarations under Article 36(2) for fear of being compelled to defend, at considerable expense and without the possibility of having recourse to the Trust Fund, applications made against them by wealthier States which also had made such declarations. These factors, coupled with the normal rule that each party bears its own costs regardless of the outcome, can serve only as discouragements in practice.

In the case of the Tribunal, there is no requirement for a special agreement. This change seeks to avoid the problems just outlined above by opening up the Fund more widely. The new approach also, and perhaps more significantly, reflects the difference between the jurisdiction of the Court when it is acting in a case under Article 36 of its Statute, where the jurisdiction is fundamentally consensual,[23] and the jurisdiction of courts and tribunals when they are acting under Part XV, where the basic rule is that contained in Article 286. By virtue of Article 286, in establishing their consent to be bound generally by the Convention, States also, at that stage, establish their particular consent, subject to section 3 of Part XV, to the jurisdiction of courts and tribunals "having jurisdiction" under section 2 in those disputes concerning the interpretation or application of the Convention where no settlement has been reached by recourse to section 1. (The term "having jurisdiction" is not circular or question-begging. In its context, the term is apt simply to identify an appropriate forum listed in Article 287.) It is perhaps a sad reflection on the present state of international relations that a State Party to the Court's Statute need never take the next step and consent to the jurisdiction of the Court – whether generally under Article 36(2), or in a treaty or in relation to a particular dispute. In contrast, States Parties to the Convention do consent to a measure of jurisdiction on the part of an international court (including the Court) or some other international tribunal (including the Tribunal and its Seabed Disputes Chamber) which is acting in respect of a dispute arising within the scope of Part XV. Accordingly, it was considered unnecessary to require a special agreement as a precondition for access to the Tribunal's Trust Fund,

---

[22] Sh. Rosenne, *The Law and Practice of the International Court 1920-1996*, 1997, 515.

[23] This point was stressed by Judge Oda in his recent article in the *International and Comparative Law Quarterly* entitled "The Compulsory Jurisdiction of the International Court of Justice: A Myth?", *ICLQ* 49 (2000), 251.

bearing in mind that the Tribunal's jurisdiction is founded on the Convention.

The purpose of the Fund is to provide financial assistance to States Parties to the Convention for expenses incurred in connection with cases submitted, or to be submitted, to the Tribunal, including the Seabed Disputes Chamber and any other Chamber. According to paragraph 5 of the terms of reference,

> "Assistance ... should only be provided in appropriate cases, principally those proceeding to the merits where jurisdiction is not an issue, but in exceptional circumstances may be provided for any phase of the proceedings".

This is an important paragraph, on which some comments may be in order. Assistance is to be provided, naturally, in accordance with the terms of reference. It is the Secretary-General who will have no doubt the recommendations of a Panel of Experts. The reference to the "merits" would include the broad substantive questions arising for decision in a normal case, as well as the more limited "merits" of applications for the prompt release of vessels and crews brought under Article 292. The existence of "exceptional circumstances" would again be determined by the UN Secretary-General. Where such circumstances exist, assistance may be provided for "any phase" of the proceedings in a case. These phases would include a preliminary objection to the jurisdiction of the Tribunal or to the admissibility of an application, an application for provisional measures of protection, the presentation of a counterclaim and an application to intervene.

3. Applications for Financial Assistance

The requirements facing an applicant for assistance are more onerous in regard to the Court's Fund than that of the Tribunal. In the former case, the State seeking assistance must be in a position to produce a copy of a special agreement submitting to the Court a specific dispute for its decision. In the latter case, the State seeking assistance (being a State Party to the Convention) must simply describe the nature of the case which is to be, or has been, brought by or against the State concerned. In other words, the applicant for assistance could be the applicant in the proceedings or the respondent. There may or may not be a special agreement. The applicant may not have taken any step in the case. For example, it may not have filed an application. Indeed, the State may wish to consult Counsel at the outset as to whether or not there is a good case to be made and, if so, whether the Tribunal is an appropriate forum in which to institute proceedings, in circumstances where there is more than one possibility.

4. Offers of Professional Assistance

The terms of reference of the Trust Fund for the Court refer only to financial assistance, without mentioning the possibility of making offers of professional assistance. In contrast, paragraph 13 of the terms of reference of the Tribunal's Trust Fund reads as follows:

> "The implementing office also maintains a list of offers of professional assistance which may be made on a reduced fee basis by suitably qualified persons or bodies. If an applicant for assistance so requests, the implementing office will make the list of offers available to it for its consideration and decision: both financial and other assistance may be extended in respect of the same case or phase thereof".

This paragraph opens the door to individuals and law firms that have a particular interest or experience in international litigation, the law of the sea or the Tribunal to make a standing offer of their services on a reduced fee basis for the purposes of the Trust Fund. It is made clear in paragraph 13 that, in every instance, the decision whether or not to take up such an offer lies with the government that is seeking assistance. It is for that government to decide whether to take up such an offer or not. This is not a question upon which the Panel of Experts advises the Secretary-General. There may be many reasons why a government is unwilling to avail itself of such an offer, without having to give reasons. On the other hand, a particular offer may prove to be attractive and discussions about the scope and terms of possible assistance could be entered into. Offers may be confined, for example, to the giving of advice or to the preparation of studies. Equally, offers may extend also to appearing before the Tribunal as counsel or as a member of the team of representatives.

Paragraph 13 concerns what has sometimes been described elsewhere as *pro bono* work,[24] although that particular expression, often inappropriate in an international context, was not used. Suitably qualified persons or bodies include individuals and law firms that have experience in appearing before international courts and tribunals, as well as professors and experts in disciplines such as international law, in

---

[24] Suggestions were made to extend the Court's Trust Fund in this direction by T. Franck, in C. Peck and R. S. Lee (eds.), see note 4, 496, as well as in the comments of the United Kingdom, see note 8.

particular the law of the sea, as well as shipping, maritime, mining and environmental law. The number of such persons and firms with international experience is growing rapidly. At one time, the pool of advocates before the Court was composed of a small number of law professors. Today, with the increase in international litigation, more and more lawyers have had the experience of appearing before an international court or tribunal. Indeed, many governments now engage the services of large law firms, usually based in Europe or North America, to assist with their representation before the Court. The same is true of other courts and tribunals. These individuals and firms may also be willing to undertake some *pro bono* work in both the national and now the international context.[25] In other words, they are now enabled, pursuant to paragraph 13, to bring their expertise to bear at the international level on a reduced fee basis.

Suitably qualified persons may also include technical experts, such as hydrographers, cartographers, historians, fisheries scientists, environmental scientists, mining engineers and the like. In the past, such experts have often been attached to legal teams in boundary or fisheries cases.

V. Some Concluding Remarks

In the case of individuals, equality before the law and access to justice are both basic human rights. Many States operate schemes of legal aid and assistance in order to avert the risk of there being a denial of justice to poor persons on account of their inability to meet the high costs of litigation, whether civil or criminal.[26]

With the recent expansion in the number of international courts and tribunals and the increase in the number of cases coming before them, international litigation today involves States from all parts of the world. In some contexts, litigation is becoming a routine element in the work of governments. A major element in international relations, world trade, is now subject to a system of dispute settlement which is very active. States in all stages of development, rich and poor, participate as applicants, respondents or interveners in cases before the Court and other international tribunals. While States are in principle equal before the law,[27] some small developing States possess only limited resources

---

[25] For a recent survey of the evolution of practice in this regard in England, see M. Zander, *Cases and Materials on the English Legal System*, 8th ed., 1999, 542.

[26] In England, the legislation has recently been reformed in the Access to Justice Act 1999.

[27] This is an aspect of the Charter Principle of the sovereign equality of States, reviewed by C. Warbrick in C. Warbrick and V. Lowe (eds.), *The United*

for the conduct of international litigation. In practice, they tend to be remote from the seats of most international tribunals. Such a State, when faced with a legal dispute with another State, may need advice and assistance from experienced counsel for a variety of reasons: for example, in order to assess the strength of its legal position, determine which are the legal procedures open to it, and, if it decides to have recourse to a court or tribunal, present its arguments in the most effective manner. By using the administrative skills and good offices of impartial public servants, such as the UN Secretary-General and his legal staff, Trust Funds have been structured, and are being managed, in ways which do not affect the impartiality of the courts or tribunals concerned. For their part, courts, both national and international, benefit from the appearance before them of expert advocates, versed not only in the relevant rules of law but familiar also with the practice and procedure of the forum. Their presence helps to ensure that relevant points of fact and law, including precedents from all international courts and tribunals (and especially those of the Court), are presented and argued before decisions are reached. In that sense, Trust Funds assist not only litigants but also the courts before which assisted litigants appear. Indeed, the administration of justice generally is enhanced thereby.

It is a good development, therefore, that each major international court and tribunal, including most recently the Tribunal, has its own facility for providing appropriate assistance upon request. Such facilities avoid the risk of access to justice being rendered illusory, or even denied in practice, for financial reasons. In short, Trust Funds and other similar forms of assistance help to advance the rule of law in international relations. It is to be hoped, therefore, that both Trust Funds, as well as the other facilities, will receive full and continuing support from governments and other contributors.[28]

Nations and the Principles of International Law, 1994, 209. In his survey, Warbrick alludes to equality in relation to the Court and to the Trust Fund.

[28] As stated by Sir Arthur Watts (United Kingdom) in the General Assembly, the best help "is money" (UN Doc. A/44/PV.43), quoted by P. Bekker, see note 2. Upon the adoption of GA Resolution 55/7, establishing the Trust Fund for the Tribunal, one delegation announced a donation and others indicated that consideration would be given to making one.

Annex I to General Assembly Resolution 55/7:
The Terms of Reference of the Trust Fund for the International Tribunal for the Law of the Sea

Reasons for Establishing the Trust Fund

1. Part XV of the United Nations Convention on the Law of the Sea ("the Convention") provides for the settlement of disputes. In particular, Article 287 specifies that States are free to choose one or more of the following means:
   (a) The International Tribunal for the Law of the Sea;
   (b) The International Court of Justice;
   (c) An arbitral tribunal;
   (d) A special arbitral tribunal.

2. The Secretary-General already operates a Trust Fund for the International Court of Justice (see A/47/444). The Permanent Court of Arbitration has established a Financial Assistance Fund. The burden of costs should not be a factor for States, in making the choices under Article 287, in deciding whether a dispute should be submitted to the Tribunal, or in deciding upon the response to an application made to the Tribunal by others. For these reasons, it was decided to create a Trust Fund for the International Tribunal for the Law of the Sea ("the Tribunal").

Object and Purpose of the Trust Fund

3. This Trust Fund ("the Fund") is established by the Secretary-General in accordance with General Assembly Resolution XXX and pursuant to the Agreement on Cooperation and Relationship Between the United Nations and the Tribunal of 18 December 1997 (General Assembly Resolution 52/251, annex).

4. The purpose of the Fund is to provide financial assistance to States parties to the Convention for expenses incurred in connection with cases submitted, or to be submitted, to the Tribunal, including its Seabed Disputes Chamber and any other Chamber.

5. Assistance, which will be provided in accordance with the following terms and conditions, should only be provided in appropriate cases, principally those proceeding to the merits where jurisdiction is not an issue, but in exceptional circumstances may be provided for any phase of the proceedings.

Contributions to the Fund

6. The Secretary-General invites States, intergovernmental organizations, national institutions, non-governmental organizations, as well as natural and juridical persons, to make voluntary financial contributions to the Fund.

Application for Assistance

7. An application for assistance from the Fund may be submitted by any State Member to the Convention. The application should describe the nature of the case which is to be, or has been, brought by or against the State concerned and should provide an estimate of the costs for which financial assistance is requested. The application should contain a commitment to supply a final statement of account of the expenditures made from approved amounts, to be certified by an auditor acceptable to the United Nations.

Panel of Experts

8. The Secretary-General will establish a panel of experts, normally three persons of the highest professional standing, to make recommendations on each request. The task of each panel is to examine the application and to recommend to the Secretary-General the amount of the financial assistance to be given, the phase or phases of the proceedings in respect of which assistance is to be given and the types of expenses for which the assistance may be used.

Granting of Assistance

9. The Secretary-General will provide financial assistance from the Fund on the basis of the recommendations of the panel of experts. Payments will be made against receipts showing expenditures made in respect of approved costs. The latter may include:
(a) Preparing the application and the written pleadings;
(b) Professional fees of counsel and advocates for written and oral pleadings;
(c) Travel and expenses of legal representation in Hamburg during the various phases of a case;
Execution of an Order of Judgment of the Tribunal, such as marking a boundary in the territorial sea.

Application of the Financial Regulations and Rules of the United Nations

10. The Financial Regulations and Rules of the United Nations will apply to the administration of the Fund, including the procedures for audit.

Reporting

11. An annual report on the activities of the Fund, including details of the contributions to and disbursements from the Fund, will be made to the Meeting of States Parties to the Convention.

Implementing Office

12. The Division for Ocean Affairs and the Law of the Sea of the Office of Legal Affairs is the implementing office for this Fund and provides the services for the operation of the Fund.

Offers of Professional Assistance

13. The implementing office also maintains a list of offers of professional assistance which may be made on a reduced fee basis by suitably qualified persons or bodies. If an applicant for assistance so requests, the implementing office will make the list of offers available to it for its consideration and decision; both financial and other assistance may be extended in respect of the same case or phase thereof.

Revision

14. The General Assembly may revise the above if circumstances so require.

The Uses and Abuses of Advisory Opinions

Sir Franklin Berman

When the International Court of Justice adopted its by now notorious advisory opinion on the *Legality of the Threat or Use of Nuclear Weapons*, it began by taking a formal vote on whether or not to give the opinion requested by the UN General Assembly. One may surmise that the vote was taken at the insistence of Judge Oda who was the only member of the Court to vote against.[1] This note will examine the

---

[1] This supposition is fortified by the fact that this is the only occasion in the history of the Court's advisory jurisdiction on which the decision to render the opinion requested was formally opposed by just one judge, in isolation. There had, on one or two occasions, been small dissenting minorities, notably in the case of the *Western Sahara* opinion, which was further distinguished by the peculiarity that *separate* votes were taken and recorded on the two questions submitted by the General Assembly, and further still by the fact that the voting figures were different on each occasion: 13-3 in favour of answering the first question *(terra nullius)* and 14-2 the second (links to Morocco). (For a possible explanation, but an implicit one, see the separate opinion of Judge Petrén.)
  In fact, the practice of the Court over taking, and recording, a formal vote on the prior question of whether to respond to the request for an advisory opinion, is very uneven and not readily susceptible to rationalization. Despite the evidently strong differences within the Court (notably for example over the *Interpretation of Peace Treaties* case in 1950), no vote on the prior question is recorded in this case or in any of the other early cases in which the Court plainly devoted much thought to whether it ought or ought not to respond. No such vote is in fact recorded until 1956, when the Court first confronts staff cases – on this occasion emanating from the *Judgments of the ILO Administrative Tribunal*. Thereafter, with the sole exceptions of *Western Sahara* (mentioned above) and the present case, all the other cases in which a vote is recorded on the prior question are the staff cases emanating from judgments of the UN Administrative Tribunal: *Fasla* in 1973; *Mortished* in 1982 (the first occasion after the practice was adopted of recording the votes of individual

nature of the Court's discretion to refuse to comply with a request that it give an advisory opinion, and offer some concluding remarks on the benefits and risks of the use of the advisory jurisdiction.

The Court had of course preceded its decision on the request from the General Assembly by declining to accept the earlier (and largely parallel) request submitted by the Assembly of the World Health Organization. This was however on jurisdictional grounds, and therefore has no direct bearing on the question presently under discussion; there was never at any stage any controversy as to whether questions to do with the legality of nuclear weapons fell within the mandate of the UN General Assembly, or more precisely within the General Assembly's *competence* to request advisory opinions under Article 96(1) of the Charter. Whether a particular point to be referred to the Court does or does not represent a "legal question" within the meaning of Article 96(1) is a different matter, and will be addressed below.

The starting point is naturally that, whereas Article 96 of the Charter establishes the competence of organs of the United Nations and other bodies to submit requests for advisory opinions to the ICJ, it is Article 65 of the Court's Statute which regulates the Court's power to respond. As is well known, paragraph 1 of Article 65 is written in permissive terms ("The Court may give an advisory opinion ... at the request of whatever body may be authorized by or in accordance with the Charter ... to make such a request".). Moreover the Court has repeatedly held that this provision does more than merely create the power to give advisory opinions, but is intended to give the Court a discretion whether to do so or not. There is of course a degree of overlap between the power conferred on the General Assembly and other bodies under Article 96 and the discretion conferred on the Court by Article 65, in that both of them apply (and apply exclusively) to

---

judges) and *Yakimetz* in 1987. The last-named stands out because, although a separate formal vote was taken and recorded, the decision so recorded was in fact a *unanimous* one.

Conversely, no separate vote was taken in respect of the *Namibia* opinion of 1971 nor in respect of the *Privileges and Immunities* opinion of 1989 *(Mazilu)*, despite the strong objections lodged before the Court in each case by a Member State directly concerned.

While it thus becomes difficult to provide any satisfactory general formula to explain when prior votes are or are not taken, the answer (so far as the present case is concerned) no doubt lies in Article 8(ii)(a) of the 1976 Resolution concerning the Internal Judicial Practice of the Court, which – despite paragraph (iii) of the same article – leads one to suppose that a judge who insists upon the especial importance he attaches to a separate vote will get his way.

"any legal question". It follows that – were the Court to find in any given instance that a request as formulated did not pose a "legal question" – the matter would at one and the same time have to be ruled outside the *competence* (in a strict jurisdictional sense) of the requesting organ and outside the competence of the Court itself to respond. The connection between the two will be gone into below; for the moment, however, let us remain with the discretionary aspect of the Court's function under its advisory jurisdiction.

To say that it is within the Court's discretion to respond or to decline to respond to a request for an advisory opinion invites the question, on what basis is this discretion exercised? Quite clearly, it cannot be a free discretion to be exercised on policy grounds. The Court, whether exercising its advisory or its contentious jurisdiction, remains a court of law, and the principal *judicial* organ of the United Nations. This is the basis on which the Court has consistently approached its function since its creation. Any suggestion that the Court might decide not to respond to a request on general policy grounds would therefore be as abhorrent to the Court itself as it would to the political organs of the institutions in the UN family. It would amount to no less than suggesting that the Court should be in a position to overrule the political judgement of the requesting organ. Nothing would be more likely to give rise to a crisis in which the judicial function would be the probable victim.

It follows that the discretion which represents the starting gate to the whole advisory process is accepted to be an essentially *judicial* discretion, in other words a discretion of a kind that is uniquely appropriate to be exercised by judges, sitting collectively as a court, and in accordance with the same approach as they apply to other questions that arise in proceedings before them. It does not, however, answer the question why the discretion exists in the first place. There is no equivalent doctrine in respect of the contentious jurisdiction. It would be startling in the extreme if the Court were to assert that, in the exercise of its discretion, it would not hear a contentious case properly brought before it by the parties. Issues as to appropriateness of exercising the judicial function do certainly arise even in respect of contentious proceedings; the most frequently cited is the *Northern Cameroons* case, and the Court deals with them *in limine* through the flexible concept of "admissibility". Rosenne, however, believes that the distinction between what he refers to as the "objective" and the "subjective" element in advisory proceedings exactly parallels what is to be found in the Court's practice in respect of contentious cases.[2]

---

[2] Sh. Rosenne, *The Law and Practice of the International Court, 1920-1996*, 1997, 987. An element of confusion is however introduced by the author's

In a recent thoughtful article, Abi-Saab questions whether the Court's supposed discretion exists at all.[3] There are however two flaws in Abi-Saab's line of reasoning.

To begin with, his question seems to originate in a linguistic confusion or at least an extreme reading of what "discretion" means. Abi-Saab's proposition is essentially that the Court cannot be thought to possess an *absolute* or *unlimited* power to decline to answer a request for an advisory opinion. But this is not how the matter has ever been approached by the International Court of Justice, or by its predecessor. Nor does it represent the sense in which the notion of a "discretion" is normally used in legal terminology in English, or understood by Common Lawyers.[4] There is in other words a recognised distinction between "absolute discretion" and "limited discretion", and "discretion" on its own does not imply the absolute rather than the limited form. Within the common law systems, a large part of the fertile modern development of the highly significant institution of judicial review consists in the general limits which the law has progressively read into (or imposed upon) the exercise of delegated powers or of generalised administrative discretion.[5]

It follows that – despite surface appearances to the contrary – there is no inherent contradiction between Abi-Saab's denial that the Court truly possesses a "discretion" (in the way in which he understands the term) whether to accept or to refuse a request for an advisory opinion, and the insistence by Common Lawyers that the nature of the Court's task in face of such a request is, precisely, not an automatic but a "discretionary" one. As it emerged in this instance, that insistence took the form of several powerfully argued national submissions that, in the very special circumstances of the case, the Court should exercise that "discretion" by declining to give an opinion at all.[6]

The second problem lies in the way Abi-Saab develops his argument. It is not so much whether he is right or wrong to read the

---

use of the terms "jurisdiction" and "competence" to refer to the "objective" and "subjective" elements.

[3] G. Abi-Saab, "On Discretion", in L. Boisson de Chazournes and Ph. Sands (eds.), *International Law, the International Court of Justice and Nuclear Weapons*, 1999, 37 et seq.; see also "Introduction by the Editors", *ibid.*, 19.

[4] An old judicial pronouncement by an English court, dating from 1792, captures the flavour neatly: "according to law, and not humour. It is to be, not arbitrary, vague and fanciful, but legal and regular" (*Wilson v. Rastell*).

[5] See the *locus classicus* of *Associated Provincial Pictures Houses Ltd. v. Wednesbury Corporation*, [1948] 1 K.B. 223.

[6] The references can be found at para. 15 of the opinion, ICJ Reports 1996 (I), 236-237.

cases as a progressive whittling-down to nothing of the Court's discretion, as the fact that his reading relies too heavily on the "classic" pronouncements of the Court, those one finds recited in all the textbooks: *Status of Eastern Carelia*[7]; *Interpretation of Peace Treaties*[8]; *Certain Expenses of the United Nations*.[9] Although he does refer to the opinion on the *Judgments of the ILO Administrative Tribunal*,[10] Abi-Saab fails to tap the rich vein of the ICJ's jurisprudence in the long line of advisory opinions on staff cases, of which the 1956 opinion was but the beginning. These cases contain the plainest possible (and repeated) reaffirmation by the Court that it does retain a discretion whether or not to accept a request for an advisory opinion. At the same time they throw clear light on why the Court considers the discretion to exist and on the basis on which it is to be exercised.

Thus, while in *ILO Administrative Tribunal*[11] the Court contents itself merely with recalling the "permissive character" of Article 65 of the Statute, in the first case with which it confronts the review of judgments of the UN Administrative Tribunal, the *Fasla* case in 1973,[12] the Court goes a clear step further. It goes out of its way to say that, although none of the statements or comments submitted to it raises a question as to its competence or as to the propriety of its giving the opinion sought, the Court is nevertheless *obliged* to examine both matters *proprio motu*. It takes a very similar approach in the *Mortished* case in 1982,[13] which must surely be the one case (until the WHO request in the present matter) in which the Court came closest to declining outright to answer a request. Indeed in that case the Court says that "there can be no question ... of any restriction" on its discretion. In other words, the Court has not only consistently taken the view that its own character as a judicial organ necessarily requires it to retain a discretion whether or not it will provide advice in any given case, but that that same judicial character obliges it *ex officio* to exercise that discretion in each given case. Admittedly, the jurisprudence has developed over the years a clear presumption in favour of the positive exercise of the discretion, which the Court has encapsulated in the well-known phrase that there must be "compelling

---

[7] PCIJ, Ser. B, No. 5 (1923).
[8] ICJ Reports 1950, 221.
[9] ICJ Reports 1962, 155.
[10] ICJ Reports 1956, 77.
[11] *Ibid.*
[12] ICJ Reports 1973, 183.
[13] ICJ Reports 1982, 325.

reasons" before it will decline; but that is far from removing the discretion for all practical purposes from the scene altogether. To read the cases in that way would hardly be reconcilable with the visible struggle the Court re-enacts in its successive opinions as to why the circumstances before it, although they differ from the paradigm of judicial process, are nevertheless sufficiently close to it that they do admit of treatment in a properly judicial way. The position is precisely summed up in another of the ICJ's famous utterances, but one the second sentence of which is often omitted: "The Court is a judicial body and, in the exercise of its advisory functions, it is bound to remain faithful to the requirements of its judicial character. Is that possible in the present case?".[14]

The citation just given from the 1956 opinion on *Judgments of the ILO Administrative Tribunal* poses a criterion of judicial propriety. It becomes not a mechanical question of whether this or that given feature is present but a judgement whether in all the circumstances of the particular case the question put can be answered judicially. That is a judgement which is very fact-specific and is a judgement of a kind that only the Court itself can make, as the ultimate guardian of its own integrity. Hence the repeated stress on propriety, which represents the guiding thread through the decisions, and which in the Court's mind is another name for compatibility with the judicial function.[15]

So, in the *Western Sahara*[16] case, the Court poses the question whether "in the circumstances of a given case, considerations of judicial propriety" should oblige it to refuse an opinion[17]; it twice says that the consent (or lack of it) of an interested State is relevant, not however for the Court's *competence*, "but for the appreciation of the *propriety* of giving an opinion" (emphasis added) and sums up by declaring that "[i]n certain circumstances" lack of consent may render the giving of an advisory opinion "incompatible with the Court's judicial character".[18] In the *Mortished* case, the Court points out that a "compelling reason" (sc. to decline to give an opinion) would be "that its judicial role would be endangered or discredited".[19]

---

[14] ICJ Reports 1956, 84.
[15] In other words, Abi-Saab adopts too narrow an approach when he limits the Court's anxieties to avoiding compulsory jurisdiction by the back door and avoiding being used for "mere doctrinal speculations". Abi-Saab, see note 3, 40.
[16] ICJ Reports 1975, 21.
[17] *Ibid.*, 24.
[18] *Ibid.*, 25.
[19] See ICJ Reports 1982, 347.

In the *Nuclear Weapons* case, the question whether the Court should exercise its discretion in a positive or negative direction had (as indicated) been actively canvassed in the written and oral submissions made to the Court by Member States under Article 66 of the Statute. Judge Oda devotes almost the whole of his dissenting opinion to it.[20] It is striking, by contrast, how few of the other judges delivering individual opinions feel it necessary to touch on the question at all. There is a very brief mention in the declaration by Judge Ferrari Bravo[21] and a short section in the separate opinion of Judge Guillaume,[22] as well as a treatment of the question in Judge Weeramantry's separate opinion in the *WHO* case[23] (which he expressly applies to the General Assembly case as well). For the rest, the majority of the judges are content to rest on the relatively brief passage in paragraphs 10 to 19 of the Opinion.[24] Given the controversy which saturated the entire proceedings, this reticence is a little surprising. It is doubly surprising in the face of the remarkable fact that the Court, having expressly decided that the question put to it, properly interpreted, was one capable of being given a legal answer, ended up concluding that it was not in a position to give a definitive legal answer to it. As Judge Shahabuddeen put it, "the contradiction between promise and performance cannot, really, be concealed".[25] So it is disappointing, in these circumstances, that in the main Judge Oda's fellow judges passed up the opportunity to join him in grappling publicly with the important issues of judicial function and technique that are involved.

Judge Oda based his lone dissent on reasons of judicial *propriety* and *economy*, and to some extent judicial *credibility*. The first of these is by far the most important, and in many respects the third rejoins it, but the question of "judicial economy" is of less interest within the scope of the present article.

What Judge Oda means by judicial propriety seems largely to be encompassed by what he refers to in the introductory part of his opinion as "The Inadequacy of the Question put by the General Assembly". This in turn refers back to his conclusion that the question was put, not so much out of a genuine wish to ascertain the Court's

---

[20] Paragraphs 1-53 out of a total of 55; the dissenting opinion is at ICJ Reports 1996 (I), 330 et seq.
[21] *Ibid.*, 282.
[22] *Ibid.*, 287-288.
[23] *Ibid*, 154 et seq.
[24] *Ibid.*, 232-238.
[25] *Ibid.*, 375.

opinion, as to seek the endorsement by the Court of an alleged legal axiom. In other words, that for those who were the driving force behind seeking the opinion, the question simply admitted of one answer, and of one answer only. To this Judge Oda adds a query about the question's imprecision (especially as regards the "threat of use") and a further query as to whether the request really did represent a true "consensus" of the General Assembly.

His palpable unease is all too understandable. It can be summed up as an anxiety lest the Court was being used as an instrument in an orchestrated political campaign. For this fear there is of course some evidence, which Judge Oda marshals in his opinion.[26] Judge Guillaume approaches the same material, but does so head on, by posing the question "de savoir si, dans ces conditions, on pouvait encore regarder les demandes d'avis comme émanant des assemblées qui les avaient adoptées, ou si, appliquant la théorie de l'apparence, la Cour ne devait pas les écarter comme irrecevables".[27] Of the two, the robust Guillaume approach is surely preferable. To go at the issue round about – as Judge Oda seems to imply – would have the Court taking upon itself either to decide that governments had to be protected from themselves (or from their representatives in plenary international organs), or else even to have us suppose that the judges of the ICJ could not be trusted to act judicially in circumstances of strong external pressure.

Understandable as this unease may be, therefore, the question remains whether it does not take us too far in the direction of a broad policy discretion of a kind which the Court ought not to possess (and which it has never regarded itself as possessing). It might be otherwise if – following Judge Guillaume – we were able to construct a properly legal criterion to distinguish a request for an opinion validly emanating from the General Assembly or other authorized requesting organ, from a request that only purports to do so and is therefore invalid. But it will be noted that Judge Guillaume, having teased us with this proposition, sagely draws back. In any case, a doctrine of this nature has more in it of *competence* (in the strict sense) than of *discretion* (once competence has been established). Moreover, to refer to issues of this kind, however troubling they may be, as issues of "judicial propriety" is far removed from the way the Court has used this term in its previous pronouncements. The Court's primary concern would seem clearly to have been the integrity of the judicial process as such. Issues as to the part the Court is being asked to play in a wider process are not

---

[26] *Ibid.*, 334-341.
[27] *Ibid.*, 287-288.

necessarily absent, but play a subsidiary role, and can often be circumvented in any case by the application of judicial techniques.[28]

If "judicial propriety" is thus the proper touchstone, how was the question approached by those States who (as indicated) urged the Court to decline both the General Assembly's request and that from the World Health Assembly?

The Court in its opinion summarizes these submissions through a brief passage from the written pleading of the United States, which mentions various matters.[29] A fuller treatment in the written pleading of the United Kingdom[30] puts the matter however squarely in the context of judicial propriety. It does so principally, not by investigating directly what the concept of "propriety" consists in, but by analyzing into categories the types of cases in which the Court had found it proper in the past to give an opinion – the purpose being, of course, to demonstrate that the request in hand fell into none of them.[31] This indirect approach seems not to have held much appeal for the Court, and it is interesting to speculate how the Court (or individual judges) might have responded to a more direct line of argument. Such an argument might have run somewhat as follows.

The judicial function in its normal sense consists in pronouncing an authoritative legal decision on a dispute referred to the court through a *process* binding the parties to the proceedings and leading to a *result* binding on those parties. It may entail findings of fact as well as of law, and will almost invariably consist in applying the law to facts authoritatively established for the purposes of the proceedings. The court is accepted as being in a better position than any other available authority to make those pronouncements. That acceptance is the product not simply of the accumulated legal knowledge and professional experience collected in the judges composing the court,

---

[28] A classic example being dispensing with oral hearings in staff cases in order to avoid creating an unacceptable inequality between the real parties to the underlying issue.

[29] See note 6.

[30] At paras. 2.23-2.36.

[31] The four categories proposed were: (a) cases where the legal question involved the interpretation of a constitutional provision which had become the subject of dispute in the organ making the request; (b) cases where the legal question involves matters on which the requesting organ or agency seeks guidance in the exercise of its constitutional functions; (c) cases where the legal question involves the interpretation of agreements between the organization and a Member State; (d) cases where the legal question concerns the obligations of Member States consequential upon decisions or resolutions of the competent organs of the organization.

but is in significant measure also the product of the judicial *process* itself. In order to achieve that result, courts over the ages and in all legal traditions have developed fundamental rules governing the process played out before them. The purpose of these rules is to ensure that the outcome of the process before the court qualifies for the authority expected to attach to judicial findings and pronouncements. The rules are thus designed to ensure that the process embodies the qualities of expertise, impartiality, and fairness as between the parties. These rules have a primordial importance and may therefore be regarded as essential guarantees for the very integrity of the judicial process. By their nature these rules are ones of which the courts themselves are the sole guardians; by the same token they are rules which the courts are under an inherent duty to apply and to enforce in all cases. The rules in question will include, for example: rules for distinguishing between assertions and factual proof; the requirement to hear both (all) parties; the equality of the parties; and the principle that the judges must decide on the case presented to them and not on the basis of prejudgments or personal predilections.

Process rules of the kind just described[32] derive naturally from the paradigm of a direct dispute between claimant and defendant parties over their mutual legal rights and obligations in respect of a given subject matter. There is thus no problem of substance in applying these process rules, with the necessary adjustments, to the circumstances of legal disputes between States. So an international court can place itself in a position to carry out the judicial function with integrity in respect of such disputes. Where modifications are necessary, they take the form of adjustments only, leaving the essence of the rule intact. Conversely, to introduce a purely advisory jurisdiction poses an inherent problem. This is precisely that a purely advisory jurisdiction asks the court to play a role which escapes from the paradigm: there is an absence of "parties" in the strict sense; there may well be no defined "dispute"; and the outcome is not designed to be an authoritative and definitive pronouncement as to the mutual rights and obligations of defined parties in respect of a defined issue. A court asked to play an advisory

---

[32] The term "process rules" is used as a convenient shorthand, but without any special technical meaning. It is important only to note that the "process rules" in view are *not* Rules of Procedure in the technical sense normally given to that expression. The "process rules" lie at a deeper level; they define what we think of as judicial process itself, as opposed to Rules which govern the conduct of particular judicial processes. They are what the PCIJ described in *Eastern Carelia* as "the essential rules guiding their activity as a Court". PCIJ, Ser. B, No. 5 (1923), 40.

role is therefore faced with a choice. It may decide that the role requires it to bring to bear its collective judicial experience and wisdom, to be sure, but nevertheless not to act as a court; so it may conceive its function as analogous instead to that of a trusted advisor, like a family lawyer or the legal counsel of a government department or international organization.[33]

It may, on the contrary, decide that the advisory role is a judicial one, requiring it still to function as a *court*. This is what the International Court of Justice and its predecessor have consistently maintained in respect of their advisory jurisdiction. If so, however, a court may be faced with modifying the essential process rules in ways that amount to more than mere "adjustments". And were that to become necessary in order to accommodate the advisory function in a particular set of circumstances, it would raise the question whether the essential guarantees the process rules are there to furnish remain valid – in a word, whether to carry out, as a court, the particular request for an advisory opinion remains compatible with the integrity of the judicial function.

Against that background (the argument might have continued) it can be seen – *via* a detailed comparison with the requests previously made to and answered by the Court – that the present case presents features all of its own. These include the following:

- less so than in any other previous case, there is nothing remotely equivalent to *parties*: the requesting organization has no rights or functions of its own at play, nor are the Member States divided in legal terms into a *pro* and a *contra* camp, but rather maintain a varied spectrum of views; as a result (in the absence of any compulsion on States to appear and argue their legal viewpoints, and failing any *amicus curiae* institution authorised to present in a representative capacity the full range of legal views) the Court can not be confident that all of the necessary legal elements would be fully, effectively and equally argued before it;
- although all issues presented for the advisory treatment must needs be to some extent "abstract questions", this question is abstract in a deeper sense, in that it covers a wide range of potential fact situations, and in circumstances in which the answer due to the question is very likely to depend upon the facts; in the absence (again) of any compulsion over the production of evidence – indeed

---

[33] Other comparisons are the (auto-limited) legislative role the most senior English judges may perform in the House of Lords, or the advisory function of the French *Conseil Constitutionnel* and similar bodies.

of any available *process* for reliable fact-finding in an area at once highly controversial and highly confidential – the Court might have every reason to fear that its members might find themselves with no option but to fall back on unverified assumptions or personal opinions;[34]

- that in the circumstances, it becomes highly probable that the "question" (or more accurately "proposition") presented to the Court for its opinion will not prove capable of the form of answer which the Court has always chosen for its response to advisory opinion requests.[35]

That reasoning might well be thought to have justified the fear (in anticipation) or the conclusion (in retrospect) that the Court would not be (was not) in a position to *answer* the question put to it – in the form in which the Court had always believed an "answer" ought to take – or at least not in a position to "answer the question" (in normal *form*) by applying the essential process rules so as to preserve the integrity of the judicial function. This is a somewhat complex proposition, but the two sides of it are clearly linked: from the one angle, the question becomes not properly a "legal question" because it is not capable of answer in the form that has hitherto been prescribed; from the opposite angle, the question should not be answered because, if it is to be "answered" in the way the law normally prescribes, it can only be done by a process which deviates from the essential rules defining the judicial process. The one objection is a fundamental one, deriving from absence of "competence"; the other objection is a mediate one, resulting from the court's overriding "discretion" to decline to answer, if to answer would

---

[34] The opinion does, of course, contain an express denial by the Court of engaging in a hypothetical or speculative exercise, involving "writ[ing] 'scenarios'", evaluating complex technical information etc. (ICJ Reports 1996 (I), 237, para. 15 *in fine*), but this is a mere affirmation that does not carry great conviction in the light of the separate and dissenting opinions taken as a whole; moreover, in paragraphs 35-36, the Court declares *ex cathedra* certain unique characteristics or consequences of nuclear weapons, which it says are "imperative" for it to take into account "in order correctly to apply [the law] to the present case", but without indicating their probative origin, whether they are agreed by all States who appeared, etc.; they are simply "material before the Court" (*ibid.*, 243-244, paras. 35 and 36).

[35] It goes without saying that the formal *determination* of whether or not this was so would not have to be made by the Court at the outset of the process, but would follow in the normal course as part of its substantive decision on the request.

involve deviating from legal propriety. The interesting finding is that at this point the two established grounds, legal and discretionary, for declining to answer seem to merge into one broader ground.[36] President Anzilotti is reported as saying, as long ago as 1935: "It is ... difficult to see how the Court's independence of the political organs of the League of Nations could be safeguarded, if it were in the power of the Assembly or the Council to oblige the Court to answer any question which they might see fit to submit to it".[37]

One remains therefore with a question of a fundamental kind: is there in truth a double barrier consisting of a *competence* hurdle (the "legal question") and then a *discretionary* hurdle (the "compelling reason of judicial propriety")? Or are these in fact no more than separate facets of a broader discretion inherent in the Court – broader thus than has often been assumed? No precise answer can be given. It might be said that Article 65 of the Statute, by specifying that the Court's facultative power applies only to "legal questions", makes the nature of the question into a preliminary issue which goes strictly to competence. On the other hand, the Statute gives no definition of what constitutes a "legal question". This has been left to the appreciation of the Court itself, and the Court has given every appearance of approaching the matter not as an ordinary question of treaty interpretation but as a question touching on the inherent nature of the judicial function, that is, as a matter uniquely within the Court's (discretionary?) appreciation.[38] There are however one or two pointers. The leading commentator on the Court's law and practice[39] says: "... the Court's advisory competence is a unitary one. Division of the material into sections dealing with jurisdiction and discretion as though they were discrete compartments of the advisory practice can be misleading".[40]

---

[36] Cf. Rosenne, see note 2, 1000: "... the difference between the Court's declining to give an advisory opinion on the ground that the question put is not a legal question, and its declining to give the opinion in exercise of its judicial discretion, is largely one of formulation and of judicial technique rather than one of substance...".

[37] PCIJ, Ser. A/B, No. 65, 61 (cited with approval by Judge Winiarski in ICJ Reports 1956, 105).

[38] Cf. ICJ Reports 1975, 18.

[39] Even while adopting the jurisdiction/competence distinction, Rosenne, see note 2.

[40] He continues: "In this respect, the general practice of the Court is first to establish that it has jurisdiction to give a requested advisory opinion. ... The Court has never, however, included the two aspects as separate paragraphs of the operative clause of an advisory opinion". *Ibid.*, 988.

A further pointer may perhaps lie in asking whether the term "any legal question"/"legal questions" carries the identical meaning in Article 96 of the Charter as it does in Article 65 of the Statute. The one defines the *power* of the requesting organ to make the request, the other the power of the Court to respond. But Article 96 does no more than define the *requesting organ's* power. It clearly cannot be the case that a decision by the requesting organ, however properly taken, that the case is one within its power to request an advisory opinion, binds the Court even as to the existence *vel non* of a "legal question". Rosenne, once again, is categorical on the point:

> "The decision of the requesting organ to avail itself of Article 96 of the Charter and to adopt a request for an advisory opinion carries an implication that it regards the question as a legal question. The requesting organ makes a political determination that to that formulated question it desires a legal answer, produced by application of the judicial techniques of the Court's advisory jurisdiction. However, that determination of the characteristic of the question is not binding on the Court. Following Article 65 of the Statute, the Court may only give an advisory opinion on 'any legal question', and therefore it has to satisfy itself that it has a legal question before it and can give a legal answer to that question. 'Legal answer' means an answer according to the applicable rules of international law, reached through the Court's deliberative techniques".[41]

He draws attention also to the way in which the Charter provision and the provision in the Statute were differentiated by the Court in the *Expenses* case,[42] and continues:

> "By putting the matter in this way, without any mention of Article 96, the Court has made it plain that the concept of 'legal question' applicable for a political organ is not necessarily the same as the concept of 'legal question' applicable for the Court".[43]

The point is surely unchallengeable as a matter of principle.[44] Of particular interest in Rosenne's analysis is however his shrewd

---

[41] *Ibid.*, 999.
[42] ICJ Reports 1962, 155.
[43] Rosenne, see note 2, 1000.
[44] An issue of a similar kind was however raised (including by one member of the Bench) in the consideration of the World Health Assembly's request in the *Nuclear Weapons* proceedings; the discussion can be found at 82-84, paras. 29 and 30.

suggestion that the issue (sc. as to whether the Court has before it a "legal question" or not) is not determined simply by the terms of the question but is also influenced by the shape of the prospective *answer*. In other words, however much a question may be written in legal categories and solicit the application of legal rules, it is still not a "legal question" (for the purposes of the advisory jurisdiction) unless it is susceptible of a legal answer – in the form which the Court has always regarded such an answer should take. With respect, Rosenne's analysis seems yet again to be exactly to the point.[45]

There is moreover room at this juncture for speculation as to whether (notwithstanding its formal vote to comply with the request for the advisory opinion) the Court did in fact "answer" the General Assembly's questions in the sense given above. Doubt on this score is an amalgam of the form in which the "answer" was cast and the manner in which its paragraph E was adopted.

As to the construction of the *dispositif*, in previous instances the practice has been to set out or refer to the question or questions posed, and opposite each to give a condensed answer, often of the "yes or no" variety, digesting the essence of the Court's response. This has become the absolutely standard form in which the Court delivers its advisory opinions. The only deviation from it is to be found in the 1971 *Namibia* opinion,[46] where the answer consists of three separate propositions developed by the Court itself. But there the *Question* as posed to the Court was in the form "What are the consequences of ... ?". It was not therefore susceptible of answer except along the lines of the answer given. Here, however, in clear contrast, the Question was cast in almost compellingly "yes or no" terms: "Is the threat or use of nuclear weapons in any circumstance permitted under international law?". Yet the Court "answered" it,[47] in sharp divergence from its standard practice, by an elaborated series of six[48] propositions of its own manufacture which, far from building up to an approximation to a "yes or no" answer, culminate instead in the famously Delphic evasion of paragraph E.[49] Not merely that, but the Court instructs its addressee

---

[45] See also the tart comment by Judge Dillard: "The notion that a legal question is simply one that invites an answer 'based on law' appears to be question-begging and it derives no added authority by virtue of being frequently repeated". ICJ Reports 1975, 117.
[46] ICJ Reports 1971, 16.
[47] Opinion, ICJ Reports 1996 (I), 265-266, para. 105.
[48] Or seven, if both limbs of paragraph 105 E are counted.
[49] With which in turn (as has been pointed out) every member of the Court expresses him- or herself dissatisfied.

expressly, before proceeding to the *dispositif* proper, "that its reply to the question put to it by the General Assembly *rests on* [emphasis added] the totality of the legal grounds set forth by the Court above",[50] adding for good measure "each of which is to be read in the light of the others".[51]

The oddity of this result becomes more troubling still if one asks the question whether paragraph E represents an "opinion" of the Court at all. The question arises, of course, out of the fact that the paragraph in question was one on which the votes in the Court (as recorded in the text) were exactly evenly divided, and the paragraph was "adopted" on the casting vote of the then President. We have no indication at all in the text of the opinion[52] on what statutory authority this was done, or what its effect was thought to be.

It is well known that the Presidential casting vote is the creature of Article 55 of the Statute, which is understood from the context to be a rule applying to the contentious procedure. It is equally well known that, for the advisory competence, Article 68 provides that the Court shall be "guided" by the provisions applicable in contentious cases, and then "to the extent to which [the Court] recognizes them to be applicable".[53] Are we then to understand that the Court has "recognized", but only in some implicit form, the Article 55 rule to be "applicable" to the advisory procedure? If so, was it well-founded in doing so? And did it follow from that that "guided" meant "bound"? The lack of transparency is puzzling. But it is also bothersome, since there is no obvious reason why the rule creating a casting vote for contentious proceedings should be "applicable" at all to the advisory procedure.

To see why this is so, one must go back to the presumed reason for having the rule. At the fundamental level, it could be said that the purpose of a contentious procedure is to *decide* cases, which inevitably means decide them either in favour of the Applicant or, if not, then in favour of the Respondent; and from that starting point it makes eminent sense that the Statute should determine precisely in which way an equality of votes should be resolved.[54] At the procedural level, it

---

[50] There follows a precise reference back to the 81 paragraphs between nos. 20 and 103.
[51] Opinion, ICJ Reports 1996 (I), 265, para. 104.
[52] Or in the declarations or the separate or dissenting opinions.
[53] For the origin of Article 68, see D. Pratap, *The Advisory Jurisdiction of the International Court of Justice*, 1972, 27-29.
[54] The Article 55 rule, let it be noted, does not in itself determine in which *direction* the President should cast his deciding vote; it does not

could be said that the rule (being a *statutory* provision) is part of what the contending parties assent to in giving consent to the Court's jurisdiction in respect of their case. But how does any of that reasoning apply to the advisory procedure? Since the Court's advice is advice only (and not a binding decision) what can the rationale be for applying a rule designed to ensure that there will always be a binding *decision*? Still more so, what possible juridical value can there be in giving greater weight to the voice of one particular judge? It is surely hard to resist the conclusion that an advisory proposition on which the judges divided equally is a proposition on which the Court as such was unable to pronounce a view. It neither accepted the proposition nor rejected it; it simply could not decide. It is impossible to see how such a proposition acquires extra persuasiveness as a statement of *law* in virtue of the fact that one of the judges voted for it twice. The issue – granted its crucial, controversial and highly visible position – is doubly puzzling in the light of the fact that Article 68 is not a general power in the Court to "decide" what provisions from the contentious procedure to apply, but allows the Court to put into operation those which it *"recognises"* to be applicable. This use of terminology,[55] problematic though it might be in certain circumstances, surely connotes the Court acknowledging the application of a rule that in some sense already notionally applies in logic. Whereas in fact, as the preceding analysis suggests, the inherent logic on this issue runs in quite the opposite direction.

The purpose of this discussion is not to be unduly critical – with the divine gift of hindsight – of how the Court was driven to extricate itself from the severe dilemma in which the General Assembly had placed it.[56] It is rather to point out that there is something puzzling and quite seriously disturbing in the procedural handling of this advisory opinion. However understandable its intentions, should not the ultimate

---

automatically follow, for example from the precedents in national legal systems, that a casting vote must necessarily be cast in the same direction as the President's personal vote, since other considerations may enter into the situation. Cf. Sh. Rosenne, "The President of the International Court of Justice", in Lowe and Fitzmaurice (eds.), *Fifty Years of the International Court of Justice*, 1996, 410-411.

[55] Identical in the French language text, "[la cour] s'inspirera en outre des dispositions ... qui s'appliquent en matière contentieuse dans la mesure où elle les reconnaîtra applicables".

[56] The author's most severe criticism is reserved for the almost frivolous lack of regard he observed within the General Assembly majority for the General Assembly's own interest in preserving the integrity of the judicial function.

outcome have led the Court to wonder whether there was something smacking of "judicial impropriety" in the process, and therefore to think again whether its discretionary power to refuse to give an opinion should be brought into play? In a way, the worst aspect of the situation is the lack of transparency. Perhaps the Court thought about some of these matters; perhaps it did not.[57] But there is no trace in the decision of whether it did, or what it thought. And that in itself must be damaging to the judicial propriety and integrity which the Court has always (and quite rightly) regarded it as its first duty to preserve.

What else might the Court have done? Again, this is a question easier to pose than to answer. Even the committed enthusiasts for the whole process concede the difficult position in which the Court was placed.[58] That said, must it in fact be accepted that there is an absolute dichotomy between answering and declining to answer a request for an advisory opinion? It would certainly have been more honest for the Court to have *replied* to the General Assembly along the following lines: "We are not in a position to deliver the advisory opinion requested, both because the extreme breadth and generality of the question posed does not admit of an answer in accordance with accepted judicial process, and because of the very sharp division of legal view on the wide range of situations and circumstances which the question potentially encompasses. The Court is not, in other words, able to say in the abstract whether 'the threat or use of nuclear weapons [is] in any circumstance permitted under international law'. The Court does however conceive it to be in accordance with its judicial vocation, and with its position as the principal judicial organ of the United

---

[57] It must surely have thought about the "casting vote", given its dramatic rarity in past practice.

[58] Cf. the "Introduction by the Editors", see note 3, 1 et seq.: "If the Court failed to answer the requests altogether, it risked incurring the wrath of those countries ... which had put them before the Court. If it answered the requests ... and did so by ruling any use of nuclear weapons to be unlawful, it risked an erosion of credibility with the four declared nuclear-weapon states that participated in the proceedings. If it answered and did so by ruling some uses lawful, it threatened its credibility with the public, or at least those sections of it which had been most active in bringing the requests to the Court, and it risked opening the door to states wishing to acquire those weapons which were not yet parties to the NPT. In the circumstances the Court followed a path which allowed all except those adhering to the most extreme of views to claim a degree of satisfaction with the outcome...".

One wonders however whether that last sentence embodies an acceptable approximation of the judicial vocation.

Nations, to offer the following propositions of law, on which it is broadly agreed, these being propositions which bear on the question posed by the General Assembly and which the Court believes may prove of assistance to the General Assembly in carrying out its future functions". In a curious way, that may have been what the Court did conceive itself to be doing, to judge by paragraph 19 of the Opinion, in which it says:

> "... An entirely different question is whether the Court, under the constraints placed upon it as a judicial organ, will be able to give a complete answer to the question asked of it. *However, that is a different matter from a refusal to answer at all*".[59] (emphasis added)

This imputed intention is however, sad to say, quite incompatible with the *dispositif* in paragraph 105, and particularly with the adoption of subparagraph E by casting vote.

It will be evident that the writer is not among those who believe that the salvation of the world lies in a wholesale extension of the advisory jurisdiction or its exercise. In a recent essay, one of the current judges has written:

> "The relatively infrequent contemporary recourse to Advisory Opinions is not necessarily a matter to be regretted. Its use most naturally occurs at certain moments in the institutional life of an international organization (in the early years), and from time to time when unforeseeable problems arise for the working of UN organs or agencies. Little purpose is served by artificially inventing new business for the Advisory Opinion.... Recourse to the Court must work with the seams of political and institutional realities and not against the grain".[60]

The experience over the two requests for advisory opinions in respect of the legality of nuclear weapons must surely have reinforced that balanced judgement. In particular, it has shown that there are strict limits on the use of the advisory jurisdiction in a further area not mentioned by Judge Higgins: the solution of controversial questions of general international law. Indeed, not merely are there limits, but serious dangers for the preservation of the integrity of the judicial

---

[59] ICJ Reports 1996 (I), 238.
[60] R. Higgins, "A Comment on the current health of Advisory Opinions", in Lowe and Fitzmaurice, see note 54, 567 et seq.

process, of which the Court itself will be acutely conscious. If the Court's rigorous barring of the way to the World Health Organization constitutes a recognition of this fact, and of the fact that Specialized Agencies in general might prove a temptingly soft route for campaigners trying to "use" the judicial process, then it is only to be welcomed.[61] Judge Oda's root and branch suspicion of the entire advisory jurisdiction as a *détournement* of the essential function of the International Court of Justice is likely to remain a distinctly minority view. But his insistence on this occasion that the Court should take a formal position on whether to respond to the General Assembly's request, by helping to focus attention on vital issues that might otherwise have been buried, is deserving both of our respect and of our gratitude.

---

[61] Judge Oda would see it as part of what he called "judicial economy".

Why Do States Take Disputes to the International Court?

Ian Brownlie

I. Introduction

As Judge Oda would readily appreciate, this essay in his honour has been prepared not from an academic standpoint, but against a background of work as counsel and advocate representing States before the International Court and other international tribunals. Because the author works as a practising member of the English Bar, a professional discipline is imposed and, as counsel, there is a certain independence from the client. Indeed, if counsel were to be a mere outwork or spokesman of the government concerned, he would be of little use. At the same time, the professional relationship imposes duties of confidentiality and, in consequence, the views expressed here are not to be attributed to any government. In any case, governments are complex creatures and motives may vary from one official to another.

As a further preliminary, though it is stating the obvious, it may be observed that an application may have several motives, and this multiplicity of purpose is usually entirely respectable.

In no particular order, the issue will now be addressed: why do States take disputes to the International Court?

II. To Assist or Induce Negotiations

Foreign ministry officials regard negotiation as the preferred mode of peaceful settlement of disputes,[1] and this for sound practical reasons. However, there are circumstances in which the Court can play a complementary role in the process of negotiation. In the *North Sea Continental Shelf* cases,[2] the Court was given a major role in the ongoing

---

[1] Sir Gerald Fitzmaurice, Institut de Droit International, *Livre du Centenaire 1873-1973*, 1973, 276-280.
[2] ICJ Reports 1969, 4 et seq.

negotiations between three riparian States concerning valuable areas of continental shelf. By means of two cases introduced by special agreements, the proceedings being joined, the Court was given the significant task of clarifying the nature of the applicable law concerning continental shelf delimitation.

In the *Nauru* case,[3] the Republic of Nauru had a dispute with Australia which dated from the time of its independence in 1968. While the Government of Nauru had sought negotiations on several occasions, the Government of Australia had always refused. It was also the case that Australia held the view that the matters in issue had been resolved by negotiations at the time of the independence of Nauru. The application of Nauru was drafted in such a way as to postpone the merits phase, if and when the Court found that it had jurisdiction, and that the case was in other respects admissible. On the principal claim presented by Nauru, the Court rejected all the preliminary objections of Australia and, in the wake of this decisive outcome, Australia for the first time embarked upon negotiations with Nauru.

III. To Obtain the Definitive and Authoritative Resolution of a Dispute

The heading may well astonish by its obviousness. In principle, all contentious cases are concerned with the settlement of legal disputes as, indeed, are some requests for advisory opinions. Moreover, the existence of a dispute (in the legal sense) is a condition of the exercise of jurisdiction by the Court. A proportion of cases before the Court proceed on the basis of a special agreement relating to a dispute, the existence and dimensions of which are agreed between the parties. Cases of this type include the *Minquiers and Ecrehos* case,[4] the *Gulf of Maine* case,[5] and the *Botswana/Namibia* case.[6] In the background of such cases brought by special agreement there will be a variety of political and practical considerations.

In the *Minquiers and Ecrehos* case, the two parties, France and the United Kingdom, evidently needed a clear and definitive resolution of a troublesome dispute which had given rise to localised breaches of the peace. Although both States were parties to the optional clause system, the vehicle of a special agreement was preferred in order that the Court could be instructed not to determine the existence of either a *terra*

---

[3] ICJ Reports 1992, 240 et seq.
[4] ICJ Reports 1953, 47 et seq.
[5] ICJ Reports 1984, 246 et seq.
[6] ICJ Reports 1999.

nullius or a condominium (by conduct), given that such outcomes would not lead to a firm resolution of the dispute.

In the *Gulf of Maine* case, the application was called for in the wake of the failure of strenuous attempts to settle the fisheries question by negotiation. The establishment of a single maritime boundary was to be the substitute for a sophisticated agreement on management of the fish stocks. Again, in the *Botswana/Namibia* case, the special agreement resulted from the failure of previous attempts at settlement of the boundary dispute.

In certain political circumstances, the applicant State is seeking to obtain an authoritative determination when the potential respondent State is either reluctant to accept the jurisdiction of the Court, or reluctant to negotiate, or both. Thus, in the *Aegean Sea Continental Shelf* case,[7] Greece sought, without success, to satisfy the Court that it had jurisdiction in a situation in which Turkey was reluctant to negotiate a special agreement.

IV. To Obtain an Authoritative Characterisation of Illegal Conduct

In a number of cases, it is probable that the applicant State is not merely seeking the resolution of a dispute, but is intent on alerting the international community to the illegal conduct of another State. This course of action is particularly attractive in situations in which, because of the veto, the Security Council is unable to make any finding by virtue of Chapter VII of the Charter. The use of the Court in this context is particularly effective in alerting the relevant national political systems to the illegal conduct, particularly when the conduct is clearly illegal in relation to both national and international legal criteria. This was true of the mining of Nicaraguan harbours in *Nicaragua* v. *United States*.[8] It was also true of the proceedings initiated by the United Kingdom against Albania in the *Corfu Channel* case.[9]

Such an approach to the Court may take place either in retrospect or *in media res*. The latter case is illustrated by the proceedings brought by Yugoslavia against certain members of NATO in respect of the bombing of Yugoslavia.[10]

---

[7] ICJ Reports 1978, 3 et seq.
[8] See the Request for Interim Measures, ICJ Reports 1984, 169. An interesting account of the crisis from the point of view of a member of Congress is provided by D. Moynihan, *On the Law of Nations*, 1990, 120-148.
[9] ICJ Reports 1949, 4 et seq.
[10] Request for Interim Measures of Protection, Order of the Court dated 2 June 1999.

V. To Obtain a Declaration

Applications to the Court intended to obtain findings of illegality may be drafted (or interpreted) in two senses. They may request a declaratory judgment, and no more, or they may request a declaratory judgment by way of an injunction, that is to say, with a forward reach, restraining the continuation of the action. This was one of the objectives of the case brought by Nicaragua against the United States. However, in the *Nuclear Test* cases,[11] the majority of the Court interpreted the applications of Australia and New Zealand in such a way as to give effect to French promises not to carry out atmospheric nuclear tests, and declared the cases to be moot, but without giving a declaration relating to future conduct. In this general context, the applicant State may meet difficulties if the Court, or a segment thereof, is convinced that in the area of law in question the rules of customary law are in a process of evolution. This factor may explain the complex and qualified response of the Court to the cases brought by the United Kingdom and Germany against Iceland[12] in respect of fishery limits.

VI. To Remove a Threat to Public Order

Particularly in relation to boundary and other territorial disputes, a subsidiary motive for approaching the Court is the removal of the chronic risk of incidents and threats to the peace. This has constituted an element both in cases brought by unilateral application and those brought by special agreement. Thus in the *Botswana/Namibia* case,[13] while the relevant area was free of tension, and the two defence forces had good relations, the difficulty over the location of the boundary had in the past given rise to exchanges of fire. In this context, it may be noted that, where such threats are prominent, and expeditious adjudication is called for, as it was in the *Red Sea Islands* arbitration[14] between Eritrea and Yemen, the slow movement of the Court's calendar, caused by certain financial stringencies, will militate in favour of resort to arbitration with accelerated exchanges of pleadings.

---

[11] ICJ Reports 1974, 253 *(Australia v. France)*; ibid., 457 *(New Zealand v. France)*.
[12] ICJ Reports, 1974, 3 *(United Kingdom v. Iceland)*; ibid., 175 *(Federal Republic of Germany v. Iceland)*.
[13] ICJ Reports 1999.
[14] *ILR* 114, 1 et seq. (Phase One).

VII. To Establish a Stable Régime for the Exploitation of Natural Resources

This is an element in cases involving maritime delimitation, including the *Tunisia/Libya* case,[15] the *Gulf of Maine* case (Canada and the United States),[16] and the *Libya/Malta* case.[17] In a number of cases, if not all, active exploitation will be delayed until the issue of attribution has been settled on an authoritative legal basis.

VIII. International Political Advantage

On occasion, it is probably the case that the existence of certain political advantages, within the compass of the internal thinking of the particular government, provides an incentive to place a question before the Court. While negotiation always remains the primary vehicle for peaceful settlement of disputes, the essence of negotiation is compromise and this may present serious internal risks, especially in the context of territorial disputes. The involvement of an authoritative and independent institution provides a means of avoiding the awful realities of compromise proposals concerning what have become major issues of principle, and the risks such circumstances involve for officials proposing compromise.

IX. Certain Related Issues

There are certain subsidiary issues which deserve consideration. In the first place, there is the often ventilated issue of the distinction between legal and political disputes. The key criterion is not whether a dispute has a political aspect, because disputes between States inevitably do. It is true, of course, that some disputes have a relatively prominent political aspect. This was true of the *Corfu Channel* case and also of *Nicaragua v. United States*. In the final analysis, as Jennings[18] has pointed out, the criterion from the practical standpoint is whether there is an identifiable legal element which can be the appropriate subject of judicial attention. It is also true to say that, in the sphere of peaceful settlement, international law is deficient in having no criteria to offer for the settlement of political disputes *as such*.

---

[15] ICJ Reports 1982, 18 et seq.
[16] ICJ Reports 1984, 246 et seq.
[17] ICJ Reports 1985, 13 et seq.
[18] R. Jennings, Josephine Onoh Memorial Lecture, University of Hull, 1986; and in *Essays in Honour of Roberto Ago*, Vol. III, 1987, 139-151.

A rather different question concerns the volunteer freelance lawyer, who may persuade a government to embark on proceedings which would otherwise not have taken place. Providing the freelance lawyer has the necessary expertise, it may be that he or she plays a helpful catalytic role in indicating useful remedies and assisting the cause of justice. In the background history of the *Nauru* case, certain Australian academic historians played a very positive role in providing Nauruan leaders with sound advice. At the same time, it may happen that the freelance lawyer has his or her own agenda, which may diverge substantially from that of the applicant State. In any event, even small States are nearly always adept at taking their own view of the interests at stake, and it is unlikely that freelance activity has much effect in practice.

X. Conclusion

While the purposes for which States take cases to the Court are varied, the primary purpose is to achieve the closure of a dispute, or, as in the *North Sea* cases (in which Judge Oda was counsel), to provide the elements contributing to a successful negotiation. In pursuing this mission, the Court has a number of significant advantages in comparison with arbitration. Its services are free and the intrusion of factors involving compromise is much less likely.

Appearing "for" or "on behalf of" a State: The Role of Private Counsel Before International Tribunals

Jean-Pierre Cot

Judge Oda has recently called our attention to the role of private counsel appearing before international jurisdictions. In the *Armed Activities on the Territory of the Congo* case *(Democratic Republic of Congo v. Uganda)*, Request for Indication of Provisional Measures, Order of 1 July 2000, Judge Oda noted that the Democratic Republic of Congo had chosen a Belgian lawyer as its agent to represent it before the Court. In recent years, this is not the only instance of problems arising with private counsel. In the *Banana* case, the panel appointed by the World Trade Organisation refused to allow Saint Lucia to have its case presented by private lawyers. In the *Grand Prince* case, the International Tribunal on the Law of the Sea experienced a similar sort of difficulty. These instances are worth reflecting upon.

Lawyers have always played an important role in international litigation, whether representing private clients before claims commissions or appearing as counsel to States in intergovernmental proceedings. But cases involving private counsel appeared to be the exception rather than the rule. In fact, access to the International Court of Justice was restricted to a small group of highly qualified professors and barristers, the "invisible" Bar of the Court.

The development of new international, but also transnational, fora of litigation has changed the picture. The number of cases has dramatically increased. Intergovernmental jurisdictions, such as Hamburg, have opened access to private parties and nongovernmental organisations. Certain jurisdictions adjudicate disputes between private parties and governments, such as the European Court of Human Rights or the International Centre for the Settlement of Investment Disputes. Others have a mixed nature, sometimes adjudicating purely intergovernmental or interinstitutional disputes, sometimes dealing with private parties as well, like the European Court of Justice.

As a result, the distinction between intergovernmental proceedings and private litigation tends to be blurred, both procedurally and in substance. I will nevertheless limit my remarks to intergovernmental jurisdictions. I will address the problems of the lawyer appearing as counsel, acting as an agent, or applying to an international tribunal "on behalf of" a State, respectively.

I. The Lawyer Appearing as Counsel

International tribunals quite often possess a heavy and impressive machinery. Fifteen judges sit in The Hague, 21 in Hamburg, and even more in Strasbourg, chosen from amongst the most highly qualified jurists in their own countries, and representing the major legal traditions of the world. Such imposing benches call for top-quality Bars. Counsel appearing before such tribunals are expected to deliver perfect or near-perfect performances. Their briefs are to be crisp, yet thorough. Their oral presentation must be exciting, but duly restrained. They are expected to be in full command of the latest developments with regard to a large amount of complex and sophisticated information.

However, there are no formal requirements for appearing as counsel for a State before an international tribunal. States are free to choose whoever they wish to represent them. It is their sovereign right and privilege to do so.

No problems arose as long as international litigation was rare and mainly confined to the World Court. The "invisible" Bar was mainly composed of a handful of eminent professors of international law, teaching at the most prestigious universities, and an occasional and highly specialised barrister, backed by a few major law firms. Counsel were very careful to stay within the guidelines, as they were putting their professional reputation at stake each time they signed a brief or delivered a speech. In such a small world, reputations were quickly established or undone. As a result, and quite predictably, the quality of the product was superb.

Globalisation has done away with all that, for better or for worse. The multiplication of international fora and the corresponding expansion in the number of lawyers called to appear before international tribunals have exacted their toll on the quality of presentations.

The development of fast track procedures has added pressure to the system. Demands for interim measures of protection tend to become systematic in the World Court. They have precedence over ordinary cases and are dealt with in a matter of weeks. One cannot expect the parties and their counsel to prepare their legal statements with the same care as if they were allowed months for each memorial. The same is true of applications for permission to intervene, which have multiplied over the last few years.

They also have priority and must be dealt with within tight schedules. Three of the four cases examined by the International Tribunal for the Law of the Sea have concerned cases of prompt release of vessels. Article 292 of the Convention on the Law of the Sea calls for a swift procedure with scarcely a month between application and judgment.

Inevitably, international tribunals have become familiar with the sort of slippage national tribunals live with daily. An eminent *avocat* of the Paris Bar, more familiar with criminal cases than with international law, forgot to establish *prima facie* competence while asking for provisional measures. An *abogado* from Vigo, well versed in maritime law, became confused as to the functioning of the French *Cour de Cassation* in a Hamburg proceeding. It all is pretty natural if you think about it. But international tribunals are used to another set of standards.

It would be unfair to blame private practice for all the mischief in the courts. The International Court of Justice recently witnessed a very serious case of falsified documents that a party had to withdraw. Yet counsel were seasoned, experienced, respectable ... and very embarrassed.

The variety of new problems calls for some sort of an answer. There are bound to be problems of ethics, misrepresentation of facts and conflicts of interest. Yet there is no minimum level of qualification, no code of ethics or of deontology, no Bar association that could admonish its members, much less deal out sanctions for misconduct.

Such problems are bound to crop up. An interesting one appeared in the *Elettronica Sicula S.p.A. (ELSI)* case. The Chamber of the Court noted:

> "Mr. Giuseppe Bisconti also addressed the Court on behalf of the United States; since he had occasion to refer to matters of fact within his knowledge as a lawyer acting for Raytheon Company, the President of the Chamber acceded to a request by the Agent of Italy that Mr. Bisconti be treated *pro tanto* as a witness. Mr. Bisconti, who informed the Chamber that both Raytheon Company and Mr. Bisconti himself waived any relevant privilege, was cross-examined by Mr. Highet".[1]

Judge Oda has pointed to the danger of law firms indulging in unnecessary litigation. The WTO panel in the *Banana* case addressed that same objection when refusing St. Lucia the right to be represented by an American law firm during the proceedings. The panel drew attention to the danger that small, underdeveloped nations could become instruments of large law firms in these sorts of cases.

---

[1] ICJ Reports 1986, 19. See also Sh. Rosenne, *The Law and Practice of the International Court, 1920-1996*, Vol. III, 1997, 1181-1182, para. III.282.

The complaining parties reargued the case before the Appellate Body. The report of 9 September 1997 sums up their argument:

> "The Complaining Parties submitted that from the earliest years of the General Agreement on Tariffs and Trade (the "GATT"), presentations by governments in dispute settlement proceedings have been made exclusively by government lawyers or government trade experts. With respect to developing countries, the Complaining Parties argued that, unlike the practice before other international tribunals, under the provisions of Article 27.2 of the DSU, developing countries are entitled to legal assistance from the WTO Secretariat. The Complaining Parties also cited certain policy reasons in support of their position. WTO dispute settlement, they argued, is dispute settlement among governments, and it is for this reason that the DSU safeguards the privacy of the parties during recourse to dispute settlement procedures. Furthermore, the Complaining Parties asserted that if private lawyers were allowed to participate in panel meetings and Appellate Body oral hearings, a number of questions concerning lawyers' ethics, conflicts of interest, representation of multiple governments and confidentiality would need to be resolved".[2]

The Appellate Body declined to pronounce on the decision of the panel, but accepted representation of St. Lucia by the law firm for just the opposite reason. It considered that a small developing country did not necessarily have the administrative and legal talents necessary to put its case efficiently before the Appellate Body and that it should be free to choose its counsel. It noted

> "... that there are no provisions in the *Marrakesh Agreement Establishing the World Trade Organization* (the '*WTO Agreement*'), in the DSU or in the *Working Procedures* that specify who can represent a government in making its representations in an oral hearing of the Appellate Body. With respect to GATT practice, we can find no previous panel report which speaks specifically to this issue in the context of panel meetings with the parties. We also note that representation by counsel of a government's own choice may well be a matter of particular significance – especially for developing-country Members – to enable them to participate fully in dispute settlement proceedings. Moreover, given the Appellate Body's

---

[2] Appellate Body Report, x, para. 9.

mandate to review only issues of law or legal interpretation in panel reports, it is particularly important that governments be represented by qualified counsel in Appellate Body proceedings".[3]

One answer to the problem of qualification would be to formalise the "invisible" Bar. The institution of a category of QCs or *avocats aux conseils* would be a guarantee of a certain qualification. There is no objection in principle to the idea.[4] Most national judicial systems have some sort of filter to select lawyers appearing before the high courts. But one can hardly imagine what the process of selection is like at an international level. In 1922, the Permanent Court was seised of a proposal that advocates should be persons admitted to practice before the highest court of their country, university professors of international law, or members of the great academies of international law. But the Court decided that no provision "limiting the right of pleading before the Court should be introduced into the Rules".[5]

National Bar associations certainly should give the matter some thought. They are responsible for the corporate image of the profession. They must at least offer adequate training in international law, maritime law, European law or whatever topics are appropriate, and ensure some control over the quality of the counsel who appear before international tribunals.

Ultimately, the decision rests with the sovereign States. As the WTO correctly put it, sovereign States are free to organise their representation and the defence of their interests as they wish. The ultimate sanction is a case lost through negligent choice of counsel.

---

[3] *Ibid.*, xi, para. 12.

[4] B. Oxman, "Observations on Vessel Release under the United Nations Convention on the Law of the Sea", *Int'l J. Marine & Coastal L.* 11 (1996), 214-215, has suggested that certain law firms in Hamburg could seize the opportunity of the newly installed International Tribunal to specialise in such litigation and examined the possibility for foreign law firms to open an office in Hamburg under German law. Such a project should be seen in long term. The short term docket of ITLOS is not so full as to justify the ambition. The precedent of The Hague is not inspiring. I know of no major Dutch law firm specialised in Hague Court litigation. The bulk of the work is done by university professors, some eminent Dutch colleagues among them. Whatever law firms are associated with the proceedings are mainly British and American. Local firms provide some useful technical facilities, but not much more.

[5] M. Hudson, *The Permanent Court of International Justice 1920-1942. A Treatise*, 1943, 537.

II. The Lawyer Acting as an Agent

Private lawyers can appear before international courts or tribunals as agents of States, even when they do not have the nationality of the State concerned. This is precisely the practice Judge Oda was regretting in the *Armed Activities on the Territory of the Congo* case.

There is no objection in principle to the lawyer appearing as an agent for a sovereign State, as long as he is not representing a private interest at the same time. I understand Judge Oda's irritation and the WTO panel's remarks on the issue. But I do not think the reasoning of the WTO panel in the *Banana* case is well founded. I do not see why a sovereign State cannot appoint as an agent a person of its choice, national or foreigner, to represent it. It may have good reasons to think a competent New York law firm will more adequately further its interests than a less experienced domestic official. States have in the past – and certainly in the present – privatised more essential sovereign prerogatives than representation before an international tribunal. I see no ground for formal exclusion of private persons acting as agents, unless the statute of the tribunal stipulates otherwise.[6]

An interesting provision to that effect is to be found in the United Nations Convention on the Law of the Sea. The Convention, addressing the issue of seabed disputes, provides the following:

"Article 190
Participation and appearance of sponsoring States Parties
in the proceedings

If a natural or juridical person is a party to a dispute referred to in article 187, the sponsoring State shall be given notice thereof and shall have the right to participate in the proceedings by submitting written or oral statements.

If an action is brought against a State Party by a natural or juridical person sponsored by another State Party in a dispute referred to in article 187, subparagraph (c), the respondent State may request the State sponsoring that person to appear in the proceedings on behalf of that person. Failing such appearance, the respondent State may arrange to be represented by a juridical person of its nationality".

---

[6] The doctrine seems to accept the appointment of private lawyers of foreign nationality as agents, at least before the World Court. Hudson, see note 5, 529; G. Guyomar, *Commentaire du Règlement de la Cour Internationale de Justice*, 1983, 263; Rosenne, see note 1, Vol. III, 1167-1168.

This provision in fact is concerned with a different issue, that of the very specific procedure within an arbitration process. The State may be represented by a juridical person, an issue that is quite different from the issue of private lawyers which is addressed here. But it is an interesting example of the delicate balance struck by the Convention on the question of representation between the interests of the private party, the sponsoring State and the defending State.

Agents play an essential role in international litigation.[7] They represent the State. They also commit the State by their actions, as the Permanent Court of International Justice noted in 1926:

> "The representative before the Court of the respondent Party, in addition to the declarations above mentioned regarding the intention of his Government not to expropriate certain parts of the estates in respect of which notice had been given, has made other similar declarations which will be dealt with later; the Court can be in no doubt as to the binding character of all these declarations".[8]

Article 42 of the Statute of the Court clearly distinguishes the functions of the agent and of the counsel or advocates:

> "1. The parties shall be represented by agents. They may have the assistance of counsel or advocates before the Court".

Professor Verzijl strongly voiced his opinion regarding the specific role that agents are to perform. Presiding at the Franco-Mexican Claims Commission in 1924, he declared:

> "Je tiens à déclarer que les agents doivent être considérés non comme de simples avocats, ayant liberté d'énoncer toutes sortes d'opinions personnelles, quand bien même ces opinions seraient en contradiction avec l'opinion de leur gouvernement, mais comme les représentants officiels de ce dernier. S'il en était autrement, l'on ne saurait jamais si l'agent expose des opinions personnelles, ou bien le point de vue officiel de son gouvernement, et les débats revêtiraient un caractère hybride et indéfinissable. Et seule cette interprétation du rôle des agents est conforme au droit international commun en matière arbitrale,

---

[7] Guyomar, see note 6, 256-266; Rosenne, see note 1, Vol. III, 1165-1182, para. III.277 - III.282; R. Monaco, "Représentation et défense des parties devant les instances internationales", E. Diez (ed.), *Festschrift fur Rudolf Bindschedler*, 1980, 373 et seq.

[8] Judgment, PCIJ, Ser. A, No. 7 (1926), 13.

tel qu'il se trouve formule a l'article 62 de la 1ère Convention de La Haye du 18 octobre 1907 pour le règlement pacifique des conflits internationaux".[9]

The main problem of private counsel acting as an agent for a State is that of his reliability as a representative. Appointed by the government, the agent is acting under instructions. Advocates and counsel act under his authority. He is the only person authorised to speak in the name of his government, file submissions, etc. The agent must be in a position to take all the necessary procedural decisions. Moreover, he is expected to state precisely the position of the government he is representing.

The agent is normally a high-ranking official, of ministerial or sub-ministerial level. He is expected to be in close contact with his national authorities and to act as a link between the international jurisdiction and the State concerned. A private lawyer may well do the job. But he normally doesn't have the same sort of permanent association with government as an official. The difficulty is compounded if the lawyer has only distant relations with the government he is supposed to represent, or only with certain branches of the government.

The International Tribunal for the Law of the Sea ran into some difficulties in the *Grand Prince* case. The agent appearing for the flag State obviously had some difficulty in communicating with the Belizean Government and producing the exact position of the Government as to the existence of a national link of Belize to the ship at the time of the application. Judge Anderson noted in his separate opinion:

> "The Agent appointed by Belize is not well placed, as a non-Belizean lawyer in private practice in Spain, to explain to the Tribunal the seeming inconsistencies in the statements of different government departments and agencies in Belize...".

The authors of the joint dissenting opinion disagreed and considered that "there was coordination on the question between the various Belize authorities involved".[10] The Tribunal was not convinced. It considered that Belize had not demonstrated that it was the flag State at the date of application and found it had no jurisdiction to hear the application.

One of the difficulties the Tribunal had to face was that of an eventual conflict of interest between the lawyer acting as an agent for Belize and as counsel for the ship-owner.[11] The lawyer obviously had contacts with the department of justice and the agency for ship

---

[9] *RIAA* V, 355-356.
[10] Dissenting opinion of Judges Caminos, Marotta Rangel, Yankov, Yamamoto, Akl, Vukas, Marsit, Eiriksson and Jesus.
[11] See ITLOS/PV/01/3, 10.

registration, but none with the ministry of foreign affairs. The difficulty is that the Statute of the Tribunal institutionalises such a conflict of interest by allowing lawyers to appear "on behalf of" the applicant State.

III. The Lawyer Applying to the Tribunal on Behalf of a State

The situation is more complicated if the lawyer represents not only the interests of the State, but also the interests of a private party. In such a case, the appointment of a lawyer to such a double function blurs the difference between intergovernmental and transnational disputes. Georges Scelle did theorise on what he called the *"dédoublement fonctionnel"*. But in practice, such a situation is not always as clear cut as one might wish.

The United Nations Convention on the Law of the Sea created such a possibility in connection with the procedure for prompt release (Article 292). A private party may be authorised by the national authorities to apply for the release of the detained vessel "on behalf of" the flag State.

The *ratio legis* of the procedure is evident. Undue detention of a vessel entails important economic and financial costs for the shipowner. It is essential that the vessel be authorised to proceed to sea and resume its activities with no delay, on the condition that the coastal State has an adequate guarantee in the form of a reasonable bond or other financial security. The Convention provides for a swift procedure before an international jurisdiction, the International Tribunal for the Law of the Sea, unless the parties otherwise agree. The decision is obligatory for both parties.

To ensure the rapidity and the efficiency of the procedure, the Convention provides the "on behalf" clause. The flag State may well be negligent of or simply uninterested in the proceedings. in addition, there is the inevitable red tape involved in such proceedings. Appointing a representative before a distant tribunal is a complicated and costly venture, not likely to be expedited in a matter of days. The development of flags of convenience, with their minimal "genuine link", leaves shipowners with little official protection. It is the price they pay for the convenience of such a flag.

The "on behalf" clause was drafted to overcome this difficulty and to give shipowners a fast-track procedure, cutting through red tape and gaining a form of direct access to the Tribunal while preserving the intergovernmental nature of the dispute and the litigation.

Authorisation to act on behalf of a State will normally be given after the arrest of the vessel. But it may be given in advance. The Tribunal will check its authenticity, as it amounts to quite an important delegation of sovereignty to a private person.

Article 292 clearly qualifies prompt-release disputes as intergovernmental disputes. The parties to the procedure are the States involved, the coastal State on the one hand, the flag State on the other. This is not a case of private litigation with the flag State appearing alongside the private party or sponsoring its action.

This situation is quite the opposite of that provided for in Part XI of the Convention on Seabed Disputes.[12] Article 187 of the Convention states that the Seabed Disputes Chamber may have jurisdiction over disputes between a State Party to the Convention or the Authority and a prospective contractor or a natural or juridical person sponsored by another State Party. In such a case, the "sponsoring State" has the right to participate in the proceedings by submitting written or oral statements (Article 190). But the case clearly opposes the private party to the State or the Authority. The sponsoring State is a third party in the procedure. Yet, the respondent State may request the State sponsoring the private person to appear in the procedure "on behalf of that person". Here, the "on behalf" clause plays the other way around. The transnational procedure is blurred by a remnant of intergovernmental litigation.

Article 292(2) is explicit as to the nature of the proceedings: "The application for release may only be made by or on behalf of the flag State of the vessel". The flag State is clearly in command of the procedure. The rules of the Tribunal provide, in the procedure for prompt release, for notification by the State Party of the name, address and office "of any person who is authorised to make an application on its behalf". They add that the State Party may also and at any time notify the Tribunal of "any clarification, modification or withdrawal of such notification" (Article 110(2)(d)).

The Tribunal proceeds to a verification of the credentials of the appointed agent. In the *Grand Prince* case, the Belizean Government did not consider an appointment by the Solicitor General of Belize as sufficient and asked for confirmation of the appointment by the Attorney General.

It has been suggested that the coastal State may also be represented by a private person acting on its behalf.[13] The Convention does not provide for anything of the sort, and the interests are of quite a different nature. They relate to the exercise of sovereign rights over the exclusive

---

[12] But *contra* Oxman, see note 4, 212-213, who insists on the similarity of the two procedures and the eminent role of the private parties in both cases. M. Nordquist, Sh. Rosenne and L. B. Sohn, *United Nations Convention on the Law of the Sea, 1982. A Commentary*, 1985, Vol. V, 60-65, do not confirm the legislative history as described by Oxman.

[13] V. Bernard, "Arrest of Seagoing Vessels and the LOS Convention; Does the New International Tribunal for the LOS Offer New Prospects", *Int'l Maritime L.* 5 (1998), 114.

economic zone and are not of a private nature. This does not bar a coastal State from choosing a private lawyer of a third nationality as an agent. But such an agent, whatever his private status, would be acting exclusively for the coastal State and certainly not at the same time for a private party, contrary to the "on behalf" situation.

Does the "on behalf" clause provide direct access to the Tribunal for private parties, and shipowners in particular, or is it just a technical convenience within the traditional law of diplomatic protection? The question was at the centre of the *Grand Prince* case.[14]

The Tribunal read Article 292 of the Convention as calling for proof of the nationality of the vessel at the date of the application for prompt release. It considered that the nationality of the ship had lapsed in the case after the arrest of the vessel and before the application was filed.

Judge Treves was explicit in his individual opinion:

"Article 292 of the Convention establishes, for limited purposes, a form of diplomatic protection. In submitting an application for release the flag State espouses a private claim of persons linked to it by the nationality of the ship. This becomes even clearer considering that the application may also be submitted directly by the interested private persons 'on behalf' of the flag State.

In cases of diplomatic protection, the nationality requirement must be satisfied at least at the time of the submission of the claim and at the time of the commission of the wrongful act".

The minority challenged that view. In their opinion, the dissenting judges considered that the *Grand Prince* was registered in Belize and that the Tribunal had jurisdiction to entertain the application. They added:

"15. A more general point of interpretation of the Convention, going beyond the scope of the present case, is raised by the fact that the decision of the Tribunal proceeded from the assumption that the applicant in a proceeding under article 292 of the Convention must be the flag State at the time the application is submitted. In the circumstance of prompt release proceedings, the flag State at the time of detention, and at the time when an allegation is made of non-compliance with the provisions of the Convention on prompt release, would ordinarily still be the flag State at the time of making an application under article 292. The reasoning of the Tribunal to justify this as a legal requirement

---

[14] ITLOS, *Grand Prince*, Judgment, 20 April 2001.

is, however, not convincing. Regrettably, the deliberations in the present case have not allowed a full treatment of the consequences of this approach in various other circumstances which could be contemplated".[15]

The problem is not only one of procedure. This amounts to challenging the applicability of the principles of diplomatic protection to the prompt release proceedings. From this perspective, the private party, once duly authorised to act "on behalf of" the flag State, has an independent right to pursue the procedure, whatever the nationality of the vessel in the following phases of litigation.[16]

The traditional rule of nationality of claims reflects the essence of diplomatic protection. It is not always an easy one to maintain in a globalised world.[17] It was inevitable that the problem should be even more acute when diplomatic protection concerns vessels. The national link is indeed a tenuous one. Article 91 of the Convention on the Law of the Sea may well affirm that there must exist a genuine link between the State and the ship. The nature of the link is not comparable to that of the link between physical persons and a human community. McDougal and Burke did have some reason to criticise the false analogy of the *Nottebohm* doctrine applied to the nationality of vessels.[18] In the case of a flag of convenience, the flag State tends to have a minimal, sometimes only financial, relationship with the vessel. The temptation is obvious for the shipowners to do away with what is left of State control and to try to apply directly to an international tribunal to seek remedy. This is the legal translation of globalisation.

The majority in the *Grand Prince* case did not accept this view. As Judge Wolfrum put it in his declaration:

"3. I would like to emphasize that the statement made by the Director and Senior Deputy Registrar that '... despite the expiration of the Patent of navigation and ship station license, the vessel *is still considered* as registered in Belize until final decision of this Administration pending the result of the court proceeding in which the vessel is engaged at the present time...' (italics supplied) cannot be regarded as registration in the

---

[15] Dissenting opinion of Judges Caminos, Marotta Rangel, Yankov, Yamamoto, Akl, Vukas, Marsit, Eiriksson and Jesus.
[16] Also Oxman, see note 4, 212-213.
[17] See *Oppenheim's International Law*, Vol. I(1), R. Jennings and A. Watts (eds.), 9th ed., 1992, paras. 150-152 on the issue, in particular on problems arising out of the *Nottebohm* doctrine and the protection of corporations in the *Barcelona Traction* case.
[18] M. S. MacDougal, *The Public Order of the Oceans*, 1987, 1008-1035.

The Role of Private Counsel Before International Tribunals

meaning of article 91 of the Convention or as an equivalent thereto. It does not conform to objective and purpose of registration... Article 94 of the Convention establishes certain duties of the flag State. Apart from that article 91, paragraph 1, third sentence of the Convention states that there must be a genuine link between the flag and the State. This means the registration cannot be reduced to a mere fiction and to serve just one purpose, namely to open the possibility to initiate proceedings under article 292 of the Convention on the Law of the Sea. This would render registration devoid of substance – an empty shell".[19]

The point Judge Oda made in his declaration thus goes far beyond procedure. It is directly related to the changing structure of international law, and the role of the State in this new arrangement. The challenge is a legal one, but also a political one, for the years ahead.

---

[19] Separate opinion of Judge Rudiger Wolfrum. The Tribunal did not take an explicit position on this issue in its judgment. The Tribunal was divided 12 to 9 in the final vote.

Deprivation of Property of Foreigners under International Law

Yoram Dinstein

I. Limitations on the Right of States to Deprive the Property of Foreigners

Customary international law requires respect for the "vested rights" of a foreigner (i.e., a non-national) and restricts the entitlement of the local State to deprive him of those rights.[1] Generally speaking, the deprivation of property of foreigners is permissible only upon the fulfilment of strict conditions. When reversed, the same conditions can be presented as the fulcrum of state responsibility for economic injury to foreigners. That is the approach of the 1986 (Third) edition of the American Law Institute's *Restatement*:

> "#712. *State Responsibility for Economic Injury to Nationals of Other States*
> A state is responsible under international law for injury resulting from:
> (1) a taking by the state of the property of a national of another state that
> (a) is not for a public purpose, or
> (b) is discriminatory, or
> (c) is not accompanied by provision for just compensation".[2]

The international legal protection of the property of foreigners was subjected – especially during the 1970s – to a cross-fire of sharp

---

[1] See the Judgment of the Permanent Court of International Justice in the *German Interests in Polish Upper Silesia and the Factory at Chorzów* (1926), World Court Reports, Vol. 1, 475 et seq., 510, 523-524.
[2] American Law Institute, *Restatement of the Law: The Foreign Relations Law of the United States*, Vol. 2, 3rd ed., 1986, 196.

criticism emanating from the USSR and countries from the Third World. It is noteworthy that earlier, in 1962, the General Assembly adopted Resolution 1803 (XVII), dealing with "Permanent Sovereignty over Natural Resources", which includes the following paragraph:

> "4. Nationalization, expropriation or requisitioning shall be based on grounds or reasons of public utility, security or the national interest which are recognized as overriding purely individual or private interests, both domestic and foreign. In such cases the owner shall be paid appropriate compensation in accordance with the rules in force in the State taking such measures in the exercise of its sovereignty and in accordance with international law. In any case where the question of compensation gives rise to a controversy, the national jurisdiction of the State taking such measures shall be exhausted. However, upon agreement by sovereign States and other parties concerned, settlement of the dispute should be made through arbitration or international adjudication".[3]

Resolution 1803 – incorporating explicit demands that the deprivation of property shall be for the public good and that appropriate compensation shall be paid in accordance with international law, as well as a reference to possible settlement of any dispute by international fora – was passed by an overwhelming margin of 87 in favour and only two against, with 12 abstentions.[4] However, over the years, the General Assembly's position underwent a radical change.[5] In 1974, in Resolution 3281 (XXIX) – adopted with a view to establishing a "Charter of Economic Rights and Duties of States" – an altogether different message was inscribed in Article 2(2)(c):

> "2. Each State has the right:
>
> ...
>
> (c) To nationalize, expropriate or transfer ownership of foreign property, in which case appropriate compensation should be

---

[3] United Nations General Assembly, Resolution 1803 (XVII): Permanent Sovereignty over Natural Resources (1962), *United Nations Resolutions, Series I: Resolutions Adopted by the General Assembly*, Vol. 9, 107 et seq., 107 (D. J. Djonovich ed.).

[4] The result of the vote is detailed *ibid.*, 46.

[5] As an illustration of the process, see United Nations General Assembly, Resolution 3171 (XXVIII): Permanent Sovereignty over Natural Resources (1973), *United Nations Resolutions, Series I: Resolutions Adopted by the General Assembly*, Vol. 14, 422 et seq., 422, para. 3 (D. J. Djonovich ed.). The vote was 108 for, one against, with 16 abstentions. *Ibid.*, 211.

paid by the State adopting such measures, taking into account its relevant laws and regulations and all circumstances that the State considers pertinent. In any case where the question of compensation gives rise to a controversy, it shall be settled under the domestic law of the nationalizing State and by its tribunals, unless it is freely and mutually agreed by all States concerned that other peaceful means be sought on the basis of the sovereign equality of States and in accordance with the principle of free choice of means".[6]

This time the text mentions neither international law nor international fora, and in fact leaves the local country unfettered discretion as to whether or not to pay appropriate compensation upon the deprivation of the property of foreigners. The Resolution was passed by 108 votes in favour, six against, and ten abstentions.[7]

Needless to say, the General Assembly is not a legislative organ. However, as pointed out by the International Court of Justice in 1996, in the *Nuclear Weapons* advisory opinion, "General Assembly resolutions, even if they are not binding, may sometimes have normative value. They can, in certain circumstances, provide evidence important for establishing the existence of a rule or the emergence of an *opinio juris*".[8] Which, if any, of the two contradictory resolutions – 1803 or 3281 – is truly declaratory of customary international law? Notwithstanding some confusion in the 1970s, there is no doubt today that Resolution 1803 – rather than Resolution 3281 – provides an accurate expression of the *lex lata*.[9] It is worthwhile to quote the Award rendered in 1977 in the *Texaco* case, where the Arbitrator R.-J. Dupuy examined these Resolutions carefully and in detail, arriving at the following conclusion:

> "Resolution 1803 (XVII) seems to this Tribunal to reflect the state of customary law existing in this field. ...
>
> While Resolution 1803 (XVII) appears to a large extent as the expression of a real general will, this is not at all the case with

---

[6] United Nations General Assembly, Resolution 3281 (XXIX): Charter of Economic Rights and Duties of States (1974), *United Nations Resolutions, Series I: Resolutions Adopted by the General Assembly*, Vol. 15, 300 et seq., 302 (D. J. Djonovich ed.).
[7] The result of the vote is detailed *ibid.*, 149.
[8] Advisory Opinion on *Legality of the Threat or Use of Nuclear Weapons*, ICJ Reports 1996, 226 et seq., 254-255.
[9] See C. F. Amerasinghe, "Issues of Compensation for the Taking of Alien Property in the Light of Recent Cases and Practice", *ICLQ* 41 (1992), 22 et seq., 32-36.

respect to ... other Resolutions. ... In particular, as regards the Charter of Economic Rights and Duties of States, several factors contribute to denying legal value to those provisions of the document which are of interest in the instant case".[10]

Subsequent cases reconfirm the view that it is Resolution 1803 – and not Resolution 3281 – which correctly mirrors customary international law.[11]

II. The Normal Hazards Faced by a Foreign Investor

It is indisputable that when a foreigner enters a local country in order to carry on economic activities, he cannot complain should his property be injured as a result of either occurrences over which the local State has no control (such as strikes or riots, let alone a revolution) or ordinary governmental activities (like taxation and devaluation). The protection of the property of foreigners by international law does not amount to an airtight insurance against all eventualities. In the words of L. B. Sohn and R. R. Baxter:

> "It is recognized, in the first place, that the incidence of taxation may deprive an alien of some of his assets and that a failure to pay taxes may lead to the seizure of the alien's property. A revaluation of the currency of a particular State, if not adopted in a manner which discriminates against aliens individually or collectively, may deprive an alien of a portion of his economic wealth, but the measure is not on that account wrongful. As examples of the taking or deprivation of property of an alien arising out of the action of the competent authorities of the State in the maintenance of public order, health, and morality may be mentioned the confiscation of goods which have been smuggled into a country and the seizure of such articles as narcotics, liquor, obscene materials, firearms, and gambling devices which are unlawfully in a person's possession".[12]

---

[10] Award on the Merits in Dispute between *Texaco Overseas Petroleum Company/California Asiatic Oil Company and The Government of the Libyan Arab Republic* (International Arbitral Tribunal, 1977), *ILM* 17 (1978), 1 et seq., 30.

[11] See particularly *Sedco, Inc.* v. *National Iranian Oil Company and the Islamic Republic of Iran* (Interlocutory Award), *Iran-United States Claims Tribunal Reports* 10 (1986), 180 et seq., 186.

[12] L. B. Sohn and R. R. Baxter, "Responsibility of States for Injuries to the Economic Interests of Aliens", *AJIL* 55 (1961), 545 et seq., 561.

Whereas the extent of ordinary governmental activities cannot be precisely defined, the emphasis must be placed on good faith and reasonableness. Even taxation, the epitome of ordinary governmental activities, is subject to an important exception spelt out by S. R. Swanson:

> "The underlying question is when does state action amount to mere regulation by the host state, in which case no true 'taking' has occurred, and when does it constitute a compensable taking. A typical example of the difficulties posed by this question can be found in the area of taxation. The state's right to impose reasonable measures of taxation is unquestioned. But what about a tax of one hundred percent or more, which deprives the investor completely of his investment? Other regulations may also prove to be so onerous that the investment loses all value. Such state regulation is often referred to as 'creeping expropriation' because a taking occurs without direct action to physically deprive the investor of his property".[13]

In a nutshell, a foreigner entering a local country for commercial purposes must take into account normal hazards and have no more than plausible expectations. He cannot rely on the local law standing still. Taxes may be increased within reason. Exchange rates may be modified. Zoning rules concerning construction may alter. However, when taxation becomes excessive, when other rules become too onerous, the issue of deprivation of property arises.

III. The Meaning of Deprivation of Property

Multiple synonyms are in vogue to connote the deprivation by a local State of the property of foreigners, from the neutral "taking" to the ideologically pregnant "nationalization", including the more or less stigmatized "confiscation", "expropriation" and "requisitioning". As the Iran-United States Claims Tribunal articulated it in the *Tippetts* case:

> "The Tribunal prefers the term 'deprivation' to the term 'taking', although they are largely synonymous, because the latter may be understood to imply that the Government has acquired something of value, which is not required.

---

[13] S. R. Swanson, "Iran-U.S. Claims Tribunal: A Policy Analysis of the Expropriation Cases", *Case Western Reserve Journal Int'l L.* 18 (1986), 307 et seq., 311.

A deprivation or taking of property may occur under international law through interference by a state in the use of that property or with the enjoyment of its benefits, even where legal title to the property is not affected".[14]

The deprivation of the property of foreigners must result from action attributed to the local State, taken by any organ of the three branches of the government: legislative, executive or judicial.[15] What ultimately counts is the *de facto* outcome of the action resorted to by the organ of the State and not its ostensible *de jure* designation. This is sharply attested in the jurisprudence of the Iran-United States Claims Tribunal:

> "Most cases brought before the Tribunal ... involve the grey area in which no formal taking is announced by the host Government, but the claimant argues that its property has been seized *de facto*; ... . In deciding such cases, the Tribunal consistently has ruled that interference by the Government with the alien's enjoyment of the incidents of ownership – typically the use or control of the property, or the income and other economic benefits derived therefrom – constitutes a compensatable taking".[16]

It should be underscored that, in addressing the issue of *de facto* deprivation of the property of foreigners, the Tribunal based its rulings on norms generally recognized in international law and not on any *lex specialis* governing the relations between the parties.[17]

A similar position was taken by the European Court of Human Rights in the 1993 *Papamichalopoulos* case, dealing with certain Greek legislation:

> "the applicants were never formally expropriated: Act 109/1967 did not transfer ownership of the land in question to the Navy Fund.

---

[14] *Tippetts, Abbett, McCarthy, Stratton* v. *TAMS-AFFA Consulting Engineers of Iran, Government of the Islamic Republic of Iran et al.*, Iran-United States Claims Tribunal Reports 6 (1984), 219 et seq., 225.

[15] See M. Pellonpää, "Compensable Claims Before the Tribunal: Expropriation Claims", in R. B. Lillich and D. B. Magraw (eds.), *The Iran-United States Claims Tribunal: Its Contribution to the Law of State Responsibility*, 1998, 185 et seq., 221.

[16] C. N. Brower, "The Iran-United States Claims Tribunal", *RdC* 224 (1990), 123 et seq., 285.

[17] See M. Mohebi, *The International Law Character of the Iran-United States Claims Tribunal*, 1999, 308-309.

Since the Convention is intended to safeguard rights that are 'practical and effective', it has to be ascertained whether the situation complained of amounted nevertheless to a *de facto* expropriation, as argued by the applicants [See, among other authorities, Sporrong and Lönnroth v. Sweden, para. 63].

...

The Court considers that the loss of all ability to dispose of the land in issue, taken together with the failure of the attempts made so far to remedy the situation complained of, entailed sufficiently serious consequences for the applicants *de facto* to have been expropriated in a manner incompatible with their right to the peaceful enjoyment of their possessions".[18]

In its earlier Judgment of 1982, in the *Sporrong and Lönnroth* case, the Court held (in the paragraph referred to in the later decision):

"In the absence of a formal expropriation, that is to say a transfer of ownership, the Court considers that it must look behind the appearances and investigate the realities of the situation complained of. Since the Convention is intended to guarantee rights that are 'practical and effective', it has to be ascertained whether that situation amounted to a *de facto* expropriation, as was argued by the applicants".[19]

It follows that the concept of deprivation of property is comprehensive enough to encompass any serious, direct or indirect, interference in the property.[20] The Iran-United States Claims Tribunal held, in 1983, in the *Starrett* case:

"it is recognized in international law that measures taken by a State can interfere with property rights to such an extent that these rights are rendered so useless that they must be deemed to have been expropriated, even though the State does not purport to have expropriated them and the legal title to the property formally remains with the original owner".[21]

---

[18] *Papamichalopoulos and Others* v. *Greece, Eur. H. Rts. Reports* 16 (1993), 440 et seq., 459-460.

[19] *Sporrong and Lönnroth* v. *Sweden, Eur. H. Rts. Reports* 5 (1982), 35 et seq., 51.

[20] See R. Higgins, "The Taking of Property by the State: Recent Developments in International Law", *RdC* 176 (1982), 259 et seq., 322-324.

[21] *Starrett Housing Corporation et al.* v. *Government of the Islamic Republic of Iran et al., Iran-United States Claims Tribunal Reports* 4 (1983), 122 et seq., 154.

Moreover, as the Tribunal pronounced in the *Tippetts* case, in 1984:

> "The intent of the government is less important than the effects of the measures on the owner, and the form of the measures of control or interference is less important than the reality of their impact".[22]

Obviously, there is no deprivation of the foreigner's property unless the government's interference with the property rights is intentional – in the sense of being deliberate rather than accidental – but it is not required to establish that "the Government had an intention to expropriate, in the technical meaning of the term".[23]

References are sometimes made to "constructive taking" (which includes excessive taxation legislation).[24] The point is that the deprivation of the foreigner's property does not rest on any formal decree.[25] There is no doubt that, even in the absence of formal measures, any far-reaching intervention by the local State in the foreigner's property will qualify as a deprivation (leading to the duty to compensate).[26] In such instances, there may be a question as to what juncture may be deemed the date of the deprivation, but usually that would be the point in time when it may be determined by an objective observer that "there is no reasonable prospect the owner will be able to resume enjoyment of his interests".[27]

IV. The Scope of the Property Subject to Deprivation

The deprivation of property need not relate to a tangible *res* (movable or immovable): it also covers contractual rights,[28] including debts.[29] In the words of R. Y. Jennings:

---

[22] *Tippetts* case, see note 14, 225-226.

[23] H. Piran, "Indirect Expropriation in the Case Law of the Iran-United States Claims Tribunal", *Finnish YBIL* 6 (1995), 140 et seq., 174-175.

[24] See I. Seidl-Hohenveldern, "Aliens, Property", in R. Bernhardt (ed.), *EPIL*, Vol. 1, 1992, 116 et seq., 117.

[25] See A. B. Avanessian, *Iran-United States Claims Tribunal in Action*, 1993, 37.

[26] See R. Dolzer, "Expropriation and Nationalization", in R. Bernhardt (ed.), *EPIL*, Vol. 2, 1995, 319 et seq., 322.

[27] See J. A. Westberg, *International Transactions and Claims Involving Government Parties: Case Law of the Iran-United States Claims Tribunal*, 1991, 111.

[28] See *Mobil Oil Iran Inc. et al.* v. *Government of the Islamic Republic of Iran et al.*, *Iran-United States Claims Tribunal Reports* 16 (1987), 3 et seq., 25.

[29] "It must be recognized at the outset that an international debt is property (a chose in action) to the lender. Like other property, debts are subject to the

"That the idea of acquired or vested rights is part of public international law is a proposition that has received the sanction of numerous authorities including the Permanent Court of International Justice and is now hardly open to question; further, there is good authority for the view that acquired rights are not confined to the notion of property in its narrowest sense, but also include rights derived from contract or concession".[30]

Differently put by B. M. Clagett:

"Customary international law has long recognized the obligation of a State to pay compensation upon the expropriation of property of foreign nationals, including contract rights and other intangibles as well as physical property".[31]

Or as stated by G. M. White:

"the legal notion of property refers to the totality of rights which one or more natural or legal persons possesses at a given moment in relation to a physical or intangible thing. The term *rights* is used here in the sense of interests recognized by a particular system of municipal law, and which will be upheld and enforced against appropriate defendants by courts applying that law. A debt, which is a chose in action, is such a right. Its origin in contract is material, in the present context, only as an indicator, albeit a crucially important one, that the creditor's first remedy is against the debtor, invoking their contract. If this remedy has been removed by an executive or legislative act of the State in whose courts the remedy would normally be

---

risk of expropriation. It is a contention of this writer that debt repudiation is analogous to expropriation by a state of any other property held by aliens within the state". K. M. Siegel, "The International Law of Compensation for Expropriation and International Debt: A Dangerous Uncertainty", *Hastings Int'l & Comp. Law R.* 8 (1984-1985), 223 et seq., 223.

[30] R. Y. Jennings, "State Contracts in International Law", *BYIL* 37 (1961), 156 et seq., 173.

[31] B. M. Clagett, "The Expropriation Issue Before the Iran-United States Claims Tribunal: Is 'Just Compensation' Required by International Law or Not?", *Law and Policy in International Business* 16 (1984), 813 et seq., 814.

pursued, the impact, both economic and legal, would seem to be indistinguishable from expropriation of a property right".[32]

Of course, contracts concluded between a foreigner and a local national – just like contracts concluded between local nationals *inter se* – are subject to the general laws of contracts, which may change in time either in favour of the foreigner or against him. But if the local State abuses its powers in order to deprive the foreigner of his contractual rights through legislation, this constitutes a violation of international law.[33] To quote F. A. Mann:

> "where a legislative measure cannot fairly be described as being adopted for the benefit of the general public ... the interference with contractual rights constitutes a taking".[34]

In the *Pressos* case of 1995, the European Court held that even the denial of a right to claim damages in torts is tantamount to deprivation of property:

> "The Court notes that the 1988 Act exempted the State and other organisers of pilot services from their liability for negligent acts for which they could have been answerable. It resulted in an interference with the exercise of rights deriving from claims for damages which could have been asserted in domestic law up to that point and, accordingly, with the right that everyone, including each of the applicants, has to the peaceful enjoyment of his or her possessions.
>
> In so far as that Act concerns the accidents that occurred before 17 September 1988, the only ones in issue in the present proceedings, that interference amounted to a deprivation of property within the meaning of the second sentence of the first paragraph of Article 1".[35]

---

[32] G. M. White, "Wealth Deprivation: Creditor and Contract Claims", in R. B. Lillich (ed.), *International Law of State Responsibility for Injuries to Aliens*, 1983, 149 et seq., 177.

[33] See M. Akehurst, *A Modern Introduction to International Law*, P. Malanczuk (ed.), 7th ed., 1997, 238-239.

[34] F. A. Mann, "State Contracts and State Responsibility", *AJIL* 54 (1960), 572 et seq., 585.

[35] *Pressos Compania Naviera S.A. and Others* v. *Belgium, Eur. H. Rts. Reports* 21 (1995), 301 et seq., 335.

V. Non-Discrimination

As the American Law Institute's *Restatement* sets forth, a State incurs responsibility for injury to a foreigner caused by the deprivation of his property when the measure is "discriminatory". The Iran-United States Claims Tribunal pronounced, in the *Amoco* case, that "[d]iscrimination is widely held as prohibited by customary international law in the field of expropriation".[36] In the 1977 Arbitral Award in the *LIAMCO* case, the requirement of non-discrimination was accentuated:

> "It is clear and undisputed that non-discrimination is a requisite for the validity of a lawful nationalization. This is a rule well established in international legal theory and practice ... . Therefore, a purely discriminatory nationalization is illegal and wrongful".[37]

Discrimination, in the language of the Court of Justice of the European Communities, means "treating either similar situations differently or different situations identically".[38] The European Court of Human Rights determined that, for the claim of discrimination to be substantiated, "it must be established that other persons in an analogous or relevantly similar situation enjoy preferential treatment, and that there is no reasonable or objective justification for this distinction".[39] More succinctly, discrimination "implies unreasonable distinction".[40] What is unreasonable has to be determined contextually by evaluating the factual circumstances of each particular case.[41]

Not every act of discrimination against foreigners is necessarily prohibited.[42] Discrimination against foreigners may exist as compared

---

[36] *Amoco International Finance Corporation* v. *Government of the Islamic Republic of Iran et al.*, Iran-United States Claims Tribunal Reports 15 (1987), 189 et seq., 231.

[37] Award in Dispute between *Libyan American Oil Company (LIAMCO) and The Government of the Libyan Arab Republic* Relating to Petroleum Concessions (Arbitral Tribunal, 1977), *ILM* 20 (1981), 1 et seq., 58-59.

[38] *Re Electric Refrigerators: The Italian Government* v. *E.E.C. Commission*, Court of Justice of the European Communities (1963), *Common Market Law Reports* 2 (1963), 289 et seq., 312.

[39] *The National & Provincial Building Society et al.* v. *United Kingdom*, Eur. H. Rts. Reports 25 (1977), 127 et seq., 174.

[40] *Restatement*, see note 2, 200.

[41] See A. F. M. Maniruzzaman, "Expropriation of Alien Property and the Principle of Non-Discrimination in International Law of Foreign Investment: An Overview", *J. Transnat'l L. & Policy* 8 (1998-1999), 57 et seq., 67.

[42] See J. S. Stanford, "International Law and Foreign Investment", in R. S. MacDonald et al. (eds.), *The International Law and Policy of Human Welfare*, 1978, 471 et seq., 474.

either to other foreigners (usually of a different nationality) or to any class of local nationals residing in the State whose action is impugned.[43] The existence of illicit discrimination is determined not in theory but in practice. As phrased by the Permanent Court of International Justice:

> "There must be equality in fact as well as ostensible legal equality in the sense of the absence of discrimination in the words of the law".[44]

The non-discrimination requirement in the sphere of deprivation of the property of foreigners may appear to be paradoxical, inasmuch as the very protection of foreigners by international law is founded on discrimination in their favour as compared to local nationals. Yet, the preferred treatment of foreigners in this regard is fully in tune with modern thinking. In the *James* case of 1986, the European Court of Human Rights held:

> "Especially as regards a taking of property effected in the context of a social reform, there may well be good grounds for drawing a distinction between nationals and non-nationals as far as compensation is concerned. To begin with, non-nationals are more vulnerable to domestic legislation: unlike nationals, they will generally have played no part in the election or designation of its authors nor have been consulted on its adoption. Secondly, although a taking of property must always be effected in the public interest, different considerations may apply to nationals and non-nationals and there may well be legitimate reason for requiring nationals to bear a greater burden in the public interest than non-nationals".[45]

Within the European system, even nationals are normally entitled to compensation for the deprivation of property, but whereas there is an "automatic requirement" of compensation to foreigners (under the general principles of international law), a total lack of compensation to nationals may be considered justifiable under "exceptional circumstances" (which apparently have not yet arisen in practice).[46]

---

[43] See G. C. Christie, "What Constitutes a Taking of Property under International Law?", *BYIL* 38 (1962), 307 et seq., 332.

[44] *German Settlers in Poland*, Advisory Opinion (1923), World Court Reports, Vol. 1, 207 et seq., 218.

[45] *James v. United Kingdom*, Eur. H. Rts. Reports 8 (1986), 123 et seq., 150.

[46] D. Anderson, "Compensation for Interference with Property", *Eur. H. Rts. L. Rev.* 6 (1999) 548-550.

VI. The Public Purpose Condition

The condition that the deprivation of the property of foreigners must be carried out "for a public purpose" is unequivocally articulated in the American Law Institute's *Restatement*, as quoted above.

The list of public purposes recognized as legitimate in the international jurisprudence is long and flexible.[47] Given the wide discretion bestowed on the local State in this respect, there are those who question the practical value of the condition.[48] Yet, as suggested in the latest edition of *Oppenheim*:

> "Perhaps the most clearly established condition is that expropriation must not be arbitrary".[49]

Arbitrariness has been defined by a Chamber of the International Court of Justice in the *ELSI* case of 1989:

> "Arbitrariness is not so much something opposed to a rule of law, as something opposed to the rule of law. ... It is a wilful disregard of due process of law, an act which shocks, or at least surprises, a sense of juridical propriety".[50]

But this formulation of the norm appears to be too stringent: suffice it to state that a legitimate deprivation of property must be effected for public purposes.

The clearest case in which deprivation of property is effected for a public purpose is that of nationalization. As the Iran-United States Claims Tribunal held, in the *Amoco* case of 1987, "nationalization is generally defined as the transfer of an economic activity from private ownership to the public sector".[51] The Tribunal went on to pronounce that the right to nationalize the property of foreigners for a public purpose "is today unanimously accepted".[52]

---

[47] See B. Cheng, *General Principles of Law as Applied by International Courts and Tribunals*, 1987, 39.

[48] See M. Pellonpää and M. Fitzmaurice, "Taking of Property in the Practice of the Iran-United States Claims Tribunal", *Netherlands YBIL* 19 (1988), 53 et seq., 62-63.

[49] *Oppenheim's International Law*, Vol. 1(2), R. Jennings and A. Watts (eds.), 9th ed., 1992, 919-920.

[50] *Case Concerning Elettronica Sicula S.p.A. (ELSI)*, ICJ Reports 1989, 15 et seq., 76.

[51] *Amoco* case, see note 36, 222.

[52] *Ibid.*

Much light has been shed on the subject of deprivation of property for public purposes by the European Court of Human Rights, in construing Article 1 of the First Protocol (of 1952) to the European Convention for the Protection of Human Rights and Fundamental Freedoms. Article 1 prescribes:

> "Every natural or legal person is entitled to the peaceful enjoyment of his possessions. No one shall be deprived of his possessions except in the public interest and subject to the conditions provided for by law and by the general principles of international law".[53]

Although the text employs the phrase "public interest", there is no material difference between it and public purposes.

The interpretation and application of the expression "public interest" were the crux of the issue before the Court in the aforementioned *James* case of 1986.[54] The proceedings revolved around an English law, which enabled long-term lease-holders of residences in Belgravia (London) to buy them from the landlord at a fixed price determined by set criteria (below the market price). There was common ground that this amounted to a deprivation of property rights, but the question was whether the deprivation was carried out in the advancement of a public interest. The Court rejected the contention that the transfer of property from one individual to another can never be in conformity with the public interest:

> "The Court agrees with the applicants that a deprivation of property effected for no reason other than to confer a private benefit on a private party cannot be 'in the public interest'. Nonetheless, the compulsory transfer of property from one individual to another may, depending upon the circumstances, constitute a legitimate means for promoting the public interest".[55]

The Court amplified:

> "a taking of property effected in pursuance of legitimate social, economic or other policies may be 'in the public interest', even if the community at large has no direct use or enjoyment of the property taken. ...

---

[53] [First] Protocol to the [European] Convention for the Protection of Human Rights and Fundamental Freedoms, 1952, *European Conventions and Agreements*, Vol. 1, 39 et seq., 40.
[54] *James* case, see note 45, 123.
[55] *Ibid.*, 140.

Furthermore, the notion of 'public interest' is necessarily extensive. In particular, ... the decision to enact laws expropriating property will commonly involve consideration of political, economic and social issues on which opinions within a democratic society may reasonably differ widely. The Court, finding it natural that the margin of appreciation available to the legislature in implementing social and economic policies should be a wide one, will respect the legislature's judgment as to what is 'in the public interest' unless that judgment be manifestly without reasonable foundation. In other words, although the Court cannot substitute its own assessment for that of the national authorities, it is bound to review the contested measures under Article 1 of Protocol No. 1 and, in so doing, to make an inquiry into the facts with reference to which the national authorities acted".[56]

After a careful examination of the factual setting, the Court arrived at the conclusion that this reform, which had been on the agenda for almost a century and was the topic of an exhaustive debate by the public and in Parliament, was indeed intended to further the public interest.[57]

This general rule applies also to the deprivation of rights derived from contracts. The European Court explicitly ruled that

"in remedial social legislation ... it must be open to the legislature to take measures affecting the further execution of previously concluded contracts in order to attain the aim of the policy adopted".[58]

VII. The Duty of Compensation

As noted, the American Law Institute's *Restatement* establishes the responsibility of the local State for injuring foreigners through the deprivation of their property when any one of three alternative conditions is met. No doubt, of the three, the most important condition is the last: the absence of "just compensation". The coinage "just compensation" is borrowed from the Fifth Amendment of the Constitution of the United States, although that clause is not really germane to the issue under consideration. Anyhow, the United States

---

[56] *Ibid.*, 142.
[57] *Ibid.*, 143-144.
[58] *Mellacher* v. *Austria*, Eur. H. Rts. Reports 12 (1989), 391 et seq., 410.

Government sharply disagrees with the proposition that customary international law mandates only "just" compensation to foreigners whose property has been deprived: its firm opinion is that there is an obligation to pay them compensation which is "prompt, adequate and effective" in every instance.[59] The framers of the *Restatement* defend their stand on the basis of international practice.[60] Others maintain that the official American position is the one gaining more support in the cases.[61]

An analysis of the case law of the Iran-United States Claims Tribunal shows that its "favourite standard of compensation" – derived from customary international law – is "full compensation", representing the "fair market value of the property taken" considered as a "going concern".[62] "'[F]ull' compensation and not anything less" was held to be the correct standard "with near unanimity".[63] Admittedly, in determining what constitutes "full" compensation, "a variety of different methods of valuation has been employed".[64]

The standard of "full" compensation was applied by the Iran-United States Claims Tribunal irrespective of the question of the lawfulness of the deprivation of the property of foreigners.[65] That is to say, compensation must be paid in full even when the deprivation of the property is deemed legitimate because it is carried out for a worthwhile public purpose. The lawfulness of the act of deprivation does not vitiate the fundamental requirement of compensation.[66] As observed by G. H. Aldrich (after surveying the jurisprudence of the Tribunal dealing with deprivation of property):

---

[59] The official position of the State Department is expressed by D. R. Robinson, "Expropriation in the Restatement (Revised)", *AJIL* 78 (1984), 176-178. The phrase "prompt, adequate and effective" is usually called the "Hull formula", after the American Secretary of State who used it in 1938 in an exchange of notes with Mexico. See G. H. Hackworth, *Digest of International Law*, Vol. 3, 1942, 659.

[60] See O. Schachter, "Compensation for Expropriation", *AJIL* 78 (1984), 121-130.

[61] See, e.g., M. H. Mendelson, "Compensation for Expropriation: The Case Law", *AJIL* 79 (1985), 414-420.

[62] R. Khan, *The Iran-United States Claims Tribunal: Controversies, Cases and Contribution*, 1990, 247.

[63] Westberg, see note 27, 239.

[64] *Ibid.*

[65] Khan, see note 62, 247.

[66] See P. M. Norton, "A Law of the Future or a Law of the Past? Modern Tribunals and the International Law of Expropriation", *AJIL* 85 (1991), 474 et seq., 489-490.

"Liability is not affected by the fact that the State has acted for legitimate economic or social reasons and in accordance with its laws".[67]

A good authority for this general proposition is the Tribunal's decision, in 1986, in the *Phelps Dodge* case:

"The conclusion is unavoidable that, as of 15 November 1980, control of the SICAB factory was taken by the Respondent, thereby depriving Phelps Dodge of virtually all of the value of its property rights in SICAB. That such deprivation has lasted for five years is undisputed, and that it is likely to continue indefinitely seems clear to the Tribunal. The Tribunal fully understands the reasons why the Respondent felt compelled to protect its interests through this transfer of management, and the Tribunal understands the financial, economic and social concerns that inspired the law pursuant to which it acted, but those reasons and concerns cannot relieve the Respondent of the obligation to compensate Phelps Dodge for its loss".[68]

Even in the *Amoco* case, in which (as indicated in Part VI *supra*) it was established that the right to nationalize the property of foreigners for a public purpose is unanimously accepted, the Tribunal hastened to add that nationalization – albeit lawful – is subject to the obligation, under customary international law, of the payment of compensation.[69]

The distinction between a lawful deprivation of property (non-discriminatory and for a recognized public purpose) and an unlawful one (either discriminatory or not for a legitimate public purpose) does have consequences regarding the amount of compensation to be assessed. The Tribunal in the *Amoco* case drew a distinction between *damnum emergens* (which must be compensated in every instance) and *lucrum ceassans* (lost profit, which is germane only to unlawful expropriations).[70] The relevance of the application to expropriation cases of the dichotomy of *damnum emergens* and *lucrum ceassans* (derived from breach of contract cases) has been questioned.[71] However, the

---

[67] G. H. Aldrich, *The Jurisprudence of the Iran-United States Claims Tribunal*, 1996, 217.
[68] *Phelps Dodge Corp. and Overseas Private Investment Corp. v. Islamic Republic of Iran* (1986), Iran-United States Claims Tribunal Reports 10 (1986), 121 et seq., 130.
[69] *Amoco* case, see note 36, 223.
[70] *Ibid.*, 248-252.
[71] See W. C. Lieblich, "Determinations by International Tribunals of the Economic Value of Expropriated Enterprises", *J. Int'l Arb.* 7 (1990), 37 et seq., 47-48.

principle of setting apart a lawful deprivation of property from an unlawful one – when compensation is calculated – transcends in its importance the precise manner in which this is accomplished.

D. W. Bowett suggests that the case law may actually be going even further, supporting three standards of compensation: "(i) for an unlawful taking; (ii) for a lawful *ad hoc* taking; and (iii) for a lawful, general act of nationalization".[72] Under this three-pronged approach, there is a difference between a general nationalization and a situation in which an individual property holder is singled out for deprivation because of a public need: that individual should be more generously compensated than those affected by a general nationalization.[73]

The European Court of Human Rights, in the already cited *James* case (see Part VI *supra*), did not confine its analysis to the question whether the deprivation of property was designed to advance the public interest:

> "This, however, does not settle the issue. Not only must a measure depriving a person of his property pursue, on the facts as well as in principle, a legitimate aim 'in the public interest', but there must also be a reasonable relationship of proportionality between the means employed and the aim sought to be realised. This latter requirement was expressed in other terms in the *Sporrong and Lönnroth* judgment by the notion of the 'fair balance' that must be struck between the demands of the general interest of the community and the requirements of the protection of the individual's fundamental rights.[74] The requisite balance will not be found if the person concerned has had to bear 'an individual and excessive burden'[75]".[76]

The Court concluded:

> "the taking of property without payment of an amount reasonably related to its value would normally constitute a disproportionate interference which could not be considered justifiable under Article 1. Article 1 does not, however, guarantee a right to full compensation in all circumstances. Legitimate objectives of 'public interest', such as pursued in measures of economic reform or measures designed to achieve

---

[72] D. W. Bowett, "State Contracts with Aliens: Contemporary Developments on Compensation for Termination or Breach", *BYIL* 59 (1988), 49 et seq., 73.
[73] *Ibid*.
[74] *Sporrong and Lönnroth* case, see note 19, 52.
[75] *Ibid*., 54.
[76] *James* case, see note 45, 144-145.

greater social justice, may call for less than reimbursement of the full market value".[77]

The *James* decision was reconfirmed by the Court later in the same year (1986) in the *Lithgow* case (dealing with nationalization).[78]

In the *Pressos* case of 1995, the Court held that the non-payment of compensation would generally undermine the "fair balance" that must exist between the interests of the individual (whose property has been deprived) and the public:

> "An interference with the peaceful enjoyment of possessions must strike a 'fair balance' between the demands of the general interest of the community and the requirements of the protection of the individual's fundamental rights. The concern to achieve this balance is reflected in the structure of Article 1 as a whole, including therefore the second sentence, which is to be read in the light of the general principle enunciated in the first sentence. In particular, there must be a reasonable relationship of proportionality between the means employed and the aim sought to be realised by any measure depriving a person of his possessions.
>
> Compensation terms under the relevant legislation are material to the assessment whether the contested measure respects the requisite fair balance and, notably, whether it imposes a disproportionate burden on the applicants. In this connection, the taking of property without payment of an amount reasonably related to its value will normally constitute a disproportionate interference and a total lack of compensation can be considered justifiable under Article 1 only in exceptional circumstances".[79]

Manifestly, compensation goes to the heart of the matter of deprivation of property. It has been suggested that the European Court

> "has made it clear that in deciding whether a deprivation of property can be regarded as 'in the public interest' it will normally regard the provision of compensation as a major consideration".[80]

---

[77] *Ibid.*, 147.
[78] *Lithgow and Others* v. *United Kingdom*, Eur. H. Rts. Reports 8 (1986), 329 et seq., 371-372.
[79] *Pressos* case, see note 35, 336.
[80] See J. G. Merrills, *The Development of International Law by the European Court of Human Rights*, 2nd ed., 1993, 217.

But on the whole it is more accurate to summarize the European case law to the effect that, notwithstanding compliance with the public interest requirement, the failure to pay fair compensation would render the deprivation of property inconsistent with the condition of proportionality.[81]

No less than the Iran-United States Claims Tribunal, the European Court of Human Rights has regarded the payment of some compensation as indispensable even when the public purpose of the deprivation of property is unassailable. As observed by I. Seidl-Hohenveldern:

> "All types of takings, however, have one thing in common. The State taking the property becomes liable to pay compensation, without regard to what this State does with the property it took. A pipeline may be laid across a foreign-owned real estate for the benefit of an oil company, latifundia may be split up into homesteads for demobilized soldiers, – nevertheless, it was the State which took the property in the public interest so the State will be responsible for the payment of compensation".[82]

In J. A. Frowein's language:

> "The taking of property without payment of an amount reasonably related to its value would normally constitute a disproportionate interference which could not be considered justifiable under Article 1. However, Article 1 does not guarantee a right to full compensation in all circumstances. Legitimate objectives of 'public interest', such as pursued in measures of economic reform or measures designed to achieve greater social justice, may call for less than reimbursement of the full market value".[83]

Compensation, according the European Court, must not only be related to the value of the property: it must also be timely. In the *Matos e Silva* case of 1996, the Court held that the "fair balance" between the competing interests of the individual (whose property is affected), on the one hand, and the community, on the other, is upset (and the individual bears an "excessive burden") when the ordinary enjoyment of the right

---

[81] P. van Dijk and G. J. H. van Hoof, *Theory and Practice of the European Convention on Human Rights*, 3rd ed., 1998, 634.

[82] I. Seidl-Hohenveldern, "International Economic Law: General Course on Public International Law", *RdC* 198 (1986), 9 et seq., 173.

[83] J. A. Frowein, "The Protection of Property", in R. St. J. Macdonald et al. (eds.), *The European System for the Protection of Human Rights*, 1993, 515 et seq., 524.

to property is hindered for more than 13 years during which time no progress has been made in legal proceedings and no compensation has been paid.[84] In the *Guillemin* case of 1997, the Court pronounced:

> "Compensation for the loss sustained by the applicant can only constitute adequate reparation where it also takes into account the damage arising from the length of the deprivation. It must moreover be paid within a reasonable time".[85]

VIII. Conclusion

Despite attempts in the 1970s to derail the international legal engine of special protection of the property of foreigners, the state of the law at the present time is probably more lucid than ever before. The deprivation of such property can be justified by noble public-spirited motivations, but compensation must be paid. The exact measure of compensation may be controversial, yet the principle today is virtually impregnable. Absent compensation altogether, the deprivation of the property of foreigners amounts to a breach of international law.

---

[84] *Matos e Silva, Lda. and Others* v. *Portugal, Eur. H. Rts. Reports* 24 (1996), 573 et seq., 601-602.
[85] *Guillemin* v. *France, Eur. H. Rts. Reports* 25 (1997), 435 et seq., 451.

Third Parties Before International Tribunals: The ICJ and the WTO

Yuji Iwasawa

I. Introduction

In 1983, Judge Shigeru Oda published an article[1] based on the separate opinion he had given in the *Tunisia/Libya Continental Shelf* case two years earlier.[2] He developed his ideas further in his dissenting opinion in the *Libya/Malta Continental Shelf* case in 1984.[3] Article 62(1) of the Statute of the International Court of Justice (ICJ) provides: "Should a state consider that it has an interest of a legal nature which may be affected by the decision in the case, it may submit a request to the Court to be permitted to intervene". The ICJ has interpreted "an interest of a legal nature which may be affected by the decision in the case" narrowly, rejecting the requests for intervention submitted by Malta and Italy respectively in the above two cases. Judge Oda was the most forceful critic of the view taken by the majority of the Court. He criticized the narrow construction adopted by the Court, arguing that "[i]ntervention

---

[1] S. Oda, "Intervention in the International Court of Justice: Article 62 and 63 of the Statue", in R. Bernhard and W. Geck (eds.), *Völkerrecht als Rechtsordnung, Internationale Gerichtsbarkeit, Menschenrechte: Festschrift für Hermann Mosler*, 1983, 629 et seq., also published in Japanese as "Kokusai Shihō Saibansho ni okeru Daisangoku no Soshō Sanka", *Kokusai Ho Gaiko Zassi*, 84 (1985), 1 et seq.

[2] *Tunisia/Libya Continental Shelf* case *(Tunisia v. Libya)*, 14 April 1981, ICJ Reports 1981, 3 et seq., 23 (separate opinion of Judge Oda).

[3] *Libya/Malta Continental Shelf* case *(Libya v. Malta)*, 21 March 1984, ICJ Reports 1984, 3 et seq., 90 (dissenting opinion of Judge Oda). While he wrote a *separate* opinion in the *Tunisia/Libya* case, he wrote a *dissenting* opinion in this case. Despite the seeming difference of opinion, his ideas remained essentially the same. The separate opinion in the *Tunisia/Libya* case was virtually a dissent. For his views on intervention, see also S. Oda, *Kokusai Shihō Saibansho (The International Court of Justice)*, 1987, 267-284; S. Oda, "The International Court of Justice Viewed from the Bench (1976-1993)", *RdC* 244 (1993), 9 et seq., 76-87.

within the meaning of Article 62 [should] be considered to have a far broader scope than the Court's Judgment allows".[4]

The World Trade Organization (WTO) has an elaborate mechanism of dispute settlement which is judicial in nature.[5] Third parties are routinely allowed to intervene in cases before the WTO. What Judge Oda advocated for the ICJ has been realized in the WTO. In fact, the extent to which the WTO allows intervention of third parties goes beyond what Judge Oda had in mind for the ICJ. This article will demonstrate that, compared with the ICJ, the WTO allows participation of third parties in its adjudicatory process much more liberally, and it will suggest that this is so because compulsory jurisdiction has been achieved in the WTO, and because the WTO adjudicatory procedures have the character of "international supervision".

II. Some Features of the WTO Dispute Settlement Procedures

If a WTO member considers that any benefit accruing to it under the WTO Agreement is being nullified by a measure of another member, it may request consultations with the other member. If no satisfactory adjustment is effected between the members concerned within a reasonable time, the aggrieved member may request the establishment of a panel. A panel, consisting of three individuals acting in a personal capacity, examines the matter in an adjudicatory manner. The parties present their case to the panel in writing and orally. After hearings, the panel compiles a report and submits it to the parties. The report of the panel includes a summary of facts, arguments of the parties, findings and conclusions of the panel. The parties may appeal the panel report to the Appellate Body established within the WTO. After review, the Appellate Body issues a report to the parties. The Dispute Settlement Body (DSB), an organ of the WTO, adopts the reports of the panel and the Appellate Body, unless it decides by consensus not to do so. The adopted reports have legal effect. Some salient features of the WTO panel procedures will be highlighted below to lay the basis for further analysis in the following sections.

---

[4] *Tunisia/Libya Continental Shelf* case, see note 2, at 23.
[5] For details on dispute settlement in the WTO, see generally Y. Iwasawa, *WTO no Funsō Shori (Dispute Settlement in the World Trade Organization)*, 1995; D. Palmeter and P. C. Mavroidis, *Dispute Settlement in the World Trade Organization: Practice and Procedure*, 1999; E.-U. Petersmann, *The GATT/WTO Dispute Settlement System: International Law, International Organizations and Dispute Settlement*, 1997; E.-U. Petersmann, *International Trade Law and the GATT/WTO Dispute Settlement System*, 1997.

1. Compulsory Jurisdiction

One of the most notable features of the panel procedures in the WTO is that the WTO has compulsory jurisdiction over a case brought by its members. In the General Agreement on Tariffs and Trade (GATT), the precursor of the WTO, consensus was required for the Contracting Parties to establish a panel. This meant that the consent of the defendant had to be obtained before a panel could be established. In contrast, the WTO establishes a panel unless there is a consensus not to do so. Therefore, the consent of the defendant State to establish a panel is no longer required. If a member requests the establishment of a panel, consensus not to establish a panel is lacking by definition, and the DSB will establish a panel. Thus, a WTO member has a right to a panel, and the WTO effectively has compulsory jurisdiction. This is significant in view of the fact that less than a third of the parties to the ICJ Statute have made declarations under the optional clause, and that the ICJ is far from having achieved compulsory jurisdiction in any real sense of the term.[6]

Other conventions that provide for compulsory jurisdiction for a court established in the convention include the UN Convention on the Law of the Sea, the Treaty Establishing the European Economic Community, the European Convention on Human Rights, and the American Convention on Human Rights. However, the jurisdiction of the European Court of Justice, the European Court of Human Rights, and the American Court of Human Rights are geographically limited to specific regions. Moreover, these courts have not dealt with many interstate disputes. The WTO is unique in that it has established a universal system of dispute settlement with compulsory jurisdiction and has dealt with numerous interstate disputes.

2. Binding Force of Reports of WTO Panels and the Appellate Body

GATT Article XXIII authorizes the Contracting Parties to make "recommendations" or give "rulings" on the dispute brought before the body. The Contracting Parties, however, commonly adopted a panel report without making "recommendations" or giving "rulings". Under these circumstances, different views were expressed as to the binding force of adopted panel reports. One may conclude, however, that the adopted reports of a panel and the Appellate Body do bind the parties to

---

[6] See, e.g., S. Oda, "The Compulsory Jurisdiction of the International Court of Justice: A Myth? A Statistical Analysis of Contentious Cases", *ICLQ* 49 (2000), 251 et seq.

the dispute for the following reasons: (1) the Dispute Settlement Understanding (DSU), which constitutes an integral part of the WTO Agreement, provides that *only in cases of a non-violation complaint* is there no obligation to withdraw the measure challenged (Article 26(1)(b)); (2) the DSU also provides that compensation and the suspension of obligations are available only *as temporary measures* in the event that the recommendations and rulings are not implemented within a reasonable period of time (Article 22(1)); and (3) the DSU further provides that an Appellate Body report must be unconditionally accepted by the parties to the dispute (Article 17(14)).[7] In the *Japan-Alcoholic Beverages* case (1997), the Appellate Body acknowledged that adopted panel reports are binding on the parties to the dispute.[8]

Nevertheless, the WTO has no doctrine of *stare decisis*. In this respect, the WTO is no different from the ICJ. In the *Japan-Alcoholic Beverages* case, the Appellate Body reversed the panel's conclusion that adopted panel reports constitute "subsequent practice" in the sense used in Article 31 of the Vienna Convention on the Law of Treaties, and stressed that they were not binding, except with respect to resolving the particular dispute, referring to Article 59 of the ICJ Statute. At the same time, however, the Appellate Body recognized that adopted panel reports are "an important part of the GATT *acquis*".[9] Thus, in reality, WTO adjudicatory bodies place considerable value on their own decisions as precedents. In this respect, they are no different from the ICJ.

3. Panel Procedures as Adjudication

What is the character of panel procedures in the WTO? Are they a form of adjudication or conciliation? Panel procedures in the WTO resemble adjudication in the following respects: an applicant has a right to be heard by a panel; the parties present their case in writing and orally in a legal manner; third parties can intervene in the proceedings; the panel makes legal findings based on law; the panel uses various legal techniques in reaching findings (e.g., burden of proof, methods of interpretation, respect for precedents); the parties may appeal the case to the Appellate Body; the report is adopted virtually automatically; and the adopted report binds the parties. Whether findings of a third party bind the parties to the dispute is often considered the decisive factor

---

[7] Iwasawa, see note 5, 135-138. See also J. H. Jackson, "The WTO Dispute Settlement Understanding – Misunderstandings on the Nature of Legal Obligation", *AJIL* 91 (1997), 60 et seq.

[8] *Japan-Taxes on Alcoholic Beverages*, Report of the Appellate Body, WTO DSR 1996, 97 et seq., 108 (adopted 1 November 1996).

[9] *Ibid.*

distinguishing adjudication from conciliation. Panel procedures are definitely adjudicatory in this sense because, as demonstrated above, reports of panels and the Appellate Body bind the parties. Among adjudications, one may distinguish between arbitration and judicial settlement. To which of these are the WTO panel procedures more closely related? A WTO panel is established on an *ad hoc* basis. In this sense, panel procedures are similar to arbitration. However, they are more akin to judicial settlement in the following respects: the applicant can refer a dispute to the WTO unilaterally and set the parameter of the dispute; working procedures and applicable law are predetermined; third parties can intervene in the case without consent of the parties to the dispute; a standing tribunal (the Appellate Body) may ultimately deal with the case; and the adjudicatory bodies value precedents.

Admittedly, panel procedures also have some elements of conciliation. A panel is expected to encourage the parties to reach an amicable agreement between themselves, and some mechanisms are provided for that purpose. For example, a panel is required to issue the descriptive sections of its draft report and an interim report to the parties, and to hold a further meeting with the parties on the issues identified in written comments submitted by the parties on the interim report.

In sum, the WTO panel procedures are a *sui generis* method of dispute settlement with elements of both adjudication and conciliation. In the beginning of their development, they were similar to conciliation. However, they gradually changed their character and have come closer to adjudication. The DSU has judicialized the panel procedures to a great extent and has made them much closer to judicial settlement. The Appellate Body and panels deal with a dispute judicially in much the same way as the ICJ does.

4. Frequent Recourse

One of the major differences between the dispute settlement in the WTO and the ICJ is the frequent use of the former compared to the latter. The extent to which GATT/WTO dispute settlement has been used is phenomenal. More than 300 requests for consultations were made during the 47-year existence of the GATT, and approximately 100 panel reports were issued. Since 1995, when the WTO came into being, disputes have increased even more. In the six years since 1995, 219 requests for consultations were made. Panels were established in 94 cases, and 53 panel reports were issued. Reports of the Appellate Body were issued in 45 cases (on 33 distinct matters). On aggregate, panels and the Appellate Body have issued more than 200 reports so far.

It is remarkable that the GATT/WTO has dealt with so many interstate disputes in accordance with defined procedures. The ICJ, of course, is the most important international tribunal. It has handled a number of important interstate disputes, and the Court is busier than ever with many disputes on its docket. However, the fact remains that the number of disputes brought to the ICJ since the Second World War is fewer than 100. Moreover, in many of those cases, the jurisdiction of the Court was denied, and the Court gave no judgment on the merits. Both the 1883 Paris Convention for the Protection of Industrial Property and the 1886 Berne Convention for the Protection of Literary and Artistic Works provide that any dispute concerning the interpretation or application of the convention may be brought before the ICJ. Nevertheless, no dispute has ever been brought to the ICJ as to the interpretation or application of these conventions.[10] In contrast, since the basic provisions of the Paris and Berne conventions were incorporated in the TRIPS Agreement, disputes concerning these conventions are now regularly adjudicated before the WTO.[11]

III. International Supervision and Dispute Settlement

1. International Supervision

International law has witnessed the growth of various institutions and procedures established by multilateral conventions for the purpose of securing States' compliance with their international obligations, especially in the areas of human rights,[12] the environment,[13] arms

---

[10] For surveys of economic disputes brought to the ICJ, see K. Wellens, *Economic Conflicts and Disputes Before the World Court (1922-1995): A Functional Analysis*, 1996; G. Jaenicke, "International Trade Conflicts Before the Permanent Court of International Justice and the International Court of Justice", in E.-U. Petersmann (ed.), *Adjudication of International Trade Disputes in International and National Law*, 1992, 43 et seq.

[11] E.g., *U.S.-Section 110(5) of the US Copyright Act*, Report of the Panel, WTO Doc. WT/DS160/R (adopted 27 July 2000) (finding, *inter alia*, that subparagraph (B) of Section 110(5) of the US Copyright Act is inconsistent with Articles 11*bis*(1)(iii) and 11(1)(ii) of the Berne Convention as incorporated into the TRIPS Agreement).

[12] See, e.g., E. Klein (ed.), *The Monitoring System of Human Rights Obligations*, 1998; N. Valticos, "Once More about the ILO System of Supervision: In What Respect Is It Still a Model?", in N. Blokker and S. Muller (eds.), *Towards More Effective Supervision in International Organizations: Essays in Honour of Henry G. Schermers*, 1994, 99 et seq.

[13] See, e.g., R. Wolfrum, "Means of Ensuring Compliance with and Enforcement of International Environmental Law", *RdC* 272 (1998), 9 et seq.

control,[14] and economic law.[15] Techniques of supervision include consideration of reports submitted by States, consideration of communications submitted by States and individuals, verification and inspection. "Non-compliance procedures", in which supervisory bodies give assistance or incentives to secure compliance, are a technique often prescribed in multilateral environmental agreements. The process of supervision entails three stages: establishment of facts, their assessment in light of legal standards, and recommendations of such corrective adjustments as may be necessary. These institutions and procedures of supervision have attracted the attention of international lawyers and have been studied under the rubric of "international supervision" or "international control".[16]

"Judicial supervision" is one technique of international supervision. Certainly, as long as an international judge makes determinations based on law, any international adjudication serves the purpose of securing States' compliance with international law – at least indirectly. "Judicial supervision", however, may be more narrowly defined than "dispute settlement". "Dispute settlement" is aimed at resolving a dispute arising between two States. It is basically a bilateral mechanism premised on the assumption that one State has injured another State and that the injured State demands relief for the injury caused by the wrongdoer. The objective of "judicial supervision", on the other hand, is to put an end to violations of international obligations by one State. For that purpose, any other State is deemed empowered to initiate judicial proceedings. The applicant State brings a claim for the protection of community interests and therefore need not allege any injury to its own interests. "Judicial supervision" is thus distinguished from "dispute settlement" in that States which have not been injured directly may initiate judicial proceedings for the purpose of putting an end to violations of international obligations by another State.[17]

---

[14] See, e.g., E. Myjer, "The Law of Arms Control and International Supervision", in M. Brus et al. (eds.), *The United Nations Decade of International Law: Reflections on Dispute Settlement*, 1991, 99 et seq.

[15] See, e.g., P. van Dijk (ed.), *Supervisory Mechanisms in International Economic Organizations*, 1984.

[16] E.g., Blokker and Muller, see note 12; W. E. Butler (ed.), *Control over Compliance with International Law*, 1991; A. Morita, *Kokusai Kontorōru no Riron to Jikkō (International Supervision in Theory and Practice)*, 2000; J. Charpentier, "Le contrôle par les organisations internationales de l'exécution des obligations des Etas", *RdC* 182 (1983), 143 et seq.; H. J. Hahn, "International Controls", in R. Bernhardt (ed.), *EPIL*, Vol. II, 1995, 1079 et seq.

[17] Morita, see note 16, 97-102, 165-74. I. Brownlie, *Principles of Public International Law*, 4th ed., 1990, 648-649 (section deleted in 5th ed., 1998).

In the *South West Africa* cases, the ICJ took the view that the right of any member of a community to take legal action in vindication of a public interest *(actio popularis)* is not known to general international law.[18] However, as the ICJ itself subsequently acknowledged in the *Barcelona Traction* case, some international conventions provide for a right of any member to take legal action in vindication of community interests.[19] Thus, judicial supervision has been provided by such multilateral conventions as minority treaties in the League of Nations, the Statute of the Memel Territory, the Versailles Treaty, the Genocide Convention, the Racial Discrimination Convention, the European Convention on Human Rights, the American Convention on Human Rights, and the Treaty Establishing the European Economic Community. The cases which have been brought under these treaties for the purpose of "judicial supervision" include the following: the *Memel Territory Statute* case (under the Statute of the Memel Territory),[20] the *Wimbledon* case (under the Versailles Treaty),[21] *Ireland* v. *UK* (under the European Convention on Human Rights),[22] and *France* v. *UK* (under the EEC Treaty).[23]

2. WTO Dispute Settlement as Judicial Supervision

In this section, it will be shown that WTO dispute settlement has the character of judicial supervision as defined above. Under GATT Article XXIII, if a member considers (1) that any benefit accruing to it under the WTO Agreement is being nullified or impaired, or (2) that the attainment of any objective of the WTO Agreement is being impeded, it may file a complaint with the WTO. The alleged consequences must result from one of the following causes: (a) the failure of another member to carry out its obligations; (b) the application by another member of a measure not inconsistent with the WTO Agreement, or (c) the existence of any other situation. Thus, in all, Article XXIII provides six different causes

---

[18] *South-West Africa* cases (Second Phase) *(Ethiopia* v. *South Africa, Liberia* v. *South Africa)*, 18 July 1966, ICJ Reports 1966, 6 et seq., 47.

[19] *Barcelona Traction, Light and Power Co.* case *(Belgium* v. *Spain)*, 5 February 1970, ICJ Reports 1970, 3 et seq., 32. For *actio popularis* in international law, see E. Schwelb, "The *actio popularis* and International Law", *Isr. Y.B. Hum. Rts.* 2 (1972), 46 et seq.; I. Seidl-Hohenveldern, "*Actio popularis* im Völkerrecht", *Comun. e Stud.* 14 (1975), 803 et seq.

[20] *Interpretation of the Statue of the Memel Territory (UK, France, Japan, Italy* v. *Lithuania)*, 11 August 1932, PCIJ, Ser. A/B, No. 47, 1932.

[21] *S.S. Wimbledon* case *(France, Italy, Japan, UK* v. *Germany)*, 17 August 1923, PCIJ, Ser. A, No. 1, 1923.

[22] *Ireland* v. *UK*, ECHR, 18 January 1978, Ser. A., Vol. 25.

[23] Case 141/78, *France* v. *UK*, ECR 1979, 2923 et seq.

of action which can trigger the dispute settlement procedures in the WTO. In reality, most complaints allege that a benefit accruing to the complaining State is being nullified – route (1). Few complaints allege that the attainment of an objective of the WTO Agreement is being impeded – route (2). Equally few are claims based on the existence of a situation – subset (c). Most complaints allege that a benefit accruing to the complaining State is being nullified by violations of the WTO Agreement by another member (route (1), subset (a) – "violation complaints") or by the application by another member of a measure not inconsistent with the WTO Agreement (route (1), subset (2) – "non-violation complaints").

The WTO Agreement explicitly authorizes a member to bring a complaint alleging that the attainment of an objective of the WTO Agreement is being impeded. A member may do so even when it has suffered no direct injury. Thus, the WTO Agreement is one of the international agreements which explicitly allow an *actio popularis*. If members of the WTO make regular use of such power and bring complaints based on community interests, one may say that WTO dispute settlement functions as judicial supervision. In reality, few complaints of such a kind are brought to the WTO.

Nevertheless, one may still conclude that WTO dispute settlement has elements of judicial supervision for the following reasons.[24] The GATT came into being on the basis of balance of tariff concessions made by the Contracting Parties. The dispute settlement procedures in the GATT were conceived as a mechanism by which the balance of concessions can be restored when it has been disturbed by a measure of one Contracting Party. They are bilateral procedures by which a State which has suffered injury brings a claim against another State which has caused the injury. The applicant State claims that the defendant State has nullified *its benefits*.

In spite of the bilateral nature of the procedures, they work, at the same time, as a mechanism to secure compliance with the WTO Agreement. Maintenance of a free and non-discriminatory world trade system serves the common interests of WTO members. Basic obligations stipulated by the WTO Agreement, such as the prohibition of quantitative restrictions, obligations to grant most-favored-nation (MFN) treatment and national treatment, are objective obligations. The world

---

[24] For characterization of the GATT/WTO dispute settlement as international supervision, see also A. Kotera, *WTO Taisei no Hō Kōzō (Legal Structures of the WTO System)*, 2000, 87-93; I. H. Courage-van Lier, "Supervision Within the General Agreement on Tariffs and Trade", in P. van Dijk (ed.), *Supervisory Mechanisms in International Economic Organizations in the Perspective of a Reconstructing of the International Economic Order*, 1984, 47 et seq.

economy has become so interdependent that trade restrictions introduced by one State will have negative effects upon the world trade system as a whole. Thanks to the MFN clause in the WTO Agreement, elimination of trade restrictions benefits not only the applicant State but all WTO members as well. Under such circumstances, even though only one member files a complaint in the WTO, alleging that *its benefits* are being nullified, the action promotes free trade and serves community interests. The applicant plays a dual role and the complaint has the effect of action on behalf of the international community.

Because WTO dispute settlement has the character of judicial supervision, the mootness doctrine is not considered applicable in the WTO. Even when the challenged trade restrictions are eliminated or terminated, panels continue their proceedings and make findings as to whether the restrictions were consistent with the WTO Agreement.[25] It is considered important to establish that the trade restrictions were inconsistent with the WTO Agreement to prevent the recurrence of such violations.

In contrast, since the ICJ is given the task of traditional bilateral dispute settlement, it refuses to give rulings on disputes which have become moot. In the *Northern Cameroons* case, Cameroon asked the Court to declare that the United Kingdom had violated the Trusteeship Agreement. The UN General Assembly had endorsed the results of the plebiscites held in Northern Cameroons and had terminated the Trusteeship Agreement. Under the circumstances, the ICJ held that it could not adjudicate upon the case because adjudication would be devoid of purpose.[26] In the *Nuclear Tests* case, Australia and New Zealand asked the Court to declare that the carrying out of further atmospheric nuclear weapon tests by France in the Pacific was inconsistent with international law. The Court held that a statement of France that it would undertake no more nuclear tests in the Pacific was binding on France. Noting that the applicants had achieved their purpose, it refused to deal with the merits of the claims.[27]

---

[25] *EC-Measures on Animal Feed Proteins*, GATT, *Basic Instruments and Selected Documents (BISD)*, 25th Supp., 49 et seq. (adopted 14 March 1978); *EC-Restrictions on Imports of Apples from Chile*, BISD, 27th Supp., 98 et seq. (adopted 10 November 1980); *U.S.-Prohibition on Imports of Tuna and Tuna Products from Canada*, BISD, 29th Supp., 91 et seq. (adopted 22 February 1982); *EC-Restrictions on Imports of Dessert Apples*, BISD, 36th Supp., 93 et seq. (adopted 21-22 June 1989); *EC-Restrictions on Imports of Apples*, BISD, 36th Supp. 135 et seq. (adopted 21-22 June 1989); *U.S.-Denial of Most-Favoured-Nation Treatment as to Non-Rubber Footwear from Brazil*, BISD, 39th Supp., 128 et seq.

[26] *Northern Cameroons* case *(Cameroon v. UK)*, ICJ Reports 1963, 15 et seq.

[27] *Nuclear Tests* cases, *(Australia v. France, New Zealand v. France)*, ICJ Reports 1974, 253 et seq., 457 et seq.

The European Court of Human Rights[28] and the European Court of Justice[29] have also proceeded to rule upon cases that have become moot. In one such case, the European Court of Human Rights stated that its function was "not only to decide those cases brought before [it], but, more generally, to elucidate, safeguard and develop the rules instituted by the Convention".[30] Like WTO panels, these tribunals reject the doctrine of mootness, because they view their tasks not only as dispute settlement but also as judicial supervision.

IV. Position of Third Parties Before WTO Panels and the ICJ

In this section, the position of third parties before WTO panels and the ICJ will be compared. The comparison will illustrate the point that WTO dispute settlement has the character of judicial supervision in addition to the traditional function of settling bilateral disputes.

1. Legal Interest

In the *South West Africa* cases, the ICJ dismissed the action brought by Ethiopia and Liberia for the reason that they did not have sufficient legal interest to bring a case against South Africa. The applicants argued that each member of the League should be deemed to have a legal interest in the performance of the Mandate for South West Africa, and the Court replied:

> "[T]he argument amounts to a plea that the Court should allow the equivalent of an *actio popularis*, or right resident in any member of a community to take legal action in vindication of a public interest. Although a right of this kind may be known to certain municipal systems, it is not known to international law as it stands at present".[31]

As the Court subsequently admitted in the *Barcelona Traction* case,[32] it is incorrect to say that *actio popularis* is entirely unknown to international law, because some international conventions, including the WTO Agreement, provide for *actio popularis*. Admittedly, however, it is debatable whether *actio popularis* is allowed under customary

---

[28] *Ireland* v. *UK*, ECHR, 18 January 1978, Ser. A., Vol. 25, 61-62.
[29] Case 26/69, Commission v. *France*, ECR, 1970, 565 et seq. (575-576); Case 7/61, Commission v. *Italy*, ECR, 1961, 317 et seq., 326.
[30] See note 28.
[31] See note 18.
[32] See note 19.

international law.³³ In any case, the majority of the Court in the *South West Africa* cases declined to interpret the relevant provisions in the Mandate to the effect that any member of the League had a legal interest in the observance of the Mandate. The majority was not prepared to see the ICJ engaged in judicial supervision.

WTO adjudicatory bodies are much more willing to be engaged in judicial supervision. Under GATT Article XXIII, if a member considers that the attainment of any objective of the WTO Agreement is being impeded, it may file a complaint with the WTO. In reality, however, all complaints brought to the GATT/WTO have alleged that a benefit of the complaining member is being nullified by the measure of another member. Thus, it may appear that the applicant has necessarily had a legal interest in all cases. It must be noted, however, that the requirement that the applicant show that its benefit is being nullified is not strenuously maintained in the WTO. Case law has established that where measures are applied in contravention of the WTO Agreement, it is *prima facie* presumed that the benefit of the applicant is being nullified.³⁴ Moreover, nullification of benefit need not be concrete and material. Even if a measure inconsistent with the WTO Agreement has no effect on the volume of trade, it still impairs the benefit of the applicant in the sense that it could lead to increased transaction costs and would create uncertainties which could affect investment plans.³⁵ Thus, if one member has taken a measure inconsistent with the WTO Agreement, other members can bring a complaint to the WTO, maintaining that a benefit is being nullified, even if the applicant has suffered no *injury* in a concrete and material sense.

The *EC-Bananas* case (1998) provides a good example. In this case, several States in Latin America brought a complaint against the EC, maintaining that the preferential treatment the EC accords to its former colonies in Africa, the Caribbean and the Pacific is inconsistent with the WTO Agreement. The United States joined the case as an applicant because US companies were engaged in the export of bananas from the Latin American countries to the EC. The United States, however, produces few bananas and its export trade in bananas is negligible. Under the circumstances, the EC questioned the *locus standi* of the

---

³³ See, e.g., J. Crawford, "The Standing of States: A Critique of Article 40 of the ILC's Draft Articles on State Responsibility", in M. Andenas (ed.), *Judicial Review in International Perspective. Liber Amicorum for Gordon Slynn*, 2000, 25 et seq.

³⁴ *Recourse to Article XXIII, BISD*, 11th Supp., 95 et seq., 99-100 (adopted 16 November 1962).

³⁵ *Treatment by Germany of Imports of Sardines, BISD*, 1st Supp., 53 et seq. (adopted 31 October 1952); *Japanese Measures on Imports of Leather, BISD*, 31st Supp., 94 et seq. (adopted 15 May 1984).

United States. The EC invoked several ICJ cases, including the *South West Africa* cases, to argue that an applicant must have a "legal interest" in the performance of obligations by the defendant. However, the Appellate Body affirmed the *locus standi* of the United States, pointing out that a member has broad discretion in deciding whether to bring a case against another member. The Appellate Body stated that "with the increased interdependence of the global economy, ... Members have a greater stake in enforcing WTO rules than in the past since any deviation from the negotiated balance of rights and obligations is more likely than ever to affect them".[36] The Appellate Body observed also that:

> "The participants in this appeal have referred to certain judgments of the International Court of Justice and the Permanent Court of International Justice relating to whether there is a requirement, in international law, of a legal interest to bring a case. We do not read any of these judgments as establishing a general rule that in all international litigation, a complaining party must have a 'legal interest' in order to bring a case. Nor do these judgments deny the need to consider the question of standing under the dispute settlement provisions of any multilateral treaty, by referring to the terms of that treaty".[37]

Thus, in the WTO, a complaint coming close to *actio popularis* is allowed in the form of a complaint that the applicant's benefit is nullified. It is submitted that such an action is allowed in the WTO because its dispute settlement procedures have the character of judicial supervision.

2. Multiple Disputes

The ICJ deals with disputes involving more than two States in several ways. First of all, one State may sue another State before the ICJ and other States may request to intervene as third parties. This method will be examined in detail in the next section. Secondly, multiple disputes may be dismantled into separate bilateral disputes and more than one dispute on the same subject matter may be submitted to the ICJ. A single

---

[36] *EC-Regime for the Importation, Sale and Distribution of Bananas*, Report of the Appellate Body, WTO DSR 1997, 589 et seq., 650-652 (adopted 13 February 1998).

[37] *Ibid.* For an early commentary of the case, see R. Bustamante, "The Need for a GATT Doctrine of *Locus Standi*: Why the United States Cannot *Stand* the European Community's Banana Import Regime", *Minn. J. Global Trade* 6 (1997), 533 et seq.

plaintiff may submit multiple disputes against more than one defendant (disputes brought by Nicaragua against the United States, Honduras and Costa Rica on military activities in Nicaragua; the *Lockerbie* cases, and the *NATO Bombing* cases). Conversely, more than one plaintiff may bring multiple disputes against the same defendant (the *South West Africa* cases, the *North Sea Continental Shelf* cases, the *Fisheries Jurisdiction* cases, and the *Nuclear Test Cases*). If the plaintiffs share interests, the Court may join the proceedings and give only one judgment (the *South West Africa* cases and the *North Sea Continental Shelf* cases). Thirdly, multiple plaintiffs may jointly bring one case against a single defendant (the *Wimbledon* case and the *Memel Territory Statute* case). There has also been a case in which one State named more than one defendant in a single case (the *Monetary Gold* case). Nevertheless, multiple disputes are rare in the ICJ. The ICJ is best suited to deal with bilateral disputes of a traditional type, and, in fact, most disputes brought to the ICJ are of that type.

In contrast, multiple disputes are a common feature in the WTO. Trade restrictions of one State affect many States and the world trading system as a whole. Given such circumstances, frequently more than one State brings complaints to the WTO on the same subject matter. In fact, since 1995, when the WTO came into being, more than a third of all disputes brought to the WTO have been multiple disputes. The DSU contains one article devoted to "Procedures for Multiple Complainants". Article 9(1) provides that "[w]here more than one Member requests the establishment of a panel related to the same matter, a single panel ... should be established to examine such complaints whenever feasible". A single panel was established in such cases as *US-Superfund* (1987),[38] *US-Customs User Fee* (1988)[39] and *EC-Bananas*.[40] Even when a single panel is established, "[i]f one of the parties to the dispute so requests, the panel shall submit separate reports" (paragraph 2), and the panel in the *EC-Bananas* case did so. Article 9(3) of the DSU also provides that "[i]f more than one panel is established to examine the complaints related to the same matter, to the greatest extent possible the same persons shall serve as panelists on each of the separate panels", and more than one

---

[38] *U.S.-Taxes on Petroleum and Certain Imported Substances*, BISD, 34th Supp., 136 et seq. (adopted 17 June 1987).

[39] *U.S.-Customs User Fee*, BISD 35th Supp., 245 et seq. (adopted 2 February 1988).

[40] *EC-Regime for the Importation, Sale and Distribution of Bananas*, Report of the Appellate Body, WTO DSR 1997, 589 et seq.; Report of the Panel on the Complaint by Guatemala and Honduras, *ibid.*, 695 et seq.; Report of the Panel on the Complaint by Mexico, *ibid.*, 803 et seq.; Report of the Panel on the Complaint by the United States, *ibid.*, 943 et seq.; Report of the Panel on the Complaint by Ecuador, *ibid.*, 1085 et seq. (adopted 13 February 1998).

panel was established in such cases as *EC-Refunds on Exports of Sugar* (1980),[41] *South Korea-Beef* (1989)[42] and *EC-Hormones* (1998).[43] One applicant may bring multiple complaints against more than one defendant. A single panel[44] or separate panels[45] may be established to examine such complaints. The fact that multiple disputes are numerous and common in the WTO provides further evidence that WTO dispute settlement has the character of judicial supervision.

In the WTO, a third party which has intervened in a case may subsequently commence another action on the same subject matter. In the *India-Patents* case, the EC initially intervened in a case brought by the United States as a third party. As soon as the panel issued a report, however, the EC requested the establishment of another panel on the same subject matter.[46] The EC had to start its own proceedings for the following reasons. If the defendant State fails to comply with the recommendations of the DSB, the DSU allows the *complaining party* to request authorization from the DSB to take countermeasures, the level of which shall be equivalent to the level of the nullification of *its benefits*. Under the circumstances, the EC felt it necessary to become a complaining party. The character of WTO dispute settlement as a bilateral dispute settlement procedure, rather than judicial supervision, predominates here.

---

[41] *EC-Refunds on Exports of Sugar*, BISD, 26th Supp., 290 et seq.; 27th Supp., 69 et seq. (adopted 10 November 1980).

[42] *Korea-Restriction on Imports of Beef*, BISD, 36th Supp., 202 et seq., 234 et seq., 268 et seq. (adopted 7 November 1989).

[43] *EC-Measures Concerning Meat and Meat Products (Hormones)*, Report of the Appellate Body, WTO DSR 1998, 135 et seq.; Report of the Panel on the Complaint by Canada, *ibid.*, 235 et seq.; Report of the Panel on the Complaint by the United States, *ibid.*, 699 et seq. (adopted 13 February 1998).

[44] *Recourse to Article XXIII*, BISD, 11th Supp., 95 et seq. (adopted 16 November 1962).

[45] *Income Tax Practices*, BISD, 23rd Supp., 127 et seq., 114 et seq., 137 et seq. (adopted with an understanding, 7-8 December 1981, BISD, 28th Supp., 114).

[46] *India-Patent Protection for Pharmaceutical and Agricultural Chemical Products*, Report of the Appellate Body, WTO DSR 1998, 91 et seq.; Report of the Panel on the Complaint by the United States, *ibid.*, 41 et seq. (adopted 16 January 1998); *India-Patent Protection for Pharmaceutical and Agricultural Chemical Products*, Report of the Panel on the Complaint by the EC, WTO DSR 1998, 2661 et seq. (adopted 2 September 1998). See also *Argentina-Footwear, Textiles, Apparel and Other Items*, Report of the Appellate Body, WTO DSR 1998, 1003 et seq.; Report of the Panel, WTO DSR 1998, 1033 et seq. (adopted 22 April 1998).

3. Intervention

WTO panels liberally allow a third party to intervene in their proceedings, in striking contrast to the ICJ, which is unwilling to allow intervention. The extent to which third parties are allowed to intervene in panel proceedings provide the best evidence that the WTO panel procedures have the character of judicial supervision.

According to the ICJ Statute, should a State consider that "it has an interest of a legal nature which may be affected by the decision in the case", it may ask the Court to be permitted to intervene (Article 62), and "[w]henever the construction of a convention to which States other than those concerned in the case are parties is in question", every State which is a party to the convention has the right to intervene in the proceedings (Article 63). Since the ICJ deals mostly with bilateral disputes of a traditional type, third parties have seldom submitted a request to intervene. So far, third parties have made only six requests for intervention under Article 62 and three requests under Article 63. Even though Articles 62 and 63 could well be interpreted liberally, the ICJ has interpreted them narrowly and has allowed intervention only under strict circumstances.[47] Greig went so far as to say that "the Court's treatment of intervention, whether under Article 62 or under Article 63, demonstrates an almost total lack of concern for the position of third parties".[48] Thus far, the Court has allowed intervention in only two cases under Article 62 (*El Salvador-Honduras Frontier Dispute* (1990), and *Cameroon-Nigeria Boundary* (1999)) and two cases under Article 63 (*Wimbledon* (1923), and *Haya de la Torre* (1951)).

The Court has interpreted "an interest of a legal nature which may be affected by the decision in the case" in Article 62 narrowly. According to the Court, it is not enough for a third party to have a general concern about the case before the Court or a general interest in principles of law to be determined by the Court. The third party must have a specific interest in the outcome of the case. In the *Tunisia/Libya and Libya/Malta Continental Shelf* cases, the Court held that a mere general interest in the principles the Court might apply in the case was insufficient.[49] In the *Frontier Dispute* case between El Salvador and Honduras, a Chamber of the Court allowed Nicaragua to intervene in the

---

[47] For details on intervention in the ICJ, see, e.g., C. Chinkin, *Third Parties in International Law*, 1993, 47-217; S. Torres Bernárdez, "L'intervention dans la procédure de la Court international de Justice", *RdC* 256 (1995), 193 et seq.

[48] D. W. Greig, "Third Party Rights and Intervention before the International Court", *Va. J. Int'l L.* 32 (1992), 285 et seq., 318.

[49] *Tunisia/Libya Continental Shelf* case, see note 2. *Libya/Malta Continental Shelf* case, see note 3.

case, but pointed out that "an interest of a third State in the general legal rules and principles likely to be applied by the decision can [not] justify an intervention".[50] In *Tunisia/Libya*, Judge Oda appended a separate opinion and put forward the view that Article 62 should be interpreted to allow intervention when a third party has concerns about the Court's interpretation of the principles and rules of international law, provided that an interest of a legal nature is present.[51] However, even Judge Oda believed it insufficient to have "a general interest in the development of international law in an abstract form", emphasizing that "the mere interpretation of principles and rules of international law [was] not at issue" in the present case.[52] As we shall see below, WTO panels allow intervention even when a third party has a general interest in the development of WTO law. Thus, the WTO allows third-party intervention more broadly than Judge Oda has advocated for the ICJ.

Article 63 of the ICJ Statute has not fared much better in the practice of the ICJ. Even though it appears to provide for intervention as of right, the Court has held that an element of discretion still exists, denying intervention in the *Nicaragua (Declaration of El Salvador)* case (1984). The ICJ is reluctant to allow third-party intervention because it has no compulsory jurisdiction. Adjudication before the ICJ rests upon consent of the parties to the dispute. It is feared that if the Court allows intervention too liberally, States will become reluctant to submit disputes to the ICJ.

WTO panels allow intervention liberally. Intervention is on the increase in the WTO. In fact, it is becoming rare to find a WTO case in which no third party intervenes. In *EC-Bananas*, the complaining parties were five States, and no less than 20 States intervened as third parties. Multiple disputes of that scale are not found in the ICJ. The WTO has even formed a practice, according to which a panel will not consider a matter which was not raised by the complaining party in its request for the establishment of a panel. One of the rationales of the practice is to safeguard intervention by third parties, for they will look at the complaining party's request for the establishment of a panel and decide whether they will intervene.

Article 10(2) of the DSU requires that a third party have "a substantial interest in a matter before a panel" to intervene in panel proceedings. This requirement resembles the requirement in Article 62

---

[50] *Land, Island and Maritime Frontier Dispute* case *(El Salvador* v. *Honduras)*, 13 September 1990, ICJ Reports 1990, 91 et seq., 124. See also *ibid.*, 126. See generally E. Doussis, "Intérêt juridique et intervention devant la Cour Internationale de Justice", *RGDIP* 105 (2001), 55 et seq.
[51] *Tunisia/Libya Continental Shelf* case *(Tunisia* v. *Libya)*, 14 April 1981, ICJ Reports 1981, 23 et seq., 30-31 (separate opinion of Judge Oda).
[52] *Ibid.*, 32.

of the ICJ Statute that a State have "an interest of a legal nature which may be affected by the decision in the case". One must note, however, that a third party in the WTO is invariably a party to the WTO Agreement. Thus, WTO cases also meet the requirement provided for in Article 63 of the ICJ Statute that "[w]henever the construction of a convention to which states other than those concerned in the case are parties is in question", and Article 63 gives a third party a *right* of intervention. Under the circumstances, the requirement of "a substantial interest" in Article 10(2) of the DSU is more loosely interpreted than "an interest of a legal nature" in Article 62(1) of the ICJ Statute.

Intervention of a third party in panel proceedings may be contrasted with the joining of a third party in consultations initiated between two members of the WTO. In order for a third party to be joined in consultations, Article 4(11) of the DSU requires that the third party have "a substantial *trade* interest" in the consultations, and the consulting members must give their consent. In contrast, a third party need only have "a substantial interest" to intervene in panel proceedings, and consent of the parties is not required. The text makes it clear that "a substantial interest" need not be a *trade* interest. A general interest in the interpretation of the WTO Agreement or a systemic interest in the dispute settlement procedures is considered sufficient. For example, even though Japan and Canada were not banana-exporting countries, they intervened in *EC-Bananas* mostly out of systemic interests in the dispute settlement procedures. A third party can even intervene in a case simply to receive the submissions of the parties to the panel and the Appellate Body. The parties to the dispute do not query whether the third party has "a substantial interest" in the matter before the panel, nor does the panel examine it. Thus, the requirement to have "a substantial interest" hardly works as a restraining factor for third-party intervention in the WTO. If a third party expresses a wish to intervene in panel proceedings, intervention is in fact allowed.[53] The liberal attitudes of the WTO towards third-party intervention may be contrasted with the restrictive attitudes of the ICJ. The MFN treatment and multilateralism

---

[53] On third-party intervention in the WTO, see also A. Rosas, "Joinder of Parties and Third Party Intervention in WTO Dispute Settlement", in F. Weiss (ed.), *Improving WTO Dispute Settlement Procedures: Issues and Lessons from the Practice of Other International Courts and Tribunals* (2001), 77 et seq.; C. Carmody, "Of Substantial Interest: Third Parties under GATT", *Mich. J. Int'l L.* 18 (1997), 615 et seq.; M. E. Footer, "Some Aspects of Third Party Intervention in GATT/WTO Dispute Settlement Proceedings", in E.-U. Petersmann (ed.), *International Trade Law and the GATT/WTO Dispute Settlement System*, 1997, 211 et seq.; F. Weiss, "Third Parties in GATT/WTO Dispute Settlement Proceedings", in E. Denters and N. Schrijver (eds.), *Reflections on International Law from the Low Countries*, 1998, 458 et seq.

inherent in the WTO create a natural inclination towards third-party intervention. Besides, the WTO need not worry that intervention will inhibit States from giving consent to its jurisdiction because it has compulsory jurisdiction.

A third party does not become a party to the case by intervening in WTO cases. The ICJ has, so far, also allowed a third party to intervene as a non-party. However, procedural rights of a third party before a WTO panel are more circumscribed than those of an intervener before the ICJ.[54] A third party can make written submissions to the panel and has an opportunity to be heard. However, it only receives the submissions of the parties to the first meeting of the panel and cannot attend the meetings with the parties. Moreover, the panel does not examine an issue raised only by a third party. The stake of a third party in the case, however, may be no smaller than that of the party to the dispute (e.g., the ACP countries in *EC-Bananas*). Thus, the panel in *EC-Bananas* expanded the procedural rights of third parties and allowed them to attend the meetings with the parties to the dispute as observers. In 1999, WTO members agreed to strengthen the procedural rights of third parties by amending the DSU.[55] The amendment did not materialize due to the failure of the WTO Ministerial Conference in Seattle.

4. Essential Parties

In *Monetary Gold* (1954), Italy brought a case against France, Britain and the United States, asking the ICJ to determine certain legal questions concerning monetary gold removed by the Germans from Rome in 1943 and found to belong to Albania. The Court declined to exercise jurisdiction because a third State's (Albania's) legal interests "would not only be affected by a decision but would form the very subject-matter of the decision".[56] According to the Court, to adjudicate upon the international responsibility of Albania without her consent would run counter to the well-established principle of international law that the Court can only exercise jurisdiction over a State with its consent. In *East Timor* (1995), the ICJ approved the principle of essential parties again, declining to exercise jurisdiction because the legal interests of a third State (Indonesia) would constitute the very subject matter of the

---

[54] See N. Covelli, "Public International Law and Third Party Participation in WTO Panel Proceedings", *JWT* 33 (No. 2) (1999), 125 et seq.
[55] Proposed Amendment of the Dispute Settlement Understanding, WTO Doc. WT/MIN(99)/8, 22 November 1999.
[56] *Monetary Gold Removed from Rome in 1943 (Italy v. France, UK and US)*, 15 June 1954, ICJ Reports 1954, 19 et seq., 32.

decision.[57] As the Court acknowledged in both cases, the rationale of the principle of essential parties lies in the bilateralism of proceedings in the ICJ and the consensual basis of its jurisdiction.[58] Rosenne stated that "[t]he existence of [the] concept, as a principle of general international law and as a feature of the law of international judicial procedure, is not ... open to question".[59]

The WTO has compulsory jurisdiction over disputes arising under the WTO Agreement and a WTO panel need not worry about a member's consent to its jurisdiction. Under the circumstances, WTO panels are more disposed than the ICJ to exercise jurisdiction when interests of a third party are implicated. In *Turkey-Textiles* (1999), India brought a dispute against Turkey, contending that Turkey's restrictions on imports of textiles were contrary to the WTO Agreement. Turkey introduced the trade restrictions pursuant to a regional trade agreement Turkey had concluded with the EC, whose compatibility with the WTO Agreement was open to question. The EC, however, was not named as a defendant and it declined to intervene in the proceedings as a third party. Under the circumstances, Turkey argued that the panel should dismiss India's claims because the EC was an "essential party". The panel referred to *Monetary Gold* and *East Timor* only in passing. Instead, the panel regarded *Nicaragua* (1984) and *Nauru* (1992) as "international practice". In these cases, the ICJ exercised jurisdiction even though interests of a third State were implicated. The panel pointed out that if a decision between the parties to the case can be reached without an examination of the position of the third State, the ICJ will exercise its jurisdiction as between the parties. The panel then declared:

> "[T]here is no WTO concept of 'essential parties' ... we consider that the European Communities was not an essential party to this dispute; the European Communities, had it so wished, could have availed itself of the provisions of the DSU, which we note have been interpreted with a degree of flexibility

---

[57] *East Timor* case *(Portugal v. Australia)*, 30 June 1995, ICJ Reports 1995, 90 et seq. Compare these cases with the other two cases in which the Court exercised jurisdiction even when the legal interests of a third State were implicated: *Military and Paramilitary Activities in and against Nicaragua (Nicaragua v. US)*, 26 November 1984, ICJ Reports 1984, 392 et seq.; *Certain Phosphate Lands in Nauru (Nauru v. Australia)*, 26 June 1992, ICJ Reports 1992, 240 et seq.

[58] See Sh. Rosenne, *The Law and Practice of the International Court, 1920-1996*, 1997, 552.

[59] *Ibid.*, 560.

by previous panels [e.g., in *EC-Bananas*], in order to represent its interests".[60]

Admittedly, the panel did not purport to examine the compatibility of the Turkey-EC customs union but only the compatibility of the Turkish measures with the WTO Agreement. And, the panel pointed out that "Panel and Appellate Body reports are binding on the parties only".[61] However, such a statement of the panel was no consolation to the EC, which undeniably had an important stake in the case.

V. Conclusion

Relying upon Article XXIII of the GATT, simply entitled "Nullification and Impairment", the GATT and the WTO have developed elaborate dispute settlement procedures. The dispute settlement procedures were conceived by the GATT/WTO as a mechanism by which the balance of concessions can be restored when it has been disturbed by a measure of one member. It has a function of settling a bilateral dispute arising under the WTO Agreement. The procedures, however, also have the character of judicial supervision. They serve as a judicial mechanism to control and secure compliance of obligations under the WTO Agreement.

The ICJ, on the other hand, is an international tribunal with a function of settling bilateral disputes of a traditional type. The procedures in the WTO and the ICJ differ conspicuously with regard to the position they assign to third parties. The WTO allows participation of a third party in its adjudicatory process more liberally than the ICJ, because the WTO procedures have the character of judicial supervision in addition to the function of settling bilateral disputes, and because the WTO has compulsory jurisdiction. The extent to which the WTO allows intervention of a third party goes beyond what Judge Oda advocated for the ICJ, because the two ostensibly similar dispute settlement procedures differ considerably in character.

---

[60] *Turkey-Restrictions on Imports of Textile and Clothing Products*, Report of the Panel, WTO Doc. WT/DS34/R, paras. 9.4-9.13 (adopted 19 November 1999), emphasis added.
[61] *Ibid.*, para. 9.11.

The Differences Between Conducting a Case in the ICJ and in an ad hoc Arbitration Tribunal – An Inside View

Sir Robert Jennings

I. Introduction

Perhaps I might begin by saying that it is a great pleasure and an honour to be allowed in this way to pay a tribute to my good friend and colleague of so many years standing. We first came together, if I remember rightly, as members of the now forgotten 1965 Committee under the chairmanship of the late Professor Friedman (the young Mr. Boutros Boutros-Ghali was also a member, and the committee resulted from an idea of the late Professor Jessup) which studied the constitution and practice of the Hague Academy and recommended important changes which seem to have stood the test of time. A high point of that committee's deliberations was, as Judge Oda will certainly remember, an evening during our session at Bellagio, when the late Professor René Jean Dupuy, with his fine baritone voice, sang Schubert *Lieder*, to the accomplished piano accompaniment of Wolfgang Friedman.

There is already ample discussion in the text books and in the articles about the differences between resort to the ICJ and resort to *ad hoc* arbitration. But most of it is written from a point of view taken from outside the tribunal, whether as an academic observer, as a client, or as an advocate. There is relatively little from the inside point of view of the judges or members of a tribunal. And yet the points which strike a member of a tribunal are also of some significance. No doubt one very good reason for this gap in the literature on the subject is that the deliberations of both kinds of tribunal are, and in the nature of things have to be, strictly confidential (although in the case of the ICJ the procedures followed in the deliberative phases are of course freely available in the officially published Resolution of 1976 on the Internal Judicial Practice of the Court[1]), and in the case of arbitrations the

---

[1] Available at http://www.icj-cij.org under Basic Documents, Other Documents.

pleadings and the hearings too, are often required by the parties to be treated as confidential. But although it would clearly be unacceptable to write about what has happened inside the deliberative chamber in any particular case, there seems to be no good reason to be unduly coy about some of one's general impressions after several years of experience and involvement in both kinds of judicial settlement.

Perhaps the first point to make about the differences between the Court and an *ad hoc* arbitration tribunal is to dispose of the myth that used frequently to be deployed in academic writings, that a permanently established court, and especially the World Court at The Hague, will be inclined to stick closely to applying the law, whereas an *ad hoc* arbitration will be perhaps more inclined to find a compromise solution. No doubt there might have been some truth in this idea in the days of the earliest experiments in arbitration – when indeed there was no international court to compare them with – but the situation for some considerable time has been rather the other way round. An *ad hoc* arbitration is strictly the creature of the *compromise* agreement between the parties and its very continued existence is dependent upon that agreement. The parties therefore control it in a fashion that finds no place in the situation of the ICJ or in any other permanently established court. No doubt the parties themselves, if they do indeed desire a compromise solution, can make this clear in the compromise, although the arbitral procedure would seem an odd one to employ for that purpose. And if it is the overwhelmingly more usual situation in international arbitration that both the parties call unmistakably for the strict application of what they regard as their legal rights, then that is surely the kind of decision the tribunal will and should deliver.

It is rather the ICJ that, on occasion, has not hesitated to be markedly innovatory. One thinks immediately of the *Nottebohm* case, the *Anglo-Norwegian Fisheries* case, and indeed the *North Sea Continental Shelf* case. In the latter case the parties accepted the Court's decision to reject the argument that the equidistance principle was in law mandatory in their particular problem, but had no great difficulty in reaching thereafter an agreed solution which virtually ignored the rest of the Court's eloquent disquisition, including their suggested boundary lines.[2]

One might note also that three very distinguished judges of the International Court – Sir Hersch Lauterpacht, Sir Gerald Fitzmaurice and Judge Jessup – have indeed declared that a bold and expansive attitude on the part of international judges, at any rate in certain kinds of cases, is actually one to be preferred. It was Sir Hersch who had first spoken of

---

[2] The Court was in fact only asked by the parties what were the applicable "principles and rules of international law", as the Court well noted. ICJ Reports 1969, 3 et seq., 13, para. 3.

what he called the "compelling considerations of international justice and of development of international law which favour a full measure of exhaustiveness of judicial pronouncements of international tribunals".[2] And indeed those three judges did all make major contributions displaying "a full measure of exhaustiveness" in their separate opinions. But whatever may or may not be compelling considerations for the judges of the Court, it would be difficult to think of anything more astutely calculated to discourage would be litigants before an *ad hoc* arbitration tribunal than the prospect of the tribunal's applying, not the law as they have heard it confidently described by their advisers, but a "developed" law which nobody had previously thought of.[3] So, for the purposes of the present discussion, the point it is desired to make is, that the notion of judges being under compelling considerations of international justice to take an opportunity to "develop" the law, should surely be a stranger to *ad hoc* arbitration, for the simple reason that it is the parties who choose, establish and control the tribunal, and no intending party to an arbitration is going to choose a tribunal that not only might "develop" the law in a new, unknown and unknowable direction, but might conceive itself to be under compelling reasons to do so. This gears onto a further matter in which there is a major difference between adjudication by a court and adjudication by an *ad hoc* arbitration tribunal, in that in the latter kind, the parties themselves choose their judges; and to the making of this choice we shall now turn.

II. The Choosing of the Members of an Arbitration Tribunal

The writer has once or twice found himself involved in the process of choosing members of a tribunal. In most cases the choice will appear initially to be of one, or perhaps of two, persons to be nominated by one of the parties. Nevertheless a party has also to have in mind the eventual composition of the tribunal as a whole, and this involves also the probable reaction of the other party to one's own choice of nominated members; and indeed one's own reactions to that other party's nominations.

---

[2] See the citations in the Separate Opinion of Judge Jessup at ICJ Reports 1970, 3 et seq., 162; see also the Separate Opinion of Judge Fitzmaurice, *ibid.*, 65.

[3] And even for the Court it is important to have in mind the warning expressed by Judge (now President) Guillaume: "I should like solemnly to reaffirm in conclusion that it is not the role of the judge to take the place of the legislator ... it is the mark of the greatness of a judge to remain within his role in all humility, whatever religious, philosophical or moral debates he may conduct with himself". ICJ Reports 1996, 71.

It seems to be usual for the advisers of parties to prepare for this stage of choosing the tribunal by undertaking thorough research into the publications, decisions and published opinions of proposed names, and by this means presumably to try to establish their likely judicial "attitudes". There may be for instance much eagerness, especially where maritime questions are involved, to find out whether so-and-so is a black-letter-law kind of person or an equity kind of person. One sometimes wonders whether the persons chosen or rejected would have appreciated the apparent assumption that their "attitudes" were already graven in stone and could not be expected to be capable of change; an assumption that does scant justice to the faith of parties in the eventual persuasive powers of their own chosen advocates. Experience suggests, however, that the kind of people who are chosen are usually determined to approach each new case with a mind fully open to consider, and if necessary reconsider, all arguments put to them.

Much more important than the past record of the proposed names – and again this is looking at it from the inside – is the question of temperament. The more important question is not what this person might have thought in the past, but rather whether he or she is a pedant, and if the answer to that question is in the that person be no longer considered. Even if the nature of the pedantry seems to be heading in what is supposed to be the right direction for that party it may still be a danger even for that party because the decisive issues in a case may in the course of the hearing turn out to be quite different from what was expected and probably anticipated; and it is surely right that this might be so, for otherwise it is difficult to see the point of the solemn confrontation of views before the tribunal. There is finally another absolutely crucial question, and one often omitted that from the considerations of the teams making the investigations: will this person get along with the likely other members of the tribunal ? Will the chosen persons be capable of working together as a team and on a friendly basis, even when they find themselves in disagreement? Will this person be prepared to discuss and argue with the others when his or her point of view is challenged, or will he or she decide on an answer early in the proceedings and take the intellectually easier course of sticking to it through thick and thin, and possibly begin preparing a separate or dissenting opinion ? The really good candidate will be more inclined, in face of disagreement, to want further strenuous argument and discussion and to have another good look at the possible merits of alternative points of view other than his or her own preliminary view. Such a discussion, if ably assisted by other members of the tribunal, could well lead, possibly to some compromise; but more desirably, and much more soundly, to the common adoption of a new and different view which the constructive argument might have given birth to. The truth is rarely simple and often emerges not from an eventual choice of the one or the other of opposing views, but as a new product arising from the confrontation of the apparently opposing viewpoints each of which are now seen to have represented an aspect of the truth.

And then there are very practical considerations which however are very seldom even looked at by the parties in choosing their possible names. This is simply whether a proposed person, however eminent and skilled, is likely to be able in fact to devote sufficient time to be available for meetings? For if the arbitration is really doing its job the meetings will almost certainly need to be frequent and sometimes to extend over several days A relevant consideration might therefore be the geographical distance between the bases of chosen members and, if the parties are not very rich, what may the bill for long-distance business class travel look like when considered alongside other costs of the operation ? Nowadays technology makes a telephone and television conference possible and it can sometimes be useful. But there is absolutely no technological substitute for the actual meeting and working together of human beings. Very often the right solution of a problem appears during the informality of a sandwich working lunch, a coffee break, or even over a supper together after the working day is supposed to be over.

The ideal tribunal membership will therefore be of persons who are likely to be able to work together, and preferably also enjoy working together, in these sorts of ways. But choosing is not an easy task for there is no end to the possible even esoteric considerations that may get introduced in this crucial task of choosing members of the tribunal. The writer remembers one case of some time ago when the English speaking side was finding it very difficult to choose a candidate for a still undecided place, who did not for one reason or another arouse the opposition of the other side, which happened to be francophone. Finally the English speaking party, almost in desperation, hit upon the idea of trying to satisfy the preoccupations of the other party, by proposing the name of a person having French as mother tongue, though also equipped with somewhat limited English, but whose juridical qualifications and experience, both technical and personal, were manifestly of the very highest order – in fact a famous and much respected name And yet back came the answer, "No". The difficulty now appeared to be that the parents of the person concerned had been refugees from their own country who fled to England during the First World War, and it so happened that this person had actually been born in England; though it was true that he was taken back to his own country in 1918 whilst still a tiny tot not having yet learned English, or indeed at that stage any other language. The situation seemed not wholly free from absurdity, and years later I had the opportunity of asking the legal adviser who had said "No" about his very odd objection. Oh yes, he said, of course the person concerned was in every way entirely acceptable to me personally, but you have to consider what my position would have been if the case had gone badly against us and a member of the opposition in parliament had got hold of the fact that a crucial member of the tribunal accepted by me had been born in England. Of course the argument would have been absurd. But that would not have prevented my position being made impossible by one of my political enemies.

III. The Size of the Tribunal

It is obvious that there is an important difference between an arbitration tribunal of three or five members in all, and the ICJ with probably fifteen members, or with *ad hoc* judges, even seventeen members. The difference is usually put in terms of the difference for counsel between addressing a small tribunal and addressing the Court, which is certainly very important to them. But the main difference by far, and one of very great practical importance, and one which affects the members of the tribunal themselves, and especially at the stage of the deliberations, is the difference between a discussion of a group of 15 or more, and a discussion in a group of three or five.

In the full Court the general rule has to be, in the terms of the Resolution on Internal Judicial Practice 1976, that "Judges will be called upon by the President in the order in which they signify their desire to speak". The inevitable outcome is a series of speeches rather than a discussion. It is true that sometimes a particular matter can be concentrated on for a short time where a judge asks leave of the President to be allowed to speak out of turn "on the same matter". But quite soon those who are still waiting their proper turn will be getting restive, and when dealing with the list in the order of "signifying their desire to speak" is resumed, it will probably be an entirely different set of issues and views that is then being raised. This kind of deliberation is different from deliberation in a smaller group, not only in degree but in kind.

Within a group of at most five judges the situation is entirely different, for it is then easily possible to have arguments across the table for as long as it seems to the chairman to be profitable to pursue that matter. Moreover, when there arises a strenuous difference of view on a particular issue, it is usually possible to have a considerable general argument and full discussion, concentrating on finding a solution for the particular matter or matters involved. This might resolve the matter one way or the other. But the important point to appreciate is that a thorough-going discussion between not only the protagonists of the opposing arguments but also the other members of the tribunal, may quite often lead to agreement on a different and new solution.

The smaller type of tribunal is also much more manageable when it comes to the very important drafting stage. Some or indeed probably all of the members can be allotted the task of drafting chapters or sections of the award according to the particular knowledge or inclination of each, and perhaps also depending on how much time each of them has available, having regard to other commitments. The president or chairperson will normally have to edit the whole. But the great advantage of a tribunal of not more than five members is that all of them can become the drafting committee, preferably going through and discussing every paragraph together, and if necessary doing so more than once. This is not only good for the drafting but also assists greatly to weld the tribunal together as a team. When the point is reached when the member who produced the original drafts is genuinely grateful when another member points out a weakness and offers a suggested solution or a better or clearer draft, then one knows that the aim of becoming an efficient team is being achieved. The other side of the coin is of course a genuine sadness and sense of loss all round when the final award is handed down and the work together is at an end.

But this kind of welding together of a team in which all members take a full part in both the decisions and the drafting, does take a great deal of time and patience. And this therefore illustrates the vital importance of what was said above about parties choosing members of the tribunal who will be prepared to make and spend the considerable time required for a deliberation which goes very much further than a mere exchange of views. If this can be done, however, there is no doubt that the small arbitral tribunal has important advantages for some kinds of cases. What it may still lack, no doubt, is the authority which comes from a broadly representative decision of the full ICJ. By the same token, however, it follows that even for the full Court, the size of the majority decision is important, and there is no doubt that when, as sometimes happens, the Court is split down in the middle this cannot but somewhat, or even in some instances gravely, weaken the authority and persuasiveness of the decision. And it must also be said that the addition of long and wide ranging separate or dissenting opinions will often weaken the authority and persuasiveness of the judgment. They must also, however much they may sometimes please, and provide materials for professors of international law, merely bewilder the layman parties who brought the case.

IV. Chambers of the Court

This is the point at which it may be convenient to look at the *tertium quid:* the use of a Chamber of the Court for a case as a possible alternative to a separate arbitral tribunal. Of the use of this kind of

Chamber of the Court there is now a considerable experience. It can be said straight away that the kind of thorough and argumentative discussion mentioned above is, or should be, also readily available in a Chamber of the Court. Indeed in some ways it is perhaps even more readily available for the simple reason that all the members, other than perhaps the *ad hoc* judges if there are such, should be available together in The Hague in any event and without having to make special arrangements about accommodation and so on, so that the travel and timetable problems are much more easily solved or avoided. And of course the considerable expense of setting up a tribunal is entirely avoided. Moreover, the Chamber members will presumably already know each other very well, so that in theory at least, it ought to be easier for a Chamber to have thoroughgoing deliberation and sufficiently longer deliberation meetings than in an *ad hoc* tribunal.

We say "in theory", because the theoretical advantage of the Chamber is, in this matter of available time and established mutual acquaintance of its members, offset by the difficulties that arise because the regular judges who are members of the Chamber are also at the same time continuously members of the full Court. And the time table of the full Court might be thought by the full Court, and not least by a hard pressed President of the Court, to have precedence. This doubtless was not a difficulty that would have been even thought of in the days gone by when the Court had so few cases that the expansion of the Chamber system was produced with the express intention of finding work for at least some members of the Court. But now that the full Court has a list of almost too many cases waiting to be dealt with, it is a real problem for the Chambers system. And then there is another difference between a Chamber of the Court and an *ad hoc* tribunal, and that is the possibility of parties before a Chamber deciding to nominate *ad hoc* judges as members of the Chamber. And this they can certainly do, and are therefore likely to do, under the ordinary Rules of the Court. This important factor requires separate consideration.

1. Ad hoc Judges as Members of a Chamber of the Court

Thus far at least the parties have been in effect allowed to decide on the membership of a Chamber of the Court to which they might bring a case; though of course there can be no guarantee that this will also be so. But the possibility of choosing the members is certainly one of the attractions of the Chamber system; and it will be remembered that in the *Gulf of Maine* case the parties made it clear to the Court that, if they were not allowed their own choice of judges, they would abandon the idea of a Chamber and resort instead to an ordinary *ad hoc* arbitration for which the formal agreement of the parties and even the agreed tribunal was, so to speak, ready waiting. This attitude, it is believed, was not wholly popular with the Court, though it is not easy to see on what juridical grounds a valid objection could be made. The Court can obviously refuse to accept the chosen list if it wants to; but that is all it can do. As a corollary of the hitherto accommodating attitude of the Court towards the choice of the members of a Chamber, it must also seem reasonable to hold that the system of *ad hoc* judges should also hold good for Chambers of the Court. And actual practice is in accord. It may no doubt be said in defence of that practice that, in the alternative of arbitration, it is normal for all the members of the arbitral tribunal, including the chairperson, to be appointed by the parties.

Nevertheless there is in actual practice a great deal of difference between the position of *ad hoc* judges as members of the full Court and *ad hoc* judges as members of a Chamber of typically five members in all. It is not only that one in five has manifestly a very different position from 1 in 16 or 1 in 17. There is the additional factor that an *ad hoc* judge of a Chamber of the Court is in a much stronger position than an *ad hoc* judge in the full Court because of the possibility in the Chamber of the kind of that much freer cross argument and discussion already described above. So the fact is that the position of an *ad hoc* judge in a Chamber is at least potentially an altogether very much stronger and more influential position than that of the *ad hoc* judge in the full Court. And where there is the possibility of the exercise of relatively great power, it may be assumed that there will be at least some *ad hoc* judges who may fall to the temptation to make use of it.

Moreover the position of the *ad hoc* judge of a Chamber of the Court is quite different from that of the members of the normal arbitration tribunal notwithstanding that, like an *ad hoc* judge, they will have been nominated in the first place by just one of the parties. (Normally only the chairperson will have been nominated by both parties or, sometimes, chosen by the agreement of other members of the tribunal, or failing that nominated by some third person or institution.) But they are all nominated for the purpose of serving in a completely impartial judicial capacity as ordinary but full members of the tribunal and not with the special position and preoccupations of an *ad hoc* judge. So also, it may be objected, should an *ad hoc* judge. An *ad hoc* judge, however, does have the recognised duty to see that the nominating State's case gets a full hearing and that its case is fully understood and not forgotten. And where this is coupled with the potentially much more powerful position of the *ad hoc* member of a Chamber of the Court, a question of balance could arise. At any rate this is a possible factor that those having to decide between a Chamber of the Court and an Arbitration Tribunal might wish in one way or another to bear in mind.

2. The Question about Questions from the Bench

Quite a lot has been written about the problem of questions posed from the bench during the course of an oral hearing in the ICJ. The impression to be gleaned from some of the writing on the subject is that there are two opposed schools of thought: the common law school which favours the asking of questions from the bench and the continental school which does not. If, however, counsel pleading before the ICJ were to assume that he need not expect searching questions from the francophone members of the Court, he might be in for a great disappointment. But there are some ways in which the common law judge may, and indeed is rather expected, to intervene which does seem to be peculiar to the common law.

The common law judge does not so much think of the matter as one of asking questions but rather of taking an active part in the whole proceeding whenever he opines that it might be useful to do so. Therefore he does not hesitate to do so whenever, for example, he feels that he had not entirely followed the argument being put to him. He will then not just ask about it but also probe it with counsel and there may be in effect a short debate between judge and counsel until the judge is satisfied that the issue is clear in his own mind. And if the judge, after understanding the point being put thinks that it is a non-point, or a waste of time, he often will not hesitate to say so there and then.

But the situation in the ICJ before 15 or more judges is a very different situation; and that not just in climate or tradition, but physically. It is obvious that it cannot be permissible for any one of 15 judges to intervene with a question, much less a discussion, just when the spirit moves him or her. That would simply produce a chaotic and impossible situation. Questions are indeed nowadays often asked from the ICJ bench; partly it may be from the pressure in that direction of common law trained judges. But the questions have to be asked at an announced time – the end of a session and just before adjournment for lunch is a favourite time, but the President will probably decide on the time for asking questions – and they will as a matter of courtesy probably have been circulated beforehand to the whole bench and therefore inescapably discussed and probably modified, typically during one of the so-called "coffee breaks"; and indeed some of the proposed questions might not have survived the coffee break. Moreover, the questions will probably be prefaced by the usual statement that the questions need not be answered immediately, and indeed may be answered in written form within a certain time even after the formal end of the oral hearings. And printed copies of the questions are normally handed to the parties immediately after their asking. This blunting of the edge of the questions is of course supposed to be necessary because one is dealing with "sovereign States" and so counsel will probably wish to seek instructions before answering the questions.

This form of asking questions, though no doubt as already mentioned probably partly the result of pressure from common law members of the Court, could hardly be more different from the common law practice. It may on occasion be very useful, especially when the question is one that the whole Court wants to put, and which will then be asked by the President or other presiding judge and in the name of the Court. But it can often also be somewhat of a waste of time, for very often, even when the question is a pertinent one, any member of the Court could make a very good shot at writing out pretty accurately, perhaps even as to matters of style as well as substance, the answers that will inescapably be given by the party concerned, or as more often is the case by each of the parties.

The discussions within the Court about questions being proposed to be asked do, however, sometimes reveal crucial differences in attitudes; and differences moreover that cannot always be explained by the differences between the common law and civil law traditions. For example the writer remembers one very distinguished judge of the Court who could be relied upon to make not just objection, but deeply shocked objection, to any question which might be thought on careful examination to reveal, or even possibly reveal, the direction in which the questioner's mind was tending. And no doubt that particular objection is the more strongly felt if that direction happens to be in the opposite of the direction in which the objector's mind is tending in the case. Other judges of course might feel that a revealing question has virtue in that it is useful to counsel on both sides to have early warning of a tendency of at any rate one, and possibly more, judges. For there may still be time to do something about it and to put in stronger, or longer, arguments on the point involved; and the very need to answer the question gives a new and separate opportunity of doing so. These differences between judges again seem to be more about differences of temperament rather than supposed differences of legal traditions.

This matter of questions is however one in which the situation in an arbitral tribunal of three or five members is completely different from that of the Court. There it is entirely possible for any member who wishes to ask a question or raise a point in the course of counsel's argument to do so, though preferably after first indicating to the chairman the intention to do so and having received his or her agreement. Such a question or intervention is not usually intended to be answered in writing much later but to be answered immediately if counsel is prepared to do that, when it might well lead to a short, or even a longer, further probing or argument. This kind of intervention, very much in the common law tradition, can be very useful and instructive and indeed productive for the tribunal as a whole. The attitude of counsel to this kind of intervention varies and is again a matter of temperament rather than tradition. Some counsel, though they will of course answer the question in some form or at any rate go through the motions of attempting to answer, are manifestly irritated by having their presentation interrupted. Other counsel plainly flourish on questions and are disappointed if they do not come. It is not really "questions" that are at issue here. It is the possibility of a fruitful investigative exchange between judges and counsel; and this, if prudently and reasonably indulged in, can be of very great value. But in its valuable form such interventions are obviously a luxury only to be enjoyed, and indeed carefully controlled, by small tribunals such as the typical arbitration tribunal.

Something also depends no doubt on whether the proceedings are in public or are confidential. And the latter possibility is pretty well confined to those arbitrations in which the compromise stipulates the privacy of the pleadings and oral proceedings. When the proceedings are open to the public, or at any rate not strictly, or at any rate effectively confidential, this is a factor to be taken into consideration in the framing of the questions. One remembers the famous *Beagle Channel* arbitration between Argentina and Chile. This was essentially about the rival claims to sovereignty over three islands in the Beagle Channel. At one point the tribunal asked the sensitive and politically charged question whether it was in the view of the parties a question of three islands or none, or whether some compromise, some splitting up of the islands might be acceptable. The apparent naivety of this question is still surprising to the observer. The tribunal must have known, or certainly should have known that, the answers would certainly be reported to the respective Governments. The tribunal might have guessed therefore that the question would be answered not by counsel but by the Agents, who would in effect be addressing the members of their own Governments at home rather than the tribunal, and would inevitably each have to say, with the maximum vehemence and obduracy of language possible, that his Government demanded its entire legal rights and could not even contemplate the idea of a compromise of those undoubted full legal rights. One doubts whether counsel on either side even bothered to listen to the answers, which anybody present on either side could have written out for the tribunal. But the writer still has an uneasy feeling that the tribunal took the answers seriously; whether through innocence or guile is still unclear.

Where one has counsel on both sides who know each other well and are both very experienced, as may happen, at least in the commercial type of international arbitration, then an element of profitable dialogue about the case may also very usefully occur between counsel even outside the confines of the court. The writer remembers one such commercial arbitration where the counsel had much experience of each other in the High Court and in particular of the Commercial Court, in London. When the arbitration had been running for some days, counsel on one side began his morning's speech by saying that he would not that morning be spending much time on a certain argument that had been put forward by the other side. This, he went on to explain, was because in a discussion he had had with his "learned friend" on the other side that morning and before the court assembled, he had gathered the impression that his learned friend might not be going to press that particular argument in its present form. His learned friend on the other side immediately intervened, not to protest, but to say that he thought it might help the court, and his learned friend, to know that he had now discussed that argument with his clients and they had come to the opinion that that argument had certain weaknesses – "would not run" was how he put it – and had therefore decided to abandon it altogether. This astonished the two international lawyers on the tribunal but not so a domestic UK Lord of Appeal, very used to cases in the London Commercial Court, and who was a member of the tribunal. He simply asked the question: Does that mean that we no longer need to look at volume X of the pleadings? Yes, Sir, replied counsel, you can now forget that volume.

Now this seems to the present writer to be a highly desirable way of conducting a case, economical of both time and money, as well as being quite astute advocacy in readily abandoning a whole argument, where the other side had so weakened it that it was better honestly and openly to abandon it than to persist. One can only hope that such enlightened behaviour from commercial courts and arbitrations may eventually infect even fully international tribunals with such civilised forensic behaviour. We have, alas, probably some time to wait before counsel will do something so sensible when pleading before the ICJ.

3. The Importance of the Registrar and Supporting Staff in an Arbitration

From the point of view of the members of an arbitration tribunal, and especially from the point of view of the president or chairperson, there is one very important difference between their situation and that of the judges of an established court like the ICJ. The arbitration agreement will usually have said something about the rules of procedure to be applied; usually some form of the UNCITRAL rules. But usually it says nothing or very little about certain big problems that have to be faced immediately: the appointment of a registrar and supporting staff; the hire of equipment such as word processors, fax machines, duplicating machines, and supplies of various kinds of paper and envelopes; telephone and computer connections; the collecting of the funds from the parties to finance the arbitration and the making of the estimates of costs and time, in order to know the sums to demand; the need to establish a bank account or perhaps several bank accounts, and some system of accounting acceptable to both to the parties and to the other members of the tribunal and staff; the amount and the basis of the remuneration of members of the tribunal and of the registrar and auxiliary staff when they are found and engaged; and finally but by no means least there is the question of suitable premises in which to hold meetings and the oral presentations, and again the question of the cost of premises and for how long they should be hired and all the dates likely to be involved. There may also be later questions about the appointment of simultaneous translators and/or those firms that provide daily a transcript of the oral proceedings with the help of those remarkable young ladies who, on little machines, take down every word with astonishing accuracy, never letting their attention be diverted or to miss a single word. And there may be a need, especially where maps or charts are involved, to appoint experts to assist the tribunal, and again the ancillary questions of pay and timetable, and the agreement of the parties. And all of these matters will have to be costed before one can determine the amounts to be added to the sums to be asked of the two parties and eventually accounted for. And many other technical problems will be met on the way. For example counsel these days do not expect just to talk. They often demand what are now called visual aids of various kinds. Screens that can be seen by both counsel on both sides and the bench are therefore necessary and this also creates problems about portable microphones and wiring. And so on and on. The possible problems are endless and many arise unexpectedly and require quick remedies. All these things cost money and some of them a great deal of money. The premises, which will include not only a large court room but also retirement room for the judges, rooms for the parties, rooms for the secretaries and transcribers, and facilities for security guards, and preferably some means of providing at least a working lunch of sandwiches for judges and staff both during hearings and during deliberations. The premises, if in a capital city will certainly cost many hundreds of dollars for every day of money which has usually to be paid without delay. The classic way for the president or chairperson to deal with these problems, or some of them, used to be to appoint some good and energetic and strong and therefore probably young-and-coming international lawyer as registrar and tell him to get on with it, no doubt offering wise and advise or admonishments from time to time; and of course to arranging to pay him relatively little in return for the honour thus bestowed upon him (it was always "him" in those days). What is mainly needed, however, is skill in organisation and management. If a person with such abilities is also an international lawyer that will be a bonus; but the essential skill required is management and the training of international lawyers does not usually include courses in management and accounting.

It is not surprising therefore that more arbitration cases are now being taken to institutions to organise, and particularly to ICSID which deals with investment disputes, or to the Permanent Court of Arbitration (PCA) at The Hague, which deals with all kinds of disputes and can provide a staff with great experience both of the management problems and of international and procedural law; and also can provide magnificent premises in the Peace Palace at costs which compare very favourably with other possibilities such as large hotels and other kinds of public buildings; although the PCA can also organise an arbitration where the place of arbitration is elsewhere than in The Hague. There is also in the Peace Palace a professional accountant who can look after the money side and the PCA will take responsibility for that side too; and this is a great relief to any person who has the charge of a tribunal. This is very important because it cannot be right for a president of a tribunal to have the responsibility of paying himself and his colleagues for his and their services. The PCA in fact in these administrative matters complements its younger partner in the Peace Palace, the ICJ, which of course does have the very great advantage of its own permanent staff and splendid premises of the Peace Palace at its disposal.

The Registrar of an arbitration is not only a manager both of the necessary staff and of the tribunal. He is also a necessary link between the president or chairperson and the parties. All correspondence between the tribunal and the parties must in principle be duple in the sense that copies of everything must be immediately available to both parties. Experienced Agents and counsel know this very well and will take great care never to communicate with the tribunal (or the president) except with a copy to the other party; and of course never correspond with the other party about the case without a copy to the tribunal. Nevertheless there are times when a discreet inquiry to one party can be useful. Suppose for example an important note is sent by one party to the tribunal with a copy to the other party. A question often arises whether the other party will wish to comment or to make a counter proposal and if so how long they think it will take to do that. A formal letter of inquiry to that party with copy to the other party would sometimes be appropriate, but more often that would not be without an element of absurdity and might waste time pointlessly. A telephone call to the Agent might be a better answer. And the registrar or his staff can do that quite properly and in accordance with normal expectations. A good registrar will also try to ensure that he or she has good and on-going relations with both parties, and will try to pick up any hints about how things are going and will hope to know about and deal with small disputes, such as those about procedures or the production of documents, before they become serious and create bad feeling. The registrar may quite properly in the ordinary course of arrangements make himself available to a party which simply wants to ask advice about the way to do some things or the way to organise its case so as to fit in with the normal arrangements and expectations. For not all parties have experience, or indeed any experience, of international litigation. In these ways the registrar, and probably also his staff, will have opportunities of getting to know the parties and assessing the climate of opinion.

The Registrar therefore may often be in a uniquely informed position to advise the President and the Tribunal on the details of the conduct of the case and help generally to avoid bad feeling arising, as it can so easily do when people are under great pressure. In these matters it is the ICJ Registrar and staff that set the pattern and have the longest experience. An important difference, in these matters of organisation, between the ICJ – whether full Court or Chamber – and arbitration is that, in an *ad hoc* arbitration, it might well be the first case that the registrar has had any experience of the scale and complexity of the organisation required. Then it is clear that putting the organisation of an arbitration case into the hands of the Permanent Court of Arbitration at The Hague, has much to commend it.

An undoubtedly great advantage of resort to the ICJ is that, in addition to the availability of the advice and services of a probably very experienced Registrar and staff, it is all free, being paid for in the United Nations budget. And of course so are the salaried members of the Court; but in an arbitration each one is to be paid, usually by the hour. So the fact is that resort to the Court is for these reasons very seriously cheaper than resort to *ad hoc* arbitration. One cannot but wonder whether the vast difference in costs is always realised by parties and taken into consideration at the appropriate time; and one wonders also whether their advisers always realise the differences in the scale of the costs. It is one of those very serious factors that the academic books on the matter seldom adequately explain.

Certainly the scale of the differences in costs is all the time getting more serious not least because of the demands of modern technology. These problems of organisation have become much more complicated than they used to be, because the pressure to use modern technology seems now to be irresistible: computers and word-processors; copying machines; the often irritating diversionary ploy of "visual aids", even though these as often as not, merely add a further layer of obscurity to what is better said in carefully chosen words; amplification to encourage mumbling; and the rest of the always expensive extras, almost certainly none of them even referred to or thought of in the agreement for arbitration. Even before the World War II, virtually none of this technology was available or even invented, and tribunals seemed to get along just as well without it. And any president of a tribunal who has tried to run an early morning preparatory meeting at the time when members are habitually busy with their laptops checking their largely pointless "e-mail", may well think there was something to be said for days when manual typewriters were the smartest available technology and visual aids, if they meant anything at all, suggested large maps hung on the wall, and long-distance telephone calls required booking several hours before the call could be expected to become available. But the contemporary situation, which is certainly not going to go away, calls unmistakably for some institution with experienced staff to see to these matters and here the institutions like the Permanent Court of Arbitration and the ICSID do readily supply the required detailed service economically and as, to them, a routine service.

In conclusion we may say that there are many differences between resort to the ICJ, or to one of its Chambers, and resort to *ad hoc* arbitration. Some of these are subtle and some are very evident. And of course there are factors that have not been considered in this paper and which might arise in certain cases, such as political prejudices of one kind or another in favour in particular circumstances of the one method rather than the other. But a major difference and factor, is the very considerable difference in the scale of costs between a system where virtually all has to be paid for by the parties and a system which is charged to the United Nations budget.

On the Legal Character of Retaliation in the World Trade Organization System

Akira Kotera

I. Introduction

The World Trade Organization (WTO) dispute settlement system,[1] in particular the panel system, is very strong in comparison with other interstate dispute settlement mechanisms. The reasons are as follows. First, when a Member State files a request to establish a panel composed of independent experts in the Dispute Settlement Body (DSB) of the WTO, a panel is established automatically unless this request is rejected by consensus (negative consensus). Before the creation of the WTO, a decision by consensus in the GATT Council was required to establish a panel when a request for a one was filed. Secondly, the negative consensus also applies to the adoption of panel or Appellate Body reports, that is, those reports are adopted by the DSB automatically unless they are rejected by consensus. Before the WTO, consensus was required for the adoption of a panel report by the GATT Council. In the GATT era, it often took a long time to establish a panel after the request was filed, because the State that was the subject of complaint objected to its establishment. But this problem was not so serious as the second one mentioned above, because a panel was finally established in practice, even in the GATT era. As regards the second problem, several panel reports were not adopted in the end because of the persistent objections of the defaulting party. A panel report could not be adopted even if only the defaulting party did not agree to its adoption. Therefore, there was no guarantee that the GATT dispute settlement system could settle a dispute quickly, especially in its final form. Such defects embedded in the GATT dispute settlement system were the main excuse of the US Government for maintaining unilateral measures based on Section 301 of

---

[1] On WTO dispute settlement, see D. Palmeter and P. C. Mavroidis, *Dispute Settlement in the World Trade Organization*, 1999.

the US Trade Act of 1974, or Super 301 of the US Omnibus Trade and Competitiveness Act of 1988. The WTO Agreement has improved this dispute settlement mechanism in a dramatic way.

However, because of these characteristics of the GATT dispute settlement system, all the panel reports were implemented by defaulting parties. Since defaulting parties agreed to the adoption of panel reports, it was natural that defaulting parties implemented the reports on their own initiative. Under the present mechanism, the defaulting parties have no such incentive to implement the reports adopted by the panels or the Appellate Body on their own. Even those panel or Appellate Body reports which could not have been adopted under the GATT system, because the defaulting parties, not being satisfied with the reports, had not agreed to their adoption, have been adopted in the WTO because of the negative consensus mechanism. As a result, several appellate body reports have not been implemented by defaulting parties, and the issue of how to cope with such non-implementation has arisen.

Compensation and retaliation, which are also known as "the suspension of concessions or other obligations" (Dispute Settlement Understanding (DSU), Article 22, paragraph 1), are the two means of responding to the non-implementation of panel or Appellate Body reports. In the GATT era, compensation was often agreed to by the parties concerned, but retaliation was not invoked except in the *US Dairy Products* case.[2] In that case, the Dutch Government obtained permission from a GATT panel to retaliate against the United States, but did not take any retaliatory measures. Thus, the retaliation issue was not so important in the GATT era. The situation surrounding the retaliation has completely changed, however, and questions have developed about what form retaliation should take, and what the legal character of the obligation of WTO agreements should be. The legal character of retaliation wholly depends upon the obligations of WTO agreements, including those of GATT.

Two opposing views on retaliation have been raised. The first is that retaliation is only a tentative measure until complete implementation of the panel or Appellate Body report. According to this view, the WTO Member States should comply with the panel or Appellate Body reports in any case, and the function of retaliation is as a sanction to force defaulting parties to implement such reports.[3] In the second view, the purpose of the retaliation is to restore the balance of trade benefits arising from concessions and other obligations under the WTO agreements between the parties concerned, which were determined at the conclusion of the latest round of multilateral negotiations, in this case

---

[2] *US – Restrictions on Dairy Products*, adopted on 8 November 1952, L/61.
[3] J. Pauwelyn, "Enforcement and Countermeasures in the WTO: Rules are Rules – Toward a More Collective Approach", *AJIL* 94 (2000), 335 et seq.

the Uruguay Round.[4] According to the second view, retaliation can be a final measure if the defaulting parties refuse to implement the panel or Appellate Body reports.

The second view is based on the premise that the purpose of the WTO system is to ensure access for Member States to other Member States' markets, and to record the results of their trade negotiations, which are the balance of export benefits achieved for the Member States. From this perspective, the most important matter is to maintain the balance of trade benefits; requiring Member States to implement the obligations of the WTO agreements is a secondary issue. The present WTO system originated from United States bilateral trade agreements, in which the balance of trade (export) benefits, which were shaped mainly by tariff concessions, occupied the central position.[5] From this viewpoint, if a prevailing party takes retaliatory measures in line with the DSU, the dispute can be settled in a final form, such as the payment of reparations in the context of state responsibility.

This controversy is not merely academic, but has practical aspects. First, the WTO dispute settlement system occupies a vital position in the WTO to ensure the implementation of the obligations of Member States. This is the main reason why so many items are being proposed for the agenda of the next round of multilateral negotiations. Secondly, to the extent that many trade-related disputes have been treated in efficient ways, politically sensitive disputes, such as those related to the environment, have also often been filed with the WTO dispute settlement system. The question how to deal with this dispute settlement mechanism is vital not only for trade diplomats, but also for international society. What the retaliation under the WTO dispute settlement system should be has become the inevitable topic of discussion in these contexts.

II. Retaliation under the WTO System

The origin of the WTO dispute settlement system lies in Article 23 of GATT. Article 23 stipulates the following two causes of action: a violation of object, and nullification or impairment of benefits. Hitherto the violation of object has not been invoked, only nullification or impairment of benefits. In order to invoke this cause of action, Member States must satisfy one of three requirements: first, the failure of another Contracting Party to carry out its obligations under the GATT (violation complaint); secondly, the application by another Member State of any

---

[4] D. Palmeter and S. A. Alexandrov, "Inducing Compliance in WTO Dispute Settlement", (mimeo) paper on file with author.

[5] J. H. Jackson, *World Trade and the Law of GATT*, 1969.

measure, whether or not it conflicts with the provisions of GATT (non-violation complaint); and thirdly, the existence of any other situation (situation complaint). Most disputes fall under violation complaints, while only a few are non-violation complaints. No report has yet been issued under situation complaints.

Compared with the regular dispute settlement mechanisms under international law, the WTO system has two remarkable features. First, according to GATT, in order to file a complaint with the dispute settlement mechanism, Member States are required to demonstrate the nullification or impairment of benefits. Under the present DSU, nullification or impairment of benefits is considered *prima facie* to constitute a case of nullification or impairment in the case of a violation complaint (Article 4, paragraph 10). However in a non-violation complaint or a situation complaint, indication of nullification or impairment of benefits is still required. Secondly, in the case of a non-violation complaint or situation complaint, the party that is the object of the complaint is not alleged to have violated any obligation under GATT. Thus, from the viewpoint of violation complaints, the central requisite of the WTO dispute settlement system is the obligation of Member States, but from the viewpoint of non-violation or situation complaints, it is the nullification or impairment of benefits.

According to Article 23 of GATT, the WTO considers "that [if] the circumstances are serious enough to justify such action, they may authorize a contracting party or parties to suspend the application to any other contracting party or parties of such concessions or other obligations" under GATT "as they determine to be appropriate in the circumstances". This is retaliation under GATT. Additional regulations are stipulated in the WTO Dispute Settlement Understanding (DSU). According to Article 22 of the DSU, if Member States do not comply with the recommendations and rulings within a reasonable time and no satisfactory compensation has been agreed between the parties, "any party having invoked the dispute settlement procedures may request authorization from the DSB [Dispute Settlement Body] to suspend the application to the Member concerned of concessions or other obligation" under any WTO agreement. We need to pay special attention to two paragraphs concerning such retaliation.

First, Article 22, paragraph 1, of the DSU stipulates that suspension of concessions or other obligations is a temporary measure available in the event that the recommendations and rulings in the adopted report are not implemented within a reasonable period of time. Furthermore, Article 21, paragraph 1, stipulates: "Prompt compliance with recommendations or rulings of the DSB is essential in order to ensure effective resolution of disputes to the benefit of all Members". These paragraphs are considered to demonstrate that panel or Appellate Body reports that include recommendations or rulings are legally binding.[6] They have been referred to as evidence that retaliation under the WTO system is a sanction to force the compliance of the defaulting party with the obligations of the WTO agreements. The first arbitral panel in the *Banana* case[7] to consider the level of suspension of concessions or other obligations in the WTO, found that "this temporary nature indicates that it is the purpose of countermeasures to induce compliance".[8]

Secondly, however, we must keep in mind the Article 22, paragraph 4, provision concerning the level of suspension of concessions or other obligations under the WTO. This paragraph provides, "The level of the suspension of concessions or other obligations authorized by the DSB shall be equivalent to the level of the nullification or impairment". Why should the level of the suspension of concessions or other obligations be equivalent to the level of the nullification or impairment although the Article 23, paragraph 2, of GATT provides that the suspension be appropriate? If the retaliation, that is the suspension of concessions or other obligations, is to be a sanction to induce a defaulting party to comply with the panel or Appellate Body reports, we cannot find any reason why its level should be the same as that of nullification or impairment. If retaliation has this character of a sanction, it may be proper that the level of retaliation should be higher than that of nullification or impairment. Rather, "appropriate" should better be considered from the viewpoint of a sanction. This level should be assessed on the grounds that the sanction is strong enough to induce a defaulting State to comply.

Thus either view has a theoretical basis to defend. Which view is appropriate should not be determined only in a formalistic way, by analysis of the words used in WTO agreements. More substantial analysis on the basic legal structure of the WTO system is required.

III. The Legal Structure of the WTO System

GATT was originally an agreement related only to trade, especially tariff concessions. Member States exchanged tariff concessions with each

---

[6] J. H. Jackson, "The Legal Meaning of a GATT Dispute Settlement Report: Some Reflections", in N. Blokker and S. Muller (eds.), *Towards More Effective Supervision by International Organizations, Essays in Honor of Henry G. Schermers*, Vol. 1, 1994, 157 et seq.

[7] *EC – Regime for the Importation, Sale and Distribution of Bananas*, Recourse to Arbitration by the European Communities under Article 22.6 of the DSU, WTO Doc. WT/DS27/ARB.

[8] *Ibid.*, para. 6.3.

other to increase the benefits of trade, and later to promote a liberal trade system. At that time, legal rules were not so important as at present. Legal rules under GATT were tools to guarantee the benefits of the Member States arising from the tariff concessions. If the trade rules had been changed arbitrarily to protect domestic industries, the tariff concessions which were agreed in international fora would have been impaired to the point that they would have become meaningless.

But in the Tokyo Round of multilateral negotiations in the 1970s,[9] non-tariff barriers such as anti-dumping measures or countervailing duties came to receive more attention than before as tariff levels for manufactured goods had been dramatically decreasing until the Tokyo Round. At the conclusion of the Tokyo Round, several agreements related to non-tariff barriers, such as the Standard Agreement and the Government Procurement Agreement, were concluded.

In the Uruguay Round, which started in 1986, the trade in services and intellectual property were included on the agenda, as well as tariff and non-tariff barriers. As a result of these negotiations, two agreements which covered new areas such as the trade in services and intellectual property rights were established, along with other agreements related to the trade in goods. These were the General Agreement on Trade in Services (GATS) and the Agreement on Trade-Related Aspects of Intellectual Property Rights (TRIPS). As a result of these negotiations, the WTO system, which covers more ground than the GATT system, came into being. At the present time, many States are proposing the start of a new round of multilateral negotiations, the agenda for which may include environmental issues, labor standards issues, competition issues, investment, etc. The scope of the WTO/GATT system has been gradually extended from only the trade area to various other economy-related areas. How are we to understand this phenomenon?

A persuasive explanation goes as follows. In order to increase export benefits and access to other Member States' markets, international regulations on national border measures such as tariff regulations alone are not sufficient. Thus, GATT Member States were led to non-tariff matters to introduce the international regulation of purely domestic matters such as subsidies or government procurement. Furthermore, when the aim of the international regulations for the trade of goods was realized to a certain extent, the question of why the trade in services was completely unregulated received more attention than previously. The WTO/GATT system then proceeded to include the trade in services. This is a course that the WTO/GATT system has followed to develop an international regulatory system step by step. Under this evolving system, is it reasonable to expect any regulation to be strictly

---

[9] See G. R. Winham, *International Trade and the Tokyo Round Negotiation*, 1986.

enforced? The question whether a certain rule which is justified from a purely trade perspective is absolutely right from another perspective is an issue that still needs to be discussed. This statement can be easily understood from the point of view of environment-related cases.

In both the *Dolphin I* and *Dolphin II* cases,[10] the GATT consistency of the general import prohibition of the United States against tuna harvested by fishing methods that result in the accidental killing of dolphins for the purpose of protecting these ocean mammals was a major issue. One of the US Government's arguments was that the measure was justified under the exceptions in Article 20, paragraph (g), of GATT. But the GATT panels found that the United States import restrictions on tuna were not justified under the exceptions. The panels considered that the measure could only be taken "if it was primarily aimed at rendering effective these restrictions",[11] but the United States restrictions did not satisfy this requirement. The panel report of the *Herring and Salmon* case mentioned the meaning of "primarily aimed" as follows: "while a trade measure did not have to be necessary or essential to the conservation of an exhaustible natural resource, it had to be *primarily aimed* at the conservation of an exhaustible natural resource to be considered as 'relating to' conservation within the meaning of Article XX(g)".[12]

Under the WTO system, the United States import restriction on shrimp for the purpose of the protection of sea turtles was complained about to the WTO dispute settlement mechanism by India, Pakistan and Thailand (the *Shrimp-Turtle* case).[13] The US Government again argued that the restriction was justified under the exceptions allowed under Article 20, paragraph (g), of GATT. In this case, the WTO Appellate Body interpreted the exception in a more liberal fashion. The Appellate Body considered that in order to satisfy the condition that "relat[es] to" the conservation of exhaustible natural resources in paragraph (g), it was required that the measure in question be reasonably related to the legitimate policy of conserving exhaustible natural resources. The Appellate Body considered that the United States measure "is a measure 'relating to' the conservation of an exhaustible natural resource within the meaning of Article XX(g) of the GATT". After this finding, the Appellate Body examined the *chapeau* standard of Article 20 of GATT

---

[10] *Tuna I: US – Restrictions on Imports of Tuna* (unadopted), BISD 39S/155; *Tuna II: United States – Restrictions on Imports of Tuna* (unadopted), ILM 33 (1994), 839.

[11] *Tuna I*, see note 10, para. 5.30.

[12] *Canada – Measures Affecting Exports of Unprocessed Herring and Salmon* (adopted on 22 March 1988), BISD 35S/98), para. 4.6), emphasis added.

[13] *US – Import Prohibition of Certain Shrimp and Shrimp Products* (adopted on 6 November 1998), WTO Docs. WT/DS58/R, WT/DS58/AB/R.

and found that the United States measure did not conform to these standards.

In comparing the interpretations of Article 20, paragraph (g), in these three cases, the interpretation in the *Shrimp-Turtle* case can be said to have been moderated. In the *Tuna* case, the measure had to be primarily directed to the purpose, but in the *Shrimp-Turtle* case it was enough that the measure be reasonably related to the purpose. Why did such a change take place? The Appellate Body took account of to the new preamble to the WTO Agreement, which states "... while allowing for the optimal use of the world's resources in accordance with the objective of sustainable development, seeking both to protect and preserve the environment and to enhance the means for doing so in a manner consistent with their respective needs and concerns at different levels of economic development", although the section of the arguments to which this referred was not concerned with the interpretation of Article 20, paragraph (g), but with the *chapeau* standard. The Appellate Body determined that "this language demonstrates a recognition by WTO negotiators that optimal use of the world's resources should be made in accordance with the objective of sustainable development". This statement indicated that the WTO would take environmental considerations into account in interpreting the provisions of GATT.

The change in the Appellate Body's attitude on the interpretation of Article 20, paragraph (g), in these environment-related cases demonstrates that the scope of the dispute settlement mechanism has become so enlarged that factors other than trade have come to be considered in the WTO dispute settlement system. In the present WTO, protection of the environment is an important factor to be taken into account. What effects have this extension of scope brought about? In the environment-related field, the legitimate basis of Article 21, paragraph 1, and Article 22, paragraph 1, of the DSU was intensified, which can be interpreted to mean that the WTO Member States will comply with their obligations under WTO agreements in all circumstances. If the sentence related to sustainable development had not been written into the preamble of the WTO Agreement, the question whether the balance of trade benefits fixed at the conclusion of the former round was impaired by environment-related measures would be the sole issue to be considered. This balance should be restored by the adjustment of the measures of Member States concerned without the defaulting party's complying with the panel's judgment, in case a measure of a Member State was found to be inconsistent with the GATT rules by a panel or by the Appellate Body. The maintenance of the balance of trade benefits was the only matter to be weighed. The previous GATT system had such characteristics. That system was based on the grounds that its sole interest was the maintenance of the balance of trade benefits. However, in the WTO system, a panel or the Appellate Body cannot take such an attitude in certain environment-related cases, such as the *Shrimp-Turtle* case, because the WTO system already contains the perspective of sustainable development. The WTO is not a pure trade system the results of which are Article 21, paragraph 1, and Article 22, paragraph 1, of the DSU. The WTO has attempted to regulate the general behavior of Member States.

What about the human health sphere? In the *Beef Hormone* case,[14] the Appellate Body, as well as another panel, considered that the European Community's prohibition of imports of meat and meat products derived from cattle to which hormones had been administered was inconsistent with the obligations under the WTO agreement, in particular the SPS Agreement (Agreement on the Application of Sanitary and Phytosanitary Measures). But the EC has not complied with the Appellate Body's Report and United States' retaliatory measures have already been invoked. At the present time, the EC has made a proposal that precautionary principles for human, animal and plant health, as well as the environment, should be introduced into the WTO system.[15] If such a proposal is adopted in the SPS Agreement in the next multilateral negotiations, a measure that was found to be inconsistent with the SPS Agreement might become consistent with the revised SPS Agreement. If such a revision is realized, how shall we consider the former reports? It frequently occurs that a measure that was previously prohibited becomes permitted after the related regulations are changed, even in domestic societies. But the protection of human health has occupied a central position in the policy of sovereign States. The purpose of the SPS agreement is "the establishment of a multilateral framework of rules and disciplines to guide the development, adoption and enforcement of sanitary and phytosanitary measures in order to minimize their negative effects on trade". From this preamble, the SPS Agreement was formed to promote trade, not human health. Therefore, the SPS Agreement could not help but declare that "no Member should be prevented from adopting or enforcing measures necessary to protect human, animal or plant life or health". But it is clear that the policy freedom of sovereign States related to sanitary and phytosanitary measures has been restricted by the agreement. In such a situation it may be unreasonable to demand that Member States comply absolutely with the obligations concerning sanitary and phytosanitary measures under the SPS Agreement, even when a Member State considers that to use a certain sanitary or phytosanitary measure which may be inconsistent with the SPS Agreement is required for human health. The reason why such a strange

---

[14] *EC – Measures Concerning Meat and Meat Products (Hormones)* (adopted on 13 February 1998), WTO Docs. WT/DS48/R, WT/DS48/AB/R.

[15] Commission of the European Communities, *Communication from the Commission on the Precautionary Principle*, COM(2000)1, 2 February 2000.

situation might occur is that the promotion of health was not fully taken into account when the SPS Agreement was formed.

It is true that the WTO/GATT system has mainly focused on trade matters, even though – recently – the emphasis has shifted from concessions to non-tariff barriers. From the perspective of non-tariff barriers, the SPS Agreement and similar agreements were drafted to promote trade by minimizing the negative effects arising from Member States' unilateral measures, and therefore the purposes of such measures themselves were not a basic policy objective to attain in making the agreements. If the EC proposal on a precautionary principle is realized in the next round of multilateral negotiations, it will be demonstrated that the present regulations of the WTO system are not perfect in dealing with health protection.

The WTO system enjoys strong legitimacy with regard to the international regulation of trade in goods and services and the protection of intellectual property rights. Does it enjoy the same legitimacy with regard to international regulation in other fields which are related to trade to some extent? Since many items do relate to trade, the WTO Agreements touch on those items. Because a certain balance of trade benefits among Member States was fixed in the WTO Agreements, it is reasonable that the Member States have to readjust trade relations amongst themselves if new trade-related national measures concerning human health protection and so on, which were not permissible under the WTO system, are adopted. From the trade perspective, such measure may impair the balance of trade interests. However, it may be unreasonable to insist that Member States change these measures on the grounds that a measure is inconsistent with obligations under GATT, because trade is not the only perspective from which every national measure should be assessed.

As Article 22, paragraph 1, and Article 21, paragraph 1, of the DSU stipulate, the obligations of GATT, and the judgments of panels and the Appellate Body, should be implemented in principle. But we must remember that the WTO system provides the proper ambit under which Member States should absolutely comply with their obligations under the WTO Agreements. This covers the trade in goods and services, and protection of intellectual property rights. Outside this sphere, the WTO system loses the legitimacy to enforce these regulations with regard to Member States, but can only direct them to readjust the balance of trade benefits among themselves. Even outside the proper sphere no one can doubt that the balance of trade benefits has been impaired by this measure. This is the reason why the defaulting party should compensate the prevailing party or suffer its retaliation.

At this point, we arrive at the final consideration for the legal character of retaliation. Within the proper ambit of the WTO system, such as the trade in goods, retaliation should be considered as forcing

Member States to comply with the panel or Appellate Body's judgment. But outside this ambit, retaliation can be a final measure to restore the balance of trade benefits among Member States.

IV. Conclusion

The WTO system is rather independent from the regular legal mechanisms in international society, and therefore may be referred to as a "self-contained régime". The WTO system, which does not make use of the International Court of Justice, has its own dispute settlement mechanism. Although the WTO dispute settlement system is similar to a judicial organ as a result of its gradual development since the period of GATT (judicialization), it has many special features, such as unique causes of action. The controversy over the legal character of retaliation has taken place in this context. From the interpretation of various conditions for retaliation under the WTO system, and the report of the arbitration on retaliation under the DSU, both views have a persuasive legal basis. In this article, I have approached this topic from the scope of the proper legal authority of the WTO. To obtain appropriate answers regarding the legal issues concerning the WTO, it is not sufficient to discuss the meanings of words provided in the agreement, or in the arguments of panel or Appellate Body reports, but it is necessary to analyze the entire structure of the WTO system.[16]

---

[16] See A. Kotera, *WTO Taisei no Houkouzou (On the Legal Structure of the WTO System)*, 2000.

Towards A More Proactive System of Dispute Settlement

Roy S. Lee

Judge Oda recently made the following engaging assessment:

> "In the light of this statistical overview of the Court's jurisprudence, I doubt whether a mere appeal ... for wider acceptance of the compulsory jurisdiction of the Court can achieve anything concrete. I am of the view that not a great deal can be expected in terms of meaningful development of the international judiciary from such an appeal or from amendments that may be made to the Court's procedures unless the parties in dispute in each individual case are genuinely willing to obtain a settlement from the Court. I wonder whether it is likely, or even possible, that States will one day be able to bring their disputes to the Court in a spirit of true willingness to settle them".[1]

I very much share this important concern. Given that very sensitive issues and complex factors are involved, one of the questions posed is whether the system of dispute settlement as a whole can be made more proactive so as to reduce States' hesitation. It thus seems appropriate that this tribute to him from a grateful friend should take the form of my thoughts on the subject.

I. Introduction

At the international level, our various means and modes for dispute settlement are all based on consent. Consent means voluntary submission of disputes and freedom of choice of means. States are

---

[1] S. Oda, "The Compulsory Jurisdiction of the International Court of Justice: A Myth? A Statistical Analysis of Contentious Cases", *ICLQ* 49 (2000), 251 et seq., 264.

therefore free to decide if they wish to settle a dispute and if so, by which means. There is therefore no compulsory adjudication *per se*, unless we mean that States have first consented to such an arrangement.

Consent is deeply rooted in States' desire to control and determine events by themselves, not through a third party. From their standpoint, adjudication is a luxury. If and when they must, they submit to it only on their own terms and conditions.

In the past, our main effort has been to devise as many as possible means and procedures, in the hope that parties to a dispute will use them. We have thus created various international or regional courts and tribunals waiting to be used. The system as a whole is not intended to procure clients.

In other fields, we have established the UNDP for development, UNICEF for children's welfare, and the WFP for food. We have also created a High Commissioner for Human Rights. All these institutions were specifically created to further and advance the intended objectives. But in the field of settlement of disputes, for instance, we have no mechanism with the task of enticing the parties to consider using a procedure of their choice or to assist them in overcoming difficulties that they may encounter. They are left on their own as adversaries. As such, they are bound to oppose each other. They therefore need help to overcome difficulties which are posed by being adversaries. A neutral third party is, however, in a position to offer middle-of-the-road suggestions and to be more proactive so as to reach out to the parties in disputes. This article is intended to address some of these issues and to make suggestions.

II. A Missing Link in the System

Article 33 of the UN Charter, for instance, lists the methods of settlement that are placed at the disposal of the parties: negotiation, enquiry, mediation, conciliation, arbitration, judicial settlement, resort to regional agencies arrangements, or other peaceful means of their own choice. The parties in dispute are given the maximum choice and they may decide if they wish to select a means and, if so, which one. It is presupposed that States have already, in some other manner, agreed to accept one or more of such methods for settling their dispute, and that they are going to take the necessary steps to implement the agreement.

But no process can even begin unless and until one of the parties takes the first step to trigger the process. But they are often unable or reluctant to do so. Taking a first step might be seen as a sign of weakness or concession by the other side, and so each side tends to adopt a waiting attitude. Meanwhile, the dispute may escalate.

Sometimes, there are no adequate channels of communication between the parties (e.g., lack of diplomatic relations).

Whatever means the parties choose, they have to meet first to discuss, for instance, the venue, agenda, language, records and communiqué. Often, these preliminary issues are as time-consuming and controversial as the substantive issues of the dispute.

The parties may also need to discuss how an enquiry, fact-finding mission, mediation, conciliation, arbitration, or tribunal may be established, what competence and composition it should have, how financial expenses are to be shared, and whether reports, decisions or awards will be issued.

The parties may also need to agree on the legal issues to be adjudicated. These issues are critical and are difficult to agree upon as they can affect the substance or outcome of the case. Neither side can afford to take them lightly.

All these matters may have to be done in writing before tackling the actual dispute. Each of these issues, as well as many others, may contain the seeds of disagreement providing ample opportunities for delays.

Difficulties may be encountered even after the parties have agreed on a specific means of settlement. For instance, disagreement may occur in the interpretation or application of the dispute settlement clause itself, particularly if the clause is ambiguous. Often, a party simply denies that a dispute exists.

In all these and other cases when the matter is left to the parties, each is more likely to insist on its own preference and wait for the other to concede. An impasse can easily be created. The process may be stalled or delayed, and the situation may deteriorate.

We need a go-between to bring the parties together, to open a dialogue, and to help them to resolve difficulties that they may encounter. The advantage of a go-between or facilitator is that he or she can fill any gap and make middle-of-the-road suggestions, making it easier for the parties to move on.

The parties may, however, find it difficult to ask for help from a third party. To do so, a formal decision may be necessary. Who is then prepared to take such a decision? To ask for assistance might also be taken as restricting freedom of action or "internationalising" a dispute, taking it out of the parties' control. Realistically, the assistance would have to come from a third party. In other words, assistance should be initiated, *proprio motu*, by a facilitator without waiting to be asked. However, it is difficult to be truly helpful without being labelled as a busybody.

There is no one institution that can fulfil this need alone. Disputes are many and varied, each involving different circumstances. Each calls for different skills and individual treatment. Many and different facilitators are needed.

It would be ideal to have an international agency performing different functions. Such an agency should be a technical body, not an intergovernmental institution. It must de-politicise its work, and it should have the technical expertise, managerial skills and adequate funding. These are, however, distant goals.

The implications drawn from the provisions of Article 99 of the UN Charter have become the broad legal basis for the conduct of the good offices of the Secretary-General in reducing and abating international tensions and conflicts.[2] As a result, an independent right of initiative of the Secretary-General has emerged that consists of pursuing investigations in situations which might contain the seeds of danger to peace, and tendering his good offices or appointing his special representatives to assist governments in the settlement of specific disputes.[3]

The General Assembly Declaration on the Prevention and Removal of Disputes and Situations Which May Threaten International Peace and Security and on the Role of the United Nations in the Field,[4] has

---

[2] This development may have started in September 1946 during a Security Council discussion of the Greek question, in which outside powers were charged with intervening in civil strife in Greece. The Secretary-General, Mr. Trygvie Lie, expressed the view that he "must reserve his right to make such inquiries or investigations as he may think necessary, in order to determine whether or not he should consider bringing any aspect of this matter to the attention of the Security Council under the provisions of the Charter". (Statement of the Secretary-General on 20 September 1946 at the Security Council, S/PV.70, 20 September 1946.) The right asserted by Mr. Lie became automatic for his successors. They have combined it with the exercise of good offices and the two functions together have led to the appointment by successive Secretaries-General of numerous special representatives to assist governments in the settlement of specific disputes.

[3] See *Handbook on the Peaceful Settlement of Disputes Between States*, United Nations Publication, 1992, 33-40. For current developments, see Report of the Secretary-General on the Work of the Organization, 2000, Official Records of the General Assembly, Fifty-fifth Session, Supplement No. 1 (A/55/1).

[4] General Assembly Resolution 43/51, Annex. Official Records of the General Assembly, Forty-third Session, Supplement No. 49. Paragraph 21 of the Declaration states that the Secretary-General "should consider approaching the States directly concerned with a dispute or situation in an effort to prevent it from becoming a threat to the maintenance of international peace and security". Paragraph 13 of the Declaration on Fact-finding by the United Nations in the Field of the Maintenance of International Peace and Security states that the Secretary-General, on his own initiative or at the request of States concerned, should consider undertaking fact-finding missions when a dispute or situation exists which might threaten the maintenance of international peace and security. He may when appropriate, bring the

specifically given the Secretary-General a right of initiative to approach the State concerned and offer his good offices with a view to preventing a situation from becoming a threat to peace. He does not need any further authorisation to use his good offices nor need he wait for the State to ask for assistance. The Secretary-General has also been encouraged to consider using "at as early a stage as he deems appropriate"[5] the right that is accorded to him under Article 99 of the Charter to call the attention of the Security Council to any matter which in his opinion may threaten the maintenance of international peace and security. The Secretary-General therefore now has a right of initiative both in law and in practice. Practice shows that he has and will use his initiative wisely, since the integrity of his high office requires that his actions must be beyond the criticism of the Member States.

III. Towards a Better Working Chapter VI

A dispute is not automatically discussed in the Security Council, unless it is brought to its attention. A victim or interested State is likely to do so, but, in other cases, it may serve no political interest for any State to bring a situation before the Council. Thus, the record shows that more cases remain outside than inside the Security Council.

The Council members decide, usually on political grounds, whether a situation should be considered. In the past, conflicts of an internal nature or associated with the Cold War were rarely taken up by the Council. Even when a situation is finally before the Council, its members may disagree on when and how to deal with it. Only when all these issues have been dealt with, may the substance of the situation be considered. Again, an agreement may or may not be reached in the end.

The Security Council is only concerned with disputes that are likely to endanger the maintenance of international peace and security. The test for that has often been the outbreak of armed conflicts. Such disputes are often "crisis situations", highly sensitive and at the peak of the controversy. Many different interests are involved and often procedures have been tried and have failed. The referral to the Council is often a last attempt at resolution. No doubt, these disputes are extremely difficult to handle.

Divergent positions may have already been taken and are publicly known. Different interests groups may have been formed and pressures exerted on the governments concerned. The parties' attitudes may thus

---

information obtained to the attention of the Security Council. See General Assembly Resolution 46/59, Official Records of the General Assembly, Forty-sixth session, Supplement No. 33 (A/46/33).

[5] *Ibid.*, paragraph 23.

have become inflexible. Time and effort are needed if there is to be any hope of persuading the parties to alter their positions.

While public debate is usually the next step after a dispute has been presented before the Security Council, consideration should be given to a quiet and discreet approach in some cases. Open confrontation in a political institution merely serves to publicise and even harden the opposing positions of the parties, providing a forum in which they condemn each other or reiterate their own position. Sometimes, an informal approach and confidential contacts conducted behind the scenes can be more helpful in preventing a situation from further deterioration.

There can be merits in approaching the parties at an early stage and on a confidential basis to allow quiet diplomacy and informal consultations to diffuse the conflict. But in other cases where public exposure is helpful, formal debate in the Council may still be an effective means of dealing with a dispute.

Under Chapter VI, the Council cannot make a specific recommendation unless both parties have so requested. At this point, that is almost impossible. The only hope is action under Chapter VII. Under this chapter, the Security Council has, as enforcement measures, created criminal tribunals and boundary commissions, applied various kinds of sanctions, and ordered the surrender of suspects to other interested States. Article 41 of the UN Charter specifically provides that the Security Council may decide what enforcement measures not involving the use of armed force are to be employed to give effect to its decision. I submit that the wording does not seem to preclude the Council from employing compulsory settlement of disputes as an enforcement measure. Thus, when circumstances so warrant and as the last resort, the parties could be ordered to submit their dispute for settlement by a specific means.

Surely, only certain disputes are amenable to this treatment. Boundary disputes that have led to severe fighting and loss of life may well be justified for compulsory adjudication by the International Court of Justice. Not the entire situation or dispute, only those elements or legal aspects that are suited for this purpose. Under existing procedure, the Council decision under Chapter VII does not automatically give jurisdiction to the Court. The parties so ordered would have to prepare a special agreement for submission to the Court. But the order would constitute a legal obligation for the parties concerned.

IV. ICJ Advisory Opinions as Building Blocks for Peacemaking Missions

The Secretariat is the only principal organ that has not been authorised to request an advisory opinion from ICJ. The last three Secretaries-

General have made known the usefulness of having such a competence for conducting peacemaking missions. The basis is that even a most complicated political situation may contain elements of a legal nature. A pronouncement on a specific legal point by the most authoritative judicial body can be used as a building block for constructing a package to provide an overall political solution.

At present, a request for an advisory opinion entails an open and complicated process and requires the adoption of a resolution by the principal organ concerned. This means not only making public the purpose of the request but also the involvement of the entire membership in the whole process from formulation to negotiation of the question or questions to be asked. Such public exposure and involvement is not conducive to the finding of a solution to a peacemaking mission.

With that competence, the Secretary-General would be able to proceed in a quiet and discreet manner without invoking public debates, and formulate the questions in the manner that would advance the search for a peaceful solution. This procedure would also help to preclude premature involvement of the principal organs and preserve their more useful role for the subsequent stages.

V. Using the Strengths of the Settlement System

Each of the means listed in Article 33 has its strengths and limitations. They are not meant to suit all needs and situations. It is important to be aware of the differing strengths and limitations so that the appropriate means may be used more effectively.

The attitudes of the parties towards the settlement of a dispute are often influenced by different considerations: the nature of the dispute (boundary question, human rights, trade, refugees), interests involved in the issue (political, military or economic), the relationship between the parties, internal politics and the objectives of the parties (e.g., to buy time, to reach a compromise, to maintain principles).

A party's decision to choose one means over another may sometimes be influenced by the degree of control given to it and the extent of its involvement in the process. A means or mode that offers more control over the outcome of the case and allows greater participation in the decision making is thus more likely to be chosen over another means that, in comparison, offers less control or participation. But in other cases, stability, established jurisprudence, recognised reputation, low costs and shorter proceedings are also important considerations that may influence the parties' decision.

For purposes of illustration, the various means listed below may be ranked in a descending order on the basis of the degree of participation in and control of the process that each of the means may offer to the parties: (1) negotiation (2) good offices (3) enquiry and fact-finding (4) mediation (5) conciliation (6) arbitration (7) judicial settlement.

Negotiation involves direct discussions between the parties and thus gives full participation and control to the parties. Virtually all other means are preceded or accompanied by some form of negotiation. Negotiation is also frequently a prerequisite to resorting to other means, for example, mediation or arbitration. Since negotiation is between the parties, each step has to be initiated by the parties themselves. When the parties experience difficulties, they may entangle themselves in arguments or procedures or face an impasse. The process may be stalled and a way out is difficult to find. Moreover, negotiation can easily be terminated unilaterally and there is therefore no guarantee of continuity.

In good offices, the services that a neutral third party may provide are not fixed but flexible, dependent on the wishes of the parties and the relationship between the principals and the intermediary.

Enquiry and fact-finding are neutral, technical means for ascertaining and clarifying facts. The parties do not have control over the outcome. But they usually have a say in the composition and the conduct of the mission. The work of the mission depends on the cooperation of the parties, particularly, when access to territories is required.

Mediation and conciliation are similar methods, consisting of negotiations between the parties with the participation of a third party. The role of a mediator or conciliator is to reconcile the claims of the parties and to make suggestions which the parties are not committed to accept. The process cannot take place unless the mediator(s) or conciliator(s) have already been agreed upon by the parties. Hence, difficulties often occur during the selection stages.

Arbitration and judicial settlement involve the rendering of a binding decision given by an *ad hoc* or standing court or tribunal. In the case of an *ad hoc* arrangement, the parties must create the institution to which the case is referred. In arbitration, the parties usually have the opportunity to choose the arbitrators and to determine the procedural rules and the applicable law. The parties therefore retain a certain degree of participation and control. But a badly crafted arbitration clause often provides excuses for the delay or failure of the arbitration. Payments by the parties for the establishment and maintenance of an arbitral tribunal is costly. A registry must also be established. Disagreement may arise on any of these matters.

Judicial settlement is more judgmental and less participatory. But the advantages offered by an established institution are significant and outweigh other considerations when an authoritative and objective legal pronouncement is required.

Being the *principal* judicial organ of the United Nations, the World Court is the only preconstituted judicial body at the global level. Its composition, jurisdiction and procedural rules are predetermined and therefore provide stability and a well-established jurisprudence, placing the Court at the pinnacle of the system. Its judgments are, therefore, the most authoritative.

There has been a steady increase in the number of cases brought to the Court and it has skilfully and successfully dealt with a variety of complex situations. Most importantly, it has provided fair and reasonable solutions to the parties in most cases brought before it. In addition, the expenses of the Court are borne by the United Nations. The parties need only to pay for their own costs (e.g., counsel fees and production of documents). This represents a considerable savings over arbitration.

VI. Trust Funds and Technical Assistance

States engaged in arbitration or judicial proceedings need highly specialised experts to prepare briefs, conduct research, collect evidence and to produce maps, charts and documents. The expenditures are substantial and the problem of costs is a real one. Not every country possesses such expertise and sometimes it may not be desirable to have all the experts from the same country, even when all necessary expertise exists. The high costs and the lack of expertise are important factors, particularly for developing countries, in deciding whether or not to resort to such proceedings. The removal of these factors can result in a positive decision when financial or technical assistance is needed from outside.

Until 1989, there had been no mechanism at the international level for helping the parties to finance their expenses in pursuing dispute settlement, even though such mechanisms are well known in municipal law. In that year, the Secretary-General of the United Nations established a Trust Fund to provide financial assistance to States in the settlement of their disputes through the International Court of Justice.[6] The Fund covers two main categories of cases: (1) cases submitted to the ICJ by a special agreement between the parties (i.e., *compromis*); (2) cases in which both parties are ready to implement a judgment of the ICJ, but one or both lack the funds or expertise. The chief reason for limiting the applications to cases of special agreement is that in such instances the Court's jurisdiction is certain and secured. However, in cases without a *compromis*, the jurisdiction is subject to challenge and until the Court has settled that question it would be better for the Fund

---

[6] See Terms of Reference, Guidelines and Rules of the Secretary-General's Trust Fund to Assist States in Their Settlement of Disputes Through the International Court of Justice, November 1989, also published in *ILM* (November 1989).

not to be involved. If the Fund were to assist one party while the other is contesting the jurisdiction of the Court, the Fund would appear to be partial. Moreover, this restriction ensures that funds are not used against a donor to the Fund without its consent. Thus when the issue of jurisdiction is resolved or when the Court's jurisdiction is not challenged, the Fund can assist. Indeed, the Fund provided financial assistance to Guinea-Bissau when the Court's jurisdiction was no longer in dispute, even though the case had not been brought before the Court by a special agreement.

The Fund may receive contributions from individuals, corporations, non-governmental organisations and international institutions. But so far States have been the only contributors. The Fund's assets are very limited. To date there have been only three applications and in each case a very small amount was awarded. This was due in part to the limited assets of the Fund. There is therefore a real need to find ways to increase the assets of the Fund, if its utilisation is to be encouraged.

The Fund is also open to contributions made in kind, apart from cash. Legal research, provision of experts (e.g., jurists, geologists, oceanographers, mineralogists) and technical assistance (translation, production of maps, demarcation of boundaries) would be very useful services. To date, however, no contributions of this kind have been made. But they should be encouraged.

VII. A Functional Approach to Dispute Settlement

Many existing disputes involve oil, gas, minerals or ground waters. When the area or the resources are under dispute, the risks involved may often deter commercial investment. Not only do the resources remain idle, but often innocent lives may also be lost.

There are instances in which it is quite feasible to develop joint projects involving utilisation or exploration of the resources under dispute. Antarctica is an excellent example: freezing the overlapping and conflicting territorial claims over the region, but at the same time allowing peaceful uses of the area. The joint development zone in the East China Sea in the disputed continental shelf area between Japan and Korea, the joint project to exploit the Red Sea's metalliferous hot brine between Sudan and Saudi Arabia, the four joint transboundary oil fields in the Gulf between Abu Dhabi, Bahrain, Kuwait, Qatar and Saudi Arabia, the joint Jan Mayen Ridge development project between Iceland and Norway, and the joint petroleum exploration zone in the Gulf of Gabes between Tunisia and Libya are all examples of this.[7] While the

---

[7] See, in general, I. T. Gault, "Joint Development of Off-shore Mineral Resources", *National Resources Forum* 12 (1988), 275 et seq.; Choon-Ho

resources or the area may still be in dispute, utilisation has taken place, notwithstanding the opposing claims of the parties.

This joint approach has the advantages of providing, on the one hand, a legal basis for dealing with a resource or an area in dispute and, at the same time, a practical way to bring economic benefit to the parties in dispute. In this approach, the contentious resources or territorial claims are frozen or set aside, but the parties instead create a joint economic interest in the development of the resources. Such a functional approach is likely to be possible when both parties have real economic interest in developing the resources and when the main contention is over the rights to explore and exploit the resources. But if human settlements are involved, the successful rate is low.

VIII. Conclusions

As long as our system is based on consent, the interests and underlying concerns of the parties will always come first in the minds of the disputants. Unless such interests and concerns can be addressed in the process, they will shy away from any method of settlement. The process must be able to permit the parties to clarify the issues and to articulate their concerns. When the issues are too large and complicated, they should be broken down into small manageable pieces. When issues are linked or interrelated, they should be combined and packaged to encourage trade-offs and integrated solutions.

Recent developments both inside and outside the United Nations have shown a widening of the traditional approaches and methods. Neutral, self-initiating, third-party assistance is growing and is being offered. This trend should be recognised and encouraged.

This article has suggested various ways to make the system more proactive towards the disputants. Specifically, it has suggested the use of the advisory opinion as a building block in peacemaking missions, compulsory adjudication as an enforcement measure and joint management for resources development. All this is easier said than done. But the more we think along these and other lines, the more we are likely to be able to deal with this extremely difficult subject of settlement of disputes in a more adequate manner.

Park, *East Asia and Thailand of the Sea*, 1983, 150; E. L. Richardson, "Jan Mayen in Perspective", *AJIL* 82 (1988), 443 et seq., 445.

Les demandes reconventionnelles devant la Cour internationale de Justice

François Rigaux

Les ordonnances prononcées par la Cour internationale de Justice le 17 décembre 1997[1] et le 10 mars 1998[2] ont donné un regain d'actualité à la faculté laissée à la partie défenderesse d'introduire une demande reconventionnelle *(counter-claim)* par un simple acte de procédure.[3] La seconde ordonnance est accompagnée d'une opinion séparée du Juge Oda : chacun sait que cet éminent juriste résiste rarement à se séparer du dispositif arrêté par la majorité ou à y adhérer avec des réserves telles que l'opinion qualifiée de séparée est très proche d'une opinion dissidente. Tel fut le cas dans l'ordonnance du 10 mars 1998 et l'auteur de la présente contribution offerte en hommage à M. Shigeru Oda, ne saurait manquer de se référer à l'opinion individuelle qui accompagne cette ordonnance.

I. Le Règlement de la Cour sur les demandes reconventionnelles

Les demandes reconventionnelles sont aujourd'hui réglées par l'article 80 du Règlement de la Cour, inséré dans la sous-section 3 de la section D (procédures incidentes) :

---

[1] *Application de la Convention pour la prévention et la répression du crime de génocide (Bosnie-Herzégovine c. Yougoslavie)*, Demandes reconventionnelles, *CIJ Recueil* 1997, 243.

[2] Affaire des *Plates-formes pétrolières* (République islamique d'Iran c. États-Unis d'Amérique), Demande reconventionnelle, *CIJ Recueil* 1998, 190.

[3] Voir notamment : F. Salerno, « La demande reconventionnelle dans la procédure de la Cour internationale de Justice », *RGDIP* CIII (1999), 329-378 ; P. H. F. Bekker, « New ICJ Jurisprudence on Counterclaims », *AJIL* 92 (1998), 508-517 ; M. Arcari, « Domande riconvenzionali nel processo di fronte alla Corte internazionale di giustizia », *Riv. Dir. Int.* 81 (1998), 1046-1064.

« Article 80
1. Une demande reconventionnelle peut être présentée pourvu qu'elle soit en connexité directe avec l'objet de la demande de la partie adverse et qu'elle relève de la compétence de la Cour.

2. La demande reconventionnelle est présentée dans le contre-mémoire de la partie dont elle émane et figure parmi ses conclusions.

3. Si le rapport de connexité entre la demande reconventionnelle et l'objet de la demande de la partie adverse n'est pas apparent, la Cour, après avoir entendu les parties, décide s'il y a lieu ou non de joindre cette demande à l'instance initiale ».

Le texte actuel, qui appartient au Règlement adopté par la Cour le 14 avril 1978, a été précédé de versions successives qui se sont efforcées de mieux préciser les pouvoirs de la Cour en la matière. L'un des commentaires qui concerne les versions intermédiaires du texte actuel émane de Dionisio Anzilotti qui fut membre de la Cour permanente dès l'institution de celle-ci, y siégea jusqu'à la fin de ses activités et la présida durant trois années.[4] Commentaire d'autant plus autorisé que son auteur avait participé en 1922 à la rédaction de l'article 40 du Règlement[5] et qu'il intervint ensuite durant les séances consacrées par la Cour en 1934 à l'élaboration du texte qui devint l'article 63 du Règlement de la Cour permanente.[6] Non moins autorisé apparaît le commentaire du même article par un juriste éminent, qui fut membre des deux Cours et conclut dans les termes suivants :

« Tout comme dans le droit interne, le système des demandes reconventionnelles devant les juridictions internationales présente des avantages et des inconvénients. Comme pour toute demande incidente, son application ne s'accommode pas de règles rigides. Elle requiert le contrôle attentif de la Cour et dépend largement des particularités de chaque espèce ».[7]

---

[4] D. Anzilotti, « La riconvenzione nella procedura internazionale », *Rivista di diritto internazionale* IX (1929), 309-327 ; « La demande reconventionnelle en procédure internationale » (traduction de l'article précédent par M. Barda), *Journal du Droit International (Clunet)* 57 (1930), 857-877.

[5] Voir notamment l'intervention du Juge Anzilotti au cours de la séance du 9 mars 1922, reproduite dans l'article cité à la note 4, 859-860.

[6] CPJI, Série D, 1936, troisième addendum au n° 2, 104-117.

[7] Charles De Visscher, *Aspects récents du droit procédural de la Cour internationale de Justice*, 1966, 113-116, 116.

Dans les deux ordonnances précitées, la Cour s'est implicitement inspirée de la conclusion finale de Charles De Visscher :

> « Considérant que le Règlement ne définit pas la notion de 'connexité directe' ; qu'il appartient à la Cour d'apprécier souverainement *(in its sole discretion)*, compte tenu des particularités de chaque espèce, si le lien qui doit rattacher la demande reconventionnelle à la demande principale est suffisant, et, qu'en règle générale, le degré de connexité entre ces demandes doit être évalué aussi bien en fait qu'en droit ».[8]

En 1997 et en 1998, la Cour a pu s'abstenir de rappeler la jurisprudence antérieure tant de la Cour permanente que de la Cour internationale,[9] eu égard à la liberté qu'elle se reconnaît face au texte de l'article 80 de son Règlement. L'ordonnance du 10 mars 1998 ne cite d'autre précédent que l'ordonnance prononcée quelques mois plus tôt, mais cette force est très grande puisque la motivation de la décision plus récente est à peu de chose près calquée sur celle qui la précède. Pareille convergence paraît démentir l'idée même d'une évaluation « compte tenu des particularités de chaque espèce », encore que le principe de cette solution mérite d'être approuvé face aux incertitudes qui entourent l'interprétation de l'article 80 du Règlement.

II. Les difficultés d'interprétation de l'article 80

Les deux premières subdivisions de l'article 80 ont pour objet la « présentation » d'une demande reconventionnelle. Le paragraphe 2 énonce une condition de forme n'appelant aucun commentaire, à savoir que la demande soit « présentée dans le contre-mémoire de la partie dont elle émane et figure parmi ses conclusions ». Le paragraphe 1er prévoit deux conditions de fond, à savoir que la demande « soit en connexité

---

[8] Ordonnance du 17 décembre 1997, *CIJ Recueil* 1997, 258, § 33 ; ordonnance du 10 mars 1998, *CIJ Recueil* 1998, 204-205, § 37. Les mots « compte tenu des particularités de chaque espèce » reproduisent textuellement la conclusion énoncée par Charles De Visscher à propos de l'interprétation de l'ancien article 63 dont l'article 80 actuel ne s'écarte que par des différences minimes de rédaction.

[9] Remontant à 1936, l'ancien article 63 du Règlement a suscité peu de jurisprudence. Voir notamment C. De Visscher, voir note 7 ; M. O. Hudson, *The Permanent Court of International Justice*, 1943, 292-293, § 322 ; 430, § 434 ; Sh. Rosenne, *The Law and Practice of the International Court, 1920-1996*, 3rd ed., 1997, vol. III, § III, 305, 1272-1277. On trouvera l'exposé le plus détaillé de la jurisprudence dans G. Guyomar, *Commentaire du Règlement de la Cour internationale de justice*, 1983, 519-525.

directe avec l'objet de la demande de la partie adverse et qu'elle relève de la compétence de la Cour ». La seconde condition exclut toute prorogation de compétence, possibilité qui est parfois admise en droit interne. Si la demande reconventionnelle satisfait à cette double condition, c'est trop peu dire qu'elle « peut être présentée », elle doit être déclarée recevable et, par conséquent, être jointe à la demande de la partie adverse.

Le paragraphe 3 revient sur la première condition de fond, celle du « rapport de connexité ». Le texte n'est pas des plus clairs. Il fait état d'une possibilité de contestation quant à la vérification d'un tel rapport, mais les deux versions ne sont sans doute pas en parfaite harmonie : le texte anglais commence par les mots « *In the case of doubt as to* . . . », alors que la même idée est reprise en français dans les termes suivants : « si le rapport de connexité . . . n'est pas apparent . . . ». Le contraste des deux expressions apparaît bien dans la traduction française d'un passage des observations écrites du Gouvernement américain, demandeur sur reconvention dans l'affaire des *Plates-formes pétrolières*. Le texte anglais qui est indubitablement le texte original est rédigé comme suit :

> « Under the Rules of Court, the only legally relevant issue now is whether there is 'doubt' as to whether the US counter-claim is 'directly connected to the subject-matter' of Iran's claim. Here there can be no such doubt . . . ».

La traduction française faite par les services de la Cour porte ce qui suit :

> « Aux termes du Règlement de la Cour, la seule question juridique pertinente pour l'instant est celle de savoir si 'le rapport de connexité' entre la demande reconventionnelle des États-Unis et l'objet de la demande de l'Iran 'n'est pas apparent'. En l'occurrence, l'existence de ce rapport n'est pas douteux ».[10]

Entre l'expression « n'est pas apparent » et les termes « *In the case of doubt* », il existe une double différence. La première, la plus obvie, sépare une affirmation d'une négation. « En cas de doute » a le premier caractère, ce qui « n'est pas apparent » le second. Faut-il en déduire que pour la mise en œuvre de l'article 80, paragraphe 3, le doute doit être établi tandis que l'apparence se laisse présumer ?

La seconde différence a pour objet les termes respectifs des deux textes qui font également foi. Ne pas être apparent et être douteux n'ont pas exactement la même signification, et la discordance est d'autant plus

---

[10] Ordonnance du 10 mars 1998, *CIJ Recueil* 1998, 200, § 22.

troublante que l'expression « en cas de doute » est familière aux juristes de langue française. Elle est liée à ce qu'on a appelé la doctrine de « l'acte clair », née en France à propos du devoir des cours et tribunaux de solliciter d'une autre autorité, le ministère des Affaires étrangères pour l'interprétation d'un traité international ou l'autorité administrative pour l'interprétation d'un acte administratif que le juge ordinaire est invité à appliquer. Pareille obligation peut être écartée quand la demande d'interprétation aurait pour objet un texte « qui ne présenterait ni équivoque ni obscurité ».[11] La doctrine fut définitivement consolidée par le vice-président du Conseil d'État, E. Laferrière, en 1896 :

> « Cela revient à dire que, pour qu'il y ait question préjudicielle, il faut qu'il y ait une question, et qu'elle préjuge en tout ou en partie le jugement du fond.
>
> Il faut qu'il y ait une question, c'est-à-dire une difficulté réelle, soulevée par les parties ou spontanément reconnue par le juge, et de nature à faire naître un doute dans un esprit éclairé ».[12]

Le même auteur s'efforce plus loin de mieux cerner la nature du « doute » :

> « Dans le doute il faut surseoir. S'il [le juge] hésite, il doit surseoir, car lorsqu'on est embarrassé de dire si une question est ou non douteuse, tout porte à croire qu'elle l'est réellement ».[13]

En d'autres termes, le doute sur le doute porte à un doute sur ce qui en fait l'objet. La « doctrine de l'acte clair » a été étendue aux questions préjudicielles réglées par l'ancien article 177 (devenu 234) du Traité CE. Dans des conclusions prises devant le Conseil d'État, Madame Nicole Questiaux, à cette époque commissaire du gouvernement, écrit ce qui suit :

> « Ce n'est que lorsque le juge n'est pas compétent pour fixer le sens d'un acte ou d'un texte qu'il lui appartient pourtant d'appliquer qu'il se trouve en présence d'une question préjudicielle. C'est le cas, selon la définition classique depuis Laferrière, d'une difficulté réelle soulevée par les parties ou spontanément reconnue par le juge et de nature à faire naître un

---

[11] Cass., civ. 13 mai 1824, cité par Laferrière (voir note 12), I, 498. Voir aussi Trib. des conflits, 20 mai 1882, *Rodier*, également cité par Laferrière.
[12] E. Laferrière, *Traité de la juridiction administrative et des recours contentieux*, t. Ier, 1896, 498.
[13] Laferrière, voir note 12, 499.

doute dans un esprit éclairé. L'application d'actes clairs au procès ne soulève pas de questions préjudicielles ».[14]

À la même époque, la doctrine de l'acte clair a été introduite à la Cour de justice des Communautés européennes par l'avocat général Lagrange,[15] encore qu'il ait émis des réserves sur « une expression d'ailleurs peu exacte et souvent mal comprise, la théorie de l'acte clair ». À la différence de Laferrière, Madame Questiaux n'a pas reculé devant cette expression mais Laferrière lui-même y conduit par la notion de « doute dans un esprit éclairé ». Ainsi le doute est lui-même qualifié, l'expression française correspondant assez bien au « *reasonable doubt* » de la jurisprudence constitutionnelle américaine.[16] Pour accueillir un grief d'inconstitutionnalité dirigé contre une loi ordinaire, le juge applique la doctrine du « *constitutional doubt* ». La loi est présumée conforme au droit international[17] ou à la Constitution et pareille présomption ne saurait être renversée si la contrariété alléguée est douteuse, le doute profitant à la conformité de la loi à la constitution. Mais ici encore les qualifications accompagnant la notion de doute indiquent qu'elle est moins limpide qu'il ne pourrait paraître : « *serious doubt* »,[18] « *grave doubt and doubt gravely* ».[19] L'un des plus récents arrêts en la matière a été prononcé à une faible majorité (5-4), étant accompagné de l'opinion dissidente du Juge Scalia, à laquelle se sont joints trois autres membres de la Cour.[20] C'est assez dire que la notion de doute n'est pas d'un maniement aisé.

Aucun commentateur de l'article 80 ni la Cour dans les ordonnances précitées ne paraissent avoir prêté attention à la différence de rédaction des deux versions du paragraphe 3. La notion de « doute » paraît préférable à celle de « non apparence » car elle est plus conforme à une terminologie traditionnelle. La Cour a certes eu raison de ne pas se laisser entraîner dans les controverses sémantiques auxquelles se prête la notion de « doute ». Dans les deux affaires, elle a donné une réponse

---

[14] Conclusions précédant Conseil d'État, 19 juin 1964, *Revue de droit public*, 1964, 1019, 1029.
[15] Conclusions précédant l'arrêt du 27 mars 1963, aff. jointes 28-62, 29-62, 30-62, *Sté Da Costa et autres c. Administration fiscale néerlandaise*, Recueil 1963, 88-89.
[16] Voir déjà : *Ogden* v. *Saunders*, 6 L Ed (12 Wheat 213) 606, 625 (1827) : « *beyond all reasonable doubt* » ; *Burgess* v. *Seligman*, 2 S Ct 10, 16 (1883).
[17] *Murray* v. *The Charming Betsy*, 6 US (2 Cr) 64, 117-118 (1804).
[18] *Blodgett* v. *Hogden*, plur. op. Holmes J., 275 US 142, 148 (1928).
[19] *United States ex rel. Attorney General of the United States* v. *Delaware and Hudson Cy*, 29 S Ct 527, 536 (1906) ; *United States* v. *Jin Fuey Moy*, 36 S Ct 658, 659 (1916) ; *Almendarez-Torres* v. *United States*, 118 S Ct 1219, 1228 (1998).
[20] *Almendarez-Torres* v. *United States*, voir note 19.

affirmative à l'existence du lien de connexité allégué par la partie demanderesse sur reconvention. Le doute qui, à une phase de la procédure et dans l'esprit d'une des parties, a pu exister sur ce point est définitivement levé par la décision prise « compte tenu des particularités de chaque cas d'espèce ». Toutefois, l'article 80, paragraphe 3, lie à la question du doute une exigence de procédure à savoir que la Cour décide « après avoir entendu les parties ». Telle est la question qu'il convient de considérer à présent.

III. La vérification d'un lien de connexité douteux

Il y aurait certes une manière de donner à la notion de doute un caractère objectif : elle consisterait à tenir pour douteuse une question de droit ou de fait qui est litigieuse. La doctrine de l'acte clair, la théorie du « *constitutional doubt* » et la jurisprudence qui en a fait application se réfèrent explicitement à l'existence d'une contestation : les points de droit ou de fait sur lesquels les vues respectives des parties concordent sont établis au-delà de tout doute. Toutefois l'inverse n'est pas toujours vrai, une partie pouvant à des fins purement dilatoires élever une contestation ou faire une réclamation qui n'a aucune « apparence » de fondement. Il faut dès lors que la contestation soit « sérieuse », ce qui ramène inéluctablement au pouvoir d'appréciation du juge.

Pour éclairants qu'ils puissent être, les parallèles jusqu'ici établis avec les doctrines de l'acte clair et du « *constitutional doubt* » ne présentent qu'une analogie partielle avec la fonction impartie à la Cour internationale de Justice par l'article 80 de son Règlement. La doctrine de l'acte clair apporte une restriction de bon sens à la répartition des compétences entre deux ordres de juridiction.[21] De même, la théorie du « *constitutional doubt* » tempère le pouvoir d'immixtion du juge dans le processus de législation. Rien de tel quand la Cour internationale de Justice se prononce sur la manière d'exercer sa propre compétence, dans les limites fixées par l'accord des États qui comparaissent devant elle. Que la connexité soit ou non douteuse, il appartient incontestablement à la Cour de se prononcer sur les demandes régulièrement formées conformément à l'article 80, paragraphe 2.

Quel est alors le sens du paragraphe 3 du même article ? Il y a quelque redondance à affirmer que « la Cour décide s'il y a lieu ou non de joindre cette demande à l'instance initiale ». Pour donner une portée à la disposition, il faut y voir une règle de procédure que la Cour s'est elle-même fixée mais qu'elle ne doit suivre qu'en cas de doute ou si le rapport de connexité « n'est pas apparent ». Ne risquant pas d'empiéter

---

[21] Voir en ce sens les conclusions de l'avocat général Lagrange citées à la note 14.

sur une compétence attribuée à un autre pouvoir, la Cour est évidemment beaucoup plus libre « d'apprécier souverainement » le degré de connexité entre les deux demandes.

Dans les deux affaires revenues devant la Cour, l'une en 1997 l'autre en 1998, le désaccord des parties ne portait pas seulement sur le lien de connexité, allégué par le demandeur sur reconvention, dénié par la partie adverse, il s'étendait à la détermination de la procédure à suivre pour vider ce différend. En effet, alors que le demandeur sur reconvention se satisfaisait des écritures échangées, la partie adverse sollicitait de la Cour un débat oral contradictoire pour que « le doute » fût levé en connaissance de cause. Pareille revendication trouvait un soutien dans le libellé du paragraphe 3 (« après avoir entendu les parties », « *after hearing the parties* »), ces expressions ayant pris la place des mots « après examen », « *after due examination* », qui figuraient dans le Règlement de 1963.[22]

Sans s'exprimer sur l'interprétation des mots « après avoir entendu les parties », chacune des deux ordonnances conclut que « saisie d'observations écrites et détaillées de chacune des Parties, la Cour est suffisamment informée des positions qu'elles défendent . . . ».[23] C'est précisément sur ce point de procédure que porte l'opinion séparée du Juge Oda jointe à la seconde ordonnance (alors qu'il a adhéré sans réserve à la première). Il estime, en effet, que la Cour n'aurait pas dû se prononcer sur l'admissibilité des demandes reconventionnelles « *without oral hearings* ».[24] Que la Cour se soit prononcée sans débat oral, dans une ordonnance réglant la procédure, paraît cependant en conformité avec la nature du dispositif prévu par l'article 80, paragraphe 3, à savoir décider « s'il y a lieu ou non de joindre cette demande à l'instance initiale ».

IV. Reconvention et connexité

À la différence du droit interne qui distingue nettement la reconvention de la connexité, la première étant une demande nouvelle greffée sur l'action initiale, la seconde justifiant la jonction de plusieurs demandes

---

[22] M. Rosenne tire argument de ce changement de rédaction pour écrire : « This means that in future there will always be some oral proceedings in the event of doubt ». Voir note 9, III, 305, 1273.

[23] Ordonnance du 17 décembre 1997, *CIJ Recueil* 1997, 256, § 25 ; ordonnance du 10 mars 1998, *CIJ Recueil* 1998, 203, § 31. Les deux considérants concluent « qu'il n'apparaît en conséquence pas nécessaire d'entendre plus avant les Parties à ce sujet », mais le texte de l'ordonnance du 10 mars 1998 inclut en outre une référence explicite « au paragraphe 3 de l'article 80 du Règlement ».

[24] *CIJ Recueil* 1998, 215-216, § 9.

introduites par des actes de procédure séparés,[25] le Règlement de la Cour fusionne les deux concepts en faisant de la « connexité directe », entre les deux actions la principale condition de fond de la recevabilité de la demande reconventionnelle. Ce n'est pas que le même Règlement ignore la jonction d'affaires introduites séparément, mais la première phrase de l'article 47 se borne à affirmer :

> « La Cour peut à tout moment ordonner que les instances dans deux ou plusieurs affaires soient jointes ».

Il n'est pas nécessaire que ces affaires soient connexes, les définitions de la connexité que connaissent certains textes de droit interne étant d'ailleurs tautologiques.[26] Toutefois, en droit interne la connexité existe souvent entre des affaires introduites devant deux juridictions distinctes, ce qui exige le dessaisissement de l'une au profit de l'autre, éventualité qui n'a évidemment aucun sens devant la Cour internationale de Justice. La jonction y est une mesure d'ordre interne, analogue à celle qui consiste à joindre plusieurs affaires pendantes devant diverses chambres du même tribunal.

En droit interne français la recevabilité d'une demande reconventionnelle est soumise à une condition qui laisse un large pouvoir d'appréciation aux juges du fond. Aux termes de l'article 70, alinéa 1er, du nouveau code de procédure civile :

> « Les demandes reconventionnelles ou additionnelles ne sont recevables que si elles se rattachent aux prétentions originaires par un lien suffisant ».

L'article 6, alinéa 3, des Conventions de Bruxelles et de Lugano précitées énonce un critère plus rigoureux :

> « Ce même défendeur peut aussi être attrait : . . .

---

[25] Sur la demande reconventionnelle, voir par exemple les articles 64 et 70 du nouveau code de procédure civile français et l'article 6, alinéa 3, des Conventions de Bruxelles et de Lugano concernant la compétence judiciaire et l'exécution des décisions en matière civile et commerciale. Sur la jonction d'affaires connexes, voir l'article 101 du nouveau code de procédure civile français et l'article 22 des Conventions de Bruxelles et de Lugano précitées.

[26] Voir par exemple l'article 101 du nouveau code de procédure civile français : « S'il existe entre deux affaires portées devant deux juridictions distinctes un lien tel qu'il soit de l'intérêt d'une bonne justice de les faire instruire et juger ensemble . . . ». L'article 22, alinéa 3, des Conventions de Bruxelles et de Lugano n'est pas moins tautologique.

3. S'il s'agit d'une demande reconventionnelle qui dérive du contrat ou du fait sur lequel est fondée la demande originaire, devant le tribunal saisi de celle-ci ».

Une commentatrice autorisée des deux Conventions a relevé la sévérité de cette condition et elle avance une interprétation qui correspondrait mieux aux intentions des rédacteurs du texte, à savoir « que la notion visée était plutôt celle, plus souple, de connexité »,[27] ce qui rejoint la condition inscrite dans l'article 80, alinéa 1er, du Règlement de la Cour et comporte l'interprétation qui y a été donnée par celle-ci.

V. Demande reconventionnelle et moyen de défense à la demande initiale

Qu'on l'appelle « connexité directe » ou « lien suffisant », l'affinité qui unit la demande reconventionnelle à l'action initiale implique-t-elle que la première puisse être tenue pour un moyen de défense à cette action tout en y ajoutant une prétention distincte ?

La définition de la demande reconventionnelle que contient l'article 64 du nouveau code de procédure civile français marque bien la dualité d'une telle demande :

« Constitue une demande reconventionnelle la demande par laquelle le défendeur originaire prétend obtenir un avantage autre que le simple rejet de la prétention de son adversaire ».

Ne peut-on déduire de cette définition que la « connexité directe » entre les deux actions requiert que la demande reconventionnelle contribue *aussi* à obtenir le rejet de la prétention initiale ? Pareille condition qui paraît assez raisonnable était mieux satisfaite dans l'affaire des *plates-formes pétrolières* qu'à propos de *l'Application de la convention pour la prévention et la répression du crime de génocide*. Dans la première affaire, l'un des moyens de défense avancé par le gouvernement de l'État défendeur était que le bombardement de ces plates-formes, dont la matérialité n'était pas déniée, était justifié par les attaques armées menées par la partie adverse contre les navires circulant dans le Golfe persique. Réclamer ensuite, par la voie d'une demande reconventionnelle, la réparation du dommage causé auxdits navires visait simultanément un double but : amplifier le moyen de défense à l'action initiale et y ajouter une prétention supplémentaire.

---

[27] H. Gaudemet-Tallon, *Les Conventions de Bruxelles et de Lugano*, 1993, n° 297.

Le raisonnement suivi par la Cour dans le litige entre la Bosnie-Herzégovine et la Yougoslavie est moins convaincant. Telle qu'elle est énoncée dans le paragraphe 34 de l'ordonnance du 17 décembre 1997, l'identité de temps, de lieu et de qualification juridique des allégations de génocide portées par chacune des parties contre l'autre, ne suffit pas à établir le lien direct de connexité prévu par l'article 80 du Règlement de la Cour. À les supposer l'un et l'autre vérifiés, le génocide de l'un n'est pas un moyen de défense au génocide de l'autre. Même si elle est isolée, l'opinion dissidente du vice-président Weeramantry paraît sur ce point plus convaincante que la décision de la majorité.[28] L'admissibilité de la demande reconventionnelle entraîne aussi un retard excessif dans le traitement de la demande initiale. Dans le cas du génocide, spécialement, il s'agit de faits complexes, indépendants les uns des autres, dont la preuve devra être séparément apportée. Le retard ainsi imposé au jugement de la demande initiale est d'autant plus regrettable que – et ceci vaut pour les deux affaires – le défendeur à l'action initiale avait d'abord soulevé une exception préliminaire, à savoir l'incompétence de la Cour, et ce n'est qu'après avoir été battu sur ce point[29] qu'il a introduit une demande reconventionnelle. C'est de bonne guerre mais il est permis de présumer qu'il n'aurait pas pris l'initiative de l'action ensuite introduite à titre reconventionnel. Dans son opinion dissidente déjà citée, le vice-président Weeramantry rappelle que l'un des éléments retenus en droit comparé pour faire obstacle à la recevabilité d'une demande reconventionnelle est la volonté d'épargner à la partie ayant introduit la demande initiale le retard qu'entraîne nécessairement la jonction des deux actions.[30]

---

[28] *CIJ Recueil* 1997, 287.
[29] *Application de la Convention sur la prévention et la réparation du crime de génocide*, exception préliminaire, arrêt du 11 juillet 1996, *CIJ Recueil* 1996, 595 ; *Plates-formes pétrolières*, exception préliminaire, arrêt du 12 décembre 1996, *CIJ Recueil* 1996, 803.
[30] Voir notamment en ce sens l'article 810 du Code judiciaire belge : « Si la demande reconventionnelle est de nature à faire subir un trop long retard au jugement de la demande principale, les deux demandes sont jugées séparément ».

Exhaustion of Domestic Remedies

Henry G. Schermers

I. Definitions

International law increasingly grants rights to individuals, but individuals cannot make use of the (still too few) means to pressure States to respect such rights. They cannot initiate diplomatic *démarches*, withdrawals of ambassadors or economic sanctions. Such measures must be taken by their States, acting on their behalf.

A State has a duty to protect the interests of its nationals. It should act on behalf of a national when his interests are illegally violated by the public authorities of a foreign State. But that foreign State is wrong only when it offers no means of redress within its own legal order. A policeman, a burgomaster, a politician or a judge may infringe the rights of a foreigner without creating state liability. That arises only when the State offers no means of redress to the foreigner. The foreigner is obliged to exhaust the domestic remedies of the State that caused harm to him before his own State has the right to act under international law on his behalf.

In the present article we will discuss the question what exhaustion of domestic remedies actually means. What kind of remedies must be exhausted? What procedure must be followed? Are there exceptions to the obligation of exhausting domestic remedies? Many of these and similar questions have been discussed in the European Commission of Human Rights which for more than 40 years has had the obligation to verify whether applicants under the European Convention on Human Rights had exhausted domestic remedies. Much of the experience of this Commission is of a general nature and will apply also in other situations where domestic remedies must be exhausted.[1]

---

[1] Though it offers the richest case law, the European Commission of Human Rights was not the only body concerned with the exhaustion of domestic remedies. The Human Rights Committee, established by the International Covenant on Civil and Political Rights, also has a growing case law in this field. Reference may be made to T. Zwart, *The Admissibility of Human*

The Commission's obligation rested on Article 26 of the European Convention on Human Rights which, until 1 November 1998, read:

> "The Commission may only deal with the matter after all domestic remedies have been exhausted, according to the generally recognised rules of international law, and within a period of six months from the date on which the final decision was taken".

After 1 November 1998, this task was taken over by the European Court of Human Rights.

Out of the thousands of decisions of the Commission a rich experience developed on which the present article is founded.[2] According to the Commission:

> "The basis of the rule of exhaustion of domestic remedies is that before proceedings are brought in an international judicial organ, the State made answerable must have an opportunity of redressing the alleged damage by domestic means. In this respect the remedies which are to be taken into account are these which are capable of providing an effective and sufficient means of redressing the wrongs which are the subject of the international claim".[3]

The purpose of exhaustion of domestic remedies is to repair the damage suffered by the individual as much as possible. Sometimes a possible conviction can be quashed, a refused authorisation can be granted or an unlawful dismissal can be withdrawn. Often, however, the violation itself cannot be wiped out with retroactive effect. *Restitutio in integrum* cannot be made. Then, only reparation can be made and may constitute the necessary redress.[4]

Rights Petitions, 1994, who devoted a chapter to the exhaustion of domestic remedies (187-219).

[2] The author has gratefully relied, to a large extent, on the summaries of the decisions of the European Commission of Human Rights prepared by the Secretariat of the Commission in *Decisions and Reports* (DR), in particular the "Summaries and Indexes" of this series. For the judgments of the European Court of Human Rights, he has also used P. Kempees, *Systematic Guide to the Case-law*, 1996, 1998, and the surveys by A. Sherlock and J. Andrews in the Human Rights Surveys of the *European Law Review*.

[3] See, e.g., Application 13427/87, *Stran Greek Refineries S.A.*, Decision of 4 July 1991, or Application 5964/72, 3 DR 57, 60. See also the European Court of Human Rights in the "*Vagrancy*" cases, A 12, § 50.

[4] Application 5575/72, *X* v. *Austria*, Decision of 8 July 1975, 1 DR 44, 45; Application 10668/83, *E* v. *Austria*, Decision of 13 May 1987, 52 DR 177, 181.

II. Domestic Remedies to be Exhausted

According to Article 26 of the European Convention on Human Rights, all domestic remedies must be exhausted according to the generally recognised rules of international law. In its case law the Commission has further qualified this rule.

1. The Remedy Must Be Effective

According to the European Court of Human Rights, international law only imposes the use of the remedies which are not only available to the persons concerned but are also sufficient, that is to say capable of redressing their complaints.[5] This means that only effective remedies must be exhausted. Whether a remedy is effective or not will largely depend on the subject of the case. An extraordinary remedy, the use of which depends on the discretionary power of a public authority, cannot be considered as effective.[6] Nor can a remedy which may merely remedy the criticised situation indirectly.[7] The length of time taken for the examination of a complaint may also be taken into account in assessing the effectiveness of the examination.[8]

When the complaint concerns the length of criminal proceedings, an application by the accused to accelerate the proceedings cannot be regarded as an effective remedy. Such an application would not have afforded redress for the violation complained of which concerned the allegedly excessive length of the proceedings.[9]

In respect of the length of civil proceedings, the question of the methods by which the applicant could have accelerated the proceedings is not one which concerns the problem of exhaustion of domestic remedies.[10]

When other redress is impossible an applicant may request financial compensation. If he does, he has to exhaust all remedies open to obtain such compensation. A request for compensation is not a remedy which needs to be exhausted, however, where the granting of compensation lies within the discretion of a public authority.[11]

---

[5] *Stögmüller* case, judgment of 10 November 1969, A 9, 42, § 11.
[6] Decision 8395/78 of 16 December 1981, 27 DR 50.
[7] Decision 17550/90, 17825/91 of 4 June 1991, 70 DR 298; Decision 13800/88 of 1 July 1991, 71 DR 94; Decision 14807/89 of 12 February 1992, 72 DR 148.
[8] Decision 13156/87 of 1 July 1992, 73 DR 5.
[9] Decision 8435/78 of 6 March 1982, 26 DR 18.
[10] Decision 8961/80 of 8 December 1981, 26 DR 200 and Decision 8990/80 of 6 July 1982, 22 DR 129.
[11] Decision 10530/83 of 16 May 1985, 42 DR 171; Decision 12810/87 of 18 January 1989, 59 DR 172.

In times of disturbances it may be more difficult to obtain effective judicial remedies. According to the Commission, the effectiveness should then be judged on the following elements:

1. the effect of police measures taken against the person attempting to exhaust;
2. impartiality of judges and juries;
3. length of proceedings before the courts;
4. cooperation of the authorities in obtaining evidence;
5. offers in settlement by the authorities to end the proceedings.[12]

Where doubts exist as to the effectiveness of a domestic remedy, that remedy must be tried.[13]

2. The Remedy Must Be Accessible

The remedy must be accessible, i.e., the person concerned must be able to institute the relevant proceedings himself.[14] A remedy, however effective it may otherwise be, is of little help if the applicant cannot invoke it.

Ms. Van Kuijk, a Dutch national, was detained in Greece for not having paid import duties. Information about the various possible steps and remedies was given to her only in Greek. As she had no contact with a lawyer until several days after her arrest, the Commission was of the opinion that possible remedies were not accessible to her.[15]

3. Financial Compensation

Whether the payment of financial compensation is an effective remedy depends on the case in question. We can distinguish between three different situations in which an applicant has requested a particular amount of financial compensation as a remedy to the violation suffered:

1. The government paid the compensation in full, be it without or after a court decision. The domestic remedy has been successful, there is no case under international law.

---

[12] Decision 5577 - 5583/72 of 15 December 1979, 4 DR 4.
[13] Decision 10148/82 of 14 March 1985, 42 DR 98.
[14] Decision 8950/80 of 16 May 1984, 37 DR 5; Decision 12604/ 86 of 10 July 1991, 70 DR 125.
[15] Decision 14986/89 of 3 July 1991, 70 DR 240, 250.

2. The case came before the domestic courts, but was not (fully) successful. As the State had the opportunity of redressing the alleged damage by domestic means, the domestic remedies have been sufficiently exhausted. The international case should be admitted.
3. The applicant did not bring an application for financial compensation before the domestic courts. His international claim will be rejected for non-exhaustion only if the content of the international claim is that he did not receive (sufficient) financial compensation. Other claims, such as a claim that persons responsible for inhuman treatment should be punished, or a claim that unacceptable restrictions of the freedom of information should be abolished, cannot be rejected for non-exhaustion if no compensation has been claimed.[16] In a case of excessive length of detention on remand, an action for damages against the State is also not a remedy which must be exhausted.[17]

Normally, adequate financial compensation constitutes an adequate and sufficient remedy. Even in the cases of killings and torture, payment of adequate financial compensation will terminate the liability of the State. However, this is not always the case. It would be wrong to suggest that torture is acceptable as long as certain financial compensation is paid. To make clear that this is not the case, the Commission held a number of times that the situation is different if an administrative practice of torture exists. In the case of *Donnelly* against the United Kingdom, the Commission considered that in cases of torture or inhuman or degrading treatment, the possibility of obtaining compensation will in general constitute an adequate remedy, since it is likely to be the only means whereby redress can be given to the individual for the wrong he has suffered. On the other hand, the Commission also considered that compensation could not be deemed to have rectified a violation in a situation where the State has not taken reasonable measures to comply with its obligation to subject no one to torture, inhuman or degrading treatment, or punishment. The obligation to provide a remedy does not constitute a substitute for or alternative to those obligations, but rather an obligation to provide redress within the domestic system. If torture or inhuman or degrading treatment were to be authorised or tolerated by domestic law, compensation would not constitute a remedy even if the law provided for the payment of such compensation.[18] This position was confirmed in a number of subsequent cases.[19]

---

[16] Decision 19092/91 of 11 October 1993, 75 DR 207; Decision 11208/84 of 4 March 1986, 46 DR 182.
[17] Decision 10868/84 of 21 January 1987, 51 DR 62.
[18] Decision 5577-5583/72 of 15 December 1979, 4 DR 4, 78, 79.
[19] See, e.g., Decision 8462/79 of 8 July 1980, 20 DR 184.

4. Settlement

The acceptance of damages in settlement of an action terminates the action which implies that claims for higher damages have been withdrawn. Such a claim for higher damages should therefore be rejected for non-exhaustion.[20]

5. Pardon

A request for a pardon is not a remedy which must be exhausted. The same goes for any act of grace which is within the discretion of a government authority.[21]

6. Censure of a Judge

Only normal remedies must be exhausted. Censuring of a judge for malpractice is not a remedy which must be exhausted.[22] Under Austrian law, a hierarchical appeal *(Dienstaufsichtbeschwerde)* can be directed against a judge. According to the Commission, this cannot be considered as an effective remedy when it is alleged that the excessive length of civil proceedings was due to an unjustified decision to suspend the proceedings.[23]

7. Appeal to the Authorities

The possibility of requesting an authority to reconsider a decision taken by it does not constitute an effective remedy.[24]

8. Appeal to the Ombudsman

Recourse to an organ which supervises the administration, such as the Ombudsman, is not an effective and adequate remedy and therefore need not be exhausted.[25]

---

[20] Decision 5577-5583/72 of 15 December 1979, 4 DR 4, 86.
[21] Decision 7705/76 of 5 July 1977, 9 DR 196, 203.
[22] Decision 8261/78 of 8 July 1981, 25 DR 157.
[23] Decision 7464/76 of 5 December 1978, 14 DR 51.
[24] Decision 11932/86 of 9 May 1988, 56 DR 199; Decision 12609/86 of 8 March 1990, 64 DR 84; Decision 14838/89 of 5 March 1991, 69 DR 286.
[25] Decision 11192/84 of 14 May 1987, 52 DR 227.

9. Extraordinary Remedies

Normally, only the normal remedies must be exhausted. Extraordinary remedies, such as requests to re-open a case, need not be exhausted.[26] This may be different, however, in some legal orders. There are legal systems where requests for revision or re-opening of the proceedings are a normal remedy with a reasonable chance of success. In such a case, these extraordinary remedies must be exhausted before the State can be held liable. Swiss domestic law establishes that a request to re-open proceedings is permitted when an authority has disregarded relevant facts or evidence, or has otherwise breached important procedures or principles. In such a case, this request constitutes an effective remedy and failure to request the re-opening of proceedings therefore constitutes non-exhaustion.[27] Also, in Turkish law, an appeal for rectification of a Court decision *(recours en rectification d'arrêt)* is a remedy so frequently used in civil proceedings that it must be considered as a remedy which should be exhausted.

10. New Remedies

The obligation to exhaust domestic remedies is not terminated by the launching of an international action. The Commission considered that it must normally decide the question whether domestic remedies have been exhausted by reference to the situation presented to it at the date of its decision on admissibility. If on that date it is clear that a domestic remedy is, or has been, available to the applicant, then the existence of that remedy must be taken into account.[28] If, therefore, a new remedy becomes available after an application under international law has been initiated but before its admissibility has been decided, the applicant is still obliged to exhaust this remedy if possible.

In another case, in which under national law an applicant had one month's time for the introduction of a constitutional appeal against the decision which he considered to be contrary to the European Convention on Human Rights, he became aware of a new remedy after that month, but before the European Commission had discussed this application. In this case, the applicant was not obliged to invoke the new remedy, as it would not have been effective under domestic law.[29]

---

[26] Decision 10431/83 of 16 December 1983, 35 DR 241.
[27] Decision 19117/91 of 12 January 1994, 76 DR 70. See also Decision 7805/77 of 5 May 1979, 16 DR 68.
[28] Decision 7878/77 of 19 March 1981, 23 DR 102, 113.
[29] Decision 8544/79 of 15 December 1981, 26 DR 55, 69.

III. Burden of Proof

When an applicant claims to have exhausted all domestic remedies, it is for the government that claims that domestic remedies have not been exhausted to demonstrate that the applicant did not make use of a remedy at his disposal.[30] I have not found any exception to this rule, which has been confirmed by the Commission in many other cases.[31]

If the government has demonstrated that a remedy exists, the applicant must prove that he exhausted it.

The government must not only prove the existence of remedies, it must also prove so in time. In November 1993, the Commission communicated the application of *Dikme* to the government, inviting them to present their observations on admissibility and merits within ten weeks. When, after several reminders, the government had still not made any observation, the Commission declared the case admissible in October 1994. Later, when the case was before the Court, the government submitted that the domestic remedies had not been exhausted. The Court did not exclude that an argument of non-exhaustion may sometimes arise after a case has been declared admissible, but it confirmed that non-exhaustion should normally be raised before a case is declared admissible. As there were no special reasons why this could not have been done in the *Dikme* case, the Court did not admit the argument of non-exhaustion.[32]

IV. Who Must Exhaust?

1. The Victim

In order to be entitled to lodge a complaint under the European Convention on Human Rights the applicant must be the victim of an actual infringement. A complaint against unlawful legislation is generally impossible. Only when individually affected by the legislation may a person bring a complaint after exhaustion of domestic remedies.

Under the European Convention on Human Rights, the victim has to complain himself. A complaint cannot be brought by an association of victims when the association itself is not a victim. No problem arises if

---

[30] Decision 7456/76 of 8 February 1978, 13 DR 40.
[31] See, e.g., Decision 9013/80 of 11 December 1982, 30 DR 96; Decision 8805/79, 8806/79, Decision of 7 May 1981, 24 DR 144; Decision 11208/84 of 4 March 1986, 41 DR 182; Decision 11932/86 of 9 May 1988, 56 DR 199; Decision 13057/87 of 15 March 1989, 60 DR 243; Decision 12686/87 of 3 October 1990, 66 DR 105.
[32] *Dikme* judgment of 11 July 2000, §§ 45, 49.

under domestic law it is also the victim himself who has to bring the complaint and who exhausts the domestic remedies. In some national legal systems, however, organisations of victims are allowed to complain for violations of law committed against their members. It may then happen that under national law an association exhausts the domestic remedies, but is subsequently not permitted to launch a complaint under the European Convention. The individual victim is not allowed to bring a complaint under the Convention either, because he has not exhausted domestic remedies.[33] This situation is rather unfair, as in a State where an association can bring a claim on behalf of its members, the members will usually leave the litigation to the association. Informed lawyers can easily avoid problems by always including an individual applicant in the domestic litigation. The actual work can of course be done by the association on behalf of an expressly mentioned individual as well as on its own behalf.

2. The Potential Victim

In some exceptional cases applications by potential victims have been admitted. Mr. Klass suspected that his telephone was tapped, but was unable to obtain information as to whether or not that was actually the case. The Commission, and subsequently the Court, accepted his application, notwithstanding the fact that the German authorities later expressly declared that Mr. Klass' telephone had not been tapped. The Court found it unacceptable that the assurance of the enjoyment of a right guaranteed by the Convention could be removed by the simple fact that the person concerned is kept unaware of its violation.[34]

Sometimes a person risks being killed or tortured after expulsion or extradition to a particular country. Permitting him to lodge a complaint only after such expulsion or extradition has taken place would be of little help. The Commission therefore permitted an application against the decision to expel or extradite. On this issue the Court held:

> "It is not normally for the Convention institutions to pronounce on the existence or otherwise of potential violations of the Convention. However, where an applicant claims that a decision to extradite him would, if implemented, be contrary to Article 3 by reason of its foreseeable consequences in the requesting country, a departure from this principle is necessary, in view of the serious and irreparable nature of the alleged suffering

---

[33] Decision 14570/89 of 1 July 1993, 75 DR 5, 17, 18.
[34] *Klass* judgment of 6 September 1978, A 28, § 36.

risked, in order to ensure the effectiveness of the safeguard provided by that Article".[35]

3. Family, Heirs

Sometimes the victim is unable to lodge a complaint, e.g., he is held incommunicado, or is so badly wounded that he cannot act. In such cases, one of his next of kin may lodge the complaint on his behalf. The same applies when a victim was unlawfully killed by the authorities.

In other cases related to the death of the victim, the right of an heir to initiate or to continue the application depends on the circumstances. If the violation of such nature that the heir also suffers from it (e.g., the honour or the possessions of the family are at stake), then the heir can act. If, on the other hand, the violation is personal (e.g., detention on remand has taken too long), then the case may be closed after the death of the applicant. In any case, the Court shall continue the examination of the application if respect of human rights as defined in the Convention and the Protocols so requires.[36] Three recent cases offer the following picture:

1. In *Kazimierczak*, the applicant had complained about the length of his detention on remand. After his death, the Court struck the case off the list, concluding that it was no longer justified to continue the examination of the application.[37]
2. In *Fatourou*, the applicant had complained about the length of the proceedings before her national court. After her death, the Court decided that her son was qualified to continue the proceedings in Strasbourg and afforded just satisfaction to the son both for moral damages suffered and for costs and expenses.[38]
3. In *Ječius*, the applicant had been unlawfully detained on remand. The Court noted the applicant's death and the wish of his widow to pursue the proceedings, and considered that the widow had a legitimate interest to maintain the case in the applicant's stead. It granted non-pecuniary damages, as well as compensation for costs and expenses.[39]

4. Legal Persons

Can legal persons be the victims of the violation of certain rights, such as the right to the peaceful enjoyment of possessions? There are not

---

[35] *Soering* judgment of 7 July 1989, A 161, § 90.
[36] Convention, Article 37.
[37] *Kazimierczak* judgment of 27 July 2000.
[38] *Fatourou* judgment of 3 August 2000.
[39] *Ječius* judgment of 31 July 2000.

many rights regarding which they can be victims. Sometimes, legal persons have invoked the right to life when particular government activities have caused their bankruptcy. The Commission has always rejected this. In the case of *Agrotexim*, shareholders complained of a violation of their right to the peaceful enjoyment of their property caused by government acts against the company that resulted in a fall in the value of the shares. The European Court of Human Rights considered that the piercing of the "corporate veil", or the disregarding of a company's legal personality, is justified only in exceptional circumstances, in particular when it is clearly established that it is impossible for the company to apply to the Convention institutions through the organs set up under its articles of incorporation or – in the event of liquidation – through its liquidators. As this was not the situation in *Agrotexim*, the shareholders could not be regarded as being entitled to apply to the Convention institutions.[40]

5. States

The rules requiring the exhaustion of domestic remedies apply not only in individual applications but also in cases brought by States. In principle, this means that remedies which are shown to exist within the legal system of the responsible State must be used and exhausted in the normal way before an international application can be brought.[41] However, in the case of state applications, the rule concerning the exhaustion of domestic remedies does not apply to applications concerned with the compatibility of legislative measures with the Convention. Nor does it apply when there is *prima facie* evidence of an administrative practice in violation of the Convention. An administrative practice exists if there is repetition of acts and official tolerance, even if it exists only at a subordinate level, and despite occasional reactions from the authorities. In the establishment of a *prima facie* case, regard must be had to the submissions and documents presented by both parties.[42]

6. More Applicants

When an application is lodged by two parties, the Commission may, in certain circumstances, examine the entire application despite the fact that only one party has exhausted domestic remedies. In its decision on application 9905/82 the Commission held:

---

[40] *Agrotexim* judgment of 24 October 1955, A 330 A, §§ 62, 66, 71.
[41] Decision 8007/77 of 10 July 1978, 13 DR 85, 150.
[42] Decision 9940-9944/82 of 6 December 1983, 35 DR 143, 163. See also Decision 9471/81 of 13 March 1984, 36 DR 49 and Decision 8007/77 of 10 July 1978, 13 DR 85.

"It is true that only the first applicant has exhausted the domestic remedies in accordance with Article 26 of the Convention. However, in view of the particular circumstances, the second applicant can be absolved from doing the same as his complaint, although presented from a different angle, is virtually the same as that of the first applicant. The Commission is therefore required to deal with the substance of the complaints in respect of both applicants".[43]

In a case involving more than one party, the exhaustion of domestic remedies by one of them will usually be sufficient. Often, an applicant may invoke the above quotation from case 9905/82, but even if that is not possible, he can normally invoke the argument that existing case law would make his exhaustion of domestic remedies ineffective.

V. When Must the Application Be Lodged?

Normally, an applicant will plead that only a restrictive number of domestic remedies are sufficiently effective to be exhausted. The government concerned will usually argue that there are more domestic remedies available. This may lead to some uncertainty. Under the European Convention on Human Rights, applications must be brought within six months after the exhaustion of domestic remedies. Also, under general international law, one may plead that an international claim must be brought within a reasonable time. It seems unacceptable that many years after the exhaustion of domestic remedies it would still be open to an applicant to launch an international action against the State concerned.

Whenever there are time-limits it may be in the interest of the applicant to plead that particular remedies are effective. When after the exhaustion of the normal domestic remedies more than six months have passed, an applicant may be tempted to raise a request for revision and plead that the six-month time-limit starts only after the rejection of that request. Understandably, the Commission has rejected claims of that nature. Extraordinary and non-obligatory remedies cannot re-open the time-limit within which a case must be brought. This may raise a serious problem for an applicant when the effectiveness of a remedy is unclear. His application may be rejected for non-exhaustion if he does not try that remedy and the international authorities subsequently hold that the remedy was effective. However, his application may be rejected as having run out of time if, having invoked the doubtful remedy, the exhaustion of which takes more than six months, the international authorities subsequently hold that the remedy was ineffective. An example of this situation is offered in Portuguese law, in which a request for the rectification of a court decision

---

[43] Decision 9905/82 of 15 March 1984, 36 DR 187, 192.

is a rather normal remedy. Mr. Neves tried that remedy, which was rejected by the domestic court. Subsequently, he brought a petition in Strasbourg which would have been out of time if the request for rectification was not accepted as an effective remedy. The Commission did not declare the case out of time.[44] As a general rule it seems advisable to accept that, in cases of doubt, an application will not be rejected for non-exhaustion, nor for being out of time.

As the proceedings before the Commission take a considerable amount of time, the Commission, wishing to avoid delays as much as possible, has permitted the introduction of a case before national remedies have been exhausted, provided that the exhaustion is completed by the time the Commission is called upon to decide on the admissibility.[45] Zwart pointed out that the French text of the Convention does not permit this interpretation. Where the English text prohibits the Commission from *dealing with* an application before exhaustion, the French text provides that the Commission *cannot be seized* ("ne peut être saisie") before exhaustion.[46] An application is inadmissible for non-exhaustion when a remedy which cannot on the face of it be regarded as ineffective is still pending,[47] but in practice the Commission usually postpones its discussion on the admissibility of the case when national remedies are still pending. However, such postponement cannot last for too long. In a case in which the Commission had been waiting for years for a decision of the Brussels Court of Appeal, it finally decided to declare the case inadmissible for non-exhaustion.[48] The question is not of great importance, as the applicant may re-introduce his case after the exhaustion of domestic remedies.[49]

VI. Domestic Remedies

1. Procedure

In the previous section we saw which remedies must be exhausted. In the present section, we want to consider how this exhaustion must be effected. Under generally recognised rules of international law, remedies

---

[44] Case 20683/92, unpublished.
[45] Decision 13319/87, unpublished; Decision 9019/80 of 7 July 1981, 27 DR 181; Decision 12850/87 of 13 March 1990, 64 DR 128; Decision 13370/87 of 7 July 1991, 70 DR 177; Decision 15530-15531/89 of 10 October 1991, 72 DR 169.
[46] Zwart, see note 1, 205.
[47] Decision 10092/82 of 5 October 1984, 40 DR 118.
[48] Decision 5024/71 of 7 October 1976, 7 DR 5.
[49] See also F. Martinez Ruiz, in M. de Salvia and M. E. Villiger (eds.), "The Birth of European Human Rights Law", *Liber Amicorum Carl Aage Nørgaard*, 1998, 264.

must be exhausted according to the provisions of the domestic legal order. National remedies have not been exhausted if an applicant's case has been declared inadmissible on the ground of a procedural mistake by the applicant (e.g., when he lodges appeal out of time). It may happen, however, that the domestic legal order is too formalistic. Assume, for example, that under domestic legislation particular documents have to be handed over to the Court in triplicate. Assume further that, after photocopying, a paragraph, or a word, has become unreadable in one of the copies, and that, subsequently, the national judge declares the case inadmissible because it has not be presented in triplicate. In this kind of case, the European Commission of Human Rights would not declare an application inadmissible for non-exhaustion. Rather, the Commission would see this as a problem of access to justice, which is guaranteed under Article 6 of the Convention. The question whether the rejection of a case on too formal a ground is contrary to the European Convention will be decided along with the merits of the case, and not under the question of admissibility.

2. Claim in Substance

In order to be admissible under international law, a claim must have been submitted to the domestic courts. It is not enough that an applicant has litigated before the domestic courts; he must also have expressly brought before the national judge the claim he later wants to bring as an international claim. What exactly must he have presented to the domestic courts? In many States the European Convention on Human Rights is part of domestic law, and the articles of the Convention can be invoked before the domestic courts. In other States, the Convention is not part of domestic law and cannot be invoked before the domestic courts, although the human rights incorporated in the Convention can be raised before the national judge. Irrespective of this difference, the European Commission of Human Rights did not require that the particular article of the Convention was invoked. It is sufficient if the complaint before the Commission had been raised in substance before the domestic courts.[50] Even in a State where the Convention is directly applicable, the applicant may, instead of invoking a precise provision of the Convention, raise equivalent arguments before the national authority.[51]

Even if a complaint has not been clearly formulated before the appeal courts, domestic remedies have been exhausted when the latter have in fact examined the complaint, albeit in an *obiter dictum*.[52]

---

[50] See, e.g., Decision 5574/72 of 21 March 1975, 3 DR 10; Decision 6878/75 of 6 October 1976, 6 DR 79 or Decision 5573 and 5670/72 of 16 July 1976, 7 DR 8.
[51] Decision 7367/76 of 10 March 1977, 8 DR 185.
[52] Decision 8130/78 of 10 May 1979, 16 DR 120.

Raising the claim before the domestic courts may not be enough. Important arguments and documents supporting the claim must also be brought in national proceedings. Someone fearing persecution in the event of his expulsion has not exhausted domestic remedies if he has not submitted to the domestic courts an important document produced to the Commission which is intended to establish that his fears are well founded.[53] This applies only to decisive information. Normally, domestic remedies have been exhausted when the applicant has limited himself to mentioning the relevant provision in the course of the proceedings and the subject matter of these proceedings clearly falls within the scope of that provision. The essential criterion is that the national authorities have the opportunity of putting right the violations alleged against them.[54]

Under Article 192 of the Greek Penal Code, it is an offence to refer to the Moslem minority in Western Thrace as "Turks". In publications in his election campaign, Mr. Ahmed Sadik referred to "the Turkish Community of Western Thrace" and wrote: "Long live the Turkish and Muslim youth of Western Thrace!" He was prosecuted on this ground. In his defence, he attacked the accusation but did not refer to Article 10 of the Convention which is directly applicable under Greek law. Nor did he invoke the provisions of Article 14 of the Greek Constitution which guarantees the right to freedom of expression. However, while basing his case on the narrower domestic criminal law provision of Article 192 of the Greek Penal Code, the applicant claimed the right to use the words "Turk(s)" or "Turkish" to identify the Moslems of Western Thrace. The Commission then considered:

> "that this indicates an issue falling within the scope of freedom of expression. In claiming a right to use these terms the applicant was formulating a complaint which was linked to the alleged violation of Article 10 of the Convention. Therefore he provided the national courts with the opportunity which is in principle intended to be afforded to Contracting States by Article 26, namely the opportunity of putting right the violations alleged against them".

Accordingly, the Commission considered that the applicant did invoke before the Greek Courts, at least in substance, the complaints relating to Article 10 of the Convention which he subsequently brought before the Commission. The Commission accepted that he had exhausted domestic remedies.[55] The Court, however, was of a different opinion. It reiterated

---

[53] Decision 12023/86 of 23 January 1987, 51 DR 232.
[54] Decision 11798/85 of 7 November 1989, 63 DR 89, 103.
[55] Decision 18877/91 of 4 April 1995, Appendix II (Decision of 1 July 1994 as to the admissibility), § 4.

that the supervision machinery set up by the Convention is subsidiary to the national human rights protection systems, and that this principle was reflected in the rules set forth in Article 26 (now Article 35), which dispenses States from answering to an international body for their acts before they have had the opportunity to put matters right through their own legal systems. As the applicant had at no time relied on Article 10 of the Convention, or on arguments to the same or like effect based on domestic law in the courts dealing with his case, the European Court considered that he had not exhausted domestic remedies.[56]

The substance of a complaint may also be the absence or the poor quality of domestic remedies. Then, non-exhaustion is not a matter of admissibility. The case will also be admissible without proper exhaustion, and all questions of domestic remedies, including their exhaustion, will be discussed along with the merits of the case.[57]

3. Appeal

It is not enough that the complaint has been raised in a national court. If the national court does not offer satisfaction its decision must also be appealed, if necessary to the Supreme Court. The appeal must be in accordance with domestic requirements, but it does not have to be introduced if it has no chance of success.[58] There is no exhaustion when a domestic appeal is not admitted because of a procedural mistake.[59]

In a case against Ireland, the applicants, who were in prison, failed to appeal to the Supreme Court. They stated that they were prevented from appealing within the prescribed time-limit of 21 days because they did not receive the High Court decision and other relevant documents until after this time-limit, allegedly because the decision and documents were withheld by the prison governor until after that date. The applicants felt that as a result of these special circumstances they were absolved from exhausting the remedy concerned. The Commission disagreed, however, because under Irish law it was open to the applicants to seek an extension of the time-limit for appeal under the circumstances claimed. The case was declared inadmissible for non-exhaustion.[60]

In a case against Austria, the Court of Appeal had rejected the submissions of the applicant because of a failure to observe the forms

---

[56] *Sadik* judgment of 15 November 1996, §§ 30-34.
[57] See, e.g., *Sibson* judgment of 20 April 1993, A 258 A, §§ 25-27; *Ekinci* judgment of 18 July 2000, § 62.
[58] Decision 7308/75 of 12 October 1978, 16 DR 32.
[59] Decision 6878/75 of 6 October 1976, 6 DR 79; Decision 13467/87 of 10 July 1989, 62 DR 269; Decision 18079/91 of 4 December 1991, 72 DR 263.
[60] Decision 8299/78 of 10 October 1980, 22 DR 51, 75.

prescribed by law. Nonetheless, the appeal had been examined. In this case, the Commission held that non-exhaustion of domestic remedies could not be held against the applicant, as in spite of the applicant's failure to observe the forms prescribed by law, a competent authority had nevertheless examined the appeal.[61]

VII. Exceptions Claimed

The Commission has always accepted that there can be special circumstances which may absolve an applicant, according to the generally recognised rules of international law, from exhausting the domestic remedies at his disposal. Even *ex officio* the Commission may study whether such circumstances exist.[62] Over the years many possible exceptions have been invoked.

1. No Reasonable Chance of Success

Remedies must be invoked only if they offer a reasonable chance of success. An applicant cannot be required to exhaust remedies which have no chance of success according to domestic case law at the time.[63] Well-established case law before the national authorities showing the futility of an appeal may dispense the applicant from raising the complaint.[64]

Normally, there is no reasonable chance of success if the violation of the international rule is caused by legislation to which the domestic court is bound. Neither does the issue of exhaustion of remedies arise when the facts of the case show that every remedy is excluded.[65]

Domestic case law may also release an applicant from his obligation to exhaust domestic remedies. To what extent this is the case may vary. Recent and consistent case law is more authoritative than a single case. In some legal orders, case law forms part of the legal system, in others cases may be more easily overruled. This alone is not decisive. Under Icelandic law, the courts may lawfully diverge from their own precedents. In a case in 1988, the Icelandic Supreme Court clearly stated that the Icelandic constitution did not prevent the legislator from

---

[61] Decision 12794/87 of 9 July 1988, 57 DR 251, 259. See also Decision 9024/80 and 9317/81 of 9 July 1982, 28 DR 138.
[62] Decision 1094/61 of 4 October 1962, *Collection of Decisions of the European Commission of Human Rights*, Vol. 9, 40 et seq., 48.
[63] Decision 10103/82 of 6 July 1984, 39 DR 186.
[64] Decision 10027/82 of 5 December 1984, 40 DR 100.
[65] Decision 11723/85 of 7 May 1987, 52 DR 250.

requiring taxicab operators by law to become or remain members of a private association. In 1989, an Icelandic taxi driver complained that a law ordering him to become a member of a private association was contrary to the freedom of association guaranteed in Article 11 of the Convention. The Government submitted that he should have raised this issue before the Icelandic courts. The Commission shared the applicant's view, recalling that if it can be established that the remedies that may exist are ineffective or inadequate, the domestic remedies rule does not apply.[66] If a person establishes that a remedy is bound to be unsuccessful, on the basis of established case law, he is not required to make use of that remedy.[67] If there exist doubts as to the prospects of success of the domestic remedy, such a remedy must be attempted.[68]

2. No Effective Remedy

Only effective remedies have to be invoked. In Section II.1 above, when defining the remedies which an applicant has to exhaust, we saw a number of possible remedies which have not been recognised as domestic remedies that need to be exhausted according to the generally recognised rules of international law.

In addition, a remedy need not be exhausted, even if it could be effective, when it would not solve the main complaint of the applicant. Under the case law of the European Court of Human Rights, the Contracting States are obliged to conduct an investigation capable of leading to the identification and punishment of those responsible for killings and torture. This obligation might be rendered illusory if in respect of complaints under those articles an applicant were to be required to exhaust an administrative law action leading only to an award of damages.[69]

In order to deprive fighters of the PKK (Workers' Party of Kurdistan) of food and shelter, hundreds of villages in Southeast Turkey have been destroyed. In one of these attacks, the home of Mr. Akdivar was burnt down. Mr. Akdivar claimed under the European Convention, *inter alia*, that he had been inhumanly treated, contrary to Article 3 of the Convention, that his home had not been respected, contrary to Article 8 of the Convention, and that he had been deprived of his

---

[66] Decision 16130, unpublished.
[67] Decision 9856/82 of 14 May 1987, 52 DR 38; Decision 14312/88 of 8 March 1989, 60 DR 284; Decision 11945/86 of 12 March 1987, 51 DR 186; Decision 15070/89 of 6 December 1990, 67 DR 295.
[68] Decision 9856/82 of 14 May 1987, 52 DR 38.
[69] *Ilhan* judgment of 27 June 2000, § 61; *Salman* judgment of 27 June 2000, § 83.

possessions, contrary to Article 1 of the First Protocol. Mr. Akdivar did not try to exhaust any domestic remedies. No allegation or claim for compensation was ever submitted to the Turkish courts. The Commission considered that no effective remedies were available and declared the case admissible. Subsequently studying the question of non-exhaustion the European Court of Human Rights held:

> "There is, as indicated above, no obligation to have recourse to remedies which are inadequate or ineffective. In addition according to the "generally recognised rules of international law" there may be special circumstances which absolve the application from the obligation to exhaust the domestic remedies at his disposal (see the *Van Oosterwijck* v. *Belgium* judgment of 6 November 1980).[70] The rule is also inapplicable where an administrative practice consisting of a repetition of acts incompatible with the Convention and official tolerance by the state authorities has been shown to exist, and is of such a nature as to make proceedings futile or ineffective (see the *Ireland* v. *the United Kingdom* judgment of 18 January 1978),[71] and the report of the Commission in the same case.[72]
>
> ... As regards the application of Article 26 to the facts of the present case, the Court notes at the outset that the situation existing in Southeast Turkey at the time of the applicants' complaints was – and continues to be – characterised by significant civil strife due to the campaign of terrorist violence waged by the PKK and the counter-insurgency measures taken by the government in response to it. In such a situation it must be recognised that there may be obstacles to the proper functioning of the system of the administration of justice. In particular, the difficulties in securing probative evidence for the purposes of domestic legal proceedings, inherent in such a troubled situation, may make the pursuit of judicial remedies futile and the administrative inquiries on which such remedies depend may be prevented from taking place.
>
> ... In the Court's view, the existence of mere doubts as to the prospects of success of a particular remedy which is not obviously futile is not a valid reason for failing to exhaust domestic remedies (see the *Van Oosterwijck* judgment[73]).

---

[70] Series A, No. 40, 18-19, §§ 36-40.
[71] Series A, No. 25, 64, § 159.
[72] Series B, No. 23-I, 394-397.
[73] *Van Oosterwijck*, see note 70, 18, § 37.

Nevertheless, like the Commission, the Court considers it significant that the government, despite the extent of the problem of village destruction have not been able to point to examples of compensation being awarded in respect of allegations that property has been purposely destroyed by members of the security forces or to prosecutions having been brought against them in respect of such allegations. In this connection the Court notes the evidence referred to by the Delegate of the Commission as regards the general reluctance of the authorities to admit that this type of illicit behaviour by members of the security forces had occurred. It further notes the lack of any impartial investigation, any offer to cooperate with a view to obtaining evidence or any *ex gratia* payments made by the authorities to the applicants".

The Court concluded that the application cannot be rejected for failure to exhaust domestic remedies.[74]

Just over a year later the *Akdivar* case was confirmed in the case of *Menteş*.[75]

3. Unknown Facts

Only after Mr. Farmakopoulos had brought an action before a domestic court did a situation creating abuse of power in his case became apparent. The Commission considered that his failure to invoke abuse of power before the domestic courts could not be seen as a failure to exhaust domestic remedies.[76]

However, applicants in similar situations may still have to exhaust domestic remedies if this is possible after they are informed of the facts.[77]

4. Waiver

No exhaustion of remedies is necessary if the respondent government has waived the right to rely on the rule of non-exhaustion.[78] The waiver

---

[74] *Akdivar* judgment of 16 September 1996, §§ 67, 70, 71, 76.
[75] *Menteş* judgment of 28 November 1997, §§ 55-61.
[76] Decision 11683/85 of 8 February 1990, 64 DR 52, 66. See also Decision 11613/85 of 16 May 1990, 65 DR 75, 88.
[77] Decision 8334/78 of 7 May 1981, 24 DR 103.
[78] Decision 1994/63 of 5 March 1964, 7 *YB ECHR* (1964), 252 et seq., 258; Decision 8919/80 of 17 March 1981, 23 DR 244.

need not be express. It is not quite clear whether the Commission is bound to respect a tacit waiver.[79]

In a case against Austria, domestic appeal proceedings had taken place after hearing the senior public prosecutor's office without the applicant or his lawyer being present. The applicant did not expressly request to be present and, on that ground, the Government submitted that he had not exhausted domestic remedies. Under those circumstances the Commission considered that it could not be found that the applicant failed to exhaust available domestic remedies.[80]

5. Government Pressure

Some Turkish applicants have been under pressure by the military or the police not to bring their cases before a court. Serious fear as a result of such pressure may be a valid reason for not exhausting domestic remedies.[81]

Pressure is in violation of Article 34 (former Article 25) of the Convention when it is exerted in order to prevent an applicant from bringing his case to Strasbourg. Article 34 obliges the Contracting States not to hinder in any way the exercise of the right of individuals to address a complaint to the European Court.[82]

6. Time

A remedy cannot be effective if it cannot be obtained within a reasonable time. The Commission considered several times:

> "according to the generally recognised rules of international law, where there are protracted domestic proceedings together with a continuing prejudicial situation an individual is absolved from the obligation to exhaust domestic remedies".[83]

Under the case law of the Commission, complaints concerning the length of procedures can be brought before it, prior to the termination of the procedures in question, if the applicant has at least made use of those remedies available to him concerning the length of the procedures.[84]

---

[79] Martinez Ruiz, see note 45, 263.
[80] Decision 11894/85 of 8 May 1989, 61 DR 156.
[81] *Salman* judgment of 27 June 2000, §§ 126-133.
[82] *Ibid.*
[83] Decision 14556/89 of 5 March 1991, 69 DR 261, 270.
[84] Decision 9816/82 of 9 March 1984, 36 DR 170, 176.

7. Cases of Killing or Torture

Sometimes, an application is lodged against a decision to expel or extradite a person to a country where he runs a serious risk of being killed or tortured. In such cases, the expulsion or extradition may constitute inhuman treatment, and is therefore prohibited under Article 3 of the Convention. Where a violation of the Convention could be brought about by putting into effect the decision to remove a person from a State's territory, a Court action without suspensive effect cannot be considered as effective and, therefore, domestic remedies may not have to be exhausted.[85]

However, this applies only to cases where irreparable damage is feared. When an expulsion is complained of as being in violation of the respect for family life required by Article 8 of the Convention, all remedies must be exhausted, even those which have no suspensive effect. Nothing would prevent the applicant from returning after his expulsion in case his complaint would be successful.[86]

Governments are under the obligation to investigate serious claims about killings and torture. When applicants have brought killing or torture to the attention of the government and no investigation has followed at all, the European Court may find a violation of the Convention even if no further exhaustion of domestic remedies has occurred.

This does not necessarily mean that in cases of killing and torture no exhaustion of remedies is needed. In the case of *Bahaddar* the Court noted that

> "although it has held the prohibition of torture or inhuman or degrading treatment contained in Article 3 of the Convention to be absolute in expulsion cases as in other cases, applicants invoking that Article are not for that reason dispensed as a matter of course from exhausting domestic remedies that are available and effective. It would not only run counter to the subsidiary character of the Convention but also undermine the very purpose of the rules set out in Article 26 of the Convention if the Contracting States were to be denied the opportunity to put matters right through their own legal system. It follows that, even in cases of expulsion to a country where there is an alleged

---

[85] Decision 7011/75 of 3 October 1975, 4 DR 215; Decision 7216/75 of 20 May 1976, 5 DR 137; Decision 7465/76 of 29 September 1976, 7 DR 153; Decision 14312/88 of 8 March 1989, 60 DR 284; Decision 17550/90 and 17825/91 of 4 June 1991, 70 DR 298.

[86] Decision 7014/75 of 21 May 1976, 5 DR 137, 141, 142. See also Decision 12097/86 of 13 July 1987, 53 DR 210.

risk of ill-treatment contrary to Article 3, the formal requirements and time limits laid down in domestic law should normally be complied with, such rules being designed to enable the national jurisdiction to discharge their case-load in an orderly manner".[87]

8. Lack of Facilities

Mrs. Johanna Airey tried to obtain a decree of judicial separation from her husband. The proceedings to reach this were so complicated that she could not litigate herself. Neither could she pay the costs of a solicitor. For lack of money she did not exhaust domestic remedies. In the *Airey* case the European Court of Human Rights did not hold this against her. It was of the opinion that legal aid should have been granted to her and therefore it did not reject the case for non-exhaustion.[88]

In a recent case, Mr. Gnahoré failed to appeal to the French Cour de Cassation, on the ground that legal aid had been refused to him. As he could have appealed without a lawyer, the European Court found that the refusal to grant legal aid did not cause a violation of the Convention. But it decided the case on the merits, not rejecting it on the ground of non-exhaustion.[89] When the possibility of access to a court is the issue of the dispute, the argument of non-admissibility for non-exhaustion will not be upheld.

In criminal proceedings one has a right to an interpreter. Perhaps this can be extended to some civil proceedings, once they are before a court. There is no right, however, to an interpreter when one wants to start proceedings. Inability to understand the language of the Court does not absolve an applicant from his obligation to exhaust domestic remedies. A foreigner, who is not detained and who alleges that he does not understand a judgment and a notice of appeal which are both in the language of the Court, is not absolved from the duty to exhaust available remedies.[90]

9. Ignorance of the Applicant

Sometimes applicants invoked their ignorance about the requirements or about the availability of remedies as an excuse for not exhausting all domestic remedies. The Commission always rejected that argument. In

---

[87] *Bahaddar* judgment of 19 February 1998, § 45.
[88] *Airey* judgment of 9 October 1979, A 32.
[89] *Gnahoré* judgment of 19 September 2000.
[90] Decision 11122/84 of 2 December 1985, 45 DR 246.

its opinion, the ignorance of an applicant as to the existence of remedies at his disposal can in no way relieve him from obligations established by the rules of international law.[91]

Notwithstanding this consistent opinion of the Commission, the skill of the applicant seems to play some role. In *Beis*, the applicant could choose one of two ways to exhaust domestic remedies. He chose the cheaper one which was ineffective. In considering that he had not properly exhausted, the Court expressly took into account that the applicant was a professor of the law of civil procedure and was therefore well placed to assess whether the way he chose was sufficient and appropriate.[92]

10. Advice of the Lawyer

Often, it is not easy to know whether a particular domestic remedy could be sufficiently effective to be exhausted. The question then arises whether an applicant may rely on the opinion of a lawyer. In a case against the United Kingdom, the applicant's counsel carefully considered the applicant's prospect of success in the light of the domestic legal situation. He concluded that the civil proceedings concerned did not offer any prospects of success. On that ground, the Commission concluded that the applicant was not required to use the remedy which was available under English law.[93] In a later case in which the applicant's counsel advised that an appeal had no prospect of success on the ground that the judgment of first instance represented an *opinio communis*, the Commission again considered that the applicant did not need to exhaust this remedy.[94] However, the opinion of the counsel is not always decisive. In a case in 1984, the Court of Appeal had refused leave to appeal to the House of Lords. The counsel of the applicants had advised that an application directly to the House of Lords for leave of appeal had no substantial likelihood of success. In that case the Commission considered that the existing doubts as to the prospects of success were not a sufficient basis for absolving an applicant from the strict requirement of exhaustion of domestic remedies. The case was declared inadmissible for non-exhaustion.[95]

In a case against Sweden, the applicant had unsuccessfully applied to the chancellor of justice. He was told by his lawyer that there was no appeal against that decision, but no case law supported that contention.

---

[91] Decision 1211/61 of 4 October 1962, 2 *YB ECHR* (1958, 1959), 224 et seq., 226.
[92] *Beis* judgment of 20 March 1997, §§ 34-36.
[93] Decision 7907/77 of 12 July 1978, unpublished.
[94] Decision 10000/82 of 4 July 1983, 33 DR 247.
[95] Decision 10789/84 of 11 October 1984, 40 DR 298.

Accordingly, the Commission concluded that the applicant could have brought an action against the State before the ordinary courts and declared the case inadmissible for non-exhaustion.[96]

11. The Authorities Should Act

We have already noted that once killings or torture have come to the attention of the government, it has to investigate. No official request is needed. If the authorities act, e.g., by prosecuting the policeman who allegedly committed the killing or the torture, the victim or his family do not have to take any action. The judicial proceedings against the policeman serve as exhaustion of remedies.[97]

If no investigation takes place, the case may be brought before the European Court of Human Rights without further exhaustion of domestic remedies. The refusal to investigate indicates sufficiently that the national authorities will not remedy the situation.

Sometimes, applicants submit that they did not have to raise particular remedies before the domestic court as the court should have looked into such remedies *proprio motu*. In this situation, we should distinguish between cases in which a court is *permitted* to act *proprio motu* and cases in which a court is *obliged* to do so. With respect to the former case the Commission held:

> "the fact that a domestic court of appeal is competent to examine *proprio motu* grounds amounting to a violation of the Convention does not absolve the applicant from the obligation of raising the complaint before the Court itself".[98]

Pobornikoff is an example of the latter case. In this case, the Austrian Supreme Court carried out an assessment of the applicant's personality and character in his absence. Under Austrian law the applicant could have requested to be present at the hearing concerned, but he had failed to do so. The Government therefore claimed that he had not exhausted domestic remedies. The European Court of Human Rights found that Austria was under a positive duty to ensure the applicant's presence in order to enable him "to defend himself in person" as required by Article 6 § 3(c), and that it followed that the applicant did not need to expressly request to be present. Accordingly, there had not been a failure to exhaust domestic remedies in this regard.[99]

---

[96] Decision 10371/83 of 6 March 1985, 42 DR 127.
[97] See, e.g., *Assenov* judgment of 28 October 1999, or *Köksal* decision on admissibility, 19 September 2000.
[98] Decision 11244/84 of 2 March 1987, 55 DR 98; Decision 15123/89 of 18 April 1991, 70 DR 252.
[99] *Pobornikoff* judgment of 3 October 2000, §§ 31-33.

12. Other Excuses

Zwart mentions the following excuses which have also been invoked, but were rejected by the Commission: the applicant's indigence, his deportation, his fear that bringing a complaint would attract publicity to the case, his position as a mental patient, his depressive state, his lack of legal knowledge, his detention, and statements made by an unidentified official.[100]

VIII. Fairness

In *Ilhan* the European Court of Human Rights held:

> "the Court emphasises that the application of the rule of exhaustion of domestic remedies must make due allowance for the fact that it is being applied in the context of machinery for the protection of human rights that the Contracting States have agreed to set up. Accordingly, it has recognised that Article 35 § 1 must be applied with some degree of flexibility and without excessive formalism. It has further recognised that the rule of exhaustion is neither absolute nor capable of being applied automatically; for the purposes of reviewing whether it has been observed, it is essential to have regard to the circumstances of the individual case. This means, in particular, that the Court must take realistic account not only of the existence of formal remedies in the legal system of the Contracting State concerned but also of the general context in which they operate, as well as the personal circumstances of the applicant. It must then examine whether, in all the circumstances of the case, the applicant did everything that could reasonably be expected of him or her to exhaust domestic remedies".[101]

One can imagine cases in which stiff rules on the exhaustion of domestic remedies would be extremely unfair. Let us imagine the following fictitious case.

Mr. Herany has demolished a large photograph of the national dictator. Subsequently, he flees the country. In his absence, he is sentenced to torture or death. Arriving at the border of a European country he is asked: "Are you a political refugee?" He does not speak English, but understands the word political and, not considering himself a politician, replies: "No, no, I am not political, I am a normal citizen".

---

[100] Zwart, see note 1, 214, 215.
[101] *Ilhan* judgment of 27 June 2000, § 59.

On the basis of this reply, the local police decide that he should be sent back to his own country. As there is no aeroplane available, he is held in detention for a couple of days. While in detention, he meets a compatriot who helps him to find a lawyer. This takes a couple of days, and by the time the lawyer has met him the time-limit for appealing the decision to send him back has passed. His appeal is declared inadmissible under domestic law. He complains to the Court at Strasbourg. Under the existing case law, his application would be held to be inadmissible because he did not exhaust domestic remedies according to the domestic rules. It is obvious, therefore, that strict application of the law would lead to unfair results.

In some situations, an applicant may be so afraid of the national authorities that he does not even dare to appeal. In particular, when national laws prescribe that an appeal to a court should be preceded by an appeal to the military or to the police, applicants may fail to do so. In very compelling situations this may be a reason not to apply the requirements of exhaustion of domestic remedies too strictly.

IX. Choice of Different Remedies

Sometimes an applicant has different domestic remedies at his disposal. Under Turkish law, for example, there are administrative, civil and criminal remedies against illegal and criminal acts attributable to the State or its agents. Must an applicant exhaust all these remedies or would one of them suffice? If one of them suffices, may he simply choose or should one remedy be preferred over the others? The Commission considered:

> "The requirement of exhaustion of domestic remedies has been complied with if the applicant has raised in substance before the highest national authority the complaint he raises before the Commission. It is therefore not necessary to examine whether the applicant disposed of other legal remedies to prevent the breach of which he complaints".[102]

Usually, criminal proceedings are more attractive for an applicant, as most of the costs are borne by the public prosecutor. If the Convention has been infringed by a government agent, and that agent is prosecuted for the wrong he has done (for example for torturing a prisoner), the victim may join a civil action for damages to the action of the public prosecutor. The victim will have exhausted domestic remedies even if in

---

[102] Decision 9186/80 of 9 March 1982, 28 DR 172.

the last instance the government agent is acquitted and the civil action which was joined to the proceedings is rejected. Problems may arise when the public prosecutor withdraws the charges before the domestic remedies are exhausted. The applicant may then have to start new civil proceedings. However, if time-limits have passed and other proceedings are no longer possible, the civil action joined to the criminal one should be accepted as a domestic remedy. The impossibility of exhausting further remedies is attributable to the public prosecutor, who is a state authority. This means that the applicant would have exhausted the remedies open to him.

An applicant who has exhausted a remedy which is apparently effective and sufficient cannot be required to have also tried others which were available but probably ineffective.

An applicant must make normal use of those domestic remedies which are likely to be effective and sufficient. When a remedy has been attempted, use of another remedy which has essentially the same objective is not required.[103]

Mr. Ilhan was running when Turkish gendarmes ordered him to stop. He did not hear the order and was shot at. Subsequently, he was beaten, severely injured and arrested. The public prosecutor was aware of his injuries but did not provide any medical care. Only several days later he was admitted to a hospital. He was charged with the offence of resistance to officers and sentenced by the justice of the peace to a suspended fine of 35,000 Turkish lira. No further legal proceedings were taken. When he claimed before the European Commission of Human Rights that he had been tortured, the Turkish Government submitted that domestic remedies had not been exhausted. The European Court of Human Rights noted that Turkish law provides administrative, civil and criminal remedies against illegal and criminal acts attributable to the State or its agents. It considered each of these possibilities.

An action in administrative law would be insufficient as it could lead only to the award of damages. An investigation capable of leading to the identification and punishment of those responsible could not follow. This consideration demonstrates that it is not enough that the domestic jurisdiction is confronted with the claim, it should also have the power to offer an effective remedy.

For a successful civil action, a plaintiff must, in addition to establishing a causal link between the tort and the damage, also identify the person believed to have committed the tort. In the case in question, the public prosecutor took no steps to identify who was present when Mr. Ilhan was apprehended or when his injuries were incurred. None of the documents provided by the gendarmes enabled such persons to be

---

[103] Decision 11471/85 of 19 January 1989, 59 DR 67.

identified. Under those circumstances, the Court considered that a civil claim had no reasonable prospect of success and therefore did not need to be exhausted.

With regard to remedies under criminal law, the Court noted that the public prosecutor had been informed that Mr. Ilhan had suffered serious injuries and was accordingly under the duty to investigate whether an offence had been committed. Given that Mr. Ilhan's circumstances would have caused him to feel vulnerable, powerless and apprehensive of the representatives of the State, he could legitimately have expected that the necessary investigation would have been conducted without a specific, formal complaint from himself or his family. The public prosecutor, however, chose not to make any inquiry as to the circumstances in which the injuries were caused. The Court was of the opinion that Mr. Ilhan had taken sufficient steps for the exhaustion of the remedies under criminal law.

As none of the possible remedies were effective, the Court rejected the Government's submission that Mr. Ilhan had not sufficiently exhausted domestic remedies.[104]

X. Conclusion

Our conclusion from the case law can be found in the *Ilhan* case, in which the Court held that the rule of exhaustion is neither absolute nor capable of being applied automatically, that it is essential to have regard to the circumstances of the individual case, and that an applicant has sufficiently exhausted domestic remedies when he has taken all steps that can *reasonably* be expected of him or her.[105]

---

[104] *Ilhan* judgment of 17 June 2000, §§ 60-64.
[105] *Ibid.*, § 59.

The 1928 General Act for the Pacific Settlement of International Disputes Revisited

Christian Tomuschat

I. Introduction

In his separate opinion[1] in the case concerning the *Aerial Incident of 10 August 1999 (Pakistan v. India)*,[2] Judge Oda, to whom this *Festschrift* is dedicated, closely examined the legal significance of the 1928 General Act of Arbitration for the Pacific Settlement of International Disputes (hereinafter, "General Act" or GA).[3] The judgment of the International Court of Justice of 21 June 2000 came to the conclusion that India, through a notification to the UN Secretary-General of 18 September 1974,[4] had implicitly made use of the right of denunciation enshrined in Article 45 of the GA so that there was no need to consider whether the General Act had in fact been binding on India before that notification could take effect. Judge Oda, while approving the overall result reached by the majority, expressed the view that the General Act had never contained a compromissory clause establishing the jurisdiction, initially, of the Permanent Court of International Justice and, after 1945, of the ICJ. He drew attention to the fact that, regarding the judicial settlement of disputes, the General Act ran parallel to the optional clause of the PCIJ's Statute and that Article 17 of the GA contained nothing new compared with the optional clause. Observing that the Assembly of the League of Nations, on the same day that it adopted the General Act, also adopted a resolution calling upon States to accede to the optional clause, he concluded that the General Act, too, was meant to serve just as an encouragement for States to accept the jurisdiction of the PCIJ under

---

[1] *ILM* 39 (2000), 1133 et seq.
[2] Judgment of 21 June 2000, *ILM* 39 (2000), 1116.
[3] Act of 26 September 1928, 93 LNTS 343.
[4] Reproduced in United Nations, *Multilateral Treaties Deposited with the Secretary-General. Status as at 31 December 1990*, 919.

that clause. Indeed, in 1924, the Assembly had already pursued a similar double strategy. On the one hand, it invited States to accept the optional clause; on the other hand, it recommended for adoption by Member States a Protocol for the Pacific Settlement of International Disputes. Article 3 of the 1924 Protocol was couched in ambiguous terms. It read:

> "The Signatory States undertake to recognise as compulsory, *ipso facto* and without special agreement, the jurisdiction of the Permanent Court of International Justice in the cases covered by paragraph 2 of Article 36 of the Statute of the Court, but without prejudice to the right of any State, when acceding to the [optional clause] to make reservations compatible with the said clause".[5]

This provision could have been interpreted to mean that any future State Party would *ipso facto* be subject to the jurisdiction of the Court. However, in view of its explicit reference to Article 36(2) of the Statute of the PCIJ it would certainly not be too far-fetched to argue that it was confined to bringing into being an obligation for States to make a declaration under Article 36(2) to which they could then also make such reservations as they deemed necessary according to their specific political circumstances. The ambiguity has never been lifted, since the 1924 Protocol did not enter into force. Apparently, no State saw any necessity to accept a conventional instrument which would contain no more than a hortatory stipulation.

In his opinion, Judge Oda has had the courage to tackle a problem which never actually arose in the practice of the PCIJ – because no claim was ever based on the General Act – and which has always been eschewed by its successor. Indeed, any observer must feel amazed by the avoidance strategy of the Court. The General Act was consistently brushed aside whenever it was invoked by one of the litigant parties. It looks as if the General Act contained a dark secret that should never be revealed. The explanation for the reluctance of the Court actually to apply the General Act seems fairly simple, however. Today, after the optimism of slogans such as "world peace through law" has evaporated, it appears almost inconceivable that reasonable governments might have agreed to submit for decision to an international tribunal "all disputes" with regard to which they were in conflict with one another (Article 17 of the GA). Therefore, a lawyer of our epoch is almost compelled to assume that the relevant provision cannot really mean what, on the surface, it seems to connote. Additionally, the Court may have shied away from having to apply the *clausula rebus sic stantibus*, the ominous Trojan horse of the law of treaties (which is found today in Article 62 of

---

[5] Text taken from Judge Oda's separate opinion, see note 1, 1135, para. 7.

the Vienna Convention on the Law of Treaties; hereinafter, VCLT), with regard to a treaty rooted in the optimistic atmosphere of the years when the League of Nations reached its climax before suffering heavy blows from the emerging European totalitarianism.

Judge Oda rightly gives expression to these doubts, which are compounded by the parallelism between the optional clause and the General Act. It will be the aim of the following reflections to find out whether Article 17 of the GA may be understood as a compromissory clause in the same vein as, for example, Article IX of the Convention on the Prevention and Punishment of the Crime of Genocide, or whether it remains at the level of an obligation which would require implementation through a supplementary declaration under Article 36(2) of the ICJ Statute. For reasons of space, other aspects have to be left aside. In particular, no consideration will be given to the question whether, in light of many more decades of experience with mechanisms for the settlement of international disputes, the distinction underlying the General Act between legal and non-legal disputes should be maintained. Likewise, this study does not attempt to clarify whether compulsory conciliation and arbitration for all kinds of disputes may be considered a reasonable method suitable to satisfy the legitimate expectations of litigant parties.

Today, Article 17 must be read in conjunction with Article 37 of the ICJ Statute. In light of the latter, its reference to the Permanent Court can be understood as a reference to the International Court. The drafters of Article 37 intended to save the compromissory clauses that had come into being before 1945 by channelling them to the new main judicial body of the international community, the International Court of Justice. It stands to reason that Article 17 of the GA could survive only thanks to that transfer clause.

II. The Practice of the ICJ

The first cases under the aegis of the ICJ in which one of the parties relied on Article 17 of the GA were carefully reviewed by J. G. Merrills more than 20 years ago.[6] It will therefore suffice to give a short account of the relevant decisions.

1. The Norwegian Loans Case

In the *Norwegian Loans* case of 1957,[7] the jurisdiction of the Court could have found an unchallengeable basis in Article 17 of the GA. But

---

[6] "The International Court of Justice and the General Act of 1928", *Cambridge Law Journal* 39 (1980), 137 et seq.
[7] ICJ Reports 1957, 9.

the Court declined to accept the General Act as the requisite legal support of the French contention that the Court was competent to entertain its claim. France had based its application on Norway's acceptance of the optional clause and had referred to the General Act only in some other connection. Consequently, the Court felt that it was not its task to search for an adequate foundation establishing its jurisdiction. It held that, if France had wished to rely on the General Act, "it would expressly have so stated".[8] In his dissenting opinion, Judge Basdevant rejected this argument, contending that it was for the Court to ascertain the law essentially in accordance with the adage *jura novit curia*.[9] He held that Article 17 of the GA indeed constituted a true compromissory clause in the sense contemplated by Article 36(1) of the ICJ Statute.

2. The Temple Case

The General Act was relied upon a second time in the *Temple* case between Thailand and Cambodia (Preliminary Objections).[10] Cambodia invoked a declaration under Article 36(2) of the Court's Statute as well as a number of treaties providing for submission of any disputes to the Court, among them the General Act.[11] However, no attention was given to this second limb of the jurisdictional bases of the dispute, since the Court focused exclusively on Thailand's declaration under the optional clause, concluding that none of the contentions of the defendant according to which the dispute was not covered by that declaration could be accepted. Having therefore established its jurisdiction under Article 36(2) of its Statute, the Court found it unnecessary to dwell additionally on the General Act.[12]

3. The Pakistani Prisoners of War Case

The third case in which the General Act was invoked as basis of jurisdiction had the same configuration of litigants as the dispute settled by the judgment of the Court of 21 June 2000, to which Judge Oda appended his separate opinion. Pakistan and India were at odds over the treatment of Pakistani prisoners of war, some of whom had been placed

---

[8] *Ibid.*, 25.
[9] *Ibid.*, 71 et seq., 74. Textually, *jura novit curia* does not appear in his opinion.
[10] ICJ Reports 1961, 17.
[11] *Ibid.*, 21.
[12] *Ibid.*, 35.

by Indian authorities under criminal prosecution on charges of genocide. Pakistan based its claim largely on the General Act, and the two parties extensively discussed the validity of the Act after the demise of the League of Nations.[13] First of all, Pakistan requested the Court to indicate as an interim measure the repatriation of all prisoners. However, given the fact that the parties had started negotiations on the controversial issues and that Pakistan asked for a postponement of consideration of its request, the Court found that there was no need for the indication of such measures.[14] Regarding the preliminary stage of the proceedings, the issue of jurisdiction had consequently become moot. Later, the Court was informed of an agreement which the two parties had reached. Thereupon, it decided that the case be removed from the list of pending matters.[15] Thus, the case disappeared from the Court's agenda. However, the proceeding had important consequences. In order to set the record straight, Pakistan addressed a notification of succession to the UN Secretary-General on 30 May 1974, in which it asserted that "the Government of Pakistan continues to be bound by the accession of British India of the General Act of 1928", while distancing itself from the reservations made by British India.[16] Recognising the threat that it might again be put on the defensive by Pakistan in another proceeding, the Government of India reacted through a communication of 18 September 1974, which was also addressed to the UN Secretary-General, stating that it never regarded itself "as bound by the General Act of 1928 since her [India's] Independence in 1947, whether by succession or otherwise".[17]

4. The Nuclear Tests Cases

The issue of the validity and applicability of the General Act came up again in the *Nuclear Tests* cases between Australia and New Zealand, on the one side, and France, on the other. Like in the *Pakistani Prisoners of War* case, the applicants sought an interim injunction by the Court according to which France would have to refrain from conducting any further atmospheric nuclear tests giving rise to radioactive fall-out. They based their requests, on the one hand, on the French declaration under Article 36(2) of the ICJ Statute and, on the other hand, on the General Act. On both counts, the Court held that the provisions invoked appeared *prima facie* to afford a basis on which its jurisdiction might be

---

[13] For a summary of that discussion see Merrills, see note 6, 142-144.
[14] Order of 13 July 1973, ICJ Reports 1973, 328 et seq., 330
[15] Order of 15 December 1973, ICJ Reports 1973, 347, 348.
[16] *Multilateral Treaties*, see note 4, 919.
[17] *Ibid.*

founded.[18] However, it opined that the continuance in force of the General Act was so doubtful that it should refrain from examining the requests under that heading. For that reason, it focused exclusively on France's declaration under the optional clause[19] although that declaration was limited in scope by a reservation referring to "disputes concerning activities connected with national defence" – a limitation which the Court did not even mention in its orders granting the requests by Australia and New Zealand. Of course, it would have been incumbent on the Court to assess in detail the legal position in a later judgment on its jurisdiction, since France had raised preliminary objections. In this connection, the Court could not have avoided to take a clear stance regarding the applicability of the General Act as an instrument either being in force or having become extinct. It is a matter of common knowledge, though, that the Court disentangled itself from this inference of juridical logic by "discovering" that as a consequence of declarations made by high-ranking French representatives not to continue atmospheric tests the case had lost its object. Hence, there no longer was a dispute before the Court.[20] Once again, the Court had circumvented the necessity to pronounce on the legal significance of the General Act in the post-war period. In a well-argued dissident opinion Judges Onyeama, Dillard, Jiménez de Aréchaga and Sir Humphrey Waldock attempted to prove that the General Act still continued in force, not being affected by the demise of the League of Nations.[21] For the four judges, the notion that Article 17 of the GA provided a jurisdictional basis seemed so evident that they did not specifically address the issue. Their understanding of Article 17 of the GA as a compromissory clause underlies the joint dissenting opinion from the first to the last of the many pages (no less than 46!) devoted to the General Act. But their line of argument is not exclusively implicit. In one paragraph, it is explicitly stated:

"Under the terms of Article 17 of the 1928 Act, the jurisdiction which it confers on the Court is over 'all disputes'".

No doubt is therefore allowed concerning the personal views of the four dissenting judges who, it should be emphasised, did not have to reject any view to the contrary expressed by the Court. The disagreement that existed related solely to the conclusion, drawn by the Court, that the case

---

[18] Orders of 22 June 1973, ICJ Reports 1973, 99 et seq., 102, and 135 et seq., 138.
[19] *Ibid.*, 103 and 139.
[20] Judgments of 20 December 1974, ICJ Reports 1974, 253 et seq., 268-272 and 457 et seq., 473-477.
[21] ICJ Reports 1974, 312 et seq., 327 ff. and 494 et seq., 509 ff.

had come to its end inasmuch as it had lost its character as a genuine legal dispute between two parties.

5. The Aegean Sea Continental Shelf Case

In the *Aegean Sea Continental Shelf* case, a new attempt was made, this time by Greece, to found the jurisdiction of the Court on Article 17 of the General Act. The proceeding evolved in two stages, following the precedent set by the *Nuclear Tests* cases. As a first step, Greece requested the Court to indicate interim measures on the basis of both the General Act and Article 41 of the ICJ Statute. It contended that through exploratory activities undertaken by the Turkish State Petroleum Company its sovereign rights in the Aegean Sea were being infringed. Instead of inquiring into its bases of jurisdiction during this first stage, however, the Court rejected Greece's request on the merits as unfounded, given that the UN Security Council, in Resolution 395 (1976), had called upon Greece and Turkey to do everything in their power to reduce the tensions existing between the two countries. It based its legal argument exclusively on Article 41 of the Statute, totally leaving aside the General Act, because Turkey had contested the continuance in force of the Act and had additionally drawn attention to the reservations which Greece had appended to its act of accession to the Act.[22]

When considering its jurisdiction two years later, [23] the Court finally had to deal squarely with the General Act, since no other opening under Article 36 of the Statute was available to Greece. But the Court chose not to examine whether the General Act was still in force, holding that a finding on that issue could also entail repercussions for other States. Instead, it sought to clarify the legal position by considering a reservation by which Greece had restricted the scope *ratione materiae* of its adherence to the General Act, excluding in particular "disputes relating to the territorial status of Greece". According to the principle of reciprocity, this reservation could also be invoked by Turkey. The Court came to the conclusion that the controversy over the Greek rights in the Aegean Sea was indeed a dispute relating to the territorial status of Greece. Proceeding from that premise, it denied its jurisdiction.[24] No word was said on whether the General Act could be deemed to be still in force.

---

[22] Order of 11 September 1976, ICJ Reports 1976, 3 et seq., 8-14.
[23] Judgment of 19 December 1978, ICJ Reports 1978, 3 et seq.
[24] *Ibid.*, 13-37.

6. The Aerial Incident Case

The case concerning the *Aerial Incident of 10 August 1999 (Pakistan v. India)* repeats the pattern set by the *Aegean* case. Faced with the Pakistani contention that the General Act was in force between the two litigant parties, the Court, as already noted, took the view that, irrespective of the General Act's survival notwithstanding the disappearance of the League of Nations, India had to be considered a non-party on account of its notification of 18 September 1974 to the UN Secretary-General in response to the earlier Pakistani notification of succession of 30 May 1974. Again, the most contentious issue could in that way be avoided.

Summarising the jurisprudence of the Court as succinctly described above, one can say that the disagreement which existed as to the relevance of the General Act in the cases at hand centred invariably and exclusively on its existence as part and parcel of the body of positive international law in force. At no time was it alleged that Article 17 of the GA did not constitute a true compromissory clause on which it would be possible to base the jurisdiction of the Court. Consequently, it would appear to be necessary to construe this central provision according to the general rules of interpretation as they are today laid down in Articles 31 and 32 of the Vienna Convention on the Law of Treaties as a reflection of the customary rules governing interpretation in international law.

III. Interpretation of Article 17 of the General Act

Any effort at interpretation must first of all start out with an elucidation of the meaning of the words used. Article 17 reads:

> "All disputes with regard to which the parties are in conflict as to their respective rights shall, subject to any reservations which may be made under Article 39, be submitted for decision to the Permanent Court of International Justice, unless the parties agree, in the manner hereinafter provided, to have resort to an arbitral tribunal".

Admittedly, the words "shall ... be submitted" are not devoid of a certain ambiguity. They could indeed, as suggested by Judge Oda, be understood in the sense that the parties to the General Act entered into a commitment to conclude a *compromis* under Article 36(1) of the PCIJ Statute whenever a conflict as to their respective rights would arise. In more recent conventional instruments, care has been taken to specify that any party may seize the Court by unilateral application. Thus, Article IX of the Genocide Convention provides:

> "Disputes between the Contracting Parties ... shall be submitted to the International Court of Justice at the request of any of the parties to the dispute".

A similar formula can be encountered in the Vienna Convention on the Law of Treaties (Article 66(a)):

> "any one of the parties to a dispute concerning the application or the interpretation of Articles 53 or 64 may, by a written application, submit it to the International Court of Justice...".

Thus, it might be argued that Article 17 does not *ipso facto* grant access to the Court, but requires an additional agreement between the parties concerned to that effect. In this connection, it may be recalled that, in the late 1920s, examples evincing a better formulation of a compromissory clause were indeed not lacking. Thus, in the *Ambatielos* case, Greece founded the jurisdiction of the Court in its dispute with the United Kingdom on a declaration from 1926 in which it was stipulated that

> "any differences which may arise between our two Governments as to the validity of such claims shall, at the request of either Government, be referred to arbitration...".[25]

Likewise, in the *Barcelona Traction* case, Belgium relied on a provision of the Hispano-Belgian Treaty of Conciliation, Judicial Settlement and Arbitration of 19 July 1927, which read:

> "... chaque Partie pourra, après préavis d'un mois, porter directement, par voie de requête, la contestation devant la Cour permanente de Justice internationale".[26]

The conclusion, however, that Article 17 of the GA did not open the gates to the ICJ would appear to be contradicted by its last clause. According to this proviso, agreement between the parties has a negative function. The parties were allowed to withdraw a dispute from the jurisdiction of the Permanent Court by agreeing upon an alternative remedy. Otherwise, however, cases were channelled to the Permanent Court, which is tantamount to saying that each party was endowed with a

---

[25] Judgment of 19 May 1953, ICJ Reports 1953, 10 et seq., 15.
[26] See Judgment of 24 July 1964, ICJ Reports 1964, 6 et seq., 27.

right unilaterally to seize the Court of any relevant dispute coming within the purview of the compromissory clause.[27]

In this regard, a comparison with Article 3 of the 1924 Protocol is striking. That Protocol would have explicitly reserved the right for any party to enter a reservation at a later stage when making a declaration under Article 36(2) of the Statute of the Permanent Court. If such was the case, the 1924 Protocol did in fact lack any binding effect. A genuine treaty commitment cannot be altered through a unilateral act of will after its acceptance, entirely at the discretion of any party wishing to modify the extent of its obligations. Consequently, it is fully justified to view Article 3 of the 1924 Protocol as no more than a political encouragement to submit to the jurisdiction of the Permanent Court, which signalled at the same time that acceptance of the optional clause was less burdensome than sometimes feared by States since Article 36(2) permitted the making of reservations (though within certain limits).

Of course, the question remains whether a true parallelism between accepting the jurisdiction of the Permanent Court either through the optional clause or by acceding to the General Act could make any sense. Rightly, Judge Oda has raised this issue. However, on reflection one may be able to detect grounds justifying a double strategy. On the one hand, it was known – and accepted – in 1928 that Article 36(2) of the PCIJ Statute left ample room for reservations of the most diverse nature without any limitation *ratione materiae*. In addition, declarations on the acceptance of the Court's jurisdiction could be limited in time. While most States engaged themselves for repeated periods of five years, which corresponded to the time-limits set forth in Article 45 for denunciation of the General Act, Ethiopia in 1932 started a practice of binding itself for only two years.[28] On the other hand, accession to the General Act created firmer obligations. Pursuant to Article 39(1) of the GA, only "reservations exhaustively enumerated in the following paragraph" could be entered. Although the list given in Article 39(2) was fairly extensive, the discipline it purported to introduce was in any event stricter than the total *laissez faire* of Article 36(2) of the PCIJ Statute.

---

[27] Curiously enough, the jurisdictional clause (Article 1) of the European Convention for the Peaceful Settlement of Disputes of 29 April 1957, UNTS Vol. 320, 243, is framed in such an imprecise way that the text leaves entirely open what its connotation may be: "The High Contracting Parties shall submit to the judgment of the International Court of Justice all international legal disputes which may arise between them...". Does it permit unilateral applications? The text does not answer this question. To date, the Convention has never been applied for the purpose of judicial proceedings before the ICJ.

[28] See *Douzième Rapport Annuel de la Cour Permanente de Justice Internationale (15 juin 1935 - 15 juin 1936)*, Sér. E - No. 12, 340.

Furthermore, denunciation could be effected only with effect to the end of successive five-year periods (Article 45). Thus, a State accepting the General Act bound itself in a more far-reaching way than a State limiting itself to making a declaration under the optional clause. The General Act strengthened the jurisdictional foundations of the Permanent Court.

IV. Legal Doctrine

It would seem highly significant, moreover, that the legal literature that commented on the scope and meaning of the General Act unanimously viewed Article 17 as a compromissory clause directly conferring jurisdiction on the Permanent Court. In his lectures given at the Hague Academy, Swiss lawyer E. Borel said that with respect to parties to the Act embroiled in a legal dispute that:

> "chacune d'elles aura la faculté de porter directement le différend devant la Cour par voie de simple requête".[29]

Similarly, A. P. Fachiri wrote in 1932:

> "... by the combined effect of Articles 17 and 41 it is open to a contracting party in the absence of agreement as to the method of settlement to bring any dispute unilaterally before the Court...".[30]

B. Schenk von Stauffenberg, an outstanding commentator on the Statute of the Permanent Court, characterised the General Act as one of the main examples for the second clause in Article 36(1) of the Statute.[31] We may also refer to M. Faraggi, who devoted an entire monograph to the General Act:

> "La Cour Permanente peut être saisie, soit d'un commun accord entre les Parties, par la ratification d'un compromis rédigé par elles, soit par requête unilatérale de l'une d'elles...".[32]

This view was shared in all other contemporary writings.[33]

---

[29] "L'Acte général de Genève", *RdC* 27 (1929-II), 499 et seq., 563.
[30] *The Permanent Court of International Justice. Its Constitution, Procedure and Work*, 2nd ed., 1932, 91.
[31] *Statut et Règlement de la Cour permanente de Justice internationale. Eléments d'interprétation*, 1934, 260-261.
[32] *L'Acte Général d'Arbitrage*, 1935, 217.
[33] See J. L. Brierly, "The General Act of Geneva, 1928", *BYIL* 11 (1930), 119 et seq., 124; X. Gallus, "L'Acte général d'Arbitrage", *Revue de Droit*

After the Second World War, interest in the General Act faded away, all the more so since the General Assembly in 1949 prepared a revised version of the instrument.[34] Therefore, few voices can be gleaned which have expressed themselves on the issue under review in this study. Nevertheless, the authors concerned are all in agreement with the earlier current in the legal doctrine.[35] The United Nations has adopted the same stance.[36]

V. State Practice

The practice of States, an essential factor in the search for the meaning of a conventional instrument (Article 31(3)(b), VCLT), buttresses the views that prevailed in legal doctrine. States which took the bold step of acceding to the General Act were generally anxious to restrict their acceptance by carefully drafted reservations.[37] These reservations can roughly be divided into three classes. In the first group, the reservations were almost identical to the corresponding reservations relating to the optional clause, *mutatis mutandis*, since a reservation to a unilateral declaration requires language different from the terms of a reservation to a multilateral treaty. In particular, the condition of reciprocity plays no role within a treaty framework, since *per definitionem* an agreement is binding only between the parties to it. With regard to unilateral declarations, however, conditioning one's acceptance by reciprocity may serve as a useful reminder, although Article 36(2) of the PCIJ Statute provided that any declaration under the optional clause was in any event subject to the condition of reciprocity. Thus, the United Kingdom, whose declaration of 21 May 1931 served as a blueprint for the corresponding reservations of most other Commonwealth countries, generally excluded from the scope of its accession to the General Act (paragraph 1):

> "(i) Disputes arising prior to the accession of His Majesty to the said General Act or relating to situations or facts prior to the said accession;

International 57 (1930), 190 et seq., 231: "une doublure de l'Article 36 du Statut de la Cour"; R. R. Wilson, "Clauses Relating to Reference of Disputes in Obligatory Arbitration Treaties", *AJIL* 25 (1931), 469 et seq., 488-489.

[34] UNTS Vol. 71, 101.

[35] K. H. Kunzmann, "Die Generalakte von New York und Genf als Streitschlichtungsvertrag der Vereinten Nationen", *Die Friedens-Warte* 56 (1961/66), 1 et seq., 11, 23; Merrills, see note 6, 137.

[36] United Nations, *Handbook on the Peaceful Settlement of Disputes Between States*, 1992, 72, para. 205.

[37] For a complete list, see *Multilateral Treaties*, see note 4, 914 et seq.

(ii) Disputes in regard to which parties to the dispute have agreed or shall agree to have recourse to some other method of peaceful settlement;
(iii) Disputes between His Majesty's Government in the United Kingdom and the Government of any other Member of the League which is a member of the British Commonwealth of Nations, all of which disputes shall be settled in such a manner as the parties have agreed or shall agree;
(iv) Disputes concerning questions which by international law are solely within the domestic jurisdiction of States; and
(v) Disputes with any Party to the General Act who is not a Member of the League of Nations".

On the other hand, the British declaration under the optional clause (of 19 September 1929) covered:

"all disputes arising after the ratification of the present declaration with regard to situations or facts subsequent to the said ratification, other than:

Disputes in regard to which the parties to the dispute have agreed or shall agree to have recourse to some other method of peaceful settlement; and

Disputes with the Government of any other Member of the League which is a member of the British Commonwealth of Nations, all of which disputes shall be settled in such manner as the parties have agreed or shall agree; and

Disputes with regard to questions which by international law fall exclusively within the jurisdiction of the United Kingdom".[38]

The congruence between the two texts is striking. Obviously, the United Kingdom felt it necessary specifically to state its reservations to the jurisdiction of the Permanent Court without relying on the fact that it had already done so in making its earlier declaration under the optional clause. There is no escaping the conclusion, therefore, that the United Kingdom regarded Article 17 of the GA quite definitely as a

---

[38] Reproduced in L. Oppenheim and H. Lauterpacht, *International Law. A Treatise*, Vol. II, 7th ed., 1952, 60; for the later declarations see M. O. Hudson, *World Court Reports*, Vol. IV, 1936-1942, 1943, 35-36.

compromissory clause which was independent from the mechanisms established by virtue of the Statute of the Court.[39]

Other countries were less disciplined. Although they had made reservations when accepting the optional clause, they refrained from amending their acceptance of the General Act to the slightest extent. Luxembourg may be mentioned as an example in point. The 1936 Annual Report of the Permanent Court carries the information that Luxembourg made its declaration under Article 36(2) of the Statute dependent on reciprocity and that additionally it excluded

> "tous différends qui s'élèveraient après la signature au sujet de situations ou de faits postérieurs à ladite signature"

as well as

> "les cas où les Parties auraient convenu ou conviendraient d'avoir recours à une autre procédure ou à un autre mode de règlement pacifique".[40]

At first glance, the ensuing discrepancy between the two acts establishing some jurisdictional link may seem to confirm that at least the Luxembourg Government – and other governments which showed a similar generosity in ratifying the General Act[41] – viewed this instrument as not permitting direct access to the Permanent Court. But again, closer scrutiny leads to a fairly simple explanation. First, it was already mentioned that reciprocity is an inherent feature of any treaty system. Whoever acceded to the General Act bound himself *vis-à-vis* all other States joining the mechanisms established by the General Act – but of course not *vis-à-vis* any outsiders. To insist on the condition of reciprocity was therefore not necessary in connection with the General Act. Secondly, it is a general rule of treaty law, today laid down in Article 28 of the Vienna Convention on the Law of Treaties, that in principle treaties lack retroactive effect and therefore do not apply to any act or fact which took place or any situation which ceased to exist before the date of entry into force of the treaty concerned. And thirdly, it is clear that a special agreement between litigating parties to the effect that recourse should be had to a specific procedure, different from judicial settlement by the Permanent Court or today the International Court, prevails over any earlier general commitment which flows from declarations made under the optional clause. Thus, the notion of

---

[39] The following Commonwealth countries are in the same basket: Australia, Canada, India, New Zealand.
[40] *Douzième Rapport Annuel*, see note 28, 344.
[41] Denmark, Finland, Ireland, Latvia, Norway, Spain and Switzerland.

generosity mentioned above must not be equated with carelessness. Legal departments authorising accession to the General Act without insisting on any of the reservations entered when the optional clause was accepted were not to be blamed if the earlier reservations were of the type just described.

Some countries, however, employed different wording, and in some instances this led to difficulties in later proceedings before the International Court. Thus, in the *Nuclear Tests* case, as already mentioned, it turned out that the French acceptance of the optional clause by a declaration of 20 May 1966 had been subject to a reservation covering "... disputes concerning activities connected with national defence" while this important element limiting the jurisdiction of the Court found no reflection in France's act of accession to the General Act. But again, this discrepancy does not prove that France evaluated the General Act simply as a political act. In fact, France appended an extensive reservation to its accession to the Act (21 May 1931).[42] This reservation was largely similar – though not identical – to the reservation it had made when, on 19 September 1929, it submitted to the jurisdiction of the Permanent Court by way of the optional clause.[43] However, the exclusion of matters of national defence appeared for the first time in the French declaration of 20 May 1966, and was still in force when the *Nuclear Tests* cases arose. It would seem to be obvious to the outside observer that someone in the Legal Department of the Quai d'Orsay had lost sight of the General Act, not recalling that instrument which had indeed fallen somewhat into oblivion. Accordingly, France defended itself in the proceeding by arguing that the General Act had become obsolete – an issue which is still open today.[44] Or else, as stressed in the joint dissenting opinion of Judges Onyeama, Dillard, Jiménez de Aréchage and Sir Humphrey Waldock in the *Nuclear Tests* cases, denouncing the General Act or complementing the reservations appearing in the act of accession similarly by way of denunciation could have meant awakening sleeping dogs and was therefore eschewed.[45]

Other countries which were less punctilious in assessing the legal position as it resulted from the coexistence of the two instruments were

---

[42] For the French text, see Permanent Court of International Justice, *Collection of Texts Governing the Jurisdiction of the Court*, 4th ed., Sér. D - No. 6, 1932, 71.

[43] *Douzième Rapport Annuel*, see note 28, 341.

[44] In response to the orders of the International Court in that case for the indication of interim measures of protection (22 June 1973), ICJ Reports 1973, 99 and 135, France denounced the General Act by a notification of 10 January 1974, see *Multilateral Treaties*, see note 4, 918.

[45] ICJ Reports 1974, 312 et seq., 339 para. 57.

Estonia, Greece,[46] Italy and Peru. Each of these divergences would deserve a consideration on its own merits. It can be said, however, that the differences are generally minimal. All essential points appear in both acts. Thus, Greece accepted the optional clause on 12 September 1929 by excluding, *inter alia*,

> "des différends ayant trait au statut territorial de la Grèce, y compris ceux relatifs à ses droits de souveraineté sur ses ports et ses voies de communication",[47]

whereas its accession to the General Act (14 September 1931) lists as not being encompassed

> "les différends portant sur des questions que le droit international laisse à la compétence exclusive des Etats et, notamment, les différends ayant trait au statut territorial de la Grèce, y compris ceux relatifs à ses droits de souveraineté sur ses ports et ses voies de communication".[48]

On the whole, the above cursory review of the practice of States inevitably leads to the conclusion that governments were of the opinion that Article 17 of the GA did indeed constitute a jurisdictional clause *lege artis juris* permitting direct access to the Permanent Court by unilateral application of any party to a legal dispute. Otherwise, it could not be explained why such care was spent on formulating appropriate reservations restricting the scope of acceptance of the General Act. In fact, in its official publications, the Permanent Court itself made known its view that the General Act was indeed to be classified as one of the many treaties on which its jurisdiction could be founded.[49]

VI. The Revised General Act

The limited space open to this short examination of the scope and meaning of Article 17 of the GA does not allow for an in-depth examination of the continuance in force of the General Act after the

---

[46] In the *Aegean Sea* case, Judgment of 19 December 1978, ICJ Reports 1978, 3 et seq., the divergence played no role because Turkey had made no declaration under the optional clause.

[47] *Douzième Rapport Annuel*, see note 28, 341.

[48] *Collection of Texts*, see note 42, 71.

[49] *Neuvième Rapport Annuel de la Cour Permanente de Justice Internationale (15 juin 1932 - 15 juin 1933)*, Sér. E - No. 9, 61; *Collection of Texts*, see note 42, 31 et seq., 70.

demise of the League of Nations. In the joint dissenting opinion of Judges Onyeama, Dillard, Jiménez de Aréchaga and Sir Humphrey Waldock in the *Nuclear Tests* case, this issue was carefully considered. However, a short glance may be shed at the Revised Act for the Pacific Settlement of International Disputes of 28 April 1949. As it emerges from the *travaux préparatoires*, the General Act of 1928 was not revised because it had become extinct, but because it was felt necessary to adapt its references to the institutional system of the League of Nations to the new realities ushered in by the Charter of the United Nations.[50] However that may be, the fact is that the Revised Act was acceded to without any reservations by the few States (eight) which it has been able to attract so far. Again, this finding raises the question whether it implies an assessment of the Act as possessing a purely political character. Indeed, any student of the ICJ system immediately thinks of the many reservations which have greatly reduced the real significance of the optional clause system. He will therefore be inclined to assume that acceptance of the jurisdiction of the ICJ without any such reservation does not belong to the world of hard and fast law. Such doubts are unfounded, however. All of the States listed as parties to the Revised General Act which have at the same time made declarations under Article 36(2) of the ICJ Statute (Belgium, Denmark, Estonia, Luxembourg, Netherlands, and Sweden) have submitted to the jurisdiction of the ICJ subject only to standard clauses regarding reciprocity, non-retroactivity or overriding agreement on another method of settlement. All these conditions are inherent in Article 17, if interpreted in a reasonable way as required under general rules of international law.

A major discrepancy may be observed only with regard to Norway. In 1996, the Norwegian Government renewed its declaration under the optional clause by restricting its scope of application in accordance with earlier declarations regarding the law of the sea.[51] Seemingly, nobody in Oslo has thought of aligning this new legal configuration with the earlier acceptance of the Revised General Act (16 July 1951), an event which has almost faded into history. Yet, Norway should be warned that even such long-forgotten acts may have the most palpable consequences, as emerged in the *Norwegian Loans* case when France all of a sudden brushed the dust off the General Act, which had lain dormant for almost three decades. The General Act cannot be brushed aside as easily as the European Convention for the Peaceful Settlement of Disputes,[52] which gives States a wide latitude if they wish to narrow the scope of their

---

[50] For a summary account of the drafting history see *Yearbook of the United Nations 1948/49*, 1950, 412-419.
[51] *ICJ Yearbook 1995-1996*, 1996, 108.
[52] See note 27.

acceptance of the jurisdiction of the ICJ. Indeed, Article 35(4) permits the making of reservations *ex post facto*, i.e., after having entered into the commitments specified in Article 1:

> "If a High Contracting Party accepts the compulsory jurisdiction of the International Court of Justice under paragraph 2 of Article 36 of the Statute of the said Court, subject to reservations, or amends any such reservations, that High Contracting Party may by a simple declaration ... make the same reservations to this Convention...".

Such extreme flexibility, which on its face runs counter to the maxim *pacta sunt servanda*, cannot be read into Article 17 of the GA.

VII. Conclusion

The conclusion reached by this short piece in honour of Judge Oda is somewhat intriguing. We cannot conceal our conviction that Article 17 of the General Act of 1929 – and the same applies to the Revised General Act of 1949 since its wording is substantially identical to that of its predecessor – constitutes a compromissory clause in the sense envisioned by Article 36(1), second sentence, of the ICJ Statute. Judge Oda came to the opposite conclusion in his separate opinion. He can indeed claim for himself the merit of having drawn attention to an issue which had never before been studied in depth. As Judge Oda has rightly pointed out, there exist many clues which seem to indicate that from the very outset the General Act was conceived of as being no more than a political manifesto. Never was Article 17 invoked by any party before the outbreak of the Second World War, and the ICJ has always avoided having to affirm or deny its applicability within the new world order that emerged with the Charter of the United Nations. The General Act may have been weakened by the extinction of the Geneva system of the League of Nations. Yet, in the view of the present writer, this weakness does not stem from a substantive flaw of the compromissory clause embodied in Article 17 of the General Act.

Bilateral, Plural and Multipartite Elements in International Judicial Settlement

Santiago Torres Barnárdez

For many years, judicial settlement has been a redoubt of bilateralism, but the incidence in international procedure of pluridimensional and multilateral elements, originating from the growing interdependence in international relations and the progressive development of international law, is becoming increasingly apparent in the judicial practice of international courts and tribunals. A few years ago it would have seemed unthinkable that a State could bring simultaneous proceedings before the International Court of Justice in one fell swoop, as it were, against ten States through a series of applications, certainly separate, but virtually identical, following a collective act by the respondents which, according to the applicant, was purportedly illegal in international law. This is not to say that this course of action was not possible according to the basic texts, but rather that this did not used to happen.

Well, this is indeed what Yugoslavia did in April 1999 when it brought proceedings against ten Member States of NATO, on the occasion of the bombardments to which Yugoslavia was subjected during the Kosovo crisis *(Cases Concerning the Legality of Use of Force)*. Moreover, when, in 1995, in accordance with a paragraph of the Court's 1974 judgment, New Zealand presented a request for an examination of the situation that had arisen as a result of various new nuclear tests carried out by France in the Pacific Ocean, no less than five Pacific-rim States formally communicated their wish to intervene in the case. These requests were in some cases made on the basis of documents common to two States and, in addition, invoked simultaneously both forms of intervention admitted by the Statute of the Court. This course of action was also unprecedented up till then. In fact, the increasing interest since the 1970s of third States in intervening in cases between other States – as proved by the examples of Fiji, Malta, Italy, El Salvador, Nicaragua, Equatorial Guinea and the Philippines – also reflect the general demand to go beyond traditional bilateral parameters in the judicial settlement of international legal disputes.

The progressive development of general international law as, for example, in the realm of the law of the sea, the increasing network of conventional multilateral instruments and of the number of States participating therein, and the recognition by States and international jurisprudence, as in the 1995 *East Timor* case, of the existence in international law of obligations having an *erga omnes* character, are indeed underlying developments which help to understand the increasing need for judicial settlement to take into account plural and multipartite elements in the treatment and resolution of interstate disputes in accordance with international law. The same factors explain the renewed doctrinal interest in the so-called "indispensable parties" theory and, since the 1980s, the reconsideration in the context of a number of cases of an issue which had remained dormant since the 1954 *Monetary Gold* case.

Learned institutions have not been insensitive to the need to focus studies regarding judicial settlement on the new requirements. For example, the Institute of International Law decided to study the topic entitled "Judicial and Arbitral Settlements of International Disputes Involving More Than Two States" and, at its last Berlin session (1999), it adopted a resolution on the subject. As much because of the provisions it contains as for the silences it includes (for example, with respect to the *amicus curiae*), the resolution constitutes an interesting doctrinal pronouncement which should be kept in mind in respect to the subject of this article. The way in which the resolution deals with the topics of "intervention" and the so-called "indispensable parties" is, in my opinion, quite correct on the whole. On the other hand, the resolution is overly short and modest regarding disputes involving more than two States "as parties", although it has the merit of recognising the actual fact that such disputes do exist and are also susceptible to resolution through judicial settlement.

***

The foregoing considerations are intended neither to deny the empirical fact that the majority of international legal disputes which States refer to judicial settlement are indeed between two States Parties only, nor to disregard the principle which provides that, from the date a case is submitted to judicial settlement by the litigant parties, or by one of the litigant parties, a particular procedural relationship is established between them until the resolution of the dispute by the court or the tribunal, quite independently of the fact that the parties concerned are "singular" or "plural" parties, that is, constituted by a single State or by a plurality of States. However, those considerations do already indicate that the particular nexus established between the parties from the institution of the proceedings has become less exclusive with the passing of time, in view of the need to protect similarly specific third-state rights or general interests or social values of the international community.

In addition, there is the category of genuine tripartite or multipartite legal disputes, disputes in which three or more States participate as parties in the case in their own right, that is, each of them as a separate party and for their own particular object and purpose. This category must be distinguished from the "plural party" phenomenon referred to above. Moreover, within a given tripartite or multipartite legal dispute two or more litigating States may indeed be in a situation of *cause commune*, but not necessarily. For example, if the Nicaraguan proposal regarding the joint institution of a new maritime case as an alternative to the maritime aspect of the *El Salvador/Honduras* case would have been accepted by the latter States, it would have resulted in the International Court of Justice entertaining a case with three States Parties without *cause commune*, given that Nicaragua's claims on the merits regarding the régime and the eventual delimitation of the waters of the Gulf of Fonseca were distinct and incompatible with the claims put forward either by El Salvador or Honduras.

Bearing this in mind, the general proposition defended in this article is that it would be wise to rethink some aspects of international judicial procedure and practice so as to make them more flexible to cope with today's real needs. The fears that until a few decades ago were provoked by the non-appearance of a party or the possibility of early withdrawal of a case should not today have the same dissuasive effects that they used to. In any event, without the implementation of the update that has been suggested, we run the risk of encouraging, in the not too distant future, the progressive reduction of the number of legal disputes among States which can be resolved through judicial settlement.

***

"International disputes" are the consequence of the existence in interstate relations of conflict situations. These conflicts, the existence of which is demonstrated by empirical data, constitute part of a number of social facts in the relations between States, which, as such, are capable of objective verification. This is why international jurisprudence talks of "the facts that generate a dispute". As a result of these "generating facts", confrontation originates between the positions or interests of the respective interested States. Once this confrontation is crystallised by the very conduct of the States concerned, an "international dispute" is born between them. If such a dispute is defined by those States in legal terms or is objectively susceptible to adjudication on the basis of, or in accordance with international law, we are in the presence of an "international legal dispute".

However, the existence at the sociological or political level of an international dispute, even if it may be qualified as "legal", is by no means the end of the matter, because, in order to be susceptible to adjudication, an "international legal dispute" must be referred by the States concerned, or by some of them, to an international court or tribunal. In doing so, those States are free to frame or formulate the referred legal dispute in whatever terms they like, both with regard to the States Parties to the instituted judicial case and with regard to what is asked from the court or tribunal. It follows that, although the existence of an "international legal dispute" as an antecedent is a requirement or condition of any international judicial process, the dispute which is the subject of the instituted judicial case does not necessarily correspond in all its terms to the one constituting its antecedent. There must be a certain connection between the two, but both disputes do not need to be identical or between the same States.

Thus, it may happen, and this frequently occurs in practice, that the adjudication of the legal dispute before the international court or tribunal cannot dispose of and void all aspects of the antecedent legal dispute. It is clear, for example, that if the antecedent legal dispute involves more than two States, and the dispute before the court or tribunal involves two States only, the judgment will not be binding for the States which did not participate in the proceedings of the case, even if they are a party in the antecedent legal dispute from a substantive standpoint.

When this occurs because the referring State or States so wish and have assumed the corresponding procedural risk, or when the jurisdictional situation is such that it does not allow all the interested States to be brought before the court or tribunal, nothing or little can be done. But, on other occasions, the difficulties in bringing all the States concerned before the court or tribunal in order to settle the entire antecedent legal dispute by a single judgment derive merely from rigid interpretations and applications of certain procedural rules such as those relating to the institution of proceedings and joinders. Difficulties can also arise because of the lack of interest shown by state practice in the possibility of novation of instituted judicial cases by common agreement of all the interested States.

Thus, it is one thing to be a substantive party in the antecedent legal dispute or to have an interest which may be affected in the instituted judicial case. It is quite another to have the status of a party in the judicial procedure initiated by filing or notifying an act instituting proceedings before an international court or tribunal. This becomes crystal clear when we compare the definition given of the eventual subjects of an "international dispute" in, for example, the *Mavrommatis* case, in which the Permanent Court talked of the general concept of international disputes, with the definition that is given of the concept of "party" in a judicially instituted case, in the following citation of the judgment of 13 September 1990 of the Chamber of the International Court of Justice in the case between El Salvador and Honduras:

> "... the pattern of international judicial settlement under the Statute is that *two or more States* agree that the Court shall hear and determine a particular dispute. Such agreement may be given ad hoc, by Special Agreement or otherwise, or may result from the invocation, in relation to the particular dispute, of a compromissory clause of a treaty or of the mechanism of Article 36, paragraph 2, of the Court's Statute. *Those States are the 'parties' to the proceedings*, and are bound by the Court's eventual decision because they have agreed to confer jurisdiction on the Court to decide the case, the decision of the Court having binding force as provided for in Article 59 of the Statute. Normally, therefore, no other State may involve itself in the proceedings without the consent of the original parties".
> (ICJ Reports 1990, 133, paragraph 95, emphasis added.)

***

The remedies to cope with the negative effects of the eventual lack of coincidence between the States Parties in the antecedent legal dispute, or the States which may be affected by the decision in the instituted judicial case, and the States Parties to the proceedings in such a case are: (1) intervention in the two generally accepted forms; (2) the personal and material relative authority of the *res judicata* principle; and (3) the so-called "indispensable parties" doctrine as applied by the International Court of Justice in the *Monetary Gold* and *East Timor* cases. All of them are relevant for the study of the topic generally described as "international disputes involving more than two States", but all of them have already been clarified satisfactorily, in my opinion, by international jurisprudence and doctrine. Moreover, in practice, international courts and tribunals, such as the International Court of Justice, exercise their competences with considerable restraint in cases involving rights *in rem* when they are aware of claims of States not party to the proceedings of the case concerned.

The protection of the rights or interests of the States not party to the proceedings of an instituted judicial case therefore generally seems to be conveniently assured for present needs. What remains, however, is what might be called the positive dimension increase by the existence of "international disputes involving more than two States" and, in particular, what should be done procedurally in order to facilitate, in the case of disputes involving more than two States "*as parties*", all those States becoming "*parties to the proceedings*" from the very moment the judicial case concerned is instituted, or subsequently (without prejudice, of course, to possible jurisdictional obstacles, which raise issues of a quite different nature).

In this connection, it should be born in mind in the first place that the States that are "party to the proceedings" of the case must be expressly identified in the instruments instituting proceedings. It is so stipulated in the Statute and Rules of the International Court of Justice and of other international courts and tribunals. The number of States, however, is not limited by any applicable statutory provision. Therefore, two States can be "party to the proceedings" of a case just as a *greater number* can be.

Where, in fact, lies the difficulty in providing that more than two States be "party to the proceedings" of a given judicial case from the very moment of its institution? Not in the principle which is admitted, but in the recognised means of instituting a case judicially, particularly when the means used is an application. To understand the situation and to envisage some possible remedies, it is convenient first to recall some general questions concerning the notion of "party" in international litigation, as well as to distinguish between the participation of a State in the proceedings as part and parcel of a "plural party" or as a "single party" of its own.

***

There are many similarities between the concept of "party" and its general function during proceedings, both in domestic civil procedural laws and in international judicial procedure. There are also general principles applicable to both procedures, such as, for example, the contentious nature of the proceedings and the equality of those who are participating therein "as parties". However, some other principles which learned authorities on civil procedure insist upon do not necessarily apply, or at least not with the same scope, to international judicial procedure. The most significant exception is the principle of "duality of parties", which refers exclusively to the respective position in the proceedings of those acting as parties and not to the number of actors that are involved in such a capacity in the same.

According to the principle of "duality of parties", the procedural action is bilateral in nature which, for those acting as parties, entails having one of the following two positions in the proceedings: applicant or respondent. In other words, those acting as parties either request something of the other or others acting as parties, or have something requested of them by the latter. The *litis* which result from the confrontation of these two positions would then be a conflict involving only two parties from a procedural point of view, independently of the number of actual litigants forming each of those two parties.

Without detracting from the veracity of this proposition in general, nor from its didactic value in assisting in the basic understanding of contentious proceedings, I have to note from the outset that in international judicial procedure this proposition could only be valid eventually if the proceedings are instituted through the filing of an application. The premise upon which the principle of "duality of parties" is based simply does not stand up to scrutiny when international judicial proceedings are instituted through the notification to an international court or tribunal of a *compromis* or special or similar agreement between the litigant States.

In this last category of cases, there is not a dual position of the parties in the resulting proceedings, but rather a *double position of each of the parties*, i.e., each of the litigants is at the same time applicant and respondent. It is therefore not possible to accept the "duality of parties" principle as being a principle of necessary application or verification in international judicial procedure in all cases. In international judicial procedure contentious cases may be: (1) cases between parties acting as applicant and respondent respectively, or (2) cases between parties acting in the double position of applicant/respondent.

***

As indicated above, the "duality of parties" principle has nothing to do with the different question of the number of litigants participating in the proceedings. The phenomenon of the "plural party" is therefore quite compatible with that principle. Applicant and/or respondent may be "single" or "plural" in civil law procedures, that is, be constituted by one or more persons.

The same applies in international judicial procedure. A party may be constituted by a single litigant State or by two or more litigant States. The "plural party" resulting from the latter situation may act in the proceedings either as applicant or respondent or as applicant/respondent. The practice of the International Court of Justice and the Permanent Court confirms this. For example, there were multiple applicants in the *Wimbledon* case (four States) and in the *Interpretation of the Statute of the Memel Territory* case (four States), multiple respondents in the *Monetary Gold* case (three States) and multiple applicants/respondents in the *Territorial Jurisdiction of the International Commission of the River Oder* case (six States).

When they constitute a "plural party", the litigant States find themselves procedurally in a situation similar to that of joint litigants in a *joint action lawsuit* in civil procedure, in that there is only one set of proceedings and one judgment which resolves the dispute through one single sentence. It allows a unitary treatment of the issues, ultimately responding to the requirement of the good administration of international justice. The figure of the "plural party" is particularly appropriate for situations in which the States concerned share the same object and purpose and the same interest or cause in the case, when they are holders of a common procedural action, or when their individual procedural actions are interconnected by virtue of a common given legal premise.

In the *Wimbledon* case, the four States constituting the applicant party joined their respective procedural causes of action under Article 386 of the Treaty of Versailles. In the *Memel Statute* case, the four States constituting the applicant party undertook a joint action pursuant to their rights under Article 17 of the 1924 Convention concerning the Statute of the Memel Territory. In the *Monetary Gold* case, the three States constituting the respondent party had agreed in the Washington Declaration of 1951 that the applicant party could cite them jointly. Finally, in the *Commission of the River Oder* case, the six States constituting one of the applicant/defendant parties were the States which constituted "one of the parties to the 1928 Agreement" by whom notification proceedings were instituted.

***

The "plural party" is therefore a figure which has been received without major difficulties in international judicial procedure, and which is treated therein in a manner adequate to current needs.

However, there are also other international legal disputes in which the possible participation in the judicial case of three or more litigant States as parties to the proceedings is justified and should be facilitated for reasons of a different nature than those underlying the figure of the "plural party". I have in mind, as already indicated, those international legal disputes which from a substantive or merits standpoint are really trilateral or multilateral in nature. The "plural party" is not a procedural answer to this kind of international legal dispute because the litigant States concerned cannot be grouped into two procedural parties in accordance with the "duality of parties" principle, although in a trilateral or multilateral legal dispute one or more parties could well be a "plural party" as well.

The Statute of the International Court of Justice and the Statute of the International Tribunal for the Law of the Sea do not turn their back on international legal disputes that are trilateral or multilateral in nature. For example, Article 40 of the Statute of the Court provides that acts instituting proceedings must indicate not only the subject of the dispute but also the "parties". It does not specify "the two parties" or "one party and the other party", but rather refers to "the parties". Disputes between more than two parties therefore come within the scope of this statutory provision. In the case that the institution of the proceedings is made by an application, the Registrar has forthwith to communicate the application "to all concerned".

Furthermore, Article 31, paragraph 5, of the Court's Statute confirms without reserve that its drafters clearly did not exclude tripartite or multipartite instituted legal disputes. According to this provision, which relates to judges *ad hoc*, if there are "several parties in the same interest, they shall, for the purpose of the preceding provisions, be reckoned as one party only". It follows *a contrario* that in a given case referred to the Court there could be more than two States litigating as parties and that some of them may *or may not* share the same interest or cause. If they do not share the same interest or cause, the instituted legal dispute in question cannot but be a tripartite or multipartite dispute in which there are three or more parties to the proceedings. The provisions set forth in Articles 17 and 24 of the Statute of the Tribunal for the Law of the Sea lead to the same conclusion as Articles 40 and 31 of the Statute of the International Court of Justice.

***

One of the two accepted methods of instituting proceedings, namely the notification of a special or similar agreement, appears *prima facie* perfectly adapted to the institution of tripartite or multipartite judicial cases. It is evident that three or more States can at any time enter into a tripartite or multipartite *compromis* and notify it to the court or tribunal concerned, or make a single notification of two or more bilateral *compromis* concluded by one of them with each of the others. The institution of the proceedings in the *North Sea Continental Shelf* cases, concerning disputes between Germany and Denmark and Germany and the Netherlands, was inspired by such considerations, but only in part, and this created certain procedural problems. However, the three Governments had also notified the Court of a protocol relating to certain procedural questions arising out of the special agreements they had concluded. Denmark and the Netherlands concluded a further agreement between them according to which they "should be considered parties in the same interest" *(cause commune).*

Thus, the parties' conduct allowed the Court to formally join the proceedings in the two cases and, consequently, the disputes were resolved by a single judgment in 1969. This is a good example of how the formal joinder of cases is a procedural institution which may remedy the insufficiencies of the act instituting the proceedings for dealing properly with a genuine tripartite or multipartite dispute, even when this is done through the notification of special agreements. It underlines also, as already indicated, that tripartite or multipartite disputes should not be confused with bilateral disputes in which one or two "plural parties" participate.

The suitability of the other way of instituting proceedings, i.e., the filing of an application, for tripartite or multipartite legal disputes seems to raise greater doubts at the practical level and in doctrine. The remedy used during the 1960s in the *South West Africa* cases consisted of the presentation of separate applications by the applicants, Ethiopia and Liberia, against one single respondent, South Africa, followed by the joinder of the cases. The applicants had informed the Court separately of their intention to designate a judge *ad hoc* each, but the Court considered – after having received the respective States' memorials and before South Africa had filed preliminary objections – that Ethiopia and Liberia shared the same interest in the case for the purposes of designating a judge *ad hoc* and formally ordered, before the filing of the preliminary objections, that the two cases be joined. The judgment of 1962 on the preliminary objections and the later judgment of 1966 therefore concern the two joined cases. In this example, as in the *North Sea Continental Shelf* cases, each litigant State participated as a separate principal party to the proceedings on the joined cases.

The *North Sea Continental Shelf* cases and the *South West Africa* cases illustrate the importance which the act instituting the proceedings has for the unified treatment and solution of tripartite or multipartite international legal disputes. If such an act is not a single act of procedure properly drafted, the remedy certainly lies in the joinder of the instituted cases, an action which may be taken at an early stage or a later stage of the proceedings. However, the joinder of cases is not effected by the parties or some of the parties, but by a Court decision the adoption of which frequently poses difficult problems to the Court.

In fact, in more recent years, the Court has not ordered formal joinders of cases despite the fact that, at least *prima facie*, it was not lacking in occasions that might have warranted such joinders. For example, in the following cases: the *Fisheries Jurisdiction* cases *(United Kingdom v. Iceland)* and *(Federal Republic of Germany v. Iceland)*, the *Nuclear Test* cases *(Australia v. France)* and *(New Zealand v. France)*, and *Questions of Interpretation and Application of the 1971 Montreal Convention (Libya v. United Kingdom)* and *(Libya v. United States)*. It is true that in the first two sets of cases, the respondent failed to appear and that, in all three sets, the applicants did not request that the proceedings be formally joined and early incidental proceedings were brought (provisional measures, preliminary objections or formal exceptions). However, it is also true that the formal joinder of cases is not subject either to the agreement of the applicants or that of the applicants and respondents.

Certainly, the Court must have had its own reasons for avoiding a formal joinder in those cases. For example, the jurisdictional issues involved, the lack of information as to the position of the parties on the merits, questions relating to the designation of judges *ad hoc*, etc. In any case, joinders should respect the requirements of a fair procedure and, therefore, they should not be taken for granted in advance. But all in all, it also appears that the joinder of cases is not a very fashionable procedural device in the Court's proceedings of recent years. This is not good news for the treatment and solution of genuine tripartite or multipartite international legal disputes, which today occur frequently in international relations, unless greater attention is paid to the means whereby proceedings may be instituted.

Recourse to the alternative method, available under Article 47 of the Rules of Court, to solve practical problems by partially applying common procedures without a formal joinder or, generally, by studying related cases in parallel, may help in the case of tripartite or multipartite disputes, but these means are far from providing the States Parties involved with sufficient procedural guarantees. In any case, such alternative methods have not escaped criticism and have indeed given rise to a cumbersome jurisprudence concerning practically identical cases. In addition, they have created a certain uncertainty with respect to questions such as the determination of parties with the same interest *(cause commune)* and the related question of the parties' right to designate a judge *ad hoc*. The criticisms that can be levelled at said alternative methods naturally increase when the dispute concerns more than three States, as it does the pending cases on the *Legality of Use of Force*, originally initiated in 1999 by Yugoslavia against ten Member States of NATO.

***

In my opinion, the best way to facilitate a proper judicial settlement of the increasing number of genuine tripartite or multipartite international legal disputes would imply – without prejudice of course to paramount jurisdictional questions – a deeper and more open-minded study of the means whereby those disputes may be instituted. More conceptual flexibility is needed in this respect than shown, for example, in paragraphs 4, 5 and 6 of the resolution on Judicial and Arbitral Settlement of International Disputes Involving More Than Two States, adopted by the Institute of International Law at its 1999 Berlin session.

In particular, under current circumstances, it is difficult to find real justification for the statement that an unilateral application to a court or tribunal by one or more States directed against more than one State as respondents requires, in principle, parallel and separate proceedings *if no previous agreement between the States involved can be reached*. Why is this so? Why deprive, under contemporary international law, international judicial procedure of the benefits of a well-known institution such as that represented by the "joining of procedural actions" by the parties (which is not to be confused with the "joinder of instituted cases" by the court or tribunal concerned)?

The principal players for the "joining of procedural actions" are represented by the subjects of the litigation and not by the jurisdictional organ. It should be up to the applicant litigating State or States to decide against whom and how the case their case(s) are brought before a court or tribunal, as it should be up to the defendants to decide whether or not to join in the submission of preliminary exceptions or counterclaims to the court or tribunal concerned. Any interference by the law or by jurisdictional organs with the exercise by litigant sovereign States of their freedom to join procedural instituting actions or to submit joint preliminary exceptions or counterclaims against the applicant or applicants should be limited as much as possible, and not the other way around. In any case, the contrary proposition is not conducive to making international judicial settlement a useful means of settlement of genuine tripartite or multipartite legal disputes among States.

If one or more States are in a position to invoke an adequate jurisdictional title or titles, why are they not able to file *one single application* against more than one State Party in a legal dispute? To give a very simple example, the issue here is whether, in the context of a dispute between States A, B and C as separate principal parties, there are well-founded substantive or procedural reasons which warrant disallowing the joinder an application of the procedural actions which State A may have instituted against State B and against State C with regard to the dispute in question and, conversely, why States B and C cannot reply to such an application by a joinder of their respective preliminary exceptions and counter-claim actions. I have not found reasons that can justify a negative response, even if no previous agreement existed in that respect between States A, B and C.

***

Joinders not being a matter of the Statue of the International Court of Justice, but of regulation by the Rules of Court, my general conclusion is that, if there is really a need to do so, those Rules should contain provisions on the "joinder of procedural actions", much as they do with respect to the "joinder of instituted cases", in the interest of a unitary judicial consideration and solution of genuine tripartite or multipartite legal disputes among States wishing to act individually in the proceedings as principal parties, without the need to put in motion unpredictable, cumbersome and less satisfactory procedural or practical devices.

IX

LAND AND MARITIME BOUNDARIES, INTERNATIONAL WATERCOURSES AND OTHER WATERS

International Maritime Boundaries for the Continental Shelf: The Relevance of Natural Prolongation

Jonathan I. Charney

I am pleased to participate in this *Festschrift* in honor of Judge Oda. As an international lawyer and judge on the International Court of Justice for many years he is one of today's pre-eminent international law jurists and scholars. During his period on the Court, of all the matters brought before it those involving the law of the sea were more common than any other single subject area. Fortunately for Judge Oda, his interest in this area preceded his election to the bench.[1] He has used this knowledge to his advantage in his many scholarly opinions in cases on the subject. Among the law of the sea cases, the largest concentration involved international maritime boundaries, especially continental shelf boundaries. Judge Oda and I share considerable interest in this subject. We also have been frustrated by the inability of the international community and the Court to refine the jurisprudence in this area. This paper will focus on an aspect of this subject.

When the ICJ issued its famous judgment in the *North Sea Continental Shelf* cases it marked the beginning of serious debate on the law for the delimitation of international maritime boundaries.[2] Of course, this was preceded by the work of the International Law Commission's Committee of Experts on Technical Questions Concerning the Territorial Sea,[3] and the 1956 Report of the International

---

[1] Judge Oda's extensive list of publications on the law of the sea is included with his Hague Lectures. S. Oda, "International Law of the Resources of the Sea", *RdC* 127 (1969-II), 357 et seq., 359-361; S. Oda, "The International Court of Justice Viewed from the Bench (1976-1993)", *RdC* 244 (1993), 9 et seq., 18-20.

[2] *North Sea Continental Shelf* cases *(Federal Republic of Germany/Denmark, Federal Republic of Germany/Netherlands)*, ICJ Reports 1969, 3 et seq.

[3] J. P. A. François, Addendum to the Second Report on the Régime of the Territorial Sea, ILC, fifth session, Annex, Report of the Committee of Experts on Technical Questions Concerning the Territorial Sea, UN Doc. A/CN.4/61/Add.1 of 18 May 1953, 6-7:

Law Commission that led to the 1958 Conventions on the Law of the Sea.[4] Those Conventions included articles on the delimitation of international maritime boundaries for the territorial sea and for the continental shelf.[5] The Court threw the doors wide open to unlimited

> "Question VII:
> How should the (lateral) boundary line be drawn through the adjoining territorial sea of two adjacent States? Should it be done:
> A. by continuing the land frontier?
> B. by a perpendicular line on the coast at the intersection of the land frontier and the coastline?
> C. by a line drawn vertically on the general direction of the coastline?
> D. by a median line? If so, how should this line be drawn? To what extent should islands, shallow waters and navigation channels be accounted for?"
> "Answer:
> (1) The Committee decided that the (lateral) boundary through the territorial sea – if not already fixed otherwise – should be drawn according to the principle of equidistance from the respective coastlines.
> (2) In a number of cases this may not lead to an equitable solution, which should be then arrived at by negotiation".

[4] Report of the International Law Commission on the Work of its Eighth Session, April 23-4 July 1956, UN, GAOR, 11th. Sess., Supp. No. 9, UN Doc. A/3159 (1956), reprinted in ILCYB 253 (1956-II) and *AJIL* 51 (1957), 154 et seq.; Report of the International Law Commission on the Work of its Eighth Session, Reference Guide to the Articles Concerning the Law of the Sea Adopted by the International Law Commission at its Eighth Session, UN Doc. A/C.6/L.378 of 25 October 1956.

[5] Convention on the Territorial Sea and the Contiguous Zone, done 29 April 1958, UNTS 1966, Vol. 516, No. 7477, 205, Article 12:
> "1. Where the coasts of two States are opposite or adjacent to each other, neither of the two States is entitled, failing agreement between them to the contrary, to extend its territorial sea beyond the median line every point of which is equidistant from the nearest points on the baselines from which the breadth of the territorial seas of each of the two States is measured. The provisions of this paragraph shall not apply, however, where it is necessary by reason of historic title or other special circumstances to delimit the territorial seas of the two States in a way which is at variance with this provision.
> 2. The line of delimitation between the territorial seas of two States lying opposite to each other or adjacent to each other shall be marked on large-scale charts officially recognised by the coastal States".

Convention on the Continental Shelf, done 29 April 1958, UNTS 1965, Vol. 499, No. 7302, 311, Article 6:
> "1. Where the same continental shelf is adjacent to the territories of two or more States whose coasts are opposite each other, the boundary of the continental shelf appertaining to such States shall be determined

considerations to be taken into account in delimiting the boundaries of the continental shelf based on the theory of "equitable principles".[6] This resulted in a series of ICJ and other third-party dispute settlement tribunal decisions.[7] Most of them allowed for the consideration of a variety of facts, at least in theory. These included:

1. the coastline,
2. equidistance/special circumstances/equity,
3. opposite or adjacent relationship,
4. proportionality,
5. earth sciences (geology and geomorphology),

---

by agreement between them. In the absence of agreement, and unless another boundary line is justified by special circumstances, the boundary is the median line, every point of which is equidistant from the nearest points of the baselines from which the breadth of the territorial sea of each State is measured.

2. Where the same continental shelf is adjacent to the territories of two adjacent States, the boundary of the continental shelf shall be determined by agreement between them. In the absence of agreement, and unless another boundary line is justified by special circumstances, the boundary shall be determined by application of the principle of equidistance from the nearest points of the baselines from which the breadth of the territorial sea of each State is measured.

3. In delimiting the boundaries of the continental shelf, any lines which are drawn in accordance with the principles set out in paragraphs 1 and 2 of this article should be defined with reference to charts and geographical features as they exist at a particular date, and reference should be made to fixed permanent identifiable points on the land".

[6] *North Sea Continental Shelf* cases, ICJ Reports 1969, 38 et seq., 46-52, paras. 85, 88-99.

[7] *The United Kingdom of Great Britain and Northern Ireland and the French Republic, Delimitation of the Continental Shelf* decision, *ILR* 54 (1979), 5 et seq.; *Dubai-Sharjah Border Arbitration* award, *ILR* 91 (1993), 543 et seq.; *Report and Recommendations to the Governments of Iceland and Norway of the Conciliation Commission on the Continental Shelf Area between Iceland and Jan Mayen* (19-20 May 1981), *ILR* 62 (1982), 108 et seq.; *Continental Shelf* case *(Tunisia/Libyan Arab Jamahiriya)*, ICJ Reports 1982, 18 et seq.; *Delimitation of the Maritime Boundary in the Gulf of Maine Area (Canada/United States)*, ICJ Reports 1984, 246 et seq.; *Guinea and Guinea-Bissau Maritime Delimitation* award, *ILR* 77 (1988), 635 et seq.; *Continental Shelf* case *(Libyan Arab Jamahiriya/Malta)*, ICJ Reports 1985, 13 et seq.; *St. Pierre and Miquelon* award, *ILR* 95 (1994), 645 et seq.; *Maritime Delimitation in the Area between Greenland and Jan Mayen* case *(Norway/Denmark)*, ICJ Reports 1993, 38 et seq.; *Arbitral Tribunal in the Second Stage* award *(Eritrea/Yemen)*, Maritime Delimitation, 17 December 1999, available at http://www.pca-cpa.org.

6. socio-economic circumstances,
7. natural resource location,
8. prior conduct and historic title,
9. equitable solution,
10. differences between sea-bed boundaries and superjacent water boundaries,
11. non-States Parties, and others.

The 1982 Convention on the Law of the Sea provided little to help the international community refine the law further.[8] The Convention contains virtually identical articles on the delimitation of international maritime boundaries for the exclusive economic zone and the continental shelf, both of which call on States to settle their maritime boundaries by "agreement on the basis of *international law*, as referred to in Article 38 of the Statute of the International Court of Justice, in order to achieve an *equitable solution...*".[9]

---

[8] United Nations Convention on the Law of the Sea, 10 December 1982, UNTS 1998, Vol. 1833, No. 31363, 3.

[9] *Ibid.*, emphasis added.

"Article 74:
Delimitation of the exclusive economic zone between States with opposite or adjacent coasts
1. The delimitation of the exclusive economic zone between States with opposite or adjacent coasts shall be effected by agreement on the basis of international law, as referred to in Article 38 of the Statute of the International Court of Justice, in order to achieve an equitable solution.
2. If no agreement can be reached within a reasonable period of time, the States concerned shall resort to the procedures provided for in Part XV.
3. Pending agreement as provided for in paragraph 1, the States concerned, in a spirit of understanding and cooperation, shall make every effort to enter into provisional arrangements of a practical nature and, during this transitional period, not to jeopardize or hamper the reaching of the final agreement. Such arrangements shall be without prejudice to the final delimitation.
4. Where there is an agreement in force between the States concerned, questions relating to the delimitation of the exclusive economic zone shall be determined in accordance with the provisions of that agreement".

"Article 83:
Delimitation of the continental shelf between States with opposite or adjacent coasts
1. The delimitation of the continental shelf between States with opposite or adjacent coasts shall be effected by agreement on the basis of

The language of these articles is even more indeterminate than that used in the 1958 Convention on the Continental Shelf for delimiting international boundaries of the continental shelf,[10] although the ICJ has construed the customary international law and the 1958 and 1982 texts as having the same meaning.[11]

---

international law, as referred to in Article 38 of the Statute of the International Court of Justice, in order to achieve an equitable solution.

2. If no agreement can be reached within a reasonable period of time, the States concerned shall resort to the procedures provided for in Part XV.

3. Pending agreement as provided for in paragraph 1, the States concerned, in a spirit of understanding and cooperation, shall make every effort to enter into provisional arrangements of a practical nature and, during this transitional period, not to jeopardize or hamper the reaching of the final agreement. Such arrangements shall be without prejudice to the final delimitation.

4. Where there is an agreement in force between the States concerned, questions relating to the delimitation of the continental shelf shall be determined in accordance with the provisions of that agreement".

[10] Convention on the Continental Shelf, see note 5, Article 6:
"1. Where the same continental shelf is adjacent to the territories of two or more States whose coasts are opposite each other, the boundary of the continental shelf appertaining to such States shall be determined by agreement between them. In the absence of agreement, and unless another boundary line is justified by special circumstances, the boundary is the median line, every point of which is equidistant from the nearest points of the baselines from which the breadth of the territorial sea of each State is measured.

2. Where the same continental shelf is adjacent to the territories of two adjacent States, the boundary of the continental shelf shall be determined by agreement between them. In the absence of agreement, and unless another boundary line is justified by special circumstances, the boundary shall be determined by application of the principle of equidistance from the nearest points of the baselines from which the breadth of the territorial sea of each State is measured.

3. In delimiting the boundaries of the continental shelf, any lines which are drawn in accordance with the principles set out in paragraphs 1 and 2 of this article should be defined with reference to charts and geographical features as they exist at a particular date, and reference should be made to fixed permanent identifiable points on the land".

[11] *Maritime Delimitation in the Area between Greenland and Jan Mayen* case, ICJ Reports 1993, 38 et seq., 58-59, para. 46. Citing with approval the reasoning of *The United Kingdom of Great Britain and Northern Ireland and the French Republic, Delimitation of the Continental Shelf* decision, ILR 54 (1979), 55-56, para. 70.

Over more than three decades of cases on this subject, most of these considerations were not used to locate the international maritime boundaries in question, with the sole exception of the contemporary coastal geography related to the areas in dispute. Although the parties regularly invoked the other considerations, the tribunals routinely found them unpersuasive, especially those that related to the natural sciences of geology and geomorphology.[12] However, the tribunals seemed never to eliminate them absolutely from potential relevance at some point in the analysis, despite Judge Oda's arguments.[13] This reached its apex in the 1982 *Tunisia/Libya* case, in which the ICJ stressed the uniqueness of each delimitation:

> "The result of the application of equitable principles must be equitable. This terminology, which is generally used, is not entirely satisfactory because it employs the term equitable to characterize both the result to be achieved and the means to be applied to reach this result. It is, however, the result which is predominant; the principles are subordinate to the goal. The equitableness of a principle must be assessed in the light of its usefulness for the purpose of arriving at an equitable result. It is not every such principle which is in itself equitable; it may acquire this quality by reference to the equitableness of the solution. The principles to be indicated by the Court have to be selected according to their appropriateness for reaching an equitable result. From this consideration it follows that the term 'equitable principles' cannot be interpreted in the abstract; it

---

[12] See, e.g., *Continental Shelf* case *(Tunisia/Libyan Arab Jamahiriya)*, ICJ Reports 1982, 18 et seq., 53-54, 58, 69, paras. 61, 67, 68, 70. While the ICJ did eliminate all but contemporary geography from consideration and, thus, eliminated substantial arguments based on geology and geomorphology, it did this based on contemporary considerations. "It is of the view that what must be taken into account in the delimitation of shelf areas are the physical circumstances as they are today; that just as it is the geographical configuration of the present-day coasts, so also it is the present-day sea-bed, which must be considered. It is the outcome, not the evolution in the long-distant past, which is of importance". *Ibid.*, 53, para. 61. See also *Maritime Delimitation in the Area between Greenland and Jan Mayen* case *(Norway/Denmark)*, ICJ Reports 1993, 44, para. 11; *St. Pierre and Miquelon* award, *ILR* 95 (1994), 665, para. 46; *Delimitation of the Maritime Boundary in the Gulf of Maine Area* case *(Canada/United States)*, ICJ Reports 1984, 373-377, paras. 45-56; *Guinea and Guinea-Bissau Maritime Delimitation* award, *ILR* 77 (1988), 687, para. 117.

[13] *Continental Shelf* case *(Tunisia/Libyan Arab Jamahiriya)*, ICJ Reports 1982, 247, 257, 267, paras. 145, 160, 176 (Oda, J., dissenting opinion).

refers back to the principles and rules which may be appropriate in order to achieve an equitable result".[14]

The consequence of this unfortunate language was the charge that the law on the subject was indeterminate. Rather than law being applied, it was the unfettered discretion of the judges sitting on the case.[15]

In reaction to this charge, the ICJ retreated in the 1985 *Libya/Malta* case by observing that, although there are no legal limits to the relevant considerations, it sought a degree of consistency in the application of the rule:

> "Thus the justice of which equity is an emanation, is not abstract justice but justice according to the rule of law; which is to say that its application should display consistency and a degree of predictability; even though it looks with particularity to the peculiar circumstances of an instant case, it also looks beyond it to principles of more general application. This is precisely why the courts have, from the beginning, elaborated equitable principles as being, at the same time, means to an equitable result in a particular case, yet also having a more general validity and hence expressible in general terms; for, as the Court has also said, 'the legal concept of equity is a general principle directly applicable as law' (ICJ Reports 1982, p. 60, para. 71)".[16]

More importantly, the Court pointed out that Article 74 of the LOS Convention established that the minimum entitlement of a State to a continental shelf is 200 nautical miles, which is the same as the breadth of the exclusive economic zone as provided for in Article 57 of the Convention. It opined that this fact influenced the analysis to be made in the delimitation of the international maritime boundary of the continental shelf in areas in which coastal States had a dispute over the

---

[14] *Ibid.*, 59, para. 70.
[15] *Ibid.*, 153, para. 20 (Gros, J., dissenting opinion); *ibid.*, 157, para. 1 (Oda, J., dissenting opinion); *ibid.*, 278, 290-91, 294, 319, paras. 12, 14 and Conclusion (Evenson, J., dissenting opinion); *Maritime Delimitation in the Area between Greenland and Jan Mayen* case *(Norway/Denmark)*, ICJ Reports 1993, 113, para. 86 (Oda, V.P., separate opinion). See J. I. Charney, "Ocean Boundaries Between Nations: A Theory for Progress", *AJIL* 78 (1984), 582 et seq., 583-585.
[16] *Continental Shelf* case *(Libyan Arab Jamahiriya/Malta)*, ICJ Reports 1985, 13 et seq., 39, para. 45. See J. I. Charney, "Progress in International Maritime Boundary Delimitation Law", *AJIL* 88 (1994), 227 et seq., 230, 233.

boundary when the 200-nautical-mile line as measured from their coastlines overlapped.

> "33. In the view of the Court, even though the present case relates only to the delimitation of the continental shelf and not to that of the exclusive economic zone, the principles and rules underlying the latter concept cannot be left out of consideration. As the 1982 Convention demonstrates, the two institutions – continental shelf and exclusive economic zone – are linked together in modern law. Since the rights enjoyed by a State over its continental shelf would also be possessed by it over the sea-bed and subsoil of any exclusive economic zone which it might proclaim, one of the relevant circumstances to be taken into account for the delimitation of the continental shelf of a State is the legally permissible extent of the exclusive economic zone appertaining to that same State. This does not mean that the concept of the continental shelf has been absorbed by that of the exclusive economic zone; it does however signify that greater importance must be attributed to elements, such as distance from the coast, which are common to both concepts.
>
> 34. ... It is in the Court's view incontestable that, apart from those provisions, the institution of the exclusive economic zone, with its rule on entitlement by reason of distance, is shown by the practice of States to have become a part of customary law; ... Although the institutions of the continental shelf and the exclusive economic zone are different and distinct, the rights which the exclusive economic zone entails over the sea-bed of the zone are defined by reference to the régime laid down for the continental shelf. Although there can be a continental shelf where there is no exclusive economic zone, there cannot be an exclusive economic zone without a corresponding continental shelf. It follows that, for juridical and practical reasons, the distance criterion must now apply to the continental shelf as well as to the exclusive economic zone; and this quite apart from the provision as to distance in paragraph 1 of Article 76. This is not to suggest that the idea of natural prolongation is now superseded by that of distance. What it does mean is that where the continental margin does not extend as far as 200 miles from the shore, natural prolongation, which in spite of its physical origins has throughout its history become more and more a complex and juridical concept, is in part defined by distance from the shore, irrespective of the physical nature of the intervening sea-bed and subsoil. The concepts of natural prolongation and distance are therefore not opposed but

complementary; and both remain essential elements in the juridical concept of the continental shelf...".[17]

The Court went on to emphasize the decision to eliminate arguments based on geology and geomorphology.

"39. The Court however considers that since the development of the law enables a State to claim that the continental shelf appertaining to it extends up to as far as 200 miles from its coast, whatever the geological characteristics of the corresponding sea-bed and subsoil, there is no reason to ascribe any role to geological or geophysical factors within that distance either in verifying the legal title of the States concerned or in proceeding to a delimitation as between their claims. This is especially clear where verification of the validity of title is concerned, since, at least in so far as those areas are situated at a distance of under 200 miles from the coasts in question, title depends solely on the distance from the coasts of the claimant States of any areas of sea-bed claimed by way of continental shelf, and the geological or geomorphological characteristics of those areas are completely immaterial. It follows that, since the distance between the coasts of the Parties is less that 400 miles, so that no geophysical feature can lie more than 200 miles from each coast, the feature referred to [by Libya] as the 'rift zone' cannot constitute a fundamental discontinuity terminating the southward extension of the Maltese shelf and the northward extension of the Libyan as if it were some natural boundary.

40. Neither is there any reason why a factor which has no part to play in the establishment of title should be taken into account as a relevant circumstance for the purposes of delimitation. It is true that in the past the Court has recognized the relevance of geophysical characteristics of the area of delimitation if they assist in identifying a line of separation between the continental shelves of the Parties. In the *North Sea Continental Shelf* cases the Court said:

'it can be useful to consider the geology of that shelf in order to find out whether the direction taken by certain configurational features should influence delimitation because, in certain

---

[17] *Continental Shelf* case *(Libyan Arab Jamahiriya/Malta)*, ICJ Reports 1985, 13 et seq., 33, paras. 33-34.

localities, they point-up the whole notion of the appurtenance of the continental shelf to the State whose territory it does in fact prolong' (ICJ Reports 1969, p. 51, para. 95)

Again, in the *Tunisia/Libya* case of 1982, the Court recognized that:

'identification of natural prolongation may, where the geographical circumstances are appropriate, have an important role to play in defining an equitable delimitation, in view of its significance as the justification of continental shelf rights in some cases'. (ICJ Reports 1982, p. 47, para. 44)

And the Court remarked also that

'a marked disruption or discontinuance of the sea-bed" may constitute "an indisputable indication of the limits of two separate continental shelves, or two separate natural prolongations' (*Ibid.*, p. 57, para. 66). However to rely on this jurisprudence would be to overlook the fact that where such jurisprudence appears to ascribe a role to geophysical or geological factors in delimitation, it finds warrant for doing so in a régime of the title itself which used to allot those factors a place which now belongs to the past, in so far as sea-bed areas less than 200 miles from the coast are concerned".[18]

Judge Oda supported the direction the Court took but sought an even clearer definition of the applicable rule of law.[19] In his opinion, the Court did not go far enough. He focused on the Court's language in paragraph 34 that "[t]he concepts of natural prolongation and distance are therefore not opposed but complementary".[20] He found the statement incoherent and would have eliminated the complementarity of natural prolongation within 200 nautical miles altogether. As usual, the judgment makes clear that the Court followed prior practice and discarded natural prolongation. Thus, it *de facto* but perhaps not *de jure* continued the long practice of discarding natural prolongation as a relevant consideration in third-party settlements of international maritime boundaries in accordance with international law.[21] Nevertheless, without a definitive statement, natural prolongation

---

[18] *Ibid.*, 35-36, paras. 39-40.
[19] *Ibid.*, 125, 128, 157, paras. 1, 6, 61.
[20] *Ibid.*, 128, para. 6 (Oda, J., dissenting opinion); *ibid.*, 33, para. 34.
[21] Charney, see note 16, 236.

remains an argument for counsel and a consideration in the international law of international maritime boundaries. Compared to the *Tunisia/Libya* case, progress was made towards refining the law of international maritime delimitation, such that the law might not be susceptible to arbitrary decisions of the Court, but the actual rule used by the Court remained vague. Despite the apparent elimination or diminution of considerations within overlapping 200-nautical-mile areas, the selection of the boundary line remained indeterminate, as Judge Oda argued.[22]

Unfortunately, natural sciences, especially geology and geomorphology, may yet present another test for the law of international maritime boundaries. In 1989, Australia and Indonesia entered into the Timor Gap Agreement.[23] This agreement settled the international maritime boundary of the continental shelf between Australia and Indonesia, but established a Joint Development Zone in the area between Australia and that part of East Timor that was governed by Indonesia. A UN-administered plebiscite in 1999 has resulted in the movement of East Timor towards independence.[24] While the United Nations Transitional Administration in East Timor agreed to continue the terms of the Timor Gap Agreement with respect to the Joint Development Zone, this is without prejudice to the ultimate disposition of the matter.[25] Persons purporting to represent the nascent State of East Timor conducted discussions with Australia on the subject and have sought to renegotiate the matter on the grounds that it never had binding international legal force on the territory of East Timor.[26] The East

---

[22] *Ibid.*

[23] V. Prescott, "Australia-Indonesia (Timor and Arafura Seas), Report Number 6-(2)(2)", in J. I. Charney and L. M. Alexander (eds.), *International Maritime Boundaries*, Vol. 2, 1993, 1207 et seq.

[24] UN Security Council Resolution 1272, UN Doc. S/RES/1272 of 25 October 1999; UN Security Council Resolution 1246, UN Doc. S/RES/1246 (1999) of 11 June 1999. See S. D. Murphy, "Contemporary Practice of the United States Relating to International Law: Deployment of a Multinational Force in East Timor", *AJIL* 94 (2000), 102 et seq., 105.

[25] Exchange of Notes constituting an Agreement between the Government of Australia and the United Nations Transitional Administration in East Timor (UNTAET) concerning the continued operation of the Treaty between Australia and the Republic of Indonesia on the Zone of Cooperation in an area between the Indonesian Province of East Timor and Northern Australia of 11 December 1989 (25 October 1999), Australian Treaty Series 2000, No. 9, available at http://www.austlii.edu.au/au/other/dfat/treaties/2000/9.html. See V. Prescott, "Australia – United Nations Transitional Administration in East Timor (UNTAET), Report Number 6-15", in J. I. Charney and R. Smith (eds.), *International Maritime Boundaries*, Vol. IV, forthcoming.

[26] P. Alford and R. Garran, "East Timor Wants New Gap Treaty", *The Australian*, 15 June 2000, 2.

Timorese and many States have refused to recognize the legality of the Indonesian conquest of East Timor in 1975 and incorporation into Indonesia as a province in 1976.[27] Consequently, they take the position that any treaty entered into by Indonesia purporting to be applicable to East Timor is illegal and not binding on East Timor, including the Timor Gap Agreement. As a practical matter, Australia is required to accept this position in the case of the Timor Gap Agreement, since the Joint Development Zone would become unworkable without the cooperation of the nascent State of East Timor. Thus, by necessity the matter became open to negotiations. While it may turn out that the Joint Development Zone is used as a basis for a legally binding agreement between Australia and the State of East Timor, another solution may be negotiated. It may even happen that the negotiations will fail, although at this time a failure seems unlikely since Australia and persons purporting to represent East Timor have recently entered into a Memorandum of Understanding that is structured on the basis of the Timor Gap Agreement.[28] The possibility of failure is not insignificant because the very reason for the Joint Development Zone in the Timor Gap Agreement between Australia and Indonesia was that in this area the differences of views were insurmountable on the basis of a simple continental shelf maritime boundary line.

In this area, Australia and East Timor face each other over a distance of less than 400 nautical miles. As opposite coastlines, the equidistant line is a natural and preferred, albeit not required, starting point for analysis. However, in this area, Australia argues that the Timor Trough marks the geological limit of its continental shelf. It is located well north of the equidistant line, thereby favoring Australia. Australia continues to maintain this position and will likely do so if the negotiations devolve to the point where they only concern the delimitation of a maritime boundary. Australia maintains that natural prolongation and the geological limit of the continental shelf are valid

---

[27] UN Security Council Resolution 374, UN Doc. S/11912 of 22 December 1975.

[28] *International Herald Tribune*, 5 July 2001, 13; *Associated Press On Line*, 5 July 2001, available at Lexis-Nexis/News; *New York Times*, 4 July 2001, sec. W, 1. The text is available from the Australian Department of Foreign Affairs and Trade and is expected to be placed on its website, at http://www.dfat.gov.au. This Memorandum of Understanding (MOU) is not binding under international law and it certainly cannot commit the yet to be established State of East Timor. Those who negotiated the MOU have no legal authority to bind that State, much less represent a democratically elected government or other interim authority of the State. Even the moral authority of this MOU is limited. Thus, the government of the new State could well disavow the MOU, regardless of whether it truly is a reasonable and fair solution to the problem.

bases for the delimitation of a continental shelf boundary even in this opposite situation in which the distance between the coastlines of the two States is such that their 200-nautical-mile zones overlap. According to Prescott, "[t]he deepest parts of the Timor Trough have water depths of 3100 m[eters] and its width between the continental shelf and slope breaks to the north and south is about 70 [nautical miles]".[29] Although the exact scientific character of the Timor Trough is debated, it is a substantially more significant feature than any other sea-bed on which natural prolongation arguments have been previously presented to a third-party tribunal.[30] This dispute may present a direct challenge to the *Libya/Malta* judgment since the case for a geologically based distinction is so strong.[31] It may further provide an opening for other States to claim natural prolongation in support of their continental shelf international maritime boundary claims as well.

There is nothing in the text of Article 83 of the LOS Convention that precludes this argument, since the article makes a general reference to a delimitation on the "basis of international law", not to speak of reaching an "equitable solution". Although none of the ICJ judgments and arbitral awards on the subject rely on geology or natural prolongation,[32] the ambiguities in the prior cases and the absence of *stare decisis* before the ICJ and other international tribunals allows this consideration to be argued.[33] On the other hand, the articles of the LOS Convention on delimiting the international maritime boundaries for the exclusive economic zone and for the continental shelf are identical, suggesting perhaps that there should be no distinction between the relevant considerations. But because there are two separate articles, one for each

---

[29] Prescott, see note 23, 1211.
[30] *Ibid.* See references to the cases in note 25. See K. Highet, "Use of Geophysical Factors in Maritime Delimitation of Maritime Boundaries", in J. I. Charney and L. M. Alexander (eds.), *International Maritime Boundaries*, Vol. 1, 1993, 163 et seq., 166-177.
[31] In the *North Sea Continental Shelf* cases, the Norwegian Trough was at issue. In the *Anglo/French* arbitration, the Hurd Deep was used as a basis for argument. In the *Libya/Malta* and *Tunisia/Libya* cases, the parties argued that geological features established the limits of the natural prolongation of the respective States' continental shelves. None of these arguments succeeded. *North Sea Continental Shelf* cases, ICJ Reports 1969, 3 et seq.; *The United Kingdom of Great Britain and Northern Ireland and the French Republic, Delimitation of the Continental Shelf* decision, *ILR* 54 (1979), 32, 68-70, paras. 12, 104-109; Continental Shelf case *(Tunisia/Libyan Arab Jamahiriya)*, ICJ Reports 1982, 58-60, paras. 70-71; Continental Shelf *(Libyan Arab Jamahiriya/Malta)*, ICJ Reports 1985, 13 et seq., 34-37, paras. 35-41.
[32] See notes 12 and 16.
[33] ICJ Statute, Article 59.

zone and the references to international law and to equitable solutions may be interpreted within the context of each régime, support can be found for the use of different considerations. Even Judge Oda suggests that the régime distinctions may influence the international maritime boundary delimitation.[34] The key to the possible argument for the use of natural sciences in this circumstance of overlapping 200-nautical-mile zones is the definition of the continental shelf régime as found in Article 76(1) of the LOS Convention:

> "Article 76
> Definition of the Continental Shelf
>
> 1. The continental shelf of a coastal State comprises the sea-bed and subsoil of the submarine areas that extend beyond its territorial sea throughout the natural prolongation of its land territory to the outer edge of the continental margin, or to a distance of 200 nautical miles from the baselines from which the breadth of the territorial sea is measured where the outer edge of the continental margin does not extend up to that distance".[35]

Although the exclusive economic zone is defined solely on the basis of distance,[36] the definition of the continental shelf in Article 76 speaks in terms of "the natural prolongation of its land territory to the outer edge of the continental margin",[37] in addition to the distance criteria (and further definitions of the continental shelf limit beyond 200 nautical miles elaborated in the remainder of Article 76). Thus, one might argue first that the 200-nautical-mile limit of Article 76 did not establish the minimum breadth of a coastal State's entitlement to a continental shelf, but rather this distance could only be reached if based on natural sciences "the natural prolongation of its land territory to the outer edge of the continental margin" actually reached the 200-nautical-mile limit. Secondly, based on the rule that all terms in a treaty should be given a meaning, to read Article 76(1) as granting the coastal State an entitlement to the 200-nautical-mile limit would give no meaning or function to the phrase "the natural prolongation of its land territory to the outer edge of the continental margin...".

Obliquely, in a recent judgment on the subject, the ICJ seemed to give credence to the view that, within the area of 200-nautical-mile

---

[34] *Maritime Delimitation in the Area between Greenland and Jan Mayen* case, ICJ Reports 1993, 109-110, paras. 70-74 (Oda, V.P., separate opinion).
[35] LOS Convention, see note 8, Article 76(1).
[36] *Ibid.*, Article 57.
[37] *Ibid.*, Article 76(1).

overlapping zones considerations other than geography might be relevant to a delimitation. In the *Jan Mayen* case,[38] it purported to engage in two separate analyses of the maritime boundary – one for the continental shelf and the other for the exclusive economic zone.[39] Although the Court ultimately merged the two analyses and used the same considerations to delimit a single maritime boundary for both zones,[40] it failed to limit the considerations used to delimit the single international maritime boundary to coastal geography considerations. Thus, it considered the distribution of access to fisheries, a consideration only relevant to the exclusive economic zone boundary, in the delimitation of the boundary for both zones.[41] This suggests that as a matter of law the considerations relevant to the exclusive economic zone and the continental shelf are the same. But through that sameness it allows that the law permits the consideration of non-geographic considerations in such delimitations. Certainly, this judgment allows one to rely upon the language of the *Tunisia/Libya* case to the effect that each case is unique and the considerations may differ,[42] despite the virtual, albeit incomplete, disposal of considerations other than coastal geography in the *Libya/Malta* case.[43] As a consequence, the apparent movement towards a limitation to geography, at least within overlapping 200-nautical-mile zones, may have been undermined.

In that light might a continental shelf boundary in such a geographic situation be influenced by geology or other natural sciences? The language of the *Libya/Malta* case and the actions of tribunals in all cases other than the *Jan Mayen* case seem to confirm that, at least within 200-nautical-mile overlapping areas, all considerations are *de facto*, if not *de jure*, irrelevant to an international maritime boundary delimitation for the continental shelf and/or the exclusive economic zone.

The *Libya/Malta* case correctly interpreted Article 76(1) of the LOS Convention as setting the minimum entitlement to a coastal State's continental shelf at 200 nautical miles, understanding the reference to "the natural prolongation of its land territory to the outer edge of the continental margin"[44] as an historic retention of the idea of the continental shelf as it existed up to contemporary developments, brought

---

[38] *Maritime Delimitation in the Area between Greenland and Jan Mayen* case, ICJ Reports 1993, 38 et seq.
[39] *Ibid.*, 56-59, paras. 41-48.
[40] *Ibid.*, 61-62, 69-70, paras. 52-53, 71, 99.
[41] *Ibid.*, 79-81, para. 92. Charney, see note 16, 246-247.
[42] *Continental Shelf* case *(Tunisia/Libyan Arab Jamahiriya)*, ICJ Reports 1982, 18 et seq.
[43] See text at notes 12-20.
[44] LOS Convention, see note 8, Article 76(2)-(6).

about by the 1982 LOS Convention negotiations. It also serves as a reference to the remainder of Article 76, which allows for the régime of the continental shelf to extend *beyond* the 200-nautical-mile limit on the basis of tests that seek to approximate in legal terms the limit of the continental margin.[45] Judge Oda reviews the evolution of this terminology during the LOS Convention negotiations, including references to the *travaux préparatoires*, in his dissenting opinion to the *Libya/Malta* judgment.[46] He concludes accurately that the language of Article 76(1) was intended to provide all coastal States an entitlement to a continental shelf of 200 nautical miles regardless of the geology and

---

[45] *Ibid.*, Article 76(2)-(6). The source of the idea that the continental shelf régime is the natural prolongation of the land territory is the Truman Proclamation of 1945 in which the United States asserted rights to the resources of the shelf and initiated the movement towards the continental shelf régime. In that proclamation he states, "the Government of the United States regards the natural resources of the subsoil and sea bed of the continental shelf beneath the high seas but contiguous to the coasts of the United States as appertaining to the United States, subject to its jurisdiction and control". "Policy of the United States with respect to the Natural Resources of the Subsoil and Sea Bed of the Continental Shelf", Presidential Proclamation 2667 (28 September 1945), *Fed. Reg.*, Vol. 10, 1945, 12303.

Although this provided the backdrop to the deliberations of the International Law Commission, see note 4, and the negotiations at the first United Nations Conference on the Law of the Sea in 1958, the idea of natural prolongation or other natural sciences did not find its way into the resulting Convention on the Continental Shelf. Article 1 of the 1958 Convention on the Continental Shelf, see note 5, defined the continental shelf in terms of the so-called adjacency and exploitability test, without any refined definition of its extent or nature:

"For the purpose of these articles, the term 'continental shelf' is used as referring (a) to the seabed and subsoil of the submarine areas adjacent to the coast but outside the area of the territorial sea, to a depth of 200 metres or, beyond that limit, to where the depth of the superjacent waters admits of the exploitation of the natural resources of the said areas; (b) to the seabed and subsoil of similar submarine areas adjacent to the coasts of islands".

Nor was the topic expressly referenced in Article 12 on the delimitation of the continental shelf maritime boundary. Convention on the Continental Shelf, see note 5, Article 12. Although the term "special circumstances" could be a basis for arguing its implicit inclusion. The *North Sea Continental Shelf* cases reintroduced the ideas of natural prolongation that was not apparent from the Continental Shelf Convention. *North Sea Continental Shelf* cases, ICJ Reports 1969, 36, 47, 50-51, paras. 57, 85(c), 94-97.

[46] *Continental Shelf* case *(Libyan Arab Jamahiriya/Malta)*, ICJ Reports 1985, 13 et seq., 148-157, paras. 46-61 (Oda, J., dissenting opinion).

geomorphology of the sea-bed and subsoil.[47] That basis for the entitlement consequently conditions the relevant considerations for delimiting maritime boundaries between States with overlapping entitlements to exclude geology and geomorphology from consideration, as Judge Oda also argued in his dissent.[48]

Certainly, this would be wise advice to follow. While all international maritime boundaries are indeed unique, rights to the resources of areas within 200 nautical miles of a coastal State's coastline are now merely a function of distance from the shore. The alleged relevance of other considerations sketched out in the *North Sea Continental Shelf* cases has no relevance today. In 1969, other than the régime of the continental shelf, the only other coastal State régimes were those of the territorial sea and the contiguous zone, as well as perhaps a limited fisheries zone – all of which extended only a short distance from the coastline. The régime of the exclusive economic zone was not part of international law, and it was especially true that the concept later found in Article 76(1) of the 1982 LOS Convention, that the continental shelf régime would extend to a minimum of 200 nautical miles from the coastline, was not even under consideration. With the distance criteria attached to the régime of the continental shelf and the establishment in international law of the 200-nautical-mile exclusive economic zone, the entire context for international maritime delimitations changed. It would indeed be an anachronism to return in the 21st century to the ideas generally laid out in the 1969 *North Sea Continental Shelf* cases.[49]

---

[47] This view is supported in M. H. Nordquist (ed.), S. N. Nadan and Sh. Rosenne (volume eds.), *United Nations Convention on the Law of the Sea 1982: A Commentary*, Vol. 2, 1993. Relying on the *travaux préparatoires*, the commentary observes that the distance criteria of 200 nautical miles assured the coastal States of entitlements to the continental shelf within that distance regardless of geological or geomorphological facts. *Ibid.*, 841, 851. It removed elements of the scientific definitions that grew out of the 1958 Continental Shelf Convention that was proposed by the ILC in its 1956 report. *Ibid.*, 873. As the commentary reports, the choice of the 200-nautical-mile minimum limit was crafted to satisfy the interests of those coastal States that did not have a continental margin, as defined by natural sciences, that even reached the 200-nautical-mile limit. *Ibid.*, 845, 851. Accordingly, the natural prolongation concept became irrelevant to the area within the 200-nautical-mile limit. The primary focus of the negotiations was on crafting a formula for the limit of the continental shelf beyond 200 nautical miles. That formula was linked to the natural science concept of the continental margin and thus the concept of natural prolongation as retained in Article 76(1). *Ibid.*, 851-890.

[48] *Continental Shelf* case *(Libyan Arab Jamahiriya/Malta)*, ICJ Reports 1985, 13 et seq., 157-163, paras. 61-70.

[49] *Maritime Delimitation in the Area between Greenland and Jan Mayen* case, ICJ Reports 1993, 99, 105, paras. 37, 58 (Oda, V.P., separate opinion).

What then is to be done with the consideration of access to fisheries in the delimitation of the continental shelf boundary in the *Jan Mayen* case? Perhaps it should be ignored as ill-conceived. Interestingly, however, the arbitration panel in the more recent *Eritrea/Yemen* case found a way to avoid fisheries considerations by resolving them in an earlier phase of the arbitration in the context of awarding title over the island territories in dispute by subjecting them to an historic regional fisheries régime.[50] Consequently, when it came to the second phase in which the international maritime boundary was to be delimited it did so strictly on the basis of coastal geography with geology and geomorphology playing no part.[51]

Contemporary fisheries considerations, however, hearken back to the language used by the ICJ in the *Tunisia/Libya* case, "it is the present-day sea-bed, which must be considered. It is the outcome, not the evolution in the long-distant past, which is of importance".[52] When natural prolongation was largely set aside in favor of contemporary coastal and maritime geography in the subsequent *Libya/Malta* case, the ICJ noted that "those factors ... now belong to the past".[53] The ICJ was addressing the conclusion that because those factors did not provide a juridical basis for rights in the continental shelf as a result of the new distance criterion, such factors had become irrelevant to the delimitation of international maritime boundaries. This focus on current utility could be construed to allow utilization of other considerations with contemporary value as opposed to arguments based on the ancient origins of the continental margin. This could lead to the argument that geology and geomorphology have contemporary relevance since they are linked to the location of currently exploitable natural resources, e.g., hydrocarbons. This view would unfortunately undermine an evolution towards a refinement of international maritime delimitation law that seemed to be developing. The *Jan Mayen* judgment places this at risk.

Australia and East Timor, when the latter becomes a State, may negotiate a Joint Development Zone based on the MOU, another solution, or an international maritime boundary on any basis they wish.

---

[50] *Arbitral Tribunal in the First Stage* award *(Eritrea/Yemen)*, Territorial Sovereignty and Scope of the Dispute, 9 October 1998, paras. 526-526, available at http://www.pca-cpa.org.

[51] *Arbitral Tribunal in the Second Stage* award *(Eritrea/Yemen)*, Maritime Delimitation, 17 December 1999, paras. 62-73, 87-112, 129-168, available at http://www.pca-cpa.org.

[52] *Continental Shelf* case *(Tunisia/Libyan Arab Jamahiriya)*, ICJ Reports 1982, 54, para. 61.

[53] *Continental Shelf* case *(Libyan Arab Jamahiriya/Malta)*, ICJ Reports 1985, 13 et seq., 36, para. 40.

Since the law of international maritime boundaries is not *jus cogens* they are free to do so. However, if a tribunal were to be charged with applying applicable international law to delimit the maritime boundary between Australia and East Timor, if it were to take into consideration the natural prolongation of the sea-bed of the area within the 200-nautical-mile zones, it would be acting contrary to a long list of international decisions by the ICJ and courts of arbitration and, consequently, contrary to international law. Such an approach would do a disservice to the settlement of international maritime boundary disputes in general. Rather than bringing consistency and coherence to the law, it would contribute to the view that the law on the subject is whatever the tribunal wishes it to be, thereby further undermining the legitimacy of the law. It is time for the international community to move towards a more coherent rule of international law on the subject, rather than moving away from it.

The Problem of Delimiting the Maritime Zone Between China (Mainland and Taiwan) and Japan

Hungdah Chiu

I. Introduction

The 1982 United Nations Convention on the Law of the Sea[1] (UNCLOS) entered into force on 16 November 1994.[2] The People's Republic of China (hereinafter, "China" or "PRC") and Japan are Contracting Parties to the Convention.[3] The Republic of China in Taiwan (hereinafter, "Taiwan" or "ROC") has been unable to adhere to the UNCLOS because of opposition from the PRC. On 12 January 1983, Taiwan's Foreign Minister, Chu Fu-sung, stated in the Legislative *Yuan* (Parliament):

> "In the past we participated in various maritime conferences, but since our withdrawal from the United Nations [in 1971], we have been unable to participate in these conferences on the Law of the Sea. Consequently, we did not participate in signing the [1982 United Nations] Convention on the Law of the Sea.
>
> ...
>
> This convention includes a lot of articles and is an important convention among general international conventions. Our country ... is located within an ocean region; therefore, after its entry into force, this Convention would have far-reaching effects on our country. ... After the Law of the Sea Convention enters into force,

---

[1] UN Doc. A/CONF.62/122 and Corr. 1 to 14.
[2] *Law of the Sea Bulletin* 25 (June 1994), 1.
[3] China ratified the Convention on 7 June 1996 and Japan did so on 20 June 1996. Multilateral Treaties Deposited with the Secretary-General, Status as of 31 December 1997 (Doc. ST/LEG/SER.E 116), 1998, 1 et seq., 799 (hereinafter, "Status of Multilateral Treaties").

our attitude is that in principle we will comply with its terms. However, because of the region in which we are located and because our political environment differs from those of other countries, we must have certain reservations and [special considerations with respect to] some aspects [of the Convention]. We are still studying the problem".[4]

Since the UNCLOS entered into force on 16 November 1994, the ROC has not announced any reservations to particular provisions of the Convention. Therefore, I will assume that the Convention is acceptable to the ROC (Taiwan). China ratified the Convention on 7 June 1996, and Japan did so on 20 June 1996.[5] China ratified the Convention with the following Declaration:

"1. In accordance with the provisions of the United Nations Convention on the Law of the Sea, the People's Republic of China shall enjoy sovereign rights and jurisdiction over an exclusive economic zone of 200 nautical miles and the continental shelf.

2. The People's Republic of China will effect, through consultations, the delimitation of the boundary of the maritime jurisdiction with the States with coasts opposite or adjacent to China respectively on the basis of international law and in accordance with the equitable principle.

3. The People's Republic of China reaffirms its sovereignty over all its archipelagoes and islands as listed in Article 2 of the Law of the People's Republic of China on the Territorial Sea and Contiguous Zone which was promulgated on 25 February 1992.

4. The People's Republic of China reaffirms that the provisions of the United Nations Convention on the Law of the Sea concerning innocent passage through the territorial sea shall not prejudice the right of a coastal State to request, in accordance with its laws and regulations, a foreign State to obtain advance approval from or give prior notification to the coastal State for the passage of its warships through the territorial sea of the coastal State".[6]

---

[4] *Li-fa Yuan kung-pao (Gazette of the Legislative Yuan)*, Vol. 72, No. 38, 11 May 1983, 109, 111.
[5] See note 3.
[6] Status of Multilateral Treaties, see note 3, 806.

When Taiwan ratified the 1958 Convention on the Continental Shelf,[7] it made the following Declaration:

> "With regard to the determination of the boundary of the continental shelf as provided in paragraphs 1 and 2 of Article 6 of the Convention, the Government of the Republic of China considers:
>
> 1. that the boundary of the continental shelf appertaining to two or more States whose coasts are adjacent to and/or opposite each other shall be determined in accordance with the principle of the natural prolongation of their land territories; and
>
> 2. that in determining the boundary of the continental shelf of the Republic of China, exposed rocks and islets shall not be taken into account"[8]

Taiwan also declared, on 8 October 1979, a 200 mile exclusive economic zone from which the territorial sea is measured.[9] As stated earlier, the PRC also stated that it had sovereign rights in an exclusive economic zone when it ratified the Convention,[10] though it has not enacted a law or regulation concerning this zone. The PRC also declared that delimitation with States with coasts opposite to China should take place on the basis of international law and in accordance with equitable principles.[11]

In delimiting the maritime boundary between Taiwan and Japan, there is no reason to believe that the ROC would oppose these principles.[12]

Therefore, the principles of delimiting the maritime zone between China (Mainland and Taiwan) should follow international law and

---

[7] UNTS Vol. 499, 311.
[8] Status of Multilateral Treaties, see note 3, 794-795.
[9] *Chinese YB Int'l L. & Affairs* 1 (1981), 151-152. On 21 January 1998, the Republic of China enacted a Law on the Territorial Sea and the Contiguous Zone, which provides that the baseline of the territorial sea "shall be determined by a combination of a straight baseline in principle and a normal baseline as an exception", but it has not defined its straight baseline. On 21 January 1998, it also enacted a Law on the Exclusive Economic Zone and Continental Shelf. *Chinese YB Int'l L. & Affairs* 16 (1987-1988), 124.
[10] See note 3 and accompanying text.
[11] *Ibid.*
[12] See, e.g., Professor Kuen-chen Fu, "Delimiting the Overlapping EEZs Between the ROC and the Philippines", 4 *Soochow Law Review* 2 (Taipei, March 1984), 1 et seq.

equitable principles, as provided in Articles 74 (Exclusive Economic Zone) and 83 (Continental Shelf) of the Convention.[13]

II. The Maritime Zone Between China and Japan and the Dispute over the Senkaku Islands / Diaoyu Tai Islets

Both China and Japan claim sovereignty over the Senkaku Islands/ Diaoyu Tai Islets.[14] Under Article 121, paragraph 3, of the Convention "[r]ocks which cannot sustain human habitation or economic life of their own shall have no exclusive economic zone or continental shelf", although paragraph 2 of the same article also references the territorial sea. Both the PRC and ROC have 12-mile territorial seas, but whether the Senkaku/Diaoyu Islets can claim an independent exclusive economic zone and continental shelf is debatable. The largest islet, Diaoyu Islet, measures 4.319 square kilometers,[15] so it certainly qualifies by size to claim a continental shelf or exclusive economic zone. China proposed to jointly explore the oil resources near the Diaoyu Islands, but no positive response from Japan has been received.[16]

The East China Sea continental shelf measures about 460,000 square kilometers. China's position is that it is the natural prolongation of Chinese territory and should be divided by an equidistance line between China and the Okinawa Trough. However, Japan considers that the equidistance line should be between China and Japan, disregarding the Okinawa Trough. China has discussed the issue with Japan, but could not reach agreement.[17] China, however, already actively explores oil resources in the East China Sea about 250 miles east of Shanghai.[18]

---

[13] As stated above, the Convention entered into force on 16 November 1994. See *Law of the Sea Bulletin* 25 (June 1994), 1.

[14] For the Chinese view, see Hungdah Chiu, "An Analysis of the Sino-Japanese Dispute over Diaoyutai Islets (Senkaku Gunto)", *Chinese YB Int'l L. & Affairs* 15 (1996-1997), 9 et seq. For the Japanese view, see Yoshiro Matsui, "International Law of Territorial Acquisition and the Dispute over the Senkaku (Diaoyu) Islands", *The Jap. Ann. Int'l L.* 40 (1997), 3 et seq.

[15] Seiho Yamaguchi, *Ishigaki Cho Shi*, Naha: Okinawa Shose Ki Kabushiki Kai-sha, 1935, 13.

[16] Tang Jia Xuan (Foreign Minister of the PRC) (ed.), "Diaoyu Dao Wen Ti" ("The Question of Diaoyu Island"), in Zhongguo Waijiao Cidian, Beijing: Shi-jie shi-shi chu ban she, 52.

[17] "Dong Hai Da Lu Jia Wen Ti" ("The Question of Continental Shelf in the East Sea"), in *Zhongguo Waijiao Cidian*, Beijing: Shi-jie shi-shi chu ban she, 52.

[18] See M. Landler, "Big Chinese Oil Company Testing Equity Markets Again", *New York Times*, 22 February 2001, W1 (including a map showing that *China National Offshore* is involved in exploration as well as production).

Since Japan has recognized the PRC as the only legal government of China, it only maintains informal relations with Taiwan,[19] and has not entered negotiation with Taiwan on the maritime boundary between them.

III. Possible Settlement Mode – Judicial, Conciliation and Negotiation

Japan accepts the compulsory jurisdiction of the International Court of Justice,[20] but China has not. Thus, the issue of the maritime boundary between them can only be settled through negotiation, although, as stated earlier, negotiations have failed to yield any agreement. With regard to the Diaoyu Islets dispute, China has already enacted the Law on the Territorial Sea and Contiguous Zone of China, which lists Taiwan and the Diaoyu Islets, *inter alia*, as Chinese territory in Article 2.[21] On this basis, it is even more unlikely that China will make any concession on the sovereignty issue.

One possible solution is to use conciliation to settle the issue. The Conciliation Commission could include among its members a representative from Taiwan, in addition to a member from China. However, whether the PRC would agree to let Taiwan participate in the Commission is not clear, because the PRC has, since its establishment on 1 October 1949, regarded Taiwan as part of its territory, and maintains that the Government of the PRC is the sole legal government of all China.

---

[19] See J. W. Morley, "The Japanese Formula for Normalization and Its Relevance for US-China Policy", in Hungdah Chiu (ed.), *Normalizing Relations with the People's Republic of China: Problems, Analysis and Documents*, Baltimore: University of Maryland School of Law, *Occasional Papers/Reprints Series in Contemporary Asian Studies*, No. 2-1978, 121 et seq.
[20] International Court of Justice, *Yearbook 1988-89*, 1989, 74-75.
[21] *Zhonghuo Renmin Gonghe Guo Guowuyuan Gong Bao (Gazette of the State Council of the People's Republic of China)* 3 (1992) (Total number, No. 688), 13 March 1992, 69. The law was promulgated on 25 February 1992.

The Role of Proportionality in Maritime Delimitation Revisited: The Origin and Meaning of the Principle from the Early Decisions of the Court

Ryuichi Ida

I. Introduction

Some rules of international law leave judgment of the legality of an act to the consideration of the specific situation of the case, and offer only a general notion of the criteria for evaluation. One of these rules is the principle of proportionality. While this principle is often referred to in various situations, it seems to play different roles and to be applied in different ways. This article will revisit the proportionality principle in maritime delimitation cases, starting from the *North Sea Continental Shelf* case and some other pertinent cases from the developmental period of law of the sea, in order to clarify its criteria and its functions as originally conceived. By limiting the scope of investigation to several decisions of primary importance, this paper provides a first step towards understanding the principle of proportionality in the international law of the sea, in which Judge Oda has constantly played and will always play a leading role.

The problem of maritime delimitation[1] between States which face each other or which are adjacent has led along a difficult road. This road began in 1958 with the adoption of four conventions of Geneva, of which the convention on the continental shelf imposed the principle of equidistance in Article 6. However, this has never brought decisive solutions in maritime disputes. In search of a reasonable solution, the International Court of Justice decided the *North Sea Continental Shelf* case. The Court's judgment, for which the so-called "façade doctrine"[2]

---

[1] For an excellent theoretical and practical analysis, see generally P. Weil, *Perspectives du droit de la délimitation maritime*, 1988, esp. 203-293.
[2] *North Sea Continental Shelf* case, Judgment, ICJ Reports 1969, para. 15 (hereinafter, "Judgment 1969").

presented by Judge Oda, at that time Germany's Counsel, played a crucial role, fixed a principle, following which the delimitation was to be effected "by agreement in accordance with equitable principles, and taking account of all the relevant circumstances, in such a way as to leave as much as possible to each party all those parts of the continental shelf that constitute a natural prolongation of its land territory into and under the sea".[3] The Court recognized a decisive role for the proportionality principle, as will be examined later in this paper.

The decision had a great impact on the drafting of the Law of the Sea Convention. Articles 74 and 83 are the compromise solution between two groups, one supporting equidistance, the other equitable principles. The proportionality principle serves as a tool in working towards an equitable settlement. The objective of this paper is to determine the original concept of the proportionality principle and its function as well as the method of its application in relation to maritime delimitation. What are the criteria of proportionality? How do we elaborate a particular line of delimitation as proportionate? Such are the questions which we will try to answer. However, in this short paper it is not opportune to make a full investigation of all maritime delimitation cases. Therefore, the focus is on some cases of primary value at the early stage of the application of this principle, i.e., from the end of 1969 to 1985.

II. Starting Point of the Principle

1. Appearance of the Principle

The *North Sea Continental Shelf* case is the first leading case of maritime delimitation between States to apply the proportionality principle. The Netherlands and Denmark claimed the equidistance line between each of them on the one hand and the Germany on the other. However, the geographical situation was so particular that the two equidistance lines drawn from the respective coasts formed a sort of closing line as the two lines came across just at the entrance of the disputed area. The Federal Republic of Germany developed the notion of "maritime façade" in order to avoid the application of the equidistance criteria[4] and argued that the purpose of delimitation was to attribute a just and equitable portion of the zone of continental shelf which was not delimited.[5] In applying its doctrine Germany argued that the length of the coast should be taken into account.

---

[3] *Ibid.*, 53.
[4] *Ibid.*, para. 15.
[5] *Ibid.*, paras. 15-17.

However, as it is well known, the Court rejected the concept of a just and equitable distribution of the area.[6] The Court took the view that each coastal State has its own continental shelf which is the natural prolongation of its territory and that the disputed area was the overlapping area of the continental shelf of each coastal State. If the area that is the overlapped area of the three continental shelves extended from the land territory of each coastal State, then there should be no area to be shared by the parties to the dispute. The Court concluded that the delimitation should be done "through the agreement of the parties to the dispute, following the equitable principle and taking into account all the relevant circumstances, in order to attribute the areas of continental shelf as much as possible which constitute a natural prolongation of the territory subjacent to the sea to each of the parties to the dispute". Here the maritime façade doctrine seems to have in fact played an important role. The Court, rejecting the plea of just and equitable share, took the substance of the notion of "façade" and introduced the idea of proportionality between the continental shelf attributed to each of the States and the length of respective coast following the general direction of the coast.[7]

In this case, proportionality was conceived as a corrective element for inequitable results to be applied in a quasi-equal geographical situation for each party and to be used in order to avoid an unreasonably inequitable result deriving from such a geographical particularity as a concavity of the coasts.[8] The starting point was the equidistance line and proportionality played the role of testing the equity[9] of the initial line or of other possible methods of delimitation. We should recognize, however, that the Court applied proportionality to a certain particular situation in which the coasts of the three States Parties to the conflict themselves form a concavity in a clearly defined region by agreements with third States.

2. Affirmation of the Principle

Once proportionality is recognized for the purpose of applying equitable principles, it is used as a convenient tool, but in various ways. The

---

[6] *Ibid.*, para. 39.
[7] *Ibid.*, para. 101, D-3.
[8] *Ibid.*, para. 91.
[9] Paul Reuter pointed out that there are three principles in the application of equity: the equivalence, the proportionality and the objective. P. Reuter, "Quelques réflexions sur l'équité en droit international", *RBDI* 15 (1980), 165.

arbitration concerning the continental shelf between the France and the United Kingdom was another example of application of the principle. The Tribunal confirmed the principle of proportionality in a case of different geographical circumstance, where the region was open to the Atlantic Ocean. According to the Tribunal, the element of proportionality may appear, in general, in the form of a relation between the extent of the continental shelf and the length of the coast of each State.

However, for the Tribunal, it may also be a question – a more general one – of elements, which allows evaluation of whether the geographical characteristics or the particular configurations have a reasonable or unreasonable, equitable or inequitable effect on the line drawn according to equidistance.[10] Therefore, proportionality seems to be a test of the results of applying the equidistance method.[11] However, the Tribunal formulated this principle in negative terms. According the Tribunal, "It is the disproportion, rather than a general principle of proportionality, that constitute the criteria. ... It can never be a question of remaking the nature totally: it is question more to remedy the disproportion and the inequitable effects deriving from geographical configurations or characteristics".[12] Therefore, also according to the Tribunal, proportionality does not constitute an independent source of rights in the continental shelf, but a criterion for evaluating the equity of a line drawn in a certain geographical situation.[13] Thus, equity is the goal and proportionality is a tool of measuring equity. If equity has such a value, the Convention on the law of the sea provides exactly, in its Articles 74 and 83, that the delimitation should be done in order to reach an equitable settlement. Then the question is, what should be the specific means to delimit the disputed maritime areas so as to find an equitable settlement? Obviously, proportionality is one such means and probably, as we will see, the most valuable one.

In these two decisions, proportionality was applied in two different situations. These cases show how the principle will work in a given context. Then the questions are "How proportionality is applied?" and "What are the functions of proportionality".

---

[10] *Affaire de la délimitation du plateau continental entre le Royaume-Uni et la France*, Décision du 30 juin 1977, para. 100 (hereinafter, "Decision UK-France").

[11] The proportionality is "a factor to be taken into consideration for judgment of the effect of geographical characteristics on the equity or the inequity of a delimitation". *Ibid.*, para. 99.

[12] *Ibid.*, para. 101.

[13] *Ibid.*

III. Application of the Principle

As the basic premise of proportionality is "to compare what is comparable",[14] the proportionality principle in maritime delimitation requires comparison of the areas attributed to each of the parties. In fact, what is compared is not the extent of the areas. The common conception of proportionality in maritime delimitation is the proportion between the extent of the maritime zone (continental shelf and EEZ) attributed to each of the parties in dispute and the length of coast corresponding to each zone. It is true that, in the *Guinea-Guinea Bissau* arbitration, Guinea claimed the proportion of the landmass of each State, but the Tribunal denied it.[15] In such a situation what are the elements of comparison?[16] How should the proportion be calculated in comparing such elements? (B)

1. Elements of Proportionality

In the *North Sea Continental Shelf* case, the Court pointed out the need to take into account "the element of a reasonable degree of proportionality, which a delimitation carried out in accordance with equitable principles ought to bring about between the extent of the continental shelf areas appertaining to the coastal State and the length of its coast measured in the general direction of the coastline".[17] A clear placement of proportionality was also given in the decision of the two Guineas' arbitration. The tribunal defined proportionality as one of the relevant circumstances, and the confirmation of the equity of results as its role. Moreover, it distinguished between the elements for the tracing operation and those for ascertaining the equity of results. The former consist of the relevant coasts, the existence of numerous islands, the length of the coasts, and the configuration and the direction of the coastal lines. The latter are the circumstances for ascertaining whether the results of the delimitation are equitable.[18] It is clear from this

---

[14] The Court used this expression twice in its 1982 judgment in the *Continental Shelf* case *(Tunisia/Libyan Arab Jamahiriya)*, ICJ Reports 1982, (hereinafter, "Judgment 1982"), paras. 104 and 130.
[15] *Sentence arbitrale sur la délimitation de la frontière maritime Guinée/Guinée Bissau* du 4 février 1985, para. 119 (hereinafter, "Sentence arbitrale").
[16] The importance of the parameters was emphasized by Reuter, see note 9, 173.
[17] *North Sea Continental Shelf* case, Judgment 1969, see note 2, para. 101, D-3.
[18] Sentence arbitrale, see note 15, paras. 91-125.

distinction that proportionality is a test of results, and the four other factors are elements for drawing the line.[19]

Thus, when we compare the maritime zone and the length of the coast of each State, we face the following four questions: 1) where does the area begin? (Starting line): 2) what is the outer limit of the area? (closing line): 3) what is the corresponding coast of the area? (Relevant coasts): 4) how should the length of the coast be calculated? (Length of the coasts – general direction).

A. Starting Line

The first question may be translated in a more precise way. Which is the starting line of the area? Is it the base line, the line of the outer limit of the territorial sea, or any other line?

It is clear that, in international law, the continental shelf and the EEZ are measured from the outer limits of the territorial sea. The continental shelf in particular is in this definition a legal concept and not a geographical or geological concept. However, the Court in the *Tunisia-Libya* case, in which the Tunisian claim for the straight base line was not accepted by Libya, took the low watermark line as starting line for each continental shelf.[20] This means that the Court based its decision on a geographical concept of the continental shelf although it had distinguished between the legal and the geographical notion of the continental shelf.[21]

If we calculate the continental shelf from the low watermark line, the extent should differ slightly from that calculated from the outer limit of the territorial sea measured from the straight base line. In this sense, the Court's choice apparently lacks coherence. Such a difference in the starting line, though it seems very small and even negligible, will amplify the difference of the extent of the area attributed to each of the States, particularly when the calculation is shown in numbers.

However, in almost all cases the starting line is not a large obstacle for the settlement. It is probably because, in almost all cases, the extent of the area and the length of the coast as well as the proportion are not presented in precise numbers. It follows that measuring the proportion is rather a rough operation. In any event, the base line, natural or straight, is the basic reference mark. It should be noted, in addition, that the relevant element of proportionality is not a straight base line, but the length of the coast.

---

[19] See, Y. Ben Achour, "L'Affaire du plateau continental tuniso-libyan", *JDI* (1983), paras. 35 et seq.
[20] Judgment 1982, see note 14, para. 104.
[21] *Ibid.*, paras. 43-44.

B. Closing Line

In applying the proportionality principle, as was the case of the starting line described above, it is necessary for the calculation of proportion to determine the outer limit of the area to be delimited. It is on this very point that Judge Oda strongly criticized the Court's findings.[22]

i. Agreement with Other States

The most effective manner of maritime delimitation is the conclusion of a bilateral agreement between the States concerned. This is the main objective of the Articles 74 and 83 of the Law of the Sea Convention of 1982. Thus, a relevant treaty may determine the closing line. In the *North Sea Continental Shelf* case, the situation of the three European countries in 1969 was relatively simple. The outer limit of the disputed area was clearly defined prior to the dispute by five agreements between the coastal States of the North Sea, namely the United Kingdom, the Netherlands, Denmark and Norway. These lines were drawn according to the median line principle. It should be noted that there was no disagreement by Germany on this point. Some other examples are found, for instance, in the cases between Libya and Malta of 1984 (with Tunisia and Italy) and between Italy and Tunisia, and between Italy and Greece[23] and, in 2001, between Qatar and Bahrain (with Saudi Arabia and Iraq).[24]

ii. Line Defined by a Multilateral Convention: EEZ

It goes without saying that, when the delimitation both of the continental shelf and of the EEZ is the goal, the closing line is the 200 mile line from the baseline of the corresponding coast. Examples of this type are the *Gulf of Maine* case and the arbitration between the two Guineas.

iii. Imaginary Line

In many cases where the disputed area is open to the outside, a hypothetical line is drawn. The two Mediterranean cases are illustrative.

---

[22] Dissenting Opinion of Judge Oda, ICJ Reports 1982, para. 164.
[23] *Continental Shelf* case *(Libyan Arab Jamahiriya/Malta)*, Judgment, ICJ Reports 1985, 3 (hereinafter, "Judgment 1985").
[24] See the judgment in *Maritime Delimitation and Territorial Questions Between Qatar and Bahrain*. A summary of the judgment is available at http://www.icj-cij/icjwww/ipresscom/ ipresscom2001.

In order to determine the area for delimitation, the Court drew an imaginary line that consisted of the median line passing through Ras Tajoura and the parallel on the level of Ras Kapoudia. The Court did not explain the precise reason of this choice, but said that such a choice was founded only on cartographical convenience.[25] However, the result of the calculation based on this imaginary closing line was crucial to indicate a possible line for equitable delimitation between the two Arabian countries.

In another occasion, in the dispute between two coastal States opposite each other, it seems the Court took into consideration an area determined by an "invisible" line in order to calculate, though implicitly, the extent of the subjacent area belonging to each of the coastal States Parties in dispute. However, according the Court's decision, "... difficulties are particularly evident in the present case, where, in the first place, the geographical context is such that the identification of the relevant coasts and the relevant areas is so much at large that virtually any variant could be chosen".[26] This phrase shows the eventual inappropriateness of proportionality test in cases of practical obstacles. That is why the Court, in the case between Libya and Malta, could take the idea of an approximate equity of the result, without trying to show it in numbers.[27]

The situation has been changing since the adoption of the Montego Bay Convention. It follows naturally that the closing line is the 200 mile line from the baseline, except in cases where the claim to the continental shelf extends beyond that limit. Therefore, the Court's efforts to provide a just proportion by drawing an imaginary line in the *Libya-Malta* case have not been followed again, at least on the closing line. However, a question remains; "Should the delimitation be identical for the continental shelf and the EEZ?"

C. Relevant Coasts

As for the appreciation of proportionality, it is also necessary to determine the coast from which the extent is measured. It was in the *Tunisia-Libya* case that the Court presented generally the idea of determining the coast to be taken into account. According to the Court, the primary criteria for the definition of the legal status of the submerged zone is the contiguity of the coastal State's land territory.[28] The Court's

---

[25] Judgment 1982, see note 14, para. 130. However, for severe criticism by Judge Oda, see his dissenting opinion. ICJ Reports 1982, paras. 163-164.
[26] Judgment 1985, see note 23, para. 74.
[27] *Ibid.*, para. 75.
[28] Judgment 1982, see note 14, para. 129.

idea in that case was that, starting form the coast of each State in dispute, it seeks to determine how far the submerged space belonging to each of the parties extended to the ocean.[29] However, it was not necessary to take into account the totality of the coasts. Each segment of the coastline of one party, whose prolongation could not meet that of the coast of the other party, should have been excluded from the evaluation of proportionality. The maps showed clearly the existence of a point beyond which the sea-bed in front of the coast could not constitute a overlapping zone of the extension of the territories of both parties and, from this fact, had no role in the delimitation.[30] In the context of the *Tunisia-Libya* case, the end-points were Ras Kapoudia and Ras Tajoura.

Yet, the variety of situations led the Court to different solutions. The *Gulf of Maine* case was an interesting example of another solution for this notion. The Chamber of the Court, without having any geographical definition of the Gulf in the *compromis* of the two parties, determined as pertinent only the coasts within the two ends of the closing line of the Gulf, which were Nantucket and Cape Sable.[31] Nevertheless, the reason was not clear why the Chamber took into account the coast of the Bay of Fundy, which had never been subject of overlapping claim from the parties.[32]

The *Libya-Malta* case was one of the most limited regions for delimitation because of the existence of the Italian claim to its own maritime zone. According to the Court, the staring point in the West was the terminus of the frontier with Tunisia, i.e., Ras Tajdir, and the Eastern limit of the area was the crossing of the Libyan coast by the meridian 15°10'E, i.e., Ras Zarruq.[33] Such an operation might have reduced its judiciary function, because, from the given geographical situation in which the both parties were facing each other, it would have seemed to be more reasonable had the Court taken the whole Libyan coast in to account.

The other extreme example was the *Guinea-Guinea-Bissau* arbitration. The Tribunal considered as relevant not only the coast of these two West African States but also all the configuration of the West African coast, in particular, the coasts belonging to Senegal and Sierra-Leone,[34] although these coasts were used for the operation of delimitation as a whole and not simply for proportionality.

---

[29] Judgment 1985, see note 23, para. 74.
[30] *Ibid.*, para. 75.
[31] *Gulf of Maine* case, Judgment, ICJ Reports 1984, (hereinafter, "Judgment 1984"), para. 223.
[32] Judge Schwebel also made a criticism on this matter. Individual opinion of Judge Schwebel, ICJ Reports 1984, pp. 353-358.
[33] Judgment 1985, see note 23, para. 68.
[34] Sentence arbitrale, see note 15, para. 110.

D. Length of the Coasts: "General Direction of the Coast"

Although the starting line of the coast is either the base line or the outer limit of the territorial sea, the measurement of the length of the coast does not precisely follow these lines. According the Court, the length of the relevant coast should be measured following the general direction of the relevant coast.

It was in the *Tunisia-Libya* case that the Court for the first time suggested the operation of measuring the length following the general direction of the coast, although it did not specifically use this term. According to the judgment of this case, it was noted that the length of the coast was measured along the coastline without taking account of small inlets, creeks and lagoons, and treating an island as if it were a promontory. The relevant coastline of Libya thus measured stood in the proportion of approximately 31:69 to that of Tunisia. The Court further noted that the coastal front represented by a straight line was also taken into account, and that the proportion of the coastal fronts was 34:66.[35] Furthermore, the Court denied to consider the Tunisian base line for the claim of the historical rights, because of the lack of a direct relationship with the delimitation of the continental[36] shelf. It follows that neither the coastal front nor the low watermark line represents the general direction of the coast.

As for the extent of the maritime area to be delimited, both the continental shelf and the EEZ are measured, by definition, from the baseline. However, as we stated above, the baseline may often be sinuous so that the law watermark line is surely not the general direction of the coast in the calculation of the length of the coast. The term "general direction" supposes naturally the abstraction of the sinuosity of the coast. The easiest solution may be to use the map of a small scale.[37]

Although in a dispute between an island State and a continental State, such as the case between Libya and Malta, it might not be very easy to determine the general direction of the relevant coasts, much greater difficulties could arise when the disputed area contains islands, promontories, peninsulas, rocks or rises. In neither in the *Gulf of Maine* case, nor in the *Guinea-Guinea Bissau* case, was the general direction of the coast easy to find because of the existence of islets, rocks, shoals and even peninsulas.

---

[35] Judgment 1982, see note 14, para. 131.
[36] *Ibid.*, para. 102.
[37] On this point, see the critical remarks of Judge Evensen on how the Court could measure the length. Individual opinion of Judge Evensen, ICJ Reports 1982, para. 23.

The most important and probably most difficult element that should be considered is the existence of islands in the proximity of the coast. The *Gulf of Maine* decision gave "a half effect" to the Seal Island,[38] and the two Guineas' arbitration, the effect of twenty percent of the mainland coastal line to islands of Bihagos, in order to adjust or rectify the provisionary line.[39] On the other hand, the *Tunisia-Libya* decision admitted "a half effect" to an island,[40] not for the purpose of measuring the length of the coast, thus not in the framework of "the general direction of the coast", but for the purpose of general "configuration"[41] of the coast. Moreover, in this case the Court obliterated an island in the framework of calculating the length of the relevant coast, and treated the island as if it were a peninsula.[42]

2. Methods of Evaluating Proportionality

The concept of proportionality needs to compare the relation between the length of each of the two relevant coasts and the relation between the extents of the two attributed zones. However, throughout the cases, the method of evaluating the proportionality has not been always the same. Two methods may be extracted: the arithmetical method and the global method.

A. Arithmetical Method

An example of the strict arithmetical method[43] was given in the *Gulf of Maine* dispute. The Chamber of the Court calculated, for the second segment of the area, the difference of the length of the coasts facing each other in this segment, and the median line initially traced was transposed following the proportion estimated from this calculation.[44] This was a pure and simple application of proportionality. Needless to say, proportionality here is not used as a test of equity, but as a criterion of

---

[38] Judgment 1984, (see note 31), paras. 222-223.
[39] Arbitral decision 1985, para. 97
[40] Judgment 1982, see note 14, para. 129. On the treatment of the Kerkennah Island, see L. Herman, "The Court Giveth and the Court Taketh Away", *ICLQ* 23 (1984), 844-845.
[41] Judgment 1982, see note 14, 93, B-2.
[42] *Ibid.*, para. 132.
[43] The United States and Canada did a very precise calculation. See, J. Schneider, "The Gulf of Maine Case: The Nature of an Equitable Result", *AJIL* (1991), 564-565.
[44] Judgment 1984, see note 31, paras. 221-223.

equity, even of decisive value, for drawing the delimitation line. However, here, the subject of calculation and comparison was only the length of the coast, and it was not a question of referring to the extent of the area, at least not in numbers. It is however easy to imagine that, in the case of the quadrangle of the second segment of the Gulf, the proportion of the length of two relevant coasts is equal to the proportion of the size of each attributed zone.

The arithmetical method seems quite easy to use, because it shows precisely in numbers the relevant proportion, i.e., the extent of the each continental shelf attributed or the length of the coastal front line. The Court thus concludes, in the *Tunisia-Libya* case that the result of calculation, taking into account all relevant circumstances, meets the requirement of the test of proportionality as an aspect of equity.[45] However, it might be questionable if the difference between the relation of 31:69 (length of the coasts) and the relation of 40:60 (attributed continental shelf) is negligible from the viewpoint of the proportionality principle. Such difference might go beyond a reasonable relation. To show the difference in numbers sometimes leads to demonstrating the unreasonableness of the solution. It is, of course, no longer a question of proportionality but of equity or of reasonableness.[46]

B. Global Method

The two decisions of 1985, showing the length of the coasts in numbers, declined to indicate the extent of each of the attributed areas in an arithmetical form. According to the Court, "it is possible for it [the Court] to make a broad assessment of the equitableness of the result, without seeking to define the equities in arithmetical terms. The conclusion was that there is certainly no evident disproportion in the areas of shelf attributed to each of the parties respectively such that it could be said that the requirements of the test of proportionality as an aspect of equity were not satisfied".[47] It is true that every application of proportionality criteria does not necessarily require a precise arithmetical calculation and that the operation of the verification of results bases itself on an approximate inference. However, the Court should have at least shown how such approximation was done, all the

---

[45] Judgment 1982, see note 14, para. 131.
[46] See, J. Salmon, "Le concept de raisonnable en droit international public", D. Bardonnet (ed.), *Mélanges offerts à Paul Reuter: le droit international, unité et diversité*, 1981, 447-478; and more generally, Ch. Perelman, "Le raisonnable et le déraisonnable en droit", *Archives de philosophie du droit*, 1978, t. 23, 35-42.
[47] Judgment 1985, see note 23, para. 75.

more because the Court transposed the preliminary median line 18' north of it, taking into account of the disparity of length of the coasts, for which the relation of the length was 8:1.[48]

The two Guineas' arbitration also suggested to not apply an arithmetical but rather a more sophisticated method. The Tribunal clearly pointed out that the proportionality rule was not a mechanical rule based only on the figures transcribing the length of the coasts.[49] Having said that, it measured the length of the coasts and presented them in numbers. Moreover, it gave to the Bijagos islands about twenty percent of the real length of the coast so as to arrive to give the same length of the coasts for both the parties.[50] Such adjustment was apparently another means of applying the theory of the partial effect of islands. Despite the Tribunal's dicta in support of a global method, its reasoning seems more mechanical than global. On this particular point, though, the Tribunal noted that proportionality should play its role in a reasonable degree, taking into account other relevant circumstances. The question was whether it was opportune or appropriate to use proportionality in the circumstances of this case, and not of a question of application of proportionality proper.

IV. Functions of the Principle

If proportionality should intervene in a reasonable way of taking other circumstances into account, it was a question of the opportuneness or relevancy of proportionality, and not a question of application. Then it becomes necessary to define, in addition to these elements and methods of proportionality, the functions of the proportionality principle.

1. A Test of the Equity of Results of Delimitation

As we saw in the *North Sea Continental Shelf* case as well as in the *UK-France Continental Shelf* arbitration, the original idea of proportionality was to use it as a corrective element for inequitable results, in order to avoid an unreasonably inequitable result deriving from geographical particularities of the coasts. The objective of the proportionality principle was to apply the equitable principle, taking into account all the relevant circumstances. In other words, the principle of proportionality is a test of the equity[51] of the results of delimitation.

---

[48] *Ibid.*, para. 73.
[49] Sentence arbitrale 1985, see note 15, para. 120.
[50] *Ibid.*, para. 97.
[51] Reuter, see note 9, 145.

However, one may ask how the Court can define the elements for calculating proportions before drawing a delimitation line, or even before drawing some provisionary lines for comparison, particularly when the Court has tried to give "nice calculations of proportionality".[52] A typical example is that of the *Tunisia-Libya* case, where the Court used a very precise calculation for testing results of the delimitation line by the proportionality principle. According to the decision, the length of the relevant coasts are, respectively, 185 km for Libya and 420 km for Tunisia, the ratio of the length of the coast being 31:69 and that of the façade 34:66. Then the ratio of the extent of each continental shelf results in 40:60 and this satisfies proportionality criteria. The reasoning of the Court is apparently logical. Nevertheless, the above-supposed question still seems valuable. The more precise the calculation is, the more the proportionality test seems to turn into one of indispensable preconditions for drawing an equitable delimitation line.[53]

Finally, the test does not always accompany a neat calculation in numbers. It is essential to establish a reasonable relation between the extent of the attributed area and the length of the coasts. However, as a test of the results of delimitation, a rough appreciation of equity by way of proportionality seems sufficient. The non-existence of apparent disproportion means that the line is equitable. The *Gulf of Maine* decision has already pointed out, in discussing the economic importance of the area, its scruple that the overall results should unexpectedly be revealed as radically inequitable as likely to entail catastrophic repercussions for the livelihood and economic well-being of the population.[54] The term "proportionality" was not explicitly used here. However, should such a "radically inequitable result" be revealed through the verification of results, proportionality as a test, that is as a tool of verification, implies inevitably qualitative appreciation of results, and not only quantitative. In case of such an inequitable result, the delimitation line would be adjusted in order to wipe out such inequity.

2. Adjustment of Tentative Lines

The *Libya-Malta* decision pointed out that proportionality was neither a principle nor a rule of international law and, thus, that it did not

---

[52] Judgment 1982, see note 14, paras. 10 and 130-132.
[53] Judge Gros pointed out, "the main thrust of the Judgment's approach seems to be that, equity being the goal, proportionality is method. ... The present Judgment drastically alters the restricted role which properly belongs to proportionality factor. Opinion dissidente de Juge Gros, ICJ Reports 1982, see note 14, para. 17.
[54] Judgment 1984, see note 31, para. 237.

constitute an independent source of rights in the continental shelf.[55] However, the Court recognized the importance of proportionality as a factor of delimitation and applied proportionality on three levels. The first was on the level of the relevant circumstances,[56] the second on that of a test for equitable results after delimitation,[57] and the third on that of adjustment of the initial tentative lines.[58] The first two levels have already been discussed in previous sections, and the third one interests us here.

According to the Court, the proportionality principle requires taking into consideration the great disparity in the length of the relevant coasts of the two parties.[59] Thus, the line was transposed 18' to the North.[60] Although the Court did not calculate the extent of the continental shelf, the disparity was translated in numbers. The way the Court had taken seems to be that it was preoccupied with calculating implicitly the extent of the areas to be attributed, and that it used proportionality as an adjusting tool of lines initially tended. Such an adjustment character of proportionality is in fact its most practically important function. But, if so, proportionality is no longer a simple test *ex post facto* of equitable results, but an independent determinative factor before tracing the line of delimitation.

The *Gulf of Maine* case revealed another possible use for adjustment of lines through the proportionality principle. Recognizing the political aspect of the given geographical situations in which the geographical criteria for equity were not totally applicable,[61] the Chamber tended to make corrections to certain unreasonable effects of the application of basic geometrical criteria, and used auxiliary criteria. Thus, the Chamber applied to the second segment such complementary criteria, which consist of giving a weight in just proportion (a fair measure of weight) to a non-negligible difference of the length of the coasts.[62] The Chamber, stressing that it did not intend to make an autonomous criterion or method of delimitation out of the concept of "proportionality", said that

---

[55] Judgment 1985, see note 23, paras. 57 and 58, citing the *Arbitral Award of the UK-France Continental Shelf Dispute*.
[56] The Court cited "dis-proportionality" among the circumstances and the factors of consideration. Judgment 1985, see note 23, para. 57, B-3. According to Prof. Chemillier Gendreau, the proportionality principle is a general guideline for giving remedies to the disproportion. Ch. Gendreau, "La signification des principes équitables dans le droit international contemporain", *RBDI* 16 (1981-1982), 516.
[57] Judgment 1985, see note 23, para. 74.
[58] *Ibid.*, para. 78.
[59] *Ibid.*, para. 68.
[60] *Ibid.*, para. 73.
[61] Judgment 1984, see note 31, para. 195.
[62] *Ibid.*, para. 196.

this did not preclude using an auxiliary criterion to correct the untoward consequences.[63]

In fact, it transposed in two stages the median delimitation line initially drawn. The first was to shift the median line in a way so as to reflect the ratio of the coastal fronts. Then, as second step, it was shifted so as to give a half effect to Seal Island.[64] Such operation clearly shows the practical importance of proportionality as a decisive criterion for delimitation process.

V. Conclusion

The proportionality principle, by its nature, requires individual appreciation of every relevant element in each given concrete situation. In maritime delimitation cases, the proportion tends to be quantitative, because it is principally a question of space, which is normally measurable in numbers. However, other elements, like economic interests or historical appurtenance, may play a role in addition to measurable indexes. There are elements which are easily calculable in numbers, and elements that are difficult to measure precisely in an arithmetical manner. Thus, proportionality may also be qualitative.

The proportionality principle has played its role, often even a decisive one, in maritime delimitation disputes, probably because of its measurability or quantifiability. Proportionality seems to occupy a place in the sources of delimitation. Certainly, the application of proportionality is not an easy task because of an extremely wide variety of relevant situations and circumstances. Its application has never been uniform. The true problem of proportionality is to find objective criteria for its application. Without such objectivism in applying the proportionality concept, the result of its application might be deemed as a solution *ex aequo et bono*, or that of a judge's subjective or arbitrary choice.

The effectiveness of the proportionality principle naturally depends on its establishment as an accurate concept and objective and scientific criteria[65] of application. Otherwise, the proportionality principle would sink to the realm of subjectivity, and thus of non-legality, where it would play the role of "the principle of non-principle", as Judge Oda quite correctly already pointed out about twenty years ago.[66] Our task is thus

---

[63] *Ibid.*, para. 218.
[64] *Ibid.*, para. 222.
[65] D. Bardonnet, "Equité et frontière terrestre", in *Mélanges Reuter*, see note 46, 43, para. 5.
[66] *Continental Shelf* case *(Tunisia/Libyan Arab Jamahiriya)*, Dissenting opinion of Judge Oda, ICJ Reports 1982, 157, para. 1.

to clarify all of the factors relating to the concept of proportionality. This paper has provided just a starting point for reflection on this goal by revisiting some important cases in the incipiency of the concept.

International Watercourses in the Jurisprudence of the World Court

Stephen C. McCaffrey

I. Introduction

It gives me great pleasure to contribute to this *Festschrift* in honor of Judge Shigeru Oda. Judge Oda, in his writings and opinions, has demonstrated a keen interest in, and an understanding of the importance of shared natural resources, in particular those of the sea. It seems appropriate, therefore, to focus in this paper on one form of those resources, fresh water.

This paper will consider decisions that have been rendered in three cases brought by States to the International Court of Justice and its predecessor, the Permanent Court of International Justice. As space constraints dictate a narrowing of scope, I will concentrate upon the implications of these decisions for non-navigational uses of international watercourses.[1]

II. The River Oder Case[2]

The *River Oder* case concerned navigation rights, a subject of more interest and importance in the early twentieth century and before than

---

[1] Thus, judgments involving boundary issues will not be dealt with. Prominent among these is the recently decided *Case Concerning Kasikili/Sedudu Island (Botswana/Namibia)*, Judgment of 13 December 1999, available on the Court's website, at http://www.icj-cij.org, in which the Court found, by 11 votes to 4, that the boundary between Botswana and Namibia follows the line of deepest soundings in the northern channel of the Chobe River around Kasikili/Sedudu Island.

[2] The *Territorial Jurisdiction of the International Commission of the River Oder* case *(Czechoslovakia, Denmark, France, Germany, Great Britain, Sweden/Poland)*, PCIJ, Series A, No. 23, 1929, 5-46 (10 September 1929) (hereinafter, the *River Oder* case); Manley O. Hudson (ed.), *World Court Reports* II (1969), 602.

today. However, navigation does remain a significant use of international watercourses in nearly all regions of the world. One need only think of such rivers as the Rhine, the Danube, the Mekong, the Nile, the Congo, and the Mississippi for proof of that fact. In *River Oder*, the Permanent Court of International Justice was asked to determine whether, under the Treaty of Versailles of 1919, the jurisdiction of the International Commission of the Oder extended to a tributary of the Oder, the Warthe (Warta), and a subtributary that flows into the latter, the Netze. The upper reaches of both of these streams were situated in Polish territory.

Article 331 of the Treaty of Versailles provides that "all navigable parts of these river systems [including the Oder] which naturally provide more than one State with access to the sea", possess international status. Article 341 of the treaty placed the Oder under the administration of an International Commission,[3] whose task it was to "define the sections of the river or its tributaries to which the international régime shall be applied".

The principal question before the Court was whether the jurisdiction of the Commission extended to the portions of two above-mentioned tributaries situated in Polish territory.[4] Poland maintained that the Commission's jurisdiction should end where they cross the Polish border, while the other members of the Commission[5] contended that it should extend to the point at which each of the tributaries ceased to be navigable, even if that point were situated within Polish territory. The navigability of the Warthe and the Netze was not disputed, but Poland claimed that the sections of those rivers situated in Polish territory provided only Poland with "access to the sea" under Article 331 of the Treaty of Versailles.

The issue in the case thus concerned the competence of the Oder Commission in particular, and navigation rights in general. The decision is of particular interest, however, because the Court, having found that it was unable to answer the question before it based solely on the provisions of the treaty, resorted to "the principles underlying the matter to which the text refers", namely, those "governing international fluvial law in general...".[6] The Court acknowledged that providing upstream States with access to the sea played an important role "in the formation of the principle of freedom of navigation on so-called international

---

[3] The members of the Commission were Poland, Germany, Great Britain, Czechoslovakia, France, Denmark and Sweden.
[4] The *River Oder* case, see note 2, 16. A second question before the Court was, if the Commission's jurisdiction does so extend, "what is the law which should govern the determination of the upstream limits of this jurisdiction?" *Ibid.*
[5] Czechoslovakia, Denmark, France, Germany, Great Britain, and Sweden.
[6] The *River Oder* case, see note 2, 26.

rivers".[7] It then went on address the converse situation, which was the one involved in the dispute: whether the principle of freedom of navigation also provided downstream States with access to portions of tributaries situated wholly in upstream States. The Court found that

> "a solution of the problem has been sought not in the idea of a right of passage in favour of upstream States, but in that of a community of interest of riparian States. This community of interest in a navigable river becomes the basis of a common legal right, the essential features of which are the perfect equality of all riparian States in the user of the whole course of the river and the exclusion of any preferential privilege of any one riparian State in relation to the others. ... If the common legal right is based on the existence of a navigable waterway separating or traversing several States, it is evident that this common right extends to the whole navigable course of the river and does not stop short at the last frontier".[8]

The Court therefore held that under the treaty the jurisdiction of the Commission extended to the sections of the Oder tributaries that are situated in Polish territory.

While the dispute before the Court concerned questions of navigation, the characteristics of a river that led the Court to draw the above conclusions also support the proposition that States sharing an international watercourse have a "common legal right" in its non-navigational uses as well. Today, the best designation for this common right with regard to such uses is probably "equitable utilization". Thus the fact that the right is "common" does not imply that all riparians have exactly the same rights of use; their respective rights will depend on the facts and circumstances of each case, and may vary according to changed conditions.

As will be seen below, the applicability of the concept of "community of interest" to non-navigational uses has recently been confirmed by the International Court of Justice, in the *Gabčíkovo-Nagymaros* case.

III. The Diversion of Water from the Meuse[9]

In the words of the Permanent Court:

---

[7] *Ibid.*
[8] *Ibid.*, 27-28.
[9] PCIJ, Series A/B, No. 70, 1937, summarized in ILCYB 1974, Vol. 2, Part 2, 187, para. 1022 et seq.

"The Meuse is an international river. It rises in France ... leaves French territory near Givet, crosses Belgium, forms the frontier between the Netherlands and Belgium ... and enters Netherlands territory a few kilometres above Maastricht. Between Borgharen (a few kilometres below Maastricht) and Wessem-Maasbracht, the Meuse again forms the frontier between Belgium and the Netherlands, then below Wessem-Maasbracht both banks of the river are in Netherlands territory".[10]

Belgium and the Netherlands had concluded a treaty in 1863 "to settle permanently and definitively the régime governing diversions of water from the Meuse for the feeding of navigation canals and irrigation channels".[11] Article 1 of the treaty provided for construction in Netherlands territory below Maastricht of an intake from the Meuse which would feed all the canals then situated below Maastricht. Belgium in 1930 had begun the construction of a canal (the Albert Canal) which was to be fed by water drawn from the Meuse in Belgian territory above Maastricht. In 1936, the Netherlands instituted this case against Belgium, claiming that certain works planned or constructed by the latter in connection with the 1930 canal project violated or would violate the 1863 treaty. For its part, the Belgian Government in its counter-memorial contended that the Netherlands had violated the treaty by constructing a barrage, and also claimed that the flow of water to a certain canal constructed by the Netherlands would be subject to the treaty.

Although the Court noted that the parties had made reference in their written and oral pleadings to "the application of the general rules of international law as regards rivers", it declared: "The points at issue must all be determined solely by the interpretation and application of [the 1863] Treaty". While it is therefore difficult to draw general principles from the decision, the Court did make several observations that may be noted. First, with regard to the Netherlands' claim that Belgium's plans to divert water from the Meuse above Maastricht violated the Netherlands' right of supervision over diversions of water from the Meuse by means of the Maastricht intake,[12] the Court stated:

---

[10] PCIJ, Series A/B, No. 70, 1937, 9 and 10.

[11] Text of the treaty reprinted in PCIJ, Series C, No. 81, 15 et seq. See also ILCYB 1974, Vol. 2, Part 2, paras. 736-740.

[12] The gravamen of the Netherlands' complaint in this regard related to the fact that a portion of the new canal being constructed by Belgium (the Albert Canal) utilized the bed of the old Hasselt Canal, which was situated in Belgian territory below Maastricht and thus, under the treaty, was to be fed by water drawn from the Maastricht intake. The Netherlands claimed, in essence, that Belgium could not, consistently with the treaty, feed that

"There can be no doubt that, so far as the right of supervision is derived from the position of the intake on Netherlands territory, the Netherlands, as territorial sovereign, enjoys a right of supervision which Belgium cannot possess". The Court went on to make the following observation concerning diversion of Meuse waters from points other than the treaty feeder into canals not expressly covered by the treaty which are situated wholly in the territory of Belgium or the Netherlands:

> "As regards such canals, each of the two States is at liberty, in its own territory, to modify them, to enlarge them, to transform them, to fill them in and even to increase the volume of water in them from new sources, provided that the diversion of water at the treaty feeder and the volume of water to be discharged therefrom to maintain the normal level and flow in the Zuid-Willemsvaart [a canal situated partly in the Netherlands] is not affected".

It would thus appear that the Court recognized that the treaty régime for the supply of water to canals below Maastricht could not legally be impaired by upstream diversions of Meuse waters at points and for purposes not covered by the treaty. This may be regarded as an application of what has come to be known as the "no significant harm" rule, which is embodied in Article 7 of the 1997 United Nations Convention on the Law of the Non-Navigational Uses of International Watercourses.[13]

IV. The Gabčíkovo-Nagymaros Case[14]

The final case to be discussed is the only one concerning the use of an international watercourse that has been brought before the ICJ. While

---

portion of the Albert Canal which would be comprised of the old Hasselt Canal with water drawn from points other than the Maastricht intake. The reason for the Netherlands' concern seems to have been that it would not be able to supervise diversions in Belgian territory in the manner that would be possible at the Maastricht feeder; this element of having some degree of control over diversions through a supervisory power was presumably an aspect of the overall agreement that was important to the Netherlands.

[13] UN Doc. A/RES/51/229 of 21 May 1997, reprinted in *ILM* 36 (1997), 700.

[14] *Case Concerning the Gabčíkovo-Nagymaros Project (Hungary/Slovakia)*, Judgment of 25 September 1997, ICJ Reports 1997, 7 et seq., reprinted in *ILM* 37 (1998), 162 et seq. Also available on the Court's website, at http://www.icj-cij.org. The many articles on this case include a Symposium in *Y.B. Int'l Envtl. L.* 8 (1997), 3 et seq. The author served as counsel to Slovakia in the case.

this case also involved a treaty, it presented the Court with an opportunity to consider the influence of the customary international law of international watercourses upon the rights and obligations of the parties.

Hungary and Slovakia brought the case to the Court on 2 July 1993 by Special Agreement entered into in that year.[15] The case concerned a large project on the Danube River which was to have been constructed pursuant to a 1977 treaty between Hungary and Czechoslovakia.[16] As foreseen in the treaty, the project consisted of a series of dams and barrages on a stretch of approximately 200 kilometers of the Danube between Bratislava, the capital of Slovakia, and Budapest, the capital of Hungary.

For much of the length of this stretch (142 kilometers), the Danube forms the border between the two countries. Upstream of the point at which it begins to form the border it is for a short distance wholly within what is now Slovak territory; downstream of the border sector it passes into Hungary. According to the treaty, the principal purposes of the project are to improve navigation, provide flood protection and produce electricity. The treaty provides for the construction of two series of locks: one at Gabčíkovo on a 31-kilometer bypass canal, constructed pursuant to the treaty in Czechoslovak (now Slovak) territory, the other at Nagymaros, in Hungarian territory. In order to fill the bypass canal and thus operate the hydroelectric power plant and ship locks at Gabčíkovo, the treaty called for the Danube to be dammed at Dunakiliti, an installation which was Hungary's responsibility under the agreement.

Citing environmental concerns, Hungary stopped work on both the Nagymaros and the Dunakiliti portions of the project in 1989 and purported to terminate the treaty unilaterally in May 1992. Among Hungary's contentions was that water in the Dunakiliti reservoir would become stagnant, resulting in serious impairment of the quality of groundwater in the "Szigetköz" aquifer in Hungary.[17] Faced with this situation, and having already constructed substantial works in fulfillment of its obligations under the treaty, including much of the bypass canal and the dam at Gabčíkovo,[18] Czechoslovakia decided to put the

---

[15] Reprinted in *ILM* 32 (1993), 1293.
[16] Treaty Concerning the Construction and Operation of the Gabčíkovo-Nagymaros System of Locks, *ILM* 32 (1993), 1247.
[17] ICJ Reports 1999, para. 40.
[18] "In spring 1989, ... the Gabčíkovo dam was 85 per cent complete, and the bypass canal was between 60 per cent complete (downstream of Gabčíkovo) and 95 per cent complete (upstream of Gabčíkovo) and the dykes of the Dunakiliti-Hrušov reservoir were between 70 and 98 per cent complete, depending on the location". Hungary itself had completed 90 per cent of the work on the Dunakiliti dam. *Ibid.*, para. 31.

Gabčíkovo, or upper, part of the project into operation. This it accomplished by damming the Danube, beginning on 23 October 1992, at Čunovo, a point at which the river is wholly in Czechoslovak (now Slovak) territory. The Čunovo dam diverted most of the flow of the Danube (80 to 90 per cent) through the bypass canal to the point at which the canal rejoins the bed of the Danube. The Čunovo dam also created a reservoir, although a smaller one than had been foreseen under the 1977 treaty. The Čunovo dam and related structures are known as "Variant C" – the designation given one of several possible "provisional solutions" considered by Czechoslovakia after Hungary's cessation of work on the project. On 1 January 1993, Slovakia became an independent State.

As the President of the ICJ has indicated, the case "proved to be compendious in terms of the range of legal issues it summoned up: the law of treaties, of state responsibility, of international watercourses, of state succession and environmental law".[19] The Court essentially held that both parties had breached their international obligations, Hungary by stopping work on the project and Slovakia, as successor to Czechoslovakia, by putting Variant C into operation. The Court also held that Hungary's purported termination of the 1977 treaty was ineffective, that the project – including the Čunovo dam – would have to be placed under a joint operational régime, and that, unless the parties agreed otherwise, they were obligated to compensate each other for the injurious consequences of their respective breaches.

The Court's opinion and some of the separate and dissenting opinions – including one by Judge Oda, to be touched upon below – contain a wealth of material pertaining to the law of international watercourses and international environmental law. I will focus here upon only some of the Court's statements relating to the former field.

The Court began by noting that the riparian States of the Danube have for centuries engaged in the construction of works for the development and utilization of the river, as well as for the protection of their territories and citizens from floods. It observed that the cumulative effects of these activities on the water and the environment had not all been favorable, and that the resulting problems could be addressed only by international cooperation. With regard to the 1977 treaty in particular, the Court noted that it contained three provisions relating to water

---

[19] Keynote address of Judge Stephen M. Schwebel, "The Influence of the International Court of Justice on the Work of the International Law Commission and the Influence of the Commission on the Work of the Court", delivered at the United Nations Colloquium on Progressive Development and Codification of International Law to Commemorate the Fiftieth Anniversary of the International Law Commission, United Nations, New York, 28 October 1997.

quality and environmental protection: Articles 15 (water quality), 19 (protection of nature) and 20 (protection of fishing interests). Furthermore, the treaty provides in Article 16 for the maintenance of the bed of the Danube. Judge Oda stated in his dissenting opinion that "the Parties are under an obligation in their mutual relations, under Articles 15, 16 and 19 of the 1977 Treaty, and perhaps in relations with third parties, under an obligation in general law concerning environmental protection, to preserve the environment in the region of the river Danube".[20] The Court did not disagree with this finding.

Some of the Court's statements concerning the environment also shed light upon its attitude towards the law of international watercourses. Hungary had contended, *inter alia*, that even if its stoppage of work was in breach of the 1977 treaty, it was justified by a "state of ecological necessity" to which the implementation of the project might have given rise.[21] In the course of addressing this argument, the Court stated:

> "The Court recalls that it has recently had occasion to stress, in the following terms, the great significance that it attaches to respect for the environment, not only for States but also for the whole of mankind:
>
> > 'the environment is not an abstraction but represents the living space, the quality of life and the very health of human beings, including generations unborn. The existence of the general obligation of States to ensure that activities within their jurisdiction and control respect the environment of other States or of areas beyond national control is now part of the corpus of international law relating to the environment.'" [22] [23]

In this passage, the Court essentially confirms that the second limb of Principle 21 of the Stockholm Declaration[24] forms part of customary

---

[20] ICJ Reports 1997, 168, para. 33, dissenting opinion of Judge Oda.

[21] This argument was based on the doctrine of the "state of necessity" as a "circumstance precluding wrongfulness". See Article 33 of the Draft Articles on State Responsibility adopted by the International Law Commission on first reading, ILCYB 1980, Vol. II, Part 2, 34; ILC Reports 1996, 137; and ICJ Reports 1997, para. 50.

[22] The *Legality of the Threat or Use of Nuclear Weapons* case, Advisory Opinion, ICJ Reports 1996, 241-242, para. 29.

[23] ICJ Reports 1997, paras. 53, 41.

[24] Principle 21 reads:
> "States have, in accordance with the Charter of the United Nations and the principles of international law, the sovereign right to exploit their own resources pursuant to their own environmental policies, and the

international law. It is noteworthy that this is as close as the Court came in its judgment to endorsing a "no harm" principle in the field of international watercourses. Indeed, it is striking that the Court did not refer to such a principle at all in the context of the allocation of shared water resources.

The Court summed up its findings on Hungary's ecological state of necessity theory as follows:

> "The Hungarian argument on the state of necessity could not convince the Court unless it was at least proven that a real, 'grave' and 'imminent' 'peril' existed in 1989 and that the measures taken by Hungary were the only possible response to it.
> ...
>
> The Court concludes ... that, with respect to both Nagymaros and Gabčíkovo, the perils invoked by Hungary, without prejudging their possible gravity, were not sufficiently established in 1989, nor were they 'imminent'; and that Hungary had available to it at that time means of responding to these perceived perils other than the suspension and abandonment of works with which it had been entrusted. What is more, negotiations were under way which might have led to a review of the Project and the extension of some of its time-limits, without there being need to abandon it. ...
>
> [A]lthough the principal object of the 1977 Treaty was the construction of a System of Locks for the production of electricity, improvement of navigation on the Danube and protection against flooding, the need to ensure the protection of the environment had not escaped the parties, as can be seen from Articles 15, 19 and 20 of the Treaty".[25]

The Court thus strongly endorsed the stability of treaty régimes and the need to work within them to address perceived problems resulting from their implementation. But the Court's holding on Hungary's state of necessity argument suggests that it will be very difficult, if not impossible as a practical matter, for a State to rely upon the state of necessity doctrine to preclude the wrongfulness of breaches committed because of the fear of threats to groundwater. Such threats often take

---

responsibility to ensure that activities within their jurisdiction or control do not cause damage to the environment of other States or of areas beyond the limits of national jurisdiction". UN Doc. A/CONF.48/14, 2 (1972).

[25] ICJ Reports 1997, paras. 54, 57.

long periods of time to materialize and thus would generally not be regarded as being "imminent" under the Court's analysis. Given the fact that groundwater as a source of water supply will increase in importance in coming decades, the question arises whether traditional rules of international law are adequate to deal with the special characteristics of this resource. Certainly the doctrine of state of necessity as a circumstance precluding wrongfulness hardly seems sufficient, given that contaminants that make their way into an aquifer may not materialize in harm to humans for years. Thus, "imminence" will be virtually impossible to prove. One wonders whether the precautionary principle could not play a role in this context, although perhaps more as a tool for interpreting treaty obligations than as a means of exculpating a State that is in breach of those obligations.[26]

The Court later turned to the legality of "Variant C". It stated that the operation of these works "led Czechoslovakia to appropriate, essentially for its use and benefit, between 80 and 90 per cent of the waters of the Danube before returning them to the main bed of the river, despite the fact that the Danube is ... an international boundary river".[27] The Court continued:

> "Czechoslovakia submitted that Variant C was essentially no more than what Hungary had already agreed to and that the only modifications made were those which had become necessary by virtue of Hungary's decision not to implement its treaty obligations. It is true that Hungary, in concluding the 1977 Treaty, had agreed to the damming of the Danube and the diversion of its waters into the bypass canal. But it was only in the context of a joint operation and a sharing of its benefits that Hungary had given its consent. The suspension and withdrawal of that consent constituted a violation of Hungary's legal obligations, demonstrating, as it did, the refusal by Hungary of joint operation; but that cannot mean that Hungary forfeited its *basic right to an equitable and reasonable sharing of the resources of an international watercourse*".[28]

---

[26] Later in the judgment the Court stated that it was "mindful that, in the field of environmental protection, vigilance and prevention are required on account of the often irreversible character of damage to the environment and of the limitations inherent in the very mechanism of reparation of this type of damage". *Ibid.*, para. 140. In referring to "vigilance" and irreversibility, the Court seems to have the precautionary principle in mind, although it does not refer to the principle by name.

[27] *Ibid.*, para. 78.

[28] *Ibid.*, emphasis added.

Here the Court reinforces the status in customary international law of the principle of equitable utilization, and, in particular, a State's "basic right" to an equitable and reasonable share of the "resources" of an international watercourse. These resources include not only the water itself, but also such benefits as electric power it could be used to produce, fisheries, and recreation. While the precise implications of describing this right as a "basic" one are not entirely clear, it at least seems evident that a State will not lightly be regarded as having waived or "forfeited" the right – even where the deprivation of the State's equitable share is based upon detrimental reliance on that State's earlier undertakings and conduct.

The Court later turned to the question whether Czechoslovakia's putting into operation of Variant C could be justified as a countermeasure.[29] The Court stated, *inter alia*, as follows:

> "In the view of the Court, an important consideration is that the effects of a countermeasure must be commensurate with the injury suffered, taking account of the rights in question.
>
> In 1929, the Permanent Court of International Justice, with regard to navigation on the River Oder, stated as follows:
>
> '[The] community of interest in a navigable river becomes the basis of a common legal right, the essential features of which are the perfect equality of all riparian States in the user of the whole course of the river and the exclusion of any preferential privilege of any one riparian State in relation to the others'.[30]
>
> Modern development of international law has strengthened this principle for non-navigational uses of international watercourses as well, as evidenced by the adoption of the Convention of 21 May 1997 on the Law of the Non-Navigational Uses of International Watercourses by the United Nations General Assembly.
>
> The Court considers that Czechoslovakia, by unilaterally assuming control of a shared resource, and thereby depriving

---

[29] That is to say, a measure that would be unlawful but for the fact that it was a legitimate response to the prior wrongful act of another State. See generally Article 47 of the complete set of Draft Articles on State Responsibility adopted by the International Law Commission on first reading in 1996, ILC Reports 1996, 125 et seq., 144.

[30] The *Territorial Jurisdiction of the International Commission of the River Oder* case, Judgment No. 16, PCIJ, Series A, No. 23, 1929, 27.

Hungary of its right to an equitable and reasonable share of the natural resources of the Danube – with the continuing effects of the diversion of these waters on the ecology of the riparian area of the Szigetköz – failed to respect the proportionality which is required by international law".[31]

Several aspects of these passages call for brief comment. First, the Court breathes new life into the powerful idea of "community of interest", originally enunciated in the *River Oder* case as we have seen, applying it for the first time to non-navigational uses of international watercourses. While its precise legal implications are not clear, the theory at least emphasizes the shared nature of interests in an international watercourse and the resulting need to cooperate with other riparians in its use, development and protection. Moreover, the Court used the community of interest theory as support for its conclusion that one riparian may not "unilaterally assum[e] control" of a "shared resource", even where the "control" that was "assumed" produced results closely akin to what the other riparian had agreed to in a treaty, and even as a measure taken in response to an internationally wrongful act by the other riparian – specifically, repudiation of a treaty and stoppage of work thereunder. Thus, *a fortiori*, such an assumption of control would not be justifiable in the absence of a breach by the other riparian. This has important implications for the relations between upstream and downstream riparians on an international watercourse. But as to the case itself, it is remarkable that the Court seemed to believe it was obvious that Slovakia's diversion was *per se* a disproportionate response to Hungary's internationally wrongful act (i.e., its breach of the 1977 treaty). And it arrived at this conclusion on the basis of the customary international law of shared water resources – *in fine*, equitable utilization – rather than the 1977 treaty.[32]

Secondly, as evidence of the modern vitality of the principle of community of interest in the field of non-navigational uses of international watercourses, the Court cites the 1997 UN Convention. This is remarkable in that the Convention had only been concluded four months earlier, and, by the date the judgment was rendered, had been signed by only three States.[33] The Court sheds no further light on exactly why it regards the newly minted treaty as evidence of the principle and "modern development of international law" in the field. The answer

---

[31] ICJ Reports 1997, paras. 85 and 86.
[32] In discussing the parties' obligations of reparation the Court again had recourse to equitable utilization. It stated that the parties' respective obligations of reparation would be fulfilled, *inter alia*, if they implement the project "in an equitable and reasonable manner". *Ibid.*, para. 150.
[33] South Africa, Syria and Venezuela.

International Watercourses in the Jurisprudence of the ICJ 1067

might lie to some extent in the process that produced the Convention: twenty years' work by the International Law Commission (ILC), culminating in a diplomatic negotiation which produced an agreement that closely tracks the ILC's draft articles.[34] Whatever the case may be, the Court's invocation of the Convention constitutes a strong endorsement of it as an authoritative instrument in the field, and seems likely to lead States to refer to it in support of their positions concerning internationally shared water resources. Later in the judgment, in discussing the re-establishment of the joint treaty régime, the Court quoted Article 5(2) of the UN Convention, using it as a kind of normative guidance as to the manner in which the parties should cooperate with regard to the project.[35]

Finally, in discussing the parties' obligations of reparation, the Court states that the consequences of the parties' wrongful acts will be wiped out "as far as possible" if they implement the "multi-purpose program" in an "equitable and reasonable manner". It does not say that the parties must return to the project as originally foreseen in the 1977 treaty, only that the entire program of "use, development and protection" must be implemented equitably and reasonably. It explains that one way of doing this would be to continue to operate the Gabčíkovo sector of the project through the works at Čunovo (Variant C) and "not to build works at Nagymaros". This Solomonic solution would give each party something it wanted, but would also call for each to sacrifice one of its key objectives: Hungary would not have to build the Nagymaros dam, but would have to live with the Čunovo dam and the Gabčíkovo sector of the project (although it would operate it jointly with Slovakia); Slovakia would be able to keep the Čunovo dam, but would have to live without the peak power that could have been produced if the Nagymaros dam had been constructed.

At the time of this writing, the parties have as yet not been able to agree on how to give effect to the Court's judgment. On 3 September 1998, Slovakia submitted to the Court a Request for an Additional Judgment Pursuant to Article 5(3) of the Special Agreement of 7 April 1993, to which Hungary responded on 7 December 1998 with a Written Statement. As of December 2000, the Court had not acted formally on these submissions, instead requesting that the parties keep it informed of the progress of their negotiations on the implementation of the judgment.

Finally, a word should be said about the dissenting opinion of our honoree, Judge Oda. In essence, Judge Oda was of the view that

---

[34] Another factor could have been the familiarity of President Schwebel, a former special rapporteur of the ILC on international watercourses, and other members of the Court that had served on the ILC (Judges Bedjaoui, Shi and Koroma), with the Convention's provisions and the law in the field.
[35] ICJ Reports 1997, para. 147.

Hungary had breached the 1977 treaty by stopping work in 1989 but that Czechoslovakia had not committed a breach by putting the Čunovo dam into operation. Czechoslovakia's action in this regard "implied nothing other than the accomplishment of the original Project". The diversion of Danube waters into the bypass canal "was the essence of the whole Project with which Hungary was in full agreement".[36] On the other hand, Judge Oda wrote that the way in which the Čunovo dam was operated by Slovakia "seems to have led to an unfair division of the waters between the old Danube river bed and the bypass canal"[37] – a division for which Slovakia could bear responsibility.

V. Conclusion

The International Court of Justice and its predecessor have not had occasion to decide a great number of cases concerning international watercourses. All of the disputes the two courts have resolved have concerned situations governed by treaties. Neither tribunal has been asked to decide a case concerning watercourses solely on the basis of rules of customary international law. This is not unusual, however. Nor is it unusual for the World Court – like any other international tribunal – to use rules of customary international law in interpreting and filling gaps in a treaty and to determine consequences of a breach.

The Court's pronouncements of the latter kind – particularly those in the *Gabčíkovo-Nagymaros* case – are of great interest to States and others interested in the field of international watercourses. It seems likely that they will be influential in relations between States concerning increasingly precious shared fresh water resources. For his part, Judge Oda has demonstrated in this field as he has in others an acutely practical approach, but one which is entirely in keeping with, and indeed reinforces, the international legal order. The international community is fortunate to have a person of his intellect, integrity and independence on the Court.

---

[36] ICJ Reports 1997, 164, para. 23, dissenting opinion of Judge Oda.
[37] *Ibid.*, 166, para. 26, dissenting opinion of Judge Oda.

On the Quasi-Normative Effect of Maritime Boundary Agreements

Maurice Mendelson[*]

It is a great pleasure to contribute to this *Festschrift* for Shigeru Oda, the significance of whose contribution to international law, over many years, needs no elaboration from me here.

This essay deals with two subjects close to Judge Oda's heart: the law of the sea and the sources of international law. Specifically, it examines the legal significance of bilateral maritime delimitation agreements. Obviously, they are binding on the parties. We also have it on good authority – though the precise reasons merit further examination – that these agreements do not contribute to the formation of rules of general customary law. But does that mean that they have no significance at all in the general legal process, that juridically it is as if they had never existed? At first sight, it seems that this must be the case, and that they are no more than historical anecdotes, of interest to those dealing with the sea area in question but otherwise of no concern to lawyers. But, if so, it is a little surprising that counsel before the International Court of Justice and arbitral tribunals, commentators on the law of the sea, and government officials engaged in negotiations, are so interested in these "precedents". The purpose of this paper is to explore, a little more closely than usual, exactly what legal status they have. In particular, the tentative suggestion is made that they may have a sort of quasi-legal significance in general international law: that although they are not "binding" outside their own confines, collectively they do nevertheless constitute signposts, as it were; or in other words,

---

[*] Judge David Anderson of the International Tribunal of the Law of the Sea, Visiting Professor of International Law at University College London, kindly commented on an earlier draft of this paper. Responsibility for the views expressed here is, however, entirely my own.

that they show us what international society as a whole considers a *reasonable* solution to various sorts of problems. This is particularly important in an area of the law where – according to the prevailing wisdom – "equitable",[1] i.e., fair and reasonable, outcomes are the goal and the only generally binding criterion.

But before we reach that issue, it is necessary to deal with some preliminary considerations. I first examine what it means to say that bilateral delimitation agreements are binding only on the parties. Next, I examine why exactly they do not constitute "building bricks" of a more general customary rule. I then briefly reflect on the nature of the equitable principles that are supposed to guide delimitation (at least beyond the territorial sea), before elaborating on my thesis of the quasi-legal normative nature of bilateral agreements in this context. Space does not permit an exploration of the related question of the precedential value of the case-law of international courts and tribunals in relation to maritime boundary delimitation: whilst there are close analogies with delimitation by agreement, there are also differences which I hope to explore on another occasion.

First, then, when we say that delimitation agreements are binding on the parties, what exactly does this mean? It certainly means that the actual parties are bound to respect the agreement in their relations *with each other*.[2] This also applies to successors of the parties.[3] Third parties are also legally bound insofar as the boundary treaty establishes an "objective régime".[4] Thus, if opposite States A and B have proclaimed exclusive economic zones and agreed the boundary between their zones, the master of a boat flying the flag of State C which is arrested for illegal fishing within the EEZ of A cannot plead that the waters were really within the jurisdiction of B, and neither can State C base a diplomatic protest on such

---

[1] For the purpose of this paper, it is unnecessary to consider what, if any, are the differences between "equity", "equitable delimitation", "equitable principles of delimitation" and "delimitation in order to achieve an equitable solution"; these and similar terms will be used interchangeably here.

[2] Whether, and to what extent, they are ever free to depart from their treaty unilaterally depends on the general law of treaties; and as is well known, in the law of treaties, both customary and conventional, agreements establishing an international boundary (whether maritime or terrestrial) are afforded a special sanctity. Cf. Vienna Convention on the Law of Treaties (1969), Art. 62(2)(a).

[3] Cf. Vienna Convention on State Succession to Treaties (1978), Art. 11.

[4] Cf. *Legal Consequences for States of the Continued Presence of South Africa in Namibia (South West Africa) notwithstanding Security Council Resolution 276 (1970)*, ICJ Reports 1971, 17 et seq., 56, para. 126.

an argument. Similarly, minerals extracted from the seabed within a zone which State X has agreed is subject to the sovereign rights of State Y "belong" to Y. This assumes that no third State has claims of its own in the area, for if it does, the régime agreed in the boundary treaty cannot, of course, prejudice its position.

But to say that a maritime delimitation agreement is binding on the parties does not mean that State A is bound to apply the principles adopted in its boundary treaties with State B in relation to *other* States. Even if that treaty were expressly to assert, for instance, that it has applied the technique of equidistance, this would not oblige State A (or B) to apply this technique in a delimitation with some third State. There would seem to be at least two reasons why this is so. First, there is the well-known rule that treaties do not benefit (or burden) third parties.[5] Secondly, the law of the sea in general, and the law relating to maritime boundaries in particular, is not *jus cogens*, but *dispositivum*, and so we cannot conclude from the making of a boundary in a particular way that the parties claimed or recognised that it was in any way obligatory for them to use that particular method.[6] Indeed, a negotiating State is perfectly entitled to take into consideration matters which have nothing whatsoever to do with what it considers to be its entitlements, such as a desire to improve political relations with its neighbour or to provide it with a *quid pro quo* for a concession in the field of economic relations. So far as I am aware, no State has ever, either in pleadings before an international tribunal or in bilateral negotiations, argued that the fact that its opponent has used a particular method or approach in a treaty with a third State obliges it to accept the same method or approach in the instant case. One might invoke the apparent inconsistency in order to cause a little embarrassment, but that is all.

Is the position different if the States concerned, as well as adopting a particular approach or method, actually asserted that it was obligatory as a matter of general law? In principle, yes. Customary international law is built up through the mutual claims and responses of States, express or implied, and whilst a single claim and a single response cannot of themselves create general international law, they can certainly contribute to its formation.[7] In addition, in certain

---

[5] Cf. Vienna Convention on the Law of Treaties (1969), Arts. 34-36.
[6] The second (but not the first) reason also means that, in a subsequent delimitation between the *same* States (e.g., the delimitation of other areas of the EEZ), neither party is bound by the type of solution reached in the first treaty.
[7] Cf. M. H. Mendelson, "The Formation of Customary International Law", *RdC* 272 (1998), 155 et seq., 189-191, 204-207.

circumstances, recognition by a State that certain conduct is required or permitted (as the case may be) by the general law may preclude it from afterwards asserting the contrary. Probably it is partly for this reason that States do not in general make assertions about the general law in their boundary agreements – indeed, they sometimes seem to go out of their way to conceal even what method was employed. But examples are not wholly unknown. Thus, the agreements on boundary delimitation between Belgium on the one hand, and the France and United Kingdom, respectively, on the other, expressly stated that they had "taken full account of the current rules of international law on international boundaries in order to achieve an equitable solution", which seemingly helped Belgium in subsequent negotiations with the Netherlands.[8] And according to Elferink, the reason why the preamble to the 1978 agreement between Turkey and the Soviet Union on the delimitation of the continental shelf in the Black Sea asserts that the boundary was "based on the principles of equity, [t]aking into consideration the relevant principles and norms of international law" was that Turkey was anxious for a statement, which could strengthen its position *vis-à-vis* Greece, to the effect that what the general law requires is equitable delimitation and that equidistance is merely one possible technique, appropriate in only some circumstances.[9]

The next question to consider is whether a succession of bilateral delimitation agreements in a similar vein can in some way constitute evidence of a rule of customary law, or (help to) generate a new one. As it happens, the leading case on the question of the legal effect of a series of bilateral agreements on the general law comes from the field of maritime delimitation: the *North Sea Continental Shelf* cases.[10] By way of background, it should be pointed out that, at the time that the Geneva Convention on the Continental Shelf (1958) was adopted, it

---

[8] See J. Charney and L. Alexander (eds.), *International Maritime Boundaries*, Vol. II, 1993, 1899 and 1909. Cf. the Declaration appended to the France-UK agreement on the delimitation of the territorial sea in the Straits of Dover in 1988, which went out of its way to assert that a right of transit passage through straits was "generally accepted in the current state of international law" – though this is not, of course, an assertion about the law relating to boundary delimitation as such. *Ibid.*, 1754.

[9] "The Law and Politics of the Maritime Boundary Delimitations of the Russian Federation: Part 1", *Int'l J. Marine & Coastal L.* 11 (1996), 533 et seq., 561-562. One might surmise that the preambular statement was particularly important to Turkey, because the technique used was in fact based on equidistance.

[10] ICJ Reports 1969, 3 et seq.

was widely supposed that Article 6[11] meant that, if the parties could not agree on the boundary, it would be the equidistance/median line, *unless* it could be established that there were special circumstances. This was also, apparently, the tacit assumption of the Court. It is true that this interpretation was subsequently undermined by a Court of Arbitration in its 1977 award in the *France-UK Arbitration on the Delimitation of the Continental Shelf*, which (whether it intended to or not[12]) lent support to the notion that Article 6 afforded no priority to equidistance and that the conventional rule, like the customary one, was simply that equitable principles should apply; it is also true that the ICJ itself adopted this interpretation in a series of later cases.[13] But this is to benefit from hindsight.

In coming to the conclusion that Article 6 was not declaratory of an existing customary rule and did not help to crystallise an emerging one, the Court did not have to deal with bilateral treaties. Neither did it have to do so when it considered (to the extent that it did) whether the Continental Shelf Convention could have generated a new rule "of its own impact".[14] It did, however, consider such treaties in dealing with the Danish-Dutch submission that the substance of Article 6 had become part of customary law subsequently to its adoption, as was demonstrated, they alleged, not only by the quantity of ratifications of the Continental Shelf Convention within a relatively short period, but

---

[11] The article reads in pertinent part as follows:
   "1. Where the same continental shelf is adjacent to the territories of two or more States whose coasts are opposite each other, the boundary of the continental shelf appertaining to such States shall be determined by agreement between them. In the absence of agreement, and unless another boundary line is justified by special circumstances, the boundary is the median line, every point of which is equidistant from the nearest point of the baselines from which the breadth of the territorial sea of each State is measured.
   2. Where the same continental shelf is adjacent to the territories of two adjacent States, the boundary of the continental shelf shall be determined by agreement between them. In the absence of agreement, and unless another boundary line is justified by special circumstances, the boundary shall be determined by application of the principle of equidistance from the nearest point of the baselines from which the breadth of the territorial sea of each State is measured".

[12] See the dissenting Opinion of Judge Oda in the *Continental Shelf (Libya/Malta)* case, ICJ Reports 1985, 123 et seq., 144-148, 163-169.

[13] See footnotes 30 and 31 below and accompanying text.

[14] ICJ Reports 1969, 3 et seq., 42, para. 70. For a detailed discussion of those aspects of the Judgment which dealt with customary international law, see Mendelson, note 7, Chapter IV.

also by a number of unilateral proclamations and bilateral agreements in which the basis of delimitation of the continental shelf was equidistance.

In dealing with this submission, the Court observed[15] that there were "some fifteen cases [which] have been cited in the course of the present proceedings, occurring mostly since the signature of the 1958 Geneva Convention, in which continental shelf boundaries have been delimited according to the equidistance principle ... or else the delimitation was foreshadowed but has not yet been carried out". The Court was not impressed with this number, but this was not its only objection. It went on:

> "But even if these various cases constituted more than a very small proportion of those potentially calling for delimitation in the world as a whole, the Court would not think it necessary to enumerate or evaluate them separately, since there are, *a priori*, several grounds which deprive them of weight as precedents in the present context.
>
> To begin with, over half the States concerned, whether acting unilaterally or conjointly, were or shortly became parties to the Geneva Convention, and were therefore presumably, so far as they were concerned, acting actually or potentially in the application of the Convention. From their action no inference could legitimately be drawn as to the existence of a rule of customary international law in favour of the equidistance principle".[16]

Pausing there, although this part of the Court's reasoning is widely assumed to be correct, even by those who criticise other aspects of the decision, it is in fact a little puzzling. What is right is the Court's unstated premise that the practice of parties to a convention, *vis-à-vis* each other, should not count towards the formation of a customary rule, because it is practice taking place under the convention, not in the arena of customary law.[17] Thus the mutual grant of tariff concessions under the General Agreement on Tariffs and Trade, for instance, cannot be relied upon by or against a non-party as a rule of customary

---

[15] After eliminating certain irrelevant cases.
[16] ICJ Reports 1969, 3 et seq., 43-45, paras. 75-81.
[17] The practice of treaty parties in relation to non-parties (and *vice versa*) is, of course, a different matter, as is the practice of non-parties amongst themselves.

law. But even if the general proposition is correct, its application to bilateral delimitation agreements between parties to the Geneva Convention was rather more dubious. Because even if, as I have suggested above, the treaty rule was assumed in 1969 to be that, in the absence of special circumstances, equidistance would apply, the agreement of the parties on their boundary would have overridden any rule of general law applicable in the absence of agreement. Therefore, if a pair of States agreed on a boundary based on equidistance, for instance, one would not have been entitled to assume that this was because the Convention obliged them to: it did not, and they would have been free to choose whatever criterion they wished (including, incidentally, disregarding relevant special circumstances).[18] Indeed, it was precisely because Article 6 did not lay down a clear rule, but rather required agreement-or-failing-this-equisdistance-unless-there-are-special-circumstances (so to speak), that the Court, in another context, held that this article did not have a "fundamentally norm-creating character".[19] Like some others, I have ventured to question that characterisation of Article 6[20]; but given that the Court made it, it should at least have done so consistently. In short, even if the Court was correct in discounting this group of bilateral agreements, it may have done so for the wrong reason. Strictly speaking, it was not so much that the conclusion of equidistance agreements by parties to the Geneva Convention constituted practice under the Convention, as that the Convention neither required nor prohibited their conclusion, and so the content of these bilateral treaties told us nothing about the obligations of the parties under even multilateral treaty law, let alone customary law.

Whether or not the point just made about the Court's discounting of agreements between parties to the Geneva Convention is correct, it may in any event have been questionable, strictly speaking, for the Court to treat in the same way the delimitation agreements of those that "shortly [afterwards] became parties to the Geneva Convention, and were therefore presumably, so far as they were concerned, acting ... potentially in the application of the Convention". Even if we assume, for the sake of argument, that Article 18(a) of the Vienna Convention

---

[18] No doubt, the belief that, if they could not come to an agreement, equidistance (in the absence of special circumstances) was the Convention norm might have encouraged them to use the same criterion bilaterally, but this is not the same as saying that they were *obliged* by the Convention to do so.
[19] ICJ Reports 1969, 3 et seq., 41-42, paras. 71-72.
[20] Mendelson, see note 7, 318-321.

on the Law of Treaties 1969[21] reflects customary law, there is nothing in that rule which would have prevented, say, States which had signed the Geneva Convention but not yet ratified it from concluding a bilateral treaty which diverged from the equidistance-special circumstances rule. This is not only because such an agreement would not be incompatible with the object and purpose of the Convention; it is also because the Convention itself envisaged, and did not prohibit, such agreements.

Be that as it may, the Court's reductive reasoning still left a few equidistance agreements where one or both of the parties were not (and were not shortly to become) parties to the Convention. Were these agreements evidence that a new rule of customary law, similar in content to that contained in Article 6, had emerged? Or, perhaps more accurately, had they contributed to the emergence of such a customary rule? The Court went on:

> "As regards those States, on the other hand, which were not, and have not become parties to the Convention, the basis of their action can only be problematical and must remain entirely speculative. Clearly they were not applying the Convention. But from that no inference could justifiably be drawn that they believed themselves to be applying a mandatory rule of customary international law. There is not a shred of evidence that they did and, as has been seen [referring to the Court's previous observations on the convenience and objective character of the equidistance method], there is no lack of other reasons for using the equidistance method, so that acting, or agreeing to act in a certain way, does not of itself demonstrate anything of a juridical nature.
>
> [E]ven if these instances ... were much more numerous than they in fact are, they would not, even in the aggregate, suffice to constitute the *opinio juris*; – for, in order to achieve this result, two conditions must be fulfilled. Not only must the acts concerned amount to a settled practice, but they must also be such, or be carried out in such a way, as to be evidence of a belief that this practice is rendered obligatory by the existence

---

[21] "A State is obliged to refrain from acts which would defeat the object and purpose of a treaty when: (a) it has signed the treaty or has exchanged instruments constituting the treaty subject to ratification, acceptance or approval, until it shall have made its intention clear not to become a party to the treaty...".

of a rule requiring it. The need for such a belief, i.e., the existence of a subjective element, is implicit in the very notion of the *opinio juris sive necessitatis*. The States concerned must therefore feel that they are conforming to what amounts to a legal obligation. The frequency, or even habitual character of the acts is not in itself enough. There are many international acts, e.g., in the field of ceremonial and protocol, which are performed almost invariably, but which are motivated only by considerations of courtesy, convenience or tradition, and not by any sense of legal duty".[22]

Departing from the received wisdom, I have ventured to suggest elsewhere that there is no *general* necessity to prove the existence of *opinio juris:* normally a constant and uniform practice of sufficient generality will suffice for the formation of a customary rule.[23] However, even I conceded that there were cases, such as those where the practice was accompanied by a disclaimer or was ambiguous, where reference to the lack of *opinio juris*, whilst perhaps a little unsubtle, could be useful. This was one such case, because the delimitation agreements were inherently ambiguous. Bearing in mind that the object of the inquiry was to determine whether these agreements had contributed to the formation of a new customary rule (or corroborated its emergence), the simple answer was that they did not. Whatever the customary rule might be, it was common knowledge that it could be departed from by agreement. Therefore, in the absence of evidence as to the understanding of the various parties about their obligations in customary law, their agreements revealed nothing about those customary obligations. Suppose that A argues that he is entitled to 75 per cent of a loaf of bread, and B only to 25 per cent, whereas B responds that he (B) is entitled to 80 per cent, and A only to 20 per cent. If they go on arguing, the loaf will become stale and useless to both of them. So each may conclude that "half a loaf is better than none", but that does not mean that either disputant concedes that he would not have been legally entitled to more. *Mutatis mutandis*, the same can be said of bilateral shelf delimitation. Oil and gas companies are generally reluctant to start drilling in contested areas, so two States may very well agree to "go halves" without this meaning that the general law would not entitle one or other to more, which could indeed be awarded if they could agree to adjudication or arbitration and were willing to undertake the expense of time and money involved. One can

---

[22] ICJ Reports 1969, 3 et seq., 43-44, paras. 76-77.
[23] Mendelson, see note 7, Chapter III.

perhaps put this in another way by saying that the bilateral agreements were referable only to themselves, not to customary law, and this is why they did not count towards the formation of the latter. But, however one puts it, clearly the Court was right to refuse to count this group of treaties.

The Court's view that the conclusion of bilateral delimitation agreements tells us nothing about the general law was reiterated shortly afterwards in the context of a different type of bilateral agreement. In the *Barcelona Traction* case (Second Phase),[24] it refused to treat as precedents (for lifting the corporate veil in favour of shareholders of the company allegedly harmed by wrongful state action):

> "the various arrangements made in respect of compensation for the nationalization of foreign property. Their rationale ... derived as it is from structural changes in a State's economy, differs from that of any normally applicable provisions. Specific agreements have been reached to meet specific situations, and the terms have varied from case to case. Far from evidencing any norm as to the classes of beneficiaries of compensation, such arrangements are *sui generis* and provide no guide in the present case".

One can generalise these propositions by saying, in the words of Section 25 of the International Law Association's London Principles on the Formation of General Customary International Law, that "[t]here is no presumption that a succession of similar treaty provisions gives rise to a new customary rule with the same content".[25] As has already been indicated, it might in theory be otherwise if the treaties concerned made statements about the *general* law and there were enough of them; but in the field of maritime delimitation such statements are, in fact, rare. Therefore, most of the relatively few commentators who have studied the legal effects of a series of maritime boundary treaties have either concluded or assumed that they have no legal force outside the bilateral obligations (and, I would add, the objective local legal régime) which they create.[26]

---

[24] ICJ Reports, 1970, 3 et seq., 40, para. 61.

[25] International Law Association, Report of 69th Conference (2000), 712-777; also available at http://www.ila-hq.org under London Conference, Committee on the Formation of Customary (General) International Law, Report and Resolution.

[26] A few seem to think that these agreements constitute state practice and thus, if sufficiently uniform and widespread, could give rise to general

Whilst agreeing with the conclusion that they do not constitute rules of customary law about delimitation, I would like to tentatively to suggest that uniform patterns of treating certain features and problems could amount, if not to binding legal rules, then at any rate to *indications* of the sort of solution that the generality of States think equitable or reasonable. Such indications are not obligatory for governments negotiating a boundary, not least because they retain their freedom to agree on whatever they wish. Neither are these indications binding as such on a third-party decision maker, such as a court or arbitral tribunal, which remains free to decide that, in the particular circumstances of the case, equity demands a different solution. But still, they do represent at least indications, and I suggest that a third party either should or will normally take them into consideration. I hesitate to go so far as to assert that there is a (rebuttable) legal presumption in favour of applying generally accepted solutions, but simply that in practical terms they represent the starting point. If this is correct, it indirectly applies also to those who negotiate boundary treaties because, even if they retain the freedom to diverge from the normal pattern, that pattern is part of the background and they run the risk that, if they do not agree a boundary, they will have to ask a third party to decide it for them, and that the third party will take these criteria into account.

The justification for this proposition is, essentially, as follows. "Equity", of itself, is a rather imprecise concept;[27] but to say that a certain potential line would produce an inequitable result presupposes some criteria for determining what is, and is not, equitable. These criteria are to be gleaned from the expectations of international society as a whole, as revealed in part by the case-law, and in part by the patterns of boundary making.

I should elaborate. First of all, is equity indeed the overriding criterion? Not in the case of the territorial sea: Article 15 of the Law of the Sea Convention (1982), reflecting a comparable provision of the

---

customary rules. For the reasons already given, such a view is incorrect. Curiously, though he is an author who has taken a keen interest in the formation of customary international law as well as the law of the sea, Charney seems at one point to admit this as a theoretical possibility. But he goes on to conclude that the requirement of uniformity, in particular, has not in fact been met. J. Charney, "Progress in International Maritime Boundary Delimitation", *AJIL* 88 (1994), 227 et seq., 228. He goes on, however, at 254, to suggest that it would be desirable for the International Court of Justice to take into account the solutions adopted in cases similar to the one before it, if only as a source of inspiration.

[27] See also note 1 above.

Geneva Convention on the Territorial Sea and Contiguous Zone (1958), provides that (in the absence of agreement) equidistance is to apply, except "where it is necessary by reason of historic title or other special circumstances to delimit the territorial seas of the two States in a way which is at variance therewith". Where the territorial sea is concerned, there seems to be little tendency, in either the jurisprudence or in practice, to blur the equidistance-special circumstances rule into one of equity: rather, there remains a presumption in favour of equidistance, which can be dislodged only if the party asserting that there exist special circumstances can persuade its interlocutor or a third party decision maker that this is indeed the case.[28] But when we come to the delimitation of the continental shelf, the exclusive economic zone, the exclusive fishing zone, or a single maritime boundary, increasing reliance has been placed on the concept of equity and its congeners. The first step was in the *North Sea Continental Shelf* cases, where the ICJ, having rejected the submission that the customary rule was identical in content to Article 6 of the Continental Shelf Convention, went on to hold that equity should govern the delimitation. The next step, as already indicated, was taken by the Court of Arbitration in the *France-UK Continental Shelf* case, holding that the equidistance-special circumstances rule in Article 6 had the same object as the customary rule, which was to achieve an equitable delimitation in all the circumstances.[29] Subsequent case-law, even before the conclusion or entry into force of the 1982 Convention, endorsed and strengthened this approach.[30] And as is well known, at the Third UN Conference on the Law of the Sea, a stand-off between the proponents of equidistance and those who favoured equitable principles resulted in the exceedingly vague compromise formula in Articles 74(1) and 83(1), which states that "[t]he delimitation of the [exclusive economic zone or continental shelf, as the case may be] between States with opposite or adjacent coasts shall be effected by

---

[28] In the case of the territorial sea, however, it is possible that a modified version of my thesis would apply, so that existing delimitation agreements (as well as the limited case-law) would provide guidelines as to what constitute "special circumstances". An example is the recent agreement between Belgium and the Netherlands. See text accompanying note 8 above.

[29] ILR 54 (1977), 6 et seq., esp. paras. 70 and 75.

[30] For example, the *Tunisia-Libya Continental Shelf* case, ICJ Reports 1982, 18 et seq.; the *Gulf of Maine* case, ICJ Reports 1984, 246 et seq.; the *Guinea-Guinea-Bissau* case, Arbitral Award, *ILM* 25 (1986), 252 et seq.; and the *Libya-Malta Continental Shelf* case, ICJ Reports 1985, 13 et seq.

agreement on the basis of international law as referred to in Article 38 of the Statute of the International Court of Justice, in order to achieve an equitable solution". This is not the place to enter into a discussion of whether this rule applies in situations not governed by the 1982 Convention, and still less to investigate the merits and demerits of equidistance *versus* equity. It will be sufficient to observe that international tribunals, whether applying the 1982 Convention or post-1982 customary law,[31] regard themselves as obliged to seek an equitable result, and the literature overwhelmingly reflects this approach, *pace* Judge Oda's eloquent reservations.

But the trouble with a concept like "equity" is that it is very vague. This is to state a fact, not necessarily a criticism. Even domestic law is riddled, to an extent not frequently noticed, with concepts such as fairness, reasonableness and equity. The explanation is not far to seek. Circumstances vary so considerably that, although there are areas where it is appropriate for the law and its application to be the same for everyone, there are others where the variety of individual circumstances requires a degree of flexibility. Criminal sentencing is one such area; the exercise of judicial discretion in divorce law is another. As Tolstoy put it, "All happy families resemble one another, but each unhappy family is unhappy in its own way".[32] Similarly, there are areas of international law, such as diplomatic immunity and the prohibition of the use of force, where the rules are, quite appropriately, the same for all States. But there are others, such as maritime delimitation (outside the territorial sea) and the non-navigational uses of international watercourses, which require the application of equitable principles. It is no accident that the two latter examples are both closely related to the geographical situation of the States in question, because whilst all diplomatic missions might be said to be essentially similar, riverine and coastal geography varies enormously.

But even if some justification can be found for the use of so vague a concept, what is the basis upon which the decision maker who has to apply concepts like "equitable" or "reasonable" can ascertain what it is that reason or equity require? Reasonableness and fairness are not self-evident, and in order to be able to form the view that a proposed boundary is inequitable we have to have some criteria (conscious or unconscious) which lead us to that conclusion. Where are we to find these criteria? Some of them, it is true, are now enunciated in the decided cases. But that cannot be the whole answer. First, there is no doctrine of binding precedent in international law, and in addition to

---

[31] For example, the *Greenland/Jan Mayen* case, ICJ Reports 1993, 38.
[32] L. Tolstoy, *Anna Karenina* (1875-77), Pt. I, Chap. 1.

this some would deny that judicial and arbitral decisions in bilateral boundary cases can constitute precedents, even in the weak sense of being merely persuasive, because each case is unique as to its facts, its specific geographical and historical context.[33] Secondly, whence did those who decided the first cases derive *their* criteria? Thirdly, the cases decided so far may still be too few to cover the many possible variations in circumstances. In short, the case-law might not be a sufficient guide.

It may be helpful to consider what a domestic judge who has to apply conceptions of reasonableness or equity would do in a similar situation.[34] He or she will first look to decided cases, not because they are binding (even in the common law precedents involving the exercise of judicial discretion are often not binding as such), but because they provide guidelines. But if there are no precedents, or if the range of solutions that they leave open is still wide, he or she will need to look elsewhere. It would be inappropriate for judges to be guided by their own prejudices or preferences. In England, the judge in such a situation will often refer to what is thought reasonable or equitable by the proverbial "man on the Clapham omnibus" – that is to say, by what the average citizen is likely to think reasonable or equitable. The judge does not take an opinion poll, but is supposed to be broadly in touch with the relevant values of his or her compatriots. What I am tentatively suggesting here is that, in international law, the equivalent of "the man on the Clapham omnibus" is the collectivity of States, whose opinions on what is an equitable solution to a particular type of boundary problem can be gleaned, with all due caution, from a consistent tendency to reach similar solutions in similar geographical situations. The pattern would have to be rather pronounced: a handful of precedents may not be enough. Nor am I claiming that, even if there were a very pronounced pattern, that would be *binding* even on a third-party decision maker.[35] This is so for a variety of reasons. First, if the

---

[33] Cf. L. D. M. Nelson, "The Roles of Equity in the Delimitation of Maritime Boundaries", *AJIL* 84 (1990), 837 et seq., and authorities cited there.

[34] My own domestic legal background is primarily the common law, but so far as I am aware the situation is not markedly different in other legal systems. It should also be mentioned, in this context, that when I speak of a common-law judge applying concepts of equity, I do not refer to the body of law known as Equity which was developed primarily in certain branches of the law, such as property, succession and trusts, and which became in time just as fixed and precedent-based as the rules of the common law proper.

[35] Parties negotiating a boundary treaty of course remain free, I repeat, to agree on whatsoever they please.

decision maker were bound, we would be in the presence of actual rules of law, which we have already seen is not the case. Secondly, the nature of the process is that there is not just one correct line: rather, there is a range of possible equitable lines, whilst other lines outside that range are inequitable – rather like valuation, where competent valuers armed with the same information should be able to come up with a range of valuations within which any figure could be said to be reasonable, whilst those outside the range would not be. Thirdly, it is always possible to argue that, though some aspects of the matter in hand are similar to those which have arisen in other delimitations, there are other aspects which are different and need also to be taken into account. But still, a consistent approach in treaty making will, I submit, provide guidelines which have, if not legal force, then at least legal relevance – a sort of quasi-normative status.

The proposition I am advancing here was touched, rather obliquely, by the ICJ in the *Continental Shelf (Libya/Malta)* case. Primarily in discussing the relevance of the concept of equidistance, the parties had referred to over 70 bilateral delimitation agreements. Malta used such agreements to support its contention that there was a certain primacy to be given to equidistance, especially where the States concerned were opposite. Its argument was not so much that they represented customary law, as that this practice "must provide significant and reliable evidence of normal standards of equity". Libya disputed their relevance as well as denying that, on the substantive point, they supported the inferences Malta sought to draw from them. Rather laconically, the Court simply observed that "it for its part has no doubt about the importance of state practice in this matter. Yet that practice, however interpreted, falls short of proving the existence of a rule prescribing the use of equidistance, or indeed of any method, as obligatory". The Court unfortunately gave no indication whether, in referring to "state practice", it was characterising these agreements as capable of constituting a rule of customary law – contrary to its previous jurisprudence; neither did it address the Maltese argument that these agreements were capable of providing evidence of "normal standards of equity".[36] Further light is perhaps thrown on the Court's views in the separate opinion of Judge *ad hoc* Valticos. Citing this passage from the majority Judgment, he said: "Like the Court, I tend to the view that the States which concluded those bilateral agreements ... did not have the impression that they were following a binding rule of law, and were not guided by any *opinio juris*. But, at the very least, they did conclude these agreements in the light of the legal

---

[36] ICJ Reports 1985, 10 et seq., 38, para. 44.

background, and in the belief that the median line was the most widespread and convenient method and that it reflected what might be called an *opinio aequitatis*".[37] He did not, however, elaborate on this idea.

In making this suggestion I am of course assuming that certain features recur in maritime boundary agreements.[38] This is not entirely uncontroversial[39]; but even if I were wrong the point might not be devoid of general interest, for two reasons. First, because even if we were to conclude that there is insufficient evidence to deduce the existence of a pattern – or, more precisely, patterns – at present, they could emerge in the future, given that a large number of maritime boundaries have yet to be settled. And secondly, because the same questions could be asked about the legal significance of recurring treaty patterns in other areas of the law calling for the application of equitable principles, such as that relating to the non-navigational uses of international watercourses. So the suggestion made here is not without more general relevance.

But, in fact, it seems that recurring patterns may indeed exist – some of which are also reflected in the case-law. First, in a very large proportion of agreements the technique of equidistance is employed, at least as a starting point.[40] Here one needs to be very careful not to jump

---

[37] *Ibid.*, 104 et seq., 107-108, para. 11. He went on to add: "It would in any events be highly unfortunate if, on a point of such importance, a divorce were to set in between the treaty practice of States, to which Article 38 of the Statute of the Court refers, and the Court's jurisprudence".

[38] At any rate, those drawing a continental shelf, exclusive economic zone, exclusive fishery zone, or single maritime boundary. As we have seen, territorial sea boundaries involve somewhat different considerations.

[39] For instance, Charney asserted in 1994 that there was such a diversity of practice, both in the method of delimitation and the factors taken into account and the weight given to them, that "no normative principle of international law has developed that would mandate the specific location of any maritime boundary line". See "The American Society of International Law Maritime Boundary Project", in G. Blake (ed.), *Maritime Boundaries*, 1994, 1 et seq., 9. However, as explained above, it is not a question of (customary law) normativity in the strict sense; and furthermore, the factors to be taken into account can never be expected to "*mandate* the *specific* location of any maritime boundary line", but (at best) to indicate a range of lines which (in the absence of specific agreement) would seem to be equitable, and another range which would seem not to be, as with valuation. See also the next paragraph of the main text.

[40] Cf. S. P. Jagota, *Maritime Boundary*, 1985, 276, who put the figure at 82 per cent in 1985. In 1991, the figure was given as 77 per cent by L. Legault and B. Hankey, "Method, Oppositeness and Adjacency, and Proportionality in Maritime Boundary Delimitation", in Charney and Alexander, see note

too hastily to conclusions. After all, a "simplified" equidistance line is not a true equidistance line; and neither is an "proximate" equidistance line, far less one which starts from an equidistant construction line and then is corrected to take into account such factors as disparities in coastal length. The choice of basepoints is also very significant. As the Arbitral Tribunal pointed out in second award in the *Eritrea-Yemen* case, "both Parties to the present case have claimed a boundary constructed on the equidistance method, although based on different points of departure and resulting in very different lines".[41] The statistics can also be misleading in another respect. A treaty may be listed by a commentator as providing for the application of the equidistance method, whereas in fact it does so only in a few sectors, but not all. But still, the frequency with which equidistance comes into treaties, either pure, simplified, approximate, or at least as a starting point, is striking. The ICJ may be thought to have addressed this point in the *Libya/Malta* case. Immediately following the extract from its Judgment just cited, it went on to say that "[e]ven the existence of such a rule as is contended for by Malta, requiring equidistance simply to be used as the first stage in any delimitation, but subject to correction, cannot be supported solely by the production of numerous examples of delimitations using equidistance or modified equidistance, though it is impressive evidence that the equidistance method can in may different situations yield an equitable result". But perhaps, by focussing on whether a "rule" existed, the Court failed to do justice to the submission that such agreements represented an *opinio aequitatis* which it ought to have taken into account.

This not the place to engage in an analysis of what, if anything, these treaties really do reveal. It is simply suggested that, if there *is* indeed a significant pattern, it *may* constitute an indication of what States regard as equitable, at least in similar situations.

Other recurrent patterns seem to be the avoidance of encroachment (that is to say, avoiding the boundary swinging across the coastal frontage of one of the parties, particularly at its landward end); not giving full effect to at least relatively minor islands or other coastal features which could skew the line disproportionately in a delimitation between adjacent States; not necessarily giving full effect to islands straddling a hypothetical equidistant line[42]; similarly in the case of

---

8, 203 et seq., 214. Cf. P. Weil, "Geographic Considerations in Maritime Delimitation", *ibid.*, 115 et seq., 127.

[41] Award of 17 December 1999, para. 131.

[42] See D. Bowett, "Islands, Reefs, Rocks and Low-Tide Elevations", in Charney and Alexander, see note 8, 131 et seq., 137-141.

islands on the wrong side of a median line; taking the proportionate length of the relevant coastlines into account, at least in order to check the equity of using a particular method (e.g., equidistance), and sometimes also to correct the line produced by the equidistance method[43]; and (perhaps) taking into account the defence and security concerns of one or both parties to secure access to the open sea from their own jurisdictional zones.[44]

Lawyers, especially those of us brought up in the positivist tradition, may understandably feel uneasy about the idea of quasi-legal norms which affect the legal system but are not formally part of it. But if space permitted, it would be easy to demonstrate that, as has been briefly adumbrated here, rules of this type are in fact to be found in all legal systems, and we should not be shocked by them, as long as we confine them to their proper role. After all, law (not least international law) is not an entirely hermetic system: it is about the real world and, in appropriate circumstances and within carefully defined limits, it is acceptable for it to take into account the values and judgments of the society it serves.

---

[43] L. Legault and B. Hankey, "Method, Oppositeness and Adjacency, and Proportionality in Maritime Boundary Delimitation", *ibid.*, 203 et seq., 217-221.

[44] B. H. Oxman, "Political, Strategic and Historical Considerations", *ibid.*, 3 et seq., 22-30.

Considerations of Equity in Maritime Boundary Cases Before the International Court of Justice

Masahiro Miyoshi

I. Introduction

A notable tendency in recent cases decided by the International Court of Justice, especially those concerning maritime boundary delimitation is the resort to equitable considerations. This is a noteworthy phenomenon in view of the general understanding that judicial settlement has been considered a mode of third-party settlement based on law, while arbitration has been described as a third-party procedure which, though producing as binding a settlement as the ICJ, may be based on principles of justice and equity as well as law.

Arbitration is conceived as a process of third-party settlement based on respect for law, rather than on law. As Article 15 of the Hague Convention on the Pacific Settlement of International Disputes of 1899 states:

> "L'arbitrage international a pour objet le règlement de litiges entre les Etats par des juges de leur chois et *sur la base du respect du droit*".[1]

This provision, if not the Convention in its entirety, is a codification of the practice of States in international arbitration. The expression "respect for law" is used in lieu of "law", and implies the admissibility of resort to considerations other than strict law. In the words of the drafters' report, "La justice arbitrale internationale règle – c'est-à-dire termine définitivement – les litiges internationaux qui lui sont soumis. Elle règle ces litiges *sur la base du respect du droit, conformément aux exigences*

---

[1] J. B. Scott (ed.), *Texts of the Peace Conferences at The Hague, 1899 and 1907: With English Translation and Appendix of Related Documents*, 1908, 28-29, emphasis added.

de la justice. Elle règle par l'organe de juges choisis en vertu de l'accord des Etats eux-mêmes. Tels sont les traits fondamentaux de la justice arbitrale".[2] Thus the concept of "sur la base du respect du droit" is linked with the notion of "conformément aux exigences de la justice". This characteristic of the basis of decision, together with the choice of judges by the parties themselves, is the "fundamental trait" of arbitral justice, i.e., arbitration.

When the Permanent Court of International Justice (PCIJ) was established under the auspices of the League of Nations, it was clearly distinguished from the traditional procedure of arbitration in two main respects: the permanent composition of the Court and the applicable law. Writers tended to stress the neutral and objective nature of the new Permanent Court.[3] In terms of the applicable law, something other than strict law was only mentioned in the form of a decision *ex aequo et bono*, which however required the agreement of the parties.[4] No change has been made in this aspect in the slightly revised Statute of the International Court of Justice. Thus the PICJ and the ICJ are understood to be the courts of justice that apply law, unless authorised by the parties to decide *ex aequo et bono*. Arbitral tribunals, by contrast, are considered to be free to apply equity unless their basis of decision is limited to law.

Things seem to have changed recently, with some ICJ judgments frequently applying equitable principles in maritime boundary cases. The watershed came in the *North Sea Continental Shelf* cases of 1969, in which equitable principles were resorted to as a matter of customary law in opposition to the rule of equidistance under Article 6 of the Convention on the Continental Shelf of 1958. Reference to equitable principles has since been made in the *Tunisia/Libya Continental Shelf* case of 1982, the *Gulf of Maine* case of 1984, the *Libya/Malta Continental Shelf* case of 1985, and the *Greenland/Jan Mayen Maritime Delimitation* case of 1993. Some arbitrations have also been conducted on similar subject matters during these years: the *Anglo-French Continental Shelf Delimitation* case of 1977, the *Guinea/Guinea-Bissau Maritime Boundary Delimitation* case of 1985 and the *Canada/France*

---

[2] Conférence internationale de la paix, La Haye, 18 mai - 29 juillet 1899, Sommaire général, tome 1, 111, emphasis added. For a more detailed discussion of the preparatory work on this particular aspect of the conference, see M. Miyoshi, *Considerations of Equity in the Settlement of Territorial and Boundary Disputes*, 1993, 22-23.

[3] See M. Miyoshi, "Recent Trends in the Jurisprudence of the International Court of Justice and International Arbitral Tribunals: With Special Reference to Territorial and Boundary Cases", *Asian YIL* 6 (1997), 5-7.

[4] Article 38, para. 2, of the Statute of the Permanent Court of International Justice.

Delimitation of Maritime Areas case of 1992. In all these cases, equitable principles or considerations were applied or taken into account. Perhaps these ICJ cases and arbitration cases affected each other in sequence. In a word, increasing reference has been made to equity in recent ICJ cases and arbitrations on maritime boundary delimitation.

Why is this so? Was the application of law not enough or not even appropriate? Does this tendency signal a change in the content of contemporary international law or in the role of the ICJ as the principal judicial organ, not only of the United Nations, but also of the entire world? If proved at all, this would present an interesting challenge to international law in general.

II. Varieties of Equitable Considerations

Although equitable principles were frequently relied upon as the basis of decision in earlier maritime boundary cases before the ICJ, it was not until the *Libya/Malta Continental Shelf* case of 1985, the fourth ICJ case on this particular subject matter, that they were mentioned in the Court's reasoning. In the *North Sea Continental Shelf* cases of 1969, equitable principles were emphatically applied in lieu of the conventional rule of equidistance, which was not applicable to the Federal Republic of Germany, a non-party to the Convention on the Continental Shelf of 1958. The Court simply mentioned equitable principles as customary law, but did not specify their substance. Nor did the *Anglo-French* arbitration, which repeated the importance of equity while admitting the applicability of the conventional rule of equidistance or median line as coinciding with customary law. Nor, indeed, did the *Tunisia/Libya* case of 1982 and the *Gulf of Maine* case of 1984 attempt to clarify the content of equitable principles, although they discussed other aspects of maritime boundary delimitation in more detail.

Then, in the *Libya/Malta* case of June 1985, the ICJ did at last try to specify what equitable principles are in the context of continental shelf boundary delimitation. It mentioned five such "well-known examples" as deducted from the previous cases.[5] But it may be wondered why the

---

[5] Those five principles are: "the principle that there is to be no question of refashioning geography, or compensating for the inequalities of nature; the related principle of non-encroachment by one party on the natural prolongation of the other, which is no more than the negative expression of the positive rule that the coastal State enjoys sovereign rights over the continental shelf off its coasts to the full extent authorised by international law in the relevant circumstances; the principle of respect due to all such relevant circumstances; the principle that although all States are equal before the law and are entitled to equal treatment, 'equity does not necessarily imply equality' (ICJ Reports 1969, 49, para. 91), nor does it seek to make

Court took the trouble to specify them in this particular case. To the best of this writer's knowledge, the Court appears to have elaborated on those principles in response to the questions posed by counsel for both sides during the oral proceedings. Elihu Lauterpacht, counsel for Malta, raised the question of equitable principles in discussing the economic aspects of the dispute, and wondered why, although the Court referred to such principles for the first time in the *North Sea Continental Shelf* cases of 1969, the Court failed to quote any particular authority on the origin of this important concept.[6] At another session some days later, Sir Francis Vallat, counsel for Libya, stated that equitable principles meant taking into account all relevant circumstances.[7] These statements by counsel might have urged the Court to elaborate in some measure on the content of equitable principles.

Of the "well-known" equitable principles, arguably the one relating to "the respect due to all the relevant circumstances" is the most important, since the equitableness of applicable principles depends on the equitable result of their application[8] – the more equitableness is sought, the more important the consideration of the relevant circumstances will be. The taking into account of *all* the relevant circumstances, however, would lead to an undue process of parties being allowed to rely on *all conceivable* factors as the relevant circumstances in their pleadings. The Court rightly restricted such considerations to "those that are pertinent to the institution of the continental shelf as it has developed within the law and to the application of equitable principles to its delimitation".[9] What are these (restricted) factors that are to be taken into account for maritime delimitation? The relevant circumstances or factors are by their nature peculiar to a given case, but broadly have been divided into geographical and non-geographical factors.

The division into geographical and non-geographical factors was made as early as in the *North Sea Continental Shelf* judgment. It lists three factors that are to be taken into consideration in the negotiations on continental shelf delimitation between the parties: (1) the coastal configurations and any special unusual features, (2) the physical and geological structure and natural resources of the continental shelf, and (3) a reasonable degree of proportionality between the extent of the

---

equal what nature has made unequal; and the principle that there can be no question of distributive justice". ICJ Reports 1985, 39-40, para. 46.

[6] Public Sitting on 27 November 1984, CR 84/23, 36.

[7] Public Sitting on 20 February 1985, CR 85/9, 21-22.

[8] ICJ Reports 1982, 59, para. 70, where the Court says: "It is ... the result which is predominant; the principles are subordinate to the goal. ... The equitableness of a principle must be assessed in the light of its usefulness for the purpose of arriving at an equitable result".

[9] ICJ Reports 1985, 40, para. 48.

continental shelf area appertaining to the coastal State and the length of its coastline.[10] Factors (1) and (3) are geographical factors. The Court's reasoning that it was necessary to clarify geographical features as the basis of the natural prolongation of the landmass into and under the sea, that application of the equidistance principle in a small sea area with unusual coastal configurations could lead to an inequity and that coastal length is the key to the calculation of proportionality does indeed imply its awareness of the importance of geographical factors.[11]

This rationale for the relative importance of geographical factors was followed in the *Anglo-French* and *Tunisia/Libya* cases, which rejected the non-geographical factors presented by the parties in favour of geographical ones.[12] Thus, in the *Gulf of Maine* case, the Chamber turned down the socio-economic factors which the parties very much emphasised in their pleadings, and sought the criteria of delimitation in geographical factors instead.[13] All the subsequent cases on maritime boundary delimitation set store by geographical factors, and when they took some non-geographical factors into account, they only gave a secondary or subsidiary role to them. It would seem, therefore, that the division of relevant circumstances into geographical and non-geographical factors with emphasis on the former has taken on a kind of normativity in the law of maritime boundary delimitation.

1. Geographical Factors

Geographical factors may be divided into two broad categories: geographical definition of the delimitation area and special geographical features within the defined area.

A. Definition of the Delimitation Area

It is of fundamental importance to define the maritime area for delimitation. In the *North Sea Continental Shelf* cases, the outer limits of the continental shelf had been determined by a series of agreements between the interested countries in the region, but in the *Anglo-French* arbitration the delimitation area was clearly defined by the *compromis*. In the *Tunisia/Libya* case, the area had to be defined through a combination of a latitude and a longitude in the light of the continental shelf area claimed by a third party. In the *Gulf of Maine* case, the

---

[10] ICJ Reports 1969, 54, para. 101(D)(1)-(3).
[11] *Ibid.*, 22, para. 19; 29, para. 39; 51, para. 96; 35, para. 53; 49, para. 89(a)(b); 52, para. 98.
[12] RIAA XVIII, 112, paras. 239, 240; ICJ Reports 1982, 63, para. 78; 86-87, para. 122-124; 93, para. 133(B).
[13] ICJ Reports 1984, 340, para. 232; 278, para. 59.

delimitation area covered the whole Gulf area and some sea areas off the Gulf to a point agreed upon between the parties. In the *Libya/Malta* case, a big difference in the length of the coastlines and a third party's claim to a continental shelf area had to be taken into account in determining the delimitation area, which eventually was defined by two parallel longitudinal lines. A big difference in the length of the coasts was likewise taken into account in the determination of the delimitation area in the *Canada/France* case, concerning St. Pierre et Miquelon, and in the *Greenland/Jan Mayen* case, although the question of a third party's claim was not relevant in either case.

B. Special Geographical Features Within the Delimitation Area

The reason why the geometrically correct equidistance/median line principle has been thought unacceptable as the norm of delimitation lies in the fact that the coastlines and insular features within the delimitation area constitute special geographical features to which the equidistance principle cannot be applied in a simplistic way. However, depending on whether the relevant coasts are opposite or adjacent, the principle's applicability varies. It has been applied basically or in the first instance, where the coasts face each other, even in sea areas with complicated coastlines or insular formations.

In the *North Sea Continental Shelf* cases, the three parties had comparable coastline lengths, although the coasts of Denmark and the Netherlands were convex, while that of the Federal Republic of Germany was concave. This particular coastal configuration was taken into account in avoiding the application of the conventional principle of equidistance. In the *Tunisia/Libya* case, the great curvature of the Gulf of Gabes and the situation of the Kerkennah Islands were the geographical features requiring special consideration. Thus the boundary line which started from the outer limit of the territorial sea off the terminus of the land boundary was turned at the point where it crosses the latitude of the westernmost spot of the Gulf to veer at an angle reflecting a "half-effect" of the Kerkennah Islands.[14]

Two factors required special considerations in the *Gulf of Maine* case: the Bay of Fundy at the back of the Gulf and the shape of Cape Cod. The long coasts of the Bay are all Canadian and were taken into account in the calculation of the relative coastline lengths of the parties.[15] But the Canadian contention that the Cape's protrusion should be ignored was rejected as being against the principle of not refashioning nature.[16]

---

[14] ICJ Reports 1982, 88-89, paras. 127-129.
[15] ICJ Reports 1984, 336, paras. 221-222.
[16] *Ibid.*, 271, para. 37.

A great difference in the length of coasts was a special feature in the *Libya/Malta* and *Greenland/Jan Mayen* cases. In the *Libya/Malta* case, the distance between the two opposite coasts was another notable feature. The disparity in coastline length was taken into account by moving the provisional boundary line northwards by some minutes in favour of Libya.[17] But with respect to the distance of coasts, although it was mentioned as one of the relevant circumstances in the operative part of the judgment,[18] there is no indication of any special consideration being given to this particular factor.

In the *Land, Island and Maritime Frontier Dispute* case between El Salvador and Honduras, some disputed islands within the Gulf of Fonseca were found to be irrelevant to the delimitation, but the pertinent waters within the Gulf were historic waters bounded by not only the two contesting countries but also by Nicaragua. By its very character as an historic bay, as decided by the Central American Court of Justice in 1917 and endorsed by the ICJ Chamber, the waters within the Gulf are a condominium of the three coastal States which requires their agreement for delimitation. Nicaragua was permitted to intervene in this case as an interested third party for the first time in the history of the ICJ.[19] However, as there was no such agreement, the Chamber refrained from effecting a delimitation.[20]

C. Islands and Other Insular Formations

Islands and other insular formations like low-tide elevations are one of the important geographical factors within a specified delimitation area that can greatly affect the delimitation. Although they may be unimportant in themselves, they may bend the delimitation line or cause it to deviate, because of their position within the pertinent sea area. In the *Anglo-French* arbitration, the Channel Islands were treated as an enclave with a belt of continental shelf on the north and west sides of it on the "wrong side" of the median boundary line,[21] while the Scilly Isles and the Kerkennah Islands were given a "half effect" in the delimitation.[22] Had they been afforded a "full effect", it was thought that

---

[17] ICJ Reports 1985, 49, para. 66.
[18] *Ibid.*, 57, para. 79(B)(2).
[19] ICJ Reports 1990, 137, para. 105.
[20] ICJ Reports 1992, 616-617, paras. 432(1), (2).
[21] *RIAA* XVIII, 93-94, para. 198.
[22] Probably the first precedent of "half effect" is that given to the Iranian island of Kharg off the Iranian coast in the Offshore Boundary Agreement between Iran and Saudi Arabia, initialled on 13 December 1965. See H. M. Al-Baharna, *The Legal Status of the Arabian Gulf States: A Study of Their Treaty Relations and their International Problems*, 1968, 310-311. But this was subsequently modified in the 1968 Agreement. R. F. Pietrowski, Jr.,

they would have unduly and, therefore, inequitably distorted the delimitation.

One continental shelf case involving a good number of islands on the "wrong side" of an hypothetical median line, submitted to the ICJ but found to be outside its jurisdiction, was the *Aegean Sea Continental Shelf* case of 1978.[23] All the islands, except for a few in the north-eastern corner of the Aegean Sea at the entrance to the Dardanelles Strait, belong to Greece. The islands, lying a short distance off the western coast of Turkey, nearly block its access to the Aegean Sea, and have remained a critical issue between the two countries. As it is, this unique situation, a result of territorial transactions following World Wars I and II, would seem unfortunately to have no resolution in sight.[24]

On the other hand, islands may be entitled to have their own continental shelf or exclusive economic zone as an independent land territory. Malta, an island State, and St. Pierre et Miquelon, Greenland and Jan Mayen, all located far from their metropolitan States, are cases in point. All these islands constitute a land territory opposite another land territory. The proportion of coastal lengths had to be reflected in some form in these cases, and in the *Libya/Malta* and *Greenland/Jan Mayen* cases a provisional median line was moved towards the island with a shorter coastline for an equitable adjustment, while in the *St. Pierre et Miquelon* case the islands were both treated as an enclave on the continental shelf of Newfoundland and were allowed a "mushroom-shaped" corridor, with the islands located in the head of the mushroom.

2. Non-Geographical Factors

A. Geological and Geomorphological Factors

The *North Sea Continental Shelf* cases linked the unity of mineral deposits as a geological factor to the application of equitable principles.[25] But the *Gulf of Maine* case rejected the geomorphological considerations which the parties took great pains to present in their pleadings. Likewise, fault zones on the seabed or similar geomorphological factors were more or less ignored in subsequent cases.

"Report Number 7-7", in J. I. Charney and L. M. Alexander (eds.), *International Maritime Boundaries*, Vol. 2, 1993, 1522.

[23] ICJ Reports 1978. For a provisional measure by the ICJ in the same case, see ICJ Reports 1976.

[24] An allegedly equitable solution might be to allow Turkey a few "fingers" of seabeds between some islands which lie comparatively apart from each other off its Anatolian peninsula. A. Wilson, "The Aegean Dispute", *Adelphi Paper* No. 155, 1979-1980.

[25] ICJ Reports 1969, 51-52, para. 97.

B. Considerations of Security

Although the *Anglo-French* arbitration placed less emphasis on security considerations in view of the importance of the English Channel as an international sea lane,[26] the *Guinea/Guinea Bissau* arbitral award held that the prime objective was to avoid the prevention of the right to development and the compromising of security off the coasts of the parties,[27] and the *Libya/Malta* judgment found that, in general, security considerations were not unrelated to the concept of the continental shelf.[28] One commentator argues that diverse circumstances and considerations converge into one point, namely, the distance of the boundary line from the coast.[29] This would include security considerations as well. But the jurisprudence would seem to show that a boundary line determined mainly in the light of geographical factors ensures security considerations as well.[30]

C. Economic Factors

More often than not, parties to a dispute go to the Court for economic reasons such as the development or management of resources, for example, fisheries. But in the jurisprudence of the ICJ and arbitral tribunals, little or no attention has been paid to economic considerations. In the *Tunisia/Libya*, *Guinea/Guinea Bissau* and *Libya/Malta* cases, claims were made for special consideration for the lesser economic power, but the tribunals denied the necessity of such considerations or of a revision of the delimitation arrived at for such a reason. The alleged economic considerations were found to be variables depending on unpredictable national fortune or calamity, and "virtually extraneous factors".[31] On the other hand, the Chamber in the *Gulf of Maine* case, while ostensibly denying such considerations in the process of delimitation, did refer to a concern to avoid catastrophic repercussions on the livelihood and economic well-being of the people concerned at the verification stage of the equitableness of the delimitation.[32] One may

---

[26] *RIAA* XVIII, 90, para. 188.
[27] *RGDIP* 89 (1985), 533, para. 124.
[28] ICJ Reports 1985, 42, para. 51.
[29] P. Weil, *Perspectives du droit de la délimitation maritime*, 1988, 284. Another commentator says that security considerations are not unrelated to the concept of the continental shelf, and that they are of legal relevance to the evaluation of the equitableness of the delimitation line. R. Versan, "Legal Problems Concerning Territorial Sea Delimitation in the Aegean Sea", in S. Tashan (ed.), *Aegean Issues: Problems – Legal and Political Matrix*, 1995, 105.
[30] See, for example, the *Libya/Malta* case. ICJ Reports 1985, 42, para. 51.
[31] ICJ Reports 1982, 77, para. 107.
[32] ICJ Reports 1984, 342, para. 237.

well wonder, in fact, whether the Chamber did not take this concern into account in the process of delimitation.

In the more recent *Greenland/Jan Mayen* case, by contrast, the capelin fisheries were taken into account in the delimitation.[33] It remains to be seen whether this apparent change in the Court's attitude towards economic factors is an exception to the general rule of the past, or whether it can be regarded as a turning point in its jurisprudence in this respect.[34] However this may be, it is worthy of note that some writers admit the importance of economic considerations.[35]

D. Oil Concessions

In the *Tunisia/Libya* case, the virtual boundary between the parties' respective oil concession areas was found to coincide with a *modus vivendi* that existed concerning the delimitation of fisheries jurisdiction during colonial days.[36] But claims for the recognition of similar oil concessions were not allowed in the *Gulf of Maine* case, in which the parties failed to show sufficient respect for each other's concessions to create a virtual boundary line.[37]

E. Conduct of the Parties – Acquiescence and Estoppel

Usually, the parties to a dispute before the tribunal contend that the opponent's conduct, including omission, has come to have certain legal effect.

---

[33] ICJ Reports 1993, 70-72, paras. 73-76.

[34] In the *Eritrea/Yemen Maritime Delimitation* arbitration (Phase II) of 1999, the historical fishing practices were much discussed, but the Tribunal did not give the matter positive consideration on the grounds that the evidence was contradictory and confused. Award Phase II: Maritime Delimitation, paras. 61, 72, text available at http://www.pca-cpa.org/ERYE2chap2.htm.

[35] A. L. W. Munkman, "Adjudication and Adjustment – International Judicial Decision and the Settlement of Territorial and Boundary Disputes", *BYIL* 46 (1972-1973), 101 et seq.; D. W. Bowett, "The Economic Factor in Maritime Delimitation Cases", in *International Law at the Time of Its Codification: Essays in Honour of Roberto Ago*, Vol. I, 1987, 45 et seq.; B. Kwiatkowska, "Economic and Environmental Considerations in Maritime Boundary Delimitations", in Charney and Alexander, see note 22, Vol. 1, 1993, 75 et seq.

[36] ICJ Reports 1982, 84-85, para. 119.

[37] ICJ Reports 1984, 310-311, paras. 150-151. See also the *Eritrea/Yemen* arbitration, Award (Phase II), note 34, paras. 81, 82, in which the Tribunal stated that the petroleum contracts "lend a measure of support to a median line".

In the *North Sea Continental Shelf* cases, it was alleged that the conduct of the Federal Republic of Germany, a non-party to the Convention on the Continental Shelf of 1958, had made Denmark and the Netherlands believe that it had accepted the rule of delimitation under the Convention, but the Court found that the German conduct did not amount to acquiescence or estoppel.[38] The Chamber in the *Gulf of Maine* case pointed out that such effect could only result from "clear and consistent acceptance".[39]

F. Interests of a Third State

Except in cases of involving only two States in a large sea area, maritime boundary delimitations frequently affect third States as well. This is a macrogeographical matter of defining the delimitation area, but logically has non-geographical aspects.

The question was pointed out as being a matter of defining the delimitation area as early as the *North Sea Continental Shelf* cases,[40] and was referred to as being of potential relevance to Ireland in the *Anglo-French* arbitration as well.[41] The concern made its appearance later in the form of a third State's request for permission to intervene in the proceedings before the ICJ. In the *Tunisia/Libya* and *Libya/Malta* cases, Malta and Italy respectively tried to intervene, but their requests were both rejected,[42] although their substance was implicitly or expressly taken into account in the merits phase.[43] But such a request was formally heard, for the first time in the ICJ's history, when Nicaragua intervened in the *Land, Island and Maritime Frontier Dispute* case between El Salvador and Honduras, with regard to the waters within the Gulf of Fonseca.[44]

In state practice, a couple of agreements have delimited maritime areas between three States,[45] bearing witness to an awareness that a

---

[38] ICJ Reports 1969, 25, para. 27; 26-27, paras. 31-33.
[39] ICJ Reports 1984, 309, para. 145.
[40] ICJ Reports 1969, 54, para. 101(D)(3).
[41] RIAA XVIII, 24-25, paras. 23-26. But this was not dealt with by the Court of Arbitration on the grounds that it was outside its jurisdiction. *Ibid.*, 26-27, paras. 27-28.
[42] ICJ Reports 1981, 20, para. 34; ICJ Reports 1984, 25, para. 41.
[43] ICJ Reports 1982, 93, para. 133(B)(1); ICJ Reports 1985, 26, para. 21.
[44] ICJ Reports 1990, 137, para. 105. Likewise Equatorial Guinea's request in 1994 for permission to intervene in the *Cameroon/Nigeria Land and Maritime Boundary* case was approved. *ILM* 38 (1999), 116, para. 18(1).
[45] See the Indonesia-Thailand-Malaysia Agreement of 1971, the India-Maldives-Sri Lanka Agreement of 1976 and the India-Indonesia-Thailand Agreement of 1978. Charney and Alexander, see note 22, Vol. 2, 1993, 1452-1454, 1407-1408, 1386-1388.

delimitation between any pair of them would inevitably affect the adjacent third State's interests, and that all of them might as well get together for a single agreement. This is a reasonable approach which, theoretically, is applicable to all cases where more than three adjacent States are involved.[46]

3. Verification of the Equitableness of Delimitation

Whether the delimitation is equitable has been tested in the jurisprudence of the ICJ as well as arbitral tribunals on the basis of proportionality between the length of coastlines and the extent of maritime areas. Since the *North Sea Continental Shelf* cases, in which it was considered as one of the factors to be taken into account in the negotiations between the parties, proportionality has been taken into account in the process of delimitation, or has been used as the test of equitableness of the result of delimitation in arbitral and judicial cases. As geographical factors are the major consideration among the relevant circumstances, proportionality has indeed been treated as the most important criterion in the process of delimitation and verification. However, it is not treated as something that has to be proved positively. The question is rather whether the delimitation is not disproportionate; if there is no great disproportion, the delimitation is found to be satisfactory.[47]

Past cases have repeatedly emphasised that delimitation is not allocation or distribution of maritime areas, but identification of the boundary line between already attributed maritime areas. On second thought, however, it may be suspected that a comparison of the length of coastlines and the sea areas between the parties for a calculation of proportionality with a view to arriving at a delimitation that is equitable and, therefore, acceptable to both parties may in fact be another expression of allocation or distribution in substance, no matter how it may be reasoned. Should this suspicion not be mistaken, could it not be a case of decision *ex aequo et bono* which the ICJ has repeatedly denied in the past? Indeed, in no case have the parties ever asked the Court to decide *ex aequo et bono*, and, consequently, the Court has never been empowered to do so. It has always denied that it has ever done so. But apart from the procedural requirement, it would seem rather difficult to deny the fact that the Court has virtually had resort to such a decision in substance.

---

[46] Cf. M. Miyoshi, "International Maritime Boundaries and Joint Development: A Quest for a Multilateral Approach", in G. Blake et al. (eds.), *Boundaries and Energy: Problems and Prospects*, 1998, 453-471.

[47] See, for example, the *Anglo-French, Gulf of Maine* and *Libya/Malta* cases. *RIAA* XVIII, 58; ICJ Reports 1984, 323, para. 185; ICJ Reports 1985, 44, para. 57.

Another question worth noting is whether proportionality should be considered in the process of delimitation or used as a test of its result. The Chamber clearly distinguished the delimitation process from the verification process in the *Gulf of Maine* case, but the subsequent cases have not necessarily followed that approach.[48] What is required in the final judgment is an equitable solution of the dispute before the Court.

In the jurisprudence of the ICJ as well as arbitral tribunals, the tests of equitableness, besides proportionality, have ranged from human and economic geography,[49] through the structure and nature of the continental shelf and economic circumstances to security considerations.[50] While placing varying weight on these factors, the tribunals generally have not considered them favourably. This is the result of the emphasis placed on geographical factors.

III. How Are Equitable Considerations Distinguished from Considerations ex Aequo et Bono?

Not only the ICJ and its chambers, but also previous arbitral tribunals have repeatedly stated in their territorial and boundary cases that they have refrained from deciding *ex aequo et bono* because they have no such powers. This is true as a matter of procedure, since in no case relating to a territorial or boundary dispute have the parties ever expressly requested the tribunal to put aside the law and decide *ex aequo et bono* instead. In such cases, the tribunals have often had recourse to considerations of equity in the application of law, on the grounds that they were applying equitable principles as part of the law. Indeed, as the ICJ stressed in its judgment in the *North Sea Continental Shelf* cases, "it is not a question of applying equity simply as a matter of abstract justice, but of applying a rule of law which itself requires the application of equitable principles...".[51]

This statement of law and equity is in accord with the definition of equity as adopted by the Institut de Droit International in its resolution of 1937: "l'équité est normalement inhérente à une saine application du droit, et ... le juge international, aussi bien que le juge interne, est, de par sa tâche même, appelé à en tenir compte dans la mesure compatible avec

---

[48] Proportionality was considered in the process of delimitation in the *Libya/Malta*, *Canada/France* and *Greenland/Jan Mayen* cases, but it was discussed in the process of verification in the *Eritrea/Yemen* arbitration, Award (Phase II), see note 34, paras. 165-168.

[49] For example, the *Gulf of Maine* case, ICJ Reports 1984, 278, para. 59; 340, para. 232.

[50] For example, the *Guinea/Guinea Bissau* case, *RGDIP* 89 (1985), 529-531, paras. 113-117; 531-532, paras. 118-120; 532-533, paras. 121-123; 533, para. 124.

[51] ICJ Reports 1969, 47, para. 85.

le respect du droit".[52] Properly put, equity in this sense would mean equity *infra legem*, rather than equity *praeter legem* or equity *contra legem*.

The question, however, is whether the application of equitable principles as in the *North Sea Continental Shelf* cases and the subsequent maritime boundary cases has not overstepped the bounds of genuine equity *infra legem*. It is true that the concept of equity cannot be strictly objective. On the contrary, it can be subjective and even elusive. There is no clear-cut borderline between its objectivity and subjectivity. Consequently, its application can inevitably be subjective to some extent. For all this nature of equity, however, a suspicion persists as to the application of equitable principles in some maritime boundary cases before the ICJ and its chambers. On this point, Sir Gerald Fitzmaurice is quoted as once having confided in a private letter to Sir Elihu Lauterpacht, who made critical comments on the way equitable principles were applied in the *North Sea Continental Shelf* and the *Anglo-French* cases, that: "where ... the Tribunal is precluded by its Statute or terms of reference from deciding *ex aequo et bono, but is in fact doing just that*, it cannot avow it, and has to take refuge in silence".[53]

IV. Conclusion

The ICJ has applied equitable principles in the settlement of maritime boundary disputes. In their application, the taking into account of the relevant circumstances has played a major part. Among the relevant circumstances, geographical factors have had a dominant place to the near exclusion of non-geographical factors. In the eyes of an outside

---

[52] Institut de Droit International, *Tableau général des Resolutions (1873-1956)*, 1957, 162.

[53] E. Lauterpacht, *Aspects of the Administration of International Justice*, 1991, 125, note 19, emphasis original. See also M. B. Feldman, "International Maritime Boundary Delimitation: Law and Practice from the Gulf of Maine to the Aegean", in Tashan, see note 29, 14, where the former Deputy Legal Adviser to the US Department of State, who played a leading role in the *Gulf of Maine* case, candidly says: "although the International Court of Justice and arbitral tribunals purport to apply equitable principles within the framework of law, the judgments themselves are comprehensible only as decisions *ex aequo et bono*...". Sir Arthur Watts, formerly the Principal Legal Adviser to the British Foreign and Commonwealth Office, basing his analysis mainly on the *Greenland/Jan Mayen* case, conceded "the measure of discretion conferred on the Court by the need to arrive at an equitable result". A. Watts, "Delimitation in the Aegean Sea: Implications of Recent International Judgments", in Tashan, see note 29, 137.

observer, however, the Court's considerations of equity would seem sometimes to have overstepped the bounds of equity *infra legem*, coming close to deciding *ex aequo et bono*. But the Court has said that this was for the sole purpose of producing an equitable result.

Inasmuch as the primordial purpose of States going to the ICJ is to have their disputes settled, it is natural that the latter should aim at an equitable solution. As the Court pointed out in the *Tunisia/Libya* case, an equitable result is an end and the application of equitable principles a means to achieve it. It thus becomes understandable that the Court has sometimes "taken refuge in silence", in the words of the late Fitzmaurice, when it almost decided *ex aequo et bono*, although it had not been authorised to do so.

Since amendments were made to Article 17, paragraph 2, of the Rules of Court in 1972 and 1978 for easier use of the chamber under Article 26, paragraph 2, of the Statute, the parties to a dispute are now able to have more of a say in the composition of a chamber. Besides, each party may have one judge *ad hoc* on the bench of a chamber. All this was designed to encourage States to make more use of the ICJ in the days when it had no cases on its docket. Putting such policy considerations aside, a chamber may now have a character fairly close to an arbitral tribunal in terms of its composition.

Arguably more interesting than the increasingly similar composition of ICJ chambers and arbitral tribunals is the increasingly similar attitude of the ICJ and arbitral tribunals towards the basis for decisions, as may be seen in some recent maritime boundary cases. The tendency of the ICJ to resort to considerations of equity is critically pointed out, for example, in the separate opinions of Judge Oda and some of his colleagues in the *Greenland/Jan Mayen* case.[54] Indeed, Judge Oda's discomfort with the delimitation line of the Court's choice may be shared by many international lawyers. One may even wonder if it constituted the right application of the applicable law by a court of law.

If one returns to the basic definition of the institution of the ICJ, however, one must nevertheless admit that it *is* a judicial organ for *settling* the disputes submitted to it. This is, of course, not to say that it is permitted to do whatever may contribute to the solution of the dispute before it. It is a court of law "whose function is to decide *in accordance with international law* such disputes as are submitted to it",[55] and it is therefore not allowed to go beyond the bounds of law. The ultimate question, then, is whether or not the ICJ has overstepped such bounds by relying on certain equitable considerations for the primordial purpose of settling the disputes submitted to it.

---

[54] Separate opinions of Vice-President Oda and Judges Schwebel and Shahabuddeen, ICJ Reports 1993, 117, para. 100; 120; 193.
[55] ICJ Statute, Article 38, para. 1.

Oil and Gas in the Caspian Sea and International Law

Kazuhiro Nakatani

The Caspian region is considered to contain, as the last bonanza, the largest undeveloped oil and gas reserves in the world.[1] In this paper, which covers developments until the end of the year 2000, the author would like to consider the following three important legal questions relating to oil and gas in the Caspian Sea very briefly, namely, (I) the legal status of the Caspian Sea, (II) the unilateral developments of oil and gas pending determination of the legal status of the Caspian Sea, and (III) the transport of Caspian oil and gas by pipelines and the Energy Charter Treaty.

I. Legal Status of the Caspian Sea

The legal status of the Caspian Sea has not been settled.[2] Although the USSR and Iran (Persia) concluded some bilateral treaties concerning the

---

[1] On the Caspian oil and gas in general, see International Energy Agency, *Caspian Oil and Gas*, 1998.
[2] On the legal status of the Caspian Sea, see A. Dowlatchahi, *La mer caspienne. Sa situation au point de vue de droit international*, Theses, Université de Paris, 1961, mimeographed; M.-R. Dabiri, "A New Approach to the Legal Régime of the Caspian Sea as a Basis for Peace and Development", *Iran. J. Int'l Aff.* 6 (1994), 28 et seq.; H.-J. Uibopuu, "The Caspian Sea: A Tangle of Legal Problems", *World Today*, May 1995, 119 et seq., S. Vinogradov and P. Wouters, "The Caspian Sea: Current Legal Problems", *ZaöRV* 55 (1995), 604 et seq. B. M. Clagett, "Ownership of Seabed and Subsoil Resources in the Caspian Sea under the Rules of International Law", *Caspian Crossroads Magazine* 1-3 (1995), 3 et seq.; B. H. Oxman, "Caspian Sea or Lake: What Difference Does It Make?", *Caspian Crossroads Magazine* 1-4 (1995/1996), 1 et seq.; S. Vinogradov and P. Wouters, "The Caspian Sea: The Quest for a New Legal Régime", *Leiden JIL* 9 (1996), 87 et seq.; A. P. Mizzi, "Caspian Sea Oil, Turmoil, and Caviar: Can They Provide a Basis for an Economic Union of the Caspian Sea", *Colo. J. Int'l Envtl. L. & Pol'y* 7 (1997), 483 et seq.; D. Allonsius, *Le régime*

Caspian Sea early in the 20th century, there was no provision stipulating its legal status and mineral resources.

Article 11 of the Treaty of Friendship (signed on 26 February 1921) provides:

> "In view of the fact that the Treaty of Turkomantchai, concluded on February 10, 1828 (old style), between Persia and Russia, which forbids Persia, under the terms of Article 8, to have vessels in the waters of the Caspian Sea, is abrogated in accordance with the principles set forth in Article 1 of the present Treaty, the two High Contracting Parties shall enjoy equal rights of free navigation on that Sea, under their own flags, as from the date of the signing of the present Treaty".[3]

Article 12 of the Treaty and Commerce and Navigation (signed on 25 March 1940) which replaced the Treaty of Establishment, Commerce and Navigation (signed on 27 August 1935) provides:

> "1. Vessels plying in the Caspian under the flag of either of the high contracting parties shall be treated in all ways in the same manner as the national vessels when in the ports of the other high contracting party, whether during arrival, stay in port, or departure.
> 2. [omitted here]
> 3. Cabotage is the exclusive right of the national vessels of the high contracting party; nevertheless, it is agreed that each of the high contracting parties shall grant the right of cabotage for the transport of passengers and cargo on the

---

juridique de la mer Caspienne. Problèmes actuels de droit international public, 1997; L. Savadogo, "Les régimes internationaux de l'utilisation des ressources minerales de la mer Caspienne", *Annuaire de droit de la mer 1997-II* (1998), 265 et seq.; S. Vinogradov, "The Legal Status of the Caspian Sea and Its Hydrocarbon Resources", *and* H. R. Huttenbach, "The Post-Soviet Conflict over Offshore Boundaries in the Caspian Sea and the New Littoral States", *both* in G. H. Blake, M. A. Pratt and C. H. Schofield (eds.), *Boundaries and Energy: Problems and Prospects*, 1998, 137 et seq. and 157 et seq.; S. Kawakami, "The Legal Status of the Caspian Sea", Parts 1 to 3, *Bulletin of the Research Institute Chuo-Gakuin University* 13-2 (1998), 57 et seq.; 14-1 (1999), 137 et seq.; and 15-1 (2000), 117 et seq. (in Japanese); R. Meese, "La mer Caspienne: quelques problèmes actuels", *RGDIP* 103-2 (1999), 405 et seq.; R. Yakemtchouk, *Les Hydrocarbures de la Caspienne. Le compétition des puissances dans le Caucase et en Asie centrale*, 1999; W. Ascher and N. Miroviskaya (eds.), *The Caspian Sea: A Quest for Environmental Security*, 2000.

[3] LNTS (1922), Vol. 9, No. 268.

Caspian Sea to vessels plying under the flag of the other party.
4. Notwithstanding the provision set forth above, each of the high contracting parties reserves for its own vessels the exclusive right of fishing in its coastal waters up to a limit of 10 nautical miles, and reserves the right to grant special reliefs and privileges in connexion with imports of fish caught by vessels plying under its own flag".

Article 13 provides:

"The high contracting parties are agreed that, according to the fundamental principles set forth in the treaty of February 21, 1921, concluded between Iran and the RSFSR, no vessels other than those belonging to Iran or to the USSR or in an equal manner to the subjects and the commercial or transport organisations of one of the high contracting parties, flying the flag of Iran or of the USSR, may exist in the whole of the Caspian Sea".[4]

With the collapse of the USSR, the number of the riparian States has increased from two to five, namely Azerbaijan, Kazakhstan and Turkmenistan as well as Russia and Iran. In the Alma Ata Declaration of 21 December 1991, which established the Commonwealth of Independent States, the States participating in the Commonwealth, which include the first four States, guaranteed in accordance with their constitutional procedures the discharge of the international obligations deriving from treaties and agreements concluded by the former Union of Soviet Socialist Republics.[5] At any rate, there is no doubt that the five States are bound by the above-mentioned two bilateral treaties of 1921 and 1940.[6]

The following points are relevant to this subject matter. First, it is the five riparian States which can determine the legal status of the

---

[4] On the 1940 Agreement, see *British and Foreign State Papers* 144 (1940-42), 419 et seq.; J. Degras (ed.), *Soviet Documents on Foreign Policy* 3 (1953), 424 et seq. On the 1935 Agreement, see LNTS (1937), Vol. 176, No. 4069.

[5] *ILM* 31 (1992), 148 et seq.

[6] Russia, in a document entitled Position of the Russian Federation Regarding the Legal Régime of the Caspian Sea, UN Doc. A/49/475 (Annex) of 5 October 1994, stated that in accordance with the principles and norms of international law, Russia and other coastal States – the former Republics of the USSR and Iran – were bound by the provisions of the 1921 and 1940 Agreements.

Caspian Sea. Other States have no say as to its legal status. If the five States agree to consider the Caspian Sea as a part of the sea, each State can claim its own territorial sea, EEZ and continental shelf in accordance with UNCLOS.

Secondly, UNCLOS is not helpful in solving this problem. The Caspian Sea does not come under "enclosed or semi-enclosed" seas defined in the Article 122 of UNCLOS because it does not connect to another sea or ocean. At any rate, as Articles 122 and 123 cover only the definition and the cooperation of States bordering enclosed or semi-enclosed seas, they cannot be good guidelines for determining the legal status of the Caspian Sea.

Thirdly, even if the Caspian Sea is considered to be an international lake rather than a sea, there is no customary rule concerning its delimitation. Although most international lakes are partitioned among the littoral States, it is possible to determine its legal status as a condominium (joint ownership) or, as in the case of Lake Constance, to partition only the adjacent waters and leave the rest of the waters subject to condominium.[7]

Fourthly, with regard to the intentions of the five riparian States, Azerbaijan and Kazakhstan have been advocating that the Caspian Sea should be divided into five sectors and each State should exercise complete jurisdiction over its sector. In a joint statement, they declared: "Two Parties consider that the delimitation of the Caspian Sea between the littoral States in one form or another is fully consistent with international practice and the principles and rules of international law and will facilitate their cooperation on the basis of equal rights and mutual advantage and help to attract investments and modern technology for the purpose of the effective, rational and safe use of the natural resources of the Caspian".[8] Azerbaijan insists that she has based its sea boundaries with Iran on the Astara-Hasangulu line, which was accepted as the boundary line separating the former Soviet Union and Iran and that all activities carried out by its State Oil Company *(SOCAR)* that relate to the exploration and the development of deposits in the Caspian

---

[7] On the legal status and delimitation of international lakes, see R. Zacklin and L. Caflisch (eds.), *The Legal Régime of International Rivers and Lakes*, 1981; F. Schroter, "La délimitation des lacs internationaux: essai d'une typologie", *AFDI* 40 (1994), 910 et seq.. On the legal status and delimitation of Lake Constance, see G. Riva, "L'exercice des droits de souveraineté sur la lac de Constance", *ASDI* 24 (1967), 43 et seq.; J. A. Frowein, "Lake Constance", in R. Bernhardt (ed.), *EPIL*, Vol. XII, 1990, 216 et seq.

[8] Joint Statement on the Caspian Sea Questions Adopted by the Presidents of Azerbaijan and Kazakhstan in Baku on 16 September 1996, UN Doc. A/51/529 (Annex) of 21 October 1996.

Sea have always been based on the sectoral division of the Caspian as determined by the Ministry of Oil Industry of the former USSR.[9]

Russia at once took the following stance:

"The Caspian Sea lacks a natural link to the world's oceans and seas and is thus a land-locked body of water. The norms of international maritime law, particularly those pertaining to the territorial sea, the exclusive economic zone and the continental shelf, are not applicable to it. ... The Caspian Sea and its resources are of vital importance to all the States bordering to it. For this reason, all utilisation of the Caspian Sea, in particular the development of the mineral resources of the Caspian seabed and the rational use of its living resources ... must be the subject of concerted action on the part of all States bordering the Caspian if the flora and fauna of this unique body of water are not to be harmed. Its ecosystem is highly vulnerable, and the chief task at hand is to prevent a regional environmental disaster. This task can only be addressed through strict respect for the legal regime of the Caspian Sea and the prevention of any unilateral action, given that the Caspian Sea, by virtue of its legal nature, is subject to joint utilisation; any questions relating to activities, including the exploitation of its resources, must be decided jointly by all the countries bordering its coast".[10]

This Russian position has been dramatically changed by the Agreement Between the Republic of Kazakhstan and the Russian Federation on the Delimitation of Bottom of the Northern Part of the Caspian Sea of 6 July 1998.[11] Article 1, paragraph 1, of the Agreement provides that the seabed

---

[9] K. B. Yusifzade, "The Status of the Caspian Sea", *Azerbaijan International* 8-3 (2000), 93.

[10] Position of the Russian Federation Regarding The Legal Regime of the Caspian Sea, UN Doc. A/49/475 (Annex) of 5 October 1994. Vinogradov points out that, contrary to a widely held belief, Russia has not invoked condominium (joint ownership) but joint management as a basis for its position. S. V. Vinogradov, "'The 'Tug of War' in the Caspian: Legal Positions of the Coastal States", Ascher and Mirovitskaya, see note 2, 194. Russia cites preservation of ecosystems as one of the main reasons why the Caspian Sea is subject to joint utilisation. However, there is no warranty that the joint management is the best way to preserve the ecosystem. It might ironically result in the deterioration of the environment if the concerned States consider that "everybody's business is nobody's business".

[11] On this Agreement, see R. N. Gaisin, "Caspian Legal Regime: Dynamics of Kazakhstan's Position", Ascher and Mirovitskaya, see note 2, 180 et seq. A French translation of the Agreement is provided in Yakemtchouk, see note 2, 162 et seq.

and subsoil of the northern part of the Caspian Sea are divided between the two States subject to a (modified) medium line. Article 2, paragraph 1, provides that the two States exercise sovereign rights over resources of the seabed and subsoil in their respective parts.[12] The reason why Russia changed its position can easily be explained, as only recently there have been indications that the Russian coasts of the Caspian Sea contain significant oil fields.[13]

Iran has been taken the position that the Caspian Sea is subject to condominium and it has been opposed to unilateral developments of oil and gas in the Caspian Sea. According to the joint statement of 8 July 1998 between the President of the Islamic Republic of Iran and the President of Turkmenistan: "The two sides reiterated that the Caspian Sea and its resources are the common heritage of the littoral States and that the completion of the legal régime and the resolution of issues on the exploitation of the resources of the Caspian Sea is exclusively the right of these five littoral States while adopting the convention on the legal status of the Caspian Sea. ... The two sides were of the view that a condominium arrangement for the common use of the Caspian Sea by the littoral States through assuming a sectoral coastal strip as the national zone is the most appropriate basis for the legal régime. The extent of national zones and the procedures for common use of the Caspian Sea will be subject to supplementary agreements".[14] It is reported that the Iranian Foreign Minister Kamal Kharrazi, in a meeting on 1 August 2000 with Victor Kalzhny, the Russian President's special envoy for Caspian Sea affairs, stated that Iran, although preferring joint ownership of the Caspian Sea, was ready to accept a complete division of the Caspian Sea, with Iran having a 20 per cent share in order to speed up resolution of the issue.[15]

Turkmenistan was ambivalent as to its position on the legal status of the Caspian Sea. Paragraph 1 of the joint statement on issues relating to

---

[12] Gaisin, *ibid.*, 180 et seq. According to Yakemtchouk, see note 2, 163 et seq. Article, 1 para. 1, reads: "Les fonds de la partie septentrionale de la mer Caspienne et son sous-sol sont partages entre les Parties suivant la méthode de la ligne médiane, telle que modifiée en vertu du principe d'équité et par accord entre les Parties, les eaux de surface restant affectées à l'utilisation commune aux fins de garantir la liberté de navigation, l'application des normes de pêche adoptées d'un commune accord et la protection du milieu marin". Article 2, para. 1, reads: "Les Parties exerceront leurs droits souveraines aux fins de la prospection, de l'exploitation et de la gestion des ressources des fonds de la parties septentrionale de la mer Caspienne et de son sous-sol dans les limites du secteur qui leur aura été attribué, et qui s'étend jusqu'à la ligne de partage".

[13] Vinogradov, see note 9, 195.

[14] UN Doc. A/53/453 (Annex) of 2 October 1998.

[15] *Tehran Globe*, 1 August 2000.

the Caspian Sea, signed by the Presidents of Kazakhstan and Turkmenistan in Almaty on 27 February 1997 provides: "The parties consider that the elaboration and conclusion of a convention on the legal status of the Caspian Sea is an urgent priority task. Until the Caspian States reach an agreement on the status of the Caspian Sea, the parties will adhere to the delimitation of administrative and territorial borders along a line running through the middle of the Sea".[16] This position is hard to make compatible with the position contained in the joint statement of 8 July 1998 mentioned above. Iran, without doubt, regarded the Kazakhstan-Turkmenistan joint statement as a step in violation of the existing legal régime and declared that it bore no legal value.[17]

To sum up, although the positions of the five littoral States concerning the legal status of the Caspian Sea are not uniform, there is an undeniable tendency towards the partition of the Caspian Sea. The concept of the common heritage of the littoral States has proved to be nothing but a tool for some littoral States whose coasts have no significant oil fields.

II. Unilateral Developments of Oil and Gas Pending Determination of the Legal Status of the Caspian Sea

Pending determination of the legal status of the Caspian Sea, some unilateral development of oil and gas has already been started by some littoral States. For example, on 20 September 1994, the State Oil Company of the Azerbaijan Republic *(SOCAR)* concluded an agreement with *Amoco*, *BP* and others on joint development and production sharing for the Azeri and Chirag Fields.[18] This agreement was called "the Contract of the Century".

---

[16] UN Doc. A/52/93 (Annex) of 17 March 1997. On 3 October 2000, Turkmenistan's presidential envoy on the Caspian stated that Turkmenistan would not change its stance that the Caspian must be divided into national sectors, and criticised Russia for only aggravating the problem. Russia proposed that the seabed and its natural resources be divided nationally, while the waters, with valuable fish stocks, be shared. See http://www.cnn.com/2000/ASIANOW/central/10/03/turkmenistan.russia.ap/ index.html.

[17] Letter dated 3 September 1997 from Chargé d'Affaires a.i. of the Permanent Mission of the Islamic Republic of Iran to the United Nations addressed to the Secretary-General, UN Doc. A/52/325 of 8 September 1997.

[18] *Barrow's Petroleum Concession Handbook Supplement* 111 (1996), 38 et seq. Art. 2.0: *SOCAR* grants to contractors the exclusive right to conduct operations within the contract area. Art. 4.1: duration is 30 years. Art. 11.6: the balance of total production after deducting operating costs and accumulated capital costs is shared between *SOCAR* and contractor according to the cumulative after-tax, real rate of return (PROF). Art. 23.1:

Unilateral developments have met with protests. For example, Iran lodged a protest with Azerbaijan for extraction and exploitation of oil from the Chirag reservoir. Iran stated that such unilateral actions contravened the Iran-USSR Agreements of 1921 and 1940, as well as the Ashgabat Declaration of 12 November 1996 (in which the Foreign Ministers of the five coastal States agreed that complementation of the legal régime of the Caspian Sea, following the dismemberment of the USSR, could only occur through the unanimous decision of the five coastal States) and that such measures, adopted without the consent of the coastal States, bore no legal value, and, as such, did not establish a basis for any right of claim. Iran also stated that full responsibility for consequences of such illegal measures and actions, including damage caused to other coastal States, rests with the States violating the legal régime of the Caspian Sea.[19] Azerbaijan, on her part, lodged a protest with Iran on the occasion of the signature of an agreement between Iran and the Shell and Lasmo Oil companies on geological and geophysical prospecting in an area of the Caspian Sea which allegedly includes part of the Azerbaijani sector of the Caspian Sea. Azerbaijan stated that the Iranian claims to 20 per cent of the territory of the Caspian Sea were contrary to the rules and principles of international law and the established practice of the littoral States of the Caspian Sea and that the Iranian unilateral and illegal actions undermined the positive trends which had marked the process of determining the legal status of the Caspian Sea.[20] On 7 July 1997, Turkmenistan lodged a protest with Azerbaijan because the Agreement between Azerbaijan and Russia on the joint exploration of the Kapaz field overlapped with Turkmenistan's Serdar deposit. Turkmenistan insisted that the historical boundary line

---

the contract is governed by the law of Azerbaijan, English law and the principles of the common law of Alberta, Canada.

[19] Declaration expressing the position of the Islamic Republic of Iran on the exploitation of the resources of the Caspian Sea, UN Doc. A/52/588 (Annex) of 25 November 1997. Therefore, according to Iran, "the inclusion of such terms as the 'Kazakhstan Part' or the 'Russian Part' ... is in contravention of the existing legal regime and of the agreement reached among the coastal States in Ashgabat on 12 November 1996." Statement of the Foreign Ministry of the Islamic Republic of Iran on the results of the Kazakhstan-Russian Consultations reflected in the statement dated 13 February 1998 of the Kazakhstan Foreign Ministry, UN Doc. A/52/913 (Annex) of 21 May 1998.

[20] Statement of the Ministry of Foreign Affairs of the Azerbaijani Republic adopted on 10 December 1998, UN Doc. A/53/741 (Annex) of 14 December 1998. Iran rebutted, saying that the legal régime of this body of water should be established by the agreement of all its coastal States, and the economic activities of Iran in that area should not be a matter of concern to the coastal States of the Caspian Sea. Statement made by the spokesman of the Ministry of Foreign Affairs of the Islamic Republic Of Iran on 16 December 1998, UN Doc. A/53/890 (Annex) of 31 March 1999.

that placed the Kapaz field within the jurisdiction of Turkmenistan was based on the principle of equidistance as determined by the Soviet Union and that until there was a new agreement, no State should do anything outside these administrative borders.[21]

Pending determination of the legal status of the Caspian Sea, it is hard to tell whether such unilateral actions are illegal or not. It is hard to say, as Iran argues, whether such unilateral actions contravene the 1921 and 1940 Agreements between Iran and USSR as well as the Ashgabat Declaration of 12 November 1996, because these instruments do not directly prohibit unilateral actions. In addition, the Ashgabat agreement is non-binding.[22] Such unilateral actions might be illegal at a later stage when the legal status of the Caspian Sea is determined. In that case, contracts on the development of oil and gas in the pending area will be null and void. In this sense, the contractors have assumed a political risk.[23] This is the reality of unilateral developments in a pending area.

Protests are lodged in order to prevent unilateral actions from obtaining opposable effects, because silence on the part of a State in the face of a unilateral claim by another State might be interpreted as acquiescence when the claim affects subjective interests of the former State. *Qui tacet consentire videtur si loqui debuisset ac potuisset*, as the ICJ also pointed out in the judgment of the *Preah Vihear* case.[24]

III. Transport of Caspian Oil and Gas by Pipelines and the Energy Charter Treaty

The most suitable means of transport of Caspian oil and gas to global markets is through pipelines. The proposed oil export pipeline routes from Baku are the Russian route (via Chechnya to Novorossiysk on the

---

[21] See website of the Embassy of Turkmenistan, Washington, DC, at http://www.dc.ifni.net/~embassy/prazeri.html. On this dispute, see A. Harris, "The Azerbaijan-Turkmenistan Dispute in the Caspian Sea", *Boundary and Security Bulletin* 5-4 (1997-1998), 56 et seq.

[22] As to the unilateral development in a disputed sea area, Miyoshi points out that pending agreement on positive cooperation, there must be a negative form of cooperation to refrain from proceeding to a unilateral act of extraction and that this is implied in Articles 74(3) and 83(3) of UNCLOS. M. Miyoshi, "The Joint Development of Offshore Oil and Gas in Relation to Maritime Boundary Delimitation", *Maritime Briefing* 2-5 (1999), 1 et seq., 5. The present writer considers that it is hard to assert that a unilateral development in a disputed sea area is illegal, although it might be illegal when the legal status of the sea area is determined.

[23] In terms of contract, this problem is dealt with in the *force majeure* provision. In the case of the contract mentioned in note 18, *force majeure* is applied (Art. 21.1).

[24] ICJ Reports 1962, 6 et seq., 23.

Black Sea), the Georgian Route (to Supsa on the Black Sea), the Turkish route (via Georgia to Ceyhan on the Mediterranean Sea), and the Iranian Route (to the Persian Gulf). On 19 November 1999, in Istanbul, the Presidents of Azerbaijan, Georgia and Turkey signed a package of agreements on the Main Export Pipeline (Baku-Tbilisi-Ceyhan). This pipeline, which will stretch for 1730 kilometres across these three countries, is expected to become operational in 2004 and ship 50 million tons of oil by the following year. The sides also signed a memorandum of mutual understanding regarding the Azerbaijani gas exports to Turkish and world markets. Azerbaijan, Georgia, Turkey and Turkmenistan also agreed on an intergovernmental declaration on the Trans-Caspian gas pipeline that is to be laid across the Caspian Sea from Turkmenistan to Azerbaijan.[25]

There has been no generally established rule regulating international pipelines on the ground. However, with regard to Caspian oil and gas, Article 7 (Transit) of the Energy Charter Treaty (signed on 17 December 1994 and entered into force on 16 April 1998) is significant.[26] Paragraph 1 stipulates that each Contracting Party shall take the necessary measures to facilitate the transit consistent with the freedom of transit (cf. Article 5 of the GATT) without discrimination as to the origin, destination or ownership of oil and gas, and without imposing any unreasonable delays, restrictions or charges. Paragraph 2 provides for the encouragement of cooperation including "measures to mitigate the effect of interruption in the supply" of oil and gas. Paragraph 3 provides for the national treatment of oil and gas in transit, unless an existing international agreement provides otherwise. Paragraph 4 stipulates that, when the transit cannot be achieved on commercial terms, the Contracting Parties shall not place obstacles in the way of new capacity being established, except as may be provided in applicable legislation consistent with paragraph 1. As to Azerbaijan and Georgia, exceptions will apply in accordance with Article 32 and Annex T. Paragraph 5 stipulates that the transit State is not obliged to permit the construction of pipelines or additional transit through existing pipelines, which would

---

[25] *Caspian Business Report* 3-22 (30 November 1999), 2.
[26] *ILM* 33 (1995) 360 et seq. Among the five littoral States, all except for Iran are signatories of the Treaty. Georgia and Turkey are also signatories. Among these six States, Georgia, Kazakhstan, Turkmenistan and Azerbaijan have ratified the Treaty. On Article 7 of the Treaty, see Legal Counsel of the IEA, *The Energy Charter Treaty*, 1994, 13 et seq.; M. M. Roggenkamp, "Transit of Network-bound Energy: The European Experience", in T. W. Walde (ed.), *The Energy Charter Treaty*, 1996, 499 et seq.; P. Vlaanderen, "Energy Transit: The Multilateral Challenge", *International Association of Energy Economics Newsletter* 3 (1998), 16 et seq.; R. Liesen, "Transit under the 1994 Energy Charter Treaty", *J. En. & Nat. Res. L.* 17 (1999), 56 et seq.

endanger the security and efficiency of its energy system, including the security of supply. Contracting parties shall, subject to paragraphs 6 and 7, secure established flows. Paragraph 6 stipulates that the transit state shall not, in the event of a dispute over any matter arising from that transit, interrupt or reduce the existing flow of oil and gas prior to conclusion of the dispute resolution (conciliation) procedures set out in paragraph 7, which is independent from Part V (Dispute Settlement) of the Treaty. Paragraph 7 stipulates that only following the exhaustion of other remedies previously agreed may a Contracting Party refer the dispute to the Secretary-General of the Energy Charter Conference for the appointment of a conciliator. The conciliator, if he fails to secure agreement within 90 days of his appointment, shall recommend a resolution to the dispute and shall decide the interim tariffs and other terms and conditions. It is said that this conciliation procedure was modelled after labour dispute settlement mechanisms. On 4 December 1998, the Energy Charter Treaty Conference adopted Rules Concerning the Conduct of Transit Disputes.

The present writer would like to mention the following two points. First, even if freedom of transit is provided in paragraph 1, new access to transit is substantively subject to the permission of the transit State, as provided in paragraph 5.[27] Whereas established flows are secured, new access is not. Once new access is permitted, most-favoured-nation treatment and national treatment are secured in accordance with paragraphs 1 and 3.

Secondly, as to the compensation for the interruption of supply and, in particular, the damage to pipelines because of civil war or other political turmoil, which could occur any day in the unstable Caspian region, the extent of the application of the Article 7 paragraph 2(c) and Article 12 should be clarified. Article 12, paragraph 1, stipulates that an investor of any Contracting Party that suffers a loss owing to war or other armed conflict, state of national emergency or civil disturbance shall be accorded restitution, indemnification or compensation. According to this provision, the State where the political turmoil arises and causes damage to a pipeline has to pay indemnification or compensation to the investor(s) even if, according to the general rule on

---

[27] Liesen writes, "Transit services can be denied when they endanger security of supply, safety or efficiency of the energy systems. In fact, under normal circumstances all of the mentioned issues can be compensated by proper pricing schemes and hence do not constitute an economic argument for denying transit. Only where ... acceptable new or additional transit results in the impossibility to compensate endangerment, the objection of transit is acceptable". Liesen, see note 26, 64.

the attribution of responsibility,[28] the State is not responsible. This does not exclude the possibility that the State will claims reparations from the insurgents responsible for the damage according to the general rule, as Article 7, paragraph 8, confirms that nothing in the article shall derogate from a Contracting Party's rights and obligations under international law. Article 7, paragraph 2(c), imposes on the transit State the so-called "best efforts" obligation[29] to restore the damaged pipeline in its territory. Once again, the transit State is requested to restore the pipeline in question. However, what the transit State owes here is only its "best efforts". This means that shortage of money for the restoration of the pipeline because of non-indemnification can be a reason for exemption from the obligation, when the State's best efforts are exhausted.

Intergovernmental negotiations on the text of the legally binding Transit Protocol began in 2000. Also, a set of non-binding Model Transit Agreements are now being developed by the Energy Charter's Transit Working Group. The Working Group believes that the Agreements should consist of (i) a Model Intergovernmental Agreement which deals with operational issues that are most commonly settled horizontally among state authorities, (ii) a Model Host Government Agreement which deals with operational issues that are most commonly settled vertically among host governments *vis-à-vis* investors and operators, and (iii) a Model Transit Transportation Agreement which deals with operational issues that are most commonly settled horizontally among investor/operators/shipper/users and state companies.[30] These Agreements will be essential for enhancing the stability of pipeline transport.

As to the Trans-Caspian gas pipeline, although the question whether the relevant provisions of UNCLOS[31] are applied or not is subject to the legal status of the Caspian Sea, it seems reasonable to apply these rules, *mutatis mutandis*, even pending the determination of its legal status.

---

[28] On the general rule on the attribution of responsibility in a civil war, see Articles 14 and 15 on the ILC Draft Articles on State Responsibility, ILCYB 1975-II, 91 et seq. The territorial (transit) State can be held responsible when a lack of due diligence on the part of the State leads to the damage.

[29] "Best efforts" obligations were considered in detail in the Heathrow Airport User Charges Arbitration of 30 November 1992. *ILR* 102 (1996), 215 et seq.

[30] See http://www.encharter.org/English/General/Explanatory.html.

[31] In UNCLOS, the relevant Articles are 58, 79, and 112-115.

Maritime Boundary Issues and Their Resolution: An Overview

*M. C. W. Pinto**

I. Introduction

Boundaries are imaginary lines that represent geographic limits to a State's jurisdiction, i.e., the State's power to make laws, and interpret, administer and enforce them. Such limits have often – but not always – to do with restrictions on the very sovereignty of the State. Where the land territories of States are *adjacent* and – as is of more concern to us here – they have maritime areas that are adjacent to each other, or where the maritime areas of States *confront* each other, a boundary between them will restrict the jurisdiction of each in relation to the other. However, when the territory of a State confronts an area not subject to sovereignty, as when the marine area of a State confronts the open sea, the State's boundary with the open sea will restrict its jurisdiction in relation to the activities of all other States in the exercise of their rights on the open sea.

Because territorial boundaries may be perceived by national political leaders to be constraints on the exercise of sovereignty, in that they may restrict the power of those who govern to give external effect to policies they believe to be in the best interest of the governed, e.g., to explore and exploit marine or land-based resources in particular areas, disputes concerning them are often of a politically sensitive nature. So sensitive are such disputes, that governments often go to considerable lengths not to engage in them; or if they must, to ensure that the process is as non-confrontational as possible, with a view to maintaining good neighborly relations. Accordingly, in modern international law, boundaries, including maritime boundaries, are to be settled by *agreement between*

---

* Over the three decades of my acquaintance with Shigeru Oda as teacher, state representative and judge, I have come to value both his profound comprehension of issues of the law of the sea, and the independence of his judgment.

or among the States directly concerned, in contrast to former times, when such matters were dealt with in the far-away capitals of imperial powers.[1] We may expect such agreements to be preceded by protracted negotiations, or possibly by recourse to agreed third-party intervention through mediation, conciliation, arbitration or adjudication.

II. Maritime Boundary Issues

If a coastal State is not confronted, or laterally confined, by another State in such a manner as to restrict its claim to a 12-mile territorial sea, a 24-mile contiguous zone, a 200-mile exclusive economic zone, or a continental shelf to the full extent contemplated by the provisions of Article 76 of the UN Convention on the Law of the Sea, neither its claims regarding baselines, nor the maritime boundaries it establishes are likely to require negotiation, or be the subject of a challenge, or give rise to disputes as to their location. In practice, however, such ideal conditions are not frequently encountered. The presence of islands, questions of sovereignty over islands, the degree of "effect" an island is to be given in the making of a boundary, historic rights of access, whether a particular physical feature is to be treated as an island, reef, rock, or low-tide elevation, whether neighbors share a common continental shelf, and ways of apportioning such a shelf between them when it is less than 400 miles broad, are among the complicating factors that might need to be considered.

One authority offers the following summary of maritime boundary delimitation problems:

> "There are three basic problems in delimiting maritime boundaries. First, it is necessary to establish the baseline to be

---

[1] For a comprehensive treatment of maritime boundaries, see D. M. Johnston, *The Theory and History of Ocean Boundary-Making*, 1988; on more specialized aspects of maritime boundaries, see: D. M. Johnston and P. M. Saunders (eds.), *Ocean Boundary-Making: Regional Issues and Developments*, 1988; G. Blake, *Maritime Boundaries and Ocean Resources*, 1987; P. Weil, *The Law of Maritime Delimitation – Reflections*, 1989; K. Kittichaisaree, *The Law of the Sea and Maritime Boundary Delimitation in South-East Asia*, 1987; S. P. Jagota, *Maritime Boundary*, 1985; H. W. Jayewardene, *The Regime of Islands in International Law*, 1990; J. R. V. Prescott, *Boundaries and Frontiers*, 1978, Chapter 5: "Maritime Boundaries"; idem, *The Maritime Political Boundaries of the World*, 1985, 45 et seq., and extensive bibliography; E. Collins, "The Gulf of Maine Case and the Future of Ocean Boundary Delimitation", *Marine Law Review* 38 (1986), 1 et seq.; The Hydrographic Society, *Maritime Limits and Baselines – A Guide to Their Delineation*, 3rd ed., 1987.

used by the coastal State. The selection of this line will immediately fix the outer edge of the State's internal waters, and then permit the mechanical determination of the outer edges of the territorial waters, the contiguous zone, and the exclusive economic zone or fishing zone, since each of them consists of a uniform distance measured seawards from the baseline. In the case of archipelagic States the establishment of the baselines will instantly determine the extent of the archipelagic waters.

The second problem concerns the determination of the outer edge of the continental margin. This boundary is not related to the baseline, except that in the case of submarine ridges, other than oceanic submarine ridges which cannot be claimed, the boundary may not be drawn more than 350 nautical miles beyond the baseline.

The third problem centres on the need for countries to draw common international limits. This need arises when countries are so close together that if each claimed the full suite of maritime zones, some of them would overlap.

All countries with coasts must face these three problems, because there is no country which is so far from neighbours that the construction of international boundaries is unnecessary...".[2]

The origins of maritime boundary disputes frequently lie in (1) the method used by a coastal State (or an island or archipelagic State) for drawing the baselines from which the breadth of its territorial sea, and essentially the breadth of each of its maritime zones, is measured; (2) the demarcation by a State of maritime zones which, in the perception of another State, exceed the entitlement of the former as prescribed by international law; and (3) the activity of foreign ships, such as fishing or mineral exploration, within a coastal State's demarcated maritime zones alleged to be inconsistent with international law or coastal State legislation, or both. Such actions may be seen as threatening or causing damage, and result in the arrest of ships and crews, or other enforcement action, a corresponding response from the flag State, leading, *inter alia*, to questioning of the coastal State's boundary-making policies and practices.

---

[2] J. R. V. Prescott, *The Maritime Political Boundaries of the World*, 1985, 45-46. In another work, the same author lists some 30 "apparently uncomplicated maritime boundaries" in the South-West Pacific Ocean, which, he observes, are found mainly in the north and east of the region, away from the submarine ridges and rises which link Australia, New Zealand, New Caledonia, Fiji and Tonga, and discusses the significance of physical features in the latter area in detail. "Maritime Boundaries and Issues in the South-West Pacific Ocean", in Johnston and Saunders, see note 1, 297.

Maritime boundary issues may arise in a wide variety of circumstances. Most often bilateral, they may well involve more than two States, as when the demarcation of boundaries involves a "tripoint" or *triplex confinium* where the territories of three States converge,[3] or where traditional fishing rights in an area are claimed by more than two States. Remarking on the variety of causes of boundary controversies generally, one author suggests that the following factors often generate conflict: (i) the non-existence of any boundary line; (ii) insufficient marking of the boundary; (iii) interpretation of inexact, unsuitable or contradictory terms in the treaty establishing the boundaries; and (iv) inconvenience and dissatisfaction associated with the manner in which a boundary actually functions.[4] While maritime boundary controversies might be fitted into such broad and general classifications, the nature of the boundary concerned is such that such controversies share certain attributes that seem to set them apart.

Maritime boundaries traverse great areas of the sea, and activity at or near such boundaries is likely to be observed only by those who make a living from the sea, and whose knowledge and skills give them the opportunity and the license to navigate in such areas. We may expect any such activity, and any consequences thereof, occurring, as it does, beyond areas of normal human habitation, to remain largely unwitnessed and unreported. If a coastal State has enacted laws and regulations governing its maritime areas and their resources, the administration of these laws and regulations, including maintenance of systems of surveillance and enforcement to ensure compliance with them, would require the support of specially trained and equipped personnel. For the majority of coastal States that are entitled to substantial marine areas and resources, effective administration of them could, given their other national economic priorities, be beyond their financial and technological capability. Surveillance to detect illicit activity, as well as to collect the evidence needed to support enforcement action, including dispute resolution, may require technology and skills to which they have no ready access, leaving such areas and their resources particularly vulnerable to unauthorized use or exploitation.

Another distinguishing feature of maritime boundary issues is the nature of the terrain to be surveyed. Establishment of a maritime boundary must take into consideration, in addition to the general characteristics of the visible coast, not only an expanse of water that is featureless to the untrained eye, but also a complex underwater configuration, invisible without the aid of specialized equipment and

---

[3] See the maritime boundary agreements between India/Sri Lanka/Maldives (31 July 1976), and between India/Indonesia/Thailand (2 March 1979).

[4] P. K. Menon, "International Boundaries – A Case Study of the Guyana-Surinam Boundary", *ICLQ* 27 (1978), 738 et seq., 741-742.

trained personnel: the geological structures beneath the seabed, occasional protrusions above the water surface in the form of islands or low-tide elevations, artificial structures on the water surface or attached to the seabed, and, most importantly, the presence of living and non-living resources in the water column, or on or beneath the seabed below it.

Such factors suggest that cooperative arrangements and agreements among States are the most appropriate and efficient approach to the establishment and administration of boundaries, as well as to resolving boundary issues, since they lead to the pooling of resources for mutual protection of maritime entitlements, avoid confrontation, and build confidence. Such a course would seem particularly appropriate for States in the same geographical area, with historical and cultural affinities, and perhaps lacking the financial and technical resources to proceed independently. Where geographical, cultural and historical characteristics have given rise to similar political and policy orientations among the States concerned, *regional* approaches to maritime issues, including boundary issues and the resolution of differences and disputes, should they occur, are clearly indicated.[5]

1. Alternative Approaches to Maritime Boundary Issues

Johnston and Saunders have suggested a list of alternative courses of action open to States confronted with maritime boundary issues, based on a "functional approach". The categories of policy options considered are: (1) direct bilateral diplomacy; (2) third-party adjudication; and (3) non-adjudicative intermediation. Within the first category, a spectrum of options are suggested, namely: (i) *do nothing* in consideration of the risk of conflict with a possibly sensitive or hostile neighbor; (ii) public admission of a boundary dispute, and *agreement to disagree*, implying an appeal to neighboring States to avoid complicating matters, and foreseeing a good faith effort at resolution when conditions improve; (iii) an *agreement designating the boundary area in dispute*, at once

---

[5] Johnston and Saunders suggest, as parameters relevant to the classification of an area as a "region", the following illustrative list: (1) offshore resource potential; (2) degree of geographic complexity increasing the difficulty of resolving conflicting claims; (3) level of convergence of regional practice with global boundary norms; (4) extent of regional organization; (5) level of ideological/economic/cultural symmetry among the States concerned; (6) social and political stability of the area; and (7) level of economic development of the area as a whole. D. M. Johnston and P. M. Saunders, "Ocean Boundary Issues and Development in Regional Perspective", in Johnston and Saunders, see note 1, 313, 316-320.

limiting the area of controversy, and narrowing the issue for future negotiation; (iv) *agreement on some limited degree of cooperation or consultation within the designated area in dispute*, e.g., joint research; (v) *agreement on certain questions of access to the area in dispute for specified purposes*, envisaging a range of possible provisional arrangements that could be negotiated pending eventual settlement; (vi) negotiation of *a preliminary joint undertaking that contemplates future joint production*, a course of action that might be motivated by the prospect of the disputed area's resource potential; (vii) *a joint venture for designated production purposes within the disputed area*, a limited cooperative strategy that might be chosen when developmental prospects for both States are favorable, as are prospects for their eventual agreement on detailed conditions of operation for the entrepreneurial and managerial agencies that will be involved, implying medium-term postponement of a boundary delimitation settlement within the area; (viii) *an agreement for the sharing of specified state services*, requiring formal and continuing commitment for an indefinite period, such as might be feasible in regard to some subject of strong mutual concern, e.g., environmental emergency, and is unlikely to arouse political sensitivities; (ix) *a full-scale but incomplete joint management system*, covering the entire area in dispute, but short of stipulating how all ocean management functions are to be assigned among state agencies; (x) *a full-scale and complete joint management system* covering the entire disputed area, and providing for assignment of all management functions among agencies of the States concerned; and (xi) *agreement on a final and formal boundary treaty* specifying the exact location of the boundary line in the disputed area.[6]

While conceding that at sea, as on land, there are advantages to installing a fence between even the best of neighbors, since "it often serves as the final answer to quarrels about space and entitlement" (option 11), Johnston and Saunders would for some regions advocate options 8, 9 or 10, arrangements that would enable neighboring States with opposite or adjacent coasts to explore and exploit their wealth of marine resources on a mutually advantageous cooperative basis:

> "In many cases, it may be in the interest of a developing coastal (or island) State to defer final boundary delimitation decisions until the region as a whole has been able to assemble an effective strategy for ocean development and management on a regional basis, with carefully thought out implications for bilateral boundary making in the region".[7]

---

[6] *Ibid.*, 331-336.
[7] *Ibid.*, 337.

III. Resolution of Maritime Boundary Disputes: Methods

1. Methods of Dispute Resolution

The methods traditionally followed in resolving interstate disputes are listed in Article 33 of the UN Charter in a provision that commits all Member States to "peaceful settlement". They are: negotiation, enquiry, mediation, conciliation (often characterized as "diplomatic" methods since they do not foresee decisions that are binding on the disputants), arbitration and judicial settlement, the latter methods implying that a tribunal or court will render an "award" or "judgment" which the disputants are legally bound to carry out.[8] Article 33 goes on to require, as possible alternatives, "resort to regional agencies or arrangements, or other peaceful means of [the disputants'] choice". These are also the methods of settlement to which the States Parties to the UN Convention on the Law of the Sea commit themselves by Article 279 of the Convention, while other provisions of Part XV of the Convention establish an elaborate compulsory dispute settlement régime, which is binding among States Parties with regard to disputes which arise after entry into force of the Convention, and relates to the interpretation and application of the Convention. Article 298(1)(a) on compulsory conciliation, which is of special relevance to boundary disputes, is dealt with below. Of the methods listed, negotiation alone does not contemplate the intervention of a trusted and agreed third party, leaving the disputants to arrange matters on their own. Each of the others foresees joint submission of the dispute to a third party, and cooperation with it in its efforts to resolve the dispute.

2. Negotiation and Consultations

Negotiation and a variant thereof, sometimes referred to as "consultations", are the methods most frequently used by States in attempting to resolve their disputes. One or the other would be the first step in any initiative to settle a maritime boundary dispute. Although usually commenced voluntarily by a disputant, negotiation may also be obligatory when it is commenced in implementation of a system of dispute settlement to which the disputants are committed, or in

---

[8] For a comprehensive overview of methods of international dispute resolution, see J. G. Merrills, *International Dispute Settlement*, 1991; United Nations, *Handbook on the Peaceful Settlement of Disputes Between States*, 1992. On maritime disputes, see generally A. O. Adede, *The System for Settlement of Disputes under the United Nations Convention on the Law of the Sea*, 1987, especially Chapter XII.

implementation of the order of a court[9] or a decision or recommendation of some other body on which the disputants have conferred a measure of authority. Negotiation presents no aspect threatening to either party. It requires no involvement of a third party, enables each party to maintain control over the procedure, and does not impair the parties' sovereign rights. It offers each party the freedom to present and emphasize any element legal, technical or political, that it considers relevant, allows confidentiality and cordial relations to be maintained, involves little expenditure, and is of proven efficiency, since it is the means by which most interstate disputes are resolved.[10] The fact that negotiation may be in progress does not form a bar to some other method of settlement, and unless the contrary is specified in an agreement, negotiation may continue in parallel with any other settlement initiative. When negotiation is successful in resolving the dispute, the parties conclude an agreement embodying the terms of settlement.

Negotiation is not only the most efficient and frequently-used method of resolving maritime boundary disputes: it is also of value in *preventing* disputes. Engaged in for the latter purpose, the process is sometimes distinguished as "consultations". Thus, when a government anticipates that a decision or proposed course of action may harm another State's interests, discussion with that State can offer a way of heading off a dispute by creating an opportunity for adjustment and accommodation.[11]

3. Good Offices and Mediation

Two "diplomatic" methods available for settling disputes involving recourse to an agreed third party (institution or individual) for the resolution of boundary disputes are good offices and mediation. They may be commenced at the invitation of or with the consent of the disputants. The disputants may also engage in these procedures in implementation of a treaty obligation, the order of an international court, or a decision or recommendation of some other body, such as a regional or international organization, on whom the disputants have, by prior or *ad hoc* agreement, conferred a measure of authority. Regional organizations are in a particularly advantageous position to undertake dispute resolution between their members by these methods.

---

[9] The *Fisheries jurisdiction* cases, ICJ Reports 1974, 32; the *Gabčíkovo-Nagymaros Project* case, ICJ Reports 1997.

[10] The UN General Assembly gave fresh emphasis to the process through Resolution 53/101, entitled Principles and Guidelines for International Negotiations.

[11] Merrills, see note 8, 3-7.

Although differing slightly from each other in technique, mediation calling for somewhat greater creative initiative, the range of functions which these "diplomatic" methods require of a mutually agreed third party (e.g., state representatives, international or regional organizations, international officials, eminent private individuals) have much in common: investigating and clarifying the facts in dispute, ascertaining the limits of a disputant's position, making proposals that appear to the third party to represent a fair and equitable resolution of the dispute, and, if a settlement is achieved, assisting the parties to embody the settlement in an agreement, and organizing and overseeing its implementation. The actual range of functions conferred on a third party entrusted with applying one of these methods, is determined by the disputants, in consultation with the chosen third party. The confidentiality and privacy inherent in these methods create conditions that help to preserve the mutual relations of the disputants, besides minimizing the costs and consequences of failure.

As with the range of their functions, the methods and aims of the two settlement techniques are similar. Both are conducted with a minimum of formality, care being taken to treat each party on the basis of equality, and to afford each of them a fair opportunity to state its position on the issues. Both aim at reducing tensions, opening or maintaining channels of communication between the disputants, offering advice and helping the disputants to understand each other's point of view, suggesting ways in which views could be reconciled and, in general, preventing aggravation of the dispute, while encouraging the parties towards the goal of agreement on a negotiated compromise.

Although the expenses of a mediator or a person using good offices must be met by the disputants, this would seem to involve only minimal expenditure. When the functions are performed by a regional or international organization of which the disputants are members, or by its representatives, all or a substantial portion of these costs may be absorbed by the organization, that is, borne by the membership as a whole.

It should be noted that the Permanent Court of Arbitration has established a Financial Assistance Fund from which it will provide financial assistance to qualifying States to enable them to meet, in whole or in part, the administrative costs involved in international settlement procedures involving recourse to third parties, including mediation and the use of good offices when carried out under its auspices.[12]

---

[12] The Administrative Council of the Permanent Court of Arbitration agreed to the establishment of a Financial Assistance Fund in 1994, and approved Terms of Reference and Guidelines for the operation of the Fund. Contributions to the Fund are made on a voluntary basis. The Fund provides financial support to qualifying States to enable them to meet, in whole or in

4. Inquiry and Conciliation

Inquiry and conciliation are two other "diplomatic" methods of dispute settlement that call for assistance and direction by agreed third parties. They may be commenced only at the invitation or with the consent of the disputants, although the latter may be required to engage in one or the other in implementation of a treaty, the order of a court, or a decision or recommendation of some international or regional body on which the disputants have conferred a measure of authority. As with good offices and mediation, they apply techniques of negotiation and persuasion to bring the disputants together. They are usually, but not necessarily, carried out by "commissions" of more than one person appointed in a manner agreed by the disputants. Although commissions of inquiry or conciliation operate in a more formal manner, and some international dispute settlement institutions have developed sets of procedural rules for them which disputants may choose to adopt, they are not empowered to make decisions that are binding on the disputants.[13] Their aim, as with good offices and mediation, is to bring about agreement between the disputants through the ascertainment of facts, clarification of issues and presentation of alternative courses of action. Both processes conclude with a report and recommendations to the disputants, which they are liberty to accept or reject.

While both inquiry and conciliation are informal and "non-adversarial" in their operation, and have been used successfully to resolve complex disputes, including maritime boundary disputes, their application in practice has been infrequent. Although the appeal of conciliation is such that it is the residual dispute settlement method most frequently included in multilateral treaties, and its potential is generally acclaimed, there has been little resort to conciliation in practice. Conciliation was the method used successfully to resolve the maritime

---

part, the costs involved in international arbitration or other means of dispute settlement provided for under the Hague Conventions of 1899 and 1901. To qualify for such support, a State (1) must have concluded an agreement for the purpose of submitting one or more disputes, whether existing or future, for settlement by one of the means administered by the PCA; and (2) at the time of requesting assistance from the Fund, be listed on the DAC List of Aid Recipients, maintained by the Organization for Economic Cooperation and Development (OECD). Annual Report of the PCA for 1995, 9.

[13] The UN General Assembly has repeatedly endorsed the use of both inquiry and conciliation. Most recently, as to inquiry, see Declaration on Fact-finding by the United Nations, Res. 46/59 of 9 December 1991; as to conciliation, see UNGA Res. 50/50 and the annexed UN Model Rules on Conciliation of Disputes Between States (1995). The Permanent Court of Arbitration has established Optional Rules for both fact-finding (inquiry) and conciliation (1996).

boundary dispute between Norway and Iceland in the *Jan Mayen Continental Shelf* case. The Conciliation Commission, having applied equitable principles and criteria as prescribed by the International Court of Justice, and "balancing the equities", recommended to the parties, not the demarcation of a boundary, but a settlement through cooperative arrangements to be included in a *joint development agreement* covering "substantially all of the area offering any significant prospect of hydrocarbon production".[14]

A recent study of the role of law and legal institutions in six Asian countries (China, India, Japan, Korea, Malaysia and Taipei, China) conclude that "Asia shows a preference for the mechanisms of mediation and conciliation available through traditional dispute settlement institutions".[15] That such a preference exists in West Asia under the influence of Islamic culture has also been noted,[16] while other commentators have suggested that

> "Both Southeast and Northeast Asia, more than any other regions of the world, seem to be culturally conditioned to de-emphasize the need to use law in 'contest'".[17]

It seems reasonable to conclude that, at the present time, except where the adversarial systems inherited from the imperial powers have been received and have put down deep roots, such a preference may be observed throughout Asia and the Pacific. It has been pointed out that where such a preference prevails, a price may have to be paid, in that important economic and institutional decisions may be delayed until legal and political issues can be circumvented through diplomatic ingenuity.[18]

5. Arbitration and Judicial Settlement

Where States acknowledge the existence of an international boundary dispute, and wish to resolve it by submitting it to an independent third

---

[14] "Report and Recommendations to the Governments of Iceland and Norway of the Conciliation Commission on the Continental Shelf Area Between Iceland and Jan Mayen", *ILM* 20 (1981), 797 et seq., 825.

[15] K. Pistor and P. A. Wellons, *The Role of Law and Legal Institutions in Asian Economic Development 1960-1995*, Executive Summary, 1996, 13; Chinkin, see note 21, 253-256.

[16] A. S. El-Khosheri, "Is There a Growing International Arbitration Culture in the Arab-Islamic Juridical Culture?", Proceedings of the 1996 Conference of ICCA, 47.

[17] Johnston and Saunders, see note 1, 327.

[18] *Ibid.*

party for a decision that would be final and binding upon them, the basic choice seems to lie between recourse to the International Court of Justice (full-court or chamber procedure) or to an arbitral tribunal constituted and functioning as agreed between them. States Parties to the UN Convention on the Law of the Sea have open to them, *in addition*, the possibility of resolving such disputes through submission to the complex system provided for under Part XV of the Convention which includes recourse to the International Tribunal for the Law of the Sea, with its Seabed Disputes Chamber (Annex VI), to arbitration pursuant to Annex VII of the Convention, or to special arbitration pursuant to Annex VIII. Several factors may influence the disputant's choice between recourse to arbitration and a judicial organ, among which (1) the relatively higher degree of "control" over the composition and procedure of an arbitral tribunal, as compared with the fixed composition and procedure of an established court; (2) the relatively lower expense of recourse to either of the established judicial organs,[19] the cost of maintaining which is spread internationally, and does not fall exclusively on the disputants, as would the cost of an arbitral tribunal; (3) the relative duration of the proceedings, an established court being able to register a case immediately while possibly being constrained by a pre-existing roll of cases, while an arbitral proceeding, which could, in principle, commence immediately and reach a decision expeditiously, might be delayed if the disputants fail to agree on the composition of the tribunal, its procedures and other operational considerations; (4) the "open" character of a court proceeding, in which oral and written pleadings and reasoned decisions would be available to the public, and in which parties other than the disputants could, in principle, and at the discretion of the court, be permitted to intervene to defend a legal interest,[20] as contrasted with the

---

[19] In 1989, the UN Secretary-General established a Trust Fund, financed through voluntary contributions, from which to provide funds to assist qualifying States to meet certain of their expenses in connection with (i) disputes submitted to the International Court of Justice by way of special agreement, or (ii) the execution of a judgment of the Court resulting from such special agreement. A similar fund has been established by the Permanent Court of Arbitration, see above at 7.

[20] In the *Tunisia/Libya* case, Intervention, ICJ Reports 1984, 20, the Court, having refused the application of Malta to intervene, said:
"The findings at which [the Court] arrives and the reasoning by which it reaches those findings in the case between Tunisia and Libya will therefore inevitably be directed exclusively to the matters submitted to the Court in the Special Agreement concluded between those States and on which its jurisdiction in the present case is based. It follows that no conclusions or inferences may legitimately be drawn from those findings or that reasoning with respect to rights or claims of other States not parties to the case".

essentially closed and private nature of an arbitral proceeding, which, in principle, would inhibit intervention by interested third States; and (5) the possibility of implementing the result of the adjudicatory process, the right of recourse to the Security Council where there has been a failure to perform the obligations imposed by a judgment of the International Court of Justice (Article 94 of the UN Charter), as contrasted with other adjudicatory and arbitral processes which, as a rule, offer no means of promoting or ensuring the implementation of their decisions.

6. Settlement of Maritime Boundary Disputes under the UN Convention on the Law of the Sea: The Option of States Parties to Choose Conciliation Pursuant to Article 298(1)(a)

A generalized preference for non-adversarial dispute resolution methods in relation to politically sensitive issues such as those touching territorial sovereignty, and appurtenant rights of a nation to "permanent sovereignty" over its natural resources, may have resulted in the Convention's provision for residual application of *compulsory conciliation* in connection with disputes over maritime boundaries. Thus, while the UN Convention on the Law of the Sea in Part XV requires compulsory submission of a wide range of maritime disputes to final and binding decision by the International Court of Justice, the International Tribunal for the Law of the Sea, the International Court of Justice, arbitration pursuant to Annex VII to the Convention, or to special arbitration pursuant to Annex VIII, all of which are governed by adversarial procedures,[21] it does permit States, by declaration made pursuant to Article 298, paragraph 1, subparagraph (a), to remove from the compulsory adjudicatory scheme, *inter alia*, "disputes concerning the interpretation or application of Articles 15, 74 and 83 relating to sea boundary delimitation, or those involving historic bays or titles":

In the *Libya/Malta* case, Merits, ICJ Reports 1985, 26, the Court, having refused the application of Italy to intervene pursuant to Article 62 of the Court's Statute, said:
"The present decision must ... be limited in geographical scope so as to leave the claims of Italy unaffected, that is to say, that the decision of the Court must be confined to the area in which, as the Court has been informed by Italy, that State has no claims to continental shelf rights".
The Court permitted Nicaragua to intervene in the *Land, Island and Maritime Frontier Dispute* case *(El Salvador/Honduras)*, Application to Intervene, ICJ Reports 1990.

[21] Article 287. For a detailed review of these procedures, see C. Chinkin, "Dispute Resolution and the Law of the Sea", in J. C. Crawford and D. R. Rothwell, *Prospects for the Law of the Sea in the Asian Pacific Region*, 1995, 237-262.

"Article 298: Optional Exceptions to Applicability of Section 2
1. When signing, ratifying or acceding to this Convention or at any time thereafter, a State may, without prejudice to the obligations arising under section 1, declare in writing that it does not accept any one or more of the procedures provided for in section 2 with respect to one or more of the following categories of disputes:
(a) (i) disputes concerning the interpretation or application of Articles 15, 74 and 83 relating to sea boundary delimitations, or those involving historic bays or tides, provided that a State having made such a declaration shall, when such a dispute arises subsequent to the entry into force of this Convention and where no agreement within a reasonable period of time is reached in negotiations between the parties, at the request of any party to the dispute, accept submission of the matter to conciliation under Annex V, section 2; and provided further that any dispute that necessarily involves the concurrent consideration of any unsettled dispute concerning sovereignty or other rights over continental or insular land territory shall be excluded from such submission;
(ii) after the conciliation commission has presented its report, which shall state the reasons on which it is based, the parties shall negotiate an agreement on the basis of that report, if these negotiations do not result in an agreement, the parties shall, by mutual consent, submit the question to one of the procedures provided for in section 2, unless the parties otherwise agree;
(iii) this subparagraph does not apply to any sea boundary dispute finally settled by an arrangement between the parties, or to any such dispute which is to be settled in accordance with a bilateral or multilateral agreement binding upon those parties. ..."

A State making a declaration under Article 298(1)(a) excluding boundary disputes from the operation of "one or more" of the "compulsory procedures entailing binding decisions" provided for in section 2 of the Part XV, would thus be bound to submit such disputes to non-binding "compulsory" conciliation provided for in the article, presumably without the need to actually declare or express its acceptance of that procedure. While the declaration may exclude *all* the "binding decision" procedures of section 2, in favor of compulsory conciliation, it need not do so: a declarant State may accept *one or more* of those procedures, in addition to compulsory conciliation. When a State declares non-acceptance of one or some, but not all section 2 procedures in relation to boundary disputes, and thus accepts submission of such disputes to compulsory conciliation under Annex V *in addition to* some section 2 procedure, difficulties of interpretation could arise as to whether or not the disputants have "accepted the same procedure" for the purposes of determining, pursuant to Article 287 paragraphs 4 and 5, to what settlement method they are mutually committed.[22]

A State declaring non-acceptance of the compulsory settlement procedures of section 2 is nevertheless required under the Convention: (1) to abide by any previously agreed boundary settlement, or settlement procedure binding on the parties; (2) to abide by the general obligation and procedures aimed at dispute settlement prescribed by Part XV, section 1 of the Convention; (3) to submit any post-ratification boundary disputes which the negotiated agreement has not resolved, "within a reasonable period of time", to the procedure of "conciliation under Annex V, section 2" of the Convention; and (4) where the conciliation effort so prescribed has failed to result in a negotiated agreement, to submit the dispute to the arbitral/judicial procedures of Part XI, section 2, although such submission is only to be made "by mutual consent", and so may not take place at all.[23]

The process of "conciliation" prescribed by Annex V, section 2, of the Convention is conceived along traditional lines, in that the Conciliation Commission is chosen by the parties, its procedures are flexible, and its function is to make recommendations to the parties with a view to their reaching an amicable settlement. The Commission's report and recommendations do not bind the disputants, and may be rejected by one or both of them. However, some provisions of the Annex are of more than usual stringency, giving the process an adversarial flavor: constitution of the Commission is assured by default-curing procedures that are time-bound, the Commission's functions are themselves time-bound, the Commission's report is to be "deposited with the Secretary-General of the United Nations", and no provision is made as to confidentiality, and Article 12 provides that

> "The failure of a party or parties to the dispute to reply to notification of institution of proceedings shall not constitute a bar to the proceedings".

---

[22] For an analysis of the complexities that could arise in determining what procedural choice must be attributed to the disputants in such situations, see T. Treves, "'Compulsory' Conciliation in the UN Law of the Sea Convention", in V. Götz, P. Selmer and R. Wolfrum (eds.) *Liber amicorum Günther Jaenicke*, 1998, 611 et seq., 623 ff.; for a comparison of these provisions with those of other conventions offering similar choices, see T. Treves, "New Trends in the Settlement of Disputes and the Law of the Sea Convention", in H. N. Scheiber (ed.), *Law of the Sea*, 2000, 61 et seq., 66.

[23] Article 298, subparagraph (a)(ii), and Article 299, paragraph 1.

The Convention seems to concede that one type of dispute is to remain wholly outside the ambit of even compulsory conciliation: "any dispute that necessarily involves the concurrent consideration of any unsettled dispute concerning sovereignty or other rights over continental or insular land territory...".[24] Since many maritime boundary disputes do "necessarily involve" unsettled disputes over islands or land territory, this provision would appear to reduce further the Convention's reach in respect of maritime boundary disputes.

With a view to discouraging exclusionary declarations under Article 298, subparagraph 1(a)(i), and encouraging States Parties to the Convention to submit maritime boundary disputes to the Convention's compulsory arbitration/judicial settlement scheme, Article 298, paragraph 3, would preclude a declarant State from submitting a maritime boundary dispute with any other State Party to the Convention's compulsory scheme without the latter's consent, while paragraph 4 of that article would nevertheless leave the declarant State open to compulsory boundary dispute settlement procedures initiated by another State Party.

IV. Resolution of Maritime Boundary Disputes: Applicable Principles and Rules

The International Court of Justice and international arbitral tribunals have several times been requested to resolve disputes concerning the maritime boundaries between opposite or adjacent States. In reasoned and authoritative decisions they have clarified and developed the law on maritime boundaries, laying a foundation for the formulation of its essential elements in the relevant provisions of the 1982 UN Convention on the Law of the Sea.

1. Boundaries Between the Territorial Seas and Contiguous Zones of States with Opposite or Adjacent Coasts

Article 15 of the UN Convention on the Law of the Sea incorporates, without substantial change, the provisions of Article 12 of the 1958 Geneva Convention on the Territorial Sea and the Contiguous Zone regarding the delimitation of the territorial sea between opposite or adjacent States. It provides for application of the "median line" or "equidistance" principle, the reference points being those located on baselines which both States should have established in accordance with

---

[24] Article 298, subparagraph (a)(i), second proviso.

Article 16 of the Convention. Provision is made for non-application of the "equidistance" principle "where it is necessary by reason of historic title or other special circumstances". Coastal States are required to show the lines of delimitation on charts of a scale or scales adequate for ascertaining their position, or to make available a list of geographical coordinates of reference points, to give due publicity to such charts of lists, and to deposit a copy of each chart or list with the Secretary-General of the United Nations.

Whereas Article 24, paragraph 2, of the 1958 Geneva Convention provided for application of the equidistance/special circumstances principle in delimiting the contiguous zones of opposite or adjacent States, the 1982 UN Convention makes no specific provision on the subject.

2. Boundaries Between the Exclusive Economic Zones or Continental Shelves of States with Opposite or Adjacent Coasts

The principles applicable to the delimitation of the continental shelf between States have been distilled gradually from the decisions of international courts and arbitral tribunals, and are now reflected in the provisions of the 1982 UN Convention concerning the delimitation of the exclusive economic zone (Articles 74-75) and the continental shelf (Articles 83-84), which are, in terms, identical. Authoritative pronouncements on the subject were made by the International Court of Justice in the *North Sea Continental Shelf* cases *(FRG* v. *Denmark, FRG* v. *Netherlands)*, the *Tunisia/Libya* case, the *Gulf of Maine* case *(Canada* v. *USA)*, and the *Libya/Malta* case, and by the arbitral tribunal in the *Anglo-French Channel* case (1977).[25] The basic principles were outlined in the judgment of the Court in the *North Sea Continental Shelf* cases (1969), when it declared that, under customary international law:

> "... delimitation is to be effected by agreement in accordance with equitable principles, and taking account of all the relevant circumstances...". (paragraph 101)

Asked to determine the applicable rules and principles of delimitation, but not to demarcate the boundary, the Court found that the equidistance/special circumstances principle, which has presumptive

---

[25] The *North Sea Continental Shelf* cases *(FRG* v. *Denmark, FRG* v. *Netherlands)*, ICJ Reports 1969, 3; the *Tunisia/Libya* case, ICJ Reports 1982, 18; the *Gulf of Maine* case *(Canada/USA)*, ICJ Reports 1984, 246; the *Libya/Malta* case, ICJ Reports 1985, 13; and the *Anglo-French Channel* case (1977), *ILM* 18 (1979), 153.

application in the case of territorial sea delimitation, is only one of several principles that might be considered for application in connection with the delimitation of a continental shelf boundary, and occupies no position of particular significance unless, indeed, it were to be conducive to an equitable solution. The Court went on to expound the sense in which it had introduced the concept of "equity" into maritime boundary delimitation in the following terms:

> "91. Equity does not necessarily imply equality. There can never be any question of completely refashioning nature, and equity does not require that a State without access to the sea should be allotted an area of continental shelf, any more than there could be a question of rendering the situation of a State with an extensive coastline similar to that of a State with a restricted coastline. Equality is to be reckoned within the same plane, and it is not such natural inequalities as these that equity could remedy... It is therefore not a question of totally refashioning geography whatever the facts of the situation but, given a geographical situation of quasi-equality as between a number of States, of abating the effects of an incidental special feature from which an unjustifiable difference of treatment could result...
>
> 92. It has however been maintained that no one method of delimitation can prevent such results and that all can lead to relative injustices... It is necessary to seek not one method of delimitation but one goal ... an equitable solution...
>
> 93. In fact, there is no legal limit to the considerations which States may take account of for the purpose of making sure that they apply equitable procedures, and more often than not it is the balancing-up of all such considerations that will produce this result rather than reliance on one to the exclusion of all others. The problem of the relative weight to be accorded to different considerations naturally varies with the circumstances of the case.
>
> 94. In balancing the factors in question it would appear that various aspects must be taken into account. Some are related to the geological, others to the geographical aspect of the situation, others again to the idea of the unity of any deposits. These criteria, though not entirely precise, can provide adequate bases for decision adapted to the factual situation.

95. ... The continental shelf is, by definition, an area physically extending the territory of most coastal States into a species of platform...

96. ... the principle is applied that the land dominates the sea, it is consequently necessary to examine closely the geographical configuration of the coastlines of the countries whose continental shelves are to be delimited ... since the land is the legal source of the power which a State may exercise of territorial extensions to seaward, it must first be clearly established what features do in fact constitute such extensions...".

The operative part of the Court's judgment is a concise rendering of the principles and rules of delimitation which, later elaborated by the Court in subsequent cases, eventually provided the basis for Articles 74 (exclusive economic zone delimitation) and 83 (continental shelf delimitation) of the 1982 UN Convention on the Law of the Sea. The Court held:

"(C) the principles and rules of international law applicable to the delimitation as between the Parties of the areas of the continental shelf in the North Sea which appertain to each of them beyond the partial boundary determined by the agreements of 1 December 1964 and 9 June 1965, respectively are as follows:
(1) delimitation is to be effected by agreement in accordance with equitable principles, and taking account of all the relevant circumstances, in such a way as to leave as much as possible to each Party all those parts of the continental shelf that constitute a natural prolongation of its land territory into and under the sea, without encroachment on the natural prolongation of the land territory of the other;
(2) if, in the application of the preceding subparagraph, the delimitation leaves to the Parties areas that overlap, these are to be divided between them in agreed proportions or, failing agreement, equally, unless they decide on a régime of joint jurisdiction, user, or exploitation for the zones of overlap or any part of them;
(D) in the course of the negotiations, the factors to be taken into account are to include:
(1) the general configuration of the coasts of the Parties, as well as the presence of any special or unusual features;

> (2) so far as known or readily ascertainable, the physical and geological structure, and natural resources, of the continental shelf areas involved;
> (3) the element of a reasonable degree of proportionality, which a delimitation carried out in accordance with equitable principles ought to bring about between the extent of the continental shelf areas appertaining to the coastal State and the length of its coast measured in the general direction of the coastline, account being taken for this purpose of the effects, actual or prospective, of any other continental shelf delimitations between adjacent States in the same region...". (paragraph 101)

The arbitral tribunal in the *Anglo-French Channel* case was asked not merely to determine the applicable rules and principles but actually to demarcate the maritime boundary between the two countries. In doing so the Tribunal confirmed that the equidistance/special circumstances principle was only one of several to be considered in arriving at a balance of the equities. It therefore felt free to vary application of the equidistance rule when special circumstances – in this case, the geographical factor of the existence of islands belonging to each State lying in the area to be apportioned by the boundary – seemed to justify such a course in order to give effect to equitable principles.

In the *Tunisia/Libya* case the Court further elaborated its emphasis on the application of equitable principles to achieve an equitable result. Requested by the parties to indicate the applicable rules and principles of delimitation, the Court considered a range of special circumstances pleaded by them, including coastal formation, geological evidence, the presence of islands, claims of historic fishing rights, and interference with future economic development. Giving preference to factors conducive to certainty, the Court defined the area it considered relevant to the case, accorded significance to a pre-existing informal agreement between the parties concerning oil exploration, focused on the general geographic relationship between the natural prolongations of the coastal fronts of the parties, and, having tested the result by reference to a "reasonable degree of proportionality" of the relevant sections of the coast of the parties (Libya 31:69 Tunisia) found its solution equitable. While the Court sought again to explain the role of equity in the context of maritime delimitation, its *dicta* seem, at times, tautologous:

> "It is, however, the result which is predominant, the principles are subordinate to the goal. The equitableness of a principle must be assessed in the light of its usefulness for the purpose of arriving at an equitable result. It is not every principle which is in itself equitable; it may acquire this quality by reference to the

equitableness of the solution. The principles to be indicated by the Court have to be selected according to their appropriateness for reaching an equitable result. The term 'equitable principles' cannot be interpreted in the abstract. It refers back to the principles and rules which may be appropriate in order to achieve that end". (paragraph 70)

In the *Gulf of Maine* case, the parties, having failed to negotiate a fisheries agreement, asked a specially constituted chamber of the Court to "decide, in accordance with the principles and rules of international law applicable. ... What is the course of the single maritime boundary that divides the continental shelf and fisheries zones of Canada and the United States of America..." in the area specified in the special agreement between them. The Chamber held:

"(1) No maritime delimitations between States with opposite or adjacent coasts may be effected unilaterally by one of those States. Such delimitation must be sought and effected by means of an agreement, following negotiations conducted in good faith and with the genuine intention of achieving a positive result. Where, however, such agreement cannot be achieved, delimitation should be effected by recourse to a third party possessing the necessary competence.
(2) In either case, delimitation is to be effected by the application of equitable criteria and by the use of practical methods capable of ensuring, with regard to the geographic configuration of the area and other relevant circumstances, an equitable result". (paragraph 112)

By applying geometrical methods and a consistent emphasis on geographical features, making corrections where the result appeared inequitable, the Chamber indicated, in three segments, the course of the single maritime boundary sought by the parties.

Article 74 (exclusive economic zone delimitation) and Article 83 (continental shelf delimitation) of the UN Convention on the Law of the Sea are consistent with the Court's judgments referred to, in that they require (1) that delimitation shall be effected (a) by agreement, (b) on the basis of international law (including existing relevant agreements between the parties), and (c) in order to achieve an equitable solution; (2) that pending agreement, "provisional arrangements of a practical nature" should be maintained; and (3) that, if no agreement can be reached "within a reasonable period of time", the States concerned should resort to third-party settlement procedures. The residual procedure indicated is recourse to *compulsory conciliation* pursuant to Article 298(1)(a) and Annex V of the Convention.

The Court's analyses and pronouncements since 1969 provide guidance as to what factors might contribute to achievement of an "equitable solution" for an agreement on a boundary. The many factors taken into account by the Court in reaching its decisions have been variously referred to as "rules", "principles", "methods", "criteria" and "circumstances", but do not lend themselves readily to precise classification on any such basis.[26] However, a tentative attempt to list them follows:

I. Equitable Principles

1. Relationship between the coastal configurations of each party.
2. Effect on that relationship of relevant circumstances.
3. Existing agreements relevant to delimitation.
4. Specific instructions or mandate of the parties as to delimitation.
5. Obligation to negotiate in good faith to bring about a positive result.
6. Conduct of the parties demonstrating consent, acquiescence, estoppel, etc.
7. Optimization of natural resource conservation and management measures.
8. Minimization of likelihood of future disputes.
9. Application of "equitable criteria":
   (a) the "land dominates the sea"
   (b) equal division of areas of overlap of water or seabed
   (c) non-encroachment on the coast of another State
   (d) no cut-off of the seaward projection of the coast of another State, or its maritime area
   (e) effect of inequalities in lengths of coastlines in the area of delimitation.

II. Relevant Circumstances

1. Geographical
   (a) Coasts: whether opposite or adjacent.
   (b) Lengths and configurations of coastlines.

---

[26] For a critique of delimitation principles as perceived in the Asian context, see S.-M. Rhee and J. MacAulay, "Ocean Boundary Issues in East Asia: The Need for Practical Solutions", in Johnston and Saunders, see note 1, 97-99. See also A. G. Oude Elferink, *The Law of Maritime Boundary Delimitation: A Case Study of the Russian Federation*, 1994, 13-14. He observes that "the debate over the contents of maritime delimitation law has been characterized by the division between proponents of a predictable law and those advocating the uniqueness of each delimitation situation, which requires a large amount of flexibility of the law".

(c) Distances of coastlines from each other, and from other relevant physical features.
(d) Location of islands.
(e) Location of deepest channel.
2. Environmental
   (a) Existence of identifiable ecological régimes.
   (b) Natural features dividing such régimes.
   (c) Nature and extent of traditional exploitation and management of resources.
3. Geological and geomorphological
4. Socio-economic factors (dependence of local communities on maintenance of environmental quality and natural resource reserves).

III. Practical Methods

1. "Median line" or "equidistance" method is only one of several methods. Statistical considerations afford no indication of the greater or lesser degree of appropriateness of any particular method, or of any trend in favor thereof discernible in customary international law.
2. A line perpendicular to the coast, or to the general direction of a coast.
3. A line prolonging an existing division of territorial seas, or the direction of the final segment of a land boundary, or the overall direction of such boundary.

In sum, *equitable principles and criteria*, acting upon *relevant circumstances*, must lead to an *equitable solution* clearly indicating the course of a boundary capable of being established by *practical methods*, and embodied in an *agreement* between the disputants.

V. Suspension or Deferment of Maritime Boundary Disputes

In the *North Sea Continental Shelf* cases, the International Court of Justice, having taken into consideration, *inter alia*, "the particular configuration of the North Sea, and ... the particular geographical situation of the Parties' coastline upon that sea", which had resulted in the overlapping of areas belonging to the parties, mentioned as alternative courses of action, that the situation should be

> "resolved either by an agreed, or failing that by an equal division of the overlapping areas, or by *agreements for joint exploitation*, the latter solution appearing particularly

appropriate when it is a question of preserving the unity of a deposit".[27]

The Court seemed thus to be endorsing a process whereby maritime boundary issues might be suspended or deferred while the States concerned tried to reach agreement on the exploitation of a "straddling" mineral resource in a mutually beneficial and sustainable manner.

According to one commentator, the first example of an agreement on "joint development" of an offshore petroleum resource as a means of resolving both boundary and oil exploitation issues, was that between the Sheikdom of Bahrain and the Kingdom of Saudi Arabia of 22 February 1958. Noting the conclusion of some 24 such joint development agreements over the following three decades (more than half of the States Parties being from Asia), the author suggests that the following parameters may be derived from them:

> "(1) use or exploitation on one side of a national boundary must take account of right and legitimate interests of the State on the other side of the boundary, implying obligations of a co-operative character; (2) where activities on one side of the boundary would result in exploitation of resources on the other side, mutual consent of the States concerned is required; (3) prior notification and consultation is the means of ascertaining that consent; (4) the solution arrived at (whether or not accompanied by actual delimitation) must be equitable; (5) mutual consent must be recorded in a written agreement; (6) the agreement must have its basis in international law; (7) the parties are obliged, pending conclusion of the agreement to enter into provisional agreements of a practical nature so as not to jeopardize or hamper reaching final agreement; and (8) the agreement reached should (a) demarcate the joint development zone, (b) apportion jurisdiction among the States concerned, (c) lay down the basic system to be followed, e.g., production sharing, profit sharing, (d) establish one or more supervisory organs with proportionate representation and appropriate provisions on decision making, (e) state its duration and/or procedure for renewal, and (f) provide for settlement of disputes".[28]

---

[27] ICJ Reports 1969, para. 99, emphasis added. See also Judge Evensen's Dissenting Opinion in the *Tunisia/Libya* case, ICJ Reports 1982, 321-323, suggesting that the parties "should include provisions on unitization in cases where a petroleum field is situated on both sides of ... the dividing line for the above proposed zone of joint exploitation".

[28] Yu Hui, "Joint Development of Mineral Resources – An Asian Solution?", *Asian YIL* 2 (1992), 87 et seq., 103. See also the joint development

"Joint development" as a policy option when negotiation of a maritime boundary offers the prospect of protracted and politically sensitive negotiations, could hardly be faulted, and has received support from many experts in the field.[29] The prescriptions of Johnston and Saunders quoted above (text accompanying note 6) tend in the same direction, while emphasizing, in such cases, the value of support that is to be gained from structured regional relationships. Deferment of territorial claims in recognition of prospective mutual benefits from scientific research and environmental protection has also been pioneered successfully among the parties to treaties applying to the Antarctic region since 1959. Nor need the benefits of "joint development" be considered only in relation to mineral deposits. The basis for cooperative arrangements for "joint exploitation" and management of *living resources* in areas of overlap may well be foreshadowed in existing international and regional fishery agreements, as well as in the provisions of the UN Convention on the Law of the Sea on "straddling stocks" (Article 63) and highly migratory species (Article 64), on the rights of land-locked States of the same region or subregion as a coastal State in the latter's exclusive economic zone (Article 69, paragraph 5), and on the rights of "geographically disadvantaged States" of the same subregion or region as a coastal State in the latter's exclusive economic zone (Article 70).

A recent authoritative analysis of policy options in relation to the complex boundary and resource questions arising in the South China Sea offers a comprehensive list of the principles of objectives of a multilateral maritime régime for cooperative regional resource management. Without presuming to summarize so detailed a study, it may be said that the parameters of a scheme for establishing and maintaining such a régime appear to include: taking account of the territorial sovereignty of every claimant; resolution of regional disputes through regional mechanisms, without internationalization or military

---

precedents referred to in M. J. Valencia, J. M. van Dyke and M. A. Ludwig, *Sharing the Resources of the South China Sea*, 1997, 183 et seq.; A. Razavi, *Continental Shelf Delimitation and Related Maritime Issues in the Persian Gulf*, 1997; and V. L. Forbes, *The Maritime Boundaries of the Indian Ocean*, 1995, 111 et seq.

[29] For example, Valencia et al., see note 28; M. J. Valencia et al., "South-East Asian Seas: Joint Development of Hydrocarbons in Overlapping Claim Area?", *Ocean Development & Int'l L.* 16 (1986), 223; W. T. Onorato, "Joint Development of Sea-bed Hydrocarbon Resources: An Overview of Precedents in the North Sea", *Energy* 6 (1981), 1315 et seq.; Zhiguo Gao, *Joint Development of Overlapping Continental Shelf Areas in International Law*, 1990. Compare M. Miyoshi, "The Basic Concept of Joint Development of Hydrocarbon Resources on the Continental Shelf", *Int'l J. Estuarine & Coastal L.* 3 (1988), 9.

intervention; resource exploitation on the basis of equity and fairness; cooperative regional exploration, development and management of resources both living and non-living so as to promote regional resource use; each State to undertake the responsibility to ensure that activities within its jurisdiction or control do not cause damage to the territory, resources or environment of another State; the "precautionary principle" to govern resource development, so as to foster "sustainable development", environmental protection being integrated into the management process; rare and fragile ecosystems, and biodiversity to be protected and managed, taking into account the needs of future generations; eventual demilitarization of the region; ensuring safe navigation and the prevention of piracy, drug smuggling, pollution and other illegal activities; and accommodation of the interests of extra-regional maritime powers for the maintenance of peace, stability, freedom of navigation and security of sea lanes. The authors also suggest the features of alternative models for an organizational structure for implementing such a régime.[30]

The difficulty of how to achieve solutions along these lines, given the multiple political sensitivities involved when dealing with maritime boundary disputes, still remain. While the UN Convention on the Law of the Sea, the International Court of Justice and distinguished arbitral tribunals, building upon the practice of States, have provided guidance as to the applicable principles and methods to be applied in resolving maritime boundary issues, the internal political hazards of engaging in dispute resolution procedures entailing binding decisions still causes States to lean away from submitting such disputes to arbitration or judicial settlement.[31] Conciliation appears to be an acceptable method, both in its traditional form as well as the "compulsory" form provided for under Article 298(1)(a) of the UN Convention on the Law of the Sea, but it has rarely been tested in regard to maritime boundary disputes.

No new processes of international dispute settlement have been added to those provided a century ago by the Hague Conventions of 1899 and 1907, and negotiation and consultation remain the preferred and most frequently used methods of interstate dispute settlement, and therefore the most likely to be used in resolving maritime boundary issues. Pioneering work has been done, however, by Indonesia, in attempting what might be seen as a refinement of the negotiation/consultation method. Focusing on areas of the sea and

---

[30] Valencia et al., see note 28, 199 et seq.
[31] Notable exceptions are the *Maritime Delimitation and Territorial Questions Between Qatar and Bahrain* case, before the International Court of Justice since 1995, and the *Sovereignty over Pulau Ligitan and Pulau Sipadan* case *(Indonesia/Malaysia)*, before the Court since 1998, both in the "merits" phase awaiting the Court's judgment.

seabeds of South-East Asia that are subject to multiple and apparently strongly held claims, the initiative would, as a first step, call on experts from each claimant country to pool their scientific knowledge of the area and its resources, and examine, as far as possible dispassionately and objectively, the political, social and legal elements of the claims of each country involved. Behind this series of essentially private and scientific "workshops" is the intention to build up a body of knowledge of the issues among groups of persons of balanced judgment and sound academic credentials, who have no commitment to any politically motivated position on the issues. The next stage, we may presume, would be to bring to bear on representatives in government the weight of each group's opinion on the issues and the ways in which they may be resolved in the interest of all countries concerned. It has also been suggested that, as an adjunct to this initiative, a committee of "eminent persons" from the area could, through multiple consultations, seek to put in place confidence-building measures that could lend support to the process.[32] Although the process may appear labored and slow, it may well prove to be a relatively expeditious method of achieving a lasting solution to problems accumulated over a period of several centuries.

While innovative thinking in East Asia seeks to develop non-adversarial processes of dispute resolution, a parallel development may be seen as utilizing existing structures of the adversarial process in novel ways. Thus, Edward de Bono, in his recent book *New Thinking for the New Millennium* (1999), mentions a process, which he reports, is already "on the statute books in many States in the USA". As he describes it,

> "The parties never meet. There is no bargaining or negotiation. Each side makes its position clear: needs and fears and perceptions. Each side sets out to 'design' an outcome which would be beneficial or fair to both sides. You are no longer defending your position but 'designing' a way forward. Both designed positions are then put before a judge or a panel. The most 'reasonable' design is then chosen and accepted by both parties. If both parties are indeed making a serious effort to take the other party's needs into account, then both outcomes will be 'fair' and it hardly matters which is chosen. If neither party makes a design, then both parties are told to go away and to try harder".[33]

Thinking along these lines, consistent with a modern trend that would favor cooperative solutions will, in due course, modify the dispute resolution concepts embodied in the 1899/1907 Hague Conventions, and

---

[32] Valencia et al., see note 28, 115.
[33] E. de Bono, *New Thinking for the New Millennium*, 1999, 105.

adopted with little change in international agreements since then. The desire to obtain early benefits from marine resources, when accompanied by an openness to the concerns of another State, and the will to preserve a cooperative relationship with it, would lead, as in the past, to a negotiated settlement incorporating a system of cooperative development, thus avoiding the expense and possible rancor associated with an "adversarial" decision on a boundary. Indeed, the establishment of a boundary, if needed for other purposes, could be the subject of a separate, later negotiation. The benefits that could accrue from a system of cooperative development have been neatly summarized thus:

> "'Joint development', or cooperative exploration, exploitation and management of offshore mineral resources pursuant to intergovernmental agreement, seems assured of a place among the strategies of good-neighbourliness which may be used to establish or maintain friendly and mutually beneficial relations among States. Applied with a view to early realization of benefits from mineral wealth which might otherwise be left undeveloped during protracted negotiations concerning the limits of national sovereignties, joint development is representative of a modern trend away from narrow autarkic nationalism, and toward recognition of the benefits of consensus, cooperation and balanced interdependence among States. 'Joint development' is a pragmatic solution capable of accomplishing the avoidance of confrontation and its wasteful consequences, through focusing on positive approaches and the initiation of productive activity from which tangible benefits accrue to all concerned".[34]

---

[34] Yu Hui, see note 28, 111.

Deutsche Seegrenzen

Walter Rudolf

I.

Die deutschen Seegrenzen sind für den Jubilar nichts Unbekanntes. Als Nestor des internationalen Seerechts ist er mit Abgrenzungsfragen maritimer Zonen bestens vertraut. Im Internationalen Gerichtshof hat er als Richter in Fällen strittiger Seegrenzen das Urteil über deren endgültigen Verlauf maßgeblich mitgestaltet. Vor seiner richterlichen Tätigkeit hat er als Vertreter der Bundesrepublik Deutschland im *Nordsee-Festlandsockel*-Fall vor dem Internationalen Gerichtshof dazu beigetragen, dass die Grenzen des Festlandsockels zwischen der Bundesrepublik und seinen Nachbarn – Dänemark und Niederlande – abweichend vom Äquidistanz-Prinzip zugunsten der Bundesrepublik festgelegt wurden. Das Urteil des IGH im Festlandsockel-Streit vom 22. Februar 1969[1] war Voraussetzung für die Verträge vom 28. Januar 1971 mit Dänemark und den Niederlanden über die Abgrenzung des Festlandsockels unter der Nordsee.[2] Mit dem Inkrafttreten dieser Verträge am 7. Dezember 1971[3] war der Streit endgültig beendet.

Deutschland ist unter den führenden Industriestaaten das Land mit den weitaus kürzesten Seegrenzen. Verglichen mit Großbritannien und Japan als insularen Staaten, dem peninsularen Italien, den großen Flächenstaaten Rußland, Kanada und USA mit langen Küsten jeweils an zwei Ozeanen und deren Nebenmeeren und Frankreich, dessen Seegrenzen doppelt so lang sind wie seine Landgrenzen, grenzt Deutschland mit einer durch Buchten und vorgelagerte Inseln stark zerfransten Küstenlinie von etwa 2000 km nur an die Nord- und die Ostsee. Die jetzige Grenze an der Ostsee wurde innerhalb der letzten 81 Jahre durch die Gebietsabtretungen nach dem Ersten und dem 1990 sanktionierten Gebietsverlust nach dem Zweiten Weltkrieg etwa um die

---

[1] ICJ Reports 1969, 3.
[2] UNTS Vol. 857, 109; BGBl. 1972 II, 881.
[3] Vgl. die Bekanntmachung vom 17. November 1972, BGBl. 1972 II, 1616.

Hälfte verkürzt.[4] Vor der Wiedervereinigung Deutschlands war die Ostseegrenze der Bundesrepublik kürzer als die der Deutschen Demokratischen Republik. An der Nordsee war vor dem Westfälischen Frieden 1648 die Grenze des Heiligen Römischen Reiches deutscher Nation ebenfalls erheblich länger als heute. Im Norden gehörte das Herzogtum Schleswig freilich nicht zum Sacrum Imperium und später auch nicht zum Deutschen Bund, so dass die Eidermündung die Grenze zu Dänemark bildete.[5] Im Westen reichte das deutsche Gebiet unter Einschluss der Niederlande und Flanderns sogar bis vor Calais. Auch gab es über das Herzogtum Krain und die Markgrafschaft Istrien, die Jahrhunderte lang zu Österreich gehörten, einen Zugang zur Adria – sowohl zum Golf von Triest als auch zum Quarnero.

Abgesehen von den seit 1648 souveränen Hansestädten war Deutschland seit dem Untergang der Hanse keine Seefahrernation. Kolonien wurden erst seit den achtziger Jahren des 19. Jahrhunderts in Afrika und im Pazifik erworben und gingen mit dem Ersten Weltkrieg wieder verloren. Die unter der Regierung des Großen Kurfürsten 1683 gegründete brandenburgische Faktorei Groß-Friedrichsburg an der Guinea-Küste wurde schon 1720 wieder aufgegeben. Österreichs maritimes Engagement vor dem 19. Jahrhundert war verhalten. Don Juan d'Austria, der Sieger von Lepanto, war nicht österreichischer Admiral, sondern Befehlshaber der mit den venezianischen und genuesischen Seestreitkräften vereinigten spanischen Flotte. Am Rande sei erwähnt, dass während der Zeit des Deutschen Bundes bei der österreichischen Marine die Kommandosprache Italienisch war. Es ist deshalb nicht verwunderlich, dass Deutschland zur Geschichte des Seerechts und damit der Seegrenzen nur wenig beigetragen hat. Zur Zeit, als in den als „Strom" bezeichneten Gewässern vor der flandrischen Küste der Uferstaat die Hochgerichtsbarkeit gegen jedermann, gleichgültig welcher Herkunft und Volkszugehörigkeit, in Anspruch nahm,[6] was zur Entwicklung des späteren Küstenmeeres beitrug, gehörte Flandern noch nicht zum Heiligen Römischen Reich deutscher Nation. Als Deutschland zu Beginn des 20. Jahrhunderts als seefahrende Nation sowohl durch eine bedeutende Handelsflotte als auch durch seine Kriegsmarine in Erscheinung trat, waren die völkerrechtlichen Regeln über die maritimen Zonen und ihre Grenzen weitgehend etabliert.

---

[4] Bis zum Inkrafttreten des Versailler Vertrages am 20. Januar 1920 reichte die deutsche Ostseeküste bis zum Dorfe Nimmersatt südlich von Polangen, nördlich von Memel (Klaipeda).

[5] Von 1864 bis 1920 verlief die Grenze weiter nördlich, da Dänemark das Herzogtum Schleswig abgetreten hatte. Art. 2 des Vertrages von Wien vom 30. Oktober 1864 zwischen Österreich, Preußen und Dänemark, Martens N.R.G., Bd. 17 II, 470 f.

[6] G. Stadtmüller, *Geschichte des Völkerrechts I: Bis zum Wiener Kongreß*, 1951, 76 f.

Obwohl die Bedeutung der „Lehrmeinungen der fähigsten Völkerrechtler" (Art. 38 Abs. 1 lit. d) des IGH-Statuts) hinter der von Gerichtsentscheidungen ohnehin allmählich zurücktraten, war das nicht geringe deutsche seerechtliche Schrifttum – am Kieler Institut für Internationales Recht z.B. Schücking, Böhmert, Fritz Münch – gleichwohl noch bedeutsam für offene einzelne Rechtsfragen im Zusammenhang mit der technischen Entwicklung von den Fluginseln bis zum Tiefseebergbau. Das Gewohnheitsvölkerrecht des Meeres, das sich im 19. Jahrhundert herausgebildet hatte, wurde insgesamt aber nur wenig durch Autoren aus Deutschland geprägt. So erlebte „Das Internationale Öffentliche Seerecht der Gegenwart" von Perels zwar 1903 eine zweite Auflage, hatte aber wenig Einfluss auf die Entwicklung der Materie. Das seerechtslastige „Völkerrecht" von Kapitän z. See a.D. Vanselow von 1931 gab eine referierende Einführung in die Praxis der Staaten, ohne diese Praxis nachhaltig beeinflussen zu können.

An den Kodifikationskonferenzen des Völkerbundes und der Vereinten Nationen hat sich Deutschland rege beteiligt, doch entsprach sein Einfluss auf die endgültige Gestaltung des gegenwärtigen Seerechts nicht seiner Bedeutung als einer führenden Industriemacht. Dies kam auch darin zum Ausdruck, dass die Bundesrepublik Deutschland von den vier Seerechtskonventionen von 1958 nur die über die Hohe See ratifizierte und dem UN-Seerechtsübereinkommen von 1982 erst kurz vor dessen Inkrafttreten knapp 12 Jahre nach seiner Unterzeichnung beitrat.

II.

Die Frage nach der *Gebietshoheit des Uferstaates* über die Küstenlinie hinaus stellte sich erst mit dem Aufkommen des modernen Territorialstaates. Allerdings gab es auch schon vorher Herrschaftsansprüche über das Meer – etwa im Altertum des römischen Kaisers,[7] und im Mittelalter des römisch-deutschen Kaisers, der den Titel „des oceani König" führte.[8] Bekannt sind die Ansprüche Venedigs auf die Seeherrschaft in der Adria,[9] Großbritanniens auf das mare britannicum,[10] Dänemarks auf die Ostsee[11] und Spaniens und Portugals über weite Teile der Ozeane.[12]

---

[7] *Digesten* 14, 2, 9.
[8] W. Schücking, *Das Küstenmeer im internationalen Rechte*, 1897, 1.
[9] O. Lenel, *Die Entstehung der Vorherrschaft Venedigs an der Adria*, 1897, 12.
[10] F. Perels, *Das internationale öffentliche Seerecht der Gegenwart*, 2. Aufl., 1903, 13 f.
[11] A. Raestad, *La mer territoriale*, 1913, 58.
[12] Vertrag zu Tortesillas vom 07.06.1494, Martens, Recueil suppl., Tome I, 372; W. Grewe, *Fontes Historiae Iuris Gentium*, Bd. 2, 110 ff.

Unabhängig von diesen weit gehenden Ansprüchen gab es seit der Antike Regelungen über die nahen Küstengewässer.[13] Bei Autoren des ausgehenden Mittelalters findet sich der Satz „*territorium etiam in aquis se extendit*".[14] Allerdings handelte es sich dabei um küstennahe Zonen für spezielle Materien mit unterschiedlichen Interessen wie Gerichtsbarkeit, Fischfang, Zoll oder Neutralität. Ob die Idee des späteren Küstenmeeres, das der Souveränität der Uferstaaten untersteht, schon von Pontanus, wie O'Connell[15] meint, oder erst von van Bynkershoek[16] vertreten wurde, mag dahinstehen. Jedenfalls ist seit dem beginnenden 18. Jahrhundert die Hoheitsgrenze zur See nicht mehr die Küstenlinie, sondern die äußere Grenze der Küstengewässer. Allerdings hat sich das Konzept eines einheitlichen Küstenmeeres erst seit Beginn des 19. Jahrhunderts durchgesetzt. Eine monographische Behandlung der Küstengewässer, in der erstmals der Begriff „mare territoriale" vorkommt, erschien erst 1847.[17] Die Rechtsnatur des Küstenmeeres blieb bis in das beginnende 20. Jahrhundert umstritten.

Die seewärtige Ausdehnung der Gewässer, die der Hoheit des Uferstaates unterlagen, war weltweit letztlich niemals übereinstimmend anerkannt. Von van Bynkershoek stammt der Gedanke der Kanonenschussweite.[18] Vorher waren allerdings an deutschen Küsten andere Abgrenzungen üblich, so z.B. die „Kennige", d.i. dass vom Lande aus sichtbare Gebiet vor der flandrischen Küste oder nur geringe Entfernungen von einer halben Meile an der pommerschen Küste[19] oder die Kuriosität der „Rittgrenze" an der mecklenburgischen Küste: Der Herzog von Mecklenburg hatte insoweit Hoheitsrechte am Wismarer Tief als er mit einem Pferd ins Wasser reiten konnte, bis es ihm die Hufe bedeckte, und er alsdann mit einem Hufeisen von sich ins Wasser werden könne.[20]

Ob sich aus der Kanonenschussweite die Dreimeilenzone entwickelt hat, ist im Einzelnen umstritten. Großbritannien, die größte Seemacht des 18. und 19. Jahrhunderts, hat sich jedenfalls zunächst nicht für drei Seemeilen entschieden, sondern die genaue Grenzziehung offen

---

[13] Vgl. etwa U. Kahrstedt, *Staatsgebiet und Staatsangehörigkeit in Athen*, 1934, 4; Stadtmüller, siehe Anm. 6, 28.

[14] G. Böger, *Immunität der Staatsschiffe*, 1928, 45; vgl. vor allem G. Gidel, *Le droit international public de la mer*, Bd. 3, 1934, 26 ff.

[15] D. P. O'Connell, *International Law*, Vol. I, 2nd ed., 1970, 456.

[16] C. van Bynkershoek, *Juris consulti de dominio maris dissertatio*, 1703, Cap. 11.

[17] B. D. H. Tellegen, *De jure in mare, inprimis proximum*, 1847, 64.

[18] van Bynkershoek, siehe Anm. 16, Cap. 2: "Potestatm terrae finiri, ubi finitur armorum vis".

[19] Schücking, siehe Anm. 8, 7.

[20] F. Rörig, *Zur Rechtsgeschichte der Territorialgewässer: Reede, Strom und Küstengewässer*, 1949, 14.

gelassen. Es hat dann aber die Zoll- und später die Neutralitätsgrenze und im letzten Viertel des 19. Jahrhunderts auch die Fischereigrenze und die Grenze der Gerichtshoheit auf drei Seemeilen festgelegt.[21] Deutschland hat im *Franconia*-Fall 1876 noch wie Großbritannien die Auffassung vertreten, es gäbe keine rechtlichen Schranken für das Küstenmeer,[22] dann aber wie alle Nordseeanlieger im Nordsee-Fischerei-Abkommen von 1882 die Dreimeilenzone für die Fischerei anerkannt.[23] Bei der Gelegenheit ist für die Nordsee bestimmt worden, dass Buchten mit einer Öffnungsbreite von bis zu zehn Seemeilen nationale Fischereigewässer sind. Dem Vertrag ist weiter zu entnehmen, dass die freie Bewegung von Fischereifahrzeugen bei der Schifffahrt und beim Ankern in den Küstengewässern erlaubt war, die Grenzfestlegungen des Vertrages sich also nur auf die Fischerei bezogen.

An der Dreimeilenzone hat Deutschland im 20. Jahrhundert festgehalten. Es gehörte zu den wenigen Staaten, die sich bis kurz vor Abschluss des UN-Seerechtsübereinkommens von 1982 gegen die Ausdehnung des Küstenmeeres wehrten und unbeirrt die Dreimeilengrenze vertraten.[24] Erst mit Wirkung vom 16. März 1985 wurde das Küstenmeer in der Nordsee teilweise erweitert[25] und in Vollzug der UN-Seerechtskonvention 1994 dann weitgehend auf 12 Seemeilen ausgeweitet.[26]

III.

Die *Basislinie* als äußere Grenze der deutschen Eigengewässer und Innengrenze des Küstenmeeres wurde zunächst nicht durch eine Rechtsnorm geographisch genau festgelegt. Ursprünglich war dem Grundsatz nach die Küstenlinie Basislinie. Auf der Haager Kodifikationskonferenz 1930 wurde vorgeschlagen, diese Linie offiziell von den einzelnen Staaten angeben und möglichst in die Seekarten einzeichnen zu lassen.[27] Deutschland ist dem nicht gefolgt. In

---

[21] O'Connell, siehe Anm. 15, 459 f.
[22] J. Hallier, *Franconia*-Fall, Strupp/Schlochauer, Wörterbuch I, 554; O'Connell, siehe Anm. 15, 459.
[23] Art. 2 des Internationalen Vertrages betreffend die polizeilichen Regelungen der Fischerei in der Nordsee außerhalb der Küstengewässer vom 06. Mai 1882 (RGBl. 1884, 25, 28).
[24] Vgl. etwa die Tabelle bei O'Connell, siehe Anm. 15, 461 ff.
[25] Beschluss der Bundesregierung über die Ausdehnung des Küstenmeeres in der Nordsee zur Verhinderung von Tankerunfällen in der deutschen Bucht vom 12. November 1984 (BGBl. 1984 I, 1366).
[26] Bekanntmachung der Proklamation der Bundesregierung über die Ausweitung des deutschen Küstenmeeres vom 11. November 1994 (BGBl. 1994 I, 3428).
[27] E. Vanselow, *Völkerrecht*, 1931, 95. Die vom Deutschen Hydrographischen Institut in Hamburg herausgegebenen Seekarten weisen die Basislinie für die

Deutschland verlief die Basislinie auf der durch den Wasserstand der niedrigsten Ebbe gebildeten Linie, d.h. dort, wo bei Niedrigwasser der Strand nicht mehr trocken fällt.[28] Das war an der Ostsee die durch „Normal Null bei Mittelwasser" an der Nordsee die durch „Mittleres Springniedrigwasser" gebildete Linie.[29] Wegen der ständigen Veränderung der Sandbänke an der Nordseeküste, änderte sich dort auch die Basislinie, während sie in der Ostsee abgesehen von der Schleimündung weitgehend stabil blieb. Von deutschem Gebiet umgebene Reeden, Flussmündungen, Buchten und Baien, deren Öffnungsbreite 6 Seemeilen nicht überstieg, wurden ebenfalls als deutsche Eigengewässer beansprucht.[30] Eine Konkretisierung der Seegrenze fand nicht statt, es wurde vielmehr ein Hinweis auf die zunächst durch Art. 4 der Weimarer Reichsverfassung und dann durch Art. 25 GG in deutsches Recht inkorporierten allgemeinen Regeln des Völkerrechts für ausreichend gehalten.[31]

Nachdem in Folge des Urteils des IGH im britisch-norwegischen *Fischereifall*[32] das Prinzip der geraden Grundlinien gewohnheitsvölkerrechtlich akzeptiert, 1958 in Art. 4 des Genfer Übereinkommens über das Küstenmeer und die Anschlusszone kodifiziert[33] und schließlich in Art. 7 der UN-Seerechtskonvention von 1982 übernommen wurde, hat auch Deutschland – außer im Ems-Dollart-Gebiet – seine Basislinien in Nord- und Ostsee durch gerade Grundlinien bestimmt. Dies geschah im Auftrage des Bundesministers für Verkehr durch die vom Deutschen Hydrographischen Institut herausgegebenen Seekarten für die Nordsee 1970, für die Ostsee 1978.[34]

Demgemäß folgt die Basislinie in der *Nordsee* im Wesentlichen dem Küstenverlauf. Die zwischen den Inseln liegenden Wasserflächen werden von Juist bis Wangerooge durch gerade Grundlinien überbrückt. Von der Westspitze Wangerooges verläuft die Basislinie bis Schärhörn Riff und von dort bis St. Peter Böll an der Westküste der Halbinsel

---

Nordsee seit 1970 und für die Ostsee sogar erst seit 1978 aus. Dieses Institut wurde am 12. Dezember 1945 in Nachfolge mehrerer ziviler und militärischer Vorgängereinrichtungen nach einem Beschluss des Alliierten Kontrollrats für alle vier Besatzungszonen als zentrale Dienststelle eingerichtet. 1990 ging das Institut im Bundesamt für Seeschifffahrt und Hydrographie auf. E. Beckert/G. Breuer, *Öffentliches Seerecht*, 1991, 286, 617.

[28] Vgl. etwa Nr. 3a der Prisenordnung vom 30. September 1909 (RGBl. 1914, 275).

[29] Vanselow, siehe Anm. 27, 94.

[30] Vanselow, siehe Anm. 27, 99.

[31] Vgl. Note vom 09. Februar 1956 des Ständigen Beobachters der Bundesrepublik bei den Vereinten Nationen, UN Doc. ST/LEG./SER.B/6, 17.

[32] ICJ Reports 1951, 116; *AVR* 5 (1955/56), 214.

[33] UNTS Vol. 516, 205.

[34] R. Wolfrum, Die Kustenmeergrenzen der Bundesrepublik Deutschland in Nord- und Ostsee, *AVR* 24 (1986), 247, 253, 268 Anm. 82.

Eiderstedt. Von hier aus folgen gerade Grundlinien über die nordfriesischen Inseln bis zur deutsch-dänischen Seegrenze zwischen Sylt und Römö. Um die Insel Helgoland sind 5 gerade Basislinien gezogen.[35] Die längste Grundlinie ist nur knapp 22 Seemeilen lang, hält sich also innerhalb der Marge, die der IGH im britisch-norwegischen Fischereifall für zulässig anerkannt hat.[36] Weser- und Elbemündung werden in einer Entfernung vom Festland von etwa 10 Seemeilen überbrückt. Die zulässige Höchstbreite einer Bucht gem. Art. 10 Abs. 4 und 5 der UN-Seerechtskonvention wird damit nicht erreicht. Das bedeutet, dass die Deutsche Bucht keine Bucht im Sinne des Völkerrechts ist. Völkerrechtliche Bedenken gegen die Festlegung der Basislinien der Nordsee bestehen jedenfalls nicht, da die Bundesrepublik Deutschland die ihr durch das UN-Seerechtsübereinkommen gewährten Möglichkeiten nicht einmal ausgeschöpft hat.

Eine Abgrenzung der Eigengewässer zu den Niederlanden ist nicht erfolgt. Da Dollart und Emsmündung von zwei Uferstaaten umgeben sind, liegt keine Bucht i. S. von Art. 10 Abs. 1 der UN-Seerechtskonvention vor. Handelt es sich beim Ems-Dollart-Gebiet nicht um Eigengewässer, sondern um Küstenmeer, verläuft die deutsche Basislinie von der insoweit nicht umstrittenen deutsch-niederländischen Grenze im südlichen Dollart entlang der Küstenlinie über die Emsmündung, dann wiederum der Niedrigwasserlinie der Küste folgend über die Einfahrten zu den Häfen bis zu dem Punkt, von dem aus eine gerade Linie an die Westspitze von Borkum gezogen werden könnte. Eine genaue Fixierung der Basislinie ist jedoch nicht möglich, da die Grenzziehung im Ems-Dollart-Gebiet zwischen Deutschland und den Niederlanden strittig ist,[37] was allerdings vor allem das Küstenmeer betrifft, aber auch das als Eigengewässer zu reklamierende Wattenmeer zwischen dem Festland und den Inseln Borkum und Juist.

Zu Dänemark wurden die deutschen Küstengewässer – Eigengewässer und Küstenmeer – gem. Art. 111 des Versailler Vertrages – abgegrenzt. Auch hier ist zwischen Eigengewässern und Küstenmeer zu unterscheiden ist. Die Abgrenzungen wurden durch gerade Linien festgelegt. Im Bereich der Eigengewässer ist diese Abgrenzung bis heute überwiegend unstritten. Hierauf ist im Zusammenhang mit der strittigen Abgrenzung des Küstenmeeres zurückzukommen.

Die deutsche Basislinie in der *Ostsee* wurde für die schleswigholsteinische Küste 1978 festgelegt.[38] Sie verläuft von der deutschdänischen Staatsgrenze weitgehend am Ufer, das durch Anlandungen an

---

[35] Vgl. die Koordinaten bei Beckert/Breuer, siehe Anm. 27, 51 f.
[36] ICJ Reports 1951, 141.
[37] Vgl. Wolfrum, siehe Anm. 34, 258 ff.
[38] Vgl. die Koordinaten bei Beckert/Breuer, siehe Anm. 27, 51 f. und bei Wolfrum, siehe Anm. 34, 268.

der Schleimündung sich leicht ostwärts verschoben hat. Nur in der Hohwachter Bucht und zwischen Staberhuk auf der Insel Fehmarn und Dameshöved sind gerade Grundlinien gezogen, die sich bis zu 4,5 Seemeilen von der Küstenlinie entfernen. Durch den Beitritt der Deutschen Demokratischen Republik zum Grundgesetz wurde die Ostseeküste der Bundesrepublik bis zur polnischen Grenze verlängert. Das Beitrittsgebiet umfasst das gesamte Gebiet der DDR, wie es zurzeit ihres Unterganges bestand, so dass die Basislinien der früheren DDR von der Bundesrepublik übernommen wurden. Diese folgen im Allgemeinen dem Küstenverlauf, schließen aber die Bodden und vorgelagerten Inseln durch gerade Grundlinien ein. Die Lübecker Bucht ist durch die Wiedervereinigung der beiden deutschen Staaten zu einer Bucht im völkerrechtlichen Sinne geworden mit einer Öffnungsbreite von etwa 21,5 Seemeilen[39] – 2,5 Seemeilen weniger als die nach Art. 10 des UN-Seerechtsübereinkommens zulässige Buchtbreite.

Zu Dänemark verläuft die deutsche Basislinie in der Flensburger Förde entsprechend den Festlegungen aufgrund von Art. 111 des Versailler Vertrages bis zur inneren Grenze des Küstenmeeres zwischen Birknack und Kegnäs-Feuer. Im Oderhaff ist die deutsch-polnische Grenze durch Art. 1 des deutsch-polnischen Grenzvertrages vom 14. November 1990 bestimmt,[40] der sich auf das Görlitzer Abkommen vom 6. Juli 1950 zwischen der DDR und der Republik Polen über die Markierung der festgelegten und bestehenden deutsch-polnischen Staatsgrenze[41] und den zu seiner Durchführung und Ergänzung geschlossenen Akt von Frankfurt/Oder vom 27. Januar 1951 über die Ausführung der Markierung der Staatsgrenze zwischen Deutschland und Polen[42] bezieht.

IV.

Die seewärtige Grenze des deutschen *Küstenmeeres* und damit die Staatsgrenze verlief im Abstand von 3 Seemeilen zur Basislinie. Nach der Fixierung der Basislinien 1970 in der Nordsee und 1978 in der Ostsee lag sie geographisch fest.

Eine Ausweitung des Küstenmeeres in der Nordsee trat mit Wirkung vom 16. März 1985 ein. Durch Beschluss der Bundesregierung[43] wurde

---

[39] Beckert/Breuer, siehe Anm. 27, 53.
[40] BGBl. 1991 II, 1328.
[41] Text: I. von Münch (Hrsg.), *Dokumente des geteilten Deutschland*, 1968, 497 ff.
[42] Text: von Münch, siehe Anm. 41, 299 f. Text der Anlagen zu Art. 2: UNTS Vol. 319, 103.
[43] Bekanntmachung vom 12. November 1984 (BGBl. I 1366)

das Seegebiet von der Linie von der Drei-Meilen-Grenze nordnordwestlich der Insel Langeoog[44] nordwärts bis westsüdwestlich der Insel Helgoland,[45] dann die Dreimeilen Zone nördlich Helgolands einschließend bis nordwestlich von Schärhörn[46] dem Küstenmeer einverleibt. Zwischen Helgoland und den drei östlichen ostfriesischen Inseln und der Wesermündung gibt es danach nicht mehr hohes Meer. Ansonsten blieb es bei der Dreimeilen Zone vor den westlichen ostfriesischen Inseln, der Elbemündung und vor der schleswigholsteinischen Westküste.

Die Grenzen des deutschen Küstenmeeres wurden erweitert, „um geeignete Maßnahmen gegen die Gefahr eines Tankerunfalls und einer Ölverseuchung des Meeres und der Küste in der Deutschen Bucht treffen zu können". Es handelt sich bei dieser „Box"[47] um ein sehr stark befahrenes Seegebiet nordwestlich der Jade- Weser- und Elbmündung, d.h. zu den Häfen von Wilhelmshaven, Bremerhaven/Unterweserhäfen/ Bremen, Hamburg und Brunsbüttel am Nord-Ostsee-Kanal. Hier liegen der Kreuzungsbereich des West-Ost- und des Nord-Süd-Verkehrs. Hier haben sich die Schiffe an die bereits westlich beginnenden Verkehrstrennungssysteme einzuordnen. Einziger Zweck der Küstenerweiterung war die Neuordnung der Schifffahrt durch die Bundesrepublik abweichend von der Seestraßenordnung und damit auch die Unterwerfung unter die zentrale Radarüberwachung.[48] In die Erweiterung einbezogen waren eine Tiefseereede sowie die Reede bei Weser-Feuerschiff und die Elbereede.

Da im westlichen Teil der „Box" das Küstenmeer bis zu 16 Seemeilen von der Basislinie ausgedehnt wurde, erhob sich insoweit die Frage nach der Zulässigkeit, da nach damals geltendem Gewohnheitsvölkerrecht und der noch nicht inkraftgetretenen UN-Seerechtskonvention nur eine Küstenbreite bis zu 12 Seemeilen erlaubt war. Innerhalb der 12 Meilen liegen die Weser-Feuerschiff- und die Elbereede, während die Tiefseereede zum Teil jenseits der 12-Meilen-Distanz liegt. Deren Einbeziehung in das Küstenmeer war allerdings durch Art. 9 der Konvention über das Küstenmeer und die Anschlusszone gedeckt, dem jetzt Art. 12 der UN-Seerechtskonvention entspricht, wonach „Reeden die üblicherweise zum ... Ankern von Schiffen dienen, ... in das Küstenmeer einbezogen" werden. Der Wortlaut „are included in the territorial sea" legt die Vermutung nahe, dass die Einbeziehung einer Tiefseereede eine Selbstverständlichkeit ist. Die Küstenmeererweiterung auf das übrige recht kleine Gebiet außerhalb

---

[44] 54° 47' 58" N/07° 24' 36" O.
[45] 54° 08' 11" N/07° 24' 36" O.
[46] 54° 01' 11" N/08° 18' 40" O.
[47] Becker/Breuer, siehe Anm. 27, 12.
[48] Wolfrum, siehe Anm. 34, 256.

einer 12-Meilen-Zone vor der Basislinie in der Deutschen Bucht und um die Insel Helgoland lässt sich bei teleologischer Auslegung der einschlägigen Vorschriften sowohl der Küstenmeerkonvention als auch des UN-Seerechtsübereinkommens damit rechtfertigen, dass die komplizierte Verkehrssituation in der Deutschen Bucht und die sich daraus ergebenden Umweltgefährdungen diese Erweiterung erforderlich macht.[49] Nach der Rechtsprechung des IGH im Britisch-Norwegischen Fischereifall[50] kann im Übrigen bei der Festlegung von Küstenmeergrenzen auf die lokalen Gegebenheiten Rücksicht genommen werden.[51] Trotz dieser Rechtslage protestierten die USA gegen diese Küstenmeererweiterung. Der Protest blieb unberücksichtigt.

Am 19. Oktober 1994 – knapp einen Monat vor Inkrafttreten der UN-Seerechtskonvention – beschloss die Bundesregierung erneut eine Proklamation über die Ausweitung des Küstenmeeres, die mit Wirkung vom 1. Januar 1995 in Kraft trat.[52] Danach beträgt der Abstand der äußeren Grenze des Küstenmeeres in der *Nordsee* im allgemeinen 12 Seemeilen – gemessen von der Niedrigwasserlinie und der geraden Basislinie, wie sie 1970 festgelegt wurde. Die Tiefseereede außerhalb der 12-Meilen-Zone blieb als Küstenmeer erhalten. Die Koordinaten wurden entsprechend der Ausdehnung des Küstenmeeres um 9 Seemeilen neu definiert.[53]

Die *seitliche Abgrenzung* des deutschen Küstenmeeres zu den Niederlanden und zu Dänemark blieb einer Entscheidung der Bundesregierung zu einem späteren Zeitpunkt vorbehalten.

Die Abgrenzung zu den Niederladen im Ems-Dollart-Gebiet zwischen beiden Staaten ist strittig. Zu diesem Gebiet gehören auch die Inseln Borkum und Juist, so dass eine gerade Grundlinie zwischen diesen Inseln nicht gezogen werden konnte.[54] In Art. 46 Abs. 1 des Ems-Dollart-Vertrages vom 8. April 1960 zwischen der Bundesrepublik und den Niederlanden[55] wurde ausdrücklich festgestellt, dass unterschiedliche Rechtsauffassung über die Staatsgrenze bestehen, die sich jede Partei vorbehält. Die völkerrechtlich stärkere deutsche Position kann sich auf historische Titel seit dem 15. Jahrhundert stützen, die

---

[49] Vgl. R. Wolfrum, „Germany and the Law of the Sea", zu: T. Treves (ed.), *The Law of the Sea*, 1997, 199, 203 f.

[50] ICJ Reports 1951, 116; *AVR* 5 (1955/56), 214.

[51] So J. Kokott/L. Gündling, „Die Erweiterung der deutschen Küstengewässer in der Nordsee", *ZaöRV* 45 (1985), 674, 690; Wolfrum, siehe Anm. 49, 204.

[52] Bekanntmachung vom 11. November 1994 (BGBl. I, 3444).

[53] 1. 54° 08' 11" N/7° 24' 36" O.
2. 54° 08' 19" N/7° 26' 59" O.
3. 54° 01' 39" N/7° 33' 04" O.
4. 54° 00' 27" N/7° 24' 36" O.

[54] Beckert/Breuer, siehe Anm. 27, 52.

[55] BGBl. 1963 III, 602.

durch den Haager Vergleich von 1603, den Westfälischen Frieden 1648 und einen deutsch-niederländischen Vertrag von 1896 bestätigt wurden.[56] Danach gehört das gesamte Mündungsgebiet der Ems bis zur Niedrigwasserlinie vor der gegenüberliegenden niederländischen Küste zum deutschen Staatsgebiet. Die Niederlande akzeptieren diese Auffassung nicht, zumal danach die Grenze teilweise am Fuße des niederländischen Seedeichs verläuft und die Zufahrt zum Hafen Delfzijl vollständig im deutschen Gebiet liegt.

Die unterschiedlichen Rechtsmeinungen über den Grenzverlauf schlossen nicht aus, dass unter Offenlassung der Grenzfrage einige deutsch-niederländische Abkommen über konkrete Gegenstände im strittigen Gebiet geschlossen wurden. Besonders hervorzuheben ist der Vertrag vom 10. September 1984 über Zusammenarbeit im Bereich von Ems und Dollart sowie in den angrenzenden Gebieten.[57] Dieser Kooperationsvertrag regelte die Modernisierung des Emdener Hafens als neuen Dollart-Hafen unter Teilung eines Teils des von beiden Seiten beanspruchten Gebietes. Dem Kooperationsvertrag vorausgegangen war ein Änderungsabkommen zum Ems-Dollart-Vertrag vom 17. November 1975.[58] Ihm folgte ein ergänzendes Protokoll vom 22. August 1996 über die Zusammenarbeit zum Gewässer- und Naturschutz in der Emsmündung.[59] Trotz der noch offenen Frage der Grenzziehung bestehen funktional keine Schwierigkeiten im strittigen Gebiet, da die pragmatische Regelung, dass deutsche Schiffe der deutschen, niederländische der niederländischen Gerichtsbarkeit unterworfen sind und für fremde Schiffe der nächste Anlaufhafen oder der letzte Auslaufhafen maßgebend ist, Konflikte vermeidet.[60] Legt man die deutsche Auffassung über die Grenzziehung im Ems-Dollart-Gebiet zugrunde, handelt es sich hierbei um ein Servitut, da fremde Jurisdiktion über im Inland gesetzte Tatbestände ausgeübt wird.

Auch zu Dänemark ist die Küstenmeerabgrenzung im Bereich Sylt/Römö strittig. Die Seegrenze wurde dort gem. Art. 111 des Versailler Vertrages durch einen Grenzausschuss festgelegt, dem Vertreter Dänemarks und Deutschlands sowie der vier alliierten Hauptmächte – Großbritannien, Frankreich, Italien und Japan – angehörten.[61] Das Ergebnis wurde am 3. September 1921 in drei übereinstimmenden Kartenwerken beurkundet. Danach verläuft die Seegrenze vom Endpunkt der Landgrenze in 8 geraden Teilstrecken bis

---

[56] Vgl. – auch zum Folgenden – Wolfrum, siehe Anm. 34, 258 ff.
[57] BGBl. 1986 II, 509.
[58] BGBl. 1978 II, 309.
[59] BGBl. 1997 II, 1702.
[60] Vgl. Wolfrum, siehe Anm. 49, 205 ff.
[61] Zum Folgenden vgl. Wolfrum, siehe Anm. 34, 265 ff. und ders., siehe Anm. 49, 208 ff.

zu einem Punkt nord-nordwestlich Feuerlistwest. Von hier aus verlief sie entlang drei weiterer geraden Linien bis zum äußeren Rande der Drei-Meilen-Zone Dänemarks und Deutschlands. Für den Bereich des Lister Tiefs folgte die Grenze der Mittellinie des Fahrwassers. Aufgrund eines dänisch-deutschen Vertrages vom 12. April 1922 wurde die Möglichkeit einer Revision der Seegrenze vereinbart.[62] Die Mittellinie des Fahrwassers im Bereich des Lister Tiefs verschob sich nämlich seit 1921 durch Wanderung des Salzsandes nach Norden. Deshalb wurde aufgrund eines deutsch-dänischen Vertrages vom 5. März 1935 nach dem Vorschlag einer deutschen Verbalnote vom 10. August 1937 die Mittellinie karthographisch neu festgelegt. Da der Salzsand sich weiter nordwärts verschob und eine trocken fallende Bank vor der Südwestküste Römös entstand, wurde aufgrund einer weiteren deutschen Verbalnote vom 5. September 1941 die Grenze mit Wirkung vom 3. Januar 1942 nochmals nach Norden verlegt. Trotz dieser einvernehmlichen Veränderung der Seegrenze ist dem deutsch-dänischen Vertrag über die Festlandsockelabgrenzung der Nordsee im küstennahen Bereich[63] nicht die Grenze von 1942, sondern die von 1921 zugrunde gelegt. Der Festlandsockelvertrag von 1971 knüpft ebenfalls an den Grenzpunkt von 1921 an.[64] Seither ist der genaue Grenzverlauf im Seegebiet Sylt/Römö bis auf die die Eigengewässer abgrenzenden acht geraden Verbindungslinien strittig.

Auch an der *Ostsee* ist das Küstenmeer mit Wirkung vom 1. Januar 1995 erweitert worden.[65] Zwischen Punkt 1, der in Äquidistanz zu den dänischen Inseln Alsen und Aerö und der deutschen Küste in Angeln liegt[66] und Punkt 31, nördlichen der Halbinsel Zingst[67] wird die Seegrenze durch 30 gerade Linien unter Einbeziehung der Lübecker Bucht bestimmt. Von dort folgt sie einer Linie im Abstand von 12 Seemeilen, gemessen von der Niedrigwasserlinie und den geraden Basislinien bis Punkt 32 östlich von Rügen.[68] Ab hier ist sie bis Punkt 37 nordwestlich Swinemünde[69] durch 5 gerade Linien wieder genau bestimmt. Diese Abgrenzung des deutschen Küstenmeeres steht im Einklang mit den Art. 3 ff. der UN-Seerechtskonvention, zumal sie in Teilgebieten unter dem völkerrechtlich zulässigen Abstand von 12 Seemeilen zurückbleibt. Die Bundesregierung hat in ihrer Proklamation insoweit ausdrücklich festgestellt, dass damit keine Aufgabe des

---

[62] RGBl. 1922, 142.
[63] BGBl. 1966 II, 207.
[64] BGBl. 1972 II, 882.
[65] BGBl. 1994 I, 3444.
[66] 54° 44' 17" N/10° 10' 14" O.
[67] 54° 44' 38" N/12° 45' 00" O.
[68] 54° 26' 30,3" N/14° 04' 45,9" O.
[69] 53° 55' 42,1" N/14° 13' 37,8" O.

weitergehenden Rechtsanspruchs verbunden ist. Dies gilt vor allem auch für die Abgrenzung mit der gegenüberliegenden dänischen Küste gem. Art. 15 der Konvention.

Die *seitliche Abgrenzung* des Küstenmeeres zu Polen ist unproblematisch. Aufgrund des deutsch-polnischen Vertrages vom 14. November 1990 über die Bestätigung der bestehenden Grenze[70] hat sich auch im Seegebiet nichts geändert.

In der Proklamation über die Ausweitung des Küstenmeeres von 1994 behielt sich die Bundesregierung die amtliche Abgrenzung des Küstenmeeres zu Dänemark auch für die Ostsee einem späteren Zeitpunkt vor. Die seitliche Abgrenzung ist zwischen beiden Staaten strittig.[71] Vom Festland bis zur Linie Birknach/Kegnäs-Feuer[72] entspricht die Grenze zwischen deutschen und dänischen Eigengewässern der Festlegung nach Art. 111 des Versailler Vertrages. Vom deutsch-dänischen Grenzpunkt der Linie Birknach/Kegnäs-Feuer aus hat Dänemark die seitliche Grenze seines Küstenmeeres durch Verlängerung der südöstlich verlaufenden Grenzziehung in der Flensburger Förde auf drei Seemeilen festgelegt, während die Bundesrepublik die Äquidistanzgrenze vertritt. Nach Auffassung Dänemarks verläuft dessen Küstenmeerzone nur 1,8 Seemeilen von der deutschen Küste bei Falshöft entfernt; nach deutscher Auffassung müsste sie um 1,2 Seemeilen nordöstlich verschoben werden. Wolfrum[73] hat vorgeschlagen, die Grenze der deutschen und dänischen Eigengewässer auf die Linie Pölshuk/Falshöft hinaus zu verlegen und von dieser gemeinsamen Basislinie aus die Küstenmeergrenzen bis zum Punkt 1 der deutschen Basislinie zu bestimmen. Dänemark müsste dann allerdings seine Küstenmeergrenze zurücknehmen.

V.

Grenzziehungen in den küstenstaatlichen Funktionshoheitsräumen *außerhalb der Küstengewässer* sind weitgehend unstreitig.

Eine *Anschlusszone*, wie sie das Helsingfors-Abkommen vom 19. August 1925 zwischen den Ostseestaaten zur Schmugglerabwehr als Zwölfmeilen Zone vorsah,[74] ist nur noch dort rechtserheblich, wo die deutsche Seegrenze nicht auf 12 Seemeilen ausgedehnt worden ist, obwohl eine solche Ausdehnung völkerrechtlich zulässig wäre, d.h.

---

[70] BGBl. 91 II, 1328.
[71] Zum Folgenden vgl. Wolfrum, siehe Anm. 49, 269 f., ders., siehe Anm. 34, 210 f.
[72] 54° 49' 13" N/9° 56' 30" O.
[73] Wolfrum, siehe Anm. 34, 269.
[74] RGBl. 1926 II, 220.

keine Kollision mit der gegenüberliegenden dänischen Küstenmeergrenze eintreten würde.

Am 25. November 1994 – kurz nach Inkrafttreten der UN-Seerechtskonvention – beschloss die Bundesregierung eine Proklamation zur Errichtung einer *Ausschließlichen Wirtschaftszone*. Die Proklamation wurde durch Bekanntmachung vom 29. November 1994 im Bundesgesetzblatt am 8. Dezember veröffentlicht.[75] Auch diese Proklamation trat wie die kurz zuvor ergangene über die Erweiterung des Küstenmeeres mit dem 1. Januar 1995 in Kraft.[76] Die seewärtige Abgrenzung der Ausschließlichen Wirtschaftszone wird in der Nordsee durch 17 und in der Ostsee durch 45 gerade Verbindungslinien zwischen geographisch fixierten Punkten festgelegt.[77] Vier Verbindungslinien in der Ostsee stehen unter dem Vorbehalt vertraglicher Vereinbarungen mit den jeweils betroffenen benachbarten Staaten. Über die endgültige Position der Punkte der seitlichen Abgrenzung zu den Niederlanden und Dänemark in der Nordsee und zu Dänemark in der Ostsee sowie die Abgrenzung landeinwärts dieser Punkte wird die Bundesregierung nach Konsultationen zu einem späteren Zeitpunkt entscheiden. Die Modalitäten der Anwendung des Art. 5 Abs. 2 des Vertrages vom 22. Mai 1989 zwischen der DDR und Polen über die Abgrenzung in der Oderbucht[78] bleiben einer späteren Regelung nach Konsultationen mit der Republik Polen vorbehalten.

Die seewärtigen Grenzen der Ausschließlichen Wirtschaftszone in der *Nordsee* entsprechen denen des deutschen Festlandsockels, wie er in den Verträgen mit Dänemark,[79] den Niederlanden[80] und Großbritannien[81] festgelegt wurde. In der westlichen *Ostsee* war die Abgrenzung zu Dänemark festzulegen, was durch Fixierung auf der Mittellinie geschah, im Übrigen aber konnte an Verträge der DDR mit Schweden vom 22. Juli 1978,[82] Dänemark vom 14. September 1989[83] und Polen vom 22. Mai 1989[84] angeknüpft werden. Insgesamt ist die Abgrenzung in Übereinstimmung mit Art. 55 ff. der UN-Seerechtskonvention erfolgt. Nur in der Nordsee wird die zulässige 200-Meilen-Grenze an einem Punkt fast erreicht, sonst ist die deutsche Ausschließliche Wirtschafts-

---

[75] BGBl. 1994 I, 3769.
[76] Nr. I der Proklamation.
[77] Vgl. die festgelegten Koordinaten BGBl. 1994 I, 3770 f. und in Wolfrum, siehe Anm. 49, 221 und in A. Werbke, „Germany's Proclamation of an Exclusive Economic Zone", *Int'l J. Marine & Coastal L* 11 (1996), 85.
[78] Gesetzblatt der DDR, 1989 II vom 28. Juli 1989.
[79] BGBl. 1972 II, 882.
[80] BGBl. 1972 II, 889.
[81] BGBl. 1972 II, 897.
[82] Gesetzblatt der DDR 1979 II, 38.
[83] Gesetzblatt der DDR 1989 II, 147.
[84] Gesetzblatt der DDR 1989 II, 150.

zone sehr viel schmäler. In der Ostsee erreicht sie teilweise nicht einmal die Zwölfmeilengrenze.[85] Was die Ausübung von souveränen Rechten und Hoheitsbefugnissen in der Ausschließlichen Wirtschaftszone betrifft, ist Deutschland zurückhaltend, wie sich aus einer Erklärung anlässlich des Beitritts zur UN-Seerechtskonvention ergibt.[86]

Die Bundesrepublik Deutschland gehörte zu den wenigen Staaten, die lange gegen die Konzeption des *Festlandsockels* opponiert und deshalb auch 1958 gegen die Festlandsockel-Konvention gestimmt hatten. Erst am 22. Mai 1964 erließ die Bundesregierung eine Proklamation,[87] der das Gesetz zur Vorläufigen Regelung der Rechte am Festlandsockel vom 24. Juli 1964 folgte.[88]

Die Grenze zu den Nachbarstaaten in der *Nordsee* erfolgte durch die Verträge vom 28. Januar 1971 mit Dänemark und den Niederlanden und den Vertrag mit Großbritannien vom 25. November 1971,[89] nachdem ein Streit um die Abgrenzung mit Dänemark und den Niederlanden vor dem Internationalen Gerichtshof vorausgegangen war.[90] Bei Zugrundelegung des Äquidistanzprinzips hätte Deutschland bei einer Küstenlänge von 273 km einen Anteil am Festlandsockel von etwa 24600 qkm. Dänemark hätte bei einer Küstenlänge von 247 km einen Anteil von 56700 qkm und die Niederlande bei einer Küstenlänge von 385 km eine Anteil von 61400 qkm.[91] Durch die Verträge mit Dänemark und den Niederlanden wurde der deutsche Anteil auf 35600 qkm erhöht.[92] Es wurde allerdings keine gerade Linien bis zum Schnittpunkt von britischem, niederländischem und dänischem Festlandsockel vereinbart, sondern Ausbuchtungen bei Dänemark und den Niederlanden belassen, dafür aber ein Seegebiet nördlich vom Schnittpunkt des britischen und niederländischen Anteils auf Deutschland übertragen, so dass Deutschland seither zwischen dem britisch-niederländischen und dem britisch-dänischen Festlandsockel eine gemeinsame Festlandsockelgrenze mit Großbritannien hat.[93]

In der *Ostsee* bildet die Mittellinie die Grenze zwischen deutschem und dänischem Festlandsockel. Die Abgrenzungsverträge der DDR mit

---

[85] Vgl. Werbke, siehe Anm. 77, 79 ff.
[86] Werbke, siehe Anm. 77, 81 f.
[87] BGBl. 1964 II, 104.
[88] BGBl. 1964 I, 497.
[89] Quellen vgl. oben Anm. 79-81.
[90] Zum Festlandsockel vgl. etwa Oda, *International Control of Sea Resources*, 1963; E. Menzel, „Der deutsche Festlandsockel in der Nordsee und seine rechtliche Ordnung", *AöR* 1965, 1 ff.; B. Rüster, *Die Rechtsordnung des Festlandsockels*, 1977.
[91] Vgl. C. Gloria in K. Ipsen, *Völkerrecht*, 3. Aufl., 1990, 720.
[92] Gloria, siehe Anm. 91, 721.
[93] Vgl. die Karte bei G. Francalanci/T. Scovazzi (eds.), *Lines in the Sea*, 1994, 237.

Schweden, Dänemark und Polen[94] wurden nach der Wiedervereinigung Deutschlands von der Bundesrepublik Deutschland übernommen. – Insgesamt ist der Anteil Deutschlands am Festlandsockel wegen seiner ungünstigen geographischen Lage sowohl in der Ostsee als auch im „nassen Dreieck" der Nordsee gering.

VI.

Die Grenzziehung bei den deutschen Küstengewässern, der Ausschließlichen Wirtschaftszone und des Festlandsockels werfen auch staatsrechtliche Fragen auf. Einmal geht es um die Rechtsform, in der Grenzen festgelegt werden, zum anderen um die Abgrenzung der Kompetenzen von Bund und Ländern und schließlich um die Grenzziehung zwischen den Ländern.

Was die *Rechtsform* bei Grenzänderungen betrifft, so handelt es sich dabei um einseitige völkerrechtliche Erklärungen, für deren Abgabe gem. Art. 59 Abs. 1 GG der Bundespräsident zuständig ist. Es handelt sich dabei aber nur um die Befugnis, Entscheidungen der verfassungsrechtlich zuständigen Organe nach außen zu verkünden.[95] Innerstaatlich zuständig sind je nach Inhalt einer Regelung die Bundesregierung allein oder wegen der Erforderlichkeit der Mitwirkung der gesetzgebenden Körperschaft gemeinsam mit Bundestag und Bundesrat.[96] Der Bundespräsident übt die völkerrechtliche Vertretungsmacht nicht unmittelbar aus, sondern überlässt sie der von ihm generell oder speziell bevollmächtigten Bundesregierung und in dieser vornehmlich dem Auswärtigen Amt. So bestimmt etwa § 11 Abs. 1 der vom Bundespräsidenten genehmigten Geschäftsordnung der Bundesregierung vom 11. Mai 1951,[97] dass Mitglieder und Vertreter auswärtiger Regierungen sowie Vertreter zwischenstaatlicher Einrichtungen nur nach vorherigem Benehmen mit dem Auswärtigen Amt empfangen werden. Abs. 2 dieser Bestimmung sieht vor, dass Verhandlungen mit dem Ausland oder im Ausland nur mit Zustimmung des Auswärtigen Amtes, auf sein Verlangen auch nur unter seiner Mitwirkung geführt werden.[98]

---

[94] Vgl. Anm. 79-81.
[95] J. A. Frowein, „Verfassungsrechtliche Probleme um den deutschen Festlandsockel", *ZaöRV* 25 (1965), 1, 2 f.
[96] Zum Träger der auswärtigen Gewalt vgl. etwa W. Grewe, „Die auswärtige Gewalt", *VVDStRL* 12 (1954), 129 ff; ders. Auswärtige Gewalt, in J. Isensee/P. Kirchhof, *HDStR* III (1988), 921 ff; E. Menzel, *VVDStRL* 12 (1954), 179 ff; H. W. Baade, *Das Verhältnis von Parlament und Regierung im Bereich der auswärtigen Gewalt der Bundesrepublik Deutschland*, 1962; W. Rudolf, *Völkerrecht und deutsches Recht*, 1967, 179 ff.; R. Wolfrum, „Kontrolle der auswärtigen Gewalt", *VVDStRL* 56 (1997), 38 ff.
[97] GMBl. 1951, 137.
[98] Vgl. § 77 GGO II.

Verwunderlich ist, dass für die Erweiterung des Küstenmeeres und die Errichtung der Ausschließlichen Wirtschaftszone und des Festlandsockels kein Gesetz erlassen wurde, das bei Änderungen von Landgrenzen jedenfalls immer dann notwendig ist, wenn davon Menschen betroffenen sind, deren Rechte und Pflichten durch Gesetz normiert werden müssen. Im Streit mit Frankreich um den Mundatwald[99] wurde argumentiert, dass ohne Bundesgesetz eine Grenzänderung hinsichtlich des Germanhofs nicht in Betracht käme, weil davon sieben Personen betroffen wären. In Großbritannien bedurfte es zur Ausübung des Juriskdition im Küstenmeer schon 1876 eines Gesetzes.[100]

Im Falle der Küstenmeererweiterung und der Proklamation über den Festlandsockel verneint die herrschende Lehre h.l. die Notwendigkeit eines Gesetzes.[101] Wolfrum meint nunmehr, dass bei einer einseitigen Erweiterung der nationalen Hoheitsrechte, wie der Ausdehnung der Küstenmeergrenzen oder der Ausweisung einer Ausschließlichen Wirtschaftszone wesentliche Gründe für eine Mitwirkung des Bundestages sprechen, denn die räumliche Ausdehnung staatlicher Hoheitsgewalt erfasse Aktivitäten Einzelner.[102] Dies führe, soweit nicht gleichzeitig Gesetze zur Regelung der Aktivitäten in dem Meeresgebiet erlassen werden, zu einer Erweiterung des Anwendungsbereichs bestehender Rechtsnormen, die allein in die Kompetenz des Bundestages fällt. Von daher liege eine analoge Anwendung von Art. 59 Abs. 2 Satz 1 GG nahe.

Der von Wolfrum angesprochene Parlamentsvorbehalt ist m.E. ein echter Gesetzesvorbehalt: Änderungen der Küstengewässergrenzen und die Errichtung einer Ausschließlichen Wirtschaftszone oder eines Festlandsockelregimes bedürfen eines Gesetzes gem. Art. 59 Abs. 2 Satz 1 GG, der nicht nur entsprechend, sondern unmittelbar anzuwenden ist. Statt eines Zustimmungsgesetzes gem. Art. 59 Abs. 2 Satz 1 GG kann freilich auch die innerstaatliche Umsetzung des völkerrechtlichen Aktes durch ein spezielles Gesetz erfolgen. Soweit die Verwaltung der Länder betroffen ist, ist die Zustimmung des Bundesrats erforderlich. Bei der Erweiterung des Küstenmeeres werden auch die Zuständigkeitsbereiche der Länder erweitert. Ein Zustimmungsgesetz ist ebenso wie bei Abtretungen von Landesgebiet deshalb notwendig, weil der räumliche Kompetenzbereich verändert wird. Obwohl das Grundgesetz diesen

---

[99] Vgl. S. Jutzi, „Mundatwald und Sequesterland", *AVR* 24 (1986), 277, 293.

[100] So der Court for Crown Cases Reserved im Falle *R.* v. *Keyn, The Franconia* (1876) 2. Ex. D. 63. Der Fall gab Veranlassung zur Verabschiedung des Territorial Waters Jurisdiction Act von 1878 (41 & 42 Vict. c. 73). Vgl. G. Gidel, *Le droit international public de la mer*, Bd. 2, 1934, 245 f.

[101] Frowein, siehe Anm. 95, 3 f.; Kokott/Gündling, siehe Anm. 51, 682 f.; Wolfrum, siehe Anm. 34.

[102] Wolfrum, siehe Anm. 96, 61. Vgl. auch den Diskussionsbeitrag von K. Doehring, *VVDStRL* 36 (1978), 147 f.

besonderen Fall der Zustimmung nicht ausdrücklich vorsieht, folgt die Zustimmungsbedürftigkeit aus Art. 84 Abs. 1 GG. Dies gilt auch für die Errichtung einer Ausschließlichen Wirtschaftszone und des Festlandsockelregimes, weil insoweit die Zuständigkeit von Landesbehörden betroffen und damit Art. 84 Abs. 1 GG einschlägig ist. Deshalb ist z.B. das Gesetz zur Ausführung des Seerechtsübereinkommens der Vereinten Nationen sowie des Übereinkommens vom 28. Juli 1994 zur Durchführung des Teils XI des Seerechtsübereinkommens vom 6. Juni 1995 mit Zustimmung des Bundesrates beschlossen worden.[103]

Damit ist bereits die zweite staatsrechtliche Frage in diesem Zusammenhang angesprochen: die Abgrenzung der Kompetenzen von *Bund und Ländern*. Die Zuständigkeit der Länder erstreckt sich auch im erweiterten Küstenmeer auf alles, was im bisherigen Küstenmeer in ihre Zuständigkeit fiel. Dies folgt schon daraus, dass das Bundesgebiet aus den deutschen Ländern besteht und trotz Wegfalls des alten Art. 23 GG landesfreies Gebiet nicht existiert.

Im Bund-Länder-Verhältnis spielen Fragen des Eigentums, der Nutzungsrechte und der hoheitlichen Zuordnung der Hoheitsgewässer und die Kompetenzverteilung zwischen Bund und Ländern in der Ausschließlichen Wirtschaftszone und im Festlandsockel eine Rolle.[104] Die Kompetenzen müssen jeweils zwischen Bund und Ländern abgeschieden werden. Es gibt sogar einen Fall der Organleihe: § 3 des Meeresbergbaugesetzes i.d.F. vom 6. Juni 1995[105] sieht vor, dass dieses Gesetz „vom Oberbergamt in Clausthal-Zellerfeld als einem für diese Aufgabe vom Lande Niedersachsen entliehenen Organ des Bundes ausgeführt" wird. Das niedersächsische Oberbergamt unterliegt insoweit der Fach- und Rechtsaufsicht des Bundes.

Von nicht geringer Bedeutung ist die Festlegung der *Landesgrenzen* in den Eigengewässern und im Küstenmeer sowie die örtliche Zuständigkeitsabgrenzung in der Ausschließlichen Wirtschaftszone und im Festlandsockel.

In der *Ostsee* sind die Länder Mecklenburg-Vorpommern und Schleswig-Holstein betroffen, deren jeweiliger Zuständigkeitsbereich sowohl in den Hoheitsgewässern als auch in der Ausschließlichen Wirtschaftszone und im Festlandsockel durch die früheren Grenzen zwischen der Bundesrepublik und der DDR bestimmt ist.[106] Allerdings

---

[103] BGBl. 1995 I, 778.
[104] Zu den innerstaatlichen Kompetenzabscheidungen vgl. Beckert/Breuer, siehe Anm. 27, 54 ff.
[105] BGBl. 1995 I, 782.
[106] Vgl. die in der 2. DVO zum Gesetz über die Staatsgrenze der DDR vom 20. Dezember 1984 (Gesetzblatt der DDR I, 441) festgelegten Koordinaten. Beckert/Breuer, siehe Anm. 27, 53.

verlängert sich die mecklenburg-vorpommern/schleswig-holsteinische Landesgrenze in der Lübecker Bucht durch die Ausdehnung des Küstenmeeres aufgrund der Proklamation der Bundesregierung über die Ausdehnung des Küstenmeeres vom 19. Oktober 1994 nach Maßgabe des Äquidistanzprinzips.

Komplizierter sind die Abgrenzungen in der *Nordsee*, da nicht nur Niedersachsen und Schleswig-Holstein Anlieger sind, sondern auch Hamburg und Bremen.[107] Aufgrund des Cuxhaven-Vertrages vom 26. Mai 1961 zwischen Hamburg und Niedersachsen[108] wurde unter Einbeziehung der Inseln Neuwerk und Schärhörn eine Hamburger Exklave im Neuwerker Watt gebildet, um Hamburg den Bau eines Vorhafens im Elbmündungsgebiet zu ermöglichen. Das Gebiet der Exklave reicht über die Basislinie und die Dreimeilenzone hinaus, aber nicht bis zur äußeren Grenze der Zwölfmeilenzone. Auch Bremen hat über Bremerhaven Anteil an den Eigengewässern, nicht aber am Küstenmeer. Für eine Reihe von Materien haben die Küstenstaaten vertragliche Vereinbarungen über den örtlichen Zuständigkeitsbereich ihrer jeweiligen Behörde im Gebiet der Eigengewässer und des Küstenmeeres getroffen,[109] so z.B. über die Zuständigkeit der Polizeibehörden. Soweit keine Grenze vereinbart ist, gilt das Äquidistanzprinzip, das für den Festlandsockel in § 137 Abs. 1 Bundesberggesetz ausdrücklich verankert ist.[110]

---

[107] R. Lagoni, Ländergrenzen in der Elbmündung und der Deutschen Bucht, 1982; ders., „Der Hamburger Hafen, die internationale Handelsschifffahrt und das Völkerrecht", *AVR* 26 (1988), 261.
[108] Hamb. GVBl. 1961, 318.
[109] Beckert/Breuer, siehe Anm. 27, 67 ff.
[110] BGBl. 1980 I, 1310.

Linking Equity and Law in Maritime Delimitation

Oscar Schachter

Shigeru Oda has been a valued confrère and friend for nearly half a century. I have benefited greatly from his writings, judicial opinions and our conversations, and I welcome the opportunity to join in this tribute to him.

When we first met at Yale in the 1950s, he had already chosen the law of the sea as his principal interest. It remained a major concern throughout his illustrious career as a judge, scholar and diplomat.

My topic for this paper – the relation of law and equity in maritime delimitation – seeks to respond in some degree to Judge Oda's concerns about the normativity of the delimitation process. It reflects my own experience as a judge in the dispute between Canada and France relating to the maritime delimitation of the islands of St. Pierre and Miquelon.[1] In that case, the two States requested the Court of Arbitration to carry out a delimitation of the maritime areas pertaining to France and to Canada, "ruling in accordance with the principles and rules of international law".[2]

The two States also agreed that the fundamental norm to be applied in the case "requires the delimitation to be effected in accordance with equitable principles, or equitable criteria taking account of all the relevant circumstances, in order to achieve an equitable result".[3] As in most delimitation disputes, the parties differed as to the significance and

---

[1] *Case Concerning Delimitation of Maritime Areas Between Canada and France*, ILM 31 (1992) 1145. The members of the Court of Arbitration were Eduardo Jiménez de Aréchaga (President), Prosper Weil (appointed by the French Government), Allan E. Gotlieb (appointed by the Canadian Government), Gaetano Arangio-Ruiz, and Oscar Schachter.

[2] See note 1, 1152 (Article 2). In an analysis of the decision of the Court, Keith Highet wrote in 1993 that "it stands for certain important points that illuminate and confirm the progressive development of the international law of maritime delimitation". *AJIL* 87 (1993) 452 et seq., 463.

[3] *Ibid.*, para. 38.

weight of the geographical features that were (as usual) at the heart of the dispute. A critical question underlying this difference was whether the issues could be decided by reference to equitable principles and rules of law or whether, at bottom, they could only be settled by judgments of "fairness" and equity treated as a non-legal concept dependent on the particular facts and subjective perceptions. Judge Oda has characterised the latter position as the "principle of non-principle".[4] His several learned dissenting opinions have deplored the "subjectivism" of equitable criteria applied in delimitation cases.

The unique character of the geography in the *Canadian-French* case might well have been seen as precluding precedents and rendering legal principles inapplicable. If a court had been asked to give an *advisory* opinion on the legal principles governing this case it might plausibly have concluded it could not do so, thus rendering a *non-liquet* decision. However, this possibility was not even raised in the *Canadian-French* case. The consensual jurisdiction conferred by the parties had imposed an obligation on the Court to settle the dispute in accordance with international law and equitable principles. On this premise, the judges had to apply suitable principles of law and equity even if no precedent or rule could be found to fit the unique facts of the case. Thus, a *non-liquet* was excluded, however unusual the facts in this case. Accordingly, the Court did identify equitable principles and relevant criteria, based on prior maritime delimitation cases and applied them as the applicable law of the case.

I. Equidistance

The Court began its analysis of criteria by questioning the French contention that equidistance should be the principal criterion of the delimitation. It observed that equidistance has always been understood in delimitation cases as qualified by special circumstances, especially geographical features.[5] In his dissent, Professor Weil emphasised the simplicity of a basic equidistance criterion which could be adjusted to meet particular problems.[6] The two positions may not seem so far apart in those terms. However, the majority of the Court decided that other equitable criteria had to be applied in accordance with the principles adopted in the case law of the International Court and arbitral bodies.

---

[4] Oda, Dissenting Opinion in the *Tunisia-Libya* case, ICJ Reports 1982, 157, and Dissenting Opinion in the *Libya-Malta* case, ICJ Reports 1985, 116, para. 1.
[5] See note 1, paras. 39-48.
[6] Weil, Dissenting Opinion, see note 1, paras. 36-37.

II. Coastal Lengths and Frontal Projections

The Court was also faced with contending positions regarding the seaward projection of the several coasts in relation to their lengths. Canada maintained that the length of a coastal sector was relevant to its projection; the longer the coast, the farther the projection. France contended that a coast, however short, could have a seaward projection up to 200 miles if no competing coasts curtailed its reach. The Tribunal accepted the latter position, going as far as to award France a 200-mile projection based on the coast of the French island that was unobstructed by any opposite or laterally aligned Canadian coast. The width of this corridor as determined by the lateral length of the "opening" was fixed at 12 miles and the seaward length at 200 miles.[7] Judge Weil, who disagreed with this decision, likened the resulting configuration to a mushroom with a long stem (the narrow corridor). He argued that prior case law and doctrine had not recognised such narrow frontal projections based on "openings" to the sea. Moreover, Judge Weil also noted that so narrow a corridor meant that France could make little use of the corridor for fishing or any other economic use.[8]

The Court did not extend this 12-mile corridor beyond 200 miles on the ground that the open sea beyond 200 miles was international waters and therefore outside the terms of reference of the Tribunal. It did suggest that any allocation of authority over the shelf beyond 200 miles would require international authority.[9]

III. The Principle of Non-Encroachment

This principle was first introduced in the *North Sea Continental Shelf* cases where the International Court of Justice decided that delimitation must be affected "in such a way as to leave as much as possible to each party of all those parts of the Shelf that constitute a natural prolongation of its land territory ... without encroachment of the land territory of the other".[10] The principle was affirmed in several other cases. In the *Canadian-French* case, the Court adopted a compromise solution that gave the French islands a 12-mile exclusive economic zone on one side, even though this would seem to be an encroachment of Canadian seaward projections (i.e., from Newfoundland). Canada argued that this was an unwarranted encroachment that did not do justice to the longer coastal length of Canada (i.e., Newfoundland). Encroachment was also

---

[7] See note 1, paras. 70-74.
[8] Weil, Dissenting Opinion, see note 1, para. 9.
[9] See note 1, paras. 78-79.
[10] The *North Sea Continental Shelf* case, ICJ Reports 1969, 21-22.

raised by Canada to object to the 12-mile French corridor running south from the frontal opening of St. Miquelon (referred to above). The Court ruled that this narrow corridor (140 miles from the Canadian coasts) would not encroach on the seaward projections of Nova Scotia and Cape Breton.

Although non-encroachment has been treated as a principle or criterion by tribunals, the term has no independent meaning; it only designates an unwarranted or excessive territorial claim, a conclusion that must be based on other factors. Even if non-encroachment is not an independent factor, the cases indicate that it serves psychologically as a "red light", drawing attention to a cut-off that would come so close to a State's coast or near-by islands "as to interfere with its right to development or put its security at risk".[11]

IV. The Relevance of Fisheries and Mineral Resources

The opinion in the *Canadian-French* case referred to the conclusion in the *Gulf of Maine* case that the Court must assure that its solution would not be likely to "entail catastrophic repercussions for the livelihood and well-being of the people most affected".[12] The opinion responded to this by noting that fishing rights would not be adversely affected, because an existing bilateral agreement gave access to the nationals of the other State in the fishing zones under its jurisdiction on the basis of full reciprocity.[13] The opinion also referred to potential hydrocarbon resources in the areas in question, but concluded that this issue was not relevant to the limitation at the present time. However, it left open the possibility that access to oil or other mineral deposits could be a relevant consideration in some circumstances based, on geographical factors and patterns of usage.

V. The Criterion of Proportionality

On its face, the idea of proportionality would seem to be an obvious equitable principle; its very meaning suggests a fair or just relation of means and ends. The International Court of Justice in the *Libya-Malta* case suggested "proportionality" as an *a posteriori* test that may be used to check on the equity of a delimitation made on other grounds.[14] In the

---

[11] Statement of Presiding Judge Lachs in the *Guinea/Guinea-Bissau* case, quoted in P. Weil, *The Law of Maritime Delimitation*, 1989, 61.
[12] The *Gulf of Maine* case, ICJ Reports 1984, para. 237.
[13] See note 1, paras. 87-88.
[14] ICJ Reports 1985, 43 et seq., paras. 55 et seq.

Canadian-French case, the Court compared the size of the areas awarded to each party to the respective coastal lengths as determined by the Tribunal. They found that the ratio of areas was about 16 to 1, and that substantially the same ratio applied to the coastal lengths of the parties.[15] In his dissent, Professor Weil objected to the criteria applied, emphasising the subjective character of the identification and measurements of the various segments. Adding a light touch, Judge Weil commented "the identification and measurement of the relevant coasts and the relevant areas have this in common with love or Spanish inns; each finds in them what he brings to them".[16]

The Court did not say what it would do if it had found an unreasonable disproportion between the ratios of coastline lengths and those of areas. That such a finding of unreasonable disproportion has never been made by a tribunal led Professor Weil to suggest that in these cases the tribunals selected the data to confirm a predetermined result. Despite this comment, Weil recognised that the Court in the *Canadian-French* case did not make proportionality the operative principle of the delimitation, nor did it claim that the delimitation adopted was a "proportionality line".[17]

VI. The Legality of Equitable Principles and Their Application in International Delimitation

Judge Oda's concern that the equitable principles applied in delimitation cases by the International Court of Justice and by arbitral tribunals involved subjective assessments was not expressly addressed by the Tribunal in the *Canadian-French* case. However, it is fair to say that all of the judges in that case would reject the charge of "subjectivity", used in the sense of unlimited discretion. Their recourse to the principles mentioned above, such as non-encroachment, frontal projection and proportionality, indicate an understanding that their judgment had to rest on normative principles accepted by international law. Moreover, no Member of the Court suggested that decisions as to boundaries, rights to fish or mineral resources should be determined by the relative economic or political needs of the States involved. It was understood that geography was not to be "refashioned" for political or economic reasons in order to fix the boundaries. The Tribunal (like the International Court) did not regard its obligation to apply equitable principles and reach an equitable result as a ground for decisions *ex aequo et bono*.[18]

---

[15] See note 1, paras. 89-91.
[16] Weil, Dissenting Opinion, *ibid.*, para. 24.
[17] *Ibid.*, para. 26.
[18] See note 2, comments of K. Highet.

All this said, it is undeniable that tribunals faced with delimitation disputes are charged with applying normative concepts of such breadth and flexibility that judges (as well as governments) may reasonably differ as to the specific conclusions. Of course, this is not unique to maritime delimitation. Equity *infra legem* is widely accepted to attenuate a legal rule when particular circumstances required an "equitable result" that would not be reached under a strict reading of the law. The International Court of Justice, in its 1982 judgment in the *Tunisia-Libya* case, held that an equitable result was required by the law and, therefore, that legal principles had to be selected to reach that conclusion. The Tribunal in the *Canadian-French* case recognised that the unique geographical features "at the heart of the case" could not in themselves determine the result. Their relevance and weight had to be determined by equitable principles and rules of law recognised by international tribunals and widely accepted practice. The decisions should not be dependent on the "eye of the judge". As the ICJ declared in the 1985 *Libya-Malta* case both equity and law required

> "a certain generality and a certain consistency; otherwise it [the decision] will not fulfil the essential functions of the law: certainty and predictability. ... Even though it [equity] looks with particularity to the peculiar circumstances of an instant case, it also looks beyond it to principles of more general application...."[19]
>
> While every case of maritime delimitation is different in circumstances from the next, only a clear body of equitable principles can permit such circumstances to be properly weighed and the objective of an equitable result as required by general international law to be attained".[20]

These comments of the International Court would appear to be broadly in accord with Judge Oda's concerns over the "subjectivity" of delimitation decisions based on principles of equity. Their actual significance will depend on their application in particular cases. We can be confident that Judge Oda, even after his retirement from the Court, will continue to keep a sharp eye on the reasoning of tribunals with the aim of upholding the role of law as a factor in achieving equitable results in delimitation decisions.

---

[19] ICJ Reports 1985, 38-41, paras. 45-50.
[20] *Ibid.*, 55, para. 76.

The Austrian Continental Shelf

Ignaz Seidl-Hohenveldern

I. Introduction

We have to begin this article with an apology to the readers and to the editor's typist, who believed she was correcting a slip of my pen when she listed the title of my contribution to this *Festschrift* as "the Australian Continental Shelf". I would not feel qualified to deal with this topic. Nor did I intend to discuss the practice of the Austro-Hungarian monarchy concerning maritime boundaries. For many years, our late lamented confrère Stephan Verosta had recruited the services of a group of retired Austrian ambassadors to examine the archives of the Foreign Office of the Austro-Hungarian monarchy for documents of significance for international law. The outcome would have filled six volumes, which, for reasons of economy, had to be reduced to two. The full fruits of this research are filed as a separate corpus in the Austrian *Staatsarchiv*. I was concerned with the review of the documents and can thus certify that these, even the ones not published in Verosta's Digest,[1] did not include any documents concerning Austrian territorial waters nor, for obvious reasons, Austria's continental shelf. The only reference we found concerned an incident in 1865, in which an Italian naval vessel tried to claim possession of the island of Pelagosa in the Adriatic as successor to the Kingdom of Naples. Italy apologised for this incident, blaming incorrect maps used by the Kingdom of Naples.[2] This near-absence of Austrian practice in matters of maritime delimitation may be due to the fact that, in general, the borders had already been clearly drawn, and that, compared with other concerns of the Dual Monarchy, such disputes were deemed to be of such little importance that they were not considered worthy of being recorded in the Ministry's files.

---

[1] S. Verosta and I. Seidl-Hohenveldern, *Die völkerrechtliche Praxis der Donaumonarchie von 1859 bis 1918*, 1996.
[2] *Ibid.*, Vol. 2, 403-404.

However, we did find in these files a request by Brazil for information on a problem which to this day is a matter of dispute between Austria, Switzerland and Germany[3]: are there state borders on Lake Constance (Bodensee) and, if so, where should they be drawn? We venture to submit this problem to the English-speaking readers of this *Festschrift*, because – apart from an article in the *Encyclopaedia*[4] – we have found no other articles in English on the subject. However, we hope that it will be of interest to experts on the law of the sea, and thus also to our eminent confrère, Shigeru Oda. Of course, nobody would venture to apply the law of the sea *tel quel* to Lake Constance. What appears to be appropriate in the case of the Caspian Sea would be ridiculous in the case of the much smaller Lake Constance (571.5 square kilometres). Yet, as we shall see, the issues concerning this lake ask for solutions, some of which may be similar to those found in the law of the sea. Roughly speaking, Lake Constance forms a rectangle, the large northern side belonging to Germany and the large southern side mostly to Switzerland. The small eastern side bulges out into Austria as the Bay of Bregenz; the small western side is framed by two fjord-like extensions: the Überlingen Lake in the north, which is completely surrounded by German territory, and the Constance Funnel in the south, which to the west of Constance becomes the Lower Lake (Untersee), forming a barrier between Switzerland and Germany.

II. Condominium or Limits on the Lake?

The riparian States of Lake Constance have discussed for centuries whether the lake is a condominium belonging to all of them or whether the territory of the lake should be divided and attributed to the several riparian States, and, if so, what criteria should be applied for this purpose.

Three theories have been advanced in support of the conflicting claims, although none of the States have upheld their claims throughout the centuries without deviations from what otherwise may be assumed to be "their" doctrines.

With this caveat, we may state that, generally, Switzerland believes that Lake Constance is, or at least should be, divided between the riparian States.[5] Germany tends to regard the lake as a condominium of the riparian

---

[3] *Ibid.*, Note of 7 June 1908, 407-408.
[4] J. A. Frowein, "Lake Constance", in R. Bernhardt (ed.), *EPIL*, Vol. III, 1997, 116 et seq.
[5] G. Riva, "L'exercise de souveraineté sur le Lac de Constance", *Schweizerisches Jahrbuch für internationales Recht* XXIV (1967), 43 et seq., 62.

States.[6] Austria stated in the above-mentioned note[7] that there is much that argues *de lege ferenda* for the establishment of a condominium on the lake. Thus Austria has failed to adhere unequivocally to the condominium theory. In any case, Austria has introduced a further element into this dispute which, in a certain way, reminds us of the continental shelf. According to this theory only the "High Lake" or "Blue Lake" is, or has become, a condominium of the riparian States. However, the lake down to the 25-metre isobath, the "White Lake", is regarded as part of the territory of the bordering riparian State. The words "blue" and "white" refer to the colour of the waters of the High Lake and the waters near the shore respectively.[8] The slope of the lake down to the isobath, before the steep descent to the normal depth of the lake, which is between 100 and 200 metres, is called the Halde.

A decision of the Austrian District Court at Bregenz of 22 November 1964[9] follows the conclusions reached by a conference held in Vienna, on 18 January 1961, between the relevant Austrian Federal Ministries and the authorities of Vorarlberg,[10] according to which Austrian sovereignty ends at the 25-metre isobath, whereas the remainder of the lake is a condominium of the riparian States and not part of Austria. However, as Tretter and Pelzl[11] point out, the authorities failed to make the corresponding changes to the "cataster line", a straight line across the Bay of Bregenz which includes, as part of the territory of the Austrian communities bordering the lake, waters deeper than the isobath.[12]

1. Origin of the Condominium

An almost philosophical dispute exists between adherents of the condominium theory as to its origin. Some believe that there should be

---

[6] Bavarian Administrative Court, 20 February 1963, with critical comments by H. Hüber, "Gebietshoheit und Grenzverlauf im Bodensee", *Schweizerisches Jahrbuch für internationales Recht*, XXI (1964), 192 et seq.
[7] Cf. note 3 *supra* and accompanying text.
[8] B. Schuster, *Die Entwicklung der Hoheitsverhältnisse am Bodensee seit dem Dreissigjährigen Kriege unter besonderer Berücksichtigung der Fischerei*, 1951, 11.
[9] Decision cited by K. Berchtold, "Die Hoheitsverhältnisse auf dem Bodensee", *Juristische Blätter* 87 (1965), 401 et seq., but see the subsequent decisions of the Austrian Supreme Court at notes 32 and 33 *infra* and accompanying text.
[10] Berchtold, see note 9, 401, 403.
[11] H. Tretter and A. Pelzl, "Streit um den Bodensee", *Österr. Juristen-Zeitung* 46 (1991), 793, footnote 1.
[12] See note 31 *infra* and accompanying text.

clearly drawn borders anywhere in the world, including on lakes.[13] The latter should be divided between the riparian States, often according to the median line. Dipla, however, denies that this rule has become part of customary law.[14] Customary international law holds that the border on the bridge across a frontier river – in the absence of special rules in a treaty – will be drawn in the middle of the bridge. Yet, according to other authors, no corresponding rule exists regarding the frontiers on a lake. These authors, referred to by Schuster,[15] thus assume that any border lake forms a condominium as long as the riparian States have not concluded a treaty drawing borders on the lake. Huber denies that the condominium is the natural status of a lake and should prevail subsidiarily in the absence of frontier treaties.[16]

Other writers defend the condominium theory on historical grounds. They claim that the borders of the various riparian entities on the lake extended merely to their parts of the shore. All these entities were parts of the Holy Roman Empire. Sovereignty over these waters was held to reside in the Emperor.[17] Actually, in 1434 and 1488, the Emperor did attribute fishing privileges on the lake respectively to Heiligenberg and Mainau,[18] thus manifesting a sovereign right over the waters of the lake. When Switzerland was detached from the Holy Roman Empire by Article VI of the Treaty of Westphalia in 1648,[19] the Treaty remained silent on the borders on the lake. Nor was this problem mentioned when the Holy Roman Empire was dissolved in 1806, nor when Napoleon I attributed former Austrian territory bordering the lake to Baden, Württemberg and Bavaria, nor when the Congress of Vienna, in 1814-1815, established the German Federation and restored to Austria only Vorarlberg, but not the rest of the Austrian possessions bordering on the lake, mainly in Swabia (Vorderösterreich). The result of this alleged negligence on the part of the drafters of these treaties was that, from then onwards, the waters of the lake were be considered to be a condominium of the riparian States.

No border treaties exist concerning the whole of Lake Constance. A treaty was concluded between 20-31 October 1854 between Baden and

---

[13] Riva, see note 5, 43 et seq., 52.
[14] H. Dipla, "Le tracé de la limite sur les lacs internationaux", *Schweizerisches Jahrbuch für internationales Recht* XXXVI (1980) 9 et seq., 52.
[15] Schuster, see note 8, 144, quoting Carathéodory and Vorwerk as holding this view.
[16] Huber, see note 6, 192, 200.
[17] Schuster, see note 8, 30.
[18] *Ibid.*, 54.
[19] Treaty of Osnabrück of 24 October 1684, in C. Parry (ed.), *Consolidated Treaty Series*, Vol. 12 (1648), 119 (hereinafter, CTS); J. Dumont, *Corps Universel Diplomatique du Droit des Gens contenant un Recueil des Traitez* VI/1 (1728), 479.

the Swiss Confederation on the border of the Untersee (Lower Lake),[20] and a further treaty was concluded on 28 April 1878 concerning the Constance Funnel (Trichter),[21] which was later explicitly accepted by Germany.[22] Moreover, local customary law attributes sovereignty on the Überlingen Lake to Baden-Württemberg.[23] If we start from the assumption that the lake forms a condominium between the riparian States, the treaties concluded by only some of them would not be binding on the others, as was shown, *mutatis mutandis*, in the case of Fonseca Bay.[24] However, none of the riparian States of Lake Constance appears to have protested against these treaties. The States claiming a condominium on the lake thus may have accepted by acquiescence[25] that the condominium does not extend to the Überlingen Lake or to the Constance Funnel.

2. Practice Concerning Sovereignty on the Lake

Frowein has rightly said that the conflicting theories on sovereignty over the waters of Lake Constance have lost nearly all practical importance.[26] Treaties between the riparian States regulate most matters likely to cause controversies between them. This is especially true as far as navigation rights[27] and fishing rights, including rules for the preservation of fish stock,[28] are concerned, but there even exists a treaty from 1880 regulating the registration of childbirths and deaths on ships navigating the lake.[29] These events are to be registered by the public authorities of the home port of the ship, paralleling the rules of UNCLOS.[30]

However, the fisheries treaty was unable to solve the problem of the *Bilgeri* claims. In 1825, the predecessor-in-title of the Austrian citizen Mr. Bilgeri had purchased fishery rights on Lake Constance from the Austrian State, entitling him to fish as far out on the lake as the rights and practice allowed. In a long drawn out dispute before Austrian and

---

[20] CTS 112 (1854-1855), 250-258.
[21] CTS 152 (1877-1878), 497-500.
[22] Agreement of 24 January 1879, CTS 155 (1879-1880), 143-144.
[23] Berchtold, see note 9, 401.
[24] *Case Concerning the Land, Island and Maritime Frontier Dispute (El Salvador/Honduras)*, Judgment of a Chamber of the ICJ, 11 September 1992, paras. 413-415, ILR 97, 522.
[25] Schuster, see note 8, 108.
[26] Frowein, see note 4, 118.
[27] Treaty of 22 September 1867, CTS 135 (1867), 373-396.
[28] Treaty of 5 July 1893, CTS 178 (1893-1894), 37-52; Schuster, see note 8, 111, but see the *Bilgeri* claims, see notes 31-34 *infra* and accompanying text.
[29] Schuster, see note 8, 109.
[30] Art. 92, para. 1.

Bavarian Courts, Mr. Bilgeri claimed that he owned an exclusive right to fish within the area forming part of the purchase. The 1825 sale of fishery rights preceded the drawing up of the Austrian cataster in 1856, i.e., the straight line across the Bay of Bregenz between the points where the Austro-Swiss frontier in the south and the Austro-German frontier in the north reach the shores of the lake.[31] The line not only includes the Halde, but also parts of the Blue Lake. The cataster marks the borders of several communities *(Gemeinden)* in Vorarlberg, including their borders on the Blue Lake. However, these outside borders were established unilaterally, without the consent of the other riparian States. Thus, the outward limits of Bilgeri's fishing grounds were not clearly defined. Bilgeri resorted to "self-help", and removed nets and fishing installations placed by Bavarian fishermen at a depth of 40 metres outside the cataster line. Such self-help was deemed illegal by the Austrian Courts, and the fishing equipment was returned to its Bavarian owners. However, the latter failed to obtain a court order to the effect that Bilgeri had to desist from claiming fishing rights outside of the cataster line in the future. The Supreme Court held that a general freedom to fish on any lake does not entitle an individual fisherman to obtain a desist order against his competitors.[32] Thus, this decision did not solve the problem of the outside limit of the Bilgeri claim. In any case, the main dispute concerned the part of Bilgeri's claim within the cataster line. The Halde, on the Bavarian shore, is very narrow. Only a few metres from the shore, the lake reaches a depth considerably deeper than 25 metres. Thus, most of the area of the Bilgeri purchase forms part of the Blue Lake, even if we were to assume that his claim does not reach further than the cataster line. Bilgeri applied to the Austrian Supreme Court to oppose the claims of Bavarian fishermen. The latter claimed to be entitled to fish in this part of the Blue Lake, just as Austrian fishermen were entitled to fish anywhere on the Blue Lake, even in waters claimed by Switzerland or Germany. The Supreme Court rejected Bilgeri's claim to exclusive fishing rights within the cataster line beyond the Halde, as he had failed to prove that the Austrian State actually held exclusive fishing rights in this area in 1825.[33] The legal costs incurred by Mr. Bilgeri far exceeded the profits he could have hoped to derive from exclusive fishery right in "his" part of the Blue Lake, yet he continues to fight for what he believes to be his right. Recently, his claim that he had acquired this right by prescription was rejected for lack of sufficient proof.[34]

---

[31] See the map reproduced in M. Schlag, "Die österreichische Bundesgrenze auf dem Bodensee", *ÖZÖR* 43 (1992), 241 et seq., 252.

[32] Austrian Supreme Court, 9 October 1991, Official Collection SZ 64, No. 137, 230, *Juristische Blätter* 114 (1992), 176 et seq., 178.

[33] Austrian Supreme Court, 28 July 1998, *Juristische Blätter* 121 (1999), 656 et seq.

[34] *Ibid.*

Apart from the *Bilgeri* case, there are other cases that treaties were unlikely to foresee. Thus, Veiter[35] mentions the case of a ban, introduced in 1962 by the Vorarlberg authorities, i.e., the authorities of an Austrian *Land* bordering on the lake, on dancing the twist, a dance regarded as immoral by these authorities. Students had chartered a ship, belonging to the Austrian state railways, in order to hold a twist dance party on the Blue Lake, outside the Halde. Disciplinary actions were brought against the captain and the crew for having neglected their duties as Austrian civil servants.

A condominium, in theory, authorises any riparian State to exercise its sovereign rights over the entire lake, at least on the Blue Lake (with the exception of the Überlingen Lake, the Lower Lake and the Constance Funnel). This state of affairs is highly impractical, even in peacetime, for the enforcement of customs controls and the pursuit of escaped criminals. Matters are likely to be even worse in wartime. Actually, customs lines have been drawn across the lake, and Switzerland, as a neutral country in the two World Wars, succeeded in having its claim to sovereignty over "its" part of the lake respected by the belligerents. German authorities ceased their pursuit of deserters and refugees at this line, which was likewise respected by German anti-aircraft boats. However, adherents of the condominium theory deny that these lines were established as frontiers. According to them, these were merely demarcation lines for specific administrative purposes,[36] a reasoning which does not appear to be very convincing. The same reasoning was used to reject the limits of the Austrian cataster line[37] as a precedent for a dividing line.

Recent developments in the field of the protection of the environment have given a new impetus to the defenders of the condominium theory, especially because some German towns on the shores of the lake acquire their drinking water by purifying water from the lake. It is true that the rules of good-neighbourliness would oblige the other riparian States to prohibit pollution of the lake. This obligation led to the adoption of a treaty on 30 April 1966.[38] However, the position of the riparian State objecting to water pollution would certainly be improved if it could rely on shared sovereignty over the waters of the lake,[39] rather than on the terms of this treaty and on the principle of good-neighbourliness.

---

[35] Th. Veiter, "Die Rechtsverhältnisse auf dem Bodensee, eine völkerrechtliche Untersuchung", *AVR* 28 (1990) 458 et seq., 461.
[36] Schuster, see note 8, 132 et seq.
[37] Austrian Supreme Court, see note 33.
[38] Treaty Regulating the Withdrawal of Water from Lake Constance, UNTS Vol. 620, 191-209.
[39] *Obiter dictum*, Austrian Supreme Court, see note 33, 656.

III. Dividing Lines on the Lake

As we have already seen, dividing lines exist on the lake, although not all riparian States regard them as frontiers. As far as Austria is concerned, these dividing lines leave it with only a very narrow part of the lake. The circumference of the lake measures 273 kilometres, of which 173 kilometres belong to Germany, 72 kilometres to Switzerland, and 28 kilometres to Austria. Austria borders on the easternmost part of the lake and would thus enjoy sovereignty over only 5.8 per cent of the lake. *Mutatis mutandis*, the situation of Austria is reminiscent of that of Germany regarding the North Sea continental shelf.[40] Understandably, Austria thus leans towards the condominium theory. Yet, the cataster, the official register showing the borders of several communities in Vorarlberg, follows the same straight line when indicating the external border of the communities bordering the sea shore.

Some writers[41] claim that these frontiers, unfavourable as they are to Austria, have to be accepted as a consequence of the Treaty of St. Germain of 1919.[42] This treaty contains a map of the new frontiers of Austria after the dismemberment of the Austro-Hungarian monarchy. On this map, the lines of the Austrian frontiers with Switzerland and Germany are continued in a straight line across Lake Constance from the above-mentioned points. However, maps attached to treaties may have very different probative value.[43] In the present case, the drafters of the Treaty of St. Germain did not intend to deprive Austria of any sovereignty which Austria hitherto might have enjoyed with regard to Switzerland and Germany. In contrast to the new Austrian frontiers with Italy, Yugoslavia, Hungary and Czechoslovakia, the lines on this map concerning Lake Constance did not intend to bring about any change. These lines therefore serve a merely informational purpose and cannot be used to confirm Swiss or German claims concerning sovereignty on the lake.[44]

As we have seen, Austria does not fully adhere to the condominium theory. According to this logic, the condominium should extend right up to the shore. However, an incident as early as 1825 led Austria to

---

[40] Dipla, see note 9, 21, points out this analogy. Oda, pleading for the Federal Republic of Germany, came out in favour of special circumstances. Cf. also Judge Oda's dissent in the *Case Concerning the Continental Shelf Between Libya and Malta*, 3 June 1985, ICJ Reports 1985, 160-161.

[41] Schlag, see note 31, 241 et seq., 253-254.

[42] *Austrian StGBL* 1920, No. 303.

[43] I. Seidl-Hohenveldern, "Landkarten im Völkerrecht", in W. Karl (ed.), *In memoriam Herbert Miehsler*, 1998, 67 et seq., 74-75.

[44] B. Simma and D.-E. Khan, "Nochmals: Der Staatsvertrag von St. Germain und die Österreichische Staatsgrenze auf dem Bodensee", *ÖZÖR* 45 (1993), 211 et seq., 248.

advance its claim of sovereignty over the Halde. A wine merchant had anchored a boat a few metres from the shore and tried to sell wine from it without paying Austrian taxes. Thus, well before the continental shelf theory was ever developed, Austria claimed that its sovereignty extended at least over shallow waters,[45] a claim accepted by the other riparian States by acquiescence. In contrast, Austria could only enforce its tax and customs claims in a roundabout way against a Swede selling tax-free butter from a ship anchored on the Blue Lake and flying a non-Austrian flag. The Swede gave up his venture when Austrian customs officers began to collect Austrian customs duties from the Swede's customers as they returned to the Austrian shore.[46]

IV. Parallels with the Law of the Sea

Like causes produce like effects. There is a basic uncertainty concerning frontiers on water, at sea as well as on lakes. When the territory of a lake or bay once belonged to a single larger State, be it the Holy Roman Empire or the Kingdom of Spain, the absence of clear dividing lines on the waters will lead to claims of the existence of a condominium. A treaty concluded between two or several, but not all, riparian States will not bind the other States, although they may be deemed to have acquiesced to it. When frontiers on the waters are drawn in an inequitable manner, leaving one of the riparian States with a disproportionately small part of the sea or lake concerned, this State will try enlarge its territory by claims which cast aside the usual rules regarding the sharing of waters. Riparian States will try to extend their rights of control at least to the areas closest to their shores, whether by the extension of territorial waters or by claims to a continental shelf. It is rather interesting to see that, in 1825, Austria developed an idea reminiscent of the continental shelf doctrine, when it argued that the shallow waters near its shores came under Austrian sovereignty. The use of the very word Halde already aims to justify this claim. Usually this word is used to designate the slope of a hill formed by rubble. On Lake Constance this rubble consists of stones carried into the lake by Austrian mountain rivers that flow into the lake. In the *Bilgeri* claim, we may even find a parallel with efforts to find a remedy for a situation in which unusual geological formations fail to provide an adequate extension to the rights of the riparian State, be it with regard to the continental shelf on the west coast of South America or to the Halde on Lake Constance.

---

[45] Schuster, see note 8, 67-68.
[46] The *Björn Sunne* case, mentioned by Veiter, see note 35, 463.

Peaceful Settlement of Boundary Disputes under the Auspices of the Organisation of African Unity and the United Nations: The Case of the Frontier Dispute Between Eritrea and Ethiopia

Bruno Simma and Daniel-Erasmus Khan

I. Introduction

As a prominent member of the World Court for more than a quarter of a century, Judge Shigeru Oda has not only taken a particularly active part in the resolution of a multitude of specific disputes of a territorial character, but also in the development and refinement of an entire body of legal rules and principles pertinent to this subject. As a distinguished expert on dispute settlement recently emphasised,[1] the rules and principles endorsed by the Court in this field and their application to the facts of particular cases amount to a significant contribution to the development of international law. Neither the States involved, nor any judicial or arbitral body called upon to decide a dispute of this kind, can and will afford to ignore this contribution. This will certainly also be valid in the forthcoming arbitral settlement of a boundary dispute that has cost hundreds of thousands of lives, namely that between Ethiopia and Eritrea.

For the third time in 50 years, the United Nations is about to play a crucial role in the resolution of a major conflict between these two East African States. If we include the various activities of the UN's predecessor, the League of Nations, we find that the organised international community has already taken a particularly keen interest in political developments in this particular region of the world since the early 1920s. While the Ethiopian Empire appears in the annals of the League for the first time in 1923, on the occasion of a somewhat

---

[1] J. G. Merrills, "The International Court of Justice and the Adjudication of Territorial and Boundary Disputes", *Leiden JIL* 13 (2000), 901.

troublesome procedure leading to its admission to the Organisation,[2] already in the 1930s[3] the League engaged in unprecedented multi-level efforts to stabilise the rapidly deteriorating bilateral relations between the then Italian colony of Eritrea and Ethiopia.[4] This development was due, in the first instance, to the expansionist colonial aspirations of (fascist) Italy, culminating in a full-scale armed aggression against the Empire launched from Italian bases in Eritrea in 1935-1936, which ultimately lead to the annexation of Ethiopia on 9 May 1936.[5] The

---

[2] The Ethiopian request for admission met with particularly strong resistance on the part of Great Britain, which alleged a lack of power of the central Ethiopian Government over the outlying provinces, resulting in the continuation of slave trade and arms traffic. See, with further references, A. Toynbee, "Tropical Africa", *Survey of International Affairs 1920-1923* (1927), 393-396; and L. Friedlander, "The Admission of States to the League of Nations", *BYIL* IX (1928), 84 et seq. at 95 et seq. However, certain observers held that, due to the strong economic and political interests of the British Empire in the region, its resistance was ultimately motivated by British unwillingness to allow Ethiopia to benefit from the protection granted by Article 10 of the Covenant: "The Members of the League undertake to respect and preserve as against external aggression the territorial integrity and existing political independence of all Members of the League...". See, e.g., *Revue de Droit International* XVI (1935), 75: "Si L'Ethiopie ne devait pas entrer dans la S.D.N. c'était pour ne pas être mise au bénéfice de l'article 10 du Pacte!"

[3] After certain "minor" involvements in the 1920s, such as the political sequel to the Anglo-Italian Agreement of 14-20 December 1925 (see *infra* footnote 11).

[4] See the exhaustive presentation by Ch. Rousseau, "Le conflit italo-éthiopien", *RGDIP* XLIII (1936), 546 et seq.; XLIV (1937), 5 et seq., 162 et seq., 291 et seq. and 681 et seq.; and XLV (1938), 53 et seq., as well as *Revue de Droit International* XVI (1935), exclusively devoted to the topic and including, *inter alia*, extensive documentary material about the conflict.

[5] The official decree of annexation is worded as follows: "Victor Emmanuel III, par la grâce de Dieu et par la volonté de la nation, roi d'Italie: [...] Art. 1er. Les territoires et les populations qui appartenaient à l'empire d'Ethiopie sont mis sous la souveraineté pleine et entière du royaume d'Italie. Le titre d'empereur d'Ethiopie est assumé par le roi d'Italie pour lui-même et pour ses successeurs. [...] Text taken from K. Strupp, "Problèmes soulevés à l'occasion de l'annexion, par l'Italie, de l'empire éthiopien", *RGDIP* XLIV (1937), 43. On 1 June 1936, Italy united Ethiopia with Eritrea and Italian Somaliland to constitute Italian East Africa. For a critical assessment of the various, and in the end unsuccessful, activities of the League of Nations in this final phase of the conflict from the Wal Wal incident (5 December 1934) until the end of the war (occupation of Addis Ababa by Italian forces on 5 May 1936), see J. Spencer, "The Italian-Ethiopian Dispute and the League of Nations", *AJIL* 31 (1937), 614 et seq. On the diplomatic activities in the aftermath of the Wal Wal incident which triggered the outbreak of the open

United Nations on its part played a key role both in the decision of the fate of the former Italian colonies after an interim period[6] of British occupation and administration during and immediately after World War II,[7] as well as in the transition process in the early 1990s which finally, in 1993, led to Eritrean sovereign statehood after one of Africa's longest struggles for independence.[8]

Unfortunately, international intervention in the Horn of Africa did not always produce the desired results: neither political pressure nor economic sanctions imposed by the League could prevent Ethiopia from being annexed by Italy in 1936, and the United Nations model of "a Swiss federation adapted to an African absolute monarchy",[9] developed in 1950-1952, turned out to be no shield against the gradual erosion of Eritrea's federal status in the late 1950s, and finally disappeared with the abrogation of the Federal Act in 1962 and the wholesale incorporation of Eritrea into Ethiopia as one of that country's provinces. However,

---

military confrontation, see in particular A. Zimmern, "The League's Handling of the Italo-Abyssinian Dispute", *International Affairs* 1 (1935), 751 et seq. Finally, on the background of the sanctions which the League imposed upon Italy, see O. von Nostitz-Wallwitz, "Die Entwicklung des Abessinienkonfliktes bis zum Beginn der Sanktionen", *ZaöRV* VI (1936), 496 et seq.

[6] See G. Trevaskis, *Eritrea. A Colony in Transition: 1941-52*, 1960.

[7] The question was put on the agenda of the General Assembly upon the request of the four Allied Powers (letter dated 15 September 1948, see UN Doc. A/645 of 16 September 1948), which led to the very controversial decision to render Eritrea an autonomous unit federated with Ethiopia under the sovereignty of the Ethiopian Crown. See GA Res. 390 (V) of 2 December 1950, effective 11 September 1952 by the Emperor's ratification of the Federal Act prepared by a United Nations Commissioner. For a summary of the whole process, see M. Shaw, *Title to Territory in Africa. International Legal Issues*, 1986, 117 et seq., with further references.

[8] The UN's role consisted in particular in the organisation and supervision of a respective referendum in Eritrea (United Nations Observer Mission to Verify the Referendum in Eritrea, UNOVER) and in a number of measures to help reconstruction and long-term development of the country. See in detail UN Department of Public Information, *The United Nations and the Independence of Eritrea*, 1996. See also R. Iyob, *The Eritrean Struggle for Independence: Domination, Resistance, Nationalism, 1941-1993*, 1995; and, with emphasis on the final phases of the struggle, D. Connell, *Against All Odds: A Chronicle of the Eritrean Revolution*, 1993 and R. Goy, "L'indépendance de l'Érythrée", *AFDI* XXXIX (1993), 337 et seq.

[9] Time Magazine, October 1952 (citing from H. Erlich, "The Eritrean Autonomy 1952-1962: Its Failure and Its Contribution to Further Escalation", in Y. Dinstein (ed.), *Models of Autonomy*, 1981, 171. See on the same subject T. Meron and A. Pappas, "The Eritrean Autonomy. A Case Study of a Failure", *ibid.*, 183 et seq.

following the amicable divorce of Ethiopia and Eritrea in 1993 and the instant admission of the latter as the United Nation's 182nd Member State,[10] there grew legitimate expectations that for the first time in history[11] the time was ripe for the establishment of normal and peaceful relations between the two neighbouring States, and that, due to a definite pacification of the region, the United Nations would be able to finally bring its mediation efforts to a successful end. To the great disappointment of the international community, which had emphatically praised the model character of the peaceful separation of the two countries, only five years later, in May 1998, heavy fighting erupted between Eritrea and Ethiopia over a border dispute.[12] This necessitated intervention by both the OAU and the United Nations in order to bring about a ceasefire and a political solution of the conflict. Notwithstanding intensive diplomatic activities to halt the fighting, armed hostilities flared up time and again for the following two years and could only be brought to an end in the summer of 2000. Leaving apart the complex and highly politicised issues surrounding the border dispute *strictu sensu*, as well as incidental matters such as the compensation question, the

---

[10] GA Res. 47/230 of 28 May 1993.

[11] Although a certain "détente" in the bilateral relations can be witnessed in the late 1920s, in December 1925, Great-Britain and Italy, by once again dividing the area in spheres of (economic) influence, not only denied Ethiopia the status of an equal partner in the region but, at least in the understanding of the Imperial Ethiopian Government, put into question the Empire's right to sovereign statehood as such. Exchange of Notes of 14/20 December respecting certain British and Italian interests in Abyssinia, Cmd. 2680 of 1926; cf. A. Toynbee, *The Islamic World since the Peace Settlement*, Survey of International Affairs 1925, Vol. I, 1927, 268 et seq. The Ethiopian Government reacted to this Anglo-Italian Agreement, which had been made over its head, by bringing it to the attention of the League of Nations and asking for the League's opinion. In a letter of protest sent to the Member States of the League on 19 June 1926, the heir to the throne of Abyssinia, Tafari Makonnen, complained, *inter alia*, that "... on our admission to the League of Nations we were told that all nations were to be on a footing of equality within the League, and that their independence was to be universally respected, since the purpose of the League is to establish and maintain peace among men in accordance with the will of god...". League of Nations Official Journal, November 1926, 1517. The entire correspondence between the Abyssinian, British and Italian Governments and the Secretary-General of the League of Nations is reproduced in *ibid.*, 1517-1527.

[12] It is an open secret that the high-intensity conflict owed its outbreak not only to the controversy over the exact drawing of the boundary line in certain frontier areas, but also to a whole number of economic and other causes. See for example M. Fielding, "Bad Times in Badme. Bitter Warfare Continues along the Eritrea-Ethiopia Border", *Boundary and Security Bulletin* 7 (1999), 91 et seq.

following presentation will limit itself to highlighting some basic elements of the international legal background of the conflict (II.), and further to briefly outlining the ongoing efforts on the part of the United Nations and the Organisation of African Unity to bring the dispute to a judicial settlement (III.).

II. A Glimpse at the Legal Background of the Boundary Dispute

At the outset of studies of this type, it is a commonplace to recall that in the colonial era a great number of boundaries in Africa were drawn "by a blue pencil and a rule",[13] in neglect, to a greater or lesser extent, of established political, social, ethnic and economic structures. Facing this colonial heritage in the early 1960s, African leaders of newly-independent States, as well as of the few already established States, among them Ethiopia, had a choice between Scylla and Charybdis: wiping out the injustices committed by the colonial powers on the territorial map of Africa by turning the existing political structure of the continent upside down in a second, almost unavoidably violent scramble for the distribution, and thus fragmentation, of African territory,[14] or accepting the territorial *status quo* embodied in the colonial boundaries as a historical *fait accompli* by adapting the principle of *uti possidetis*[15]

---

[13] Sir Claude Mac Donald in a lecture before the (British) Geographical Society on the boundary making between the British Colony of Lagos and the German Protectorate of Cameroon, The Geographical Journal (January to June 1914), 649. According to Shaw, see note 7, 50, some 30 per cent of the total length of borders in Africa follow geometrical lines.

[14] From the vast literature on the initial partition of Africa culminating in the decades before World War I, for which the term "scramble" was coined, see T. Pakenham, *The Scramble for Africa: 1876-1912*, 1991 and M. Chamberlain, *The Scramble for Africa*, 2nd ed., 1999. Highly interesting insights on the impact of this development on the area under discussion are offered by H. Erlîk, *Ethiopia and Eritrea During the Scramble for Africa: A Political Biography of Ras Alula, 1875-1897*, 1982. Although using the term in a much broader way, observers speak of a second round in the scramble for Africa. See, for example, the contributions in K. Ndeti (ed.), *The Second Scramble for Africa: A Response and a Critical Analysis of the Challenges Facing Contemporary Sub-Saharan Africa*, 1992.

[15] Although the Roman law principle *uti possidetis ita possideatis* literally means "as you possess you shall continue to possess", it changed its meaning when it entered the domain of international law in the early years of the 19th century. First adopted by the Spanish-American States after their independence, it was not intended to solve boundary problems between them on the basis of factual possession, but rather on the basis of *uti possidetis juris*, that is to say, according to "the rule that the boundary between them [Colombia and Venezuela] must be identical with the boundaries as laid

to the post-colonial structure of the African continent. At their first summit conference following the creation of the Organisation of African Unity, the African Heads of State, in their well-known Cairo Resolution of July 1964, decided in favour of the second option by declaring that they "pledge[d] themselves to respect the frontiers existing on their achievement of national independence".[16] Since then, as the International Court of Justice emphasised in 1986, the principle of the intangibility of frontiers inherited from colonial powers or any other territorial predecessor has surpassed the status of "a special rule which pertains solely to one specific system of international law", and has grown into "a general principle, which is logically connected with the phenomenon of the obtaining of independence, wherever it occurs".[17] Already in the late 1970s, Ian Brownlie could rightly note that, with regard to the African continent, the *uti possidetis* principle had become a "rule of regional customary law binding upon those States which have unilaterally declared their acceptance of the principle of the *status quo* at the time of independence".[18] In the context of the present study, it deserves to be mentioned that, unlike Morocco and Somalia,[19] neither Ethiopia nor – after its independence in 1993 – Eritrea ever objected to the emergence of this legal rule. In the wake of its application in Europe in the process of the emergence of new States in the years 1990-1993, it has been said that the *uti possidetis* principle has meanwhile "acquired the character of a [truly] universal, and peremptory norm".[20] Thus, it is of no importance that, since Ethiopia was never subjected to colonial

---

down by the Spanish authorities between the respective territorial units prior to the establishment of the independence of the Latin-American Republics...". Arbitral Award by the Swiss Federal Council in the *Boundary Dispute Between Colombia and Venezuela* (1922), 1 Annual Digest 84. In effect, this doctrine conflated boundary and territorial questions by assuming as a governing principle that boundaries must remain as they existed in law at the declaration of independence in 1810 and 1822 respectively.

[16] OAU Doc. AHG/Res. 16 (1).
[17] *Frontier Dispute*, Judgment, ICJ Reports 1986, 565. For a valuable analysis of today's state of the law, see S. Ratner, "Drawing a Better Line: Uti Possidetis and the Borders of New States", *AJIL* 90 (1996), 590 et seq.; and M. Shaw, "The Heritage of States: The Principle of Uti Possidetis Juris Today", *BYIL* 67 (1996), 75 et seq.
[18] I. Brownlie, *African Boundaries. A Legal and Diplomatic Encyclopaedia*, 1979, 11.
[19] See Shaw, see note 7, 186.
[20] A. Pellet, "The Opinions of the Badinter Arbitration Committee. A Second Breath for the Self-Determination of Peoples", *EJIL* 3 (1992), 180. For a detailed analysis of recent state practice in Europe and elsewhere, see also C. Antonopoulos, "The Principle of Uti Possidetis in International Law", *Revue Hellénique de Droit International* 49 (1996), 35 et seq.

rule, the issue of the boundary between Ethiopia and Eritrea does not really fit into the decolonisation context proper. What might therefore turn out to be of utmost importance in the resolution of the boundary dispute between Eritrea and Ethiopia is that the *uti possidetis* principle is not confined to mere application of legal title deriving from international (colonial) treaties. Rather, it encompasses the entire range of legal titles over territory which may exist at the time of the achievement of independence. In fact, limiting the scope of the principle to treaty-based titles would have been quite useless, since – at least in the original American context – there existed almost no such titles, but only former administrative delimitations between the various Spanish colonies in the great majority of cases. The situation was the same with regard to the vast territories of French West Africa. For this reason, in the case of the *Frontier Dispute*, the International Court of Justice did not confine the applicability of *uti possidetis* to colonial treaties, since it had to determine the course of an international boundary line on the basis of mere administrative acts of a colonial power dividing two former French colonies. In this case, the International Court of Justice applied the principle in such as way as to "photograph" the territorial situation at the moment of independence, thus freezing the territorial title, of whatever nature it might have been. *Uti possidetis* has thus been equated with "colonial heritage", meaning "the photograph of the territory at the critical date".[21] This "photograph" may reproduce a treaty settlement alone, but it may also display any other type of a legal delimitation or contain elements such as effective possession or acquiescence. The wide scope thus attributed to the principle has a simple explanation. The very aim and purpose of *uti possidetis* is to guarantee to the greatest extent possible the intangibility of frontiers inherited from colonisation. Thus, the principle plays a prominent role in the realisation "one of the primary objects" of boundary settlements, namely "to achieve stability and finality".[22] In order to achieve this primary purpose, various arguments may be advanced, provided, however, that they constitute, or give evidence of, the legal *status quo* at the time of independence. *Uti possidetis juris* can thus be regarded as an expression of the fundamental and more general principle of continuity and stability of boundaries.

---

[21] ICJ Reports 1986, 568.

[22] Cf. in particular the judgment of the ICJ in the *Temple of Préah Vihéar* case: "... when two countries establish a frontier between them, one of the primary objects is to achieve stability and finality". ICJ Reports 1962, 34. This was taken up in the Arbitral Award of 18 April 1977 in the *Beagle Channel* case *(Argentina/Chile)*, *ILR* 52 (1979), 131. See also, with further references, A. O. Cukwurah, *The Settlement of Boundary Disputes in International Law*, 1967, 119 et seq.; and the Separate Opinion of Judge Shahabuddeen, *Territorial Dispute* case *(Libya/Chad)*, ICJ Reports 1994, 44 et seq.

Today's boundary between Ethiopia and Eritrea owes its existence to the colonial occupation from 1863 onwards by Italy of a land strip on the coast of the Red Sea, a territory which was later to become known as Eritrea. However, it was only when Italy extended her colonial claims from the southern seaport of Assab northwards along the coast, and then further into the *hinterland* of Massawa, that the practical necessity of an alignment between the respective territories of the Ethiopian Empire and the then Italian colony of Eritrea arose. From the late 1880s onwards, a number of international treaties were concluded to this effect.[23] Due to her colonial interests in the area (Anglo-Egyptian Sudan), Great Britain also played a vital role in the drawing of the north-western sector of the boundary line in question. At the beginning of the 20th century, a set of pertinent treaty instruments had become legally binding upon the parties, covering the common frontier over its total length of 912 km,[24] from its tripoint with the Sudanese boundary in the north-west to the tripoint with the boundary of Djibouti in the south-east. Having never since been subjected to any fundamental changes in substance by subsequent treaty instruments,[25] the legal régime thus established consists of the following three treaties: (1) the Frontier Delimitation Agreement between Ethiopia and Italy of 10 July 1900,[26] (2) the Treaty between Ethiopia, Great Britain and Italy for the Delimitation of the Frontiers between Ethiopia and Eritrea and the Sudan of 15 May 1902,[27] and (3) the Convention between Ethiopia and Italy settling the Frontier between the Italian Colony of Eritrea and Ethiopia of 16 May 1908.[28]

Treaty No. 1 covers the north-westernmost sector (Sector I) of the boundary in the following terms: "Commencing from the junction of the Khor Um Hagar with the Setit, the new frontier follows this river to its junction with the Maieteb, following the latter's course so as to leave Mount Ala Tacura to Erythraea, and joins the Mareb at its junction with

---

[23] On the treaty-making activities in the early phase of Italian colonial expansion in the area (1882-1896), see T. Scovazzi, *Assab, Massaua, Uccialli, Adua: Gli strumenti giuridici del primo colonialismo italiano*, 1996.

[24] This figure stems from the CIA World Factbook, available at http://www.cia.gov/ cia/publications/factbook/fields/ land_boundaries.html – last visited on 13 May 2001. However, due to a number of uncertainties surrounding the course of the boundary, this figure can only be considered as approximate.

[25] This observation of principle appears to be shared by Ethiopian and Eritrean officials.

[26] C. Parry (ed. and annotator), *The Consolidated Treaty Series* 189 (1900/1901), 6.

[27] *Ibid.*, 191 (1902), 180.

[28] *Ibid.*, 207 (1908), 28.

the Mai Ambessa".[29] The adjacent, central sector (Sector II) of the boundary is legally defined in a twofold way. First, by the verbal description "Maereb-Belesa-Muna" in Article I of the Treaty of 10 July 1900 and, second, by the depiction on the map annexed to the same legal instrument.[30] Finally, the course of the almost 700 km long south-easternmost sector (Sector III) of the boundary line, running mainly through arid and rugged desert terrain, is governed by the simple terms: "... running to and at a distance of 60 kms from the coast", stipulated in Article 1 of the treaty of 16 May 1908.

The mode in which the various boundary sectors are defined differs considerably. Sub-segment 1 of Sector I, as well as Sector II, of the boundary essentially follow the natural features of watercourses, thus sharing some basic issues and problems relating to their delimitation and demarcation. In sharp contrast to this, Sector III of the boundary is defined by a purely geometrical criterion. Although this criterion – if accepted by the parties as constituting the legal alignment between the two countries – is a very clear and objective one, it will almost inevitably lead to difficulties when it comes to applying the line thus defined on the ground at the demarcation stage. Finally, the way in which the line is drawn with regard to Sector II of the boundary (the Badme area) is not free of ambiguities. Although the pertinent treaty provisions do not provide for a geometrical line in this sector, but rather resort to topographical and ethnical criteria, all cartographical material published since the early years of the 20th century depicts the former criterion, e.g., a straight line. The three treaties just mentioned do, however, share the following general features. First, they are bilingual and plurilingual (Amharic/Italian or Amharic/Italian/English) – a factor to be taken into account for the purpose of interpretation. Secondly, the verbal description of the boundary line in the treaties does not go into great detail. However, this is a feature common to many, if not most, legal instruments from colonial times drawing boundaries on the African continent. The mere superficiality of boundary definitions as such should therefore not constitute a barrier to considering the respective instruments as valid and full-scale delimitations. Thirdly and finally, the drafters of the treaties were not fully aware of all the topographical and other relevant peculiarities of the terrain. This is once again a not uncommon feature of boundary treaties of that time.

---

[29] Article I, para. 2, of the Treaty.
[30] The provision is worded as follows: "The line Tomat-Todluc-Mareb-Belesa-Muna, traced on the map annexed, is still recognized by the two Contracting Parties as the boundary between Erythrae and Ethiopia".

The only instrument that is accompanied by a map is the Treaty of July 1900.[31] The depictions on this map – though not very precise – constitute an integral element of the consent reached between the parties and have therefore to be taken into account in the process of treaty interpretation.[32] The various other maps published before, during and after the drafting of this and the other relevant treaties of 1902 and 1908[33] do not share the specific status of such a "treaty map" and are therefore of limited probative value only, at least for the purpose of treaty interpretation. In its judgment in the case of the *Frontier Dispute*, the International Court of Justice summarised the established legal doctrine on the significance of maps in boundary disputes in the following terms:

> "Whether in frontier delimitations or in international territorial conflicts, maps merely constitute information which varies in accuracy from case to case; of themselves, and by virtue solely of their existence, they cannot constitute a territorial title, that is, a document endowed by international law with intrinsic legal force for the purpose of establishing territorial rights. Of course, in some cases maps may acquire such legal force, but where this is so the legal force does not arise solely from their intrinsic merits: it is because such maps fall into the category of physical

---

[31] The great number of maps that exist with regard to Sector I and III of the boundary are merely of a "unilateral" character and the depiction of frontiers contained therein has never been made the object of any legal accord between the Parties.

[32] For a more precise and differentiated look at the legal significance of maps within the framework of a specific treaty settlement with extensive references, cf. D. Khan, *Die Vertragskarte. Völkerrechtliche Untersuchung zu einem besonderen Gestaltungsmittel in der internationalen Rechtsetzung*, 1996, 122 et seq. The empirical research undertaken there demonstrates that the legal significance of incorporated maps can vary considerably, reaching from maps with a merely illustrative function to maps serving as the sole means of defining the object of the treaty. Thus, generalisations can never grasp the entire reality, or, as India argued before the Arbitral Tribunal in the *Rann of Kutch* case: "A key to the understanding of maps is that there is no democracy in the world of maps. All maps are not equal...". *RIAA* XVII, 88.

[33] It is to be expected that the Boundary Commission established to resolve the present frontier dispute will not escape the fate of being overwhelmed, as was the ICJ in the *Frontier Dispute*, with "an abundant and varied collection of cartographic materials, consisting of a series of maps and sketch-maps differing as to date, origin, technical standard and level of accuracy". ICJ Reports 1986, 582. See also D. V. Sandifer, *Evidence Before International Tribunals*, 2nd ed., 1975, 229: "Parties frequently have employed a strategy, in fact, apparently of overwhelming the tribunal by sheer force of numbers [i.e., of maps]".

expressions of the will of the state or states concerned. This is the case, for example, when maps are annexed to an official text of which they form an integral part. Except in this clearly defined case, maps are only extrinsic evidence of varying reliability or unreliability which may be used, along with other evidence of a circumstantial kind, to establish or reconstitute the real facts".[34]

In this highly authoritative statement, the Court thus reaffirmed once again that depictions on a map cannot as such create a territorial title, and a valid territorial claim can therefore not be based on cartographic evidence as such, unless the maps concerned "fall into the category of physical expressions of the will of the State or States concerned".

In the case of the Ethiopian/Eritrean boundary, with the exception of the two tripoints (Ethiopia/Eritrea/Sudan and Ethiopia/Eritrea/Djibouti), the boundary has not been demarcated anywhere up to the present day. Although the treaties of 1902 and 1908 both expressly provide for such demarcation to be undertaken in due course,[35] no such operation on the ground has ever taken place, nor has any subsequent treaty instrument ever been concluded denoting "the means by which the described alignment is marked, or evidenced on the ground by means of cairns of stones, concrete pillars, beacons of various kinds, cleared roads in scrubs, and so on".[36] The only attempt undertaken in this respect ended in complete failure. The Joint Italo/Ethiopian Commission of 1904, instructed to demarcate the boundary sector eastwards of the Belesa river, was not able to carry out its task, *inter alia*, because of the "incertezza nell'applicazione pratica della formula geografica Mareb-Belesa-Muna adoperata nel trattato".[37]

---

[34] ICJ Reports 1986, 32, para. 54.
[35] See Article II of the 1908 Treaty: "The two Governments undertake to fix the above-mentioned frontier line on the spot by common accord, and as soon as possible, adapting it to the nature and variation of the ground". Article I, para. 3, of the 1902 Treaty provides: "The line from the junction of the Setit and Maieteb to the junction of the Mareb and Mai Ambessa shall be delimited by Italian and Ethiopian Delegates, so that the Cunama tribe belong to Erythraea".
[36] *Ibid.*
[37] "Uncertainty of the practical application of the formula Mareb-Belesa-Muna used in the Treaty of 10 July 1900", Ministero degli Affari Esteri (ed.), *Eritrea. Il confine fra lo Scimezana e l'Agame,* Monografie e Rapporti Coloniali, No. 6, 1912, 3, 1913. The difficulty of applying the treaty line on the ground was reconfirmed and dealt with *in extensu* in the early 1930s by an Italian expedition exploring the area to the east of the Belesa River: See the report on the results of the Italian expedition of May 1930 by C. Zoli, "Un territorio contestato tra Eritrea ed Etiopia. Escursione nel paese degli

Without questioning the predominant role that boundary treaties play in the international legal discourse on boundary régimes in general, and in the present case in particular, the boundary between Ethiopia and Eritrea demonstrates – at least with regard to certain segments of it – that in order to render the conventional line operational for delimitation and/or demarcation purposes, additional legal considerations are necessary. A lack of knowledge of geographical features,[38] merely superficial textual descriptions not always free of ambiguities, as well as imprecise cartographical material used at the time of the drafting of the respective conventions, will make recourse to legal considerations going beyond the narrow limits of pure treaty interpretation inevitable. In particular, the element of "effective control" *("effectivités")* over boundary regions in the course of the last 100 years or so might turn out to be a crucial issue, at least with regard to areas in which the boundaries are insufficiently defined. In this respect, the statement of Max Huber in the *Island of Palmas* case *(Netherlands/USA)* might be recalled: "If ... no conventional line of sufficient topographical precision exists or if there are gaps in the frontiers otherwise established, or if a conventional line leaves room for doubt ... the actual continuous and peaceful display of

---

Iròb", *Bolletino della R. Societá Geografica Italiana* Serie VI – Volume VIII (1931), 715 et seq. With regard to the crucial geographical feature referred to in the 1908 Treaty, the expedition came to the conclusion that a river denominated "Muna" simply did not exist: "Le vere difficoltà e le contestazioni più aspre sorgevano quando si trattava di decidere quale fosse il corso d'acqua indicato come 'Muna' nel testo dell'art. 1 del Trattato 10 Luglio 1900 e come 'Mai Muna' nello schizzo annesso al Trattato stesso; perchè effettivamente in tutto il territorio dell'Eritrea e dell'Etiopia – manco a farlo a posta! – non esiste alcun corso d'acqua nè piccolo nè grande che porti tal nome". A Mixed Commission constituted in 1911 to undertake a similar operation with regard to the boundary line between Ethiopia and the then Italian colony of Somalia, delimited by a treaty dated 16 May 1908 (not to be confused with the treaty of the same date mentioned above), was likewise unsuccessful. On the latter undertaking, see C. Citerni, *Ai confini meridionali dell'Etiopia: Note di un viaggio attraverso l'Etiopia ed i paesi galla e somali*, 1913.

[38] This was, for example, openly admitted by the British Foreign Minister Lord Salisbury: "We have been engaged ... in drawing lines upon maps where no white man's feet have ever trod; we have been giving away mountains and rivers and lakes to each other, but we have only been hindered by the small impediment that we never knew exactly where those mountains and rivers and lakes where". *The Times*, 7 August 1890, cited by Judge Ajibola, Individual Opinion in the *Territorial Dispute* case, ICJ Reports 1994, 53. See also Brownlie, see note 18, 20: "In any case lines were commonly drawn on maps at a stage when there was no great knowledge of the region concerned".

State functions is in case of dispute the sound and natural criterium of territorial sovereignty".[39]

---

[39] *RIAA* II, 840. Probably the most authoritative statement on the relation between, on the one hand, effective possession and acquiescence, and, on the other hand, title based on boundary treaties or alignments by administrative acts of former colonial powers, is to be found in the 1986 judgment of the International Court of Justice in the *Frontier Dispute* case:
"[T]he Parties have invoked in support of their respective contention the 'colonial *effectivité*', in other words, the conduct of the administrative authorities as proof of the effective exercise of territorial jurisdiction in region during the colonial period. ... The role played in this case by such *effectivités* is complex, and the Chamber will have to weigh carefully the legal force of these in each particular instance. It must, however, state forthwith, in general terms what legal relationship exists between such acts and the titles on which the implementation of the principle *uti possidetis* is grounded. For this purpose a distinction must be drawn among several eventualities. Where the act corresponds exactly to law, where effective administration is additional to the *uti possidetis iuris*, the only role of *effectivité* is to confirm the exercise of the right derived from a legal title. Where the act does not correspond to the law, where the territory which is the subject of the dispute is effectively administered by a state other than the one possessing the legal title, preference should be given to the holder of the title. In the event that the *effectivité* does not co-exist with any legal title, it must invariably be taken into consideration. Finally, there are cases where the legal title is not capable of showing exactly the territorial expanse to which it relates. The *effectivités* can play an essential role in showing how the title is interpreted in practice". ICJ Reports 1986, 586 et seq.
In its 1992 Judgment in the *Land, Island and Maritime Frontier Dispute*, the International Court of Justice expressly confirmed that the possibility of a consideration of *effectivités* in order to ascertain the *uti possidetis* at the time of independence is not confined to effective control during colonial times but might also include so-called "post-colonial" effectivités. ICJ Reports 1992, in particular 351 and 399. The present state of legal doctrine on this very issue may thus be summed up as follows. The starting point for the examination of questions of boundaries is the "photograph" of the territory at the moment of independence. This is the very essence also of the 1964 OAU Cairo Resolution, calling for the intangibility of the frontiers inherited from colonial powers. However, the International Court of Justice, in some recent decisions, has taken the matter further by implying that the intangibility of former colonial boundaries sanctioned by the *uti possidetis* principle means that at the moment of independence it represents a *de facto* position of boundaries. Thus, the actual "photographic" position of the boundary may – according to the circumstances of the individual case – represent colonial *de jure* or *de facto* possession. The Court has furthermore pronounced quite specific guidelines as to whether it is the title to sovereignty as evidenced by the colonial *jus* or rather the effective possession which is to prevail in a

Regrettable as deficits in boundary making through colonial treaties may be, it should nevertheless be underlined that, from the perspective of international law, there can be no doubt that the terms of the colonial treaties concluded in the first decade of the 20th century must be regarded as the starting point for each and every legal analysis of the present boundary régime. Only if the meaning and scope of the terms of the applicable treaty instruments cannot satisfactorily be clarified by the various methods of treaty interpretation can the issue of "effective control" or other kinds of "auxiliary" evidence come into play. This basic principle is also mirrored in the Preamble to the Framework Agreement adopted by the OAU in late 1998,[40] which recognises as key elements for the solution of the present boundary conflict the "respect for the borders existing at independence as stated in Resolution AHG/Res. 16(1) adopted by the OAU Summit in Cairo in 1964 and, in this regard, [i.e., with regard to their determination] on the basis of pertinent Colonial Treaties and applicable international law...". Although rightly underlining and confirming the particular importance of the *uti possidetis juris* principle in the African context in general, and in the present case in particular, the above-quoted statement agreed upon by Ethiopia and Eritrea also makes it possible to draw upon additional legal considerations ("*and* applicable international law"), including, *inter alia*, the element of "effective control".

---

specific case. In the first place, the Court has reaffirmed earlier practice. See, for example, the territorial dispute between Equatorial Guinea and Gabon over sovereignty over the islands of Mbane and the Île des Cocotiers, in the settlement of which in 1972 the "mediation" of an *ad hoc* commission of the OAU played a crucial role (cf. Ch. Rousseau, *RGDIP* 77 (1973), 1216 et seq.), and according to which in the case of effective possession conflicting with legal title "preference should be given to the holder of the title". However, the Court has limited the application of this rule to cases in which the legal title is "capable of showing exactly the territorial expanse to which it relates". This latter qualification might obviously turn out to be of the utmost importance in the present case.

[40] UN Doc. S/1998/1223, Annex. This document was approved by the Central Organ Summit of the Organisation of African Unity (OAU) Mechanism for Conflict Prevention, Management and Resolution on 17 December 1998, and served as the basis for all further steps in the peaceful resolution of the conflict. It was, however, only on 18 June 2000 by signing the Agreement on Cessation of Hostilities (UN Doc. S/2000/601) that Ethiopia and Eritrea finally accepted the OAU Framework Agreement and the Modalities for its Implementation, which had been endorsed by the 35th ordinary session of the Assembly of Heads of States and Government, held in Algiers from 12 to 14 July 1999.

III. Towards a Peaceful Settlement of the Boundary Dispute

Soon after fighting erupted between Eritrea and Ethiopia in May 1998, the United Nations joined earlier calls for peace,[41] including in particular the prompt mediation efforts of the Organisation of African Unity based on a prior peace initiative by the Rwandan and US Governments. The objective of the joint US-Rwandan facilitation effort, launched within days of the outbreak of hostilities, was to promote a peaceful and durable settlement of the dispute and to prevent a war which could, and unfortunately did, cost many lives and undermine regional stability. Within two weeks of intensive diplomatic activities, on 30-31 May 1998, the so-called "facilitators" presented to both parties recommendations on a "practical principled basis for peaceful resolution of this conflict",[42] together with a detailed implementation plan, and asked them to confirm their acceptance of both the substantive as well as the procedural elements of this peace plan. Immediately endorsed by the OAU and the UN Security Council,[43] the basic principles enshrined in these recommendations became the cornerstone of all further mediation efforts and were shortly after translated into the so-called "OAU Framework Agreement".[44] Deplorably enough, it took several rounds of further armed hostilities, involving heavy casualties and economic losses,[45] until both sides, after extremely difficult negotiations on "Technical

---

[41] As a very first step, the Secretary-General immediately contacted the leaders of both countries, urging restraint and offering assistance in resolving the conflict peacefully. For other early peace initiatives, see, *inter alia*, the Resolution of the European Parliament of 18 June 1998; the Declaration of the Co-Presidents of the ACP-EU Joint Assembly of 8 July 1998; the Resolution of the Union of African Parliaments at its 21st Conference in Niger (18-20 August 1998); and Resolutions 233 and 234 of the Non-Aligned Movement's summit in South Africa (29 August 1998 - 3 September 1998).

[42] US Department of State, Office of the Spokesman, Statement by James P. Rubin of 3 June 1998, "The Dispute Between Ethiopia and Eritrea".

[43] At a Special Session on 5 June 1998, the OAU Council of Ministers adopted a resolution in which it paid tribute to the work of the facilitators and requested them to continue and "urgently appeal[ed] to the two parties to, at the same time and simultaneously, put an end to all hostilities, accept and implement the recommendations of the facilitators..." (UN Doc. S/1998/485). The Security Council, in a resolution of 26 June 1998, "commend[ed] the efforts of the OAU and of others, in cooperation with the OAU, to achieve a peaceful settlement of the conflict" (SC Res. 1177 (1998)).

[44] See note 40.

[45] For the most recent and hopefully last round of armed confrontation in May/June 2000, cf. J. Pearce, "Facts on the Ground. War and Peace in the Horn of Africa, May-June 2000", *Boundary and Security Bulletin* 8 (2000), 74 et seq.

Arrangements for the Implementation of the OAU Framework Agreement and its Modalities", were finally prepared to accept the legal principles for a peaceful settlement of the dispute.[46] With regard to the final disposition of the common border, agreement could be reached that this issue was to be settled "on the basis of pertinent Colonial Treaties and applicable international law". This formula does thus not limit the legal basis for the settlement of the boundary issue to a narrow *uti possidetis* approach, but leaves room for the consideration of factors such as effective control and other kind of auxiliary evidence.

It was the continuous and patient mixture of mediation efforts,[47] on the one hand, and growing international pressure,[48] on the other, which proved to be most helpful in finally convincing the parties that the only possible solution of the conflict was its peaceful settlement, and that such a solution would not result in a loss of face by either of the parties. Without in any way trying to diminish the leading role played by the OAU and its Member States in this regard, it is apt to observe that the United Nations also made a substantial contribution to the peace process. A whole series of Security Council resolutions not only increased the pressure on the parties, but at the same time demonstrated the continuous concern of the international community of States in general, and of the great powers in particular.[49] Tacit diplomacy, close monitoring

---

[46] Talks resumed in Algiers on 30 May 2000 culminated in the signature, on 18 June 2000, of the Agreement on Cessation of Hostilities between Ethiopia and Eritrea together with the "reaffirmation" of their acceptance of the OAU Framework Agreement and its Modalities. Finally, further talks facilitated by President Bouteflika of Algeria, resulted in the signature, on 12 December 2000, in Algiers of a comprehensive Peace Agreement between Ethiopia and Eritrea. Text available in *Boundary and Security Bulletin* 8 (2000/2001), 58 et seq.

[47] Apart from the manifold initiatives of the OAU and its Member States and the ongoing efforts of the facilitators, one could mention the Security Council mission to the region, which took place on 8 and 9 May 2000 and was headed by the US Permanent Representative to the United Nations, Richard Holbrooke. See Report of the Security Council Special Mission Visit to Eritrea and Ethiopia (9 and 10 May 2000) of 11 May 2000 (UN Doc. S/2000/413).

[48] On 17 May 2000, sanctions were imposed on both countries by Security Council Resolution 1298, aimed at preventing the supply of weapons or arms-related assistance to the two countries.

[49] Cf. SC Res. 1226, 1227 (1999); 1297, 1298, 1312 and 1320 (2000); as well as number of statements of the Council's President. The most recent one, of 15 May 2001, UN Doc. S/PRST/2001/14, concludes with the words: "The Security Council remains vigilant and expresses its intention to take appropriate measures if the situation between Eritrea and Ethiopia again threatens regional peace and security. The Security Council will remain seized of the matter".

of the situation and the "good offices" provided by the Secretariat and the Secretary-General himself were indispensable elements on the way to a peaceful solution of the conflict. Finally, the deployment of a Peacekeeping Force (UNMEE) of up to 4,300 troops was decided upon,[50] the function of which is to reduce to a minimum the danger of a resumption of the military confrontation. During the signing ceremony of the Peace Accord in Algiers in December 2000, the Secretary-General emphasised that "[t]he United Nations and the international community are determined to work closely with the parties to ensure the implementation of both the 18 July Agreement and the one signed today, so that lasting peace can be achieved and that reconstruction can begin".[51] Even after the conclusion of the peace agreement, the United Nations remain seized of the dispute. The Boundary Commission established in January 2001 in accordance with the Algiers Agreement will be serviced by the Cartographic Section of the UN Secretariat,[52] the Commission itself being funded by the Trust Fund established pursuant to Security Council Resolution 1177 (1998).[53] On-site technical work for the demarcation of the boundary line will be carried out by the UN Cartographic Section with the logistical support of UNMEE. Finally, the Organisation itself claims certain informal supervisory functions in order to "guarantee" the continuation of the peace process within the agreed time schedule.[54]

Unlike many other recent cases, the multi-level involvement of the United Nations in the peaceful settlement of the boundary dispute

---

[50] Established on 31 July 2000 by SC Res. 1312 (2000) as a small mission consisting of up to 100 military observers, the mission was expanded to its actual size by SC Res. 1329 of 15 September 2000. Detailed information about the mandate of the ongoing mission and its legal and factual background are to be found at http://www.un.org/Depts/dpko/unmee/body_unmee.htm – last visited on 17 May 2001.

[51] Press Release SG/SM/7659AFR/290 of 12 December 2000. See also the letter of the Secretary-General to the President of the Security Council of 14 December 2000, regarding the signing of the agreement between the Government of the State of Eritrea and the Government of the Federal Democratic Republic of Ethiopia (UN Doc. S/2000/1194).

[52] See Article 4, para. 7, of the Algiers Peace Agreement of 12 December 2000: "The UN Cartographer shall serve as Secretary to the Commission and undertake such tasks as assigned to him by the Commission, making use of the technical expertise of the UN Cartographic Unit, *ILM* 40 (2001), 261.

[53] See Article 4, para. 17, of the Algiers Peace Agreement. *Ibid.*, 262.

[54] In official communications (UN Docs. S/2000/627 and S/2000/612) both Governments requested the United Nations' assistance in the implementation of the Agreement on Cessation of Hostilities of 18 June 2000. The United Nations is actively responding to this request on a permanent basis. See, e.g., the Statement by the President of the Security Council of 9 February 2001 (UN Doc. S/PRST/2001/4).

described will not be crowned by a judgment of the International Court of Justice. However, the World Court's rich jurisprudence on various pertinent legal aspects, such as, e.g., the probative value of maps, the scope of the *uti possidetis* principle, the role of *effectivités*, the evaluation of facts,[55] and finally the way in which ethnic, historical and other incidental elements are to be considered, will certainly have a direct and major impact on the decisions of the Boundary Commission.

---

[55] For a case study on this issue, see P.-M. Dupuy, "Fact-Finding in the Case Concerning the Frontier Dispute (Burkina Faso/Republic of Mali)", in Lillich (ed.), *Fact-Finding Before International Tribunals*, 1992, 81 et seq.

Judge Shigeru Oda and Maritime Boundary Delimitation

Jon M. Van Dyke

Shigeru Oda has played the role of the canary in the mine shaft, providing warnings when his colleagues on the International Court of Justice have strayed too far from the moorings of traditional customary international law. His contributions – both as a scholar and as a judge – cover a vast array of topics in sophisticated and sensitive detail, and the rest of us owe a deep debt of gratitude to him for his hard work in providing sensible solutions to many of the unresolved problems in this evolving field. He has served on the Court for three terms, during a time of transition and challenge for his home country of Japan, as well as for the rest of the world. His opinions are often harshly critical of the views of his colleagues, they sometimes seem stubbornly resistant to change, and they are sometimes inconsistent with each other. But they are always provocative, carefully written, thoughtful, and instructive. His views have had a significant impact on the directions that the Court has taken, they have provided guidance for a generation of scholars, and they will continue to be of importance for coming generations.

This brief comment will focus on Judge Oda's opinions concerning maritime boundary disputes. Judge Oda has assumed the important role of being the "conscience" of the International Court of Justice in these boundary cases, because he always insists on ensuring that the procedural posture of the case is appropriate to enable the Court to reach the substantive issues,[1] and on exploring the substantive issues that the Court's majority ignores.

---

[1] Judge Oda always looks for a State's true expression of consent to the jurisdiction of the Court before he feels comfortable adjudicating a dispute involving that State. See, generally, S. Oda, "The Compulsory Jurisdiction of the International Court of Justice: A Myth?", *ICLQ* 49 (2000), 251 et

Judge Oda has taken a particularly active role in the maritime boundary cases, because much of his work as an academic scholar and consultant for the Japanese Government focused on ocean issues. He was a prolific scholar before joining the Court, and participated as a member of the Japanese delegation at the negotiations that produced the 1982 United Nations Law of the Sea Convention until he began his first term as an ICJ judge in 1976.

One of his efforts during those negotiations was to discourage the formation of what has recently become the International Tribunal for the Law of the Sea.[2] He was concerned that the proliferation of international tribunals would lead to a confusion regarding the principles of international law, and has always worked to strengthen the role of the International Court of Justice as the primary body to interpret and apply international law, and thereby ensure uniformity.[3]

In the boundary context, Oda has always favoured the equidistance approach as the primary and central means to resolve disputes.[4] His instinctive preference would be always to use this approach, but he also recognises situations where adjustments are appropriate, and so he calls the test he prefers the "equidistance/special circumstances" rule.[5]

Judge Oda has been frustrated with the notion of "equitable principles", which assumed a dominant position in Articles 74 and 83

---

seq. In the *Jan Mayen* case, for instance, he argued that Denmark's unilateral application presenting the dispute to the Court should be dismissed, even though both Denmark and Norway had filed declarations accepting the Court's compulsory jurisdiction under Article 36(2) of the ICJ Statute, because international law does not mandate that any particular maritime boundary be drawn and hence no international law "dispute" can exist regarding the matter. *Case Concerning Maritime Delimitation in the Area Between Greenland and Jan Mayen (Denmark* v. *Norway)*, ICJ Reports 1993, 18 et seq., Separate Opinion of Judge Oda, paras. 2 and 47-89 (hereinafter, *Jan Mayen* separate opinion).

[2] Shigeru Oda, "Colloque – Droit de la Mer et Pêche Responsible", Brest, 19 May 2000, 1; B. Kwiatkowska, "Judge Shigeru Oda's Opinions in Law-of-the-Sea Cases: Equitable Maritime Boundary Delimitation", *GYIL* 36 (1993), 225 et seq., 241-242.

[3] S. Oda, "Queries Relating to the Viability of the UN Convention on the Law of the Sea in the Light of State Practice", in M. H. Nordquist (ed.), *Proceedings of the 14th Annual Seminar of the Center for Oceans Law and Policy*, University of Virginia, Cascais, Portugal, 19-22 April 1990, 1991, 320-321.

[4] The *Continental Shelf* case *(Tunisia/Libya)*, ICJ Reports 1982, Dissent of Judge Oda, 157-277, paras. 176, 180-181 and 188 (hereinafter, *Tunisia/Libya* dissent).

[5] See, e.g., the *Continental Shelf* case *(Libya/Malta)*, Dissent of Judge Oda, ICJ Reports 1985, 123-171, para. 7 (hereinafter, *Libya/Malta* dissent).

of the 1982 United Nations Law of the Sea Convention,[6] because he views this amorphous term as being devoid of content. In his *Jan Mayen* opinion, Judge Oda questioned whether Articles 74 and 83 "have any validity as customary international law".[7] He also noted in that opinion that the language in Articles 74 and 83 "may indicate a frame of mind, but it is not expressive of a rule of law".[8] In a subsequent paper, he argued that the expression "in order to achieve an equitable solution", as used in those articles, fails to afford "any objective criteria other than the concept of equity. The maritime boundary cannot be clearly defined in legal terms and nor therefore is the matter of division of the geographical area subject to judicial determination".[9] Later in the same paper, he presented his view that "[e]quity comprises no objective legal criteria and varies in each circumstance".[10] Others have argued that the concept of "equitable principles" does have a content to it, based on ICJ decisions, and that it can provide predictability and stability to maritime boundary delimitations.[11] Even Judge Oda has noted that "relevant" or "special circumstances" that have always guided maritime boundary delimitation would normally be considered in any effort "to arrive at an 'equitable solution'".[12] But Judge Oda still insists on the traditional view that "equity" is something different from "law", and that it is not a proper source on which to base a judicial determination. In his *Jan Mayen* separate opinion, he argued that because Articles 74 and 83 allow neighbouring countries to agree upon a maritime boundary line "among infinite possibilities",[13] any disagreement they might have regarding the proper line "cannot constitute a 'legal dispute'".[14]

---

[6] UN Convention on the Law of the Sea, 10 December 1982, UN Doc. A/CONF.62/122 (1982), reprinted in *The Law of the Sea: Official Text of the United Nations Convention on the Law of the Sea with Annexes and Index*, UN Sales No. E.83.V.5 (1983) and *ILM* 21 (1982), 1261 et seq.

[7] *Jan Mayen* separate opinion, see note 1, para. 64.

[8] *Ibid.*, para. 66.

[9] Oda, see note 1, 3.

[10] *Ibid.*, 5.

[11] See, e.g., J. I. Charney, "Progress in International Maritime Boundary Delimitation Law", *AJIL* 71 (1977), 641 et seq.

[12] *Jan Mayen* separate opinion, see note 1, para. 66.

[13] But see *ibid.*, para. 76, where Judge Oda acknowledges that the "infinite number of ways" that a delimitation line can be drawn must be "within a certain range".

[14] *Ibid.*, para. 68; see also para. 85 ("there are in fact no rules of law for effecting a maritime delimitation in the presence of overlapping titles"). Judge Oda views the ICJ decision delimiting the boundary between Malta

Although Judge Oda served as counsel for the Federal Republic of Germany in the 1969 *North Sea Continental Shelf* case,[15] which established the "natural prolongation" idea,[16] he has played a significant role in ensuring the demise of this idea in subsequent opinions. His long dissenting opinion in the 1982 *Tunisia/Libya* case[17] explained that the concepts of the continental shelf and exclusive economic zone had become parallel,[18] that it would always be logical to draw a single line to delimit the boundaries between those of opposite and adjacent States,[19] and that "the notion of natural prolongation ... has greatly lost its significance".[20] In this elaborate opinion, he analysed the evolution of customary international law during the "laboratory" conditions of the Law of the Sea negotiations, and concluded, among other things, that the concept of the exclusive economic zone, with its uniform limit of 200 nautical miles, "has rapidly been accepted in the realm of international law".[21]

This important 1982 *Tunisia/Libya* dissent also stresses that boundary delimitations should always be based upon actual geography and that factors such as population, location of resources, and economic development should be ignored in any delimitation.[22] But a decade later, in his *Jan Mayen* separate opinion, Judge Oda seems to be struggling to give meaning to the concept of "equity", and he indicates that any search for equity would include non-geographical factors. This more recent opinion has some internal conflicts, because Part II argues that "equity" is something that has little or no substantive content and

---

and Libya as having been made *ex aequo et bono* rather than based on any legal principles. *Ibid.*, para. 86.

[15] The *North Sea Continental Shelf* cases *(FRG/Denmark, FRG/Netherlands)*, ICJ Reports 1969, 1-54.

[16] Kwiatkowska, see note 2, 227.

[17] *Tunisia/Libya* dissent, see note 4.

[18] Indeed, in his *Jan Mayen* separate opinion, see note 1, at para. 71, Judge Oda says that the régimes of the continental shelf and the exclusive economic zone "should have been amalgamated in the new law of the sea", and he criticises the 1982 Law of the Sea Convention for presenting an "immature" concept of the exclusive economic zone.

[19] But in his *Jan Mayen* separate opinion, see note 1, at para. 69, Judge Oda points out that the delimitation of the exclusive economic zone could produce a different line from that delimiting the continental shelf because "the 'special' or 'relevant' circumstances to be taken into account when defining a delimitation line may well be different in each case".

[20] *Tunisia/Libya* dissent, see note 4, para. 107.

[21] *Ibid.*, para. 108. Judge Oda remained cautious, however, about whether the detailed rules governing the EEZ régime in the Law of Sea Convention had become part of customary international law. *Ibid.*, para. 120.

[22] *Ibid.*, para. 157.

cannot be the basis for resolving a dispute,[23] while Part III includes ideas that should logically be included in an equitable determination.[24] Judge Oda says, for instance, that "I can accept that, just as in the North Sea Continental Shelf cases where the 'natural resources of the continental shelf areas involved' were suggested as a factor to be taken into account, the 'fishing resources may well be a relevant factor in the delimitation of the exclusive economic zone'".[25] But then he proceeds to criticise sharply the Court's opinion for failing to offer "the slightest hint in its reasoning" regarding why it had adjusted the median line in the manner that it did.[26]

Several paragraphs later, Judge Oda offers an elaborate description of the factors that might be taken into account in the search for equity:

> "In *my* concept of equity, it is not merely the simple disparity of opposite coastlines which must be taken into account but also disparity of geographical (natural or socio-economic) situations, for example, population, socio-economic activity, existence of communities behind the coastline and the distance of an uninhabited island from the nearest community of the mainland or main territory. The existence, quality and quantity of marine resources (either fishery or mineral) are relevant, but equity surely requires that any decision as to how these resources should be allotted to each party should take account not only of such relatively objective ecological facts but also of their relative significance, perhaps amounting to dependence, to the communities appertaining to either party. Certainly it is impossible to calculate and balance up the elements mathematically in order to draw a line with total objectivity. Thus the drawing of a line must depend upon the conscientious but infinitely variable assessment of those drawing it".[27]

This complex description of "*my* concept of equity" (emphasis added) illustrates the tension in Judge Oda's thinking. While his traditional roots reject the notion that equity can be quantified and utilised as a predictable mode of decision making, his logical analysis understands that even an amorphous and ephemeral concept such as equity can be broken down into its component parts and given shape and content. He

---

[23] *Jan Mayen* separate opinion, see note 1, paras. 47-89.
[24] *Ibid.*, paras. 90-100 (especially paras. 94 and 98).
[25] *Ibid.*, para. 94.
[26] *Ibid.*, para. 95.
[27] *Ibid.*, para. 98, emphasis added.

thus appears willing to undertake an "equitable" determination to draw a maritime boundary, but remains frustrated with the way his colleagues on the Court have undertaken the task. His closing paragraph acknowledges the "conscientious" effort of the Court "towards the finding of an equitable solution", but nonetheless he denounces their effort as "arbitrary", and "unsupported by any sufficiently profound analysis".[28]

Offshore islands provide another example of some uncertainties in approach. Judge Oda has been reluctant to give much importance to islands in certain maritime boundary delimitations. He joined the Court in ignoring the tiny islet of Filfla in the *Libya/Malta* delimitation,[29] he sharply criticised the Court for giving "half effect" to the much more substantial Kerkennah Islands in the *Tunisia-Libya* continental shelf delimitation and argued that they should be wholly disregarded,[30] and he criticised the Court's treatment of islands again in his *Libya/Malta* dissent.[31] But he has also noted that islands are a factor that may or may not be relevant in boundary delimitations, depending on demographic or economic factors.[32]

Judge Oda has criticised the Court for being arbitrary in defining the "relevant area" that is to be affected by a boundary delimitation.[33] Although he has recognised that "a division of the area concerned in proportion to the length of the relevant coast of each State facing that area will, in principle, satisfy the requirement of equity",[34] he has also been impatient with the Court's *ad hoc* utilisation of the proportionality of coasts as a means to adjust a boundary determination. He criticised the ratios utilised by the Court in the *Tunisia/Libya* case,[35] and in the *Malta/Libya* case.[36]

In the complicated *Gulf of Fonseca* case, Judge Oda criticised the Court for its decision to recognise "condominium" ownership of the waters of the Gulf (and those extending into the open ocean) among the three neighbouring countries – Nicaragua, El Salvador, and Honduras – arguing that international law does not contain the concept of a pluristate bay.[37] Judge Oda always worries about the impact of any

---

[28] *Ibid.*, para. 100.
[29] *Libya/Malta* dissent, see note 5, para. 80.
[30] *Tunisia/Libya* dissent, see note 4, paras. 39-45, 72-73, 149-150, and 170-173.
[31] *Libya/Malta* dissent, see note 5, paras. 28 and 39-45.
[32] *Tunisia/Libya* dissent, see note 4, paras. 173 and 176.
[33] *Libya/Malta* dissent, see note 5, para. 8.
[34] *Tunisia/Libya* dissent, see note 4, para. 176.
[35] *Ibid.*, para. 164.
[36] *Libya/Malta* dissent, see note 5, paras. 13-15.
[37] *Land, Island and Maritime Frontier Dispute (El Salvador/Honduras: Nicaragua Intervening)*, Dissent of Judge Oda, ICJ Reports 1992, 732-761.

boundary decision on third parties not involved in the litigation, and has frequently criticised the Court for ignoring the interests of neighbouring concerned countries.

The innovations and evolutions that we will see in the international law field in the coming century will be built upon the shoulders of the giants who have laboured during our time, and Judge Shigeru Oda is certain to be one of those whose views and writings will be consulted repeatedly for ideas and analysis. His attention to detail and precision leaves an important legacy that will greatly facilitate the work of all those who follow. His historical analyses of the importance of the negotiations producing the 1982 Law of the Sea Convention will always be examined for guidance on that seminal international gathering. His internal struggles in deciding what factors to utilise in drawing boundaries will provide guidance for those who follow, just as his constructive criticism of the opinions of his colleagues have helped to bring clarity to their views.

X

THE LAW OF THE SEA

A Few Remarks on the Theoretical Basis of the New Law of the Sea

Milenko Kreća

I. The Concept of the International Law of the Sea[1]

1. The Concept of the Law of the Sea in a Formal Sense

Since the law of the sea is a component part of public international law,[2] there is no special need (except epistemological) to give it a separate and independent definition, for logically such a definition would either be a paraphrase of the general definition of public international law or an enumeration of the topics regulated by the rules which we refer to as the

---

[1] As follows from the formal concept, the term "international law of the sea" should be regarded as shorthand for all the rules of public international law which regulate maritime affairs. Such a description seems a fair one since this term (typical of a widespread tendency in the theory of international law to talk, for instance, about space law, treaty law, humanitarian law, etc.), besides having positive, symbolic implications in a system which cannot boast an advanced level of codification, carries the risk that the symbolic meaning be taken as a basis for other, less than logical extrapolations. A good example is the title: Convention on the Law of the Sea. Grammatically construed, this phrase suggests that a convention has been concluded on a body of existing law that is merely reshaped legislation as already in force. Such an interpretation would obviously be wrong, for the Convention has, to a great extent, created new law. If the reasoning behind the adoption of this wording were to be applied by analogy to other legal instruments, then the Law on Contracts could be called the Law on Contract Law, or the Law on Criminal Proceedings the Law on the Law of Criminal Proceedings. The same criticism could be levelled at the wording of similar titles (e.g., the Convention on the Law of Treaties).

[2] In a chapter entitled, "The Sources and Development of the International Law of the Sea", one of the authorities on the law of the sea talks about *"international law, of which the principles which govern maritime intercourse, naval warfare and neutrality form a substantial part..."*. See C. J. Colombos, *The International Law of the Sea*, 1967, 7, emphasis added.

law of the sea.[3] Therefore, we can conceptually define the law of the sea as that part of public international law which, *ratione materiae*, regulates maritime affairs.

2. The Concept of the Law of the Sea in a Substantive Sense

In a substantive sense, the law of the sea represents an institutionalisation of demands by States to extend their jurisdiction over maritime regions. Here we can distinguish between two aspects of institutionalisation:

1. A positive aspect, which is seen in recognising the right of coastal States to extend their jurisdiction over various parts of the sea. It is manifested in permissive norms such as, for example, the rule that entitles a coastal State to establish a contiguous zone bordering on its territorial sea. Permissive norms may be complete (when a right and the modalities of exercising it depend on the will of the State, as in the case of the contiguous zone, since it is left up to the coastal State to decide on the establishment and breadth of this zone within permitted limits) or incomplete (when the possession of certain rights is inherent and does not depend on the will of the State,[4] but the spatial modalities of exercising the given right is a matter of the will of the coastal State).
2. A negative aspect, which is seem in the prohibition of the assertion by States of sovereignty or jurisdiction over certain parts of the sea. This aspect is expressed in prohibitive norms, of which the principal one forbids States to lay claim to sections of the open sea.

Institutionalisation is carried out by dividing the sea as a physical entity into zones, which differ according to the number of rights acknowledged for the coastal State. In this respect, the normative logic of the international law of the sea is that the rights of a coastal State progressively diminish in each successive zone moving from the shore seawards. In effect, the legal partitioning of the sea accommodates the ambitions of coastal States to extend sovereignty over areas where it does not conflict with the sovereignty of other States and they raise no objections. However, it would be an exaggeration to think that the international law of the sea is a simple or mechanical projection of the

---

[3] An example of an epistemological definition is found in D. Rudolf, *Međunarodno pravu mora (The International Law of the Sea)*, 1985, 5.

[4] A State possesses a territorial sea *ipso facto*, by virtue of the fact that its territory borders on a sea. By analogy, a coastal State acquires rights over the continental shelf *ipso facto* and *ab initio*.

individual demands of States. In such event, it would merely represent externalised municipal law or a rationalised aggregate of the *de facto* relations of littoral States. Institutionalisation has been carried out in the parameters of the basic principles of universal international law at its present stage of development, expressing the prevailing balance between individual interests and the general interests of the international community with regard to the sea.

With the exception of the period when the leading maritime powers extended their dominion over entire seas and oceans, which may be referred to as a prelegal state of affairs, institutionalisation has been carried out in the manner of *jus strictum*, on the basis of a strict separation between the waters which are of national interest (the territorial sea) and the waters which are of international interest (the high seas). As time goes on, the international law of the sea is acquiring the attribute of *jus aequum* by satisfying the national interests (primarily those of an economic nature, but also to a degree the interests of security), while at the same time protecting international interests by imposing limitations on coastal States or articulating the rights of third States. However, in the entire history of the law of the sea, the protection of international interests has not been carried out directly through international machinery (except to some extent in cases on the high seas when a warship under one flag takes measures against a merchant vessel of another flag, acting in the capacity of an agent of the international community), because such machinery simply did not exist, as the law of the sea represented a distribution of jurisdiction, primarily to the coastal States. The first elements of direct international jurisdiction over the sea are only to be found in the international seabed area (as the common heritage of mankind).

II. Forms of Institutionalisation in the New Law of the Sea

1. The Positive Aspects of Institutionalisation

If we compare positive forms of institutionalisation in the traditional and the new law of the sea, we note differences which can be categorised as qualitative and quantitative.

Qualitative differences are manifested in the extension of the maritime zones set up prior to the adoption of the new Convention. In general, the UN Convention on the Law of the Sea (1982) displayed a high degree of tolerance for demands for existing maritime belts to be extended.[5] Admittedly, the concrete arrangements envisaged include

---

[5] According to Article 33 of the Convention, the contiguous zone may not extend beyond 24 nautical miles from a State's baselines and the territorial

elements which express the broader interests of the international community or individual groups of States.[6]

Qualitative differences are seen in the setting up of new zones to benefit coastal States. There are archipelagic waters and the exclusive economic zone. It is interesting that even though the new law of the sea has taken institutionalisation to a higher level in comparison with traditional law, the corresponding rules of law suffer from insufficient precision, which as the history of the law of the sea teaches us, will provide a springboard for coastal States to expand the prerogatives given to them over parts of the sea. In a sense, the spirit of John Selden's *Mare clausum* continues to haunt the law of the sea.

A. Archipelagic Waters

One of the notions concerning the sea around archipelagos (the so-called Asian variant), namely, that it is a component part of national waters, has, under the regulations of the new Convention, developed into a new maritime zone: archipelagic waters. This institutionalisation was carried out *in concreto* to benefit the particular interests of archipelagic States.[7] Since the Convention recognises sovereignty over archipelagic waters (the area between islands and the territorial sea in the event that an archipelagic State does not have inland waters, or between inland waters and the territorial sea), archipelagic States have sovereignty over a larger area than do coastal States, thanks to their geographical status.[8]

The Convention has taken a liberal approach from the standpoint of the methodology used to define archipelagic waters.[9] The definition of an archipelagic State is essentially based on a subjective criterion

---

sea may not extended beyond 12 nautical miles from the baselines. In addition, the new definition of the continental shelf states that the continental shelf extends to the outer edge of the continental margin or to a distance of 200 nautical miles (Article 76).

[6] Generally speaking, this category includes the rules on innocent passage, which have been laid down in great detail, the rights of landlocked and geographically disadvantaged States in the exclusive economic zone, the right of archipelagic sea lane passage, the rules on marine scientific research and the protection and conservation of the marine environment, the rule on transit passage through straits, etc.

[7] The Convention defines an archipelagic State as a State with a territory that is constituted wholly by one or more archipelagos which are so closely interrelated that they form an intrinsic geographical, economic and political entity, or which historically have been regarded as such (Article 46).

[8] Only independent States, but not dependent territories, can be granted the status of an archipelagic State.

[9] See Articles 46(b) and 47(1) and (2) of the Convention.

bolstered by way of compensation by some objective criteria. The section of the definition which states that an archipelagic State is "constituted wholly by one or more archipelagos and may include other islands" is a broad one, for the basic concept ("constituted wholly") is expanded with the additional "and may include..." in order to cover the subjective situations of some archipelagic States. Furthermore, the definition of an archipelago shows evident signs of a compromise expressed in elements of a different character. On the one hand, it is required that these islands, their "interconnecting waters and other natural features" should be closely interrelated, a clause which could be interpreted primarily as a matter of distance or physical links, while, on the other hand, it is specified that these features should form a "geographical, economic and political entity", a wording which lays primary emphasis on interdependence and *de facto* links. Particularly ambiguous is the alternative condition accorded to those groups of islands which "historically have been regarded as such".

Of an objective nature are the lengths of the straight baselines drawn for the purposes of enclosing the waters of an archipelagic State (Article 47(2) of the Convention) and the ratio between the area of water and the area of land, which should be between 1:1 and 9:1.

Regardless of the fact that the definition incorporates the reasonable and justified demands of archipelagic States, which for the most part are underdeveloped, there is no doubt that with the establishment of archipelagic waters, large portions of the open sea are being brought under national jurisdiction. Thus we have a Copernican twist in that geographical handicaps are transformed into an advantage. The national territories of archipelagic States have been increased many times over. However, as regards limits to the sovereignty of archipelagic States (the right of innocent passage by foreign vessels or archipelagic sea lane passage – and the right of neighbouring States), it would seem that the existence of these rights, particularly the right of passage, will not escape differences of interpretation,[10] the resolution of which will greatly depend on such factors as the balance of power and expediency.

B. The Exclusive Economic Zone

The exclusive economic zone is a qualitatively new category in which the tendency to expand national jurisdiction has found scope. In the Convention, it is revealed in a hybrid régime, a régime *sui generis*, which is somewhere between the régime of the territorial sea and the

---

[10] See, for instance, the view of the delegations from the Philippines and Sao Tome and Principe on the innocent passage of warships. UNCLOS, Official Records, XIV, pp. 58-59 and XVI, p. 34.

régime of the high seas. The sovereign rights and jurisdiction of a coastal State in this zone[11] have their corollaries in the rights of third States, either coastal or landlocked, which derive from the general principle of freedom of the high seas.[12]

However, it is to be presumed that the coastal States do not consider these rights as satisfying their demands completely. They are encouraged in this feeling by the fact that in addition to the aforementioned, expressly defined rights, coastal States also have residual rights in the exclusive economic zone, as the Convention stipulates that they have "other rights and duties provided for in this Convention" (Article 56(1), subparagraph (c)). The vague, generalised wording of the residual rights leads to the "expansion of the rights of a coastal State in the zone, over and above this sphere of resource exploitation and economic rights, leaving only freedom of intercourse on the high seas".[13]

However, unlike the régimes in the other zones, the régime of the exclusive economic zone contains regulations which express the idea of solidarity and the new international economic order. These elements are discernible in the provisions of Article 62(2) of the Convention, which envisage the right of other States, particularly developing countries, under the conditions and modalities laid down by the Convention, to have access to the surplus of the allowable catch on equal terms. Belonging to the same category are the provisions of Article 70, which regulate the rights of geographically disadvantaged States to participate, on an equitable basis, in the exploitation of an appropriate part of the surplus of the living resources of the exclusive economic zones of coastal States of the same subregion or region.

How should the significance of the institution of the exclusive economic zone be assessed? There is no doubt that the establishment of an exclusive economic zone is a form of institutionalisation of the demands of maritime countries to extend their jurisdictional rights over the sea. In this sense, the zone, regardless of the elements of the new international economic order incorporated in its régime, is basically an expression of the concept of an individual distribution of jurisdiction, but for reasons other than those which motivated the establishment of bays and estuaries, etc., as inland waters and the territorial sea. Therefore, there are no justified arguments to back up the claim that the exclusive economic zone gives developing countries special benefits, for

---

[11] Article 56 of the Convention.
[12] Article 58 of the Convention.
[13] Z. Perišić, "Isključiva ekonomska zona. Prinosi za porodbeno proučavanje prava i međunarodno pravo" ("The Exclusive Economic Zone. Contributions for a Comparative Study of Laws and International Law"), XV *Novo pravo mora* 17 (1982), 76.

the most elementary indicators prove the folly of such an assertion.[14] As a form of extending jurisdictional rights over sections of the sea which, according to the Geneva Conventions, used to be under the régime of the high seas, the exclusive economic zone is a general gain for coastal States and, what is more, according to statistics, the advanced countries reap the greatest benefits from it. The overriding interest of the developing countries was to establish an international area (the common heritage of mankind) for the exploitation of the resources of the seabed and ocean floor beyond the zone of three nautical miles, and since this idea could not come to anything, the interests of small and underdeveloped countries with regard to the exploitation of the living and mineral resources of the sea were defended to some extent by the establishment (or confirmation) of the exclusive economic zone. In short, the exclusive economic zone, from the standpoint of the interests of developing countries, is a type of defensive response,[15] for without it they would have found themselves relegated to the position of observers of a *de facto* appropriation of the resources of the seabed and subsoil by the developed countries.

2. The Negative Aspects of Institutionalisation

In terms of the substantive concept of the law of the sea, the international seabed area represents a negative form of institutionalisation, for *ex definitione*, States cannot exercise sovereignty or sovereign rights over any part of the international area of its resources. However, this is just one prohibitive aspect which does not differ from the interdictory rule forbidding occupation of the open sea. What makes the international seabed area special and, historically, the first expression of a qualitatively different negative form of institutionalisation is a second, "positive" aspect, the thrust of which is to designate the international seabed area as the common heritage of mankind, and it is expressed in a number of specific provisions

---

[14] See the table in D. Rudolf, *Terminologija međunarodnog prava mora (The Terminology of the International Law of the Sea)*, 1980, 54. The table shows that of the seven countries with the largest economic zones, which constitute 45.03 per cent of the total surface of all economic zones in the world, only one is a developing country (Indonesia).

[15] Mention is made of the "defence value of the concept of the exclusive economic zone" by E. M. Borgese, *Pacem in Maribus. Convocation*, Malta 23-26 June 1973, International Ocean Institute, *viii-ix*. Z. Perišić talks about a "reserve function" in the case of disputes over the system and régime of exploration and exploitation of the international seabed area, Z. Perišić, see note 13, 78-79.

regulating activities undertaken for the purpose of exploring and exploiting the resources of this zone.

The international seabed area is the common heritage of mankind. The rights in this Area are vested in mankind as a whole,[16] on whose behalf the International Seabed Authority, an independent international organisation based on the principle of the sovereign equality of its members, acts as an organised institution. Activities in the international seabed area are carried out for the benefit of the whole of mankind, regardless of the geographic location of the States, regardless of whether they are coastal or landlocked, and particularly taking into consideration the interests and needs of developing countries and nations which have not yet gained full independence. In a nutshell, the concept of the international seabed area represents an application of the idea of the new international economic order.

We can assess the implications of the international seabed area from two standpoints: normative-theoretical and practical. From the normative-theoretical standpoint, the international zone represents a revolutionary innovation in the law of the sea and in international law in general. The direct international jurisdiction which is asserted over the seabed and subsoil of the high seas is in itself an historical advance based on the philosophy of collectivity and solidarity. The logic of *laissez faire*, of equal opportunity for *de facto* unequal subjects, is replaced by the logic of fair and effective participation in a division of the resources of the sea. Such a model of relations among States is a milestone in the international community's movement towards a genuine *genus humanum*.

From a practical point of view, however, the implications of the concept of the international seabed area are swallowed up in legal *de facto* lacunae.

The *de facto* lacunae are the fact that the setting up of an exclusive economic zone has drastically reduced the economic potential of the international area, since the most valuable living and non-living resources have come under national jurisdiction.[17] The legal drawbacks

---

[16] According to Dupuy, the concept of "mankind" has two connotations: spatial, since mankind includes all human beings alive today, and chronological, in the sense that it includes not just living human beings but also those who are yet to be born. See R. J. Dupuy, *La gestion des ressources pour l'humanité: Le droit de la mer*, Colloque, The Hague, 29-31 October 1981, 11.

[17] According to statistics, the greatest part of biological and mineral resources lies precisely in the parts of the sea that are under the national jurisdiction of coastal States. For instance, 87.5 per cent of the oil and gas (see *Economic Significance in Terms of Sea-Bed Mineral Resources of the Various Limits Proposed for National Jurisdiction*, UN Doc. A/AC.138/87, 1973, 17) and 96 per cent of the catch (see A. W. Koers, *International Regulation of Marine Fisheries*, 1973, 25) come from this zone.

are that with the so-called parallel system of exploiting the international seabed area, direct international jurisdiction has been impaired, namely, the exploitation of the resources in the international seabed area by natural and legal persons, regardless of the corresponding restriction contained in the Convention and in Annex III, cannot be regarded as exploitation for the benefit of mankind, because essentially it is a case of exploitation by individuals.

III. The Trichotomous Structure of the New Law of the Sea

The erosion of the dichotomous structure of the law of the sea began with the establishment of new zones in the sea which, from the standpoint of the rights of littoral States, lie midway between the régime of the high seas and the régime of the territorial sea. It is a general feature of the legal régimes of these zones that the coastal States are given either restricted sovereign rights or jurisdiction to an extent less than that which they exercise in internal waters or the territorial sea, but in areas which enjoyed the status of open sea in the traditional law of the sea. For instance, the contiguous zone, a belt of water on the high seas adjacent to the territorial sea, has been set up to enable coastal States to carry out the necessary controls to enforce their customs, fiscal or sanitary regulations. In addition, coastal States have been granted "sovereign rights for the purpose of exploring [the continental shelf] and exploiting its natural resources", independently of effective or notional occupation or of any express proclamation (Article 2(1) and (3) of the 1958 Convention on the Continental Shelf). Attention is drawn to the fact that, in this Convention, the models of dichotomous reasoning have been retained, as both the contiguous zone and the continental shelf are described as parts of the high seas, although clearly the rights that are granted to coastal States in the aforementioned zones affect the exercise of rights pertaining to the concept of the high seas. The dichotomous mimicry has lost all rational connection with the creation of the exclusive economic zone as "an area beyond and adjacent to the territorial sea, subject to the specific legal régime established in this Part..." (Article 55 of the 1982 UN Convention of the Law of the Sea), since the rights which are expressly acknowledged for a coastal State in the exclusive economic zone, combined with residual rights, make the exclusive economic zone a zone *sui generis*.

Therefore, the dichotomous structure of the law of the sea has been formally done away with, and in its place a trichotomous structure has been established. The dichotomy of sovereignty/anti-sovereignty has been replaced with the trichotomy of sovereignty/semi-sovereignty/anti-sovereignty. The opposite poles of the dichotomy (sovereignty and anti-sovereignty) are expressed in terms of positive law in the same zones as in the traditional law of the sea (the territorial sea and the high seas), except for the fact that the belts of water making up the territorial sea have been extended, while the exclusive economic zone as an institutional form of semi-sovereignty has been introduced in a section of the sea which traditionally fell under the régime for the high seas. Not only has the area of the high seas been quantitatively reduced to a great extent with the establishment of the exclusive economic zone and the expansion of other zones, but it has also undergone a qualitative transformation with the creation of the international seabed area. In a real sense, the international seabed area is the most worthwhile achievement of the new Convention on the Law of the Sea, not only because it is the only zone which escapes the logic of the extension of the territorial rights of coastal States, but also because the régime of this Area supersedes the previous system of distributing jurisdiction in favour of direct international jurisdiction.

Changes in Japan's Policy on the Law of the Sea

Chiyuki Mizukami

I. Introduction

Japan is a small island country with few natural resources. As an industrialised country, in need of great quantities of raw materials, its industries depend heavily on raw materials from abroad. Japan is also one of the most developed distant-water fishing countries. In the 1960s and 1970s, Japanese fishing vessels were engaged in fishing in all of the world's oceans. The tonnage of the long-range fleets of Japan and the Soviet Union in 1976 (without support vessels), for example, represented 50 per cent of the total world gross tonnage of fishing vessels over 100 GRT each,[1] although the tonnage of Japan has been reduced thereafter due to a decrease in quotas in fisheries agreements.

For these reasons, i.e., the need for raw materials to support its domestic industries and the need for a catch to support its domestic fish industry and consumption, sea transportation is very important to Japan. This is why Japan has adhered to freedom of the high seas.

Until the 1970s, Japan supported and complied with the three-mile territorial sea; it did not have straight baselines nor did it establish a contiguous zone even after it became a party to the 1958 Convention on the Territorial Sea and Contiguous Zone (hereinafter, the "Territorial Sea Convention"). However, Japan's policy on the law of the sea began to change greatly in tandem with the general trends of the law of the sea. For example, Japan extended its territorial sea to 12 miles and established a 200-mile fishing zone in 1977. When it ratified the UN Convention on the Law of the Sea in 1996, Japan adopted the straight baseline system and established a contiguous zone and an exclusive economic zone.

This article will discuss changes in Japan's policy regarding the law of the sea.

---

[1] V. Kaczynski, "Distant Water Fisheries and the 200 Mile Economic Zone", The Law of the Sea Institute, Occasional Paper No. 34, 1983, 9.

II. The Territorial Sea

1. Breadth of the Territorial Sea

Before Japan enacted the Law on the Territorial Sea in 1977, it had no legislation on this subject.

During the Prussian-French War, Japan, which had just opened its door to the world in 1868 after its long self-imposed national isolation policy, established a three-mile neutral zone in 1870 by the Order of the *Dajokan* (predecessor of the Cabinet). This proclamation stated, in part, that "States shall not be engaged in acts of belligerency in that part of the open sea which lies within the limit of about three miles (the distance of the reach of cannons from the land) ... though the mere passage shall be kept free as it has been both to warships and to merchant ships". The Order was intended to prohibit Prussia and France from fighting a battle in the neutral zone, but was also the first time that Japan officially made public its position on the three-mile rule.[2]

In practice, Japan thereafter adopted three miles as the outer limit of the territorial sea. For example, the prize courts of Japan during the Russo-Japanese War applied the three-mile limit. After the Russian Revolution, Japan appealed to the Soviet Union not to arrest Japanese vessels beyond the three-mile limit.

After World War II, Japan continued to follow the three-mile limit for the territorial sea.

Japan enjoyed the freedom of the sea and the limited breadth of the territorial sea was consonant with the national interests of Japan. The three-mile limit was, of course, a practice followed by Western countries such as the United States, the United Kingdom, France, Germany and others, although the Scandinavian countries adopted a four-mile limit and the Soviet Union adopted a 12-mile limit.

Japan attended the First and Second United Nations Conferences on the Law of the Sea, which were held in Geneva in 1958 and 1960, with a basic position of support for the three-mile limit.

Throughout the First Conference, Japan supported the three-mile limit. When a proposal was submitted at the Second Conference by the United States and Canada which would allow a six-mile limit for the territorial sea and, beyond that, another six-mile limit for the fishery zone, Japan favoured the proposal at the committee stage, but abstained from the voting for a slightly amended proposal at the plenary meeting.[3]

---

[2] Minutes of the Foreign Affairs Committee, House of Representatives, 58th Session, No. 4, 15 March 1968; S. Oda and H. Owada, *The Practice of Japan in International Law 1961-1970*, 1982, 144.

[3] S. Oda, *International Control of Sea Resources*, 1989, 101 et seq.

After the Conference, the Deputy Director-General of the Treaties Bureau, Ministry of Foreign Affairs, stated on 15 March 1968 in the House of Representatives of the Diet:

> "At the [Second] Conference [on the Law of the Sea] Japan supported, by way of compromise, a proposal for six miles as the breadth of the territorial sea as a matter of *lex ferenda*. However, the representative of Japan at the Conference made it clear in his statement that he was going to support the proposal by way of compromise and from the view point of *de lege ferenda*, and that therefore if this proposal were to be defeated, Japan would revert to the principle of three miles which was the only rule of international law on this point at present.
>
> Since the proposal was defeated, Japan has since reverted to its traditional position of three miles as the breadth of the territorial sea".[4]

After the Second United Nations Conference on the Law of the Sea, many States established a 12-mile fishery zone, and Japan protested to those countries, stating that it could not accept the unilateral establishment of the zone. At the same time, Japan concluded fisheries agreements with New Zealand, Mexico and others in order to guarantee the continued operation of Japanese vessels in the fishery zone of those countries.[5]

Especially after 1970, the number of States adopting the 12-mile limit for the territorial sea increased. Japan's position on this issue changed slightly, although at this stage it still adhered to the three-mile limit. For example, the delegate of Japan (Professor Shigeru Oda) at the meeting of the Asian-African Legal Consultative Committee, which was held in Colombo, Ceylon (now Sri Lanka) in 1971, said that Japan was ready to accept the 12-mile limit on the conditions of freedom of passage in straits and the granting of preferential fishing rights to the coastal States beyond their territorial seas.[6]

At the 1971 summer session of UN Sea-Bed Committee, the delegate of Japan said:

> "My delegation is of the view that the figure of twelve miles presently claimed by more than forty-five States, represents the

---

[4] Minutes of the Foreign Affairs Committee, House of Representative, 58th Session, No. 4, 15 March 1968; Oda and Owada, see note 2, 143.

[5] C. Mizukami, *Nihon to Kaiyoho (Japan and the Law of the Sea)*, 1995, 11.

[6] S. Oda, "Proposal Regarding a 12-Mile Limit for the Territorial Sea by the United States in 1970 and Japan in 1971: Implications and Consequences", *Ocean Development & Int'l L.* 22 (1991), 192.

best possible compromise as the maximum breadth of the territorial sea".[7]

At the 1973 spring session of the UN Sea-Bed Committee, the delegate of Japan said:

"My delegation can support the 12-mile limit as the maximum limit for the territorial sea, with all the normal attributes recognized traditionally within the territorial sea".[8]

However, Japan, at this stage, did not intend to implement a 12-mile territorial sea, but supported it by way of compromise for making an international rule.

By 1977, some of the developed States such as Canada, France and Italy had extended their territorial seas to 12 miles. Japan, likewise, extended its territorial sea to 12 miles in 1977. The reason why Japan did so was to follow the worldwide trend of extending the territorial sea to 12 miles and, more importantly, to protect its fishermen from Russian fishing vessels operating just off the coast of Japan.

Russian fishing vessels, especially since 1968, operated in the Pacific Ocean off Japan. Large trolling vessels and purse seine vessels operated off Hokkaido and northern Honshu (the main island). In 1974 Russian fishing vessels even travelled down to the area near the Izu Islands where mackerel fishing grounds are located. Russian vessels often broke the fishing nets of Japanese vessels, causing great damage. Korean vessels also operated off Hokkaido and caught Alaskan pollock and other fish.

To avoid disputes between Japanese and Russian vessels, Japan and the Soviet Union concluded an Agreement on Fishing Operations between the Governments of Japan and the Soviet Union, which provided for signs, lights, signals of vessels, signs of fishing nets and navigation and operation of vessels, and set up committees in Tokyo and Moscow to deal with compensation problems.

Meanwhile, the domestic fishing industries requested the Japanese Government to extend the breadth of the territorial sea to 12 miles.

On 26 October 1976, the Committee on Agriculture and Forestry of the House of Representatives adopted a resolution, entitled Matters Concerning Early Realisation of the Twelve-Mile Territorial Sea and Procurement of Food from the Sea. The resolution said it was important, *inter alia*, to take measures to extend the territorial sea to 12 miles,

---

[7] Law of the Sea Office, Ministry of Foreign Affairs, Statement Delivered by the Delegation of Japan at the Third United Nations Conference on the Law of the Sea, 74 et seq. (date of publication unknown).
[8] *Ibid.*, 143.

responding to the long-standing aspirations of fishermen and paying attention to national interests.[9]

The Japanese Government decided on 26 January 1977 to extend the territorial sea to 12 miles. On 2 May 1977, Japan enacted the Law on the Territorial Sea.[10] This Law provides that the breadth of the territorial sea is 12 miles, except for five designated areas where the breadth remains three miles.[11] The five designated areas are areas of the Soya Strait, the Tugaru Strait, the eastern channel of the Tushima Strait, the western channel of the Tushima Strait, and the Osumi Strait. The reason for keeping three miles in these areas was explained by the Government with respect to the trend of the régime of passage through straits used for international navigation. The trend in the Third UN Conference on the Law of the Sea was towards recognising freer passage through international straits than for passage through the ordinary territorial sea. Thus, the Government decided for the time being that the extent of the territorial sea should be maintained as it was in the case of waters of international straits. The Government considered, in particular, that it was desirable, in view of Japan's overall national interest, to wait until this question was resolved internationally in line with such a trend.[12] However, it was also true that if Japan recognised passage of ships carrying nuclear weapons in the territorial sea of international straits, it would contradict the non-nuclear principles of Japan, a subject which will be discussed later in this article.

2. Baselines

Before 1977, Japan had adopted and maintained the low-water line as the baseline of the territorial sea.

The Law on the Territorial Sea provided that the baseline shall be the low-water line and the straight line drawn across the mouth of or within a bay, or across the mouth of a river, except for the Seto Naikai (Inland Sea of Seto), where the baseline shall be the line prescribed by Cabinet Order as the boundary with other adjacent areas.[13]

---

[9] Minutes of the Agricultural and Forestry Committee, House of Representatives, 78th Session, No. 6, 26 October 1976.
[10] For this Law, see S. Yanai and K. Asomura, "Japan and the Emerging Order of the Sea – Two Maritime Laws of Japan", *Jap. Ann. Int'l L.* 21 (1977), 48 et seq.
[11] Article 1.
[12] Minutes of the Budget Committee, House of Representatives, 80th Session, No. 12, 23 February 1977, 5.
[13] United Nations, National Legislation and Treaties Relating to the Law of the Sea, UN Doc. ST/LEG/SER.B/19.56.

The Law on the Territorial Sea did not adopt the straight baseline system. This was unlike the Law on the Territorial Sea of the Republic of Korea of 1977, which adopted the straight baselines along its coast. When the Law on the Territorial Sea was drafted, it was considered unnecessary to adopt such a system.[14]

The Law on the Territorial Sea was amended in 1996. The title of the Law was also changed to the Law on the Territorial Sea and Contiguous Zone. The amended Law[15] adopted the straight baseline system in accordance with Article 7 of the UN Convention on the Law of the Sea. A government official commented in an article that the introduction of baselines was expected to bring about straighter and clearer outer limits of the territorial sea and benefit both foreign ships navigating near the Japanese coast and the law enforcement authorities of Japan, and put Japan on an equal footing with her neighbours such as the Republic of Korea, Russia and China.[16]

In 1997, several incidents occurred in which fishing vessels of the Republic of Korea were seized by Japanese patrol boats in Japan's enlarged territorial sea based on straight base lines or in hot pursuit beyond the territorial sea. In some of them, Korean fishermen were fined on summary verdicts. In two cases the Supreme Court upheld the High Court decisions. It ruled that the Korean vessels were operating in Japan's territorial sea and that the seizures were legal in spite of the defendant's arguments. The defendants had argued under the Japan-Korea Fisheries Agreement of 1965, which was in force at the time of the seizures, that only the flag State could enforce its laws beyond the 12-mile limit and that the seizures, therefore, were illegal.

3. Innocent Passage of Foreign Warships

As for the right of innocent passage of foreign warships in the territorial sea, Japan's position has been that foreign warships also have the right of innocent passage in the territorial sea of Japan. However, Japan has not regarded the passage of warships carrying nuclear weapons as innocent passage.

On 17 April 1968, the then Foreign Minister Takeo Miki summarised the position of Japan at the Foreign Affairs Committee of the Diet as follows:

---

[14] E. Sato, "Ryokai, 12 Kairi" ("The Territorial Sea: 12 Miles"), *Toki no Horei* 974 (1977), 9.
[15] United Nations, *Law of the Sea Bulletin* 35 (1997), 76.
[16] T. Takata, "The Conclusion by Japan of the United Nations Convention on the Law of the Sea (UNCLOS) and the Adjustment of the Maritime Legal Régime", *Jap. Ann. Int'l L.* 39 (1996), 127.

"... submarines of the Polaris type, or those normally equipped with similar nuclear weapons, cannot be regarded as falling within the category of ships whose passage can be innocent in the sense that it is not prejudicial to the peace, good order, or security of the coastal State".[17]

Japan's position that passage of warships carrying nuclear weapons cannot be regarded as innocent derives from Japan's three non-nuclear principles. In November 1971, a resolution was adopted at the Plenary Meeting of the House of Representatives of the Diet, entitled Resolution on Non-nuclear Weapons and Reduction of the US Military Bases in Okinawa. In the resolution, three principles were proclaimed, according to which Japan would neither manufacture nor maintain any nuclear weapons, nor would it allow them to be introduced into its territory. By this resolution, Japan hardened its principles, which have guided Japan's policy ever since. Its position on the passage of warships carrying nuclear weapons relates to the third of the non-nuclear principles.[18]

When the Diet considered the ratification of the UN Convention on the Law of the Sea, a government official said on behalf of the Government:

"... Article 19, paragraph 1 (of the UN Convention on the Law of the Sea), refers to non-prejudicial passage, and paragraph 2 of the same Article lists concrete activities which are considered prejudicial to the peace, good order or security of the coastal State.

Cases which are considered prejudicial to peace, good order or security are not limited to activities listed in Article 19, paragraph 2".

He continued as follows:

"Japan has taken the position that the passage of warships carrying nuclear weapons is not considered to be innocent passage, based on the provisions (Article 14, paragraph 4) of the Convention on the Territorial Sea and the Contiguous Zone. As the provision of Article 19, paragraph 1, of the UN Convention on the Law of the Sea is the same as that of the Convention on the Territorial Sea and the Contiguous Zone, Japan continues to take such a position".[19]

---

[17] Minutes of the Foreign Affairs Committee, House of Representatives, 58th Session, No. 12, 17 April 1968, 13; Oda and Owada, see note 2, 145 et seq.
[18] Yanai and Asomura, see note 10, 61 et seq.
[19] Minutes of the Foreign Affairs Committee, House of Representatives, 136th Session, No. 7-1, 14 May 1996, 28.

This means that after ratification of the UN Convention on the Law of the Sea, Japan's position concerning the passage of warships carrying nuclear weapons through the territorial sea is the same as its previous position. On this issue, Japan made its position clear in the 1960s, and its position has remained unchanged.

III. The Contiguous Zone

Japan did not at first establish a contiguous zone. When Japan became a party to the Territorial Sea Convention, it was not considered necessary to exercise jurisdiction concerning a contiguous zone in accordance with the Convention, although Japan had many smuggling and illegal immigration cases.

When the Law on the Territorial Sea was amended in 1996, a provision on the contiguous zone was added. Article 4 of the Law provides:

> "There is hereby established the contiguous zone, as a zone in which Japan takes necessary measures to prevent or punish infringement of its customs, fiscal, immigration or sanitary laws and regulations within its territory in accordance with Article 33, paragraph 1, of the UN Convention on the Law of the Sea".

According to a comment in an article by an official of the Japan Coast Guard, establishing a contiguous zone made it possible, in cases involving the smuggling of guns, drugs, etc., as well as illegal immigration, to spot, deal with and control the situation swiftly and properly before the suspects could enter the territorial sea of Japan.[20]

Thus, Japan changed its position concerning the contiguous zone. Japan had long thought it unnecessary to establish a contiguous zone. Now, like many other countries, Japan has a contiguous zone in accordance with the UN Convention on the Law of the Sea.

IV. The Exclusive Economic Zone

At the UN Sea-Bed Committee and the early sessions of the Third United Nations Conference on the Law of the Sea, Japan was not willing to accept the idea of a 200-mile resources jurisdiction or an exclusive economic zone. For example, at the 1973 spring session of the Sea-Bed Committee, the delegate of Japan said:

---

[20] K. Shimaya, "Ryokaiho no Ichibu o Kaiseisuru Horitsu" ("A Law to Amend the Law on the Territorial Sea"), *Toki no Horei* 1531 (1996), 12.

"We have difficulty with the exclusive zone proposals, first because a recognition of exclusive rights of the coastal States in a very extensive area of the Sea to a maximum of 200 miles, deprives other States of the right to participate in fishing in that area, and thus fails to take into account the legitimate interest of these States...

Secondly ... our difficulty with the exclusive zone proposals is that they would have a beneficiary effect for a rather limited number of countries which are in the advantageous geographical position of having very extensive and long coastlines or which adjoin the rich fishing grounds...

Thirdly, recognition of exclusive rights over fishery in an extensive area offshore would mitigate against sound conservation and management of the living resources of the sea...".[21]

When Japan began drafting a Law on the Territorial Sea in 1976, it had no intention of establishing a 200-mile zone. In the course of drafting the Law on the Territorial Sea, the Soviet Union established a 200-mile fishery zone. Japan then decided to establish a 200-mile fishing zone to be able to negotiate on fisheries talks on the same footing as that of the Soviet Union.

Japan established a 200-mile fishing zone in the Law on the Provisional Measures Relating to the Fishing Zone of 1977. However, Japan aimed at controlling only Russian fishing vessels. The Republic of Korea and China had not established a 200-mile zone at that time and fisheries relations with these countries were controlled by bilateral agreements which were based on flag State jurisdiction. Thus, Japan did not establish a 200-mile zone in the area of the Sea of Japan west of 135° East Longitude, the East China Sea, a part of the Pacific Ocean adjoined thereto and the Yellow Sea. Moreover, fishing control within the fishing zone was directed at Russian vessels. Korean and Chinese vessels were exempted from control.

When Japan ratified the UN Convention on the Law of the Sea, it established an exclusive economic zone around all the coasts of Japan in the Law Relating to the Exclusive Economic Zone and the Continental Shelf. By that time, many fishing vessels of Korea and China had come to the area just off the territorial sea of Japan and had had trouble with Japanese fishing vessels. In 1983, Korean trolling vessels, sea eel vessels, etc., were spotted off the San-in area (the Sea of Japan side of western Honshu).[22] Chinese vessels also operated in the East China Sea off Kyushu. Fishermen from the Sea of Japan side and the East China

---

[21] Law of the Sea Office, see note 7, 145 et seq.
[22] *Asahi Shimbun*, 25 June 1987.

Sea side had requested that the Government establish an exclusive economic zone all around Japan.

The Law Relating to the Exclusive Economic Zone and the Continental Shelf provides for the establishment of an exclusive economic zone in which Japan would exercise its sovereign rights and other rights as a coastal State as prescribed in Part V of the UN Convention on the Law of the Sea. Whereas in the case of the fishing zone, Japan exercised jurisdiction according to the Law on Provisional Measures Relating to the Fishing Zone, in the exclusive economic zone, Japan exercised "sovereign rights and other rights", although, in fact, even jurisdiction in the fishing zone was said to mean sovereign rights.[23]

The Law Relating to the Exclusive Economic Zone and the Continental Shelf provides that the laws and regulations of Japan shall apply with respect to exploring and exploiting, conserving and managing the natural resources, the establishment, construction, operation and use of artificial islands, installations and structures, and the protection and preservation of the marine environment and marine scientific research in the exclusive economic zone (and continental shelf).

This time, control of fisheries in the 200-mile zone is directed not only at Russian vessels but at other countries' vessels as well. Japan concluded fisheries agreements with China and the Republic of Korea in 1997 and 1998 respectively.

The Law on the Exercise of Sovereign Rights Concerning Fisheries in the Exclusive Economic Zone provides necessary measures for exercising sovereign rights over fisheries in the exclusive economic zone. This Law superseded the Law on the Provisional Measures Relating to the Fishing Zone. One of the important changes between these Laws is the deletion of the highly migratory species exception. Whereas the Law on Provisional Measures Relating to the Fishing Zone exempted foreigners from needing to obtain permission to catch highly migratory species, the new Law now claims jurisdiction over these species within the 200-mile zone, like other countries, including the United States, which included these species within its jurisdiction in an amendment, effective 1 January 1992, to the Magnuson Fishery Conservation and Management Act.

The Law Relating to the Preservation and Management of Marine Living Resources, which was enacted at the same time, has the purpose of (a) preserving and managing marine living resources in the exclusive economic zone through a programme to preserve and manage the said resources, (b) taking measures to control the allowable catch, along with

---

[23] S. Yoshida, "Nihyakkairijidai, Wagakuni no Gyogyousuiikinimo" ("An Age of 200-mile Zones – Japan Also Sets a 200-mile Zone"), *Toki no Horei* 974 (1977), 14.

the traditional measures, and (c) ensuring (at the same time) proper implementation of the UN Convention on the Law of the Sea.

Japan introduced the TAC system in this Law, in accordance with Article 61 of the UN Convention on the Law of the Sea.

Thus, Japan, which was against the idea of a 200-mile zone in the 1970s, now has an exclusive economic zone in accordance with the UN Convention on the Law of the Sea. One of the reasons is the world trend in establishing 200-mile zones. A second reason is the development of the fisheries of the Republic of Korea and China. A third reason, finally, is that Japan's fishing vessels have been expelled from other countries' 200-mile zones. It has therefore become important to Japan to preserve and manage fisheries resources in its own 200-mile zone.

V. The Continental Shelf

In April 1953, Japan and Australia resumed negotiations regarding a dispute over pearling on the continental shelf in the Arafura Sea, which had been suspended during World War II. In August 1953, Australia suspended negotiations and issued a continental shelf proclamation in September 1953, in which it claimed sovereign rights over the sea-bed and subsoil of its continental shelf for the purpose of exploring and exploiting the natural resources.[24]

Thereafter, Japan and Australia agreed to submit the dispute to the International Court of Justice, but in view of the global trend to exercise sovereign rights over exploring and exploiting resources of the continental shelf, Japan gave up the submission of the dispute to the Court.

Japan was not a party to the 1958 Convention on the Continental Shelf. The reason was that the Convention included sedentary species in the continental shelf resources, whereas Japan's position was that some crabs, such as king crabs, were not continental shelf resources but high seas resources.

However, with the development of the continental shelf régime in international law, Japan had been engaged in the exploration and exploitation of the mineral resources of the continental shelf, such as oil and gas, on the understanding that the coastal State had exclusive rights to explore and exploit mineral resources of the continental shelf.

It is also on this understanding that Japan concluded, in 1974, an agreement with the Republic of Korea concerning the joint development of oil and gas resources in the continental shelf of the East China Sea.[25]

---

[24] S. V. Scott, "The Inclusion of Sedentary Fisheries Within the Continental Shelf Doctrine", *ICLQ* 41 (1992), 798.

[25] For this agreement, see M. Miyoshi, "The Joint Development of Offshore Oil and Gas in Relation to Maritime Boundary Delimitation", in: International

In a case concerning the imposition of tax under the Corporation Tax Law on a foreign corporation that was drilling for oil in the continental shelf of Japan, the Tokyo High Court, supporting the decision of the Tokyo District Court of 22 April 1982, said on 14 March 1984:

> "According to established rules of customary international law, sovereign rights as a natural extension of Japan's territorial sovereignty extend over the continental shelf adjacent to Japan's coast as far as exploration and exploitation of mineral resources of the seabed and subsoil are concerned, even though Japan had not yet ratified the Convention on the Continental Shelf during the years concerned. It is a matter of course that exploration and exploitation of mineral resources and related activities are under Japanese jurisdiction and control, and sovereign rights include the power to impose tax on income from those activities. Accordingly, the continental shelf adjacent to Japan's coast has become 'an enforcement area' of the Corporation Tax Law according to the established rules of customary international law".[26]

In this decision, the High Court approved the imposition of tax under the Corporation Tax Law on foreign corporations.

The Law on the Exclusive Economic Zone and the Continental Shelf of 1996 provides that the continental shelf over which Japan exercises its sovereign and other rights as a coastal State in accordance with the UN Convention on the Law of the Sea comprises the sea-bed and subsoil of zones (1) ... (2) ... .[27] It seems to follow from this provision that Japan accepted Part VI of the Convention, which contains a provision concerning sedentary species, although in practice it does not matter within the 200-mile zone whether maritime living resources belong to exclusive economic or continental shelf resources.

VI. Control of Marine Pollution

Japan, as a maritime State, has been very active in the control of marine pollution.[28]

Boundaries Research Unit (University of Durham, England), *Maritime Briefing* 2 (1999), 12 et seq.

[26] Judgment of 14 March 1984, Tokyo High Court, Case No. *(gyo-ko)* 43 of 1982, Appeal from Tokyo District Court (Third Division of Civil Affairs), Judgment of 22 April 1982, *Jap. Ann. Int'l L.* 28 (1985), 202 et seq.

[27] Article 2.

[28] C. Mizukami, "The Law and Practice Relating to Marine Pollution in Japan", *Jap. Ann. Int'l L.* 29 (1986), 65 et seq.

Japan acceded to the 1954 International Convention for the Prevention of Pollution of the Sea by Oil, as amended, on 21 August 1967. To implement the Convention, the Law for the Prevention of Pollution of the Sea by Oil Discharged from Ships was enacted on the same day.

In 1970, the Law was replaced by the Marine Pollution Prevention Law. The 1976 amendment of this Law after the Yuyo Maru incident contained provisions concerning the prevention of maritime disasters and changed the title of the Law to the Law Relating to the Prevention of Marine Pollution and Maritime Disasters.

On 6 April 1971, Japan acceded to the 1969 International Convention Relating to Intervention on the High Seas in Cases of Oil Pollution Casualties and amended the Law Relating to the Prevention of Marine Pollution and Maritime Disasters.

On 9 June 1983, Japan acceded to the 1978 Protocol Relating to the International Convention for the Prevention of Pollution from Ships of 1973 (MARPOL 73/78). To implement the provisions of this Convention, the Law Relating to the Prevention of Marine Pollution and Maritime Disasters was amended. Japan has revised the Law in accordance with amendments to the Protocol.

Japan acceded to the 1972 London Convention on 15 October 1980. The Law Relating to the Prevention of Marine Pollution and Maritime Disasters and the Waste Disposal and Cleaning Law were amended to give effect to the Convention.

As for civil liability for oil pollution damage, Japan ratified the 1969 International Convention on Civil Liability for Oil Pollution Damage (Private Law Convention) on 3 June 1976, and the 1971 International Convention on the Establishment of an International Fund for Compensation for Oil Pollution Damage (Fund Convention) on 7 July 1976. To give effect to them, Japan enacted the Law on Liability for Oil Pollution Damage.

On 24 August 1994, Japan acceded to the 1992 International Convention on Civil Liability for Oil Pollution Damage, which, *inter alia*, sets a higher limit for ship owner's liability than the 1969 Convention. On that same day, it also acceded to the 1992 International Convention on the Establishment of an International Fund for Compensation for Oil Pollution Damage, which sets a higher limit of compensation than the 1971 Fund Convention.

As for coastal State jurisdiction, Japan's position has changed during its membership of the UN Sea-Bed Committee. At the 1973 summer session of the Committee, Japan submitted a proposal on enforcement measures by coastal States for the purpose of preventing marine pollution. According to the proposal, a coastal State could investigate and prosecute in relation to the discharge or dumping of any harmful substance which is proven to be in violation of international rules and standards and which occurs within a zone beyond the outer limit of the territorial sea and extending a certain number of miles from shore. In view of Japan's previous position, which was very passive with regard to the adjacent water concept, this proposal was indeed unprecedented.[29]

The Law on the Exclusive Economic Zone and the Continental Shelf of 1966 established the exclusive economic zone in accordance with the UN Convention on the Law of the Sea and provides that the laws and regulations will apply with respect to certain matters, including the protection and preservation of the marine environment in the exclusive economic zone and on the continental shelf. Thus Japan now controls marine pollution in the framework of the UN Convention on the Law of the Sea. Application of the Law Relating to the Prevention of Marine Pollution and Maritime Disasters in the exclusive economic zone is adjusted by the Cabinet Order Relating to the Adjustment of the Application of the Law Relating to the Prevention of Marine Pollution and Maritime Disasters.

VII. Scientific Research

Japan has supported the freedom of scientific research on the high seas, like the United States, the United Kingdom and other western countries. At the 1972 summer session of the UN Sea-Bed Committee, the delegate of Japan (Professor Shigeru Oda) said:

> "How could research within the national jurisdiction be detrimental to the interests of the coastal State? On the contrary, it would seem that such research could only benefit the coastal State. The inclination of the coastal States to require consent for research within its jurisdiction probably finds its roots in the suspicion that research by foreigners will lead to exploitation of the resources. But this suspicion is probably groundless, since it is hardly possible to exploit either mineral or living resources under the guise of research.
>
> I am convinced that in such a régime where too many permissions, approvals, inspections and other constraints are imposed upon the liberal activities of scientific research, what we would be losing would be very much larger than what we would be gaining".[30]

---

[29] S. Oda, *The Law of the Sea in Our Time – I. New Developments 1966-1975*, 1977, 222.
[30] Law of the Sea Office, see note 7, 131 et seq.

Before the establishment of the exclusive economic zone, Japan's position concerning marine scientific research on its continental shelf and in its 200-mile fishing zone was that foreigners were free to undertake scientific research, except when its purpose was to explore or exploit non-living natural resources on the continental shelf,[31] or when it involved the catching and removal of marine animals and plants in the fishing zone.[32] Japan requested the research States to submit information concerning their programme for marine scientific research in its surrounding waters in accordance with the provisional guidelines.

The Law on the Exclusive Economic Zone and the Continental Shelf provides in Article 3 that laws and regulations are applied to certain activities in the exclusive economic zone, including scientific research. However, Japan thus far has no laws or regulations concerning scientific research in the exclusive economic zone. Japan only has guidelines for conducting marine scientific research in areas under national jurisdiction.[33]

The guidelines provide that foreign nationals or foreign institutions, whether private or public, are requested to obtain the consent of the Government of Japan prior to conducting marine scientific research in the territorial sea, in the exclusive economic zone or on the continental shelf of Japan. The guidelines also provide that those foreign nationals or institutions planning to conduct marine scientific research are requested to submit to the Ministry of Foreign Affairs of Japan, through diplomatic channels, a formal request for consent together with the application form provided in the annex at least six months before the expected starting date of research activities.

The guidelines further provide that when the research project involves catching, removal or exploration of marine animals and/or plants in the exclusive economic zone of Japan, a separate approval from the Ministry of Agriculture, Forestry and Fisheries of Japan shall also be necessary, in accordance with the Law on the Exercise of Sovereign Rights Concerning Fisheries in the Exclusive Economic Zone.

These guidelines are in line with the relevant provisions of Part 13 of the UN Convention on the Law of the Sea. They are only guidelines and not a law and have no provisions for sanctions in cases of infringement.

Chinese research ships have come to Japan's exclusive economic zone to conduct maritime activities. The Japanese Government has repeatedly urged China to exercise restraint over increasing instances of Chinese research and naval activities conducted in Japan's zone. On 29

---

[31] The Mining Law, Article 7. In practice, this Law is interpreted to apply to the continental shelf of Japan.
[32] The Law on Provisional Measures Relating to the Fishing Zone, Article 9.
[33] United Nations, *Law of the Sea Bulletin* 33 (1997), 38 et seq.

August 2000, the Foreign Ministers of Japan and China agreed to create a mutual notification system to resolve disputes over Chinese research activities in Japan's exclusive economic zone and the agreement was confirmed at working-level talks in September.

Thus, Japan exercises jurisdiction concerning marine scientific research in the exclusive economic zone and on the continental shelf, unlike the United States, which does not assert jurisdiction in its exclusive economic zone over marine scientific research.[34] However, Japan, for the moment, is not ready to fully control it.

VIII. High Seas Fisheries

Japan has made the most of fishing on the high seas, but at the same time, it has paid attention to preservation and management of fisheries resources on the high seas. In fact, Japan has joined many high seas fisheries treaties. It recently took a position which attached more emphasis to the preservation and management of fisheries resources than to the maintenance of the freedom of fisheries on the high seas.

For example, in the negotiations which led to the conclusion of the Convention on the Conservation and Management of Pollock Resources in the Central Bering Sea (1994), Japan at first took a position giving priority to maintenance of freedom of fishing on the high seas. However, before the second meeting, Japan proposed active fisheries' controls based on each country's own judgment, thus changing its high seas fisheries policy to one of attaching importance to preservation and management of high seas resources based on scientific research.[35]

Japan cooperated with international efforts to manage and preserve fisheries resources on the high seas. For example, it responded swiftly to the recommendation of the General Assembly of the United Nations concerning large-scale drift net fishing in the Pacific Ocean. Japanese fishing vessels had been engaged in drift net fishing of neon flying squid in the North Pacific region since the end of the 1970s, and had been fishing for albacore in the South Pacific Ocean since the 1980s. Drift net fishing by Japan and other countries caused concern, especially in Canada and the United States and the island countries in the South Pacific Ocean. In December 1989, the General Assembly of the United Nations adopted a resolution recommending a moratorium on "large-scale pelagic drift net fishing" by 30 June 1992, applicable to all areas of

---

[34] The reason is that the United States has an interest in encouraging marine scientific research and promoting its maximum freedom while avoiding unnecessary burdens. See J. Ashley Roach and Robert W. Smith, *United States Responses to Excessive Maritime Claims*, 2nd ed., 1996, 437.

[35] *Nihonkeizai Shimbun*, 22 July 1991.

the high seas, and cessation of large-scale pelagic drift net fishing activities in the South Pacific region by 1 July 1991. On 17 July 1990, the Japanese Government decided to suspend drift net fishing for the 1990-1991 season. In 1990, Japan joined the United States to propose a resolution recommending a complete end to drift net fishing of the world's oceans and seas by 31 December 1992. Japan stopped all large-scale drift net fishing on the high seas by the end of 1992. Another example occurred when the UN Food and Agriculture Organization adopted an Action Plan in 1999 which, among other things, recommended that members of the FAO reduce the number of long-distance tuna long liners. Japan swiftly responded by reducing the number of the liners.[36]

IX. Conclusion

Japan's policy on the law of the sea has placed importance on the freedom of the sea. When Japan ratified the Territorial Sea Convention and the Convention on the High Seas in 1968, its policy on the law of the sea changed little. In the 1970s, Japan's position began to change gradually in concert with the general change in the law of the sea. During and after the Third UN Conference on the Law of the Sea, Japan greatly changed its policy. The main reasons for the changes were (a) the worldwide trend of extending the territorial sea to 12 miles and establishing exclusive economic zones, and (b) the fact that fishing vessels of neighbouring countries, namely, the Soviet Union (since 1991 the Russian Federation), the Republic of Korea, and China, began to operate off Japan's coast. Now Japan's policy on the law of the sea is in line with the UN Convention on the Law of the Sea.

---

[36] *Gyogyo Hakusyo (White Paper on Fisheries) 1999*, 2000, 96.

The Continental Shelf: Interplay of Law and Science

L. D. M. Nelson

I. Definition

The 1982 Convention on the Law of the Sea defines the continental shelf as comprising "the seabed and subsoil of the submarine areas that extend beyond its territorial sea throughout the natural prolongation of its land territory to the outer edge of the continental margin, or to a distance of 200 nautical miles from the baselines from which the breadth of the territorial sea is measured where the outer edge of the continental margin does not extend up to that distance" (Article 76, paragraph 1). This definition embraces two elements: the first based on the notion of natural prolongation and the other on the distance criterion.

The continental margin has been described as comprising "the submerged prolongation of the land mass of the coastal State", and as consisting "of the seabed and subsoil of the shelf, the slope and the rise" (natural features well known to geo-scientists.) The continental margin, the Convention proceeds to declare "does not include the deep ocean floor with its oceanic ridges or the subsoil thereof" (Article 76, paragraph 3). Here the Convention seeks to make a distinction between what may be termed "continental territory" and "oceanic territory", a distinction which, in not a few cases, is far from clear-cut. The point has been made that "there is not in fact a neat division between the two features but there is rather a zone where it is quite difficult to say with certainty 'this' is oceanic, and 'that' is continental".[1] Thus this

---

[1] A. J. Kerr and M. J. Keen, "Hydrographic and Geologic Concerns of Implementing Article 76", *International Hydrographic Review* 62 (1985), 139 et seq., 141. The Commission on the Limits of the Continental Shelf itself noted that "continental crust is compositionally distinct from oceanic crust, but the boundary between these two crustal types may not be clearly defined. Simple subdivision of margins into shelf, slope and rise may not always exist owing to the variety of geological and geomorphological continental margin types resulting from different tectonic and geological

apparently simple provision masks a fundamental problem with respect to the interpretation and application of the provisions of the Convention dealing with the continental shelf beyond the 200-mile limit.

A coastal State may establish the outer edge of the continental margin when it extends beyond the 200-mile mark by either:

"(i) a line delineated in accordance with paragraph 7 by reference to the outermost fixed points at each of which the thickness of sedimentary rocks is at least 1 per cent of the shortest distance from such point to the foot of the continental slope; [Irish formula or Gardiner line] or

(ii) a line delineated in accordance with paragraph 7 by reference to fixed points not more than 60 nautical miles from the foot of the continental slope [Hedberg formula]".[2]

These two complex formulae have brought a certain measure of precision to the process of determining the boundary between maritime areas under national jurisdiction and the international seabed area. But, as we shall see, the interpretation and application of these formulae are not free from difficulties, especially with regard to the location of the foot of the slope and the complex issues arising from the implementation of the sediment test. These relate, *inter alia*, to the identification of the sediment/basement interface, the calculation of sediment thickness and the variability of sediment distribution.

Paragraphs 5 and 6 of Article 76 impose certain constraints on the operation of paragraph 4.[3] Under paragraph 5 the fixed points drawn in

---

settings". Scientific and Technical Guidelines of the Commission on the Limits of the Continental Shelf, CLCS/11 of 13 May 1999, 45, para. 6.2.4.

[2] Article 76, para. 4(a). Hedberg considered the creation of a boundary zone an "essential feature" of his formula, since the foot of the slope was not defined sharply enough to serve as the boundary. He had suggested that the standard width of the boundary zone should be no less than 100 kilometres (54 nautical miles). H. D. Hedberg, "Ocean Boundaries and Petroleum Resources", *Science* 191 (1976), 1009 et seq., 1011.

[3] In 1985, Chile and Ecuador both made claims to areas of continental shelf beyond 200 miles. Chile declared that its sovereignty over the continental shelves of Easter Island and Sala y Gomez Island in the Pacific Ocean extended up to a distance of 350 nautical miles, measured from the baselines from which their respective territorial seas were measured. Chile's claim is based on Article 76, para. 6, which stipulates, *inter alia*, that "notwithstanding the provisions of paragraph 5, on submarine ridges, the outer limit of the continental shelf shall not exceed 350 nautical miles". Ecuador relied on the operation of the criterion in Article 76, para. 5, delineating the limits of its continental shelf up to a distance of 100 miles from the 2,500 metre isobath in the seabed and subsoil located between its continental territorial sea and its insular territorial sea around the Galapagos

accordance with the Irish formula or the Hedberg formula may not extend beyond 350 nautical miles from the baseline of the territorial sea nor go beyond 100 nautical miles from the 2,500 metre isobath which is "a line connecting the depth of 2,500 metres". The first constraint is based purely on a distance criterion, whereas the second is based on a depth-cum-distance criterion.

By virtue of paragraph 6 of Article 76 the depth-cum-distance criterion contained in paragraph 5 may not be used on submarine ridges – the limit on such ridges being fixed at 350 nautical miles. This exception does not apply to "submarine elevations" that are natural components of the continental margin, such as "its plateaux, rises, caps, banks and spurs". It is generally acknowledged that this provision does not lend itself to easy interpretation. For instance, what is the distinction, if any, between the ridges referred to in paragraph 3 and the ridges and submarine elevations mentioned in paragraph 6?[4]

II. The Commission on the Limits of the Continental Shelf

The Commission on the Limits of the Continental Shelf is the institution most directly concerned with the business of establishing the limits of the continental shelf beyond 200 miles, i.e., the limits of the outer continental shelf. Its broad mission is "to act as a watchdog to prevent excessive coastal state claims".[5]

Islands. Both these claims drew protests from the United States on the grounds that Article 76, para. 5, or Article 76, para. 6, may, "only be invoked if the conditions ... precedent in Article 76(4) are fulfilled". In both cases, the US protests made clear that these conditions cannot be fulfilled. The Federal Republic of Germany also protested these claims in similar terms. However, it introduced two further points which are of some interest: (i) that the 1982 Convention on the Law of the Sea was not yet in force, and (ii) that the Commission on the Limits of the Continental Shelf had not yet been established. T. Treves, "Codification du droit international et pratique des états dans le droit de la mer", *RdC* 223 (1990), 96 et seq.; J. A. Roach and R. W. Smith, *United States Responses to Excessive Maritime Claims*, 1996, 205 et seq.

[4] It has been observed that paragraph 6 had no known nor "accepted applicability". R. W. Smith and G. Taft, "Legal Aspects of the Continental Shelf", in P. J. Cook and C. M. Carleton (eds.), *Continental Shelf Limits – The Scientific and Legal Interface*, 2000, 20. See also text accompanying footnotes 34-39.

[5] P. R. R. Gardiner, "Reasons and Methods for Fixing the Outer Limit of the Legal Continental Shelf Beyond 200 Nautical Miles", *Revue iranienne des relations internationales* 11-12 (1978), 145 et seq., 161; J.-F. Pulvenis, "The Continental Shelf Definition and Rules Applicable to Resources", in R.-J.

1. Membership

The Commission consists of 21 members who are "experts in the field of geology, geophysics, or hydrography". They are elected by States Parties from among their nationals, "having due regard to the need to ensure equitable geographical representation". They serve in a personal capacity. It must be noted that the Convention does not provide for the participation of legal experts in the Commission, a seemingly intentional omission.[6] In defence of this omission, Oxman has observed that "the Commission is not a court, and legal expertise is not included among the express qualifications of its members".[7] This omission was, in this writer's view, unwarranted, given the fact that one of the cardinal functions of the Commission must necessarily be to interpret or apply the relevant provisions of the Convention – an essentially legal task.[8] There has been some criticism of the absence of legal expertise in the Commission. Brown has, for example, stated that "a further cause for concern is that the Commission's membership does not include a lawyer. Given the fact that its principal task is to make recommendations on the basis of a complex legal instrument, this seems rather unfortunate".[9]

It is the State Party that has nominated a member of the Commission which will defray the expenses of that member while in performance of his duties (Annex II, Article 2, paragraph 5). It has been suggested that the fact that the expenses of members of the Commission are defrayed by their governments may compromise the independence of the members

---

Dupuy and D. Vignes (eds.), *A Handbook on the New Law of the Sea*, Vol. I, 1991, 359 et seq.

[6] The Draft United Nations Convention on the International Seabed Area, submitted as long ago as 1970 by the United States to the Seabed Committee, likewise did not provide for the participation of legal experts in what was then called the International Seabed Boundary Review Commission. See Article 45 of UN Doc. A/AC.138/25 of 3 August 1970, reproduced in S. Oda, *The International Law of the Ocean Development: Basic Documents*, 1972, 83.

[7] B. H. Oxman, "Third United Nations Conference on the Law of the Sea: The Ninth Session (1980)", *AJIL* 75 (1981), 211 et seq., 230. See also UN Division for Ocean Affairs and the Law of the Sea, Office of Legal Affairs, *Definition of the Continental Shelf: An Examination of the Relevant Provisions of the United Nations Convention on the Law of the Sea*, UN Sales No. E.93.V.16, 29, para. 87 (hereinafter, "*Law of the Sea, Definition of the Continental Shelf*").

[8] See Section III *infra*.

[9] See E. D. Brown, *Sea-bed Energy and Minerals: The International Legal Regime*, Vol. I: *The Continental Shelf*, 1992, 31; S. Karagiannis, "Observations sur la Commission des Limites du Plateau Continental", *Espaces et ressources maritimes* 8 (1994), 163 et seq., 167-168.

of the Commission.[10] This may not be an unfair criticism. The 21 members of the Commission were elected at the Sixth Meeting of the States Parties held in March 1997.

2. Functions

The prime function of the Commission, as set out in the Convention, is to consider "the data and other material" submitted by coastal States concerning the outer limits of the continental shelf where those limits extend beyond 200 miles and to make recommendations on their establishment. "The limits of the shelf established by a coastal State on the basis of these recommendations shall be final and binding".[11] What if the coastal State disagrees with the recommendations of the Commission? The Convention specifically provides that in such a situation the coastal State shall, within a *reasonable* time, make a revised or new submission to the Commission (Annex II, Article 8). The next step is not specified in the Convention and is as a consequence uncertain. Presumably this process can and perhaps should be continued until some accommodation is reached. It is certain however, that the Commission has not been granted the power to submit any dispute concerning the outer limit of a coastal State's continental shelf to any court or tribunal. Questions regarding a possible appeals procedure and the relationship between the Commission with "the proposed dispute settlement procedures under the new Convention" were raised in the Evensen Group but the matter never seems to have gone further.[12]

It may be useful here to recall that during the course of the prolonged and difficult negotiations on Article 76, the expression "taking into account these recommendations" was replaced by the words

---

[10] See Uruguay, *Off. Rec. Third UN Conference on the Law of the* Sea, XIII, 35-36, para. 50; Karagiannis, see note 9, 169-170.

[11] Article 76, para. 8; Annex II, Article 3, para. 1(a).

[12] Fourth Revision of the Evensen Text on the Continental Shelf, dated 6 May 1975, on which see *The United Nations Convention on the Law of the Sea 1982: A Commentary*, Vol. II, 1993, 850. It may be remembered that the 1970 Draft United Nations Convention on the International Seabed Area had indeed given the International Seabed Boundary Review Commission the power to form such a submission. Article 45 reads in part as follows: "The International Seabed Boundary Review Commission shall: (a) Review the delineation of boundaries submitted by Contracting Parties in accordance with Articles 1 and 26 to see that they conform to the provisions of this Convention, negotiate any differences with Contracting Parties, and if these differences are not resolved initiate proceedings before the Tribunal...". Oda, see note 6.

"on the basis of these recommendations" (Article 76, paragraph 8). Several wide-margined coastal States were unhappy with this amendment.[13] It was viewed as an attempt to erode the sovereign rights of coastal States. Indeed the United Kingdom had proposed a formal amendment seeking to reverse the change.[14] That amendment was later withdrawn. The Commission has not been given the power to impose its recommendations on the coastal State thus determining the outer limit of the continental shelves of coastal States.[15] That is why it seems important that both the Commission and the coastal State concerned should strive to the maximum extent possible to reach some accommodation.

Besides making recommendations to coastal States on the establishment of their continental shelves, the Commission has other functions which are of no less importance. It is also the business of the Commission to provide scientific and technical advice, if requested by the coastal State concerned during the preparation of the relevant data (Annex II, Article 3, paragraph 1(b)).[16] It will also, in the nature of things, provide useful guidance on the interpretation and application of the technical requirements to be found in Article 76, which would surely influence the practice of both parties and non-parties.

---

[13] Canada, for instance, declared that: "... The commission is primarily an instrument which will provide the international community with reassurances that coastal States will establish their continental shelf limits in strict accordance with the provisions of Article 76. It has never been intended, nor should it be intended, as a means to impose on coastal States limits that differ from those already recognized in Article 76. Thus to suggest that the coastal States limits shall be established 'on the basis' of the commission's recommendations rather than on the basis of Article 76, could be interpreted as giving the commission the function and power to determine the outer limits of the continental shelf of a coastal State". *Off. Rec. Third UN Conference on the Law of the Sea*, XIII, 102, para. 15. To the same effect, see also Australia *ibid.*, 33, para. 13; Pakistan *ibid.*, 21, para. 140; United Kingdom *ibid.*, 25, para. 15; Venezuela *ibid.*, 20, para. 136, and Uruguay *ibid.*, 35-36, para. 50. The Canadian statement seems, to this writer, to disregard a significant function (though unexpressed) of the Commission – to interpret and apply the provisions of Article 76, discussed in Section III *infra*.

[14] For the amendment, see UN Doc. A/CONF.62/L.126 of 13 April 1982, *Off. Rec. Third UN Conference on the Law of the Sea*, XVI, 233.

[15] *Law of the Sea, Definition of the Continental Shelf*, 29, para. 86.

[16] A trust fund has been established to help developing countries in the preparation of relevant data and information for submission to the Commission on the Limits of the Continental Shelf. See Annex II to GA Resolution A/55/L.10 of 20 October 2000.

III. The Role of the Commission in the Interpretation and Application of the Convention

An important function of the Commission which is not expressly referred to in the Convention has to do with the interpretation or application of the Convention. Recent developments in the Commission have thrown this role into high relief. It can be argued that the Scientific and Technical Guidelines of the Commission on the Limits of the Continental Shelf adopted on 13 May 1999 (CLCS/11) represents in essence an interpretation of the Convention – in particular of Article 76.

The Commission is aware of its role in the interpretation and application of the relevant provisions of the Convention. It has made the following observations with respect to the Guidelines and they are quoted *in extenso*:

> "... the Commission aims also to clarify its interpretation of scientific, technical and legal terms contained in the Convention. Clarification is required in particular because the Convention makes use of scientific terms in a legal context which at times departs significantly from accepted scientific definitions and terminology.[17] In other cases, clarification is required because various terms in the Convention might be left open to several possible and equally acceptable interpretations. It is also possible that it may not have been felt necessary at the time of the Third United Nations Conference on the Law of the Sea to determine the precise definition of various scientific and technical terms. In still other cases, the need for clarification arises as a result of the complexity of several provisions and the potential scientific and technical difficulties which might be encountered by States in making a single unequivocal interpretation of each of them".[18]

This may be termed a comprehensive exposition of what the Commission is all about.

The Commission is engaged in clarifying the ambiguities in Article 76 and giving a precise definition of the meaning and scope of that provision. The well-known definition of interpretation given by the Permanent Court of Justice will be recalled: "... the Court is of the opinion that the expression 'to construe' (*interprétation* in French) must be understood as meaning to give a precise definition of the meaning and scope" of a legal instrument.[19] The process involves an interplay

---

[17] As an example, see text accompanying footnotes 24 ff.
[18] CLCS/11, 7.
[19] "Interpretation of Judgments Nos. 7 and 8 (The Chorzów Factory)" PCIJ, Series A, Judgment No. 11, 10.

between law and science. It is all the more surprising that legal expertise was deliberately not included in the membership of the Commission.

Of course, in the nature of things, differences or conflicts may arise over the interpretation or application of certain provisions of Article 76 between the Commission and the coastal State concerned. It must be stated that these Guidelines are not legally binding but they come close to being an authoritative interpretation of the technical provisions found in Article 76. It was to be expected that conflicts of interpretation will arise over such difficult issues as (i) the location of the foot of the slope; (ii) the sediment thickness test; (iii) the selection of the 2,500 metre isobath; and (iv) ridges. Some reactions to the Draft Provisional Scientific and Technical Guidelines prepared by the Commission have already revealed differences emerging on these questions.

1. The Location of the Foot of the Slope

The foot of the slope is a primary feature in the delimitation of the continental shelf beyond the 200-mile limit. Both the Irish and the Hedberg formulae use the foot of the continental slope as "the reference baseline" to set the limits of the outer edge of the continental margin (Article 76, paragraph 4(a)(i) and (ii)). The location of the foot of the continental slope is therefore of prime importance.

Paragraph 4(b) of Article 76 deals with this question. It states that "in the absence of evidence to the contrary, the foot of the continental slope shall be determined as the point of maximum change in the gradient at its base". The Commission interprets this provision as consisting of (i) a general rule whereby the foot of the slope "coincides" with the point of maximum change in the gradient at its base[20]; and (ii) an exception to this general rule contained in the phrase "in the absence of evidence to the contrary...".[21] In the implementation of the general rule set out in paragraph 4(b), the Commission stated that it "will be guided by bathymetric, geomorphologic, geologic and geophysical sources of evidence".[22]

It is noteworthy that the Commission construes the exception as complementing the general rule. "The complementary character of this provision [i.e., the determination of the foot of the continental slope

---

[20] The French version of para. 4(b) reads as follows: "Sauf preuve du contraire, le pied du talus continental coïncide avec la rupture de pente la plus marquée à la base du talus".

[21] CLCS/11, 43, para. 6.1.1

[22] *Ibid.*, 37, para. 5.1.4.

when evidence contrary to the general rule is invoked] is emphasised by the fact that in addition to bathymetric and geomorphological evidence, all other necessary and sufficient geological and geophysical evidence must also be included as part of a submission by a coastal State".[23]

In other words, the Commission appears to be requiring from the coastal State not only bathymetric and geomorphological evidence, but also geological and geophysical evidence. Such a requirement may well result in an increased financial burden on States, especially developing States, preparing data and information on the definition of the continental shelf.

The Commission itself has pointed out that Article 76 makes use of scientific terms in a legal context which at times departs significantly from accepted scientific definitions and terminology. The evolution of the concept of the continental margin provides an example. "Current scientific knowledge about the nature and extent of the continental margin has evolved greatly from its original definition". That definition at its inception was based on a geomorphological concept. It is now broadened to include "many additional geological and geophysical concepts within the framework provided by plate tectonics".[24]

This raises a general question concerning the interpretation and application of Article 76. Must the scientific and technical terms contained in Article 76 be interpreted in the light of today's scientific and technical developments (which have advanced so rapidly since the adoption of the Convention) or must regard be had to the state of the science some 20 or more years ago? The concepts embodied in these provisions were *not intended* to be static but by their very nature evolutionary.[25] The Commission itself has noted that "it is also possible that it may not have been felt necessary at the time of the Third United Nations Conference on the Law of the Sea to determine the precise definition of various scientific and technical terms".[26] It seems right that these Guidelines should necessarily take the evolution of these scientific and technical terms into account.[27]

---

[23] *Ibid.*, 43, para. 6.1.3.
[24] CLCS/11, 44, para. 6.1.6.
[25] Cf. *Legal Consequences for States of the Continued Presence of South Africa in Namibia (South West Africa) notwithstanding Security Council Resolution 276 (1970)*, ICJ Reports 1971, 16 et seq., 31, para. 53.
[26] CLCS/11, 7.
[27] It must also be borne in mind that though new technological developments will give a more accurate description of the sea floor, they are costly techniques which may be beyond the financial capacity of some States, especially those which have already based their claims on less accurate standards.

2. Sediment Thickness

The Irish formula contains the rule regarding sediment thickness. It provides that the limits of the outer continental shelf may be delineated by a line connecting fixed points at each of which the thickness of sedimentary rocks is at least one per cent of the shortest distance from such point to the foot of the continental slope.

The Commission has invoked a principle of continuity in the interpretation or application (implementation) of this provision. It consequently requires "documentation of the continuity between the sediments" at each of the fixed points and "the sediments at the foot of the continental slope". The Commission is not prepared to accept a calculation of average distribution of sediments as a solution to the problem of ragged topography.[28]

This principle of continuity seems to presuppose a fairly even deposit of sediments forming an ideal sedimentary wedge, but as has been remarked a "serious ambiguity arises from the assumption that the thickness of sediment tapers seaward, and that it eventually diminishes to 1% of the distance back to the foot of slope. In fact, many seismic records portray a rugged basement where there are several locations at which the sediment thickness equals 1% of the distance back to the foot of the slope".[29] In short, sediment thickness depends on the topography of the basement surface.

Thus, this principle of continuity may pose a problem in the implementation of the Irish formula in the case of coastal States with complex and irregular topography.

3. The Selection of the 2,500 Metre Isobath

Paragraph 5 of Article 76 imposes a constraint on the operation of paragraph 4 in the sense that it limits the extent of the continental margin to 350 nautical miles or to a distance of 100 miles from the 2,500 metre isobath. Where there are multiple repetitions of the 2,500 metre isobath (created by geological and tectonic processes) the Guidelines recommend, unless there is evidence to the contrary, "the use of the *first 2,500 m isobath* from the baselines from which the breadth of the territorial sea is measured that conforms to the general configuration of the continental margin".[30]

---

[28] CLCS/11, 67, paras. 8.5.3-8.5.4.
[29] Ron MacNab, "Initial Assessment", in P. J. Cook and C. M. Carleton (eds.), *Continental Shelf Limits – The Scientific and Legal Interface*, 2000, 259.
[30] Emphasis added. CLCS/11, 36, para. 4.4.2.

The selection of the first 2,500 metre isobath has been criticised. It has been contended that:

"some coastal States may feel this introduces an additional requirement or restriction not supported by article 76, and that it may result in a loss of parts of their natural prolongation that are a natural component of their margin".[31]

"An example of this situation is provided by the Joey Rise off western Australia, which is part of the volcanic margin province that formed around the margins of the Exmouth Plateau during breakup. A small, isolated closure of the 2500-m isobath on the Joey Rise associated with a volcanic buildup produces a significant protrusion of the isobath cutoff. It can be readily argued that the use of such isolated closures is within the meaning of article 76 as long as they occur on features that are natural components of the continental margin".[32]

There is much to be said for treating such situations on a case-by-case basis, given the uniqueness of each particular situation.

4. Ridges

Paragraph 6 of Article 76 was intended to act as a brake on the extension of the continental shelf, in that the distance-cum-depth criterion (i.e., the 100 miles from the 2,500 metre isobath) did not apply on submarine ridges and as a result only the distance criterion applied.[33] As has already been pointed out, the difficulty with this paragraph lies in the fact that the relationship between "submarine ridges" referred to in paragraph 6 and the "oceanic ridges" referred to in paragraph 3 is not clear.

The Guidelines of the Commission relating to ridges have stirred some controversy. The Commission is of the opinion that the provisions

---

[31] The addition to the text of the words "natural components of the continental margin, such as its plateaux, rises, caps, banks and spurs" was in fact based on an Australian proposal. See *The United Nations Convention on the Law of the Sea: A Commentary*, Vol. II, 1993, 868; R. Platzöder (ed.), *Third United Nations Conference on the Law of the Sea: Documents*, IV, 524; UN Doc. A/CONF.62/L.51 of 29 March 1980, *Off. Rec. Third UN Conference on the Law of the Sea*, XIII, 82-83, paras. 4-7.

[32] P. A. Symonds et al., "Characteristics of Continental Margins", in Cook and Carleton, see note 29, 25 et seq., 58.

[33] A list of the various categories of ridges can be found in *Law of the Sea, Definition of the Continental Shelf*, at 21-22.

in paragraphs 3 and 6 may cause some difficulties in defining ridges on the basis of the origin of the ridges and the composition. It remarked that "some ridges (including active spreading ridges) may have islands on them. In such cases it would be difficult to consider that those parts of the ridge belong to the deep ocean floor".[34]

With respect to this statement the following questions have been posed: "How can an island on an oceanic ridge of the deep ocean floor change the character of the ridge from either a legal or a scientific perspective? ... Can it be argued that the ridge is a submarine ridge to the 350-M limit and an oceanic ridge beyond?"[35]

The question of ridges was the subject of particularly intense consultations and negotiations and indeed was only resolved during the latter stages of the Conference by a compromise put forward by the Chairman of the Second Committee.[36] It may be useful to recall here that certain States made what can best be described as interpretative statements with respect to these proposals on ridges. These statements were as follows:

Denmark:

> "His delegation interpreted the concept to mean submarine elevations that belong to fundamentally the same geological structure as the land territory of the coastal State in question and would support paragraph 5*bis* only if that interpretation applied".[37]

---

[34] CLCS/11, 54, para. 7.2.8.

[35] P. A. Symonds et al., "Ridge Issues", in Cook and Carleton, see note 29, 285 et seq., 303; V. E. McKelvey, "Interpretation of the UNCLOS III Definition of the Continental Shelf", in D. M. Johnston and N. G. Letalik (eds.), *The Law of the Sea and Ocean Industry: New Opportunities and Restraints*, 1984, 465 et seq., 469-470. *Per contra* see P. A. Verlaan, "New Seafloor Mapping Technology and Article 76 of the 1982 United Nations Convention on the Law of the Sea", *Marine Policy* 21 (1997), 425 et seq., 427-428.

[36] The compromise proposals read, *inter alia*, as follows: "Amend the last sentence in paragraph 3 of Article 76 to read as follows: 'It does not include the deep ocean floor with its oceanic ridges or the subsoil thereof'. Add a new paragraph 5*bis* [now paragraph 6] to read as follows: 'Notwithstanding the provisions of paragraph 5, on submarine ridges the outer limit of the continental shelf shall not exceed 350 miles from the baselines from which the breadth of the territorial sea is measured. This paragraph does not apply to submarine elevations that are natural components of the continental margin, such as its plateaux, rises, caps, banks and spurs". A/CONF.62/L.51 of 29 March 1980, *Off. Rec. Third UN Conference on the Law of the Sea*, XIII, 82.

[37] *Ibid.*, 17, para. 96.

Iceland:

> "He understood that the new provision regarding submarine ridges meant that the 350-mile limit criterion would apply to ridges which were a prolongation of the land mass of the coastal State concerned".[38]

United States:

> "His delegation's support for the proposal on the continental shelf contained in the report of the Chairman of the Second Committee (A/CONF.62/L.51) rested on the understanding that it was recognized – and, to the best of his knowledge, there was no contrary interpretation – that features such as the Chukchi plateau situated to the north of Alaska and its component elevations could not be considered a ridge and were covered by the last sentence of the proposed paragraph 5*bis* of article 76".[39]

The interpretative statements made during plenary sessions of the Third United Nations Conference, with no dissenting voices, seem to form part of the *travaux préparatoires* of Article 76, especially of paragraphs 3 and 6 of that article. They must at least play a role in the interpretation of these difficult provisions. Of course it was not to be expected that such considerations, being of a primarily legal nature, could have been raised by the Commission.

---

[38] *Ibid.*, 36, para. 58. On 9 May 1985 Iceland issued regulations concerning the delimitation of the Icelandic continental shelf to the west, south and east. These regulations which covered continental shelf areas beyond the 200-mile limit were stated to be based on Article 76 of the 1982 Convention, where applicable. For instance, a certain segment on the limits of the continental shelf reads as follows: "[S]egment FHG is defined by the 350 nautical mile distance limit from Iceland. The Continental Shelf boundary, if defined on the basis of the Foot of the Slope, extends beyond 350 nautical miles. But since in this area it lies on the Reykjanes submarine ridge the boundary of the Continental Shelf is limited to the 350 nautical mile distance from Iceland by Article 76". Regulation No. 196 of 9 May 1985 concerning the delimitation of the continental shelf to the west, south and east, reprinted in UN, *National Legislation on the Continental Shelf*, 1989, 127 et seq., 128. On 19 June 1985, the United Kingdom protested these regulations on the grounds that they had "no basis in international law", being "a misrepresentation of the United Nations Convention on the Law of the Sea, and of existing law". See *BYIL* 56 (1985), 494.

[39] *Off. Rec. Third UN Conference on the Law of the Sea*, XIII, 43, para. 156.

IV. Submission of Data and Information by Coastal States

A coastal State which intends to establish, in accordance with Article 76, the outer limits of its continental shelf beyond 200 nautical miles is under an obligation to submit data and information on such limits within ten years of the entry into force of the Convention for that State (Annex II, Article 4).[40] States Parties to the Convention are evidently bound by this provision. Such an obligation will also bind those States Parties which, before the entry into force of the Convention and therefore before the establishment of the Commission on the Limits of the Continental Shelf, had begun implementing the terms of Article 76. These would include (i) States which have made unilateral claims to continental shelves beyond 200 miles[41] and (ii) those that have concluded maritime boundary agreements covering seabed areas which extend beyond the 200 nautical mile limit.[42] These agreements extend the delimitation lines

---

[40] At the Eleventh Meeting of States Parties held in May 2001, it was decided that "(a) In the case of a State Party for which the Convention entered into force before 13 May 1999, it is understood that the ten-year time period referred to in Article 4 of Annex II to the Convention shall be taken to have commenced on 13 May 1999; (b) The general issue of the ability of States, particularly developing States, to fulfil the requirements of Article 4 of Annex II to the Convention be kept under review". (SPLOS 72) It should be noted that 13 May 1999 was the date of adoption of the Scientific and Technical Guidelines.

[41] For example, Chile: Declaration of 15 September 1985, reprinted in UN, *National Legislation on the Continental Shelf*, 1989, 62; Ecuador: Proclamation of 19 September 1985, *ibid.*, 82 (see also note 3 *supra*); Iceland: Regulation No. 196 of 9 May 1985 concerning the delimitation of the continental shelf to the west, south and east, *ibid.*, 127 et seq. (see also note 38 *supra*); Republic of Madagascar: Ordinance No. 85-013 Determining the Limits of the Maritime Zones (Territorial Sea, Continental Shelf and Exclusive Economic Zone) of 16 September 1985, *ibid.*, 151 et seq.

[42] Examples are: (i) Agreement on Maritime Delimitation between the Government of Australia and the Government of the French Republic (New Caledonia) of 1982, reprinted in J. Charney and L. M. Alexander (eds.), *International Maritime Boundaries*, Vol. I, 1993, 911-913; (ii) Agreement on Maritime Delimitation between the Government of Australia (Heard/McDonald Islands) and the Government of the French Republic (Kerguelen Island) of 1982, *ibid.*, Vol. II, 1192-1194; (iii) Agreement between the Government of Solomon Islands and the Government of Australia Establishing Certain Sea and Seabed Boundaries of 1988, *ibid.*, Vol. I, 983-984; (iv) Agreement between the Government of the United Kingdom of Great Britain and Northern Ireland and the Government of the Republic of Ireland Concerning the Delimitation of Areas of the Continental Shelf between the Two Countries of 1988, *ibid.*, Vol. II, 1774-1779; (v) Treaty between the Republic of Trinidad and Tobago and the Republic of Venezuela on the Delimitation of Marine and Submarine Areas of 1990,

beyond 200 miles and they seem to take account of the requirements of Article 76. Take for instance these observations on the UK - Ireland Agreement: "Article 4 keeps open the question of the outer limits of the respective shelves. Point 94 was chosen according to the criterion of the foot of the slope plus 60 n.m.; Point 132 according to that of the 2500 meters isobath plus 100 n.m., thereby taking into account Article 76 of the UN Convention on the Law of the Sea".[43]

It has been observed that the 1990 Treaty between Trinidad and Tobago and Venezuela "may be the first agreement in the world whereas [sic] the edge of the margin was calculated on the basis of the thickness of the sedimentary rocks as equal to 1 per cent of the shortest distance from the slope".[44] This Agreement treats the 350-mile mark as limiting the continental margin claims of Trinidad and Tobago and Venezuela (Article 2 of the Agreement).

A question may arise whether non-parties are under a similar obligation. There is probably no such obligation. This matter was in fact raised in the Commission. The fact that the term "a coastal State" is used in Annex II, Article 4, has led the Commission to seek clarification from the meeting of States Parties whether the term "coastal State" as specified in Annex II, Article 4, included a non-State Party. This clarification was considered necessary for the application of Rule 44 of the Rules of Procedure dealing with a submission by a coastal State.[45] It was felt that the Meeting of States Parties did not have the competence to provide such a legal interpretation and it was preferable not to pursue the matter any further. The Commission should request the Legal Counsel for an advisory opinion only if the actual need arises. The Commission took note and accepted this recommendation.[46]

---

ibid., Vol. I, 685-689; (vi) Agreement between the United States of America and the Union of the Soviet Socialist Republics on the Maritime Boundary of 1990, *ibid.*, 455-460; (vii) Treaty between the Government of the United States of America and the Government of the United Mexican States on the Delimitation of the Continental Shelf in the Western Gulf of Mexico beyond 200 Nautical Miles of 9 June 2000 (entered into force on 17 January 2001). This Treaty covers enclaves of high seas in the Gulf of Mexico, the seabed of which extends beyond the 200-mile limit. It has been suggested that "the coastal States concerned will probably have to confirm with the Commission that these areas form part of their respective continental shelves in accordance with Article 76". See C. M. Carleton, "Delimitation Issues", in P. J. Cook and C. M. Carleton (eds.), *Continental Shelf Limits – The Scientific and Legal Interface*, 2000, 312 et seq., 317.

[43] J. Charney and L. M. Alexander (eds.), *International Maritime Boundaries*, Vol. II, 1993, 1770.

[44] *Ibid.*, Vol. I, 681.

[45] Report of the Eighth Meeting of States Parties, SPLOS/31 of 4 June 1998, 12, paras. 51-52.

[46] Report of the Secretary-General, A/53/456 of 5 October 1998, 13, para. 67.

However, as has been pointed out, a non-party should be able to submit its data to the Commission for "confirmation".[47]

V. Disputes

The procedure for the consideration of the submission made by a coastal State is designed to ensure that the recommendation made by the Commission is acceptable to the coastal State concerned. For example, the coastal State which has made a submission is entitled to "send its representatives to participate in the relevant proceedings without the right to vote".[48] Disputes may nevertheless arise when the Commission and the coastal State cannot, in spite of all efforts, reach agreement on the extent of that State's continental margin.[49]

How can such disputes be resolved?[50] As has already been noted, the Commission has not been granted the power to submit any dispute concerning the determination of the limit of the outer continental shelf to the dispute settlement procedures provided for in the Convention.

The drafters of the Convention did not give the Authority any right to participate in the determination of the common boundary between the international seabed area and the outer limit of the continental shelf.[51]

---

[47] See further T. A. Clingan, Jr., "The Law of the Sea in Perspective: Problems of States not Parties to the Law of the Sea Treaty", *GYIL* 30 (1987), 101 et seq., 112.

[48] Annex II, Article 5 and Rule 51 of the Rules of Procedure of the Commission on the Limits of the Continental Shelf. Oxman has remarked "it is assumed that, in practice, both the coastal state and the subcommission will do their best to ensure that the Commission will not approve, even initial recommendations, that are likely to be rejected by the coastal state". Oxman, see note 7, 231.

[49] For issues that may give rise to disputes, see Section III(1-4) *supra*.

[50] The writer has dealt with this matter in more detail in his article "Claims to the Continental Shelf Beyond the 200-mile Limit", in V. Götz, P. Selmer, R. Wolfrum (eds.), *Liber amicorum Günther Jaenicke – zum 85. Geburtstag*, 1998, 573 et seq.

[51] "The process is not adversarial, and the International Seabed Authority plays no part in determining the outer limit of the continental shelf. Ultimate responsibility for delimitation lies with the coastal State itself, subject to safeguards against exaggerated claims". United States: President's Transmittal of the United Nations Convention on the Law of the Sea and the Agreement relating to the Implementation of Part XI to the US Senate with Commentary, 7 October 1994, *ILM* 34 (1995) 1393 et seq. 1427. See F. H. Paolillo, "The Institutional Arrangements for the International Seabed and their Impact on the Evolution of International Organizations", *RdC* 188 (1984), 139 et seq., 190 et seq.; S. Mahmoudi, *The Law of Deep Seabed Mining*, 1987, 77 et seq.; A. Pardo, "Before and After", *Law and Contemp. Probs.* 46 (1983), 95 et seq., 103; A. Pardo, "The Convention on the Law of

The business of the Authority is to organise and control activities in the Area,⁵² i.e., the exploration and exploitation of the resources of the Area. The competence of the Authority is limited. It does not possess the *locus standi* to mount a legal action with respect to any dispute concerning the outer limits of the continental shelf.

As to the Seabed Disputes Chamber, since its jurisdiction is confined to disputes with respect to activities in the Area (Article 187, preamble), it will most probably be beyond its competence to give an advisory opinion on matters relating to the establishment of the limits of the outer continental shelf.⁵³

Does a third State Party have the right to take legal action challenging another State Party's outer continental shelf claims which are not based on the recommendations of the Commission? Has each State Party a right to institute proceedings to vindicate the international community interest in the preservation of the international seabed area which forms part of the common heritage of mankind – a kind of *actio popularis* or *actio communis*?⁵⁴

It will be recalled that in 1966 the International Court of Justice famously responded to this question. It stated as follows: "the argument amounts to a plea that the Court should allow the equivalent of an *'actio popularis'*, or right resident in any member of a community to take legal action in vindication of a public interest. But although a right of this kind may be known to certain municipal systems of law, it is not known to international law as it stands at present".⁵⁵ It seems doubtful to this writer whether any State Party has the competence to take action to vindicate this international collective interest unless it can be shown that its interests had been harmed by these claims. Such a collective interest

---

the Sea: A Preliminary Appraisal", *San Diego L. Rev.* 20 (1982/1983), 489 et seq., 499; Vincente Marotta Rangel, "Le plateau continental dans la Convention de 1982 sur le droit de la mer", *RdC* 194 (1985), 273 et seq., 364.

⁵² Annex, section 1, of the 1994 Implementation Agreement.
⁵³ The Seabed Disputes Chamber is empowered to give advisory opinions under Articles 191 and 151, para. 10.
⁵⁴ See, among others, R. Jennings and S. Watts (eds.), *Oppenheim's International Law*, Vol. I 1993, 5; I. Brownlie, "General Course of Public International Law", *RdC* 255 (1995), 13 et seq., 88; C. Gray, *Judicial Remedies in International Law*, 1990, 211-215; M. Ragazzi, *The Concept of International Obligations Erga Omnes*, 1997, 210-214.
⁵⁵ ICJ Reports 1966, 6 et seq., 47. It may be remarked, however, that international law seems to be ineluctably moving towards the acceptance of this concept. The observation has been made that "there can be little doubt that there is an incremental progression towards an *actio popularis*". I. Brownlie, "General Course of Public International Law", *RdC* 255 (1995), 13 et seq., 88.

is left to be protected by the Authority and the Commission neither of which, as has been stated above, appears to have the capacity to take legal action with respect to the determination of the limits of the outer continental shelf.

There seems to be a need to give judicial protection to régimes such as the deep seabed régime which have been created, in the words of the Convention, "for the benefit of mankind as a whole" (Article 140). Thus States Parties which are not directly affected by the outer continental shelf claims of other States Parties may be accorded the right to take public action (*actio popularis*) to protect the integrity of the international seabed area. It may also be considered whether the International Seabed Authority should not be granted the right to institute legal proceedings in such matters on behalf of the international community.[56]

VI. The Commission and Delimitation Disputes

The Convention states that "the actions of the Commission shall not prejudice matters relating to the delimitation of boundaries between States with opposite or adjacent coasts".[57] The Commission must therefore steer clear from dealing with submissions that may prejudice matters relating to the delimitation of boundaries between States.

The Commission has drafted rules of procedure precisely to deal with this problem.[58] First, the Commission expressly recognises that the competence with respect to matters regarding disputes which may arise in connection with the establishment of the outer limits of the continental shelf rests with States. Secondly, where there is a dispute on the delimitation of the continental shelf between States with opposite or adjacent coastlines, or in other cases of unresolved land or maritime disputes, the coastal State making a submission must inform the Commission of the dispute and must ensure to the extent possible that the submission will not prejudice matters relating to the delimitation of boundaries between States. Thirdly, two or more States are entitled to make joint or separate submissions by agreement. Fourthly, perhaps most importantly, "in cases where a land or maritime dispute exists, the

---

[56] Lauterpacht has suggested (in the context of international environmental protection) that a suitable international organisation be accorded "the duty and capacity to commence legal proceedings on behalf of the international community generally against the wrongdoing State". See E. Lauterpacht, *Aspects of the Administration of International Justice*, 1991, 62.

[57] Annex II, Article 9. See also Article 76, para. 10.

[58] Rule 44 and Annex I of the Rules of Procedure of the Commission, adopted on 4 September 1998 (CLCS/3/Rev.2).

Commission shall not examine and qualify a submission made by any of the States concerned in the dispute. However, the Commission may examine one or more submissions in the areas under dispute with prior consent given by all States that are parties to such a dispute".[59] Of course, there remains the basic duty of the Commission itself that its own actions do not prejudice matters relating to the delimitation of boundaries between States.

The work of the Commission will nevertheless have effects, albeit indirect, on the delimitation of the outer continental shelves between States with opposite or adjacent coasts. International tribunals dealing with such maritime boundary disputes will surely be guided by the Scientific and Technical Guidelines of the Commission with respect to the interpretation and application of the relevant provisions of the Convention. In this regard the actual practice of the Commission in dealing with submissions will also be a source of guidance for international tribunals. Of course it is reasonable to expect that States – parties and non-parties to the Convention – will rely on the work of the Commission in negotiating outer continental shelf boundaries.

VII. Conclusions

The Commission, in preparing the Scientific and Technical Guidelines, has had to operate at the interface between law and science. It was faced with a legal provision (Article 76) utilising a host of scientific and technical terms. The task of the Guidelines was to interpret and apply these scientific terms within the context of the law. In that sense the Commission has been engaged in interpreting and applying a key provision of the Convention on the Law of the Sea – a provision which has as its object the establishment of the common boundary between areas falling under national jurisdiction and the international seabed area. The Commission is now ready to receive submissions from coastal States and to provide scientific and technical advice that States preparing submissions may require. The Commission, in applying its Guidelines, should take very much into account the need for "negotiating differences" which may arise between itself and the coastal State concerned so as to avoid disputes for which there is as yet no clear forum for their resolution.

---

[59] Annex I, para. 5(a), Rules of Procedure of the Commission.

Waves Within and Outside the Law of the Sea: Traversing Gaps, Ambiguities and Priorities

Jordan J. Paust

I. Introduction

The United Nations Convention on the Law of the Sea[1] is a fairly comprehensive treaty addressing numerous aspects of the law of the sea, but, like nearly all international legal instruments, it is not without gaps and ambiguities. The preamble implicitly admits this common aspect of international agreements while affirming: "matters not regulated by this Convention continue to be governed by the rules and principles of general international law." Of course, customary international law is a necessary background for interpretation of any treaty[2] and will be useful in determining whether and how a given matter is regulated by the Convention. Further, it would not be unusual to discover unavoidable clashes between the Convention and other international agreements. Relevant international agreements, like customary international law, will be useful for identification and clarification of the content of the 1982 Convention,[3] but in some instances a clash between agreements may be unavoidable. In the latter case, there is a need to identify which agreement will have priority. Article 311(2) is not helpful in this regard, since it merely states that the Convention "shall not alter the rights and obligations ... which arise from other agreements compatible with this Convention,"[4] and does not address the circumstance of an unavoidable clash between agreements. In a sense, gaps and ambiguities in the Convention can be supplemented by waves of meaning outside and

---

[1] UN Doc. A/CONF.62/122 of 10 December 1982 (hereinafter, UNCLOS).
[2] See, e.g., Vienna Convention on the Law of Treaties, Art. 31(3)(c), UNTS Vol. 1155, 331; J. J. Paust, J. M. Fitzpatrick and J. M. Van Dyke, *International Law and Litigation in the United States*, 2000, 38 et seq., 58-59, 131, 395, 487, 500.
[3] Vienna Convention on the Law of Treaties, see note 2, Art. 31(3)(a).
[4] See also UNCLOS, Art. 293(1).

without the Convention, and some normative waves within and without the Convention will have a priority as overriding waves.

II. Gaps, Ambiguities and Normative Waves

1. Ports and Vessels in Distress

One gap in coverage under the Law of the Sea Convention involves entry of vessels in distress into a port of a foreign coastal State. Article 18(2) of the Convention recognizes the right of foreign flag vessels in distress to stop and anchor in another State's territorial sea, but most ports will be in internal waters of a coastal State and, thus, implicitly within the sovereign control of the coastal State,[5] subject to limits in customary or treaty-based international law. Since customary international law recognizes the right of foreign flag vessels in distress to enter a coastal State's port,[6] the gap or ambiguity in the Convention concerning entry in distress is supplemented by such a customary norm.

2. Pollution Controls in Territorial Seas

Another interpretive problem involves coastal State competence to regulate foreign vessel pollution within its territorial sea. Article 19(1) contains the general standard with respect to non-innocent passage, i.e., such passage is not "innocent" if in a given context it is "prejudicial to the peace, good order or security of the coastal State." Article 19(2)(h) expressly recognizes that passage "shall be considered to be prejudicial" if a foreign flag vessel engages in "[a]ny act of willful and serious pollution, contrary to this Convention." Does paragraph 2(h) set the limit concerning coastal State competence to regulate foreign vessel pollution, i.e., must such pollution be "willful and serious"? Does paragraph 1 provide a potentially even greater competence (e.g., covering other pollution when it is prejudicial to the "good order" or "security" of a coastal State), while the activities listed in paragraph 2 are merely examples of conduct that under all circumstances "shall be considered to be prejudicial"? Further, does Article 21(1)(f) provide guidance when it recognizes a broad competence of coastal States to adopt laws and regulations for the "preservation of the environment of the coastal State and the prevention, reduction and control of pollution thereof," given

---

[5] See *ibid.*, Art. 2(1).
[6] See, e.g., Hoff, Administratrix *(United States v. United Mexican States)*, General Claims Commission, 1929, in 4 UN Rep. Int'l Arb. Awards 444; M. S. McDougal and W. T. Burke, *The Public Order of the Oceans*, 1962, 110.

that paragraph 1 of that article contains a conditioning phrase "in conformity with the provisions of this Convention and other rules of international law"? Does Article 21(1)(h), recognizing broad coastal State competence to create "sanitary" regulations, supplement coastal State competence concerning pollution controls when relevant pollution interferes with "sanitary" controls? Does Article 2(1)-(2), which recognizes general "sovereignty" of the coastal State in its territorial sea, air space over the territorial sea, and the bed and subsoil thereof, provide a general norm that provides guidance in case of any ambiguity?[7]

If coastal State competence to control pollution within its exclusive economic zone (or beyond) is broader (e.g., if it is not limited to cases of "willful and serious" pollution[8]), it seems logical to conclude that in areas closer to its shores, where the coastal State generally has greater competence, a coastal State must have at least the competence that it possesses in areas farther from its shores, unless such competence is expressly excluded. If so, should a principle of inclusion govern interpretation of various articles to allow inclusion in areas closer to shore of competencies that pertain in areas farther from a coastal State's shores?[9] Concerning coastal State criminal enforcement competence within the territorial sea, it is stated in Article 27(5) that the coastal State may not arrest any person in connection with any crime committed before the ship entered the territorial sea if the ship is merely passing through, "[e]xcept as provided in Part XII or with respect to violations of laws and regulations adopted in accordance with Part V." Part V concerns rights and obligations with respect to the exclusive economic zone, including coastal State jurisdiction with regard to "protection and preservation of the marine environment,"[10] and Part XII concerns rights and obligations more generally with respect to protection and preservation of the marine environment. Thus, some resource-related or environmentally-related enforcement powers within the territorial sea pertain with respect to violations of appropriate laws and regulations applicable outside the territorial sea. Article 211(4), in Part XII, allows coastal States, "in the exercise of their sovereignty within their territorial sea, [to] adopt laws and regulations for the prevention, reduction and control of marine pollution from foreign vessels, including vessels exercising the right of innocent passage," but such laws and regulations "shall ... not hamper innocent passage of foreign vessels." Thus, even application of Part XII shifts one back to the question of what constitutes

---

[7] Cf. UNCLOS, Art. 2(3) (such "sovereignty ... is exercised subject to this Convention and to other rules of international law").
[8] See Section 7 below; UNCLOS, Arts. 56(1)(a), (b)(iii), 73(1), 220(5)-(6), 230(1).
[9] See also section 7 below.
[10] UNCLOS, Art. 56(1)(b)(iii).

non-innocent passage with respect to pollution within the territorial sea. Apparently, Article 27(5) still conditions Article 211(4) in the sense that coastal States retain the right to exercise criminal enforcement jurisdiction over vessels polluting sea areas beyond the territorial sea. Also, under Article 1(1)(4), pollution of the marine environment covered, for example, in Part XII includes "the introduction by man, directly or indirectly, of substances or energy ... which results or is likely to result in" certain deleterious effects. Thus, the standard concerning controllable pollution found within Article 1 can involve negligent pollution and is not limited to "willful and serious" pollution addressed in Article 19(2)(h).

Article 220(2) complicates the general issue. It states that where there are clear grounds for believing that a vessel

> "navigating in the territorial sea of a State has, during its passage therein, violated laws and regulations of that State adopted in accordance with this Convention or applicable international rules and standards for the prevention, reduction and control of pollution from vessels, that State, without prejudice to the application of the relevant provisions of Part II, section 3, may undertake physical inspection of the vessel ... and may, where the evidence so warrants institute proceedings, including detention of the vessel..."[11]

Apparently, a coastal State could adopt laws and regulations addressing pollution covered under the standard set forth in Article 1(1)(4), but enforcement would have to be "without prejudice" to the relevant provisions in Part II, section 3, which include Articles 19 and 25 regarding innocent and non-innocent passage, 21(1)(f), and 27. Perhaps the broader standard regarding regulable pollution set forth in Article 1 at least conditions the meaning of Articles 19(1) and 21(1)(f) to allow regulation within the territorial sea of less than willful and serious pollution.[12] This seems to follow from Article 230(2), which allows "[m]onetary penalties only" regarding pollution of the marine environment committed by foreign vessels in the territorial sea "except in the case of a willful and serious act of pollution in the territorial sea." Article 230(2) thereby recognizes that controllable pollution within the territorial sea does not have to be "willful and serious," although penalties are limited to "[m]onetary penalties only" with respect to negligent or less than serious pollution.

---

[11] Unless section 7 applies, concerning bonding or other appropriate financial security and the need to allow vessels covered to proceed.

[12] See also *Restatement of the Foreign Relations Law of the United States*, 3rd ed., 1987, §604, comment d.

3. Drug Smuggling in or through Territorial Seas

An example of supplementation of Article 19 with other articles of the Convention involves interpretation of coastal State competence to regulate drug smuggling within its territorial sea. Article 19 is generally silent concerning drug smuggling, although drug smuggling will normally be contrary to the "good order" if not "security" of a coastal State within the meaning of paragraph 1 and, thus, vessels engaged in such conduct alone or with others would be in non-innocent passage. Further, paragraph 2(g) will cover some smuggling (as non-innocent passage) when it involves "[t]he embarking or disembarking of any commodity, currency or person contrary to the customs ... regulations of the coastal State." In both cases, enforcement efforts against such non-innocent passage will be permissible under Article 25(1) when steps taken are "necessary" under the circumstances.[13] Additionally, Article 21(1)(h) recognizes coastal State prescriptive competence over "customs, fiscal, immigration or sanitary" matters, and "customs" competence can certainly include regulation of drug smuggling. It is evident, therefore, that the list of non-innocent passage activities in Article 19(2) is not exclusive, that a general standard is contained in Article 19(1), and that the general standard can be supplemented by reference to other articles in the Convention, perhaps also including the general retained "sovereignty" of the coastal State recognized in Article 2(1).

Article 27(1)(d) is even more specific with respect to enforcement jurisdiction, recognizing coastal State criminal enforcement competence with respect to measures "necessary for the suppression of illicit traffic in narcotic drugs or psychotropic substances ... in connection with any crime committed on board the ship during its passage." In fact, Article 27(1)(d) provides an enforcement competence broader than that concerned with actual or attempted smuggling, since the crime involving illicit traffic does not have to involve an attempt to smuggle relevant drugs into the coastal State. A principle of inclusion would also recognize coastal State enforcement competencies in its territorial sea like those listed within Article 33 with respect to the contiguous zone and, thus, that, within its territorial sea, a coastal State can enforce controls necessary to prevent and punish infringement of its customs regulations within its territory or territorial sea. Apparently, Article 27(1)(b) will also allow coastal State criminal enforcement competence

---

[13] Cf. L. Sohn and K. Gustafson, *The Law of the Sea*, 1984, 99, stating that Article 25 arguably includes a "unilateral right to verify the innocent character of passage." If so, is the right to stop a vessel for verification governed by a "reasonable" suspicion, "clear grounds for believing" (see Arts. 220(2), 226), or "necessary steps" (see Art. 25) standard?

with respect to all matters listed under Article 33, since crimes in violation of a coastal State's customs, fiscal, immigration or sanitary regulations would be "of a kind to disturb the peace of the country or the good order of its territorial sea."[14]

Civil enforcement competencies within the territorial sea under Article 28 are more limited[15] and perhaps could be supplemented by inclusion of competencies under Article 33 under a principle of inclusion. With respect to civil enforcement of fiscal, immigration or sanitary regulations, however, Article 19(1)'s "good order" and "security" provision should cover the interests listed in Article 33. Article 19(2)(g) should also cover such interests whenever there is an "embarking or disembarking of any commodity, currency or person contrary to the customs, fiscal, immigration or sanitary regulations of the coastal State." In both cases, "necessary" steps could be taken by the coastal State under Article 25(1) to prevent such forms of non-innocent passage.

Some smuggling into the coastal State may not even involve "passage" within the meaning of Article 18, since one definition of "passage" in Article 18(1)(a) involves traversing the territorial sea "without entering internal waters or calling at a roadstead or port facility outside internal waters." Yet, smuggling could involve "passage" and the use of other boats to transfer drugs to shore, "passage" and the dropping of drug packets into territorial waters for others to pick up and bring to shore, and other forms of smuggling while "proceeding to or from internal waters or a call at such roadstead or port facility" within the meaning of Article 18(1)(b).

4. Nuclear-Powered Ships and Ships with Nuclear Weapons in Territorial Seas

A few coastal States have claimed a competence to exclude nuclear-powered ships or ships with nuclear weaponry from their territorial sea as ships engaged in inherently non-innocent passage.[16] It seems obvious, however, that the Convention recognizes that passage of such vessels is not inherently non-innocent. Article 22(2) of the Convention allows coastal States to create sea lanes within their territorial seas for the

---

[14] UNCLOS, Art. 27(1)(a) also applies "[i]f the consequences of the crime" actually "extend to the coastal State."

[15] Article 28(2) allows certain enforcement powers against a foreign ship with respect to the "obligations or liabilities assumed or incurred by the ship itself," some of which may arise from coastal regulations permitted under Article 21(1)(h) and (4).

[16] For example, Paust, Fitzpatrick, Van Dyke, see note 2, 758 (New Zealand).

passage of "nuclear-powered ships and ships carrying nuclear or other inherently dangerous or noxious substances or materials" and to confine such ships to such sea lanes, but the obvious implication is that such ships cannot be excluded altogether from a territorial sea. Article 23 is nearly explicit on this point, since it recognizes the duty of such ships to "carry documents and observe special precautionary measures established for such ships by international agreements ... when exercising their right of innocent passage." Clearly, the latter phrase assumes that such ships can engage in innocent passage under appropriate circumstances and, thus, that their passage is not inherently non-innocent. This implication is also affirmed, for example, in the Netherlands declaration upon ratification and in a 1989 US-USSR agreement.[17]

5. Crimes Committed Prior to Entry into Territorial Seas

Article 27(5) of the Convention on the Law of the Sea seems to preclude the exercise of coastal State enforcement jurisdiction over foreign flag vessels passing through a coastal State's territorial sea with respect to "any crime committed before the ship entered the territorial sea," since the coastal State "may not take any steps on board" such vessels "to arrest any person or to conduct any investigation in connection with" such crimes.[18] Sovereign enforcement jurisdiction within the territorial sea recognized under Article 2(1) thus seems limited under Article 27(5) whether or not the ship is engaged in innocent passage and whether or not a clash in concurrent jurisdictional competencies between a coastal State with territorial jurisdiction over its territorial waters and a flag State with the equivalent of territorial jurisdiction over a ship flying its flag[19] should generally be resolved in favor of the coastal State. However, are certain coastal State territorial enforcement competencies assured by overriding customary international law or by other treaties with respect to certain crimes even when the crimes were committed prior to entry of a foreign flag vessel in a territorial sea?

---

[17] *Ibid.*, 773.
[18] But see McDougal and Burke, note 6, 294. From a policy perspective, "[w]hen criminal acts have been committed outside the territorial" sea, "the importance of the alleged offense seems to be critical. If the crime is generally regarded as a serious one ... or one for which, perhaps, any State might be competent to apply authority, the coastal State should be authorized to act despite the interference with navigation." *Ibid.*, 300, 302 ("crimes of considerable importance").
[19] Concerning flag territorial jurisdiction, see Paust, Fitzpatrick and Van Dyke, note 2, 69, 404, 427, 467-471, 493, 626, 744-751 (including attention to *The SS Lotus* case *(France* v. *Turkey)*, PCIJ Series A, No. 10 (1927) ("assimilated to ... territory")), 758, 792-793, 804-814.

The question is significant in the case of alleged international crimes.[20] Under customary international law, States have universal jurisdiction[21] and universal responsibility *aut dedere aut judicare*[22] to

---

[20] See *ibid.*, 773-774.
[21] See *ibid.*, 447-460; J. J. Paust, M. C. Bassiouni et al., *International Criminal Law*, 2nd ed., 2000, 132-134, 157-176.
[22] See, e.g., Paust, Fitzpatrick and Van Dyke, note 2, 129, 133, 447, 453, 456-458, 730-731; Paust, Bassiouni et al., see note 21, 132-136, 140-147, 170-171, 175, *passim*; M. C. Bassiouni, E. M. Wise, *Aut Dedere Aut Judicare: The Duty to Prosecute or Extradite*, 1995; J. J. Paust, *International Law as Law of the United States*, 1996, 405-412; *US Dep't of Army Field Manual 27-10, The Law of Land Warfare*, 1956, 181, para. 506(b) (re: customary responsibility concerning war crimes, including violations of the Geneva Conventions); British War Office, *Manual of Military Law, The Law of War on Land*, Vol. 3, 1958, 95 n. 2 (no refuge for war crimes is possible in any State bound by the Geneva Conventions); W. Winthrop, *Military Law and Precedents*, Washington, DC: US GPO, 1920, 796 (duty to punish war crimes); *Kent's Commentary on International Law*, 1866, 3, 427 (same); L. A. Steven, "Genocide and the Duty to Extradite or Prosecute: Why the United States is in Breach of Its International Legal Obligations," *Va. J. Int'l L.* 39 (1999) 425 et seq.; Opinion of the Magistrate in *Ex parte Pinochet*, Belgian Tribunal of First Instance (8 November 1998), quoted in Paust, Bassiouni, et al., see note 21, 140-41; *Case Concerning Questions of Interpretation and Application of the 1971 Montreal Convention Arising from the Aerial Incident at Lockerbie (Libyan Arab Jamahiriya v. United States of America)*, Provisional Measures, Order of 14 April 1992, ICJ Reports 1992, 114 (separate opinion of Weeramantry, J.); *United States v. Arjona*, 120 US 479 (1887) (duty to prosecute international crime of counterfeiting foreign currency); *United States v. Klintock*, 18 US (5 Wheat.) 144, 147-148 (1820) (universal jurisdiction and duty to punish piracy); *Ex parte dos Santos*, 7 F.Cas. 949, 953 (C.C.D. Va. 1835) (No. 4016) (quoting E. de Vattel: "duty to punish or surrender"); 1 Op. Att'y Gen. 68, 69 (1797) ("duty of every government to punish"); UNSC Res. 1261, para. 3, UN Doc. S/RES/1261 (25 August 1999); Principles on the Effective Prevention and Investigation of Extra-Legal, Arbitrary and Summary Executions, UNGA Res. 1989/65 (1989); Declaration on the Protection of All Persons from Enforced Disappearances, UNGA Res. 47/133 (1992); Principles of International Cooperation in the Detection, Arrest, Extradition, and Punishment of Persons Guilty of War Crimes and Crimes Against Humanity, UNGA Res. 3074, 28 UN GAOR, Supp. No. 30, 78, UN Doc. A/9030 (1973); UNGA Res. 2840, 26 UN GAOR, Supp. No. 29, 88, UN Doc. A/8429 (1971); UNGA Res. 96, UN Doc. A/64, 189 (1946) (duty to prosecute genocide); Human Rights Commission, Res. 32, Torture and other cruel, inhuman or degrading treatment or punishment, E/CN.4/RES/1999/32 (23 April 1999). See also *Henfield's Case*, 11 F.Cas. 1099, 1108 (C.C.D. Pa. 1793) (No. 6360) (Wilson, J., on circuit). A 1985 resolution of the International Law Association affirms: "States must try or extradite *(aut judicare aut dedere)* persons accused of acts of international terrorism. No

take into custody all persons within their territory or equivalent territorial bases of jurisdiction who are reasonably accused of having committed crimes under customary international law. Since a coastal State's territorial sea is an area equivalent to its land territory,[23] especially for purposes of jurisdiction, and an area within which it generally exercises sovereign controls,[24] the issue arises whether coastal State jurisdictional enforcement competencies and duties concerning customary international crimes apply within its territorial sea with respect to persons on board foreign flag vessels. In a very general sense, the answer must be yes. There is general concurrent enforcement jurisdictional competence for both the coastal State and the flag State. The question shifts to whether there is any priority or exclusivity in enforcement jurisdiction either under customary international law or a relevant treaty, such as the Law of the Sea Convention. The latter question shifts attention back to the meaning of Article 27(5). From a policy perspective, it makes sense to recognize that a coastal State's universal and customary obligations *aut dedere aut judicare* concerning international crimes, being of great significance to the international community,[25] should override the Law of the Sea Convention and, in particular, language in Article 27(5) concerning "any crime."[26]

Also relevant are relatively new international criminal law treaties that expressly obligate signatories in whose "territory" a relevant accused is "present" or "found" to "establish its jurisdiction," to take the accused into custody or take measures to ensure his presence for criminal or extradition proceedings, and, "if it does not extradite him, ... without exception whatsoever and whether or not the offence was

---

State may refuse to try or extradite a person accused of an act of terrorism, war crime, or a crime against humanity." Res. No. 7, Report of the Sixty-First Conference, 1985, 7. Such an obligation can be traced to 1625. See H. Grotius, *De Jure Belli ac Pacis*, Book II, Chap. XXI, in J. Brown Scott (ed.), *The Classics of International Law*, F. W. Kelsey trans., CEIP, 1925, 526-528. Newer international criminal law treaties frequently expressly include the obligation, thus reflecting its continual affirmation and mirroring consistent patterns of *opinio juris*. See, e.g., note 27; see also *United States v. Yunis*, 681 F.Supp. 896, 900-901 (D.D.C. 1988). The duty is absolute, but in the alternative.

[23] For example, Sohn and Gustafson, see note 13, 94 ("coastal State has the same sovereignty ... as it has with respect to its land territory" subject to applicable treaty limits); *Restatement*, see note 12, §§511, comment b, and 512, comment a ("like land territory," "rights and duties of a State ... and its jurisdiction ... are the same in the territorial sea as in its land territory").

[24] UNCLOS, Art. 2.

[25] See notes 21-22.

[26] See also note 18.

committed in its territory," to initiate prosecution.[27] Again, since the coastal State's territorial sea is an area equivalent to its land territory, it is obvious that duties under new international criminal law treaties noted above are applicable in its territorial waters and that such obligations apply without exception whatsoever and whether or not the offense occurred abroad. Thus, if Article 27(5) is interpreted to cover "all crimes," including international crimes, there will be a clash between obligations under the new international criminal law treaties and the Law of the Sea Convention concerning international criminal law enforcement. That clash should be resolved in favor of the primacy of international criminal law treaties, especially since they address matters of great significance to the international community implicating universal jurisdiction and responsibility.

6. Civil Enforcement in Territorial Seas

Although Article 28(1) appears to provide a significant limit to coastal State civil enforcement jurisdiction within its own territorial waters, stating that the coastal State "should not stop or divert a foreign ship passing through the territorial sea for the purpose of exercising civil jurisdiction in relation to a person on board the ship," there may be ambiguity in use of the word "should"[28] as opposed to "shall" (a more mandatory term used elsewhere in the Convention). This ambiguity may be heightened by the fact that under Article 2(1), the coastal State generally has "sovereignty" over its territorial sea, and sovereign control certainly includes civil enforcement competence for a variety of purposes.

An interesting question arises as to whether a coastal State can stop a foreign flag vessel passing through its territorial sea when a person on board the foreign flag vessel is being denied his or her right to seek

---

[27] See, e.g., International Convention for the Suppression of the Financing of Terrorism, Arts. 7(4), 9(2), 10(1), adopted by UNGA Res. 54/109 (9 December 1999); International Convention for the Suppression of Terrorist Bombings, Arts. 6(4), 7(2), 8(1), adopted by UNGA Res. 52/164 (9 January 1998); Convention for the Suppression of Unlawful Acts Against the Safety of Maritime Navigation, Arts. 6(4), 7(1), 10(1), IMO Doc. SUA/CON/15/Rev.1 (10 March 1988); Convention Against Torture and Other Cruel, Inhuman or Degrading Treatment or Punishment, Arts. 5(2), 6(1), 7(1), UN Doc. A/39/51 (10 December 1984); International Convention Against the Taking of Hostages, Arts. 5(2), 6(1), 8(1), UNTS Vol. 1316, 205 (1979).

[28] See generally McDougal and Burke, note 6, 299-300 (use of the word "should" in connection with criminal enforcement competencies might mean that coastal State competencies are partly but "not wholly discretionary").

asylum, freedom from arbitrary detention, freedom from slavery or involuntary servitude, or freedom from torture or cruel or inhuman treatment in accordance with human rights law. Does the coastal State retain an overriding "sovereignty" under Article 2 concerning civil enforcement within its territorial sea and can such a sovereign competence be enhanced by obligations or competencies under human rights treaties?[29] If a retained sovereignty under Article 2 is limited or obviated by Article 28(1),[30] will an unavoidable clash between the Convention and human rights treaties be resolved by recognition of a primacy for human rights law?

Since human rights are generally *obligatio erga omnes*, human rights law should have a higher status than more ordinary provisions of the law of the sea.[31] Further, some human rights norms are *jus cogens*[32] and, as recognized by the Vienna Convention on the Law of Treaties, norms *jus cogens* must prevail over more ordinary obligations under an international agreement.[33] Additionally, customary human rights are protected by and through Articles 55(c) and 56 of the United Nations Charter,[34] and Article 103 of the Charter assures that Charter-based rights, obligations and competencies of States will prevail over other international agreements, such as the Convention on the Law of the Sea. For these reasons, the obligations and competencies of States to respect and ensure respect for customary human rights, especially those of a *jus cogens* nature, must override any seeming limits to coastal State civil enforcement competence under Article 28(1) of the Convention on the Law of the Sea.

It may also be the case that a given human rights violation on board a foreign flag vessel in a coastal State's territorial sea will constitute a crime "of a kind to disturb the peace of the country or the good order of the territorial sea" within the meaning of Article 27(1)(b) of the Convention and thus provide coastal State criminal enforcement

---

[29] See also UNCLOS, Art. 2(3) ("subject to ... other rules of international law").

[30] See also *ibid.*, Art. 2(3) ("sovereignty ... is ... subject to this Convention").

[31] See generally Paust, Fitzpatrick, Van Dyke, see note 2, 47-48, 774.

[32] See *ibid.*, 49-51.

[33] See *ibid.*, 49; Vienna Convention on the Law of Treaties, note 2, Arts. 53, 64.

[34] Article 56, coupled with Article 55(c), of the UN Charter contains the duty of States to take joint and separate action to assure universal respect for and observance of human rights of all persons. See also *Filartiga* v. *Pena-Irala*, 630 F.2d 876, 881 (2nd Cir. 1980): "'[H]uman rights and fundamental freedoms' [are] guaranteed to all by the Charter" and the prohibition of torture "has become part of customary international law, as evidenced and defined by the Universal Declaration of Human Rights... The General Assembly has declared that the Charter precepts embodied in this Universal Declaration 'constitute basic principles of international law. GA Res. 2625 (XXV), Oct. 24, 1970."

jurisdictional competence under Article 27.[35] Crimes involving violations of customary human rights, thus implicating universal prescriptive and enforcement jurisdiction in all States,[36] clearly should meet the test set forth in Article 27(1)(b).[37]

7. Pollution Controls Within the Contiguous Zone

Article 33 of the Convention on the Law of the Sea does not expressly address the competence of coastal States to regulate pollution within their contiguous zones. Perhaps the closest competence expressed in Article 33 involves the conferral of enforcement "control necessary to ... [p]revent infringement of its ... sanitary regulations," although it may be that not all pollution will implicate "sanitary" concerns. Since the coastal State clearly has "sovereign rights for the purpose of exploring and exploiting, conserving and managing the natural resources" within its exclusive economic zone[38] and such rights include "jurisdiction ... with regard to ... the protection and preservation of the marine environment,"[39] it seems obvious that the coastal State should have at least the same rights relevant to pollution and other controls in its contiguous zone. As noted above in section 2, in general, coastal State competencies closer to its shores are greater and it would be logical and policy-serving to adopt a principle of inclusion recognizing that, at a minimum, competencies of a coastal State within its exclusive economic zone must also pertain within its contiguous zone and its territorial sea. If so, a coastal State can take measures within its territorial sea or

---

[35] See Paust, Fitzpatrick and Van Dyke, note 2, 774.

[36] See, *ibid.*, 447-460. The UN Charter and the Universal Declaration of Human Rights are not limited to territorial or similar jurisdictional competencies, neither is any customary human right over which there is universal jurisdiction. Some human rights treaties contain duties of each signatory to respect and ensure to all individuals within its territory and subject to its jurisdiction the human rights set forth therein. See, e.g., International Covenant on Civil and Political Rights, Art. 2(1), UNTS Vol. 999, 171 (9 December 1966). Since the territorial sea of a coastal State is generally equivalent to its territory (see note 23) and its "sovereignty" "extends beyond its land territory ... over ... the territorial sea" (UNCLOS, Art. 2(1)), it is logical and policy-serving to recognize that such treaty-based human right duties extend to persons within the territorial sea of a coastal State signatory.

[37] By analogy, consider the broad test involving the phrase "of a character to disturb the peace and tranquility of the country" associated with the peace of the port doctrine and *Wildenhus's Case*, 120 US 1 (1888).

[38] See UNCLOS, Art. 56(1)(a).

[39] See *ibid.*, Art. 56(1)(b)(iii); see also *ibid.*, Arts. 1(1)(4), 192, 194, 211, 220, 230(1).

contiguous or exclusive economic zones that are necessary to ensure compliance with its laws and regulations concerning conservation and management of natural resources, including relevant pollution threatening such resources and the protection and preservation of the marine environment.[40]

With respect to the contiguous zone, the answer is provided in Article 55, since the article's definition of the exclusive economic zone demonstrates that it "is an area beyond and adjacent to the territorial sea subject to the specific legal régime established" in Part V of the Convention concerning the exclusive economic zone. Thus, coastal State competence within its contiguous zone is enhanced by coastal State competence concerning its exclusive economic zone, since both zones reach towards the coast at least to the outer edge of the territorial sea and thus competencies concerning an exclusive economic zone are overlapping.[41] The same point pertains with respect to coastal State competencies with respect to its continental shelf, since the shelf area addressed in Part VI of the Convention also starts at the outer edge of the territorial sea.[42]

8. High Seas Enforcement Competencies

Another area of ambiguity involves enforcement competencies over foreign flag vessels on the high seas with respect to vessels engaged in the transport of slaves or illicit traffic in narcotic drugs or psychotropic substances. Articles 99 and 108 of the Convention are most relevant. Does either article authorize the boarding and seizure of foreign flag vessels on the high seas? Article 99 declares a broad duty: "Every State shall take effective measures to prevent and punish transport of slaves in ships authorized to fly its flag and to prevent the unlawful use of its flag for that purpose... ." The article is silent regarding a similar duty or competence with respect to foreign flag vessels on the high seas. In view of such silence, are other States necessarily precluded from enforcing customary international criminal law proscribing the transport or trafficking of slaves? Article 110(1)(b) allows warships to board a foreign flag vessel when there are reasonable grounds for suspecting "[t]hat the [foreign] ship is engaged in the slave trade." Is a warship in such a circumstance only justified in boarding the foreign flag vessel and engaging in reasonable searches, or can the warship seize the foreign vessel and arrest relevant crew? Sohn and Gustafson are emphatic that

---

[40] See *ibid.*, Art. 73; see also *ibid.*, Arts. 1(1)(4), 211, 216-217, 219-221, 230(1).
[41] Cf. *ibid.*, Arts. 33(1) and 55.
[42] *Ibid.*, Art. 76.

neither article authorizes seizure of such a foreign vessel and that seizure is not authorized "unless specially provided for by treaty,"[43] although it is clear that lawful seizure could also occur if the State of registry (or flag State) provides *ad hoc* consent to seizure,[44] or if customary international law allows seizure under such circumstances. Article 92(1) seems to assure the Sohn-Gustafson approach, since it declares that ships under the flag of a State, "save in exceptional cases expressly provided for in international treaties or in this Convention, shall be subject to its exclusive jurisdiction on the high seas." Yet, it does not seem logical or policy-serving that silence regarding seizure, either in connection with Article 99 or especially Article 110, necessarily precludes the seizure of a vessel engaged in an international crime under customary international law over which there is at least a universal prescriptive jurisdiction in all States. Decision makers should consider relevant patterns of *opinio juris* and practice. In this regard, the 19th century practice of boarding and seizing ships with piratical international criminals on board may still be relevant.[45]

A similar problem involves interpretation of Article 108 of the Convention.[46] Article 92(1) seems to provide exclusive flag State enforcement jurisdiction unless otherwise "expressly provided for" in the Convention. Article 108(2) seems consistent, recognizing flag State competence to "request the co-operation of other States to suppress" illicit drug trafficking involving vessels flying its flag. However, does Article 108(1) provide enforcement competence for other States? The duty is very broad: "All States shall co-operate in the suppression of" such illicit traffic "by ships on the high seas contrary to international conventions." The duty is mandatory for all States. It is a duty to cooperate in "suppression." Does the article require flag State consent to cooperate in suppression or does the duty to suppress exist independently of special flag State consent? Is the duty to cooperate in suppression a duty to cooperate with "all States" in assuring the suppression of such illicit trafficking whether or not the flag State provides *ad hoc* consent or is it merely a duty to cooperate with a flag State? The article seems unavoidably ambiguous.[47]

---

[43] See Sohn and Gustafson, note 13, 20.
[44] See Paust, Fitzpatrick and Van Dyke, note 2, 825-826, *passim*.
[45] See, e.g., J. J. Paust, "Responding Lawfully to International Terrorism: The Use of Force Abroad," *Whittier L. Rev.* (1986), 711 et seq., 727-728, n. 57 (addressing *The Mariana Flora*, 24 US (11 Wheat.) 1, 40, 43 (1826); *The Antelope*, 23 US (10 Wheat.) 66, 119 (1825); and other materials).
[46] For example, Paust, Fitzpatrick and Van Dyke, see note 2, 824-826.
[47] But see Sohn and Gustafson, note 13, 20.

9. Use of Force and Self-Defense

Various provisions of the Law of the Sea Convention generally mirror the three forms of prohibition of the threat or use of force set forth in Article 2(4) of the United Nations Charter. Article 19(2)(a) of the Convention assures that such forms of force are non-innocent with respect to passage through the territorial sea, thus implicating coastal State enforcement competence set forth in Article 25(1) to take "necessary" steps to prevent such passage as well as that set forth in Article 27(1)(a) or (b) with respect to relevant criminal sanctions. Article 39(1)(b) contains a similar duty of ships engaged in transit passage through straits, Article 54 assures that the same duty is applicable with respect to passage through archipelagic waters, and Article 301 contains a similar duty with respect to the exercise of all rights and duties of States under the Convention, thus assuring application in all sea areas.[48] However, in view of Article 103 of the United Nations Charter, obligations of States under Article 2(4) of the Charter apply even in case of an unavoidable clash with rights and duties under the Law of the Sea Convention, so inclusion of the prohibited forms of force recognized in Article 2(4) of the Charter would not have been necessary to assure that such forms of force cannot lawfully occur in various sea areas. However, inclusion seems useful to stress permissible enforcement responses that coastal States can take in addition to those evident under the United Nations Charter – for example, the general enforcement responses noted above under Articles 25 and 27 of the Law of the Sea Convention.[49]

More generally, Article 51 of the UN Charter, coupled with Article 103 of the Charter, assures that a coastal (and any other State) has the right to engage in proportionate measures of self-defense in case of an armed attack against it, its military vessels or aircraft, and its nationals in or above any sea area. Despite the impropriety of foreign flag vessels engaging in ordinary police or enforcement measures in another State's territorial sea or strait without consent,[50] a foreign flag vessel should be able to use proportionate force to destroy mines within its immediate path, since such imminent harm posed by a mine is similar to a process

---

[48] See also UNCLOS, Art. 88 ("The high seas shall be reserved for peaceful purposes").

[49] See also *ibid.*, Arts. 30 (a coastal State can "require" a relevant warship "to leave the territorial sea immediately") and 225; Paust, Fitzpatrick and Van Dyke, see note 2, 809-811 (addressing the *Saiga* case *(Saint Vincent and the Grenadines* v. *Guinea)*, ITLOS, paras. 155-159 (1 July 1999)).

[50] See, e.g., *Corfu Channel* case *(United Kingdom* v. *Albania)*, ICJ Reports 1949, 1; *The Apollon*, 22 US 362, 370-371, 376-379 (1824); Paust, Fitzpatrick, Van Dyke, see note 2, 460-461, 774-775.

of armed attack on the vessel.[51] The UN Charter Article 103 override will also assure that permissible regional peace or security action undertaken by regional organizations (such as NATO) in accordance with Articles 52 and 53 of the Charter are not inhibited by inconsistent provisions of the Law of the Sea Convention. Similarly, measures lawfully decided upon by the UN Security Council[52] under Articles 39 and 41 or 42 of the Charter will prevail over inconsistent provisions of the Convention. Thus, Article 103 of the Charter can provide several types of waves over the law of the sea.[53]

III. Conclusion

Hopefully this brief excursion through selected aspects of the law of the sea will provide useful approaches to the solution of problems concerning interpretation and application of the Law of the Sea Convention. Gaps and ambiguities are evident, as well as overriding norms within and outside the Convention. Relevant normative waves that one might ride will be influenced not merely by text and purpose, but also by undulating currents of *opinio juris* and practice in the unending sea of humanity. Both seas have known the professional influence of "sea captain" Oda.

---

[51] See Paust, Fitzpatrick, Van Dyke, see note 2, 775.
[52] I use the phrase "lawfully decided upon" because of the limitation on the authority of the Security Council evident in Article 24(2), which requires that the Security Council "act in accordance with the Purposes and Principles of the United Nations," which are themselves evident in the preamble to and Article 1 of the Charter and elsewhere within the Charter. Further, the obligation of States to accept and carry out decisions of the Security Council is limited either to decisions "in accordance with the present Charter" or to state actions "in accordance with the present Charter," depending on one's interpretation of Article 25. See UN Charter, Art. 25.
[53] See also section 6 above.

Coastal State Jurisdiction over Refugees and Migrants at Sea

Natalino Ronzitti

I. Introduction

Mass migration has become a permanent feature of our time, with migrants moving from different areas and for different reasons. Mass migration is very often illegal migration. Migrants try to penetrate the frontiers of wealthy countries and their routes inevitably follow one direction: from less-developed countries to Western countries. European Union Member States, Canada and the United States are the most sought after destinations. Economic motivations are generally coupled with other reasons: people escape their countries not only to flee famine and poor living conditions, but also to avoid countries ravaged by war or persecution. Unfortunately, third world countries are often an example of poverty and their record of civil liberties and the rule of law is not always outstanding. People escaping their countries because of persecution are asylum seekers. However, poverty often accompanies a lack of rule of law and it may be difficult to distinguish migration from asylum seeking.

Since the 1980s, a new phenomenon has appeared: migrants and refugees have begun to flee their country of origin by sea. Vietnamese refugees were rescued by the British and Australian navies in the South China Sea in 1979 and 1980.[1] The United States has been and is confronted with migration and refugees from Cuba and Haiti, trying to reach US shores on board of precarious boats. After the end of Cold War, given the proximity of the Italian coast, Italy immediately became the country of choice for resettlement for a multitude of people coming from Albania. This situation was aggravated by the fall of the former Yugoslavia and the beginning of the Balkan wars. The Mediterranean is crossed by people coming from the southern shores, but most of them

---

[1] R. P. Schaffer, "The Singular Plight of Sea-borne Refugees", *Austrian Yb. Int'l L.* 8 (1983), 213 et seq.

started their journey in Central and Eastern Africa. Turkish ports are often the point of departure by sea for people coming from Asian countries, some of them – as the Kurds – fleeing a lack of freedom and the absence of elementary human rights.[2]

Migrants and asylum seekers have become the object of exploitation by criminal organisations which transport them from their country of origin. Transport by sea has increased and this phenomenon poses new challenges to maritime countries. As we shall see, the international community has provided a first response by drafting and opening for signature, on 12 December 2000, a Protocol Against the Smuggling of Migrants by Land, Sea and Air, supplementing the United Nations Convention Against Transnational Organized Crime.[3] Yet, no new instrument specifically dealing with refugees at sea has been drafted. The purpose of this article is to see how the law of the sea applies in cases in which refugees and migrants at sea are involved, to examine the rights and duties of the coastal State and to assess whether current law is satisfactory or whether new norms should be considered.

II. Internal and Adjacent Waters (Territorial Sea and Contiguous Zone)

The coastal State has full jurisdiction over ships anchored in its internal waters. The only limits are related to foreign warships lawfully present, regardless of their nationality. Consequently, the coastal State may exercise its control over commercial ships in its internal waters carrying migrants and/or refugees. The same is true, *mutatis mutandis*, for the territorial sea. In principle, the limitation stemming from innocent passage does not apply if the ship is directed towards or coming from internal waters, since it is assumed that the ship has not obtained the consent of the coastal State. Passage is innocent provided that it does not prejudice the peace, good order or security of the coastal State. Mass exodus certainly does not comply with that requirement.[4] Illegal migration, strictly speaking, falls under the conduct exemplified under Article 19, paragraph 2(g), of the UN Convention on the Law of the Sea, prohibiting "the loading or unloading of any commodity, currency or *person* contrary to the customs, fiscal, *immigration* or sanitary laws and regulations of the coastal State".[5]

---

[2] See, generally, F. Pastore, "La politica migratoria", *L'Italia e la politica internazionale* (a cura di R. Aliboni, F. Bruni, A. Colombo, E. Greco), 2000, 83 et seq., and bibliography.
[3] See UN Doc. A/55/383.
[4] Asahikawa District Court, 19 February 1954, *Japan* v. *Kulikov*, in *Jap. Ann. Int'l L.* 1 (1957), 66 et seq.
[5] Emphasis added.

Illegal migration is also a ground for establishing criminal jurisdiction under Article 27 of the UN Convention on the Law of the Sea, as smuggling migrants is a crime, the consequences of which extend to the coastal State, disturbing its "good order" (Article 27, letters a and b).

It is open to question whether a ship carrying illegal migrants can be subject to the jurisdiction of the coastal State when in transit passage. In effect, it is difficult to say that the ship threatens the territorial integrity or political independence of the coastal State or is in contravention of the principles of international law embodied in the Charter of the United Nations.

However, the jurisdiction of the coastal State is clearly established in waters belonging to its contiguous zone. Article 33 of the UN Convention on the Law of the Sea unequivocally affirms that the coastal State may exercise control in order to prevent and punish infringement of its immigration laws within its territory or its territorial sea.

III. The High Seas

On the high seas, ships are entitled to navigate free from interference from any foreign flag State. This liberty is the consequence of freedom of navigation stemming from a norm of customary international law. Boarding a foreign vessel is permitted if the coastal State's warship is proceeding in hot pursuit from a zone under the jurisdiction of the coastal State in which the vessel has broken the law. Consequently, if a foreign vessel has entered the territorial sea or the contiguous zone of the coastal State to disembark illegal immigrants, it can be boarded on the high seas, provided that the pursuit began in the territorial sea or the contiguous zone of the coastal State.

The right of a foreign warship or other duly authorised ship on government service to visit a foreign vessel that has not entered waters under the coastal State's jurisdiction, is permitted in the following circumstances by Article 110, paragraph 1, of the UN Convention on the Law of the Sea:

- piracy
- slave trade
- unauthorised broadcasting
- absence of nationality
- uncertain nationality.

Trafficking and transporting illegal migrants is not contemplated by the UN Convention on the Law of the Sea as a specific ground for visiting a foreign vessel. Can trafficking of illegal migrants be included among the classical grounds admitted both by customary and conventional international law to exercise the right of visit? It can be, depending on the circumstances.

The case of uncertain nationality apart, the instances closest to migrant trafficking are slave trade and absence of nationality of the ship carrying illegal migrants.

According to Article 110(1)(b) of the UN Convention on the Law of the Sea, ships engaged in slave trade may be visited and, according to Article 99 of the same Convention, any slave taking refuge on the boarding ship shall *ipso facto* be free. It is, however, difficult to equate migrant trafficking with slave trade. Article 1 of the 1926 Geneva Convention defines slavery as "the status or condition of a person over whom any or all of the powers attaching to the right of ownership are exercised" (paragraph 1), while slave trade encompasses any act "involved in the capture, acquisition or disposal of a person with intent to reduce him to slavery; all acts involved in the acquisition of a slave with a view to selling or exchanging him; all acts of disposal by sale or exchange of a slave acquired with a view to being sold or exchanged, and, in general every act of trade of transport in slaves" (paragraph 2). The 1956 supplementary Convention reinforces the prohibitions contained in the 1926 Convention forbidding practices similar to slavery. Illegal migrants are not slaves or the object of any practise similar to slavery, since they are not the property of anybody. Isolated instances of slavery (e.g., forced prostitution) are sometimes prosecuted under domestic criminal law. The case law is limited, however, and does not yet involve migrant trafficking.

The case of stateless vessels engaged in migrant trafficking is far more important. Trafficking is often carried out using non-registered small vessels, without name or flag, manned by people who do not admit to command. The coastal State may visit such a category of ships without committing any wrongful act.

IV. Consent of the Flag State

If the flag State consents to its vessels being visited on the high seas, the principle *volenti non fit iniuria* applies, and the boarding vessel will not commit a wrongful act by visiting a foreign vessel. The consent may be given orally whenever a visit is requested by the coastal State or may be the object of an agreement between the coastal and the flag State. Examples of treaties of this kind include the exchange of notes between Haiti and the United States of 23 September 1981,[6] and the exchange of notes between Albania and Italy of 25 March 1997 and the Protocol thereto of 2 April 1997.[7]

---

[6] See M. N. Leich, "Contemporary Practice of the United States Relating to International Law", *AJIL* 76 (1982), 374 et seq.

[7] *Gazzetta Ufficiale della Repubblica Italiana*, Suppl., No. 163, 15 July 1997.

The exchange of notes of 23 September 1981 allows the US Coast Guard to board Haitian vessels on the high seas suspected of transporting illegal migrants. Detention of vessels on the high seas is also permitted if an infringement of US immigration laws has been or is being committed. Detained vessels and persons on board are returned to Haiti, upon prior notification. US ships engaged in repressing illegal migrants transported under the Haitian flag have a Haitian navy representative on board, acting as a liaison.

The exchange of notes between Albania and Italy follows, to a certain extent, the one concluded between Haiti and the United States. The Italian navy is given the right to stop Albanian vessels and send them back to Albanian ports. The Italian navy is also given the right to carry out its mission in Albanian territorial waters. In this case, the Italian navy operates as if it were the Albanian navy; it may stop any private vessel irrespective of the flag it flies. The Protocol of 2 April 1997 establishes the technical modalities to be followed when the mission is carried out and sets out the principle that Albanian representatives must be present on Italian vessels operating in Albanian internal and territorial waters. The activity of the Italian navy includes the following measures:

- *vérification du pavillon*
- arrest
- visit
- diversion.

The Protocol sets out the modalities for carrying out these activities. Offensive use of weapons is ruled out unless force is exerted in self-defence or as a warning measure.

V. The Application of Article 33 of the 1951 Refugees Convention on the High Seas

The 1951 Convention Relating to the Status of Refugees establishes, in Article 33, the basic principle of *non-refoulement*. An asylum seeker cannot be expelled or returned "in any manner whatsoever" to a country where his life or freedom is in danger. This right is conferred only on refugees and not on migrants. However, in order to determine whether an individual is a refugee or a migrant, the authorities of the coastal State should screen all persons on board a ship. Migrants are often mixed with refugees, and without proper screening, it is impossible to identify refugees correctly. The practice of naval interdiction, which excludes any screening of persons on board, would be in violation of

Article 33 of the 1951 Convention if this provision were applied on the high seas.

However, Article 33 applies to a refugee while in the territory of a State Party to the 1951 Convention. This conclusion may be inferred both from the letter of Article 33 and from its drafting history.

Article 33 of the 1951 Convention states

> "1. No contracting State shall expel or return ('refouler') a refugee in any manner whatsoever to the frontiers of territories where his life or freedom would be threatened on account of his race, religion, nationality, membership in a particular social group or political opinion.
> 2. The benefit of the present provision may not, however, be claimed by a refugee whom there are reasonable grounds for regarding as a danger to the security of the country in which he is, or who, having been convicted by a final judgement of a particularly serious crime, constitutes a danger to the community of that country".

Paragraph 2 presupposes that a refugee be in the territory of a State Party to the Convention (the country in which he is seeking asylum). Paragraph 1 should be read in the light of paragraph 2: expulsion or return implies that a potential refugee is within the border of the country where the injunction of expulsion is taken.[8]

As far as the *travaux préparatoires* are concerned, it is worth recalling the statement by the Swiss delegate, supported by the Netherlands, affirming that the prohibition of expulsion and the duty of *non-refoulement* applies to individuals present within the country.[9] This interpretation is in keeping with US practice and with a judgment by the US Supreme Court.

Since 1981, the United States has interdicted Haitian and Cuban migrants at sea, putting an end to the previous practice of screening migrants on board and performing a *prima facie* determination of their status. The US Coast Guard is entitled to interdict the Haitian and Cuban boats by Executive Order 12,807 of 23 May 1993. The Executive Order was challenged before the Supreme Court as contrary, *inter alia*, to conventional international law. However, the Court held that the Order

---

[8] See G. W. Palmer, "Guarding the Coast: Alien Migrant Interdiction Operations at Sea", in M. N. Schmitt (ed.), *The Law of Military Operations, Liber Amicorum Professor Jack Grunawalt, International Law Studies* 72 (1998), 162.

[9] For the statement of the Swiss and Dutch delegates see GAOR, UN Doc. A/CONF/SR. 16, 23 November 1951, 6 and 11.

was not contrary to the 1951 Convention and the 1967 Additional Protocol, since they do not apply on the high seas.[10]

The contrary has been maintained not only by writers, but also by the UN High Commissioner for Refugees, who stated that Article 33 of the 1951 Convention has no territorial limitation and applies anywhere.[11] If this view were to be accepted, the agreement on maritime interdiction concluded by the United States with other States would be contrary to the 1951 Convention. Another possible consequence is to consider those agreements void, provided that the principle of *non-refoulement* is embodied in a peremptory norm of international law. In this connection, one may quote the decision of the Swiss Federal Council of 14 May 1996 declaring void a proposal for a referendum on asylum. The Swiss Federal Council restated a decision of the *Conseil des Etats* of 16 March 1995 and affirmed that the principle of *non-refoulement* stems from customary international law *"à caractère contraignant ('jus cogens')"*.[12]

It seems to us that the correct position has been expressed by Goodwin-Gill, who makes a distinction between denial of entry into the coastal State's territorial waters, and interdiction followed by "the actual, physical return of passengers to the country of origin". Only in the latter case might a problem of infringement of Article 33 of the Refugee Convention be posed, since the obligation of *non-refoulement* does not imply that the coastal State is obliged to admit asylum seekers to its territory. Its only obligation is not to send those people back to their country of origin.[13]

VI. Naval Interdiction on the High Seas

Naval interdiction for containing illegal immigration consists of measures aimed at preventing a ship from entering the territorial and

---

[10] See *Sale v. Haitian Centers Council, Inc.* 113 S. Ct. 2549 (21 June 1993), in *AJIL* 88 (1994), 114 et seq.; *ibid.*, the critical comment by T. D. Jones.

[11] See, for instance, UN General Assembly, 9 September 1991, Executive Committee of the High Commissioner's Programme, Forty-Second session, Note on International Protection (submitted by the High Commissioner), para. 27; General Assembly, UN Doc. A/AC.96/898, 3 July 1998, Executive Committee of the High Commissioner's Programme, Forty-Ninth session, para. 10

[12] SZIER/RSDIE 6 (1996), 419.

[13] G. S. Goodwin-Gill, *The Refugee in International Law*, 2nd ed., 1996, 166. He also quotes the comment by the *Ad Hoc* Committee on the Refugees Draft Convention, which is very clear in stating that "the obligation not to return a refugee to a country where he was persecuted did not imply an obligation to admit him to the country he seeks refuge. The return of a refugee-ship, for example, to the high seas could not be construed as a violation of this obligation". *Ibid.*

internal waters of a coastal State. As a rule, naval interdiction does not imply the arrest and visit of the intruding vessel, but only harassment by navigational means and warning shots aimed at diverting the ship from its route. Naval interdiction has two aspects: whether it is lawful to interdict foreign ships from their route on the high sea; and the mode in which such activity should be carried out.

1. Legality of Naval Interdiction

As a rule, freedom of navigation on the high seas implies that a foreign ship cannot be the object of any measure aimed at impeding its navigation and diverting it from its normal route. Only the flag State is entitled to interfere with its own vessels. However, this rule is not without exceptions.

Measures may undoubtedly be taken if the foreign ship is anchored on the high seas but boats commute between the ship and shore to unload migrants. In this case, the rule of constructive presence applies and the measures that might be taken include the arrest of the ship and not only its diversion.

The same is true for a ship without nationality. Migrant carriers are often not registered ships or can be equated to ships without nationality, since they are without a master, without flag, and without a name or other identifying document.

Naval interdiction could also be employed to prevent the entry of a foreign ship carrying illegal migrants into the territorial waters of a coastal State (assuming that a contiguous zone has not been established). In this case, the actions involved are the normal activities of state border guards. The right of innocent passage to enter the port of the coastal State cannot be claimed since the ship is not allowed to enter the coastal State's internal waters without its consent.

If the ship is unequivocally directed towards the coast of a foreign State in order to unload illegal migrants, the coastal State is entitled to take the necessary steps to prevent illegal activity from taking place on its shore. The US navy did this in accordance with Presidential Proclamation 4865 on High Seas Interdiction of Illegal Aliens of 29 September 1981 and Executive Order 12,324, which entitled the Coast Guard "to enforce the suspension of the entry of undocumented aliens and interdiction of any defined vessel carrying such aliens".[14] The same practice was followed by Italy in the waters between the Apulian and Albanian coast between 1994 and 1997.

A number of writers also agree that the coastal State is entitled to take measures on the high seas to prevent the violation of its laws if the

---

[14] *AJIL* 76 (1982), 376.

ship is heading towards the coast. This opinion is shared by Conforti, among others, even though this was in a different context (drug smuggling).[15] Churchill and Lowe recognise that "even in peacetime States do unilaterally take exceptional measures of enforcement jurisdiction on high seas, any opposition from other States being insufficient to deter them".[16]

2. Modes of Naval Interdiction

The second aspect of naval interdiction, as mentioned above, is the mode in which naval activities are performed. The aim of interdiction is the diversion of the vessel from its established route in order to prevent its entry into the coastal State's territorial waters. Vessels engaged in naval interdiction are not allowed to sink the ship that is the object of the interdiction, or to put the life of those on board in danger. However, naval interdiction is a police activity and implies limited coercion. The use of weapons is not prohibited, provided that human life is not seriously endangered and that firearms are used as last resort. O'Connell, for instance, states that "every device, including, presumably, harassment by navigational means, must be employed and for a sufficient period of time before force is justified".[17] Firing without warning and without proved necessity is forbidden if the use of weapons creates danger to human life, as has been stated by the international commission of enquiry appointed to review the Red Crusader incident.[18] However, the use of weapons to intimidate the vessel (warning shots) is not forbidden. The US navy advocates more severe measures for arresting a vessel, such as disabling fire, i.e., fire into the steering gear or engine room of the targeted vessel.[19] The international practice, according to O'Connell and Shearer, even legitimates small calibre weapons fire against the pursued vessel following a warning shot.[20]

A vessel engaged in naval interdiction is not obliged to abide by the rules preventing collisions at sea enshrined in the 1972 International Rules for Preventing Collisions at Sea (Colregs). The 1972 Colregs apply to warships or other vessels in government service, but they do not apply when a warship is engaged in hot pursuit or in crime prevention

---

[15] B. Conforti, *Diritto internazionale*, 5th ed., 1997, 256.
[16] R. R. Churchill and A. V. Lowe, *The Law of the Sea*, 3rd ed., 1999, 217.
[17] D. P. O'Connell, *The Influence of Law on Sea Power*, 1975, 67.
[18] *ILR* 35 (1967), 485 et seq.
[19] Annotated Supplement to the Commander's Handbook on the Law of Naval Operations, Part I: Law of Peacetime Operations, 1997, Rule 3.11.5.2.
[20] D. P. O'Connell, *The International Law of the Sea*, I. Shearer (ed.), Vol. II, 1984, 1072.

service. The 1972 Colregs apply to ships with precise sea routes, while crime prevention activity may imply boarding the pursued vessel. The Colregs include a number of rules which are difficult for a ship engaged in police activity to abide by strictly, for instance, speed security, collisions risks, or manoeuvring to prevent collisions. This is not to say that, if an incident occurs, migrants at sea are not under the protection of the rules on rescuing people at sea. For instance, Article 98 of the UN Convention on the Law of the Sea applies.[21]

VII. The IMO 1997 Resolution and the 1998 Circular on Unsafe Practices Associated with the Trafficking or Transport of Illegal Migrants by Sea

The International Maritime Organisation (IMO) was one of the first international organisations to take action against the smuggling of migrants at sea.[22] In 1997, the IMO Assembly adopted a Resolution on combating unsafe practices associated with the trafficking or transporting of migrants by sea. This resolution was followed, in 1998, by a Circular of the Maritime Safety Committee containing interim measures for combating unsafe practices associated with the trafficking or transport of migrants at sea.[23]

The aim of the IMO Assembly resolution is to suppress the smuggling of migrants by ships, by calling upon governments to take action against ships that do not possess the necessary navigational requirements. Governments are requested to take action in order to detain unsafe ships used for migrant smuggling and are called upon to report them to the flag State. In this connection, it is worth recalling that Article 94 of the UN Convention on the Law of the Sea obliges the flag State to ensure that ships under its flag are seaworthy and observe all regulations concerning the safety of life at sea. Article 94 does not provide the port State with the power to inspect and detain the ship. However, this power derives from the general principles. The port State might also apply its provisions on safety of ships to foreign vessels, as port waters are under its exclusive jurisdiction.[24] The IMO Assembly

---

[21] See also Article 11 of the Convention for the Unification of Certain Rules of Law Relating to Assistance and Salvage at Sea (23 September 1910) and Article 10 of the 1989 International Convention on Salvage.

[22] The efforts of the UNHCR have been concentrated on the rescue of asylum seekers at sea. See the Report submitted by the High Commissioner on Problems Related to the Rescue of Asylum Seekers in Distress at Sea", 26 August 1981, UN Doc. EC/SCP/18.

[23] GAOR, Oceans and Law of the Sea Report of the Secretary-General, 20 March 2000, UN Doc. A/55/61, paras. 109-114.

[24] Churchill and Lowe, see note 16, 276.

resolution sets out a duty to inform both the flag State and the IMO regarding unsafe ships and incidents occurred at sea.

The 1998 IMO Circular is very specific and the forerunner of the UN Protocol on Migrant Smuggling, which will be examined later. Its importance is not diminished by the conclusion of the Protocol, since the date of entry into force of the latter is not certain.[25] The Circular, embodied in an IMO Assembly resolution, is not legally binding.

Its main object is the safety of life at sea and the suppression of migrant smuggling. Usually ships lacking seaworthiness and other requirements that make a ship safe are used for transport of migrants. The Circular establishes a duty of information. IMO members are invited to report to the Organisation and to States Parties all relevant information concerning migrant trafficking. States are also encouraged to apply the existing instruments, pending the conclusion of an *ad hoc* convention. The port State is authorised by the 1974 SOLAS Convention (International Convention for the Safety of Life at Sea) to detain a foreign ship that does not comply with the standards required for navigation and to report it to the flag State, which is obliged to ensure the safety of ship. In this connection, Article 94, paragraph 6, of the UN Convention on the Law of the Sea also comes into consideration: the port State may report the fact to the flag State, which is obliged to act.

VIII. The UN Protocol Against the Smuggling of Migrants by Land, Air and Sea

In order to combat transnational crime, the UN General Assembly adopted a Global Action Plan Against Organized International Crime in 1994, the implementation of which generated, *inter alia*, a Convention and two Protocols annexed thereto,[26] opened for signature in Palermo on 12 December 2000. The Convention is entitled United Nations Convention Against Organized Crime, while the two Protocols deal, respectively, with trafficking in persons (Protocol to Prevent, Suppress and Punish Trafficking in Persons, Especially Women and Children, supplementing the United Nations Convention Against Transnational Organized Crime) and migrants (Protocol Against the Smuggling of Migrants by Land, Sea and Air, supplementing the United Nations Convention Against Transnational Organized Crime). The purpose of the

---

[25] The Protocol on migrant smuggling needs 40 ratifications to enter into force (Article 22).

[26] GAOR, Report of the *Ad Hoc* Committee on the Elaboration of a Convention Against Transnational Organized Crime on the work of its first to eleventh sessions, UN Doc. A/55/383, 2 November 2000.

second Protocol is to combat migrant smuggling.[27] For this reason, the Protocol obliges States Parties to treat the smuggling of migrants as a criminal offence. However, illegal migrants are not criminalised and the States Parties are not obliged to make illegal migration an offence under their penal code. States are obviously not prevented from doing so.[28]

The Protocol on migrant smuggling contains a special provision on trafficking by sea that was drafted using Article 17 of the 1988 Convention Against Illicit Traffic in Narcotic Drugs and Psychotropic Substances as a model.[29] Language has also been drawn from the IMO Circular 896. The Convention on organized crime and the two additional Protocols are accompanied by interpretative notes, adopted by the *Ad Hoc* Committee which negotiated them.[30] The pertinent provisions are contained in Articles 7-9 of the Protocol on migrant smuggling.

Article 7 establishes a generic duty of cooperation for preventing and suppressing smuggling of migrants by sea[31] and is reminiscent of other provisions aimed at eradicating sea crimes such as piracy[32] and traffic in narcotics.[33] The duty of cooperation is implemented "in accordance with the international law of the sea", to make clear that States cannot take measures that are not allowed by the law of the sea.[34]

The vessels and aircraft allowed to take measures against migrant smuggling are the customary category of vessels and aircraft engaged in policing the seas: warships and military aircraft, or ships and aircraft clearly marked and identifiable as being in government service. Unlike warships and military aircraft, which are presumed to be authorised to take police measures, vessels or aircraft in government service need to be authorised to do so. It can be inferred from Article 3(d) that measures cannot be taken against those means of transport that enjoy immunity according to international law, such as warships and naval auxiliary

---

[27] Article 2 of the Protocol.
[28] Article 5 of the Protocol.
[29] See W. C. Gilmore, "The 1988 United Nations Convention Against Illicit Traffic in Narcotic Drugs and Psychotropic Substances", *Marine Policy* 15 (1991), 183 et seq.
[30] UN Doc. A/55/383/Add.1.
[31] "States Parties shall cooperate to the fullest extent possible to prevent and suppress the smuggling of migrants by sea, in accordance with the international law of the sea".
[32] Article 100 of UNCLOS.
[33] Article 108 of UNCLOS.
[34] This has been restated in the interpretative notes of the Protocol, where it is said that "the *travaux préparatoires* should indicate that it is understood that the measures set forth ... cannot be taken in the territorial sea of another State except with the permission or authorization of the coastal State concerned. This principle is well established in the law of the sea and did not need to be restated in the Protocol". UN Doc. A/55/383/Add. 1, para. 98.

vessels, or other government vessels in non-commercial government service. Measures may be taken against those vessels which no longer qualify as warships. This is for instance the case with warship employed in migrant trafficking, manned by crews that are not under military discipline and not under the command of an officer duly commissioned and listed in the appropriate service list.[35] For instance, the Albanian ship *Kates I Rades*, which sank in the Adriatic after colliding with an Italian warship on 28 March 1997, cannot be qualified as a warship, since it had lost all the required qualifications.[36]

Article 8 of the Protocol establishes, in paragraph 1, that a coastal State that is party to the Protocol may request the assistance of other States Parties when it takes action against vessels under its jurisdiction or against vessels that, according to the law of the sea, might be under the authority of the coastal State. The States requested have the duty to render assistance. The vessels covered by this provision are those that fly the same flag as the coastal State, vessels without nationality and vessels which fly a false flag, but in reality have the same nationality as the coastal State. The provision does not specify what duty of assistance entails. The cogency of the norm is reduced by the caveat that the assistance shall be rendered "to the extent possible" and "within the means" of the State requested. For instance, the duty of assistance is fulfilled if the requested State allows migrants to disembark at its ports, pending repatriation. At the same time, the requested State can object that it is unable to give such assistance, as it lacks structures serving as a temporary shelter for migrants.

Unless otherwise provided, coastal States are not allowed to take measures against foreign vessels exercising their freedom of navigation. It is therefore supposed that the vessels are beyond the territorial sea and that the coastal State does not have grounds for jurisdiction (for instance since the vessel has not penetrated its contiguous zone). In such a case, the *volenti non fit iniuria* (consent) principle applies. The coastal State, which suspects that the vessel is engaged in migrant smuggling, is entitled to take measures only under proper authorisation from the flag State. Article 8, paragraph 2, of the Protocol lists the measures that might be taken in a non-exhaustive manner, as is evidenced by the words *"inter alia"*. Such measures may consist, first of all, of boarding and searching the vessel. If the inspection reveals that the vessel is engaged in migrant smuggling, the boarding State can take any measure agreed with the flag State with regard to the vessel, the persons on board and its cargo. The measures taken should be notified to the flag State. As said,

---

[35] Article 29 of UNCLOS.
[36] See F. Caffio, "L'accordo tra l'Italia e l'Albania per il controllo e il contenimento in mare degli espatri clandestini", CXXX *Rivista Marittima* 6 (1997), 124.

the measures to be taken are part of an agreement between the boarding State and the flag State. No other measure outside the agreement can be taken, except for those required by an imminent danger to the lives of persons. Should the two countries be parties to a bilateral or multilateral agreement, they are allowed to refer to measures therein established. For instance, this is the case, as noted above, between Albania and Italy, which have concluded an agreement to this effect.

As in other conventions, the Protocol establishes a mechanism for facilitating communication with the flag State. Each State Party is requested to designate an authority for dealing with the Protocol, which shall be communicated to other parties through the UN Secretary-General. This allows the flag State to discharge its duties of cooperation quickly, in particular as regards the true flag of the suspected vessel or the authorisation to board it.

The Protocol also regulates cases in which a suspicion that the vessel is engaged in migrant smuggling proves to be unfounded and the measures taken cause loss or damage. Article 9, paragraph 8, of the Protocol states that the vessel should be compensated. This provision is similar to Article 110, paragraph 3, of the UN Convention on the Law of the Sea relating to the exercise of right of visit. The only problem here is to identify the State responsible for compensation. Under the Convention on the Law of the Sea, the flag State of the boarding vessel is responsible. Under the Protocol, this conclusion is questionable, since the boarding State is authorised to do so by the flag State. Is the responsibility joint? The flag State could always make its consent conditional upon payment of compensation by the boarding State should the search prove the innocence of the boarded vessel.[37]

IX. Duties of the Boarding State under the UN Protocol

Article 9 of the Protocol establishes two categories of duties for the boarding vessel: a precautionary obligation and a duty to minimise the interference with the flag State by the boarding State.

---

[37] Article 17, para. 6, of the 1988 UN Convention on Illicit Traffic in Drugs stipulates that the flag State may establish "conditions relating to responsibility", when it authorises a foreign flag State to board its vessels. The 1995 Council of Europe Agreement on Illicit Traffic by Sea is even more specific. Article 26, para. 2, affirms "... The intervening State shall also be liable to pay compensation for any such loss, damage or injury if the suspicions prove to be unfounded and provided that the vessel boarded, the operator or the crew have not committed any act justifying them". See W. C. Gilmore, "The 1995 Council of Europe Agreement", in *Marine Policy* 20 (1996), 10.

The first set of duties encompasses obligations of result, as well as less intense obligations. As to the former, the boarding State should ensure the safety and human treatment of the persons on board as well as the environmental soundness of the measures taken against the vessel. While the obligation to ensure the lives of persons on board is paramount, the cogency of the second obligation is less stringent, since the boarding vessel is obliged to abide by it "within available means". Less intense obligations relate to the need not to endanger the security of the vessel and its cargo and not to prejudice the commercial or legal interests of the flag State. In effect, the boarding State is only requested to take "due account" of those needs and interests. The provision is vague and one has to consider that the interests of any other interested State should also be taken into account.

The second set of duties has been inserted to protect the sovereignty of the flag State and does not add much to the duties already established. It is a type of saving clause, stipulated to reaffirm the rights which the law of the sea confers on the flag State. In particular, it states that the measures taken should not interfere with the rights, obligations and exercise of jurisdiction by the coastal State and with the right to exercise jurisdiction and control in all administrative, technical and social matters involving the vessel. Since the powers given to the boarding State depend on the consent of the flag State, the usefulness of this provision is open to question.

X. Conclusion

The UN Convention on the Law of the Sea does not contain any specific provision on migrants and refugees at sea. As has been noted, the provisions on slavery (Articles 99 and 110, paragraph 1(b)) can hardly be applied to migrant trafficking. Hence, the need to regulate a phenomenon which has assumed giant proportions. However, while migrant trafficking has been the object of a regulation, the status of refugees at sea remains uncertain as the application of the 1951 Convention on the high seas is still controversial.

There tends to be no clear distinction between smuggled migrants and refugees fleeing their country, and experience shows that ships and boats manned by smugglers transport both migrants or persons who qualify as migrants and refugees. This may make the return of smuggled migrants to their country of origin dramatic. The Protocol on migrant smuggling contains a saving clause, in Article 19, according to which the principle of *non-refoulement* established under the 1951 Refugees Convention is not impaired. This means that before sending back a person to his country of origin, the boarding State must screen migrants in order to determine whether they can rightly claim the status of refugee. If this is not the case, Article 18 applies and the States Parties, of which migrants are nationals or permanent residents, should accept their return.[38] This duty is set out in Article 18, paragraph 1, which obliges each State Party "to facilitate and accept, without undue or unreasonable delay, the return of a person who has been the object of conduct set forth in Article 6 of this Protocol [i.e., smuggling] and who is its national or who has the right of permanent residence in its territory at the time of return". The implementation of this provision should be made smooth by the numerous bilateral agreements that countries of immigration have concluded with countries of origin of migrants.

For the time being, insofar as the law of the sea is concerned, the Protocol cannot be considered a point of departure for creating new customary international law. It follows the practice of other conventional instruments, such as the 1988 Convention on Drug Trafficking, aimed at extending the jurisdiction of the coastal State with the consent of the flag State. It is therefore an implementation of the principle *volenti non fit iniuria*.

---

[38] Article 18 of the Protocol on migrant smuggling restates and expands the duty, stemming from a norm of customary international law, concerning the obligation of a State to readmit its own citizens. See, generally, K. Hailbronner, "Readmission Agreements and the Obligation on States under Public International Law to Readmit their Own and Foreign Nationals", *ZaöRV* 57 (1997), 1 et seq.

The Intertemporal Character of International Law Regarding the Ocean

Sompong Sucharitkul

I. The Evolving Rules of International Ocean Law

"The theme of Grotius that the common interest of mankind in the great oceans must prevail over the selfish mercantile interests of a few nations, is a foundation of all international law. ... [T]o promote newly found mercantile interests at the risk of prejudicing other common interests in the seas, is to invite anarchic disregard of all international law. It is Grotius and not Selden who teaches us what is in our national interest".

Philip C. Jessup, 1981[1]
From a message to the conference
commemorating the 400th anniversary of
the birth of Hugo de Groot (1583-1645)

At the height of his accumulated wisdom, Philip C. Jessup was actually preaching to his compatriots that, in the course of the lively debate between the Dutch and the British at the dawn of the European Thirty-Years War (1618-1648), it was the Dutch and not the British who expressed what was truly in the national interest of the United States, as opposed to the newly found mercantile interests of a few nations.

Clearly, among these few nations, Judge Jessup did not exclude his own fatherland, where selfish mercantile interests of the unenlightened, privileged few in the influential private sectors might have beclouded the vision of the administration to turn a blind eye to

---

[1] Frederick Tse-Shyang Chen (ed.), Proceedings of Conference on Deep Seabed Mining and Freedom of the Seas, UN Association of Connecticut and University of Bridgeport School of Law, Grotius Society of International Law, 1981.

the glaring longer term common interests of the country and also of mankind. As far as Judge Jessup was concerned, the national interests of the United States had necessarily to coincide with those of the international community and humanity as a whole, and not reflect only sectarian interests of some exclusive private quarters.

The contention between the Dutch and the British advocates was advanced by Hugo Grotius in his *Mare Liberum* (1609),[2] and was responded to by John Selden (1584-1654) a quarter of a century later in his *Mare Clausum* (1635).[3] Few today would have imagined that it was Hugo de Groot and not Selden who advocated freedom of the ocean. The Grotian thesis endorsed the principle that the oceans were open to all, for all purposes, and could not be appropriated, nor enclosed for exclusive use or exploitation, by anyone, any entity or any country, European or otherwise, to the exclusion of others.[4] As such, the oceans were to be shared by all nations and every living thing for the benefit of all creatures, especially mankind.[5] If *Mare Liberum* was originally rejected by the Portuguese and the British for reasons that may today

---

[2] *Mare Liberum* appeared anonymously in 1609, as Chapter XII of the manuscript *De jure praedae*, in an attempt to justify the Dutch warfare in Asia in defence of the Dutch right to trade with the East Indies, in the light of Portuguese claims – after Vasco de Gama's discovery of the sea route to Asia around the Cape of Good Hope – to monopolise the sea lane, especially after the succession of Filippe II to the Portuguese crown in 1580. The Spanish kings ruled Portugal until 1640, steadfastly upholding Portuguese pretensions. For an English translation, see H. Grotius, *Freedom of the Seas*, Magoffin, New York, 1916.

[3] See the English translation by M. Nedham, *Of the Dominion of the Sea*, 1659, reprinted in New York, 1972. Grotius rejected not only the Portuguese pretensions, but all claims to sovereignty over the sea, whether it was Venice referring to the Mediterranean, and especially to the Adriatic, as *Mare nostrum*, or Genoa with regard to the Ligurian Sea, or England and Scotland with regard to the "narrow seas". T. W. Fulton, *The Sovereignty of the Sea*, Edinburgh, 1911.

[4] One contemporary commentator, C. G. Roelofsen, said of Grotius: "The publication of *Mare Liberum* owed much to the particular circumstance of the moment. However, even if there is some truth in the conclusion that Grotius stumbled upon the doctrine that was to make him famous, his terse reasoning certainly lent it much of its force. Grotius's opponents, relying on long-winded historical expositions, found themselves in the long run beaten by the naturalist catch-words of *Mare Liberum*", L. W. van Holk and C. G. Roelofsen (Eds.), *Grotius reader: a reader for students of international law and legal history*, 1983 15.

[5] In 1609, it was true that the division of the seas could serve no useful purpose, "since their fishing resources are inexhaustible and navigational restrictions are needless". Grotius, *Freedom of the Seas*, see note 2, 27.

appear obscure, it would have stood to reason had the positions of the Dutch and the Portuguese or the British been reversed in the contest for the supremacy of the oceans in the first half of the 17th century.[6] It did not become apparent till the time of the Napoleonic War,[7] and ever since that both the British and the French, as well as the Spaniards, also on behalf of the Portuguese, have come to embrace freedom of the ocean as a means of levelling the playing field in the race for supremacy as a maritime power. The United States was not to be left out of this worldwide venture.

For a long while, nations harboured lingering doubts and hesitations with regard to the freedom of the open sea, which was initially opposed by the British in respect of the "narrow seas" and the British channels, only to welcome its application two centuries later. The eventual dissipation of these doubts may well illustrate the intertemporal character of international law regarding the ocean. The

---

[6] Thus, in the Anglo-Dutch negotiations of 1613-1615 concerning the commercial rivalry over the Moluccas spice trade, the English quoted *Mare Liberum* against Grotius, then the leading Dutch delegate. Grotius replied that the Dutch United East Indies Company neither curtailed freedom of navigation nor purported to monopolise the free trade of the region. See G. N. Clark and J. van Eysinger, "The Colonial Conferences Between England and the Netherlands", *Bibliotheca Visseriana* 15 (1940), 120 et seq. In point of fact, the Grotian principle was not at all observed at the time, since the Dutch insisted on their rights by treaty to the monopoly of the spice trade. See Boxer, *The Dutch Seaborne Empire*, 1965, 94. See also the Trade Agreement Between the Netherlands and Thailand, 12 June 1617, Colonial Archive inventory No. 44464 L., 122 – verse 125, concerning the sale of stag skins by the United East Indies Company from Soupattrou, opra a Japanese buyer in the service of the King of Siam and an agent in charge of retail sales in this branch of the King's monopoly trade. A subsequent trade agreement was concluded in 1664, and was renewed in 1688. State Papers of Kingdom of Siam, London; William Ridgway, 169, Piccadilly, 1886, 237; *Collection of Treaties of the Ministry of Foreign Affairs of Thailand*, Vol. I: 1617-1869, 1-9; see Treaty Between the King of Siam with the Honourable Company, confirming the sole and exclusive privilege to buy up the tin, granted on 14 November 1668, *ibid.*, 9.

[7] The contesting claims of various sea powers to exclusive jurisdiction have been limited to the territorial sea where coastal States still enjoy exclusive fishing rights. Even this narrow strip of maritime belt or the territorial sea was subject to the right of innocent passage to freedom of navigation. Beyond the territorial waters were the high seas over which all flags enjoyed complete freedom of navigation as well as freedom to fish. In the critical 19th century, after Napoleon, it was enforced by the fleets of Spain, the United Kingdom and France, and today by the United States, Canada, Russia and China.

reversal of the British position, endorsed by the French and other European powers, distinctly mirrored the increasing strength and considerable improvement in the formidable might of their respective naval forces, which were ultimately deployed in the 19th and early 20th centuries in the search for virgin lands and the relentless quest for the expansion of their respective empires in Asia and Africa. Meanwhile, in North and South America, European-dominated empires started to crumble, giving way to newly emerging nations of northern, central and southern American origins, and more recently the resurgence of Caribbean island nations.

Further fundamental changes of circumstance occurred after the two World Wars, especially following the advent of the United Nations and the adoption and implementation of General Assembly Resolution 1514 of December 1960. An irreversible trend has since been put into effect, with non-self-governing territories successively attaining independence by leaps and bounds, flocking into the United Nations as the organisation is fast turning truly global.

It is not unnatural that the law of the ocean, as international law, which was the invention and conception of the European world or the Christian world, has failed to respond to the needs of non-European communities. European imperialism has now been replaced by European integration, European community and European union. The world of Grotius is not a contemporary world any more, although many of the principles formulated then still appear to apply without much satisfaction from the perspective of newer and much older nations which remain non-European in their thinking, in their culture and in their traditions.

Law has to evolve if it is to continue to exist. Unless time stands still or the clock stops, international law must go on, not always in its original formulation. Its rules must continue to evolve and transform to suit the changing needs of the growing global community.

If law is necessarily intertemporal in character, it follows that the international law of the ocean cannot be otherwise but intertemporal. As amply demonstrated by the debate between Grotius and Selden from centuries long passed, there is little doubt today that the oceans are not susceptible to appropriation by anyone or any nation.

Keeping in mind the intertemporal character of international law regarding the ocean, the problem confronting us today is no longer whether the sea should or should not be free. While freedom of the ocean must by now be taken for granted, the trend in the contemporary world appears to be directed towards the problem of how to manage the resources of the oceans in the best sustainable manner: so that not only a few technologically advanced multinationals in a handful of highly developed countries can monopolise the benefits and reserve for themselves the resources derived from the oceans in the area beyond the limits of national jurisdiction, but that mankind as a whole can benefit. This area clearly forms an integral part of the common heritage of mankind.

It is proposed in this brief cursory study to demonstrate by way of specific examples the inevitably intertemporal character of international law regarding the ocean. A selection of instances, most illustrative of this particular characteristic of world ocean law is readily visible within the last 50 years.

II. The Unsettling Notion of the Territorial Sea

Few notions of international law are more illustrative of the intertemporal character of the law of the ocean than the breadth or width of the territorial sea. This is indispensable to any meaningful structure of the law of the sea or ocean law.

As is generally known, the classic formulation of the definition and delimitation of the territorial sea or territorial waters, otherwise known as the maritime belt, was three nautical miles, being the distance of a canon shot. This "canon shot" rule, placing the limit of territorial waters at three nautical miles, was indeed a realistic appraisal at the time of the potency and capacity of a coastal State to defend its coastlines, to ward off invaders from the sea and thereby effectively to protect the hinterland against unwelcome intruders and hostile invaders.

This traditional notion was settled and remained unquestioned for centuries ever since the advent of gun powder and the use of artillery from shore batteries. Conversely, it also served to notify coastal seafarers not to come too close to the coasts, as three miles would seem to be within the range of a gunboat to start bombarding the shore, or perform acts to enforce a blockade of a coastal city or area. With the exception of the right of innocent passage, coastal trade is otherwise reserved by the coastal State for its inhabitants engaging in the transport of goods and passengers along the coastline from one point to another on the seashore.

This solid notion of a three-mile limit of territorial sea, dormant for centuries, had to undergo rapid and sudden changes and modifications after the end of World War II with the advent of the United Nations and the establishment of the International Court of Justice as one of its principal organs.

World War II in effect destroyed a considerable amount of tonnage of ships, for transport as well as for fishing. The fishing industry had to resume in earnest to feed the population of various parts of the world suffering from the scourges and spoliation of the war. Fishing grounds were narrowed and far too few. The first attempt at restraining or containing the expanding width of territorial seas within the three-mile limit became unsuccessful when Norway was allowed to justify the extension and widening of its territorial waters by one extra mile beyond the straight base line it purported to draw from one headland to another to close the indentation of its fjords.[8] The international bench endorsed the Norwegian claims on several grounds: historic titles and self-preservation for fishing communities whose means of livelihood depended solely upon the catches from the seas.[9] Thus, the maritime belt of Norway was widened to four miles with considerable implications in terms of outward or seaward adjustments.[10]

This landmark decision also constituted a point of no return in the legal status and width of territorial waters. Other coastal States also sought to extend their exclusive fishing grounds beyond the three or four-mile limit to reserve for themselves and their national fishing fleets and enterprises, at least within this enlarged maritime belt, a form of exclusive fishing rights or monopoly in fishing.[11]

---

[8] *United Kingdom* v. *Norway*, Judgment, ICJ Reports 1951, 116 et seq. See M. D. Johnsson, *The International Law of Fisheries*, 1964, 163, with reference to the doctrine of *mare adjacens*.

[9] On this case, See J. Evensen, "Anglo-Norwegian Fisheries Case and its Legal Consequences", *AJIL* 46 (1952), 609 et seq. See also a commentary by C. H. M. Waldock, *BYIL* 28 (1951), 114 et seq.; D. H. N. Johnson, "The Anglo-Norwegian Fisheries case", *ICLQ* 1 (1952), 145 et seq., and T. Kobayashi, *The Anglo-Norwegian Fisheries Case of 1951 and the Changing Law of the Territorial Sea*, 1965.

[10] The snow-ball effect of the *Anglo-Norwegian Fisheries* decision by the International Court of Justice was irresistible and not opposable in the course of the United Nations Conferences on the Law of the Sea, UNCLOS I in 1958 and UNCLOS II in 1960. The ripples were also noticeable during UNCLOS III. The evolution of *mar patrimonial* since the San Diego Declaration of 1947 into the 200 miles of exclusive economic zone has been perceptible. The concept finally crystallised in the integrated text of the 1982 United Nations Convention on the Law of the Sea (hereinafter, LOS Convention).

[11] See B. G. Heinzen, "The Three-Mile Limit: Preserving the Freedom of the Seas", *Stanford Law Review* 11 (1959), 597 et seq. Immediately before UNCLOS I in 1958, at least 27 States attending the Conference made unilateral declarations claiming jurisdiction over coastal waters beyond three miles:

Intertemporal Character of Int'l Law Regarding the Ocean 1293

This trend, which at first appeared attractive to most coastal States alike, except for the land-locked countries which found themselves handicapped without landing facilities for their fishing fleets, could not be followed with equal enthusiasm by distant-water fishing nations, such as Japan, Norway, Poland, Russia and Thailand.[12] These distant-water fishing nations stood to lose out on their habitual fishing grounds if foreign territorial waters were to expand beyond the three-mile limit. Each mile of extended or widened territorial sea would imply a mile of lost ground for fishing by their national fleets.

Thus, the First Law of the Sea Conference of the United Nations opened in Geneva in 1958 with some optimism. However, no agreement was reached on the extended width of territorial waters. It could be anywhere between three and twelve miles. Major sea powers and distant-water fishing nations entertained different aspirations. A formula of six plus six, that is, six miles of territorial sea and six miles of contiguous zone was all but adopted for want of a single vote to obtain the required two-thirds majority.

The Second United Nations Conference on the Law of the Sea closed in 1960 without a compromise agreement. This unsettling character – the deviating but expanding breadth of territorial waters – fluctuated between the extreme positions of two groups: those that wished to extend their exclusive jurisdiction as far as possible, and those enjoying the fuller freedom of the ocean that wished to contain the exclusive coastal jurisdiction within the narrowest possible margin. An acute debate followed, culminating in the adoption of the widest

---

- 6 miles: Sri Lanka, Greece, Haiti, India, Iran, Israel, Italy, Libya, Spain and Yugoslavia
- 9 miles: Mexico
- 10 miles: Albania
- 12 miles: Bulgaria, Colombia, Ethiopia, Guatemala, and Indonesia, Romania, Saudi Arabia, USSR, USA and Venezuela
- up to 200 miles: Chile, Ecuador, El Salvador, Korea and Peru

[12] For instance, the following table (from *Statistical Abstract of the United States*, US Department of Commerce, 1977-1983) indicates the actual catches by the United States, USSR and Japan in the US Fishery Conservation Zone (FCZ) in 1977-1981 (in millions of pound live weight):

Table of Actual Catches					
	1977	1978	1979	1980	1981
USA	4,853	1,414	1,771	2,003	2,320
USSR	844	823	628	128	xxx
Japan	2,487	2,610	2,485	2,602	2,559

possible margin of exclusive territorial jurisdiction, 12 plus 12, that is, 12 miles of territorial sea and another 12 miles of contiguous zone for health, immigration, customs, fiscal and other policy enforcement by the coastal State.

As the maritime belt has widened, the definition of the right of innocent passage appears to have eroded. So also does the ever-receding area known as the high seas. The moving frontier of the high seas and the expanding exclusive fisheries zone of coastal States clearly illustrate the intertemporal character of contemporary ocean law. An awareness of these fast-moving maritime boundaries, noticeable since the *Anglo-Norwegian Fisheries* case in 1951, came to a peak during the subsequent cod wars between various maritime powers and Iceland. The latter had extended its exclusive fishing zone to 50 miles and concluded bilateral agreements with some European fishing nations to that effect. In the *Iceland Fisheries Jurisdiction* case, although the Court did not find the existence of a fundamental change of circumstances to justify Iceland's repudiation of its acceptance of the compulsory jurisdiction of the Court under the Optional Clause (Article 36(2) of the Statute of the Court), significant lip service was paid to the modernisation of fishing equipment and methods and the progressive evolution of international law of the sea on the merit of the question under examination.[13]

III. The Exclusive Economic Zone of 200 Miles

The conflicts of interest between coastal States and maritime powers took a nasty turn when fishing fleets had to be escorted by warships, whether in the cod wars off the coast of Iceland or in the central and southern American oceans. This was especially so in the Pacific, where no fewer than five Latin American nations declared a *mar patrimonial* of 200 miles to protect and enforce conservation measures for their tuna fishing industries.[14] While, in the 1950s, the notion of exclusive

---

[13] See *Fisheries Jurisdiction* cases *(UK* v. *Iceland)*, Jurisdiction, ICJ Reports 1973, 3 et seq.; Merits, ICJ Reports 1974, 3 et seq.; *FRG* v. *Iceland*, Jurisdiction, ICJ Reports 1973, 49 et seq.; Merits, ICJ Reports, 1974, 175 et seq.

[14] See the historic Santiago Declaration of the 200-mile zone of *mar patrimonial* by Chile and Peru in 1947, El Salvador in 1950, Honduras in 1951, and Ecuador in 1960. J. Carroz, "Les problèmes de la pêche dans la convention sur le droit de la mer et la pratique des Etats" in *Le nouveau droit international de la mer* (RGDIP, direction D. Bardonnet), 1983, 128-229, Annexes at 221-229. The United States sought to enforce its position in regard to highly migratory species in foreign waters by adopting the Fishermen's Protective Act of 1954, Pub. L. Nos. 90-482, 82 Stat. 729

fisheries in a 200-mile zone would have been considered far-fetched, if not outright absurd, it would appear that what seemed to be an absurdity was soon to become a living reality once the concept of the exclusive economic zone (EEZ) of the coastal State was incorporated in the negotiating text of the composite draft of the third United Nations Conference on the Law of the Sea.[15]

Even before the final adoption of the text of the Convention in 1982, the distance of 200 nautical miles was a foregone conclusion for the limit of national jurisdiction for several purposes. These included fisheries as well as marine scientific research, which were to be subject to national regulation by the coastal States to the exclusion of all other flags.

It was an impossible task for a coastal State that was at the same time a maritime power and a distant-water fishing nation to reconcile its contradictory or inconsistent interests. Every such State would prefer to see all its national interests fully protected, or even overprotected. The stronger the maritime power, the more effective and aggressive appeared to be the means at its disposal to implement self-contradictory policies. Ideally, coastal States would have liked to ensure the best of both worlds. Such an untenable position nonetheless appeared irresistibly attractive. The United States was no exception. As a coastal State, the United States proclaimed its exclusive economic zone on 10 March 1983,[16] once it was clear that many if not most other coastal States had proceeded with such notifications and declarations with regard to own exclusive economic zones. In the face of this Proclamation, it would be awkward for the United States to escort or protect its distant-water tuna fishing fleets operating or fishing within the exclusive economic zones of other coastal or inland nations, especially when these highly migratory species appeared to have congregated in foreign waters or areas within the limits of foreign EEZs. It would seem most tempting if these archipelagic nations were poor, weak and defenceless. Indeed their only means of subsistence was the harvest of migratory species within these zones.

The Reagan Proclamation of 1983 established sovereign rights over resources and jurisdiction over activities within a maritime area encompassing 3.9 billion acres, significantly more than the 2.3 billion

---

(1968), especially section 2(a), providing compensation for United States tuna fishermen where vessels have been seized for illegal fishing in the fisheries zones of foreign countries.

[15] Informal Composite Negotiating Text, UN Doc. A/CONF.62/WP 101, Rev. 3 and Add. 1 (Geneva), Rev 3 (NY); UN Doc. A/CONF. 62/L 67 and Add. 1-16 (1981).

[16] Proclamation No. 5030 of 10 March 1983, reprinted *Va. J. Int'l L.* 23 (1983), 600; see also President Reagan's statement, *ibid.*, 598.

land acres of the United States and its territories.[17] The exact size of the exclusive economic zone proclaimed by the United States depends on whether some or all Pacific island territories are included, each with its own exclusive economic zone.[18]

Very little has been done by way of executive decrees or otherwise to explore this vast expanse of virtually unknown aquatic wilderness. Only a few per cent of the newly acquired acreage of water columns and submarine areas has been explored in search for mineral resources such as oil and natural gas. Thus, the Reagan Proclamation served to extend the maritime jurisdiction and sovereign rights of the United States as a coastal State over the 200 miles of its newly declared exclusive economic zones. On the other hand, as a fishing nation, the United States Proclamation disclaimed jurisdiction or sovereign rights over highly migratory species such as tuna within its EEZs. It hoped thereby not to appear to be acting inconsistently with its desire to protect United States distant-water tuna fishing fleets operating within the exclusive economic zones of other coastal States. Such a disclaimer could not possibly create fishing rights for the United States to fish freely and without prior authorisation or licence from the competent authorities of other coastal States. Such an inconsistent and self-contradictory position could not be sustained for long and soon gave way to transparency and consistency in the overall national views and attitudes towards international law regarding the ocean.[19]

The practice of States tends to demonstrate that not a single nation, however powerful, was able to resist the force of logic of the progressive development of the rules of international law of the ocean, which time and again will not forsake its basically intertemporal

---

[17] See W. O. McLean and S. Sucharitkul, "Fisheries Management and Development in the EEZ: The North, South and Southwest Pacific Experience", *Notre Dame Law Review* 63 (1988), 492 et seq.

[18] See P. R. Ryan, "The Exclusive Economic Zone", *Oceanus* (Winter 1984-85), 3, Annex I. See the LOS Convention, Part V: Exclusive Economic Zone, Articles 55-75.

[19] Past records indicates that only about 1 per cent of the total tuna catch by fishing fleets of US flags has come from the United States EEZs. To implement the United States position, the Magnuson Fishery Conservation and Management Act of 1976 authorised the US Government to impose an embargo on the importation of fisheries products from any country that seizes a US fishing vessel harvesting tuna without a licence. Following two disputes with Pacific Island Forum States (Papua New Guinea and Solomon Islands), a Treaty of Fisheries was concluded between the United States and the Pacific Island States, thereby ending an otherwise unwarranted conflict.

character. In the ultimate analysis, sound legal reasons will prevail against short term sectarian self-interest.

IV. The Continental Shelf and Non-Living Resources of the Seabed: Disputed Submarine Areas

States finally settled for a 200-mile exclusive economic zone beyond their coastlines, primarily to extend their coastal jurisdiction and sovereign rights over living and non-living resources of the sea in the water columns within their national jurisdiction.[20] But the submarine areas including the seabed and subsoil beneath the sea waters also required final settlement.

The Geneva Convention on Continental Shelf of 1958[21] did not put an end to the controversies over the régime of the seabed and ocean floor. Scientific advancement in technology for the exploitation of resources in the deep seabed served to extend the definition and the limits of continental shelf beyond the 100-fathom isobath (200 metres), and following the alternative criterion of exploitability, a coastal State could claim exclusive sovereign rights over the seabed and subsoil that constitute a natural prolongation of its coast as far as its 200-mile exclusive economic zone, and even well beyond. At a minimum, whether or not a coastal State actually possesses a geophysical continental shelf, its submarine areas coincide with the limits of its exclusive economic zone. In general, this area can measure 200 nautical miles from the coastlines, subject to delimitation by similar exclusive economic zones and continental shelves of adjacent or opposite coastal States.

In reality, a few coastal States, such as Australia and the Russian Federation, at some points, can claim an actual continental shelf almost 450 miles from their coastlines. A compromise reached during the course of the negotiations at the Third Conference on the Law of the Sea led to the adoption of a limit of 350 miles in cases where there is an actual continental shelf and a limit of 200 miles regardless of geophysical conditions.[22]

Unlike the régime of the exclusive economic zone, the continental shelf is inherently attached to the coastal State without the necessity to make any proclamation. But not unlike the exclusive economic zone, the outer limits of the continental shelf and the lines of delimitation are

---

[20] See Part V of the LOS Convention.
[21] *Ibid.*, Part VI, Articles 76-85.
[22] *Ibid.*, Article 76: Definition of the Continental Shelf.

required to be shown on charts of a scale or scales adequate for ascertaining their position.[23] The problems of delimitation exist for exclusive economic zones and the continental shelf for adjacent and opposite coastal States as well as for the seas and for the seabed and ocean floor beyond national jurisdiction. While the delimitation of the exclusive economic zone is relevant for exclusive fishing rights, fisheries management and scientific researches, the delimitation of the continental shelf has become a necessity for the exploitation of non-living resources, such as oil, gas and other mineral resources.

The richness of the non-living resources of the seabed and subsoil within national jurisdictions has given rise to a number of disputes between adjacent and opposite coastal States. These disputes or differences of opinions regarding overlapping claims have been negotiated, arbitrated, adjudicated or otherwise settled by peaceful means. Notable among these are, for example, the *Gulf of Maine* case between the United States and Canada,[24] and the agreement between Thailand and Malaysia on the Joint Development Area.[25]

In a few areas relating to islands, disputes and conflicts of interest have taken a toll, including the loss of life and the use of force, as in the Falkland Island dispute[26] and the current multiple overlapping claims relating to the Spratly Islands in the South China Sea.[27] Problems relating to the delimitation of exclusive economic zones and the continental shelf have also been compounded by islands which, if habitable and capable of sustaining lives, can lead to the extension of exclusive economic zone for fisheries by 200 miles, and the extension of the continental shelf for the exploitation of natural resources by 200-350 miles.

---

[23] *Ibid.*, Article 84: Charts and List of Geographical Coordinates.

[24] *Gulf of Maine* case, Special Chamber, ICJ Reports 1984, et seq. 246. See also the *Aegean Sea Continental Shelf* case, ICJ Reports 1978, 3; and the *North Sea Continental Shelf* cases, ICJ Reports 1969, 3 et seq.

[25] Memorandum of Understanding on the Establishment of a Joint Authority for the Exploitation of the Resources of the Seabed in a Defined Area of the Continental Shelf, dated 21 February 1979. See also Thailand-Malaysia Joint Authority Act, B.E. 2533, 29 August 1990, *Government Gazette*, Vol. 108, Part II, dated 22 January B.E. 2534 (1991).

[26] For the régime of islands, see Part VIII of the LOS Convention. See also C. H. M. Waldock, "Disputed Sovereignty in the Falkland Island Dependencies", *BYIL* 25 (1948), 311 et seq.

[27] See, e.g., J. M. van Dyke and D. L. Bennett, "Islands and the Delimitation of Ocean Space in the South China Sea", *Ocean Yearbook* 10 (1993), 54 et seq.; V. L. Forbes "Seeking a Solution of the Spratly Dispute", *Indian Ocean Review* 7 (1994), 14 et seq.

Legal developments in the law of the sea since 1967, 1982 and 1994 have introduced new possibilities for coastal States, particularly for island States and islands whose status has never been the same. Earlier disputes relating to sovereignty over islands, such as the *Island of Palmas* arbitration,[28] are not comparable to the problems of the contemporary world. Thus the Spratly Islands, spread over a large area in the South China Sea, have now been claimed, in various parts, by China, Taiwan, the Philippines, Vietnam, Malaysia and Brunei. The traditional theories of *occupatio*, i.e., occupation after discovery, or *usucapio, possessio longi temporis, nec vi, nec clam, nec precario*, or even *uti possidetis* seem to have little or no relevance, so the overlapping claims could be advanced from entirely different dimensions of time, concept and new legal, economic and philosophical theories. Contiguity, continuity, consistent display of sovereign authority, adverse peaceful possession and other such theoretical legal jargon have been utilised in support of competing and overlapping national claims, all because of new technologies, new discoveries of resources of the ocean floor and the new law of the sea extending the domain of coastal States and islands as far as 200 miles for exclusive economic zones, as well as even further for the exclusive sovereign right to explore and exploit natural resources on and under the seabed and ocean floor.

None of these claims could have been advanced before 1967, at least not on historic grounds. Applying the temporal dimension, claims advanced today would have been out of place and out of time if presented to Max Huber in the 1920s. Contemporary publicists giving advice on claims to parts of the Spratly Islands would do well to confine the arguments to the current practices and more recent theoretical bases. Computerisation of proportionality and equitable principles may afford a new mathematical foundation. Whatever the final results, patience and cooperation among the nations directly and indirectly involved will provide a more encouraging basis for more plausible and acceptable solutions, so that peace and prosperity in the region and elsewhere can be preserved.

V. Diminishing Freedom to Fish in the Ever-Receding Open Seas

Article 87 of the 1982 United Nations Convention on the Law of the Sea[29] confirms the principle of freedom of the high seas subject to conditions laid

---

[28] See the award rendered by M. Huber, Sole Arbitrator, UN Report, International Arbitration Awards 829; W. B. Scott, Hague Court Reports 2d 83 (1932), Permanent Court of Arbitration 1928, *RIAA* II, 829.

[29] See Part VII of the LOS Convention, High Seas, Articles 86-120, especially Article 119: Conservation of the Living Resources of the High Seas.

down by the Convention and by other rules of international law. Among the various forms of freedom of the high seas, freedom of fishing is expressly subject to conditions laid down in section 2: Conservation and Management of the Living Resources of the High Seas.[30]

The very notion of the high seas, which used to denote the open sea beyond the three-mile limit of territorial waters, is now redefined and confined by Article 86[31] to all parts of the sea not included in the exclusive economic zone, in the territorial sea or in the internal waters of a State, or in the archipelagic waters of an archipelagic State. This is done without abridgement of the freedom enjoyed by all States in the exclusive economic zone in accordance with Article 58 on the Rights and Duties of Other States in the Exclusive Economic Zone,[32] and consistently with the duty to comply with the laws and regulations adopted by the coastal State in accordance with the Convention and other rules of international law. Various multilateral treaties and conventions have since been adopted with further restraining effect on the freedom of the seas, especially in the management and conservation of stocks and fisheries.

It may soon be out of fashion to refer to freedom of the seas as it was understood in the Grotian era and for almost 400 years thereafter. What was law yesterday is no longer law today. The intertemporal character of the freedom of the sea, in particular freedom of fishery, must be understood in the light of the new law in the contemporary world, where the resources of the sea, especially living resources, are no longer inexhaustible. Biodiversity is not to be taken for granted. Endangered species may not always be saved, recycled, resuscitated or regenerated amidst continuing erosion of their natural habitats and eradication of certain components of the existing ecosystem. For the ocean, the receding frontier lines between national jurisdiction of the coastal States and the open seas are not conducive to the maintenance or even sustenance of the traditional freedom of the sea. No one is free to fish anywhere at all times regardless of the species caught.

As such, freedom of the seas must be seen and understood in the light of the intertemporal nature of the ocean law that regulates its exercise.

VI. Deep Seabed Mining and the Common Heritage of Mankind

Another area of legal development in ocean law is the regulation and management of the common heritage of mankind for the benefit of all, especially in the operation of deep seabed mining, which may only be

---

[30] *Ibid.*, Articles 116, 118 and 120.
[31] *Ibid.*, Article 86: Application of the Provisions of Part VII.
[32] *Ibid.*, Article 58: Rights and Duties of other States in the Exclusive Economic Zone.

conducted by the Seabed Authority, its Enterprise or with its authorisation and approval.

Before the Third United Nations Conference on the Law of the Sea (UNCLOS III) was called by General Assembly Resolution 2750 (XXV),[33] Resolution 2749 (XXV) was adopted[34] with 108 votes for, none against and 14 abstentions, presenting a Declaration of Principles, according to which (1) the International Seabed Area should be reserved for peaceful purposes; (2) international machinery for regulating seabed exploration should be developed; and (3) the seabed and its resources were declared "the common heritage of mankind" and not to be subject to national appropriation.

In the face of these avowed principles, the United States proceeded to enact the Deep Seabed Hard Mineral Resources Act in 1980.[35] This was almost in the same tone as earlier United States legislation regarding fisheries, tending to encourage United States citizens to conduct activities in the seas in violation of generally accepted principles of international law of the ocean. This time it was done to undermine and pre-empt a future international régime that had been designed to preserve the common interests of humanity as a whole.

The concept of the common heritage of mankind clearly, enshrined in the 1982 Convention on the Law of the Sea, was accepted without opposition,[36] notwithstanding slight discrepancies in the interpretation of its regulation, especially with regard to the exploitation of non-living resources of the seabed in the area beyond national jurisdiction. If the method of implementation was unclear in 1982, much less in 1970, there cannot be any lingering doubt today about its validity, given the establishment of the Authority and the Law of the Sea Tribunal following the entry into force of the 1982 United Nations Convention in November 1994.[37]

---

[33] UN General Assembly Resolution 2750 (XXV) of 17 December 1970. UNCLOS III was convened in New York in December 1973 pursuant to this Resolution.

[34] UN Doc. A/8097 (1970) of 17 December 1970.

[35] 30 USC 1441(1) (A) of 28 June 1980, with an explicit disclaimer of any sovereign rights to the seabed or its resources.

[36] Article 136 declares the Area and its resources to constitute the common heritage of mankind. Article 137 forbids any State or natural or juridical person to appropriate any part thereof and vests all rights in the resources of the Area in mankind as a whole, on whose behalf the Authority shall act. These resources are not subject to alienation, minerals recovered from the Area may only be alienated in accordance with Part XI on the Rules, Regulations and Procedures of the Authority.

[37] See S. Sucharitkul, *The Entry into Force of the 1982 United Nations Convention on the Law of the Sea and Their Perspective*, SEAPOL Publication, Selected Papers, 1994, 78 et seq., 84-85.

Rules of international law regarding the ocean are further crystallised and reinforced by the entry into force of a general codification convention such as the United Nations Convention on the Law of the Sea of 1982. As time progresses, obligations incumbent upon States Parties to the Convention could also be accepted by non-parties in their practice, particularly in areas of international law where once a point of no return has been reached, it becomes exceedingly difficult to oppose or delay the hardening process of law-making. No one single State can possibly withstand the collective will of the rest of the global community of nations. Once again, Philip Jessup has been clear and convincing: "It is Grotius and not Selden who teaches us what is in our national interest".[38]

VII. Conclusion

It is hoped that the selection of samples from the law of the ocean discussed in these few pages have sufficiently and succinctly illustrated one salient feature of the ocean law, namely, its intertemporality. As ever, the law is moving – and fast – in the direction of the general interest of every nation and every person on earth. As the Law of the Ocean also progresses swiftly, it is becoming all the more transparent each day "that the common interests of mankind in the great oceans must prevail over the selfish mercantile interests of a few nations... . To oppose a modern restatement of the law of the sea in order to promote newly found mercantile interests at the risk of prejudicing other common interests in the seas, is to invite anarchic disregard of all international law".[39]

---

[38] Jessup, see note 1.
[39] *Ibid.*

The Definition of the Law of the Sea

Budislav Vukas

In selecting the subject of my contribution to this *Festschrift* in honour of Judge Shigeru Oda, I wanted to find a topic that would be related to the most important aspects of Judge Oda's activities – his highest reputation for expertise in the field of the law of the sea and his outstanding achievements as a Member of the International Court of Justice. I therefore decided to discuss the notion and definition of the law of the sea, in view of the fact that the jurisdiction of some international tribunals has been limited to disputes concerning the law of the sea.

I. The Law of the Sea and the Law of Naval Warfare

Classic textbooks on international law deal with the rules that form the various régimes at sea in the framework of "the objects of international law". For the rules on the uses of the sea in times of peace they use the term "law of the sea"[1] or "international law of the sea".[2] They deal with naval warfare and maritime neutrality separately from the "law of the sea".

Nguyen Quoc, Daillier and Pellet deal with the various régimes at sea together with other "international régimes of the spaces", and define the law of the sea as "the amalgam of legal régimes applicable to various spaces".[3]

Although he deals with all the régimes at sea under the title "The Law of the Sea", Malcolm Shaw does not engage in defining this expression.[4]

---

[1] R. Jennings and A. Watts (eds.), *Oppenheim's International Law*, Vol. I: Peace, 9th ed., 1992, Parts 2 to 4, 599 et seq.
[2] J. Andrassy, *Međunarodno pravo*, 6th ed., 1976, 159 et seq., 576 et seq.
[3] D. Nguyen Quoc, P. Daillier and A. Pellet, *Droit International Public*, 6th ed., 1999, 1102.
[4] M. N. Shaw, *International Law*, 4th ed., 1997, Chap. 11, 390 et seq.

A general characteristic of the textbooks on international law is their use of the term "law of the sea" only for the rules on the uses of the sea in times of peace. The law of naval warfare is treated separately, and is not covered by the expression "law of the sea". It is in this spirit that Vladimir Ibler defines the law of the sea as:

> "the part of international law containing norms on the delimitation of sea areas, their use and the legal régime applicable to each of them, as well as the norms on the rights and duties of States in respect of the uses of the sea".[5]

However, I will always recall the first book on the law of the sea I consulted 40 years ago, in which John Colombos regarded the international law of the sea as including not only international law of the sea in times of peace, but also that part of international law that deals with maritime neutrality and naval warfare.[6]

Even the International Law Commission, in presenting its 1956 Draft Articles on the Law of the Sea, pointed out that their draft covered only one part of the law of the sea: "The draft regulates the law of the sea in time of peace only".[7] As a consequence thereof, the 1958 United Nations Conference on the Law of the Sea dealt only with the régimes and activities at sea in times of peace; it did not provide rules on the law of the sea in times of war.

In listing the issues for the Third United Nations Conference on the Law of the Sea (UNCLOS III), the Committee on the Peaceful Uses of the Sea-Bed and the Ocean Floor beyond the Limits of National Jurisdiction left out not only the law of naval warfare, but also the issues concerning disarmament and denuclearisation of the ocean space.[8]

The United Nations Convention on the Law of the Sea (LOS Convention) does not contain a definition of the law of the sea. In the Convention's preamble, there is only an indication of the goals the United Nations wanted to achieve by convening UNCLOS III, which to a certain extent reflect the scope of the contents of the LOS Convention.

---

[5] V. Ibler, *Rječnik međunarodnog javnog prava*, 2nd rev. ed., 1987, 170.
[6] C. J. Colombos, *The International Law of the Sea*, 4th rev. ed., 1959, V and 7.
[7] United Nations, Report of the International Law Commission covering the work of its Eighth Session, April 23-4 July 1956, GAOR: Eleventh Session, Supplement No. 9 (A/3159), 1956, 4.
[8] As a consequence, the Law of the Sea Terminology (Terminology Bulletin No. 297/Rev.1/Add.1, UN Doc. ST/CS/SER.F/297/Rev. 2 of 27 July 1976), prepared by the United Nations Secretariat Department of Conference Services for the Law of the Sea Conference, does not contain titles, terms or expressions dealing with naval warfare, such as belligerence, blockade, capture and neutrality.

The States participating in UNCLOS III recognised the desirability of establishing "a legal order for the seas and oceans which will facilitate international communication, and will promote the peaceful uses of seas and oceans, the equitable and efficient utilisation of their resources, the conservation of their living resources, and the study, protection and preservation of the marine environment" (fourth preambular paragraph). The Convention contains not only provisions dealing directly with these activities, but also the basic rules establishing the "legal order of the seas and oceans", in the framework of which it will be possible to achieve the mentioned goals.

In contemporary practice, this attitude of the United Nations has resulted in the use of the term "law of the sea" only for the part of international law that regulates maritime relations of States in times of peace. The comprehensive *Handbook on the New Law of the Sea*, edited by René-Jean Dupuy and Daniel Vignes, devotes only ten pages, written by Théodore Halkiopoulos, to "The Interference Between the New Law of the Sea and the Law of War", in which the author does not go further than indicating the link between the two parts of international law concerning the seas and oceans:

> "It is ... quite natural that these two sets of rules (the law of the sea in time of peace and the law of armed conflicts at sea), which belong to the same juridical system, are by no means separated by an impermeable wall; they co-exist and complement each other in order to attenuate as much as possible the perturbation of the juridical order of the sea that is caused by an armed conflict".[9]

II. The Scope of the Law of the Sea in Times of Peace

The law of the sea, as presented in the textbooks, consists of rules on various legal régimes (territorial sea, contiguous zone, etc.) and on some major activities at sea (fishing, protection of marine environment, etc.). However, not every such rule deals directly and exclusively with the sea. Particularly in the treaties on the law of the sea, norms relating to the sea are inevitably supplemented by rules on other, more "technical" issues of international law, such as the settlement of disputes concerning the interpretation or application of the treaties, the signing and ratifying the treaties, the functions of the depositary, etc. As the LOS Convention also

---

[9] T. Halkiopoulos, "The Interference Between the Rules of the New Law of the Sea and the Law of War", in: R. J. Dupuy and D. Vignes (eds.), *A Handbook of the New Law of the Sea 2*, Hague Academy of International Law, 1991, 1321.

provides for the establishment of some international institutions (the International Seabed Authority, the Commission on the Limits of the Continental Shelf, the International Tribunal for the Law of the Sea), it had to include many provisions regulating the composition and the work of these institutions, the privileges and immunities of their members, etc.

For historical and other reasons, the LOS Convention and the law of the sea in general also include some provisions belonging to other distinctive parts of international law. Thus, for example, there are in the Convention some well-known rules on human rights. States have the duty to take effective measures to prevent and punish the transport of slaves in ships authorised to fly their flag (Article 99). Warships are justified in boarding foreign ships on the high seas if there are reasonable grounds for suspecting that they are engaged in the slave trade (Article 110, paragraph 1(b)).

In regulating some major issues of the ocean space, the LOS Convention simultaneously deals with other spaces. Thus, for example, the Convention states that the sovereignty of a coastal State "extends to the air space over the territorial sea" (Article 2, paragraph 2), and that the rights of the coastal State over the continental shelf "do not affect the legal status of the superjacent waters or of the air space above those waters" (Article 78, paragraph 1). The Convention goes even further, granting to all aircraft the right of transit passage in straits used for international navigation (Article 38, paragraph 1), and the freedom of overflight in the exclusive economic zone (Article 58, paragraph 1) and on the high seas (Article 87, paragraph 1(b)). In order to prevent, reduce and control the pollution of the marine environment from or through the atmosphere, the Convention requires States to adopt laws and regulations applicable to the air space under their sovereignty (Article 212, paragraph 1). The provisions on anadromous stocks (Article 66) and catadromous species (Article 67) concern the treatment of these species not only in sea waters, but also in the rivers of States Parties to the Convention.

As the sovereignty and jurisdiction of a coastal State over marine areas is based on its sovereignty over its land territory, many of the law of the sea rules include references to the land. Some of them, although relevant for the adjacent régimes at sea, require reasoning from quite different disciplines. Thus, the LOS Convention contains a rule according to which "[r]ocks which cannot sustain human habitation or economic life of their own shall have no exclusive economic zone or continental shelf" (Article 121, paragraph 3). In order to be able to interpret and apply this provision correctly, *inter alia*, the following questions need to be answered: (a) What is the difference between "rocks" and "islands", which are the subject of the first two paragraphs of Article 121? (b) What is the basis for the claim that a rock can or cannot sustain human habitation? (c) Are there any fixed conditions for the conclusion that a rock can sustain economic life of its own? None of these questions belong to the law of the sea, but they are relevant for the right of the State to which the rocks belong to control some sea areas off the coasts of such small islands. The relevance of those questions for the related law of the sea problems can be compared with that of the land frontier between two coastal States for the determination of the initial point for their sea boundary delimitation.

Article 121, paragraph 3, is not the only provision which requires consideration of issues concerning the land territory of coastal States. There are many provisions concerning baselines for measuring the breadth of the territorial sea (Articles 5 to 14) which require an evaluation of some facts on the coast: the presence of low-tide elevations or installations similar to lighthouses (Article 7, paragraph 4); the existence of economic interests peculiar to the region where the method of straight baselines is to be applied (Article 7, paragraph 5), etc.

Related to the sea, but sometimes concerning distant land territories, are the provisions on the right of access of land-locked States to and from the sea and freedom of transit. These provisions, drafted in favour of land-locked States, impose duties not only on coastal States, but even on transit States without a sea coast! Even for such non-coastal transit States, the Convention regulates the freedom of transit of land-locked States (Article 125), including the problems of customs duties, taxes and other charges (Article 127).

III. The Law of the Sea and Maritime Law

As a part of the system of public international law, the law of the sea is closely interlinked with all other segments of the system. It has also various links with some other branches of law.

The extension beyond the traditional borders of public international law is particularly evident in the LOS Convention, which contains an enormous quantity of rules on the area, which proved to be a rather hasty "progressive development" of international law (Part XI and Annexes III and IV).

However, even in some traditional areas regulated in the LOS Convention there are rules which *stricto sensu* do not belong to the law of the sea, but are closely related to that part of international law. Thus, the rule according to which every State shall fix the conditions for granting its nationality to ships (Article 91, paragraph 1) can be considered as belonging to the law of the sea. Yet, many of the rules on the duties of the State which has granted its nationality to a ship – the flag State – belong to maritime law. Such as, *inter alia*, the duty of States to take such measures as are necessary with regard ships flying their flag to ensure safety at sea with regard to the construction, equipment and seaworthiness of ships, the manning of ships, labour conditions and the training of crews, the use of signals, the maintenance of communications and prevention of collisions (Article 94, paragraph 3).[10]

Many of the provisions on the protection and preservation of the marine environment contained in Part XII are closely linked to international rules and standards belonging to maritime law that have been adopted in the framework of the International Maritime Organization, such as, for example, the provisions on the measures to prevent, reduce and control pollution of the marine environment contained in Article 194.

Article 221, paragraph 1, permits States "to take and enforce measures beyond the territorial sea ... to protect their coastline or related interests, including fishing, from pollution or threat of pollution following upon a maritime casualty or acts relating to such a casualty, which may reasonably be expected to result in major harmful consequences". This principle was adopted at UNCLOS III under the influence of the International Convention Relating to Intervention on the High Sea in Cases of Oil Pollution Casualties, adopted in Brussels on 29 November 1969, which contains more detailed rules on the issue.[11]

According to the LOS Convention, States are responsible for the fulfilment of their international obligations concerning the protection and preservation of the marine environment, and they are liable in accordance with international law (Article 235, paragraph 1). This reference to international law in the case of civil liability for oil pollution damage is to the principles and rules contained in the International Convention on Civil Liability for Oil Pollution Damage (Brussels, 29 November 1969).[12]

IV. Courts and Tribunals Dealing with Law of the Sea Cases

Four courts and tribunals have been entrusted with the "settlement of disputes concerning the interpretation or application" of the LOS Convention: the International Tribunal for the Law of the Sea, the International Court of Justice, arbitral tribunals constituted in accordance with Annex VII to the Convention, and special arbitral

---

[10] Because of the close link between the "international law of the sea" and "maritime law", Davorin Rudolf suggested using the term "law of the sea" to cover both bodies of law dealing with the sea. D. Rudolf, *Enciklopedijski rječnik međunarodnog prava mora*, 1989, 309.

[11] A. Ch. Kiss (ed.), *Selected Multilateral Treaties in the Field of the Environment*, 1983, 230 et seq.

[12] *Ibid.*, 235 et seq.

tribunals constituted in accordance with Annex VIII (Article 287, paragraph 1). Notwithstanding the generally identical competence of the four procedures, they differ amongst themselves in many respects. The purpose of this diversity of procedures was to make it easier for States Parties to submit their disputes to one of the compulsory procedures entailing binding decisions. In dealing with these disputes, all the courts and tribunals listed in Article 287 apply the Convention itself "and other rules of international law not incompatible with this Convention" (Article 293, paragraph 1).

However, the jurisdiction of various courts and tribunals differs *ratione personae* and *ratione materiae*. While both types of arbitration and the International Tribunal for the Law of the Sea (ITLOS) are also open to entities other than States Parties, only States may be parties in cases before the International Court of Justice (ICJ).

In the context of this paper, special attention should be paid to the differences *ratione materiae*. Notwithstanding the general competence of all the four courts and tribunals to settle disputes concerning the interpretation or application of the Convention, only the competence of arbitral tribunals (Annex VII) is determined solely by this provision. The competence of special arbitral tribunals is limited: they may deal only with disputes concerning the interpretation or application of the articles of the Convention relative to: "(1) fisheries, (2) protection and preservation of the marine environment, (3) marine scientific research, or (4) navigation, including pollution from vessels and by dumping" (Annex VIII, Article 1).

The competence of ITLOS is also generally determined by Article 287, paragraph 1. Therefore, as well as arbitral tribunals (Annex VII, Article 1), ITLOS can deal with all the disputes concerning the interpretation or application of the Convention. However, its jurisdiction also comprises "all matters specifically provided for in any other agreement which confers jurisdiction on the Tribunal" (Annex VI, Article 21). Notwithstanding the apparent latitude given to such agreements, the jurisdiction of ITLOS on the basis of agreements conferring jurisdiction on the Tribunal can include only disputes in the field of the law of the sea. That is why it is important to be able to differentiate this part of international law from other fields of public international law. The limitation of ITLOS to disputes on the law of the sea is not only clear from its name, its competence under the Convention and its Statute (Annex VI), but also from the required competence of its members. In addition to having the "highest reputation for fairness and integrity", the members of ITLOS must be "of recognized competence in the field of the law of the sea" (Annex VI, Article 2, paragraph 1).

The strict determination of a dispute as falling within the field of the law of the sea is not relevant only for the ICJ. Of all the courts and tribunals listed in Article 287, paragraph 1, of the LOS Convention, only the ICJ can deal with disputes involving issues beyond the limits of the law of the sea, due to its general jurisdiction (Article 36, paragraph 2, of the Court's Statute). Therefore, for example, disputes concerning sea boundary delimitation that involve the concurrent consideration of issues concerning sovereignty or other rights over continental or insular land territory could be submitted to the ICJ, but not to the other procedures mentioned. However, for dealing with a dispute involving questions beyond the scope of the law of the sea, the jurisdiction of the Court should be established on the basis of its Statute (Article 36). For such disputes, the Court's competence on the basis of the LOS Convention (Article 287) would not suffice.

The vague limits of the law of the sea will in the future probably cause dilemmas with regard to the categorisation of some disputes as belonging to the law of the sea. This paper has indicated some of the Convention's provisions that provide grounds for disputes not belonging to the law of the sea. Agreements concluded between the States Parties to the Convention also allow for the submission to ITLOS of disputes not strictly confined to the law of the sea. Some time ago, having scarce information concerning a case brought before a human rights court, I wondered what my opinion might be regarding the competence of ITLOS, had the dispute regarding alleged violations of human rights in an armed conflict at sea been brought to that Tribunal.

XI

THE ENVIRONMENT AND THE LAW OF THE SEA

The Common Heritage of Mankind:
From Non-living to Living Resources and Beyond

Elisabeth Mann Borgese

This essay celebrates the life and work of Shigeru Oda, one of the great scholars in international law and the law of the sea of the 20th/21st century. I have chosen, in this context, a subject that has always been dear to him, and that is the extension of the concept of the common heritage of mankind from the mineral resources of the international seabed area – the Area – to which it now legally applies, to include the ocean's living resources.

On 25 November 1998 Shigeru Oda wrote to me:

> "The inaugural session of *Pacem in Maribus* was held in Malta, and I, as a member of the planning and organising committee and of the Bureau in Malta, have fond memories of the significant and enjoyable start of the work in Malta, where we discussed the broad aspects of the new Regime of the Ocean on the basis of the challenging concept which was proposed by Ambassador Arvid Pardo – the common heritage of mankind to apply to the deep ocean floor. *You may remember that you gave sympathy and understanding to my suggestion that this concept, originally applied to the seabed, would eventually surface to apply to the fisheries resources in a few decades*". (emphasis added)

This, indeed was a prophetic statement.[1] The thesis of this essay is that, "in the few decades" since our early common work in Malta, the issue has matured. The time has come to reconsider it in the light of contemporary developments. The purpose of this essay is to bring some supporting evidence to the attention of the international community.

---

[1] Oda's thinking on this subject goes back a very long way. In his address to the Coordinating Committee of the Economic and Social Council, on 10 July 1970, Arvid Pardo recalled that "as far back as 1957, Professor Oda identified the central problem of modern fisheries as the equitable sharing among States of limited desirable living marine resources". Equitable sharing is one of the attributes of the common heritage concept.

I. The Ocean Régime

While this essay focuses on the evolution of public thinking, it is perhaps interesting to mention in passing that my own model convention, *The Ocean Regime*, published in 1968 by the Center for the Study of Democratic Institutions, Santa Barbara, California, included living resources, in fact, all ocean resources and services, in the common heritage and provided mechanisms for their management. Article III of this model stated that: "Ocean space is an indivisible whole. Geological structures extend, currents and waves move, species migrate, across the high seas and the ocean floor regardless of political boundaries. The law of the seas and the seabeds must accord with this reality". And: "The natural resources of the High Sea and on or below the seabed as defined by this Statute are the common heritage of mankind. They must be developed, administered, conserved and distributed on the basis of international cooperation and for the benefit of all mankind".

Fishing organisations, fish processors and merchants, unions of seamen serving on fishing vessels and consumers, as well as representatives of regional fishing commissions, were to compose one of the chambers of the Maritime Assembly, the supreme organ of the Régime. The Maritime Assembly was to consist of five chambers: a political chamber, elected by the General Assembly of the United Nations on a regional basis; a fisheries chamber; a chamber representing international mining corporations; a chamber representing shipping companies, cable companies and other organisations providing services or communication on or under the oceans; and a chamber representing scientists in oceanography, marine biology, meteorology, etc. A majority vote of two chambers, i.e., of the political chamber and the chamber competent in the matter voted upon would be required for the adoption of any decisions or recommendations. If the two competent chambers failed to agree, they would have to discuss the matter in a joint session and vote in common. A simple majority vote of the two joint chambers would suffice for the adoption of a decision or recommendation. The full participation of those responsible for the management of living resources, in the broader context of integrated ocean management, was thus assured.[2]

---

[2] The model of the five-chamber assembly is adapted from the Yugoslav Constitution of 1958, which was based on the two fundamental principles of social ownership, a concept of non-ownership similar to that of the common heritage of mankind, applicable to resources which cannot be appropriated either by States or individuals, and self-management, including bottom-up participation in decision making. See, J. Djorjevic, "The Social Property of Mankind", in EMB (ed.), *Pacem in Maribus*, 1972, 170-174.

II. The Common Heritage of Mankind: Arvid Pardo and the Definition of the Concept

In his historic address of 1 November 1967, Ambassador Pardo proposed the following definition for the concept of the common heritage of mankind:

> "(1) The seabed and the ocean floor are a common heritage of mankind and should be used and exploited for peaceful purposes and for the exclusive benefit of mankind as a whole. The needs of poor countries, representing that part of mankind which is most in need of assistance, should receive preferential consideration in the event of financial benefits being derived from the exploitation of the seabed and ocean floor for commercial purposes".[3]

And he suggested that the following, among other principles, should be incorporated in the proposed treaty:

> "(1) The seabed and the ocean floor, underlying the seas beyond the limits of national jurisdiction, as defined in the treaty, are not subject to national appropriation in any manner whatsoever.
> (2) The seabed and the ocean floor beyond the limits of national jurisdiction shall be reserved exclusively for peaceful purposes.
> (3) Scientific research with regard to the deep seas and ocean floor, not directly connected with defence, shall be freely permissible and its results available to all.
> (4) The resources of the seabed and ocean floor beyond the limits of national jurisdiction, shall be exploited primarily in the interests of mankind, with particular regard to the needs of poor countries.
> (5) The exploration and exploitation of the seabed and ocean floor beyond the limits of national jurisdiction shall be conducted in a manner consistent with the principles and purposes of the United Nations Charter and in a manner not causing unnecessary obstruction of the High Seas or serious impairment of the marine environment".[4]

---

[3] First Statement to the First Committee of the General Assembly, 1 November 1967, in A. Pardo, *The Common Heritage: Selected Papers on Oceans and World Order 1967-1974*, 1975 IOI Occasional Papers No. 3 (Malta University Press).
[4] *Ibid.*, 40.

This definition, ready as early as 1967, has basically survived, incorporated, first, in Resolution 2749 (XXV), then in Part XI of the Convention itself. The basic attributes of the concept are:

1. The common heritage cannot be appropriated: neither by persons, nor by companies, nor by States. Areas or resources which are part of the common heritage are non-property. This concept is articulated in Article 137.
2. The common heritage must be managed, for the benefit of mankind as a whole, with particular consideration for the needs of poor countries (Article 140).
3. The common heritage is reserved exclusively for peaceful purposes (Article 141).
4. The common heritage must be managed with due consideration for the conservation of the environment (Article 145).

Marine scientific research in the Area shall be carried out for exclusively peaceful purposes and for the benefit of mankind as a whole. It is to be coordinated by the Authority, which may also engage in research itself, and research results are to be shared, which puts it effectively under a common heritage régime (Article 143).[5]

While the four attributes above only apply to mineral resources in the Area, marine scientific research is not thus restricted and presumably includes environmental, biological and genetic research.

Pardo was of course fully aware, from the very beginning, that it was inadequate to apply the common heritage concept only to the non-living resources and restrict its application to the international seabed area. From the very beginning, he recognised the essential unity of ocean space which he considered to be the common heritage of mankind. His arguments in favour of this new concept which he proposed to introduce into international law, namely, that neither high seas freedom nor national appropriation could solve the contemporary problems of pollution, resource depletion, and conflict, are even more applicable to living than to non-living resources. In his Ocean Space Draft Convention of 1971, he designed a régime applying the common heritage concept to all resources anywhere in ocean space. Article 6(2) of that Draft Convention asserts that "All States have the duty to co-operate with the competent international institutions in the adoption and enforcement of such measures as may be necessary for the conservation of the living resources of the seas".

---

[5] Arvid Pardo considered marine scientific research as "a public interest of the international community". As a public interest "it would enjoy special protection throughout ocean space...". Statement to Subcommittee III of the Committee on the Peaceful Uses of the Seabed and the Ocean Floor, 2 April 1973.

Part IV of the Draft Convention, dealing with the international ocean space, adapts all the basic principles established by Resolution 2749 (XXV) for the seabed to the international ocean space as a whole, and extends the common heritage status from non-living to living resources. The international ocean space, comprising the seabed, the ocean floor and its subsoil, as well as the water column and the atmosphere above it, is the common heritage of all mankind (Article 66). It cannot be appropriated (Article 68). The administration of the international ocean space and the exploration and exploitation of its resources is exclusively for peaceful purposes and shall be carried out for the benefit of mankind as a whole, taking into particular consideration the needs of developing countries (Article 71) and with due consideration of the conservation of the environment (Article 72).

With regard to ocean space as a whole, the Ocean Space Institutions play the same role that the International Seabed Authority plays with regard to the seabed. There is an Assembly, a Council, and a Maritime Court, and a number of subsidiary organs. What is most impressive, and way ahead of its time, is the mandate of these ocean institutions: to maintain international law and order in ocean space; to safeguard the quality of the marine environment; to harmonise the actions of nations in ocean space; to encourage marine scientific research, international cooperation, and strengthening the capacities of technologically less advanced nations; to promote the development and practical application of advanced technologies for the penetration of ocean space and for its peaceful use by man and to disseminate knowledge thereof; to develop in an orderly manner and to manage rationally the international ocean space and its living and non-living resources and to ensure the equitable sharing by all States in the benefits derived from the development of the natural resources of the international ocean space, taking into particular consideration the interests and needs of poor countries; to promote the harmonisation of national maritime laws and the development of international law relating to ocean space; to undertake in ocean space such services to the international community and such activities as may be consistent with the provisions of this Convention.[6]

All this is further detailed in the respective mandates of the Assembly, the Council, and the Court.

In his statements between 1967 and 1974 to the Seabed Committee and elsewhere, Pardo went even further. He saw the common heritage concept not only applicable to both non-living and living resources, he considered it basic and essential for sustainable development and peace in the modern world.

---

[6] Shortly before his untimely death in 1981, H. Shirley Amerasinghe, President of UNCLOS III, said to me, "Had we really looked at Arvid's Draft in 1971, we could have spared ourselves ten years of work!"

"For my delegation the common heritage concept is not a slogan, it is not one of a number of more or less desirable principles, but it is the very foundation of our work, the key that will unlock the door of the future. It is a new legal principle which we wish to introduce into international law. It is a legal principle which, we feel, *must* receive recognition if the international community is to cope constructively and effectively with the ever more complex challenges which will confront us all in the coming decades. We cannot deal effectively with the accumulating and increasingly serious problems of the total environment in which we live ... on the narrow, outdated basis of traditional international law; new concepts must be introduced, new solutions sought to enable us all, from the greatest world powers to the smallest society, to cope intelligently with new problems".[7]

III. The Law of the Sea Convention (1982) and the Straddling Stocks Agreement (1995)

If a common heritage resource (1) cannot be appropriated; (2) is reserved for peaceful purposes; (3) must be managed on the basis of equity for the benefit of mankind as a whole, with special consideration to the needs of poor countries; and (4) must be conserved for future generations, then it is clear that, in spite of its ardent assertions with regard to the freedom to fish in the high seas and the sovereign rights of coastal States in their exclusive economic zones, the Law of the Sea Convention goes some way towards making the ocean's living resources a common heritage resource. The sovereign rights of coastal States are somewhat limited by the imposition of responsibilities. Management and conservation are prescribed in Article 61. Paragraph 2 states: "The coastal State, taking into account best scientific evidence available to it, shall ensure through proper conservation and management measures, that the maintenance of the living resources in the exclusive economic zone is not endangered by over-exploitation".

Some awareness of the need for equity, and the duty of sharing is expressed in Articles 62, 69, and 70. Article 62 imposes the duty to share – if only the surplus, i.e., the difference between the total allowable catch and the harvesting capacity of the coastal State. Articles 69 and 70 assert the rights of land-locked and geographically disadvantages States "to participate, on an equitable basis", in the exploitation of an appropriate part – even if only "of the surplus of the living resources of the exclusive economic zones of coastal States...".

---

[7] Statement to the First Committee of the General Assembly, 29 October 1968.

Among other criteria, the nutritional needs of the States involved are to be taken in consideration; and even when the "surplus" disappears, because the coastal State has reached the capacity to harvest its total allowable catch, it still has the duty "[t]o cooperate in the establishment of equitable arrangements ... to allow for the participation of developing land-locked States", etc. This duty, however, is abrogated (Article 71) "in the case of a coastal State whose economy is overwhelmingly dependent on the exploitation of the living resources of its exclusive economic zone".

Management and Conservation are responsibilities that limit the Grotian freedom to fish on the high seas (Articles 116-119), even though references to the duty of equitable sharing are conspicuously absent. Reservation for peaceful purposes is implicit, both with regard to the exclusive economic zone and the high seas, insofar as this part of ocean space itself is reserved for peaceful purposes (Article 88). This indeed is a remarkable departure from the 1958 Convention on the High Seas. One might therefore come to the conclusion that three of the four criteria determining a common heritage resource are applied by the Convention to the living resources of the sea. However, this application appears to be weak, half-hearted and lacking concreteness, compared to the detail lavished on the management of non-living resources in Part XI of the Convention!

The Straddling Stocks agreement[8] has made important contributions to strengthening the management provisions, and thus brought the living resources even closer to a common heritage status. It has done this in three ways. First, it has strengthened the management role of regional and subregional fisheries commissions.[9] Secondly, it has greatly strengthened the *enforcement mechanism* on a regional basis.[10] And

---

[8] Agreement for the Implementation of the Provisions of the United Nations Convention on the Law of the Sea of 10 December 1982 Relating to the Conservation and Management of Straddling Fish Stocks and Highly Migratory Fish Stocks, 12 October 1995.

[9] See, e.g., Part III, Article 8. Only States which are members of the competent regional fisheries commission or undertake to apply the management measures adopted by these commissions are allowed to fish on the high seas within the region concerned. Article 13 mandates the "strengthening of existing organizations and arrangements". Article 10 lists the duties and responsibilities of regional fisheries commissions or arrangements. Where no competent regional commission or arrangement exists, it has to be established (Articles 8, 9).

[10] Articles 20, 21. Any State Party to the Agreement may board and inspect any ship of any other party, regardless of whether it is a member of the fisheries commission concerned. The inspector then has to notify the flag State, but if the flag State fails to respond, the inspecting State can take all necessary

thirdly, and perhaps most importantly, it has imposed the duty on coastal States to harmonise their conservation measures with those agreed upon within the regional Commission for the High Seas beyond the jurisdiction of the coastal State.[11] This was a hotly debated question, and that it could be resolved signifies progress in the direction of a common heritage régime for living resources.

The fourth criterion determining a common heritage resource, that is, that it cannot be appropriated, is missing altogether, both in the Convention and in the Agreement. Indeed, the voices of those who advocate the establishment of "property rights" over this "common resource", for instance, in the form of "individual transferable quotas" (ITQs), are getting stronger. They claim that the "privatisation" of the fishing industry and its governance by the invisible hand of the market is the only way to save it from extinction. It is, however, not too hard to counter their arguments. Quite apart from ethical and equity considerations, the proposition does not make sense from a resource-economic perspective. Thus Partha Dasgupta:

> "Now in many cases of externalities it may be impossible, or at any rate difficult, to define property rights, let alone establishing them legally and then enforcing them...".

And he points out:

---

enforcement measures, including taking the ship into one of its ports. Basic procedures for boarding and inspecting are detailed in Article 22. Article 23 lists the rights and the duty of port States with regard to inspecting and detaining ships voluntarily in their ports. This is an extension of the port State régime in the Convention, relating to violations of environmental and fisheries regulations. It should also be noted that, in the Convention, the port State has the right to take enforcement measures. In the Agreement it has not only the right, but also the duty.

[11] Article 7 establishes that coastal States and States whose nationals fish for straddling or highly migratory stocks in the adjacent high seas shall seek, either directly or through the fisheries commissions or other mechanisms, to agree upon the measures necessary for the conservation of these stocks both within the EEZ and in the adjacent high seas. Compatibility of conservation and management measures is of course essential both for the coastal State and for the conservation of the stocks on the high seas. Article 7 stipulates that "Conservation and management measures established for the High Seas and those adopted for areas under national jurisdiction shall be compatible in order to ensure conservation and management of the straddling fish stocks and highly migratory fish stocks in their entirety". If there is no agreement, the dispute is subject to mandatory peaceful settlement in accordance with Part VIII of the Agreement, which is in accordance with Part XV of the Convention.

"[T]here are many circumstances in which market solutions do not sustain an efficient allocation of resources. Many such situations can be described by saying that certain essential markets do not exist. Sometimes they just happen not to exist for accidental or historical reasons; sometimes there are logical reasons why they cannot exist; sometimes the nature of the physical situation keeps them from existing, or makes them function wrongly if they do exist. *It happens that industries producing (or using) renewable and non-renewable resources are especially vulnerable to these difficulties...*".[12]

If, by some misfortune, the establishment of property rights such as ITQs, privatisation and resource allocation entrusted solely to market mechanisms were to succeed, this most certainly "would make the system function wrongly". It would concentrate fishing power in the hands of large companies, forcing the artisanal fisher out of business. While reducing the number of fishers, it would not reduce fishing effort. Cut-throat competition between large companies and among fishing States would continue or increase, continuing the present trend of making the rich richer and the poor poorer and accelerating the ineluctable depletion and extinction of species after species. Rather than ignoring the question of property rights over the resource, it would seem necessary and urgent to deal with it in the sense that the living resources of the oceans cannot be owned, but must be managed for the good of mankind as a whole, including future generations, and with particular consideration for the needs of the poor. Thus complementing what is already there, this would make the living resources part of the common heritage of mankind.

IV. The Holy See and the Common Heritage of Mankind

On 19 May 1978, the Delegate of the Holy See, Msgr. Silvio Luoni, made an important statement before the Second Committee of UNCLOS III, which dealt, among other subjects, with the conservation, utilisation and management of living resources. Msgr. Luoni justified the participation of the Holy See in UNCLOS III on the basis of its character as a universal institution. "As such", he said, "the Holy See anxiously looks towards the adoption of measures capable of guaranteeing the common good as such, that is the peace of the international community".[13] This requires the abolition of presently existing injustice

---

[12] P. S. Dasgupta and G. M. Heal, *Economic Theory and Exhaustible Resources*, 1979, 190-191, emphasis added.
[13] Statement by the Delegate of the Holy See before the Second Committee of UNCLOS III, 7th Session, 19 May 1978. Original in French. Copy on file

and the suppression of the root causes of possible future injustices. The Holy See did not intend to contribute technical solutions to the Conference, but rather to deal with principles which could guarantee just and equitable solutions for the whole international community, and, in this context, "first of all, the statement of that tenet universally accepted, at least in theory, that the sea is 'a common heritage of mankind'".[14]

"Moreover", he continued, "this principle is part of the wider concept of the 'universal purpose of creation', it has already been applied to the States in regard to their own national territory, not as restriction to their sovereignty, but for the exploitation and use of their natural resources in such a fashion as to take into account the needs of all mankind and, especially, of that part of mankind belonging to States with limited resources".[15] He expressed his deep concern about tendencies he could not fail to observe at this conference: national ambitions with regard to the uses of ocean space, which were in flagrant contradiction with this principle, and he endorsed instead the calls for a New International Economic Order which should not aim at the grabbing of natural resources for the benefit of a few privileged and, above all, geographically advantaged States, but that these resources should be shared equitably among all peoples in accordance with their real needs. He pointed out that the sovereignty of coastal States would necessarily be subject to important restrictions, and that this was generally recognised even for the territorial sea, and even more so for the exclusive economic zone, especially with regard to the migration of fish. "It would seem logical to affirm that such of restrictions on sovereignty apply also the resources living in an economic area. That means that the coastal States with an abundance of living resources have the duty to share them with other States, particularly the less fortunate, and therefore that the latter acquire some rights on these resources. This must be an assured right, which means that the criteria for this sharing and its effective implementation cannot be left to the discretionary power and the good will of the coastal States, but that measures and regulations must be laid down to give effect to the implementation of this right".

The Apostolic Delegate could ground his proposal in a tradition as old, almost, as the Church itself. In a brilliant essay, entitled "The Common Heritage of Mankind – A Roman Catholic View", delivered at *Pacem in Maribus* in Moscow, June 1989, Father Peter Serracino Inglott of Malta cited St. John Chrysostom, a fourth-century bishop of a city in Asia. He said:

> with the author, courtesy of the Permanent Observer Mission of the Holy See to the United Nations. For the full text in French, see also R. Platzoeder, *Kommentar zum Seerechtsübereinkommen der Vereinten Nationen*, 2001, 76.

[14] *Ibid.*
[15] *Ibid.*

"Mark the wise dispensation of God! ... He has made certain things common, such as the sun, air and the ocean.... Their benefits are dispensed equally to all as brethren.... It is as if nature itself becomes indignant ... when we seek to divide and separate ourselves by appropriating such things.... Therefore, the (opposite) condition is rather our heritage and more in agreement with nature".[16]

As Father Peter pointed out, this statement "indeed literally contains almost all the key words in Ambassador Pardo's famous 1967 motion at the United Nations".[17] Dozens of quotes, from other Church Fathers could be added to this.

Father Peter drew attention to the distinction between *res nullius*, unowned resources capable of either private or national appropriation, *res communis*, as "not liable to private ownership, and *res communis omnium*, which "is also not to be subject to national sovereignty".[18]

It is this latter concept that the common heritage concept proposed by Ambassador Pardo was to replace. Father Peter sees the common heritage as

"a totality of resources not necessarily material which, because of their very nature, should not only *not* be appropriated – neither by individuals, nor by groups, nor by States, nor by groups of States; but also should be *managed* – on behalf of mankind as a whole (including future generations) and managed appropriately (i.e., participatively) through legally constituted institutions. The crucial difference here is that the universality of destination of the resources is respected through the right to share in the management of the resource rather than through unimpeded physical access to it, and through the right to share in the benefits of the use of the resource, rather than through directly picking up bits and pieces of it. It is because of this positive requirement – the need of appropriate (participatory) management rather than mere non-appropriability – that such resources constitute a specific category". (emphasis original)

In the tradition of the Catholic Church, Father Peter points out, "by nature *all* earthly resources have a universal destination, that is, they are intended for the good of mankind as a whole", a theme taken up by the Holy See's Representative, Msgr. Antonio del Giudice at the first

---

[16] Unpublished, copy on file with author.
[17] *Ibid.*, see also note 3 above.
[18] *Ibid.*

working session of UNCLOS III in Caracas, 12 July 1974.[19] The real importance of the International Seabed Authority would be the fact that it gave a legal and institutional form to the concept of the common heritage of mankind, which, in Catholic doctrine, comprises all the earth's resources, obviously including the living resources of the seas and oceans. "This concept has the strong support of the Catholic Church", Msgr. del Giudice stated, "as may be seen from her social teaching".[20] In his encyclical *Mater et Magistra*, Pope John XXIII stated, "According to the plan of creation, the goods of the earth are above all destined for the worthy support of all human beings".[21]

V. Food and the Common Heritage of Mankind

The ocean's living resources make a major contribution to world food, especially in poor countries where, in many cases, they provide the largest part of animal protein consumed by coastal populations. Food, in any case, is made of the earth's resources and thus, in accordance with Catholic doctrine, should be considered part of the common heritage of mankind. There is, however, at least one author, Judge Mohammed Bedjaoui of Algeria, who considers food as a whole, food as such, a common heritage of mankind. In various essays,[22] he deals with the basic human right to food as part of the right to development, which he considers *jus cogens*.

> "There is no place in such an analysis for charity, the 'act of mercy', considered as being a factor of inequality from which the donor expects tokens of submissiveness or political flexibility on the part of the receiving State. The concept of charity thus gives place to that of justice".[23]

This is entirely in line with Pardo's thinking, who repeatedly stated that the common heritage concept changes the relationship between rich and

---

[19] Copy on file with the author, courtesy of the Permanent Observer Mission of the Holy See to the United Nations.
[20] *Ibid.*
[21] *Ibid.*
[22] See, for example, "The Right to Development", in M. Bedjaoui (ed.), *International Law: Achievements and Prospects*, 1991; "Propos libres sur le droit au développement", in *Le droit international à l'heure de sa codification. Etudes en l'honneur de Roberto Ago*, 1987; "Les ressources alimentaires essentielles en tant que 'patrimoine commun de l'humanité'", *Revue algérienne des relations internationales* 1 (1986); "Are the World's Food Resources the Common Heritage of Mankind?", *Indian JIL* 24 (1984).
[23] Bedjaoui, "The Right to Development", see note 22, 1196.

poor countries, as it implies that there are no "donors" or "recipients" because both have an inherent right to their equitable share of the common heritage.[24]

And Bedjaoui continues:

"What belongs to the international community and is 'the common heritage of mankind' should be shared among all States in accordance with the maxim *to each according to his needs*'. This therefore implies an element of *jus cogens*".[25]

Like Pardo, the Fathers of the Church and the founders of other great religions, Bedjaoui emphasises the universal importance of the common heritage concept:

"There can be no denying that this innovative concept, the common heritage of mankind, is capable of giving world-wide solidarity a wealth of practical expression. It might prove especially productive for the future of world relations and be applied not only to the resources of the sea-bed and of space (those of the moon and celestial bodies), as is already the case, but also to the land, the air, the climate, the environment, inert or living matter and the animal and vegetable genetic heritage, the wealth and variety of which it is vital to preserve for future generations. It might also provide insights and suggest attractive solutions to questions such as those concerning the cultural and artistic property of the globe, just as it could, and even should, apply in the first place to the human race, the first common heritage of mankind, and to mankind itself, a new subject of international law and the supreme heritage that must at all costs be saved from the threat of mass destruction".[26]

Bedjaoui is fully aware that his proposal to declare "basic world food stocks" to be part of the common heritage of mankind will be considered utopian in these rough times. He is confident, however, that this will change.

---

[24] See, e.g., A. Pardo and E. Borgese, *The Law of the Sea and the New International Economic Order*, 1977.
[25] Bedjaoui, "The Right to Development", see note 22, 1192.
[26] Bedjaoui, *ibid.*, 1196-1197, cites the 16th-century Spanish lawyer Vitoria asserting that the Christian Holy Scriptures intended "the goods of the earth" for "the whole of the human race", "for common use" and for "a universal purpose". He also refers to "the spirituality of the seventh century when the Koran announced to all mankind that 'all wealth, all things, belong to God and thus to all members of the human community'".

> "It is dangerous to write off the aspirations of four billion human beings by dismissing them too readily as being no more than fevered incantations. What we are advocating is that 'the world food stocks' essential to life, that is to say principally *grain* stocks, be declared to be the 'common heritage of mankind' so as to guarantee every people the vital minimum of a bowl of rice or loaf of bread in order to eradicate the monster which kills fifty million human beings a year".[27]

He even goes so far as to suggest an "immediate and provisional first step" to bring this "new world food order" into being. This should be the establishment of a universal agency, provided with an operational administration, which might be called the International Fund for Food Stocks (IFFS). It would have a budget with funds provided from a tax levied in each State on manufactured products of high added value made of raw materials from third-world countries, and/or by a one-per cent tax on military budgets. This agency would function as an equalisation fund, subsidising the purchase of food stocks or buying them from the food-producing countries and making them available to countries with a food deficit at a token price, which might later on be replaced by a general system of food distribution without charge.

In the present context, we might suggest an alternative "immediate and provisional first step" towards creating this new world food order by replacing grain with the ocean's living resources as an initial component of the world food stock to be declared a common heritage of mankind. Since the oceans are already regarded as such and a system of governance of this common heritage is already emerging – has largely emerged already – it might be less "utopian" to complete this system of governance than starting a new one.

VI. Genetic Resources and the Common Heritage of Mankind

Genetic resources are evidently living resources, but while, in general, living marine resources consist of fish, crustaceans, molluscs, marine mammals, turtles and birds, as well as seaweed and algae, genetic resources include all of the above, as well as, in particular, the aquatic microfauna and the myriad of bacteria of the deep sea, on and under the sea floor, which have been discovered only in recent years. They form the basis of very peculiar, quite unearthly ecosystems, driven by chemosynthesis rather than photosynthesis, and may hold the key to our understanding of the origin of life on this planet. They flourish in

---

[27] *Ibid.*, emphasis original.

submarine areas of volcanic activity, in conditions of darkness, extremely high temperatures and pressures. Hence they have also been called "extremophiles" or "hyperthermophiles". The unique resistance they have developed against heat and pressure makes them particularly useful for the development of a number of bioindustrial and pharmaceutical processes, and bioprospecting for them has become part of big business.

The industries utilising these genetic resources are quite diversified. They include the pharmaceutical, waste treatment, food processing, oil and paper processing industries, as well as mining applications. The potential market for industrial uses of hyperthermophilic bacteria has been estimated at $3 billion per year.[28]

In a carefully documented paper Glowka points out:

"Hyperthermophilic bacteria are just one example of the commercial potential of microbial genetic resources from the Area; as research continues, other commercially interesting organisms may also be discovered. For example, there may be organisms that orchestrate processes for minerals transport and bioaccumulation of metals. These could be useful in bioremediation of hazardous waste. Other organisms could be useful in biomining applications. Viruses associated with the organisms of the Area, in particular hyperthermophilic bacteria, may provide new vectors useful in biotechnological applications. Researchers may also be able to isolate potential anticancer and antibiotic compounds from deep seabed bacteria or fungi associated with other macro-organisms, as they have in more accessible areas of the ocean. In short, the biodiversity of the seabed has hardly been explored, and we simply do not know what may exist".[29]

The Biodiversity Convention (1992), which entered into force on 29 December 1993, affirms that the conservation of biological diversity is a common concern of humankind (Preamble). It does not explicitly apply the concept of the common heritage of mankind to genetic resources. Its provisions, however, imply several of the attributes of the concept. The very purpose of the Convention is the conservation of biological diversity for future generations,[30] through management and the equitable

---

[28] L. Glowka, "The Deepest of Ironies: Genetic Resources, Marine Scientific Research, and the Area", *Ocean Yearbook* 12 (1996).

[29] *Ibid.*, The World Conservation Union has estimated that the deep sea may be home to 10 million species.

[30] "Determined to conserve and sustainably use biological diversity for the benefit of present and future generations". Preamble.

sharing of benefits derived from their use.³¹ Cooperation in the management of these resources is to enhance peace and friendly relations among States.³² It is only "non-appropriability" that is lacking among the attributes determining the common heritage status of the resource. The Convention indeed is based on the assumption of sovereign rights of States over their genetic resources and the right of industrial companies to acquire ownership through the controversial patenting of living resources. All these rights, however, are limited by considerations of the common good.³³

The Biodiversity Convention is a remarkably land-oriented document. It assures the rights of developing countries, of local, especially indigenous, communities, and their participation in the conservation and utilisation of genetic resources in areas under national jurisdiction. Marine resources are given consideration not so much by the Convention itself as by the Jakarta Mandate on the Conservation and Sustainable Use of Marine and Coastal Biological Diversity.³⁴ The conservation of biodiversity in international waters, including, in particular, the microfauna of the deep seabed and its subsoil, remains, for all practical purposes, a legal lacuna.

Clearly, something will have to be done

1. to protect the bioprospectors from conflicts with other users of the international area (it should be noted that the International Seabed Authority has the mandate to coordinate its own activities with other activities in the Area³⁵);

---

³¹ Article 19(2) provides: "Each Contracting Party shall take all practicable measures to promote and advance priority access on a fair and equitable basis by Contracting Parties, especially developing countries, to the results and benefits arising from biotechnologies based upon genetic resources provided by these Contracting Parties. Such access shall be on mutually agreed terms".

³² "Noting that, ultimately, the conservation and sustainable use of biological diversity will strengthen friendly relations among States and contribute to peace for humankind...". Preamble.

³³ Article 16 contains an amazingly strong provision to this effect: "(1) The Contracting Parties, recognizing that patents and other intellectual property rights may have an influence on the implementation of this Convention, shall cooperate in this regard subject to national legislation and international law in order to ensure that such rights are supportive of and do not run counter to its objectives". In other words, conservation, community and equity interests take precedence over private property interests.

³⁴ First version, October 1998, issued by the Secretariat of the Convention on Biological Diversity for the CBD Roster of Experts on Marine and Coastal Biological Diversity.

³⁵ Article 147.

2. to protect and study the resources, which is within the mandate of both the Biodiversity Convention[36] and the International Seabed Authority[37]; and
3. to live up to the spirit of partnership and benefit and technology sharing that pervades all recent conventions, laws and regulations intended to save our environment in order to save ourselves. Politicians, the business world, academia, non-governmental organisations, all agree today that it is impossible to attain this goal without the cooperation of the developing countries, that is, the vast majority of humankind. If they are to cooperate, however, they have to have the necessary technologies, and must be fully included in the new phase of the industrial revolution in which genetic resources will play a major role.

Since the protection of genetic resources is the responsibility both of the International Seabed Authority and the Biodiversity Convention, it can be achieved only through a joint undertaking of both régimes. In his report to the meeting of the parties to the Law of the Sea Convention in April 1996, the UN Secretary-General exhorted States,

> "particularly States Parties to the Law of the Sea Convention which are also parties to the Convention on Biological Diversity, to coordinate their activities particularly with respect to the conduct of reviews of the relationship between the two conventions, the identification of additional measures that may need to be taken, including the possible development of new or additional international rules".[38]

And in his report to the General Assembly the same year, he stressed

---

[36] "Each Contracting Party shall, as far as *possible and as appropriate, cooperate with other Contracting Parties, directly, or, where appropriate, through competent international organizations, in respect to areas beyond national jurisdiction an*d on other matters of mutual interest, for the conservation and sustainable use of biological diversity. Article 5, emphasis added. The "competent international organization", in this case, is clearly the International Seabed Authority; the "area beyond national jurisdiction" is the international seabed area.

[37] Article 145(b) mandates the Authority with "the protection and conservation of the natural resources of the Area and the prevention of damage to the flora and fauna of the marine environment".

[38] United Nations Convention on the Law of the Sea, Meeting of States Parties, Fifth Meeting, New York, July 24-2 August 1996, *Report of the Secretary-General under Article 319 of the UNCLOS*, SPLOS/6, 11 April 1996, para. 46.

> "The general subject of marine and coastal diversity, as well as the specific issue of access to the genetic resources of the deep seabed, raise important questions....
>
> The specific issue of access points to the need for the rational and orderly development of activities relating to the utilisation of genetic resources derived from the deep seabed area beyond the limits of national jurisdiction".[39]

The rules and regulations of a joint law of the sea/biodiversity régime should not be burdensome for the industry. They could be formulated as a protocol and adopted by the parties to both Conventions. By way of a preamble, it might be recalled that, in accordance with the United Nations Convention on the Law of the Sea, the seabed and ocean floor and subsoil thereof beyond the limits of national jurisdiction, and its resources, are the common heritage of mankind. All rights in such resources are vested in mankind as a whole, on whose behalf the International Seabed Authority acts. The objectives of these regulations would be:

1. the conservation of biological diversity in the Area
2. the sustainable use of its components
3. the fair and equitable sharing of benefits arising from the use of genetic resources
4. participation of developing countries in the bioindustries
5. the precautionary approach and intergenerational equity, and
6. international cooperation in technology development in a sector likely to be of primary economic importance in the 21st century.

It might also be stipulated that the use of genetic resources from the Area for purposes of biological warfare is prohibited.

The first part of a preamble of this kind is taken from the Regulations on Prospecting and Exploration for Polymetallic Nodules in the Area, the second part summarises the purposes of the Biodiversity Convention, and the final provision is taken from the Andean Pact Common Régime on Access to Genetic Resources.

The most important substantive point should be that bioprospectors should notify the Authority of their intention to engage in bioprospecting, with an exact description of the area in which they intend to work, a clear statement of the aims and objectives of the project, of the time frame and methodology and, if applicable, a

---

[39] United Nations General Assembly, Fifty-first Session, Agenda Item 24(a), Law of the Sea, *Report of the Secretary-General*, A/51/645, 1 November 1996, para. 231.

statement on how local communities possessing and traditionally utilising the same or similar genetic resources will participate in the project.

Such guidelines have already been elaborated by the University of the South Pacific and could easily be adapted to the requirements of the international seabed.

Prospecting for minerals in the international seabed area is subject to licensing, but without cost. Bioprospecting, of course, is different from prospecting for minerals. It is not followed by "exploration" and "exploitation", which would be subject to payment of a fee. All subsequent testing and developing is undertaken on land, under national jurisdiction. It would be fair, therefore, if, in return for the supervisory and coordinating activity of the Authority, guaranteeing the safety of the bioprospector, the bioprospector would be required to pay a modest fee upon the conclusion of an access agreement.

Provisions for the protection and preservation of the marine environment should be harmonised with those contained in the Authority's mining code, adopted in 2000.

The Authority's participation in scientific research, including biotechnological research activities based on genetic resources, fair and equitable sharing of research and development results and commercial and other benefits derived from genetic resource use, and access to and transfer of technology making use of genetic resources, should be determined in accordance with the provisions of the Biodiversity Convention, in particular Articles 15, 16 and 19. There appears to be a general consensus that joint ventures in R&D and technology co-development, funded partly by the partners (of which the Authority should be one) States, and international (GEF, UNDP, etc.) or bilateral funding agencies, are the most suitable instrument to achieve these goals.

It would seem that this kind of régime would serve the best interests of the industry as well as those of the International Seabed Authority and the parties to the Biodiversity Convention. It most certainly would enhance progress in exploring the living creatures in international waters, including the seabed which still has countless secrets to disclose. The fact is that only two to three per cent of the deep seabed has been explored thus far! Evidently, more than commercial interests are involved. Genetic resources, more than anything else, are our common heritage. Their exploration, bringing us face to face with the origin of life, should be the concern of all countries and people.[40]

---

[40] For all the foregoing, see International Ocean Institute. *The International Seabed Authority: New Tasks*, Proceedings, Leadership Seminar, Jamaica, 14-15 August 1999.

VII. Sustainable Development and the Common Heritage of Mankind

"Sustainable development" is a term that has been used, overused and abused in various ways to cover the most diverse intentions and activities. In the worst case it is a tautology or oxymoron. Development that is not sustainable, in the sense that it destroys its own resource and/or the environment, natural or social, in which it is supposed to take place, is no development at all. In the best case it is a concept of considerable complexity. In her Sir Peter Scott Lecture, delivered in Bristol on 8 October 1986, Gro Harlem Brundtland gave it the following definition, which is preferable to the oversimplified version in the "Brundtland Report".[41] In Bristol she said:

> "There are many dimensions to sustainability. First, it requires the elimination of poverty and deprivation. Second, it requires the conservation and enhancement of the resource base which alone can ensure that the elimination of poverty is permanent. Third, it requires a broadening of the concept of development so that it covers not only economic growth but also social and cultural development. Fourth, and most important, it requires the unification of economics and ecology in decision-making at all levels".[42]

In this perspective, "sustainable development" has environmental, economic, ethical (equity), legal and institutional implications. This may have a familiar ring, because it takes us back to the opening pages of this essay, to the definition of the concept of the common heritage of mankind. The "attributes", "aspects" or "dimensions" are identical in both cases.

What, one may ask, about the disarmament dimension – the reservation for peaceful purposes? Principle 25 of the Rio Declaration holds the answer to this question: "Peace, development and environmental protection are interdependent and indivisible". Sustainable development rests, depends, on peace and security. Without peace and security there can be neither economic development nor protection of the environment. At the same time, there can be neither peace nor security without equitable economic development, including the elimination of poverty, and without environmental conservation or environmental security.

Unfortunately, Agenda 21 ignores this interdependence and indivisibility, and the whole structure of the UN system is still too sectoral to take up the challenge. There are, however, new beginnings

---

[41] UNCED, *Our Common Future*, 1987.
[42] See E. M. Borgese, *The Oceanic Circle*, 1998.

which can be developed so as to transcend the sectoral approach and consider the closely interrelated problems of the ocean space as a whole. The most important of these is the General Assembly's newly established Consultative Process (UNICPOLOS) which has in fact already begun to look at the enforcement and security aspects of sustainable development, especially in a regional context.

What about the non-appropriability aspect of the common heritage concept? The sustainable development concept does to the Roman Law construct of private property or ownership what the common heritage concept does to the Grotian construct of sovereignty. Both sovereignty and ownership – the sovereignty of the individual – are being transcended and transformed.

It is the conclusion of this essay that the whole sustainable development process will either come to naught, or will have to be based on the concept of the common heritage of mankind: not only in the oceans, that great laboratory for the making of a new world order, but globally. In accordance with the cultures of the vast majority of humankind, its application must be extended from the wealth of the oceans to wealth in general, not to be owned by humankind, whether individually or collectively, but to be held in trust, and to be administered on the basis of cooperation between civil society and the institutions of governance, at local, national, regional and global levels, with special consideration for the needs of the poor.

VIII. Conclusion

Shigeru Oda was right and prophetic, three years ago, 30 years ago, 50 years ago, when he predicted and advocated that the living resources of the oceans must be declared a common heritage of mankind. Our present, market-based economic system is failing us miserably, giving all the wrong incentives and leading, ineluctably, to conflict, degradation and extinction. This is a market failure of the first magnitude. One might mention, in this context, that our Western economic system as a whole must be considered a war system: both historically and ideologically. Historically, because it developed in an era of aggressive expansionism, the conquest of the world by Western Europe; ideologically, because it is based on competition and conflict rather than on cooperation. Thus it is part of a "culture of war".

What we are striving to build today is a "culture of peace". The new law of the sea is at the vanguard of this effort, and the concept of the common heritage of mankind is fundamental to it.

Rome was not built in a day. It would not be realistic to think the common heritage concept could be applied universally tomorrow. If we are not to catapult into Utopia, we must envisage a step-by-step process, without, however, losing sight of the whole. The ocean's living resources, constituting part of the world's basic food stock as well as of its biodiversity, are the most obvious next candidate. The time has come, Shigeru Oda, and we do not really have any choice. For if we do not act, these resources are doomed.

The next step, not into Utopia, but into the future, is greatly facilitated by what has already been achieved. Willy-nilly, *nolens volens*, the international community has already gone more than half way. Mechanisms, such as community-based co-management of fisheries, codes of conduct, regional fisheries commissions, are already in place and do not need to be invented. What is needed now is a new Protocol or Agreement, building on everything that has been built, putting it together in a consistent architecture in the context of the emerging "culture of peace" of the 21st century, which will have to comprise an "economics of peace" based on the concept of the common heritage of mankind. The world should give you such a Protocol, Shigeru, on the occasion of your next birthday!

Das Seerecht als Motor des internationalen Umweltrechts

Gedanken zu neueren Entwicklungen im Bereich des Tiefseebergbaus

Michael Bothe

Das Seerecht, dem ein so großer Teil des Lebenswerks des Jubilars gewidmet ist, stellt seit Grotius einen zentralen Bereich des Völkerrechts dar, von dem entscheidende Impulse ausgingen.[1] Seit der frühen Neuzeit war eben die See der Raum, in dem und um den sich ein Gemeinschaftsinteresse der Staaten entwickeln konnte. Bestand dieses gemeinsame Interesse in den vergangenen Jahrhunderten zunächst in der gleichen Freiheit der Nutzung, so trat im 20. Jahrhundert der Gedanke in den Vordergrund, daß die See und die in ihr enthaltenen Naturschätze Gemeinschaftsgüter sind, die nicht nur zu nutzen, sondern auch zu bewahren sind. Dies ist die Grundlage für die Frage, welche Impulse im einzelnen vom Seerecht für die Grundkonzepte des modernen Umweltvölkerrechts ausgegangen sind und noch ausgehen. Dieser Frage sei der vorliegende Beitrag gewidmet. Aktueller Anlaß dieser Fragestellung ist die Tatsache, daß am 30. Juli 2000 die Versammlung der internationalen Meeresbodenbehörde „Regulations on Prospecting and Exploration for Polymetallic Nodules in the Area" angenommen hat, deren Stellenwert in der Entwicklung des heutigen Umweltrechts zu bestimmen ist.[2]

I. Das Seerecht als Wegbereiter

Noch bevor der Begriff Umweltrecht zum Bestandteil eines nationalen und internationalen rechtspolitischen und rechtlichen Diskurses wurde, entwickelten sich im Seerecht Regeln, die heute ein wesentlicher Bestandteil des Umweltrechts sind, nämlich die Regeln über den

---

[1] W. G. Grewe, *The Epochs of International Law*, 2000, 257 et seq., 403 et seq., 551 et seq., 689 et seq.
[2] Doc. ISBA/6/A/18.

Schutz der See vor Ölverschmutzung.³ Die Konvention über die Verhütung der Verschmutzung der See durch Öl von 1954 ist ein wichtiger Vorläufer des heutigen Umweltvölkerrechts. Der Torrey Canyon Unfall von 1967 war dann eines der Schockerlebnisse, der die internationale Debatte um den Schutz der Umwelt entscheidend motiviert hat.⁴ Die daraus entstandenen neuen Verträge vom 29. November 1969, nämlich die Konvention über die Intervention auf Hoher See im Fall von Ölverschmutzungsunfällen⁵ und die Konvention über Haftung für Ölverschmutzungsunfälle⁶ waren beide für Entwicklungstendenzen im internationalen Umweltrecht richtungsweisend. Die erstgenannte Konvention mußte mit dem traditionellen, aus dem Gedanken staatlicher Souveränität entwickelten Grundsatz der ausschließlichen Zuständigkeit des Flaggenstaates eines Schiffes aufräumen. Sie führte damit, und dies kennzeichnet das heutige Umweltrecht, zu einem neuen Ausgleich zwischen Souveränitätsinteressen unterschiedlicher Staaten. Die Haftungskonvention ist die erste einer Vielzahl völkerrechtlicher Regelungen, die, so würde es heute formuliert, Haftung als ökonomisches Instrument des Umweltschutzes vorsehen. Zukunftsweisend war, daß die Haftung des privatrechtlich Berechtigten, und nicht eine Staatenhaftung der Kernpunkt der Regelung war. In der Tat ist in der Folgezeit das Recht der Staatenhaftung für den Bereich des Umweltschutzes weitgehend bedeutungslos geworden, während die transnationale privatrechtliche Haftung einen wichtigen Stellenwert besitzt.⁷

Ein anderer wichtiger, für das heutige Umweltvölkerrecht ausschlaggebender Leitgesichtspunkt ist die Entwicklung des Konzepts oder genauer der Grenzen der Freiheiten der Nutzung der Hohen See. Art. 2 Satz 3 der Genfer Konvention von 1958 über die Hohe See

---

[3] M. Bothe, „Oil Pollution Conventions", in: R. Bernhardt (Hrsg.), *EPIL*, Vol. III, 772 et seq.

[4] R. H. Stansfield, „The Torrey Canyon", in: R. Bernhardt (Hrsg.), *EPIL*, Vol. IV, 868.

[5] International Convention Relating to the Intervention on the High Seas in Cases of Oil Pollution Casualties, UNTS Vol. 970, 212.

[6] International Convention on Civil Liability for Oil Pollution Damage, UNTS Vol. 973, 3.

[7] L. Boisson de Chazournes, „La mise en œuvre du droit international dans le domaine de la protection de l'environnment: enjeux et défis", *RGDIP* 99 (1995), 37 et seq., 50 et seq., M. Bothe, „The Evaluation of Enforcement Mechanism in International Environmental Law", in: R. Wolfrum (Hrsg.), *Enforcing Environmental Standards: Economic Mechanisms as Viable Means?*, 1996, 13 et seq., 32; P. H. Sand, *Transnational Environmental Law*, 1999, 87 et seq.

formuliert den Gedanken der Gemeinverträglichkeit der Ausübung dieser Freiheiten.

„Diese sowie die sonstigen nach den allgemeinen Grundsätzen des Völkerrechts anerkannten Freiheiten werden von jedem Staat unter angemessener Berücksichtigung des Interesses ausgeübt, dass die anderen Staaten an der Freiheit der Hohen See haben".

Die weitere Entwicklung dieses Prinzips kann man als einen wesentlichen Beitrag des Seerechts zur Entwicklung des internationalen Umweltrechts betrachten.

II. Gemeinverträglichkeit, Nachhaltigkeit und Vorsorgeprinzip

Das Prinzip der Berücksichtigung der Interessen anderer Staaten wird bereits in den Konventionen von 1958 in vielfältiger Weise konkretisiert, etwa bei der Frage der Rücksichtnahme auf von anderen Staaten verlegte Tiefseekabel oder der Konkurrenz zwischen der Nutzung des Festlandsockels und Nutzungen der darüber liegenden Gewässer.

Am deutlichsten werden die Interessenkonflikte aber dort, wo Staaten eine Ressource durch Entnahme nutzen und die Erschöpfung dieser Ressource durch exzessive Entnahme droht. Dies ist vor allem das Problem der Fischerei. Mit dem Problem der Erschöpfung von Fischbeständen hatte sich die seevölkerrechtliche Entwicklung schon seit der ersten Hälfte des 20. Jahrhunderts zu beschäftigen.[8] Wenngleich ein großer Teil dieser Entwicklung auf regionaler Ebene geschieht, nehmen sich auch weltweite Abkommen des Problems an. Die Genfer Konvention von 1958 über Fischerei und Erhaltung der lebenden Naturvorkommen der Hohen See formuliert den Grundgedanken in allgemeiner Weise. Sie verpflichtet die Staaten, zur Erhaltung der Naturvorkommen der Hohen See zusammenzuarbeiten. Diese Erhaltung wird in Art. 2 der Konvention wie folgt umschrieben:

„As employed in this Convention, the expression, 'conservation of the living resources of the High Seas' means the aggregate of the measures rendering possible the *optimum sustainable yield* from those resources as to secure a maximum

---

[8] Vgl. eingehend zur Geschichte R. Wolfrum, „Die Fischerei auf Hoher See", *ZaöRV* 38 (1978), 659 et seq., 666 et seq.

supply of food and other marine products. Conservation programmes should be formulated with a view to securing in the first place a supply of food for human consumption".[9]

Wenn man diese Bestimmung aus heutiger Sicht auch wegen ihres streng anthropozentrischen Charakters kritisieren muß, so ist doch hervorzuheben, daß hier im Umweltvölkerrecht der Begriff der Nachhaltigkeit *(sustainability)* eingeführt wird, lange bevor der Brundtland Bericht das Prinzip des *sustainable development* zu einem Zentralbegriff des internationalen Diskurses machte. Die Auseinandersetzung um die Frage, wie denn die zulässige nachhaltige Nutzung von Fischbeständen genauer zu bestimmen sei, welche Faktoren zur Bestimmung eines *optimum sustainable yield* zu berücksichtigen seien, war ein wichtiges Thema der Debatten der dritten Seerechtskonferenz.[10] Das Ergebnis ist ein ökologisch und ökonomisch modifizierter Ansatz des *maximum sustainable yield* (Art. 61 Abs. 3 und 119 SRÜ). Insbesondere sind ökologische Interdependenzen in Betracht zu ziehen.[11]

Der Gedanke der Nachhaltigkeit bedeutet einen Ausgleich der Interessen heutiger mit denen zukünftiger Nutzung. Er ist damit Bestandteil des Nachweltschutzes, der *intergenerational equity*.[12] Die Bestimmung dieses Ausgleichs ist eine der schwierigsten Fragen der Umweltrechtspolitik. Die Erhaltung von Freiräumen im Interesse zukünftiger Nutzung ist eine wesentliche Funktion des Vorsorgeprinzips *(precautionary principle)*.[13] Damit wird das Vorsorgeprinzip ein wesentliches Instrument zur Sicherung von Nachhaltigkeit, von nachhaltiger Entwicklung.[14] Das Vorsorgeprinzip hat sich zunächst im Umweltrecht der Staaten entwickelt, wobei unterschiedliche Staaten diesen Ansatz zunächst ganz unterschiedlich verfolgt oder auch eher beiseite gelassen haben. Der Begriff ist in der deutschen Debatte entstanden, wenngleich der Sache nach schon zu Beginn der 70er Jahre eine Reihe von Staaten Ansätze folgten, die in

---

[9] Hervorhebung des Verf.
[10] Wolfrum, siehe Anm. 8, 684 et seq.
[11] R. Wolfrum/V. Röben/F. L. Morrison, „Preservation of the Marine Environment", in: F. L. Morrison/R. Wolfrum (Hrsg.), *International, Regional and National Environmental Law*, 2000, 225 et seq., 233.
[12] Dazu grundlegend E. Brown Weiss, *In Fairness to Future Generations: International Law, Common Patrimony and Intergenerational Equity*, 1989.
[13] Vgl. dazu D. Freestone/E. Hey, „Origins and Development of the Precautionary Principle", in: D. Freestone/E. Hey (Hrsg.), *The Precautionary Principle and International Law*, 1996, 3 et seq., 12 et seq.
[14] Zu diesem Zusammenhang vgl. A. Epiney/M. Scheyli, *Strukturprinzipien des Umweltvölkerrrechts*, 1998, 87 et seq.

Deutschland unter diesem Begriff diskutiert wurden.[15] Das Vorsorgeprinzip fordert Maßnahmen zum Schutz der Umwelt, die nicht nur der Abwehr und dem Schutz vor bekannten Gefahren für Umweltgüter dienen, sondern auch dort und soweit greifen, wo solche Gefahren noch nicht unmittelbar nachweisbar sind. Es gibt eine Reihe typischer Instrumente zur Umsetzung des Vorsorgeprinzips, die in der Praxis der Staaten entwickelt wurden.

Bestandteil des völkerrechtlichen Diskurses wurde das Vorsorgeprinzip auf dem Weg über das Seerecht. Soweit ersichtlich taucht der Begriff zum ersten Mal in der Erklärung der zweiten Konferenz zum Schutz der Nordsee vom 25. November 1987 auf.[16] Seitdem wird er in einer Reihe von völkerrechtlichen Verträgen zum Schutze von Meeresressourcen verwandt.[17] Seinen Siegeszug in der internationalen und nationalen Praxis trat die ausdrückliche Formulierung dieses Prinzips allerdings durch die Anerkennung in Prinzip 15 der Rio-Erklärung an. Zwei auf der Rio-Konferenz angenommene Verträge (die Klimarahmenkonvention[18] und die Konvention über biologische Vielfalt[19]) anerkennen das Prinzip.

Auch durch die Rechtsprechung des Internationalen Seerechts-Gerichtshofs wurde das Prinzip der Sache nach in dem Bluefin Tuna-Fall anerkannt[20]:

„79. Considering that there is scientific uncertainty regarding measures to be taken to conserve the stock of southern bluefin tuna and there is no agreement among the parties as to whether the conservation measures taken so far have led to the improvement in the stock of southern bluefin tuna.

80. Considering that, although the Tribunal cannot conclusively assess the scientific evidence presented by the parties, it finds that measures should be taken as a matter of urgency to preserve the rights of the parties and to avert further deterioration of the southern bluefin tuna stock".

---

[15] Zu diesen Entwicklungen E. Rehbinder, *Das Vorsorgeprinzip im internationalen Vergleich*, 1991.
[16] Freestone/Hey, siehe Anm. 13, 5.
[17] Vgl. etwa Art. 2 (1)(a) des Übereinkommens zum Schutz der Meeresumwelt des Nordost-Atlantik vom 22. September 1992, *ILM* 32 (1993), 1069 (sog. OSPAR-Konvention).
[18] Art. 3 (3).
[19] Obwohl der Begriff in diesem Vertrag nicht gebraucht wird, ist das Prinzip der Sache nach in der Präambel formuliert.
[20] Order vom 27 August 1999, *ILM* 38 (1999), 1624.

Als Verhaltensmaßstab ist das Prinzip in hohem Maße konkretisierungsbedürftig. Denn das Prinzip beinhaltet zwei Schwellenprobleme. Zum einen erfordert es nicht blind jede nur mögliche Maßnahme, sondern Maßnahmen im Hinblick auf Risiken, die nicht ausgeschlossen werden können. Maßnahmen der Vorsorge setzen also eine gewisses Besorgnispotential voraus. Das Prinzip 15 der Rio-Erklärung formuliert dies so:

„Where there are threats or serious or irreversible damage...".

Dies gilt es im Einzelfall zu bestimmen. Zum zweiten unterliegen die zu ergreifenden Maßnahmen dem Verhältnismäßigkeitsprinzip, d.h. es muß ein angemessenes Verhältnis zwischen dem Besorgnispotential und den Belastungen bestehen, die die Maßnahmen für die betroffenen Akteure mit sich bringen. Auch das ist im Einzelfall zu bestimmen.

Es gibt freilich eine Reihe typischer Instrumente zur Umsetzung des Vorsorgeprinzips. Dazu gehören zum einen umfassende Methoden der Bewertung von Planungen und Vorhaben im Hinblick auf ihre Umweltauswirkungen (Umweltverträglichkeitsprüfung) und der Einsatz der jeweils verfügbaren besten Technologie (auch soweit dies zur Abwehr bekannter Gefahren nicht nötig wäre) und der besten Umweltpraxis.[21] Ihren Eingang in das Völkerrecht haben beide Instrumente im wesentlichen seerechtlichen Entwicklungen zu verdanken. Erste Ansätze für eine vertragliche Verankerung der UVP finden sich in den Abkommen von Oslo und London über die Abfallbeseitigung auf See aus dem Jahr 1972.[22] Den wesentlichen Durchbruch erzielte jedoch UNEP im Rahmen des *Regional Seas Programme*, beginnend mit der Konvention von Barcelona über die Reinhaltung des Mittelmeers und die folgenden Regionalabkommen zum Schutz der Umwelt des Persischen Golfs (1978), der west- und zentralafrikanischen Region (1981), des Südostpazifiks (1982), des Roten Meeres (1982), der Karibik (1983) und des Südpazifiks (1986). In politischen Dokumenten der Vereinten Nationen erscheint dann die Umweltverträglichkeitsprüfung etwa seit 1979.[23] Auch insofern sind

---

[21] Freestone/Hey, siehe Anm. 13, 13; Epiney/Scheyli, siehe Anm. 14, 126 et seq.

[22] Zu Einzelheiten vgl. M. Bothe/L. Gündling, *Neuere Tendenzen des Umweltrrechts im internationalen Vergleich*, 1990, 171 et seq.

[23] Ausgangspunkt waren die vom UNEP erarbeiteten „Principles of Conduct in the Field of the Environment for the Guidance of States in the Conservation and Harmonious Utilization of Natural Resources Shared by Two or More States", *ILM* 17 (1978), 1097: vgl. zum Ganzen Bothe/Gündling, siehe Anm. 22, 173 et seq.

zwei seerechtliche Dokumente bedeutsam, nämlich die von Arbeitsgruppen des UNEP erarbeiteten „Aspects Concerning the Environment Related to Offshore Drilling and Mining within the Limits of National Jurisdiction" und die „Montreal Guidelines on Protection of the Marine Environment against Pollution from Land-Based Sources.[24]

Was die Verpflichtung zur Verwendung der besten verfügbaren Technologie oder ähnliche Vorschriften angeht, so erscheint der erste Vertrag, der eine solche Begrifflichkeit vorsieht, wohl die Pariser Konvention von 1974 über Meeresverschmutzung vom Land aus zu sein. Art. 4 Abs. 3 dieser Konvention verpflichtet die Mitgliedstaaten, Regeln zu treffen, die „the latest technical developments" berücksichtigen. In der Folgezeit erscheint der Begriff der best available technology schwerpunktmäßig im Abkommen zur Luftverschmutzung, insbesondere im Genfer Abkommen 1979 über weiträumige Luftverschmutzung in Europa (Art. 6).

Moderne Definitionen des Begriffs der besten verfügbaren Technologie finden sich jedoch schwerpunktmäßig wiederum in seerechtlichen Abkommen, nämlich in Anhang I zum Übereinkommen zum Schutz der Meeresumwelt des Nordostatlantiks (1992) und in der Anlage 3 des revidierten Übereinkommens zum Schutz der Meeresumwelt des Ostseegebiets aus dem gleichen Jahr.

Es ist also festzustellen, daß das Seerecht nicht nur bei der Anerkennung des Vorsorgeprinzips im allgemeinen, sondern auch bei seinen beiden wesentlichen Instrumenten, nämlich der Umweltverträglichkeitsprüfung und der Verpflichtung auf die beste verfügbare Technologie, eine Pionierfunktion besessen hat.

III. Probleme der Rechtsdurchsetzung im Seerecht

Eine Konsequenz des Grundsatzes der Freiheit der Hohen See und insbesondere der Schiffahrtsfreiheit auf Hoher See ist es, daß kein Staat die Nutzung dieser Freiheit durch andere Staaten und deren Angehörige behindern darf. Dies führt zu einer ausschließlichen Jurisdiktion des Flaggenstaates über alle Schiffe, die seine Flagge führen. Dieses Prinzip wurde bereits in Art. 6 der Genfer Konvention von 1958 über die Hohe See kodifiziert und ist nunmehr in Art. 92 SRÜ bestätigt. Unter dem Gesichtspunkt der Durchsetzung völkerrechtlicher Pflichten und Beschränkungen hinsichtlich der Ausübung der Freiheit zur Nutzung der Hohen See ist dies in doppelter

---

[24] Vgl. Bothe/Gündling, siehe Anm. 22, 177 et seq.

Hinsicht problematisch. Zum einen hat der Flaggenstaat allein aufgrund der räumlichen Entfernung von dem Ort, an dem Rechtsverletzungen geschehen, kaum eine Möglichkeit, darauf angemessen zu reagieren. Zum anderen haben manche Flaggenstaaten (sog. Billigflaggen) auch ein wirtschaftliches Interesse, die Aufsicht über die Einhaltung relevanter Regeln nicht streng zu handhaben, weil eben dies die finanzielle Attraktivität ihrer Flagge ausmacht. Ansätze zur Lösung dieses Problems gehen in drei unterschiedliche Richtungen: verstärkte regionale Kooperation, Stärkung der Befugnisse der Küsten- und Hafenstaaten sowie eine verstärkte Rolle für internationale Institutionen. Die regionale Kooperation nimmt im wesentlichen zwei Formen an, nämlich einmal Vertragsregime zwischen den Anliegerstaaten bestimmter Meereszonen, zum anderen Vertragsregime zwischen Staaten, die bestimmte abgrenzbare Fischereiressourcen nutzen. Ersterer Typ wurde insbesondere von UNEP systematisch im *Regional Seas Programme* verfolgt.[25] Diese regionalen Abkommen können zwar gegenüber Nichtmitgliedern rechtlich gesehen das Flaggenstaatprinzip nicht aufheben, jedoch gehen offenbar von der Tatsache, daß die an einem Meeresgebiet und an einer bestimmten Ressource hauptsächlich interessierten Staaten zusammenarbeiten, wesentliche Impulse aus.[26] Bei Fragen wie der Meeresverschmutzung von Land aus ist dieses natürlich offensichtlich. Rechtlich gesehen wird die ausschließliche Hoheitsgewalt des Flaggenstaates dann ausgeschaltet, wenn dies in besonderen Abkommen vorgesehen ist, wie zum Beispiel in dem Abkommen über *straddling fish stocks.*[27] Art. 21 dieses Abkommens gibt den Teilnehmerstaaten von Regionalabkommen die Möglichkeit, im von einem Regionalabkommen erfassten geographischen Bereich auch die Schiffe von Drittstaaten zu kontrollieren, was eine bedeutsame Ausnahme vom Flaggenstaatprinzip darstellt.[28]

Die wichtigste Verstärkung der Rechte des Küstenstaates liegt zweifelsohne in der Entwicklung des Konzepts der Ausschließlichen Wirtschaftszone, wodurch ein nicht unbeträchtlicher Bereich, der zuvor zur Hohen See gehörte, nunmehr in Fragen des Umwelt- und Ressourcenschutzes der Hoheitsgewalt des am meisten betroffenen,

---

[25] Wolfrum/Röben/Morrison, siehe Anm. 11, 280 et seq.
[26] Vgl. dazu Wolfrum/Röben/Morrison, siehe Anm. 11, 247.
[27] Agreement for the Implementation of the Provisions of the United Nations Convention of the Law of the Sea of 10 December 1982, Relating to the Conservation and Management of Straddling Fish Stocks and Highly Migratory Fish Stocks vom 4. August 1995, *ILM* 34 (1995), 1542.
[28] Vgl. dazu Wolfrum/Röben/Morrison, siehe Anm. 11, 243 et seq.

nämlich des Küstenstaates unterworfen wird. Daneben hat die Seerechtskonvention wohl hauptsächlich eine Verdeutlichung der Rechte der Hafenstaaten mit sich gebracht.[29]

Dagegen ist die Rechtsdurchsetzung durch echte internationalen Organisationen (etwa die Inspektoren internationaler Überwachungskommissionen) noch deutlich unterentwickelt.

Die Frage, wie denn Umweltinteressen in bezug auf die See vertreten werden, findet damit alles in allem noch keine ganz zufriedenstellende Antwort. Insbesondere fehlt es weitgehend an unterstützenden Mechanismen, die die Zivilgesellschaft in diesen Prozeß einbezieht. Schutz der Meeresumwelt ist weitestgehend eine Sache staatlicher Bürokratien und der Kooperation zwischen ihnen. Dies unterscheidet den Meeresumweltschutz immer noch weitgehend von dem Umweltschutz zu Lande.

Was letzteren angeht, so wurden schon zu Beginn der umweltrechtlichen Entwicklungen Ende der 60er Jahre die Bedeutung des interessierten Bürgers als Promoter von Umweltinteressen entdeckt. Ein erstes Ergebnis dieses Befundes war die Entwicklung der amerikanischen Gesetzgebung über Bürgerklagen im Umweltrecht.[30] Streit herrschte immer noch über die Frage, ob und inwieweit der interessierte Bürger nur als „Nachbar", weil er in seinen eigenen rechtlich geschützten Interessen betroffen ist, oder ob er bzw. ob Verbände auch als Vertreter des Allgemeininteresses in diesem Zusammenhang agieren sollten. Dieser Ansatz, beide Varianten offen lassend, ist nunmehr durch die Aarhus-Konvention von 1998 jedenfalls für den Bereich der UN-Wirtschaftskommission für Europa völkerrechtlich abgesichert.[31] Die Frage der Bürger- oder Verbandsklage stellt sich überall dort, wo ein individuelles Recht an Umweltgütern nicht gegeben ist. Gerade hier ist die altruistische Bürgerklage besonders wichtig. In diesem Bereich ist sie auch in Deutschland, wo vielfach die Verbandsklage als allgemeines Instrument immer noch abgelehnt wird, inzwischen weitgehend anerkannt.[32] Auch die Umweltgüter der See bedürfen eines solchen Schutzes durch Vertreter der Zivilgesellschaften, die im Allgemeininteresse handeln. Dies ist bislang nur in lückenhafter Weise

---

[29] Art. 218 SRÜ. Vgl. dazu A. C. Kiss, „La pollution du milieu marin", ZaöRV 38 (1978), 902 et seq., 920 et seq.
[30] Vgl. Bothe/Gündling, siehe Anm. 22, 195 et seq.
[31] Convention on Access to Information, Public Participation in Decision-Making and Access to Justice in Environmental Matters vom 25. Juni 1998, ILM 38 (1999), 517.
[32] R. Schmidt, Einführung in das Umweltrecht, 4. Aufl. 1995, 176 et seq.

der Fall.³³ Allerdings sind es immer wieder Fischereiinteressen, die dazu eingesetzt werden, auch egoistische Interessenvertretungen für den Meeresumweltschutz dienstbar zu machen.³⁴

Ein wesentlicher Bereich in der Durchsetzung von Umweltrecht durch nicht staatliche Akteure ist allerdings der transnationalen privatrechtlichen Haftung, für dessen Entwicklung, wie bereits oben festgestellt, das Seerecht entscheidende Impulse gegeben hat. Die Regeln über die Haftung für Schäden, die durch Unfälle beim Transport von Öl entstanden sind, wurden in den 90er Jahren revidiert und durch einen Vertrag für die Haftung von Schäden im Zusammenhang mit der Beförderung sonstiger gefährlicher Güter ergänzt.³⁵ Haftung für andere Arten von Umweltschäden in der See ist bislang noch nicht vertraglich geregelt. Ein Übereinkommen zur Haftung für offshore-Aktivitäten von 1977 ist bislang nicht in Kraft getreten.³⁶ Auch für Schäden, die durch Abfallverbringung entstehen können, gibt es bislang keine völkerrechtliche Haftungsregelung, obwohl die Haftungsfrage vor nationalen Gerichten eine Rolle gespielt hat.³⁷

IV. Die neuen Regelungen im Tiefseebergbau

Vor dem Hintergrund dieser Trends ist nun ein Blick auf den Stellenwert und den Standort einer neuen Entwicklung zu werfen, nämlich die eingangs erwähnten Regelungen für das Aufsuchen und Erkunden von polymetallischen Knollen auf dem Boden der Tiefsee. Die Umweltauswirkungen solcher Erkundungen können erheblich sein,

---

[33] Zum begrenzten Zugang von NGOs zu Verfahren der Vertragsdurchsetzung im marinen Umweltrecht vgl. M. Bothe, „Compliance Control Beyond Diplomacy – The Role of Non-Governmental Actors", *EPL* 27 (1997), 293 et seq., 296.

[34] Zur Anfechtung von Genehmigungen zur Abfallbeseitigung auf See durch betroffene Fischer L. Gündling, „Abfallbeseitigung auf See", in: *Dokumentation zur 4. Wissenschaftlichen Fachtagung der Gesellschaft für Umweltrecht e.V.*, Berlin 1980, 67 et seq., 79 et seq.

[35] Ölhaftungsabkommen 1992, BGBl. 1994 II, 1152 und 1169; Fondsübereinkommen 1992, BGBl. 1996 II, 671 und 686; International Convention on Liability and Compensation for Damage in Connection with the Carriage of Hazardous and Noxious Substances by Sea of 1996, *ILM* 35 (1996), 1415.

[36] Bothe, siehe Anm. 3, 774; U. Beyerlin, *Umweltvölkerrecht*, 2000, 280.

[37] „Affaire des boues rouges" zwischen Frankreich und Italien 1975, dazu Sand, siehe Anm. 7, 92 et seq.

aber Teil XI des SRÜ enthält dazu keine Vorschriften. Freilich ist Teil XII über die Erhaltung der Meeresumwelt auch auf diese Aktivitäten anwendbar. Die dort formulierten Verpflichtungen sind jedoch sehr allgemeiner Art. Sie bedürfen der Ergänzung und Ausführung durch zusätzliche Regeln. Was das Gebiet, die Zone des Tiefseebergbaus, angeht, wird dem durch Art. 209 SRÜ Rechnung getragen, der auf die Regelungszuständigkeiten der Behörde nach Teil XI verweist. Nach Art. 160 Abs. 2 (f)(ii) in Verbindung mit Art. 162 Abs. 2 (o)(ii) bezieht sich diese Regelungsbefugnis unter anderem auf die „Prospektion, Erforschung und Ausbeutung im Gebiet". Dies ist die Rechtsgrundlage der genannten Regeln. Der wesentliche Inhalt dieser Regeln soll im folgenden im Hinblick auf die zuvor aufgezeigten Trends und Probleme analysiert werden.

1. Die Prüfung von Umweltauswirkungen

Die Regeln sehen die Sammlung umweltrelevanter Informationen umfassend vor. Der Antrag auf Erteilung einer Genehmigung zur Prospektion und Erforschung (Billigung eines Betriebsplans, *approval of a work plan*) muß sich in unterschiedlicher Weise mit den Umweltkonsequenzen der geplanten Tätigkeiten beschäftigen. Zunächst muß der Antrag eine schriftliche Verpflichtungserklärung enthalten, daß das Prospektionsunternehmen[38]

> „comply with the Convention and the relevant rules, regulations and procedures of the Authority concerning: ...
>
> protection and preservation of the marine environment;"

Ferner muß der Antrag darlegen, daß der Antragsteller

> „[is] financially and technically capable of carrying out the proposed plan of work".[39]

Für diesen Zweck muß der Antrag enthalten:

> „general description of the applicant's financial and technical capability to respond to any incident or activity which causes serious harm to the marine environment".[40]

---

[38] Reg. 3(4)(d)(i).
[39] Reg. 12(1).
[40] Reg. 12(7)(c).

Diese Anforderungen bedeuten, daß der Antragsteller eine kritische Selbstprüfung hinsichtlich der Umweltauswirkungen seines Vorhabens und hinsichtlich seiner Fähigkeit, die selben zu bewältigen, vornehmen muß.

Ferner muß der Antrag spezifische Informationen über die Umwelt enthalten und zwar die folgenden:

„(a) a general description and schedule of the proposed exploration programme, ... such as studies to be undertaken in respect of the environmental, technical, economic and other appropriate factors that must be taken into account in exploration;

(b) a description of the programme for oceanographic and environmental base line studies in accordance with these regulations and any environmental rules, regulations and procedures established by the authority that enable an assessment of the potential environmental impact of the proposed exploration activities taking into account any recommendations issued by the legal and technical commission;

(c) a preliminary assessment of the possible impact of the proposed exploration activities on the marine environment;

(d) a description of proposed measures for the prevention, reduction and control of pollution and other hazards, as well as possible impacts, to the marine environment;"[41]

Das bedeutet der Sache nach eine Umweltverträglichkeitsprüfung, die der Antragsteller mit seinem Antrag vorzulegen hat. Die Liste der erforderlichen Angaben enthält daß alle diejenigen, die zu diesem Zweck gemeinhin verlangt werden,[42] mit Ausnahme der Prüfung von Alternativlösungen und Angaben über mögliche Lücken im relevanten Wissen.

---

[41] Reg. 18.
[42] Vgl. etwa Anhang III der Umweltverträglichkeitsrichtlinie der EG (Richtlinie 85/337/EWG vom 27. Juni 1985); Anhang II zum Übereinkommen über die Umweltverträglichkeitsprüfung im grenzüberschreitenden Kontext vom 25. Februar 1991.

Die nächste Phase des Verfahrens ist die Prüfung des Antrags durch die Rechts- und Fachkommission. Dieses Gremium muß einige Feststellungen treffen, die die Grundlage seiner Empfehlung an den Rat darstellen. Dieser trifft die endgültige Entscheidung. Dazu gehört die Feststellung, daß der Betriebsplan

„(will) provide for effective protection and preservation of the marine environment...".[43]

Das bedeutet, daß die vom Antragsteller gelieferten Informationen jedenfalls hinreichend gehaltvoll sein müssen, um die Kommission in den Stand zu setzen, diese Feststellung zu treffen. Die Regeln enthalten jedoch keine Vorschrift darüber, ob die Kommission weitere eigene Untersuchungsbefugnisse besitzt. Das ist nicht unproblematisch, da keineswegs sicher ist, daß die durch die Angaben des Antragstellers gelieferte Informationsbasis wirklich hinreicht. Es gibt keine ausdrückliche Vorschrift darüber, daß die Rechts- und Fachkommission Zugang zu zusätzlichen Informationen besitzt, die die Richtigkeit der Angaben des Antragsstellers in Frage stellen könnten.

In den folgenden Phasen der Aktivität, nämlich der tatsächlichen Exploration, wird die Informationssammlung fortgesetzt. Dies wird im einzelnen durch die Vorschriften des Vertrags festgelegt, der zwischen der Behörde und dem Vertragsnehmer, d.h. dem Unternehmen, das die Tätigkeit ausführen will, zu schließen ist. Über den Inhalt dieses Vertrages enthalten die Regeln das Folgende:

„Each contract shall require the contractor to gather environmental base line data and to establish environmental base lines ... against which to assess the likely effects of its programme of activities under the plan of work for exploration on the marine environment and a programme to monitor and report on such effects...".[44]

„Contractors, sponsoring States and other interested States or entities shall cooperate with the Authority in the establishment and implementation of programmes for monitoring and evaluating the impacts of deep seabed mining on the marine environment".[45]

---

[43] Reg. 21 (4)(b).
[44] Reg. 31 (4).
[45] Reg. 31 (6).

"If the contractor applies for exploitation rights, it shall propose areas to be set aside and used exclusively as impact reference zones and preservation reference zones...".[46]

Diese Zonen werden besonders überwacht und dienen als eine Grundlage für die genaue Feststellung der Umweltauswirkungen der Tätigkeit. Diese Vorschriften sind in der Tat wesentlich als Voraussetzung für eine wirklich aussagekräftige Bewertung der Umweltfolgen der Prospektion und Exploration.

Abschließend kann also gesagt werden, daß hier eine wirklich umfassende und auch aussagekräftige Prüfung und Bewertung von Umweltfolgen gesichert ist, die im internationalen Vergleich als durchaus vorbildlich angesehen werden kann. Man darf sie als einen weiteren positiven Beitrag des Seerechts zur Entwicklung des Umweltvölkerrechts ansehen.

2. Die Intransparenz der Entscheidungsprozesse

Oben wurde auf die Grundregel guter Umweltpolitik hingewiesen (die auch Eingang in das Umweltvölkerrecht gefunden hat), daß Entscheidungsprozesse offen sein müssen. Dazu gehört öffentlicher Zugang zu Umweltinformationen und die Beteiligung der Öffentlichkeit an Entscheidungsverfahren, an die sich dann die Möglichkeit gerichtlichen Rechtsschutzes anschließt.

In diese Richtung enthalten die Regeln nichts, ganz im Gegenteil. Der Entscheidungsprozeß ist nach außen abgeschlossen. Der Antragsteller liefert die bereits dargestellten Informationen an die Behörde. Dies ist die Entscheidungsgrundlage des Rates, der den Betriebsplan billigt. Die Rechts- und Fachkommission trifft die entscheidenden Feststellungen, ohne daß die Rede von irgendwelchen zusätzlichen Informationsquellen wäre. In den weiteren Phasen der Tätigkeit muß der Vertragsnehmer, der der Aufsicht der Behörde unterliegt, weitere Informationen über die Umweltauswirkungen der Aktivität liefern und sammeln. Von einer Öffnung des Informationsflusses ist nicht die Rede.

Der geschlossene Charakter des Entscheidungsprozesses wird noch dadurch verstärkt, daß es strenge Regeln über die Vertraulichkeit der an die Behörde gelieferten Informationen in allen Phasen des Verfahrens gibt.[47] Dies macht jede Beteiligung Dritter an dem Prozeß

---

[46] Reg. 31 (7).
[47] Regs. 6, 20 (1)(b), 35, Annex 4 sec. 12.

der Informationssammlung schwierig. Zwar unterliegt es keinem Zweifel, daß Betriebs- und Geschäftsgeheimnisse sowie Urheberrechte bezüglich der beantragten und durchgeführten Aktivität gewahrt werden müssen. Inzwischen gibt es aber durchaus Regeln im innerstaatlichen Recht und im Recht der Europäischen Gemeinschaft, die Zugang zu Umweltinformationen unter Wahrung dieser Rechte ermöglichen.[48]

Bezüglich des weiteren Prozesses der Exploration gibt es freilich eine Bestimmung, die möglicherweise einen etwas offeneren Informationsprozeß gestattet. Vertragsnehmer, befürwortende Staaten und andere interessierte Staaten und Einheiten

„shall cooperate with the Authority in the establishment and implementation of programmes for monitoring and evaluating the impacts of the seabed mining on the marine environment".[49]

Das könnte bedeuten, daß möglicherweise betroffene Staaten (zum Beispiel fischereibetreibende Staaten) oder gar interessierte und betroffene Unternehmen in den Prozeß der Informationssammlung und Bewertung hinsichtlich der Umweltauswirkungen einbezogen sein könnten.

Insgesamt bleibt aber der Eindruck, daß das Entscheidungsverfahren zwischen der Behörde und dem Vertragsnehmer geschlossen, für Außenstehende intransparent ist. Damit bleiben diese Regeln weit hinter den Grundsätzen zurück, die sich für umweltrelevante Entscheidungen auf nationaler Ebene entwickelt haben und die heute auch als völkerrechtlich abgesichert angesehen werden dürfen, jedenfalls auf der Grundlage der bereits erwähnten Aarhus-Konvention. Dem kann man nicht entgegenhalten, die Situation in bezug auf die Gemeinschaftsgüter der Meeresumwelt sei mit der Lage bei umweltrelevanten Entscheidungen zu Lande nicht vergleichbar. Die „zu Lande" gewährten Beteiligungsrechte haben nicht nur den Zweck, nachbarliche Interessen zu sichern. Sie dienen auch und gerade einer besser informierten Entscheidungsfindung zugunsten des Gemeinwohls. Außerdem fehlt es auch beim Schutz der Meeresumwelt, das wurde bereits erwähnt, nicht an betroffenen egoistischen Interessen, insbesondere im Bereich der Fischerei. Bei dieser Frage der Öffnung

---

[48] Art. 3 Abs. 2 der EG-Richtlinie über den Zugang zu Umweltinformationen (90/313/EWG vom 7. Juni 1990), Art. 4 Abs. 4 der Aarhus-Konvention (siehe Anm. 31).
[49] Reg. 31 (6).

von Entscheidungsverfahren bleibt das Seerecht leider weiter hinter dem terrestrischen Standard zurück.

3. Materielle Umweltstandards für Entscheidungen

Es wurde bereits erwähnt, daß als materielle Leitlinien für die Entscheidung vorgeschrieben ist, daß der Betriebsplan einen wirksamen Schutz der Meeresumwelt vorsehen müsse. Die Frage ist nur, was dies genau heißt. Es ist Aufgabe der Behörde, dies durch weitere Rechtssetzung zu konkretisieren. Darüber hinaus gibt es jedoch einen wesentlichen materiellen Maßstab, und das ist das Vorsorgeprinzip:

> „In order to ensure effective protection for the marine environment from harmful effects which may arise from activities in the Area, the Authority and sponsoring States shall apply the precautionary approach as reflected in Principle 15 of the Rio Declaration to such activities. The Legal and Technical Commission shall make recommendations to the Council on the implementation of this paragraph".[50]

Mit dieser Bestätigung des Vorsorgeprinzips beweist das Seerecht einmal mehr seine Vorreiterrolle in dieser Frage. Dies ist wohl gerade bei Tätigkeiten des Tiefseebergbaus, wo Neuland betreten wird, besonders wichtig. Gerade hier ist von Bedeutung, daß das Vorsorgeprinzip Maßnahmen zum Schutz der Umwelt gerade auch in den Fällen verlangt, wo die Möglichkeit eines Schadenseintritts nicht hinreichend sicher prognostiziert werden kann.

Wenn, wie oben bereits dargelegt, die Regeln in vorbildlicher Form eine Umweltverträglichkeitsprüfung verlangen, so ist dies, wie bereits oben dargelegt, ein wesentliches Verfahrenselement des Vorsorgeprinzips. Als materielles Gegenstück wird verlangt, daß der Vertragsnehmer die beste verfügbare Technologie *(the best technology available to it)* einsetzen muß, um Verschmutzung und andere Gefahren für die Umwelt zu vermeiden. Diese Verpflichtung wird allerdings durch den Zusatz „*as far as it is reasonably possible*"[51] eingeschränkt. Dies ist freilich auch im innerstaatlichen Recht nicht unüblich, wo z.B. Formulierungen wie „*best available technology not entailing excessive cost*" und ähnliche gebraucht werden. Das Vorsorgeprinzip steht immer unter dem Vorbehalt des

---

[50] Reg. 31 (2).
[51] Reg. 31 (3).

Verhältnismäßigkeitsprinzips. Das bedeutet, daß die wirtschaftliche Belastung, die durch Umweltanforderungen hervorgerufen werden, nicht unverhältnismäßig in bezug auf den zu erwartenden Umweltnutzen sein dürfen. Dies erfordert ein Abwägen von ökologischen und ökonomischen Interessen, das nicht immer unproblematisch ist. An der praktischen Lösung dieses Problems kommt man bei der Anwendung des Vorsorgeprinzips aber letztlich nie vorbei.

4. Maßnahmen der Durchsetzung und Haftung

Das Regime des Tiefseebergbaus stellt in einer Hinsicht einen ganz entscheidenden Fortschritt gegenüber den Regeln für sonstige umweltbelastende Tätigkeiten im Meeresbereich dar: Die Durchsetzung aller anwendbaren Standards erfolgt durch eine eigene internationale Behörde. Die Problematik mangelnder Durchsetzung der Rechtsregime, die sich immer noch aus dem Vorrang der Durchsetzung durch den Flaggenstaat ergibt, ist damit in einer Weise gelöst, die Vorbildfunktion auch für andere Bereiche übernehmen sollte.

Für die Durchsetzung der Verpflichtungen des Vertragsnehmers sind der Behörde auch wirksame Instrumente in die Hand gegeben. Das erste Instrument dieser Art ist die Beendigung oder Suspendierung des Vertrags. Diese kann vom Rat ausgesprochen werden, wenn

„ ... the Contractor has conducted its activities in such a way as to result in serious persistent and wilful violations of the fundamental terms of this contract, Part XI of the Convention, the Agreement and the rules, regulations and procedures of the Authority...".[52]

Die Regeln über den Umweltschutz sind im Lichte des bisher Gesagten wohl in der Tat als wesentlich anzusehen. Ihre Verletzung kann also zu einer Suspendierung oder Beendigung des Vertrags führen. Zusätzlich kann der Rat auch Geldbußen gegenüber den Vertragsnehmern verhängen, und zwar sowohl anstelle einer Suspendierung oder Beendigung des Vertrags als auch im Falle von Verletzungen, die nicht in gleicher Weise schwerwiegend sind.

Schließlich rundet ein Haftungsregime das Instrumentarium der Rechtsdurchsetzung ab.[53] Der Inhalt dieses Regimes ergibt sich wiederum erst aus dem Vertrag, den der Vertragsnehmer mit der

---

[52] Annex 4 sec. 21.1(a).
[53] Annex 4 sec. 16 und 17.

Behörde schließt und dessen Klauseln im Anhang zu den Regeln näher festgelegt werden. Danach haftet der Vertragsnehmer für

„any damage, including damage to the marine environment".

Letzterer schließt ein

„the costs of reasonable measures to prevent or limit damage to the marine environment, account being taken of any contributory acts or omissions by the authority".

Dies ist ein wesentliches Element des Ersatzes von Umweltschäden. Wichtig ist dabei zunächst zu betonen, daß der zu ersetzende Schaden nicht nur Schäden an Rechtsgütern umfaßt, die einem Rechtsträger zuzuordnen sind, sondern eben auch Umweltschäden als solche. Dies führt jedoch zu zwei weiteren Fragen. Die erste geht dahin, wer denn die notwendigen Maßnahmen zur Begrenzung und zum Ausgleich von Umweltschäden trifft. Dies kann der Sache nach nur die Behörde sein, die hier im Allgemeininteresse zum Schutz der internationalen Gemeinschaftsgüter, der *global commons* tätig wird. Dabei muß man sich, und das ist das zweite Problem, mit einem möglichen *a contrario*-Argument auseinandersetzen. Wenn die Kosten für Maßnahmen der Schadensbegrenzung einen erstattungsfähigen Umweltschaden darstellen, so könnte man schließen, daß ein Umweltschaden, für den solche Maßnahmen nicht getroffen werden oder gar nicht möglich sind, auch nicht zu ersetzen ist. Solches sollte wohl nicht gemeint sein. Wenn es aber nicht gemeint ist, stellt sich das schwierige Problem einer Bezifferung von Umweltschäden durch andere Maßstäbe als die Kosten von Maßnahmen der Bekämpfung.

Ein weiteres wichtiges Element des Haftungsregimes ist, daß es eine Haftung ohne Verschulden gibt. Lediglich höhere Gewalt ist ausgeschlossen.[54]

So fortschrittlich dieses Haftungsregime ist, es teilt den Fehler, der bereits hinsichtlich der Entscheidungsprozesse festgestellt werden mußte: Es handelt sich um nach außen abgeschlossene Rechtsbeziehungen zwischen der Behörde und dem Vertragsnehmer. Das wird dadurch deutlich, daß diese Haftungsregeln nicht etwa Bestandteil einer allgemein verbindlichen Regelung sind. Sie werden vielmehr zum Bestandteil eines bilateralen Vertrages zwischen Behörde und Vertragsnehmer gemacht. Man kann nicht davon ausgehen, daß es sich dabei um einen Vertrag zugunsten Dritter

---

[54] Annex 4 sec. 17 (1).

handelt. Drittgeschädigte stehen also außerhalb dieses Vertragsregimes. Hier ist wohl dringend Abhilfe geboten.

5. Zusammenfassende Bewertung

Die neuen Regeln sind ein systematischer und bedeutsamer Versuch, hinsichtlich der Tätigkeiten im Tiefseebergbau einen ausreichenden Umweltschutz zu gewährleisten. Sie stellen damit sicher einen wichtigen Schritt dar, mit dem das Seerecht seine Leitfunktion für das internationale Umweltrecht beweist. Denn dieser Versuch ist durchaus erfolgreich. In dem Bereich der Informationssammlung und Bewertung sowie in den Entscheidungsmaßstäben, dem Bekenntnis zum Vorsorgeprinzip, aber auch und hauptsächlich in dem strikten Regime der Durchsetzung in der Hand einer internationalen Behörde ist das neue Regime vorbildlich. Allerdings darf nicht verkannt werden, daß es durch seinen abgeschlossenen, für Außenstehende intransparenten Charakter, seinem Mangel an Beteiligungsmöglichkeiten und auch den Mangel der Berücksichtigung der Interessen Dritter hinter den Entwicklungen des modernen Umweltvölkerrechts zurückbleibt.

Evolution in the Fisheries Provisions of UNCLOS

William T. Burke

It is now two decades since the adoption of the United Nations Convention on the Law of the Sea (UNCLOS) and a quarter century since the agreement within the conference on the principal provisions on the EEZ and fisheries and on the high seas. A great deal has happened over this period of time to implement these provisions in national practice and to develop them through interpretations associated with or implied by such practice, as well as through judicial pronouncements and new agreements. While in the last decade, most of the attention on the fisheries provisions of UNCLOS focused upon high seas fisheries, particularly in connection with straddling and highly migratory fish stocks, some developments directly concern coastal State authority over its EEZ fisheries.

In extending national jurisdiction to an exclusive economic zone of 200 nautical miles, Part V of UNCLOS establishes a duty for coastal States to provide for the conservation and management of the living resources found within that area. Taking into account "the best scientific evidence available to it", the coastal State is to ensure through "proper conservation and management measures" that the maintenance of the living resources of the zone is not endangered by overexploitation (Article 61(2)). Article 61(3) declares that "[s]uch measures shall also be designed to maintain or restore populations of harvested species at levels which can produce the maximum sustainable yield, as qualified by relevant environmental and economic factors, including the economic needs of coastal fishing communities, and the special requirements of developing States...". The same sentence continues with "and taking into account fishing patterns, the interdependence of stocks and any generally recommended international minimum standards, whether subregional, regional or global".

Since this is the primary language establishing coastal State authority over living marine resources in the exclusive economic zone, its interpretation and application are of some importance. For this reason it is unfortunate that this portion of Article 61(3) is sometimes treated loosely and without careful attention. Sometimes, for example, it is only partially quoted, omitting the critical phrase "as qualified by relevant environmental and economic factors". It is obvious that the phrase is not intended to qualify the reference to "maximum sustainable yield" (MSY), which is a biological concept that does not itself embrace economic considerations. Instead, reasonably interpreted, the qualifying phrase refers to the population level that the coastal State may wish to maintain or restore in accordance with its scientific understanding of the stocks concerned. Article 61 states that the coastal State may determine the level of abundance of an exploited fish population by reference to factors of environmental or economic significance for its interests, but it cannot allow a rate of exploitation which endangers the "maintenance" of that population. At the other end of the spectrum of population abundance, this language appears to mean that it is not required to maintain the largest possible population levels consistent with yielding a maximum catch, but it may do so. Indeed, under this language the coastal State could justify returning a population to its original unexploited condition if that were necessary for environmental reasons.

This interpretation of Article 61's reference to MSY reflects the understanding of its drafters that economic considerations can be important determinants in the fisheries management policies of coastal States, and that they should be given the flexibility to pursue those policies free of rigid notions of biological maximums.

Since the adoption of Part V of UNCLOS, this provision gains in importance from the understanding, not as common 25 years ago as it is today, that exploitation of fish populations at a high MSY level risks unintended excessive reduction of population levels due to the limited understanding of actual population abundance. Accordingly, maintaining fish populations at levels higher than those that yield the highest maximum sustainable yield is not only economically desirable but prudent for avoiding an unintended reduction of population abundance over time. To the extent it is capable and equipped to do so, a coastal State may attempt to maintain a fish population at whatever level of abundance it believes will satisfy its national interests; this level is more likely than not to be above the level which would produce the highest sustainable yield.

Another question, however, is also important and that is whether, in compliance with UNCLOS Article 61, the coastal State may also decide to maintain a fish population at a level lower than those that produce the highest MSY. In this scenario, the stock abundance has been significantly reduced below the level associated with the MSY. In such circumstances, the question raised is whether the coastal State is required by Article 61 to take measures to increase the population towards the level that will produce the MSY. This issue has been litigated in New Zealand in *New Zealand Fishing Industry Association et al. v. Minister of Fisheries et al.*,[1] although in the final outcome of the case the issue did not have to be resolved. The contrasting positions of the parties were clearly presented at the trial court phase of the case.

The dispute arose over the determination by the New Zealand Minister of Fisheries that the diminished abundance of snapper (a fishery conducted wholly within New Zealand jurisdiction by recreational and commercial fishermen) required the Minister to reduce the total allowable catch significantly in order to build the population to its MSY level over a 20-year period. This decision was based on the New Zealand statute which drew upon language from Article 61 of UNCLOS and, therefore, part of the justification for this decision was the view that moving to the MSY was an obligation imposed by UNCLOS.

The contrasting views, as seen by the Government, were as follows: "The fishing industry and the Treaty of Waitangi Fisheries Commission are of the view that the Minister of Fisheries can allow the stock to be 'sustainably' managed at a level less than that which will produce the MSY. They believe this is appropriate if pursuing the MSY goal would impose significant economic costs which in their view are unreasonably disproportionate to the gains in yield available at B_{msy}.[2] The Ministry, while very aware of the economic implications of TAC reductions, does not believe the legislation can be interpreted in this manner. The Ministry's interpretation is that there is an implicit obligation under the Fisheries Act and the United Nations Convention on the Law of the Sea to determine TACs that will ultimately produce the MSY in fisheries".[3]

The industry view, in contrast, was that Article 61 did not impose an obligation for coastal States to manage their fisheries to attain the MSY, and that the coastal State had the discretion to maintain population levels lower than would be required to attain the highest MSY.

The condition of the snapper fishery was indeed interesting. It was accepted by the Court of Appeal that "[w]hile the relevant biomass is at only 50 percent of the level required for MSY, it appears that the fishery itself is at 92 percent of MSY. Thus a 39 percent cut was seen as necessary to obtain an 8 percent increase from 92 percent to 100 percent MSY over 20 years".[4] The Court did not note that the reduced biomass

---

[1] The opinions in this case have not been reported, but are available through the *Maori Law Review* on the Internet. The fact that they are unreported does not affect their authority as precedent in the New Zealand system.

[2] B = biomass (or stock abundance), B_{msy} means the stock level at its highest rate of growth. This enables the MSY.

[3] High Court of New Zealand, Judgment of Justice McGechan, 24 April 1997, 71-72 (unpublished opinion on file with the author).

[4] Decision of Court of Appeal, 22 July 1997, 22 (unpublished opinion on file with the author).

had been in place for over two decades and the yield from the fishery persisted over this period in close approximation to the MSY biomass despite that fact.

The High Court of New Zealand ultimately concluded, mainly based on the specific wording of the New Zealand legislation (the Fishery Act of 1983) which differed in significant respects from Article 61, that the Minister was required "to set total allowable catch at levels specifically maintaining stock at or above MSY". The Minister did have the discretion to determine the timing of the move to MSY, which was informed by the environmental and economic factors mentioned in Article 61.

This decision of the High Court was appealed to the Court of Appeal, which reversed it on other grounds, namely, that the Minister had failed to take steps mandated by law for consideration before reducing the TAC. This failure invalidated the earlier orders for such reduction. The Court of Appeal also noted that new legislation unambiguously required the Minister to move stock abundance towards or above the MSY level, but with the discretion to determine the timing of that move.

Although not necessary for its decision, and without analysis or discussion, the Court of Appeal expressed the view that the definition of the total allowable catch in the New Zealand legislation of 1983 (no longer in effect following the introduction of the 1996 fishery law) "both alone and informed by the relevant articles of the United National Convention on the Law of the Sea (UNCLOS) cast on the Minister the *prima facie* duty to move the fishery towards MSY, if not already there, by such means and over such period of time as the Minister directed. That *prima facie* duty was subject to the so-called qualifiers, i.e., those factors introduced by the words 'as qualified by'. Those qualifiers were matters, which the Minister was required to address when considering how to implement his *prima facie* duty and, if the qualifiers were cogent enough, whether the *prima facie* duty was for the moment overtaken by one or more of those factors. Thus the qualifiers were relevant to whether, and if so, by what means and over what time the *prima facie* duty should be implemented. That in our judgment is the correct way of looking at the matter rather than saying ... that if any of the qualifiers applied the Minister had a discretion rather than an obligation to move to MSY".

In the result, therefore, the New Zealand Court decided that while the Minister of Fisheries is obligated to move a stock towards a level which can produce the MSY, the decision on when and how to do so must "consider all relevant social, cultural and economic factors". The final decision on fishery regulation is therefore not to be governed by biological concerns alone. This result is consistent with UNCLOS, if the latter is interpreted to mean that coastal State discretion is limited under

Article 61 to determining how and when to regulate a fishery stock to move it to an MSY level when it is below that level, but that it does not extend to choosing to maintain abundance below the MSY level.

If this interpretation of Article 61 is generally accepted, it could be unfortunate for coastal States that aim to maximize the economic benefits of a fishery. As the Court of Appeal noted, such a decision can be very costly to the fishing industry or, for a developing State dependent on fishing production, for the national budget.[5] The New Zealand Court of Appeal's opinion only noted on this point that "[t]he argument concerned whether the Minister's decision to make a cut of 39 percent in the TAC was unreasonable in the relevant sense, which may well be captured by the word irrational", but it said that "no ultimate conclusion is required for the purposes of this case and it is preferable not to express one". But the Court then went on to note that the Minister appeared to have paid little or no attention to the cost and benefits of his decision, or to the alternatives that might have changed the substantial effect it would have on fishing interests. It does not take much reading between the lines to discern a significant level of concern at the obliviousness of the Minister to the drastic consequence of his decision. A few sentences later the Court advised: "All we wish to say for the future is that the Minister would be wise to undertake a careful cost/benefit analysis of a reasonable range of options available to him in moving the fishery towards MSY. If the Minister ultimately thinks that a solution having major economic impact is immediately necessary, those affected should be able to see, first, that all other reasonable possibilities have been carefully analyzed, and, second, why the solution adopted was considered to be the preferable one".

The interpretation of Article 61 by the Court of Appeal, to the extent that one can be read into the decision, places a substantial burden on coastal State officials to exercise discretion with consideration and care. Under this interpretation the goal of coastal conservation and management must be to regulate stocks to reach an abundance level associated with MSY, but it has discretion regarding how and when this is to be accomplished. If that discretion is exercised without adequate regard to economic impact, local fishermen and other groups may pay a

---

[5] For an illuminating account of the significance of a lowered TAC in the face of uncertainty, see D. C. Butterworth, "Science and Fisheries Entering the New Millennium", paper presented at the 23rd Annual Conference of the Center for Ocean Law and Policy and the UN Food and Agriculture Organization, Rome, 15-17 March 2000, Rome, 7-8. The paper discusses the dramatic impact of such a decision on Namibia if such an action had been taken. Instead, on the basis of a quantitative risk assessment, the TAC was raised slightly, abundance increased, catches doubled over a period of two years, and very significant adverse financial consequences were avoided.

high price. It can be argued that it would be better to interpret Article 61 as providing that the coastal State has no obligation to set the ultimate goal of its fishery policy in accordance with a biological goal that may be divorced entirely from economic and social needs. Certainly the injunction to move towards the MSY level has little to do with conservation. In fact, it may be that the more successful the coastal State is in achieving a fishery mortality at the MSY level, the greater the risk of overfishing the resource.

The flexibility of coastal State authority established by Article 61(2) appears to be underscored by the tentativeness of the terminology "taking into account", which is used to refer both to "the best scientific evidence available" and to "fishing patterns, the interdependence of stocks and any generally recommended international minimum standards, whether subregional, regional or global". This terminology is consistent with the desired flexibility of decentralized management authority over living resources of enormous variability in almost every conceivable characteristic. To attempt to dictate absolute conformity to the unknown details of, for example, "fishing patterns", stock interdependencies, and "recommended standards" could impede, if not eliminate, the exercise of coastal discretion in realizing the benefits of its resources. Article 61 enjoins that these matters are to be taken into account in deciding what to do, but not necessarily given decisive weight in determining coastal State conservation and management measures.

In this connection it should not be overlooked that the formulation "taking into account" differs a great deal from the more demanding formula "on the basis of", which is used in other contexts. Those who wish to force restrictions on the coastal State's determination of its fisheries policy under Article 61 ought to bear the burden of explaining how the former phrase can be converted into the latter.

The reference to any generally recommended international minimum standards is also suggested as a means of restricting coastal State discretion in managing the resources of the zone. This point has been made in connection with the observation that the UNCLOS treaty was formulated largely without regard to consideration for ecosystem management. While there are some provisions which make partial reference to ecosystem considerations, the zonal structure of authority established by the treaty, that indeed dominates its provisions allocating decision-making authority among coastal and flag States, is obviously inconsistent with a comprehensive ecosystem approach by any one State or perhaps even a group of States. The omission of any provision for international standards applicable to the territorial sea – an area of great significance for the ecosystem of the coastal region and beyond – precludes any required obligation under this treaty for ecosystem management. It goes without saying, of course, that a coastal State can adopt an ecosystem management system which encompasses the entire water area subject to its national jurisdiction, although this might still not include the entire ecosystem.

But the question that has been raised is whether such a system can be imposed on the coastal State which does not adopt such a system voluntarily. It is here that at least one influential observer invokes the reference to "international minimum standards" in Article 61(3).[6]

Professor Wolfrum suggests that the failure of UNCLOS to provide for ecosystem management can be remedied by using the reference to "international minimum standards" in Article 61(3) to import a code of conduct, which is voluntary, such as that promulgated by the FAO.[7] In fact, the FAO Code of Conduct for Responsible Fisheries makes only modest reference to ecosystem management. But irrespective of its particular content, the Code is specifically voluntary and, as Professor Wolfrum mentions in a subsequent footnote reference to an article by W. R. Edeson, this "... enabled it to cover much more than could possibly have been covered in a document intended to be of a binding international agreement". It seems a bit strange to borrow concepts from a document specifically intended not to impose legal requirements and incorporate them into one that does, and to do that in an agreement negotiated more than a decade before. It at least needs to be questioned whether the UNCLOS treaty should now be interpreted to embrace concepts not considered for inclusion at the time, indeed which probably could not have secured agreement if they had been considered, and which even in 1995 might not have been included in a Code if it were intended to be legally binding.

The issue here is not, of course, the desirability of the ecosystem approach, but the mechanism that ought to be employed for increasing its use. There is still such a thing as seeking an explicit amendment to an international agreement, particularly one which has more than one means of accomplishing that end. It is worth mentioning in this regard that the capability of applying an ecosystem approach is not widely distributed. It is a daunting task even for developed States with substantial resources

---

[6] R. Wolfrum, "The Role of the International Tribunal for the Law of the Sea", conference paper, see note 5, 5.

[7] The exact statement is as follows: "One cannot but summarize that the commitment of the Convention on the Law of the Sea concerning an ecosystem approach is not strong and does not meet the standards set by subsequent international agreements. The standards can be improved, though, through international minimum standards as referred to in Article 61, para. 3, of the Convention. This does not necessarily require the adoption of another international agreement. The establishment of code of conducts implemented voluntarily may be quite sufficient. This is where FAO is playing an important role. If such code of conducts may be considered to reflect 'international minimum standards' they may equally be applied by the International Tribunal for the Law of the Sea". Wolfrum, see note 6.

for management undertakings. Expecting adoption and implementation of this approach on a wide scale among developing States that even now are unable to determine an allowable catch of their EEZ living resources does not seem entirely realistic, nor is it likely to improve fishery management. For that matter, even rich developed nations, such as the United States, are unable to assess the abundance of a significant portion of exploited fish stocks.

Of course, no one would object if an individual coastal State chooses voluntarily to embrace a particular generally recommended standard for employment in its management program. But such a decision by coastal authorities differs a great deal from a decision by an international tribunal, such as ITLOS, that such a choice is mandated by UNCLOS Article 61(3).

The Common Heritage of Mankind in the Present Law of the Sea

Vladimir-Djuro Degan

The name Shigeru Oda became known to me first through his voluminous books on the law of the sea. In 1969, I formed a visual impression of him during his lectures at the Hague Academy where I was a student. His no less voluminous dissenting and separate opinions attached to verdicts of the Hague Court have at times been a source of inspiration to me. After my election to the Institute of International Law, we soon became close friends. His office at the Peace Palace is a place I regularly visit during my stays in The Hague.

The topic of this paper was at another time the subject of numerous writings throughout the world. The idea of a common heritage of mankind had a large number of passionate advocates in the 1970s. Later on, some of them became zealous critics, but I have never read any legal arguments for this shift in their opinions. What seems to be unusual, after the UN Law of the Sea Convention finally entered into force in 1994, is that this topic has been almost entirely neglected in the doctrine. In an attempt to partly fill this gap, I have found it appropriate to select this topic for my contribution to the *Festschrift* in honour of my friend, one of the best and the finest experts on the law of the sea of the 20th century.

***

The concept of the common heritage of mankind in modern times is a reflection of the same idea in Roman law according to which the air, the water streams, the sea and the coast are *res communis omnium*, hence *extra commercium*.[1]

---

[1] *Et quidam naturali jure communia sunt omnium haec: aer et aqua profluens et mare et per hoc litora maris. Institutiones*, 2, 8, 1.

Georges Scelle had already advocated the same idea in respect of the continental shelf in 1955. The continental shelf being the seabed and subsoil of the high seas at that time, he pleaded for international management of its exploitation. This implied an international mechanism very much similar to that provided by the 1982 Convention for the International Seabed Area.[2]

At that time, this idea was rejected altogether. In its final report, the International Law Commission stressed that: "In the present circumstances ... such internationalization would meet with insurmountable practical difficulties, and would not ensure the effective exploitation of natural resources necessary to meet the needs of mankind".[3]

What was refused for the continental shelf itself was later on adopted as the basic principle for all of space law. Hence, Article 1(1) of the Treaty of 27 January 1967 provides that:

> "The exploration and use of outer space, including the Moon and other celestial bodies, shall be carried out for the benefit and in the interest of all countries, irrespective of their degree of economic and scientific development, and shall be the province of all mankind".[4]

However, for the practical implementation of this principle no agency of the international community relating to the outer space is necessary yet.

When various options for a future legal régime of the deep seabed area under the high seas were discussed, it was the US President Lyndon Johnson who declared in July 1966 "that the deep seabed and the ocean bottoms are, and remain, the legacy of all human beings".[5]

Only a year later, the representative of Malta at the United Nations, Arvid Pardo, proposed that the seabed and ocean floor beyond national jurisdictions be reserved exclusively for peaceful purposes and that their resources be declared "the common heritage of mankind".[6]

---

[2] Cf., G. Scelle, "Plateau continental et droit international", *RGDIP* 1 (1955), 5 et seq.

[3] ILCYB 1956, Vol. II, 296, para. 3.

[4] The matter was of the Treaty on Principles Governing the Activities of States in the Exploration and Use of Outer Space, including the Moon and other Celestial Bodies. The same wording was already adopted in General Assembly Resolution 1962 (XVIII) of 13 December 1963.

[5] Cf., *A Quiet Revolution, The United Nations Convention on the Law of the Sea*, 1984, 5.

[6] *Ibid.*

On 17 December 1970, the General Assembly adopted Resolution 2749 (XXV), the Declaration of Principles Governing the Seabed and the Ocean Floor, and the Subsoil thereof, Beyond the Limits of National Jurisdiction, with 108 votes in favour, none against and 14 abstentions. Among numerous other Member States, the United States voted in favour of this Declaration. Its text contains some provisions that throughout the 1970s were believed to be *jus cogens*. In any case, they were couched in these terms. Here are the essential provisions that reflected the substance of the idea of the common heritage of mankind:

"1. The sea-bed and ocean floor, and the subsoil thereof, beyond the limits of national jurisdiction (hereinafter referred to as the area), as well as the resources of the area, are the common heritage of mankind.
2. The area shall not be subject to appropriation by any means by States or persons, natural or juridical, and no State shall claim or exercise sovereignty or sovereign rights of any part thereof.
3. No State or person, natural or juridical, shall claim, exercise or acquire rights with respect to the area or resources incompatible with the international régime to be established and the principles of this Declaration.
4. All activities regarding the exploration and exploitation of the resources of the area and other related activities shall be governed by the international régime to be established.
5. The area shall be open to use exclusively for peaceful purposes by all States whether coastal or land-locked, without discrimination, in accordance with the international régime to be established...".

Nobody was sure at that time about the extent of this Area, because the rules concerning the outer limits of the continental shelf of coastal States were not certain anymore. But almost nobody contested the fact that the Area itself and its natural resources were the common heritage of mankind in the sense of the Declaration of 1970.

The mandate of the Third UN Law of the Sea Conference was, *inter alia*, to establish an international régime based on the above principles.[7]

---

[7] In Resolution 2750C (XXV) of 1970, the General Assembly decided to convoke the Conference that would deal with the question of an equitable international régime – including international machinery – for the Area and resources of the seabed, the ocean floor, with a precise definition of the Area and a broad range of related issues.

During the Third Conference, a major disagreement arose between the participating States with regard to the system of exploration and exploitation of the resources of the Area. Developing countries supported the establishment of an International Authority with strong powers. This Authority would act in the name of the international community and would directly exploit the resources of the Area through an International Enterprise for the benefit of the mankind. It is true that their proposals were the closest to the concept of the common heritage of mankind as defined above. But, as with the continental shelf in the 1950s, the viability of such an international machinery was quite another question.

The industrial States, which pretended to be in sole possession of the necessary technology (a fact they never actually proved), held quite a different view. According to them, the natural resources of the Area ought to be open to the exploitation of the most capable private enterprises. The role of the International Authority would be to issue permits and to collect revenues from the recovery of hard mineral resources. In the light of all experiences which ensued, this proved to be the most rational solution. But in human law-creating in general, solutions based on reason seldom prevail.

A compromise was reached in 1976, on the basis of a proposal by Henry Kissinger, then US Secretary of State, in the form of a "parallel system" of exploration and exploitation of the Area.

According to the text of the 1982 Law of the Sea Convention, all activities in the Area should be organised, carried out and controlled by the Authority on behalf of mankind as a whole, in accordance with its Part XI. As with the Declaration of 1970, the principles governing the Area were couched in general and peremptory terms as being directly applicable to all States and all natural and juridical persons. However, other sections of Part XI concerning activities in the Area, as well as Annexes III and IV to the Convention, were couched in terms of contractual provisions, providing rights and obligations only for States Parties to the Convention. These activities were expressly reserved for the international Enterprise, the States Parties or state enterprises, or natural or juridical persons of the States Parties.

Within the structure of the Authority and the Enterprise, a huge and costly bureaucratic machinery was provided for, referred to at the time by many as "the United Nations of the Seas and Oceans". In addition, quite a few representatives of the participating States at the Conference saw in it profitable employment for themselves, and perhaps even for their offspring, for several generations to come. None of this was realised in practice.

***

It was with respect to the activities in the Area that unilateralism appeared in the law-making of the States pretending to possess the necessary technology. Their national laws were followed by a number of mutual agreements. This happened before the future Convention was adopted and signed. Hence, since 1980, some industrialised States have adopted national laws concerning the exploration and exploitation of the deep seabed and ocean floor.

The first act of this kind was promulgated by the United States on 28 July 1980, with subsequent regulations.[8] It was followed by a German law on 22 August 1980[9] and by a British law on 28 July 1981.[10] These three States refused to sign the Convention on 10 December 1982. The United States representative explicitly announced that his country would not become a party to its actual content.

Only after the adoption of their own national legislation did several other industrialised States formally sign the 1982 Convention. They did so in order to be entitled to participate in the Preparatory Commission for the International Seabed Authority and the International Tribunal for the Law of the Sea, and in order to take advantage of Resolution II of the Final Act. These States were France with its Law of 23 December 1981 and subsequent regulations,[11] the former Soviet Union with the Decree of the Supreme Soviet of 17 April 1982,[12] and Japan with its Law of 20 July 1982.[13] Italy enacted its Law on 20 February 1985,[14] after signing the 1982 Convention.

---

[8] Deep Seabed Hard Mineral Resources Act, *ILM* 19 (1980), 1003 et seq.; Deep Seabed Mining Regulations for Exploration Licences, *ILM* 20 (1981), 1228 et seq.; Deep Seabed Mining Regulations for Exploration Licences: Procedures for Pre-enactment Explorer Applications and New Entrant Applications, *ILM* 21 (1982), 867 et seq.

[9] English translation, *ILM* 20 (1981), 393-398. See the amendments to the Law of 12 February 1982, *ILM* 21 (1982), 832 et seq.

[10] Deep Seabed Mining (Temporary Provisions) Act 1981, *ILM* 20 (1981), 1217 et seq.

[11] Loi n° 81-1135 du 23 décembre 1981: Sur l'exploration des grands fonds marins, *Journal Officiel, (JO),* 24 décembre 1981, 3499 et seq., English translation, *ILM* 21 (1982), 804 et seq.; Décret n°82-111 du 29 janvier 1982: Pris pour l'application de la loi du 23 décembre 1981, *JO*, 31 janvier 1982, 431-432; Arrêté du Ministre de l'Industrie du 29 janvier 1982: Contenu des demandes de permis d'exploration et d'exploitation des grands fonds marins, *ibid.*, 433-434; Conseil Constitutionnel: Décision du 16 décembre 1981, *JO*, 18 décembre 1981, 3448.

[12] English translation, *ILM* 21 (1982), 551 et seq.

[13] English translation, *ILM* 22 (1983), 102 et seq.

[14] English translation, *ILM* 24 (1985), 983 et seq.

The most exhaustive and most advanced act was the one adopted by the United States. It inspired the legislation of other States which, generally, is less detailed. All these laws were based on the regulations of respective States concerning mining on land, which were transposed to the marine environment.

These pieces of national legislation have some common features, but there are some subtle and even considerable differences on less important issues.

Each of these acts provides for the issuing of a licence for exploration and a permit for commercial recovery of minerals from a location within the Area. Therefore, licences for exploration of mineral resources do not empower their holders to exploit a location. Permits for commercial exploitation will be issued, subject to certain conditions, either to the natural or juridical person who explored the location, or such persons will enjoy priority among other applicants.

Some of these laws provide time-limits for the validity of licences issued, which can be extended by the national administration. According to the US Act, a licence for exploration is issued for a period of ten years and can be extended for further five-year periods. Exploitation permits are issued for an initial period of 20 years, and for as long as hard mineral resources are annually recovered in commercial quantities. Nevertheless, the permit of any permittee who is not recovering these resources in commercial quantities at the end of ten years may be terminated. Each of these laws provides a system of reciprocal recognition of licences and permits that are issued by other States.

On the other hand, the legislators of all of the above States wanted to at least partly satisfy some of the objectives of the 1970 Declaration and of the 1982 Convention, denying at the same time the obligatory character of these acts in their regard. To summarise:

1. All national laws were adopted as transitional, pending the conclusion of the Law of the Sea Convention, or its subsequent entry into force. However, the respective States did not assume any obligation to become its parties.
2. It is asserted in all of these acts that the respective State does not appropriate sovereignty, or sovereign rights, or ownership over a locations in the Area. It is furthermore asserted that all these activities do not affect the freedoms of the high seas, an allegation difficult to prove.
3. Almost all of these acts provide for taxes to be paid to a special state fund for the removal of hard mineral resources from the deep seabed. The sums to be paid are much lower than those provided for in the 1982 Convention.

4. Finally, only France and the former Soviet Union tried to harmonise their legislation with the parallel system of exploitation, as provided in the 1982 Convention. A holder of a licence for exploration can obtain a permit for exploitation of only half the location explored. Only the Soviet Act expressly provided that the other half will be reserved for exploitation by a future international organisation for the seabed.

***

Four of the above-mentioned States (France, Germany, the United Kingdom and the United States) concluded the Agreement Concerning Interim Arrangements Relating to Polymetallic Nodules of the Deep Sea Bed on 2 September 1982.[15] The Agreement provides for an exchange of information on the applications filed for the authorisation of activities in the Area between its parties, as well as the relevant coordinates of the areas claimed or approved. In addition, this information is to remain strictly confidential. The Agreement also provides for an arbitral procedure for conflict resolution in case of the overlapping of attributed areas.

On 3 August 1984, the Provisional Understanding Regarding Deep Seabed Mining was concluded, with two appendixes and a memorandum, by six States. At that time, the States Parties were Belgium, France, Germany, Italy, Japan, the Netherlands, the United Kingdom and the United States, but not the Soviet Union.

This Provisional Understanding provides that authorisations granted by each of its parties be recognised by all others. In case of an application filed in one of the States Parties in conformity with its national law and this Agreement, other parties will not issue authorisation for the Area in question.

***

By means of all the above-mentioned national laws and the two Agreements, a separate system of exploration and exploitation of hard mineral resources in the international seabed area was established. It was at variance not only with Part XI of the 1982 Convention and its annexes, but also with the basic principles set out in the 1970 Declaration.

Before the 1994 Agreement was concluded, the system was even in conflict with the legal régime of the high seas. "Freedoms" of durable appropriations of some locations of the seabed, even for limited periods, but excluding all other persons, have never existed as "freedoms" of the high seas. Never before in modern times did States

---

[15] For the text, see *ILM* 21 (1982), 950 et seq.

venture to extend the effect of their national laws on the exploitation of mineral resources – which are not renewable – beyond the limits of their territorial sea and continental shelf. Conventions having extra-territorial effect, such as those concerning the high seas, the conservation of living resources in the high seas, outer space and Antarctica, forbid such unilateral acts by States and in their national legislation. The matter was, therefore, initially *ultra vires*.

***

Nevertheless, towards the close of the Third UN Law of the Sea Conference, a concession was granted to the claims of industrial States and to licences issued by their organs, but subject to their signing of the Convention.

As an annex to the Final Act of the Conference, Resolution II "Governing preparatory investment in pioneer activities relating to polymetallic nodules" was adopted. On the basis of this Resolution, the Preparatory Commission was authorised to register "pioneer investors" and to grant them permits for their initial activities in exploration of polymetallic nodules in the Area. Paragraphs 8 and 9 awarded priority to pioneer investors over all applicants other than the Enterprise, in the approval of plans of work and exploitation of the sites explored, after the entry into force of the 1982 Convention.

The pioneer investors from the following States have registered their applications with the Preparatory Commission: Japan, the Soviet Union, France, China, India, Eastern Europe and South Korea. However, the locations registered by Japan, France and the USSR are not large enough to reserve an area through the Enterprise. However, the locations registered by Japan, France and the USSR are not large enough to reserve an area in favour of the Enterprise. Out of that procedure of authorisation remained entities which obtained permits from the national authorities of the States that did not sign the Convention in 1982. Hence three entities obtained licences from the US administration.

***

The 1994 Agreement relating to the implementation of Part XI of the UN Law of the Sea Convention of 1982 (hereinafter, "the 1994 Agreement") was adopted by the UN General Assembly Resolution 48/263 of 28 July 1994, pending the entry into force of the 1982 Convention set for 16 November of that year.

Article 4 of the Agreement forbids accession of new States to the 1982 Convention without their consent to be bound by the Agreement. Article 5 provides a simplified procedure of accession to the Agreement for the Contracting States of the 1982 Convention. All States that have given their consent to be bound by the Agreement shall be considered parties without its ratification, unless they notify the depository that they do not avail themselves of this simplified procedure.

That way, the procedure for the amendment of the 1982 Convention as provided in its text was avoided. Section 4 of the Annex to the 1994 Agreement suppressed paragraphs 1, 3 and 4 of Article 155 of the Convention on the Review Conference for its Part XI.

The 1994 Agreement has deeply affected the whole of Part XI of the Convention. According to Article 1 of the 1994 Agreement: "The States Parties to this Agreement undertake to implement Part XI in accordance with this Agreement. ... The Annex forms an integral part of this Agreement". And according to Article 2(1): "The provisions of this Agreement and Part XI shall be interpreted and applied together as a single instrument. In the event of any inconsistency between the Agreement and Part XI, the provisions of this Agreement shall prevail".

The States that have consented in the General Assembly to the adoption of the 1994 Agreement, but have not acceded so far to the 1982 Convention itself (the United States in particular), obtained the privilege of its provisional application, but only until 16 November 1998. It was expected that by that date they would have become parties to the Convention. In this provisional capacity, these States become entitled to all rights as provided, including the right to be members of all organs of the Authority. Section 1(12)(c) of the Annex provides the following in their regard:

> "States and entities which are members of the Authority on a provisional basis in accordance with subparagraph (a) or (b) shall apply the terms of Part XI and this Agreement in accordance with their national or internal laws, regulations and annual budgetary appropriations and shall have the same rights and obligations as other members, including:
>
> (i) The obligation to contribute to the administrative budget of the Authority in accordance with the scale of assessed contributions;
>
> (ii) The right to sponsor an application for approval of a plan of work for exploration...".

The law of the sea is an exemple of a domain in which the unilateral acts of some States, which are initially in breach of positive law, become lawful at a latter stage by means of subsequent conventions on codification and the progressive development of general international law. This is exactly what happened with regard to licences and permits for activities in the Area issued by the national administrations of some industrialised States.

On the basis of Section 1, paragraph 6, of the Annex, and on the basis of a recommendation of the Legal and Technical Commission, the Council of the Authority is authorised to approve applications of plans of work for exploration. That procedure is expressly provided for the plans of work of pioneer investors approved by non-States Parties to the 1982 Convention, plans previously registered with the Preparatory Commission, and plans of new entities in accordance with the principle of non-discrimination.

Understandably, the new composition of the Council of the Authority according to the 1994 Protocol will approve the richest mine sites of polymetallic nodules, which are already reserved under internal procedures of industrial States.

The 1994 Agreement has profoundly modified Part XI of the 1982 Convention in other important aspects:

First of all, the transfer of technology requirement for carrying out activities in the Area has been eliminated. Initially, beneficiaries of this transfer were to be the Enterprise and developing States. According to the 1982 Convention, this transfer was the condition for the approval of any plan of work. Section 5 of the Annex determines that the Enterprise and developing States will obtain this technology "on fair and reasonable commercial terms and conditions on the open market, or through joint-venture arrangements".

The United States does not conceal its wish to suppress the International Enterprise altogether. An intermediate solution was found according to which the Secretariat of the Authority will perform the functions of the Enterprise until it begins to operate independently. However, Section 2, paragraph 3, suppressed the obligation of States Parties to fund one mine site on behalf of the Enterprise. Under paragraph 5, a contractor that has contributed a particular area to the Authority "shall be entitled to apply for a plan of work for that area provided it offers in good faith to include the Enterprise as a joint-venture partner".

But should the Enterprise become capable of operating independently in the future, States Parties will not be allowed to subsidise its activities in the Area. The Enterprise will be bound to function according to sound commercial principles, like all other entities. In spite of these concessions, the United States did not give up the idea of liquidating the Enterprise in the future.[16]

The obligation of the Authority to render economic assistance to developing land-based producer States that suffer serious adverse effects as a result of the exploitation of minerals from the deep seabed was also virtually abolished. According to Section 7 of the Annex, the Authority shall establish an economic assistance fund from a portion of its funds exceeding its own administrative expenses.

It is therefore obvious that the first concern in collecting the revenues from the activities in the Area will be to cover the expenses of a very reduced administration of the Authority. The 1994 Agreement does not foresee any other necessities of mankind.

The United States succeeded in obtaining, to all intents and purposes, a permanent seat in the Council of the Authority, with much greater competencies than previously provided, as well as an effective right to veto all its decisions.[17] It also enjoys the right to veto any amendments to the text of the 1994 Agreement. Finally, in accordance with Section 9, a new Financial Committee was established with a strong influence on the management by the Authority.

In spite of all these far-reaching concessions, the United States has still not became a party to the 1982 Convention.

***

What can be concluded with regard to the principle of common heritage of mankind in the law of the sea?

In the 1970s there was a consensus among all States on the existence of a set of legal principles, set out in the 1970 Declaration of the UN General Assembly, as peremptory norms of general international law. The United States itself, as well as most other industrial States, participated in the *communis opinio juris*.

---

[16] In the commentary to the 1982 Convention and the 1994 Agreement, which is a part of a letter from the Secretary of State to the President of 23 September 1994, it was stated that: "... if a decision is ever made to make the Enterprise operational, it will only be on the basis that the United States would find acceptable... It is equally possible that, by the time commercial mining takes place, developing States as well as industrialized countries will recognize the Enterprise as a relic of the past and not seek to make it operational". *ILM* 34 (1995), 1436.

[17] It is stressed in the above-mentioned commentary that: "... this chambered voting arrangement will ensure that the United States and two other consumers, or three investors or producers acting in concert, can block substantive decisions in the Council". *ILM* 34 (1995), 1433.

There is little doubt that the Hague Court would at that time have ascribed the character of *jus cogens* to the rules on the legal régime of the Area, on the same basis as it later established in its 1986 judgment in the *Nicaragua* case: the principle of non-intervention. In that respect, the Court concluded:

> "... *opinio juris* may, though with all due caution, be deduced from, *inter alia*, the attitude of the Parties and the attitude of States towards certain General Assembly resolutions... . The effect of consent to the text of such resolutions cannot be understood as merely that of a "reiteration or elucidation" of the treaty commitment... . On the contrary, it may be understood as an acceptance of the validity of the rule or set of rules declared by the resolution by themselves".[18]

But in the 1970s nobody requested such a verdict from the Hague Court.

Nevertheless, the régime as established in Part XI of the 1982 Convention, together with Annexes III and IV, deserves serious critical appraisal. Being part of a political compromise not fully achieved, the system adopted was too rigid, complex and costly for potential producers. In addition, a huge, expensive and largely unnecessary administration for the Authority and the Enterprise was conceived.

After 1980, the United States and other industrial countries began to obstruct this system in the name of the supreme principle of free market economy. This tendency finally led to the adoption, by a simple resolution of the General Assembly, of the 1994 Agreement relating to the implementation of Part XI of the UN Law of the Sea Convention of 1982.

The repetition in the preamble of this Agreement of the wording that "the seabed and ocean floor and subsoil thereof, beyond the limits of national jurisdiction ... as well as the resources of the Area, are the common heritage of mankind", is mere lip-service to this principle. Its operational clauses, together with the Annex, have destroyed its substance as defined in the 1970 Declaration.

Nobody cares now that the primary objective of the exploitation of hard mineral resources in the Area should be carried out for the benefit of mankind. Potential beneficiaries of the resources of the Area will be private enterprises[19] and a very reduced bureaucracy of the Authority.

---

[18] ICJ Reports 1986, 99 et seq., para. 188.

[19] In respect of the principle of common heritage of mankind, it was said in the above-mentioned commentary by the United States, see note 14, that:

It cannot be claimed that the international community, whatever the meaning of this term may be, has no other needs.

All this considered, nobody knows when the commercial exploitation of hard mineral resources of the Area will commence, or whether it will commence at all, but the mine sites which are the richest in polymetallic nodules have been reserved for ever.

Bearing this in mind, the atmosphere throughout the Third UN Law of the Sea Conference now seems surreal. Developing States impatiently expected revenues from the exploitation of the Area in the very near future in order to settle all their urgent economic and social needs, and in order to fill the gap separating them from the industrial States in their economic development. These excessive expectations were at the root of the deficiencies in solutions adopted in Part XI of the 1982 Convention.

***

If the principle of common heritage of mankind had constituted a *jus cogens* during the 1970s, one may pose the question how it evolved later. Some authors still doubt the accuracy of the qualification of *jus cogens* in Article 53 of the 1969 Vienna Convention on the Law of Treaties. As known, it provides that "... a peremptory norm of general international law is a norm from which no derogation is permitted and which can be modified only by a subsequent norm of general international law having the same character". There is a contradiction in that provision, because any new treaty which is in conflict with the existing *jus cogens* is, according to Article 53, void at the time of its conclusion.

The 1994 Agreement has proved the fact that as the result of a shift in *opinio juris* within the "States whose interests were specially affected", an actual rule of *jus cogens* can vanish.

***

Finally, the most recent development have nearly closed the second stage in the evolution of the law of the seas and oceans. First from *res*

"This principle reflects the fact that the Area and its resources are beyond the territorial jurisdiction of any nation and are open to use by all in accordance with commonly accepted principles... It is worth noting that the Agreement, by restructuring the sea-bed mining regime along free market lines, endorses the consistent view of the United States that the common heritage principle fully comports with private economic activity in accordance with market principles". *ILM* 34 (1995), 1429.

communis omnium in Roman law to appropriations by States of entire oceans at the time of the great geographic discoveries, and then from the *Mare Liberum* of Hugo Grotius in 1609 to the 1982 Convention and the 1994 Agreement: these are two phases of the same spiral which has continuously lasted for two millennia.

From Confrontation to Cooperation on the High Seas: Recent Developments in International Law Concerning the Conservation of Marine Resources

L. Yves Fortier[*]

I. Introduction

> "... And God said, Let us make man in our image, after our likeness. And let him have dominion over the fish of the sea, and over the birds of the air and over the cattle, and over all the earth...". (Genesis, 1:26)

The general topic addressed in this brief paper is the conservation of marine resources on the high seas. The perspective from which the topic is analysed is at once timely and significant: the historic evolution of international law as it relates to the conservation and management of high seas fisheries.

If the subject has been introduced in biblical terms, the reader should rest assured that it is not my intention to trace the evolution of the relationship between mankind and the environment since "the Beginning". Yet, the passage from Genesis cited above refers to one of the most pressing problems bedevilling mankind in its constant struggle simultaneously to develop and to conserve the earth's resources – the concept of "dominion"; the dominion of mankind, generally, and the competing dominions of States, specifically, over the earth and its precious resources.

---

[*] The author wishes to acknowledge the invaluable contribution of his colleague, Stephen L. Drymer, in the preparation of this text.

It could also be said that the strong biblical encouragement to humanity to "be fruitful and multiply" has not helped the problem. Increasing populations, employing ever more efficient technology, have hastened the depletion of many of the world's once superabundant fish stocks, among other natural resources, animal, plant and mineral. Mankind, however, was not left without any guidance in this regard. According to the Old Testament, Adam and Eve were enjoined not merely to "subdue the earth", but also to "replenish" it. Surely, this sounds as though it were an early reminder that the natural environment is a gift to be enjoyed by humanity without being squandered, and an early evocation of the concept of sustainable development.

It remains as true today as in the age of Eden that the moral duty to replenish is an integral element of the right to dominion over the environment. What has changed is the urgency. It is precisely the evolving global appreciation of the duties of cooperation and conservation inherent in the concept of dominion – in particular, as regards high seas fisheries – that I wish to address.

In the area of high seas fisheries, as in so many other fields, the development of international law is largely a history of States' increasing recognition of the *obligations* which supplement, qualify and, ultimately, impart meaning to their *freedoms*. Recent developments in this area of international law demonstrate the emergence of an historic consensus. States have come to recognise that the benefits of their traditional freedom to fish are attainable only to the extent that nations are able to cooperate to ensure sustainable harvest levels.

This evolution in the law of the sea – from unrestricted exploitation to increasing conservation, from freedom to responsibility, from confrontation to cooperation – is discernible in the 1982 UN Convention on the Law of the Sea[1] (the "1982 Convention"). On 4 August 1995 it was given clear expression in the Agreement concluded during the sixth and final session of the United Nations Conference on Straddling Fish Stocks and Highly Migratory Fish Stocks (the "UN Agreement on Straddling Stocks").[2]

---

[1] United Nations Convention on the Law of the Sea, opened for signature 10 December 1982, *ILM* 21 (1982), 1261 (entered into force 16 November 1994).

[2] Agreement for the Implementation of the Provisions of the United Nations Convention on the Conservation and Management of Straddling Fish Stocks and Highly Migratory Fish Stocks, opened for signature 4 December 1995, *ILM* 34 (1995), 1542.

II. State Jurisdiction over Maritime Areas and Resources

It is said that the most salient characteristic of contemporary law of the sea has been the expanding exercise of jurisdiction by coastal States over maritime areas and resources. As the traditional freedom of States to fish on the high seas has given way to a broad redefinition and strengthening of the interests and rights of coastal States, it has become subject to a persistent pattern of regulation or restriction.

1. Early History of State Jurisdiction over the High Seas

Initially, and for centuries, activities such as navigation and fishing on the high seas were open to everybody. That was, however, before the desire of States to exercise dominion over the high seas manifested itself. In the 15th and 16th centuries, the periods of great maritime discovery by European navigators, numerous claims were made by the powerful maritime States, principally European, to the exercise of sovereignty over specific portions of the high seas. Thus, Portugal claimed maritime sovereignty over the whole of the Indian Ocean and much of the Atlantic, Spain declared herself sovereign over the Pacific Ocean and the Gulf of Mexico, and Great Britain laid claim to the North Sea. It was as a result of such claims, and largely in opposition to them, that the law of the sea developed, and with it modern international law itself. Hugo Grotius, often referred to as the father of international law, was also one of the first writers strenuously to attack such extensive claims to maritime sovereignty. In contrast to the principle of dominion, or sovereignty, the principle of "freedom of the high seas" was conceived.

The development of the concept of high seas freedom in this context was not mere sophistry. It was, rather, grounded in very pragmatic concerns. The "Age of Discovery" was also very much an age of commerce. States began to appreciate that, too often, conflicting claims were being made on the same parts of the open sea, to no practical end and to the great inconvenience of all. The principle of freedom of the high seas thus came to be seen as serving the general interest of all States.

2. The Expanding Jurisdiction of Coastal States

In 1702, Cornelius Van Bynkershoek, a judge at the Supreme Court of Appeal of Holland, enunciated the principle: *terrae potestas ubi finitur armorum vis* (territorial sovereignty extends as far as the power of arms carries).[3] Perhaps because of the increasing range of shore batteries, coupled with the willingness of States to use them to protect maritime areas and resources, the width of this figurative "cannon shot range", or territorial sea, gradually increased over ensuing centuries to three, then eventually 12 nautical miles.

The enlarged breadth of the territorial sea and the compromises leading to the establishment of the exclusive economic zone (EEZ) in the 1982 Convention[4] were the product of the age-old clash of interests between coastal States and distant-water nations. A similar pattern of confrontation also provided the backdrop to the development of the law of the sea in relation to fisheries. While coastal States pressed for increasing jurisdiction and control over key fishing grounds, others sought to rely upon the traditional rules protecting the freedom of the high seas, including the freedom to fish.

III. The 1982 UN Convention on the Law of the Sea – and Its Lacunae

The 1982 Convention, which entered into force on 16 November 1994, signalled the birth of a new legal era. The 1982 Convention formally established the concept of the 200-mile EEZ, extending dramatically the jurisdiction of coastal States and limiting considerably the geographical extent of the area previously referred to as the high seas. As a result, the traditional freedoms of the high seas, including the right of all States to fish, were significantly restricted. The EEZ represented the recognition of the primary interest of coastal States over the resources within their 200-mile zones.[5]

As regards the effective conservation and management of marine resources on the high seas, the 1982 Convention envisaged a régime based on *international cooperation*.[6] Unfortunately, the provisions of the 1982 Convention relating to high seas fisheries, and specifically to straddling stocks and highly migratory stocks, left much to be desired. Such stocks are unusual in that their straddling or migratory nature means that they can be found both within a given State's EEZ, as well as on the high seas adjacent to that zone or within the EEZ of a neighbouring State. Thus, they are subject to different legal régimes

---

[3] *De Dominio Maris*, Chap. 2.
[4] Arts. 55-75.
[5] Art. 56(1)(a).
[6] Art. 118.

and different fishing pressures. Clearly, cooperation is essential if such stocks are to be properly managed. However, the rights of coastal States with respect to straddling and highly migratory stocks within their EEZs, and the obligations of States whose nationals fish for them on the high seas, are not clearly spelled out in the 1982 Convention. Contemporary problems relating to high seas fishing, at least in respect of straddling and highly migratory stocks, are in large part due to that legal lacuna.

It is not by chance that the provisions of the 1982 Convention relating to the conservation and management of high seas fish stocks, in general, are so few: it is estimated that the creation of 200-mile EEZs placed as much as 95 per cent of the world's fisheries resources under the jurisdiction of coastal States.[7] Moreover, the uncertain relationship between the rights of coastal States and those of distant-water nations, and the clash of interest between these two groups of States, may help to explain why certain issues were left unresolved in 1982.

In sum, while the 1982 Convention established clear rights and obligations for coastal States in respect of resources within their EEZs, the high seas remained, essentially, an area free from effective conservation or management. As a result, many of the world's important fish stocks continued to be virtually unprotected.

IV. Towards the UN Conference on Straddling Stocks

Several countries, including Canada, set out to improve the 1982 Convention and to find a global solution to the problem of high seas overfishing. Their aim was to establish a set of clear, binding rules which would clarify the ambiguous provisions of the 1982 Convention and infuse meaning and substance into the general obligation of States to cooperate in management and conservation. This noble objective was laid down in the document known as "Agenda 21", unanimously adopted by the 188 States that participated in the 1992 UN Conference on Environment and Development (UNCED) in Rio de Janeiro.[8] According to that document, all States are bound to adopt measures to ensure that high seas fishing is managed in accordance with the

---

[7] See Report of the FAO World Conference on Fisheries Management and Development, Rome, June 27-6 July 1984, App. D., 1.

[8] See Report of the United Nations Conference on Environment and Development, Rio de Janeiro, 3-14 June 1992, UN Doc. A/CONF.151/26, Vol. II.

provisions of the 1982 Convention.⁹ In particular, the signatories committed themselves to convene, as soon as possible, an intergovernmental conference, under the auspices of the United Nations, with a view to promoting the effective implementation of the provisions of the 1982 Convention with respect to straddling stocks and highly migratory stocks.¹⁰

The groundwork for the UN Conference on straddling and highly migratory stocks was thus laid.

V. The Agreement on Straddling Stocks and Highly Migratory Stocks

1. Context and Background

The present stage in the evolution of international law as it relates to the conservation of high seas fisheries is fundamentally different from that of prior eras. Increasingly, the question to be resolved is no longer whether coastal States are entitled to devise new maritime areas for the exercise of given forms of jurisdiction, including in respect of fisheries. Rather, the central issue is whether, in view of serious conservation problems which require urgent solutions, those solutions should be provided by coastal States unilaterally, or whether they should be negotiated by interested parties for the international community as a whole.

Two extremely important implications, evidenced by state practice, follow from this reorientation of the age-old question of maritime jurisdiction. In the first place, the issue is not *whether* some fisheries activities should be regulated, but rather *who* shall undertake the appropriate regulatory functions, *how* and *to what extent*. Secondly, the high seas can no longer be considered an area free from regulation, just as coastal States' maritime areas can no longer be regarded as the sole source and extent of their jurisdictional authority.

Over the years, there have been several cases of unilateral state action aimed at solving questions relating to the regulation and conservation of high seas fisheries by means of varying degrees of coastal State intervention. In each instance, coastal States have, by one means or another, sought to extend their jurisdiction *(ratione materiae)* beyond their 200-mile EEZ, so as to protect and conserve certain fish stocks. One illustration is particularly apposite. In 1994, Canada

---

⁹ *Ibid.*, Article 17.49.
¹⁰ *Ibid.*, Article 17.49 (e).

amended its Coastal Fisheries Protection Act[11] so as to provide for an extension of Canadian enforcement jurisdiction beyond the 200-mile limit, for conservation purposes, in certain defined instances. In March 1995, Canada relied on this amendment to arrest a Spanish trawler, the *Estai*, on the high seas, after it was accused of violating both international and Canadian conservation rules. The incident provided a concrete demonstration of the traditional dispute between coastal and distant-water States, and underscored the need for international rules governing certain high seas fisheries.[12]

This incident is perhaps the most vivid such event in recent years, and it continues to be the subject of commentary regarding the numerous issues raised by high-seas fishing and the conservation of marine resources. While the case was referred by Spain to the International Court of Justice,[13] it is the facts "on the water" as opposed to the decision of the ICJ (which found that the Court did not have jurisdiction over the dispute[14]) that provide the most significant lesson. The arrest of the *Estai* eventually led to bilateral negotiations between Canada and the European Community.[15] On 16 April 1995, an agreement was reached[16] obligating Canada and the European Community, in the future, to submit their intended total allowable catches (TACs) of various marine species to scrutiny by the Council of

---

[11] RSC 1985, chapter C-33.

[12] On this dispute, see M. Keiver, "The Turbot War: Gunboat Diplomacy on Refinement of the Law of the Sea", 37 *Cahiers de Droit* 2 (June 1996), 543; Y. Song, "The Canada-European Union Turbot Dispute in the Northwest Atlantic: An Application of the Incident Approach", 28 *Ocean Development & Int'l L.* 3 (1997), 269.

[13] The *Fisheries Jurisdiction* case *(Spain v. Canada)*, application of Spain of 28 March 1995.

[14] Canada opposed the jurisdiction of the Court on the grounds that, on 10 May 1994, prior to the modifications it introduced in its Coastal Fisheries Protection Act, Canada had deposited with the Secretary-General of the United Nations a new declaration of acceptance of the compulsory jurisdiction of the Court. This included a new reservation providing that "disputes arising out of or concerning conservation and management measures taken by Canada with respect to vessels fishing in the NAFO Regulatory Area ... and the enforcement of such measures" were excluded from the jurisdiction of the ICJ. It was decided by 12 votes to 5 that the *Spain* v. *Canada* dispute came within the terms of this reservation.

[15] Now the European Union.

[16] The agreement is reprinted in D. Freestone, "Current Legal Developments, Canada/European Union – Canada and the EU Reach Agreement to Settle the Estai Dispute", *Int'l J. Marine & Coastal L.* 10 (1995), 397, appendix, and in *ILM* 34 (1995), 1260.

the Northwest Atlantic Fisheries Organisation (NAFO), a regional entity founded by the NAFO Convention signed in Ottawa on 24 October 1978 to regulate harvesting in the Northwest Atlantic.[17] A protocol to strengthen the NAFO conservation and enforcement measures was also agreed upon.[18] Canada then repealed the controversial amendments to its Coastal Fisheries Protection Act under which its authorities had seized the *Estai*, on 9 March 1995.[19]

According to one author, the dispute between Canada and the European Community, and the settlement ultimately reached, hastened the conclusion of the Straddling Stocks Agreement.[20] Whether or not that was in fact the case, there is no doubt that the *Estai* incident illustrated the need to strengthen the functions and competence of international and regional fisheries organisations, as it became clear that only effective *multilateral* mechanisms could solve the problem of overfishing on the high seas. The incident also occurred in an era when divergent views and opinion could be said to have coalesced into the general, if still inchoate, recognition that conservation is in the interests of all States and requires active cooperation. Indeed, the very fact that the UN Conference evolved from compromises reached at UNCED is significant. It situates the evolution of international law squarely in the context of contemporary environmental concerns, and demonstrates the extent to which those concerns have influenced the development of new approaches to the problem of high seas fisheries.

2. The Agreement

In August 1994, the Chairman of the UN Conference, Ambassador Satya Nandan of Fiji, outlined what he considered to be the five key elements of a successful outcome.[21] Each of the elements described by

---

[17] Convention on Future Multilateral Cooperation in the Northwest Atlantic Fisheries, reprinted in Council Regulation 3179/78, OJ (1978) No. L 378, 1 (entered into force 1 January 1979).

[18] Agreement, see note 16.

[19] Items 1 and 2 in Table IV in section 21 of the Coastal Fisheries Protection Regulations of 3 March 1995 applying the Coastal Fisheries Act to Spain and Portugal were repealed. See SOR/95-136, *Canada Gazette* (1995), Pt. II, 650.

[20] M. S. Sullivan, "The Case in International Law for Canada's Extension of Fisheries Jurisdiction Beyond 200 Miles", 28 *Ocean Development & Int'l L.* 3 (1997), 241.

[21] Third session of the UN Conference on Straddling Fish Stocks and Highly Migratory Fish Stocks, held at UN Headquarters in New York, 15-26 August 1994. See the Chairman's Guide (A/CONF.164/13/Rev. 1).

Ambassador Nandan are found in the Agreement concluded on 4 August 1995.[22] Each comprises an essential, integral component of an effective system of international control for the conservation of high seas fishery resources. These elements are reviewed briefly below.

A. First Element: Minimum International Conservation Standards

In adopting and enforcing conservation and management measures, States are required to apply what is known as the precautionary approach, so as to ensure that overfishing does not occur. In essence, that approach requires States to ensure that their management measures are developed and implemented so that they err – as is human, and thus inevitable – on the side of conservation.

B. Second Element: Compatibility and Non-Negative Impact

One of the most significant aspects of the Agreement, from the point of view of Canada and other coastal States, is the principle of compatibility of the conservation and management measures established for the high seas, on the one hand, and for areas under national jurisdiction, on the other. This is the second key element. Specifically, States are required to ensure that measures established in respect of relevant stocks for the high seas do not undermine the effectiveness of measures adopted by States within their EEZs.

C. Third Element: The Duty to Cooperate in Fisheries Conservation

The breakthrough in the negotiation of the Agreement occurred with the elaboration of a solution which could bridge the gulf between the competing claims of coastal and distant-water States. Rather than

---

[22] For a detailed discussion of the Agreement, see M. Collett, "Achieving Effective International Fishery Management: A Critical Analysis of the UN Conference on Straddling Fish Stocks", *Dalhousie Journal of Legal Studies* 4 (1995); H. Gherari, "L'accord du 4 août 1995 sur les stocks chevauchants et les stocks de poissons grands migrateurs", *RGDIP* 2 (1996), 367; L. Juda, "The 1995 United Nations Agreement on Straddling Fish Stocks and Highly Migratory Fish Stocks: A Critique", 28 *Ocean Development & Int'l L.* 2 (1997), 147; T. L. McDorman, "The Dispute Settlement Regime on the Straddling and Highly Migratory Fish Stocks Convention", *Canadian YIL* XXXV (1997), 57.

focusing on *creating additional rights* for one group, which could only have worsened the divide, the participants came to realise that the key to success would be the strengthening and creation of shared *obligations* designed to ensure effective conservation.

As noted above, the obligations of States in respect of the conservation of marine resources, as enunciated in the 1982 Convention, rested on a general duty to cooperate. Unfortunately, the parameters of that duty were not adequately defined, and it quickly became apparent that this shortcoming prevented effective conservation. The Agreement was conceived and drafted so as to give meaningful effect to States' duty to cooperate under international law generally, and under the 1982 Convention in particular.

According to the terms of the Agreement, States that fish for straddling stocks or highly migratory stocks are obliged to establish international organisations or arrangements to govern their management and conservation. Where such arrangements exist in respect of a particular stock or region – for example, NAFO – all States that fish on the high seas for the stocks in question are required to join the existing organisation or to apply the management measures established by it. Only those States that are members of such organisations, or that have agreed to apply the conservation measures established by them, are permitted access to the relevant fisheries.

D. Fourth Element: Compliance and Enforcement

If obligations such as the foregoing put flesh on the bones of the duty of States to cooperate in conserving marine resources, it is the Agreement's compliance and enforcement provisions that gave it its teeth. The Agreement establishes a comprehensive and rigorous system of duties in respect of States' compliance with, and enforcement of, conservation measures.

As in other multilateral agreements, and in accordance with the principle of sovereignty, in most areas it is up to the flag State to enforce those measures. However, some of the strongest enforcement provisions of the Agreement, and arguably the most innovative from the perspective of international law, are the rules which provide for the boarding and inspection of vessels *on the high seas* by the authorities of the *non-flag State*.

E. Fifth Element: Dispute Resolution

Finally, the Agreement incorporates the provisions of the 1982 Convention regarding dispute resolution, in relation to the interpretation and application of the Agreement, and the procedures for doing so.

In sum, the obligations imposed by the Agreement are onerous and far-reaching. Together, they comprise the structure of what will, hopefully, prove to be a workable, effective system of rules for the conservation of high seas fishery resources.

VI. Implementation

Implementation will require some time – not only for the Agreement to be ratified and to enter into force, but for the principles and norms that it establishes actually to be applied by the world's fishing nations. As of October 2000, 27 of the 37 States required by the instrument had ratified or acceded to the Agreement.

The fate of other multilateral instruments is also illustrative of the current transitional phase in this area, and may hold a clue as to the fate of the Agreement.

The 1993 Agreement to Promote Compliance with International Conservation and Management Measures by Fishing Vessels on the High Seas[23] (the "1993 Compliance Agreement") is also still not in force. The 1993 Compliance Agreement sets forth a broad range of obligations for parties whose fishing vessels operate on the high seas, including the obligation to ensure that such vessels do not undermine international fishery conservation measures. It also creates a form of international registry of high seas fishing vessels in order to assist in monitoring fishing on the high seas. Ratification by 25 States is necessary to bring this Agreement into force, whereas, as of the time of writing, 18 have done so.

In May 1994, Canada became the first party to ratify the 1993 Compliance Agreement. At approximately the same time, the Government of Canada decreed regulations requiring domestic-flagged vessels to obtain a high seas fishing licence when fishing outside Canadian waters. These regulations were introduced to ensure that Canadian vessels respect the obligations imposed by international conservation régimes.[24]

---

[23] FAO Conference, 27th Session (November 1993), Resolution 15/93. The text is available at http://www.fao.org/legal/treaties.htm. Eighteen of the 25 acceptances required for entry into force have taken place.

[24] 1996 Report of Canada to the United Nations Commission on Sustainable Development, available at http://www.ec.qc.ca/agenda21/96/part 3-2.html.

On 31 October 1995, the UN Food and Agricultural Organisation (FAO) unanimously adopted a global Code of Conduct for responsible fisheries[25] (the "FAO Code"). Although non-binding and voluntary in nature, the FAO Code establishes principles and standards applicable to the conservation, management and development of all fisheries. It provides a framework for national and international efforts to ensure sustainable exploitation of marine resources. This framework is in fact a list of factors that should be taken into consideration when managing all types of fisheries, fishing operations and post-harvest practices. It is not only directed at governments, but at all parties engaged in high seas fishing, including individual fishermen and companies.

The gradual implementation of these instruments suggests that the international community is struggling to come to grips with the environmental obligations newly found to be inherent in their traditional rights. The negotiation of agreements demonstrates a willingness to recognise and respect these obligations; it is only by their implementation, however, that the world's commitment will be gauged. While progress is slow,[26] developments are not entirely negative. For example, the 1999 announcement by the United States and Canada of a truce in their long-running bilateral dispute over implementation of the Pacific Salmon Treaty,[27] in the form of a ten-year agreement on abundance-based harvest levels, has been described as a "real world implementation of the concept of precautionary fisheries management as articulated in the United Fish Stocks Agreement and other global instruments".[28]

VII. Conclusion

Fish, unlike humans, know no frontiers. They have yet to learn the concept of dominion. Yet, they can, and do, arouse otherwise sensible

---

[25] FAO, Rome, 1995, text available at http://www.fao.org/waicent/faoinfo/fishery/agreem/codecond/ codecon.htm.

[26] See, for example, in J. Hyvarinen, E. Wall and I. Lutchman, "The United Nations and Fisheries in 1998", 29 *Ocean Development & Int'l L.* 4 (1998), 332. The authors criticise the UN General Assembly for its failure to provide a critical forum for debate and development on the issue of high seas fisheries.

[27] Concluded in Ottawa 25 January 1985, entered into force 18 March 1985, available at http://www.psc.org/treaty/treaty.htm. The author served as Canada's Chief Negotiator for Pacific Salmon during key stages of the resolution of the dispute from 1994 to 1998.

[28] R. Tucker Scully, 1 October 2000, text available at http://www.state.gov/www/policy_remarks/1999.

nations to territorial rage. They are an unlikely irritant between States, but one which, as history has shown, has the power to galvanise, radicalise and polarise. "Fish wars" may sound like a quaint concept, yet their impacts are real and their causes cannot be ignored: dramatically increased fishing capacity, an ever-increasing global population that needs to be fed and vastly depleted fisheries. The problem of high seas fishing is therefore a pressing issue both for States which harvest or otherwise rely upon the resource, as well as for the resource itself – and hence for the community of nations at large.

The main reason for overfishing is one perhaps familiar to economists and other students of human nature. If fishermen restrained themselves, each would benefit from enlarged stocks in the future. But no individual boat, fleet or State sees any gain from holding back unless it knows that its competitors are doing the same. Recently, States have come to realise that it is to their advantage to cooperate in the establishment and enforcement of rules designed to promote the effective conservation and management of stocks which, because they are found in the high seas, have heretofore been subject to only the barest restrictions. The establishment of the UN Conference on Straddling Fish Stocks and Highly Migratory Fish Stocks represents an acknowledgement that the existing state of international law and practice with respect to high seas fisheries was inadequate. The Agreement of August 1995, and other measures, reflect a growing consensus regarding the need for an effective, workable system of international conservation principles and rules.

The Agreement also reflects, clearly, a compromise between the interests of coastal States and those of distant-water States. As a result, the focus of the Agreement is not the rights of any given State, or group of States. Rather, the Agreement builds upon existing international law, including the 1982 UN Convention on the Law of the Sea, by reaffirming, clarifying and particularising the duty of all States to cooperate in the conservation of marine resources.

This reliance upon an international duty to cooperate constitutes far more than a practical compromise required in order to secure agreement among the various States involved in the 1995 UN Conference. It bespeaks a profound recognition of the nature of the problem to be resolved. Simply put, cooperation is the *sine qua non* of effective conservation of marine resources on the high seas. In this, as in so many areas of international law and relations, cooperation is not merely a pretty catchphrase, or simply an expression of the realities of international negotiation. It is the embodiment of enlightened self-interest.

To the extent that the Agreement breathes life into the principle of cooperation, and seeks to ensure that the principle is effectively implemented, it truly represents a proud accomplishment. It is the product of an extraordinary evolution. We have come a long way from the Beginning.

Civil Liability and Compensation for Vessel-Source Pollution of the Marine Environment and the United Nations Convention on the Law of the Sea (1982)

Thomas A. Mensah

I. Introduction: Civil Liability and Compensation under the Convention on the Law of the Sea

The provisions of the United Nations Convention on the Law of the Sea on the protection and preservation of the marine environment are, in general, addressed to States. The articles of the Convention set out obligations and responsibilities to be fulfilled, and powers and rights to be exercised, by States in various areas within and outside their jurisdictions in order to prevent, reduce and control pollution of the marine environment from the potential sources of pollution listed in the Convention.[1]

The Convention imposes responsibility on States to fulfil their international obligations concerning the protection and preservation of the marine environment and states that they "shall be liable in accordance with international law" for failure to discharge this responsibility.[2] According to Article 297, disputes that may be submitted to the dispute settlement "procedures entailing binding decisions" under Section 2 of Part XV include:

> "when it is alleged that a State, in exercising the freedoms, rights or uses of the sea under the Convention, has acted in contravention of the Convention or the laws and regulations adopted by the coastal state in conformity with the Convention or other rules of law not incompatible with the Convention,[3] and

---

[1] The various channels for the pollution of the marine environment are itemised in Article 194 of the Convention, namely, from land-based sources (para. (a)); from or through the atmosphere or by dumping (para. (a)); from vessels (para. (b)); from installations and devices used in exploration or exploitation of the resources of the seabed and subsoil (para. (c)); and from other installations and devices operating in the marine environment (para. (d)).

[2] Article 235, para. 2.

[3] Article 297, para. 1(b).

when it is alleged that a coastal State has acted in contravention of specified international rules and standards for the protection and preservation of the marine environment which are applicable".[4]

But the Convention does not provide just for the traditional remedies and sanctions of "state responsibility". It recognises that, for some damage to the marine environment, it might not be possible or appropriate to address the claim for compensation against a State Party to the Convention in its capacity as a State, but rather to an entity, governmental or otherwise, whose acts or omissions were the cause of the damage. In such a case, the "inter-State" dispute settlement procedures under Part XV of the Convention may not offer a feasible or even an appropriate mechanism for dealing with the claim or dispute in question. The Convention, therefore, makes provision for an alternative approach to deal with such cases. This is the "civil liability" approach under which a person who suffers damage may seek compensation from the person responsible for the activity that caused the damage. An essential feature of this approach is that liability to pay compensation does not necessarily arise because of non-compliance with the law on the part of the person responsible for the damage. In other words, a person who suffers damage as a result of an injurious act not prohibited by law may, nevertheless, be able to obtain compensation for the damage.[5] Thus Article 229 of the Convention states that nothing in the Convention "affects the institution of civil proceedings in respect of any claim for loss or damage resulting from pollution of the marine environment". It is pertinent to note that claims under this provision relate simply to "loss or damage resulting from pollution of the marine environment". No reference is made to non-compliance with any obligations (breach of duty). This contrasts with the claims under Article 297 which must be based on allegations that there has been "contravention" of a law, regulation or standard conforming to the Convention or not incompatible with it.

Under the Convention a State may be liable for damage or loss resulting from pollution of the marine environment. In addition to possible liability resulting from a failure to fulfil its international responsibilities concerning the protection and preservation of the marine environment pursuant to Article 235,[6] a State may incur liability if it acts beyond its powers under the Convention. Article 232 of the Convention

---

[4] Article 297, para. 1(c).
[5] On the civil liability approach in general see R. Wolfrum and C. Langenfeld (eds.), *Environmental Protection by Means of International Liability Law*, 1999.
[6] Article 235, para. 1.

provides that States shall be liable "for damage or loss attributable to them arising from measures taken pursuant to [the Convention] when such measures are unlawful or exceed those reasonably required". A related obligation is that States "shall provide for recourse in their courts for actions in respect of such damage or loss".[7] Such recourse to national courts does not, of course, displace the right of States to seek redress under international law by means of the procedures for dispute settlement established under Part XV of the Convention.

But the Convention also recognises that damage to the marine environment can result from the activities of entities other than States. As a rule, damage caused by such non-state entities will not be the subject of claims under international law, except where it is alleged that the damage resulted wholly or partly from the failure of a State to discharge its international obligations. In such a case, the claim would be against the State concerned and not the non-state entity. To ensure that there are appropriate mechanisms for dealing with liability and compensation for damage to the marine environment resulting from the activities of non-state entities, the Convention imposes an obligation on States Parties to make provision for such liability in their domestic legal systems. Thus Article 235, paragraph 2, states that "States shall ensure that recourse is available in accordance with their legal systems for prompt and adequate compensation or other relief in respect of damage caused by pollution to the marine environment by *natural or juridical persons* under their jurisdiction". (emphasis added) Paragraph 3 of the same article states: "To ensure prompt and adequate compensation in respect of all damage caused to the marine environment, States shall cooperate in the implementation of existing international law and the further development of international law relating to responsibility and liability, for the assessment of and compensation for damage, and the settlement of related issues as well as, where appropriate, development of criteria and procedures for payment of adequate compensation such as compulsory insurance or compensation funds".

Although the objectives set out in the Convention in relation to an international system of civil liability for damage to the marine environment apply to pollution from all the sources itemised in Article 194, progress has so far been made only in respect of one of these sources, namely marine pollution from vessels.[8] The régime of civil

---

[7] Article 232.

[8] No régime of civil liability has so far been developed for pollution damage from land-bases sources or from dumping. Although a Convention on Civil Liability for Damage Resulting from Exploration and Exploitation of Seabed Mineral Resources was adopted in 1977, it has so far not entered into force. In contrast, régimes on liability and compensation for damage have been established in many areas. These include the 1960 Paris Convention on Third

liability and compensation for damage resulting from vessel-source pollution of the marine environment has been developed largely under the auspices of the International Maritime Organization (IMO), the Special Agency of the United Nations with the mandate, *inter alia*, "to encourage and facilitate the general adoption of highest practicable standards in matters concerning maritime safety, efficiency of navigation, and prevention and control of marine pollution from ships".[9] Beginning from 1969 with pollution damage from oil transported in tankers, the IMO has systematically developed a régime that now covers nearly all ship-borne substances that have a significant potential to cause damage to the marine environment. And although the process actually commenced well before the negotiation and adoption of the 1982 Convention, the underlying objectives and principles of the IMO régime are the same as those finally incorporated in Part XII of the Convention. This régime establishes rules and principles for the attribution of responsibility and liability for damage resulting from pollution of the marine environment; mechanisms through which persons suffering damage may have recourse to national courts for compensation; requirements in respect of compulsory insurance to ensure the availability of compensation; and compensation funds that guarantee prompt and adequate compensation.

II. The Traditional Approach to Liability and Compensation for Damage Arising from Maritime Transport

Until 1969, there was no international régime dealing specifically with liability and compensation for damage caused by pollution of the seas resulting from the operation of ships. An international convention adopted in 1954 had, for the first time, addressed the question of pollution of the sea by oil transported in ships, but the purpose of the 1954 Convention was to prevent pollution and its scope was restricted to pollution resulting from "operational activities", such as the discharge of

---

Party Liability in the Field of Nuclear Energy (1960), the Vienna Conventions on Civil Liability for Nuclear Damage (1963); the ECE Convention on Civil Liability for Damage Caused During Carriage of Dangerous Goods by Road, Rail and Inland Navigation Vessels (1989); the 1993 Convention on Civil Liability for Damage Resulting from Activities Dangerous to the Environment (Lugano Convention, 1993); and the Basel Protocol on Liability and Compensation for Damage Resulting from the Transboundary Movements of Hazardous Wastes and Their Disposal (1999).

[9] Article 1 of the Convention on the Intergovernmental Maritime Consultative Organization (IMCO), 1948, as amended, UNTS Vol. 289, 3.

oil or "oily mixtures" into the coastal waters.[10] While the 1954 Convention, and a number of national laws, prescribed penalties for non-compliance with regulations on the discharge of oils into the sea, they did not directly or indirectly address the question of compensation or other relief for persons who suffered damage from the discharge of oil from ships, whether deliberately or accidentally.

The general rule was that persons who suffered damage from the operation of ships had to seek remedy by recourse to the normal laws of tort liability. Since the transport of goods was not considered to be an "inherently hazardous activity", a claim for compensation for damage resulting from the operation of a ship could only succeed if it was based on some proven fault on the part of the shipowner or the person against whom the claim was brought.[11]

At that point, the only international instrument dealing with the question of liability and compensation was the 1957 Convention relating to Limitation of Liability of Owners of Sea-going Ships.[12] But this Convention was not concerned with the question of liability as such. The main objective of the 1957 Convention was to establish a maximum limit to the liability of the shipowner *vis-à-vis* third parties. Under the convention, the liability of the shipowner for damage arising in connection with the operation of his ship was limited to the "liability tonnage" of the ship.[13] The governing principle of the Convention was that the liability of the shipowner should not exceed a predetermined limit. This approach was adopted even by States which were not parties to the Convention, sometimes with different criteria for setting the limit of liability. But in all cases, the principle was the same: the liability of the ship was limited to a specified amount for each incident; and this represented the only compensation available to all persons who suffered damage from the ship in respect of any one incident, regardless of the extent of damage actually suffered by them, individually or collectively.

---

[10] International Convention for the Prevention of Pollution of the Sea by Oil, 1954; UNTS Vol. 327, 3.

[11] "The principle of fault is broadly recognised in international maritime law". Comments of the USSR on the proposed International Convention on Civil Liability for Oil Pollution Damage, in *Official Records*, International Legal Conference on Marine Pollution Damage, 1969, IMO Publication, 1973, 510.

[12] The International Convention Relating to the Limitation of the Liability of Owners of Sea-going Ships, 1957, UNTS Vol. 1968, 52, N. Singh, *International Maritime Conventions*, Vol. 3, 1983, 269.

[13] The limitation figure of a ship was calculated at 3,100 gold (Poincaré) francs (approximately US$60-70) per ton of the ship's tonnage. Of this slightly more than two-thirds (2,100) was reserved for "personal claims" and the remainder for property damage. The "gold franc" unit has in many jurisdictions been replaced with the Special Drawing Rights (SDRs) of the International Monetary Fund.

And neither the 1957 Convention nor any national laws had addressed even indirectly the issue of compensation for damage which might be caused as a result of a serious accident involving a ship laden with large quantities of oil or other potentially hazardous cargoes. The assumption was either that such an accident would not occur or that, if it did happen, any questions of liability and compensation would be resolved by reference to the normal principles and rules of the applicable national law.

III. The Torrey Canyon Accident of 1967

The grounding of the tanker *Torrey Canyon* off the southern coast of England in the spring of 1967 drew the attention of the international community to a number of previously unremarked aspects of the maritime transport of oil in bulk. Among these was the issue of what measures States could take to protect themselves against damage from accidents which threatened major pollution damage. But an even more difficult question was how to ensure that States and persons who suffered damage from such accidents would be able to obtain adequate compensation for the damage suffered.[14]

In the discussions following the accident, it was relatively easy to reach agreement on the measures that States could take to protect themselves against pollution damage from incidents involving oil tankers. For this purpose, an international treaty was adopted in 1969.[15] This instrument recognised the right of a coastal State to take reasonable and proportionate measures of intervention in the event of an accident which threatened to cause major pollution damage to its coastal or related interests.[16]

The problem of liability and compensation was, however, not so easy to resolve. There were many reasons for this. In the first place, the problem affected not only the interests of States and private persons who might suffer damage but it also touched on the concerns of society at large with regard to the health and integrity of the marine environment and the amenities that it afforded to the coastal States and to humanity as a whole. In addition, the *Torrey Canyon* accident had raised issues on which there were conflicting views between the parties traditionally

---

[14] On the Torrey Canyon disaster and its impact on the law relating to marine pollution in general, see E. Gold, *Maritime Transport: The Evolution of Marine Policy and Shipping Law*, 1981, Chapter 7.

[15] The International Convention Relating to Intervention on the High Seas in Cases of Oil Pollution Casualties, 1969; UNTS Vol. 970, 216, *ILM* 9 (1970), 24.

[16] Article 1 of the Convention.

involved in the marine transportation industry. As a result, sharp differences of approach surfaced, first between these parties themselves and, secondly, between them as a group, on the one hand, and the environmentally concerned sections of the international community, on the other hand.[17]

The traditional law on liability for damage caused in the course of the carriage of goods by sea dealt with two main areas, namely, obligations due to the parties directly involved in the venture and liability towards third parties, i.e., persons who have no such involvement. The parties directly involved in the maritime venture are the carrier and the persons with an interest in the cargo (the consignor or the consignee). As between them, the general assumption has been that they are partners in a common enterprise (the maritime venture) and that they agree, or are deemed to have agreed, to share the risks inherent in that venture. The terms and conditions under which each of the parties is to bear its part of the risks are usually set out in the contract of carriage, or are otherwise regulated by national and international admiralty law.[18]

With regard to damage to third parties, the general view was that, while these parties were entitled to compensation for damage suffered by them, compensation should not be such as to undermine the integrity and viability of maritime transport as the main medium of international trade. Hence, the idea was accepted that the liability of the shipowner for any damage to third parties should be limited to a predetermined level. This limit was set by reference to the tonnage of the ship, as in the 1957 Convention on Limitation of the Liability of the Owners of Sea-going Ships.[19] Any damage beyond the prescribed limit would not be compensated by the shipowner, but would have to be accepted by the victims, presumably as part of the price to be paid for living in society dependent on international sea-borne trade.

Relying on this "ancient principle", shipowners initially reacted unfavourably to the suggestion that persons suffering damage as a result of a major accident involving a tanker should be treated differently from other victims of damage from ships. As far as they were concerned, the situation did not warrant any departures from the well-established principles and rules regarding liability and compensation for damage

---

[17] For a detailed account of the issues and parties in the negotiations on this and other aspects of marine pollution from ships, see R. M. MacGonigle and M. L. Zacher, *Pollution, Politics and International Law: Tankers at Sea*, 1979.

[18] Such as the 1957 Limitation of Liability Convention, see note 12, or the United Nations Convention on the Carriage of Goods by Sea, 1978 (Hamburg Rules), UN Doc. A/Conf.89/13.

[19] See notes 12 and 13.

arising in the course of the operations of ships engaged in international maritime commerce.[20] They accepted that a new régime was needed to deal with the new kind of damage, but they were unwilling to accept any "new burden" in terms of an obligation to compensate for damage caused by oil spills from ships. They argued that the cause of any damage from pollution was not the ship but rather the cargo – the oil. Hence, any compensation for the damage should come from the owner of the cargo or other persons with an interest in the cargo.[21]

These views were not well received in the new climate of widespread environmental awareness. Coastal States, which would have to spend large sums of money on preventive or clean-up measures in the wake of a major accident resulting in massive pollution of coastal areas, could not accept the suggestion that they would be entitled to no compensation for the expenses incurred by them, or that the compensation due should be limited to the value of the vessel or its cargo. This was particularly the case since the environmental damage which could be caused by a laden oil tanker might be far in excess of the value of the ship or even the value of the ship and its cargo put together.[22] The position of the shipowners was also rejected by the newly awakened environmental movement, as well as specific persons and groups whose property or commercial interests could be severely damaged by a maritime casualty that released large quantities of oil. They were unable to accept the proposition that "innocent third parties" that suffered damage should have their entitlement to compensation

---

[20] The position of the Government of the USSR was that while an exception to the general principle of fault liability was "called for" in the case of nuclear liability because of "exceptional causes", there were no such special causes in the case of oil pollution to warrant a departure from the traditional principle of fault liability. *Official Records*, see note 11, 510. A similar position was taken by the Liberian Government, which stated that it was "of the firm opinion that liability should be only on the basis of fault... This was necessary to prevent too drastic a change in the existing maritime law". *Ibid.*, 508. The Irish Delegation proposed that liability should be placed on the cargo because it wished to "preserve intact the existing maritime law". *Ibid.*, 447.

[21] The Irish Delegation contended that "it is the cargo which causes the type of pollution at which the Convention is aimed". *Official Records*, see note 11, 449.

[22] "The *Torrey Canyon* incident had shown that pollution caused by the escape of crude oil affects many interests, not only causing material damage but also creating the necessity for public authorities to take preventive measures at very high cost. The parties affected cannot be expected to bear the risks inherent in maritime traffic". Delegate of Netherlands, *Official Records*, see note 11, 439.

determined by reference to principles and rules that were developed for application between parties engaged in the business of transport for profit.[23]

Furthermore, the cargo interests were unwilling to accept the suggestion that they should bear liability for damage from maritime accidents. They contended that since the cargo was entirely and at all material times under the control of the shipowner, his employees or agents, any accident involving the ship or cargo could only result from the faulty handling of the ship or cargo. It was, therefore, both logical and equitable that responsibility for any damage resulting from the accident should rest on the shipowner.[24]

These differences of view emerged at the very outset of the international discussions of issues arising from the *Torrey Canyon* incident. The discussions took place mainly within the International Maritime Organization (IMO), the United Nations specialised agency with responsibility for promoting safety and efficiency of maritime transport and the prevention of marine pollution. The issues were extensively examined by representatives of governments and various sectors of the maritime industry, first in the Council of the IMO and, subsequently, in the newly established Legal Committee of the IMO. At the request of the Legal Committee, the Comité Maritime International (CMI) also gave consideration to the "private law aspects" of the matter. The conclusions emerging from these discussions were subsequently submitted to an international diplomatic conference convened by the IMO in Brussels in November 1969.

IV. The International Legal Conference on Marine Pollution Damage

1. Purpose of the Conference

In spite of the differences of opinion on the solutions to be adopted, there was general acceptance of the need for an international agreement to establish a uniform international régime to deal with the issues of

---

[23] "The notion of limited liability ... was detrimental to a victim's rights ... and implied that anyone who had suffered damage as a result of pollution was not entitled to adequate compensation but only to limited compensation". Government of Canada, *Official Records*, see note 11, 85.

[24] "It is the operator who was in a position to preclude or reduce to a minimum the risks arising out of the carriage of goods and, moreover, it was always easy to identify him or at least to identify the owner". US Delegation, *Official Records*, see note 11, 627.

liability and compensation for pollution damage arising from accidents such as the *Torrey Canyon*. All governments and parties involved recognised that the only alternative to such an international régime would be a system of unilateral national legislation; and it was agreed that such a system would create serious difficulties for the shipping industry by subjecting ship and cargo owners and their insurers to different and uncoordinated rules and procedures in different countries. In the absence of uniform regulations and criteria, there was always the possibility that some of the national laws would impose even heavier burdens on the maritime industry. For an industry as global as shipping this was clearly undesirable.

A system in which questions of liability and compensation are regulated by unilateral national law was also considered undesirable from the point of view of the potential victims of pollution damage. For it would mean that victims of the same type and extent of damage might receive widely differing levels of compensation, depending on the law applicable in the State where the damage was caused or in which a claim for compensation was dealt with. Even more seriously, persons suffering damage from the same incident might be treated differently if their claims for compensation were to be dealt with under different national laws. A major tanker accident could cause pollution damage in more than one country and it was possible, therefore, that claims for compensation might be brought before the courts of several countries in respect of that same incident. If there were no uniform international rules on liability and compensation, claimants in these States might be obliged to have recourse to different procedures to claim compensation. This could lead to a situation in which a claimant in one country receives a greater or lesser amount of compensation than a claimant with the same loss or damage in another country.

To avoid such anomalous situations, the representatives of States and industries at the 1969 Conference were united in the view that every effort should be made to develop an acceptable uniform international system. This would provide a reasonable measure of assurance to all concerned – governments, ship and cargo owners and their insurers, as well as potential victims of pollution damage – that issues of liability and compensation for damage resulting from maritime casualties would be settled by reference to commonly agreed rules and procedures, regardless of where the damage occurred.[25]

---

[25] "It was necessary to avoid a Convention which would not receive the signatures necessary for the creation of a genuine, uniform maritime law". Report of the Chairman of Committee of the Whole II, *Official Records*, see note 11, 99.

2. The Issues to be Resolved

In the Legal Committee of the IMO, as well as in the International Sub-Committee of the CMI and at the diplomatic conference, a number of basic questions were identified as requiring to be answered in any agreement to establish an international régime. The main issues raised and discussed included the following.

A. The Party to be Held Liable to Pay Compensation for Damage Caused by Oil Discharged from a Ship

There was disagreement whether liability should be placed on the owner of the ship, or on some other entity. On behalf of the shipowners, it was argued that the damage involved was not caused by the ship but rather by the cargo, and any compensation due should be claimed from the owner of the cargo or any other person that had title to, or an interest in, the cargo.[26] Another suggestion was that it would be more appropriate to place liability on the "operator" of the ship, who was always in charge of the operations and management of the ship. It was noted that the owner of the ship might have little or nothing to do with the ship at the time of the accident. This would be so, for example, where the ship is chartered by the registered owner to an operator on a "demise (bare-boat) charter" or a "time charter".[27]

Other delegations contended that the main purpose of a liability and compensation scheme should be to ensure prompt and adequate compensation for those suffering damage from maritime casualties resulting in pollution damage. For that purpose, it would be most convenient to place liability for the damage on the owner of the ship. They pointed out that it was always easy to identify and approach the registered owner of the ship, whereas it might be difficult, and sometimes impossible, to ascertain who was the owner of the cargo or the "operator" of the ship at the time that the damage was caused.[28]

---

[26] See note 21.
[27] "The burden of liability must induce a person to take all measures for the "prevention of pollution and for minimising a loss when pollution has occurred. Such measures can be taken only by the operator as the person exercising control of operation and management of the ship. On the other hand the owner of the ship in many cases (when the ship is under demise charter etc.) has no control over the operation and management of the ship". USSR Delegation, *Official Records*, see note 11, 510.
[28] "Liability on the registered owner of the ship would probably make it easier for the plaintiffs to discover the party responsible". US Delegation, *Official Records*, see note 11, 457.

Another suggestion made was to place liability on the operator of the ship, but with the proviso that the registered owner of the ship would be held liable where the operator could not be identified or if the owner did not provide information to identify the operator.[29]

B. The Basis on Which the Liability to Pay Compensation is to be Determined

It was suggested that, in line with the traditional approach in maritime law, liability for pollution caused by oil carried in ships should be based on fault, i.e., the wrongful conduct on the part of the shipowner. While accepting that the victim of damage should be entitled to compensation, those in favour of this approach maintained that it would not be justifiable to require the shipowner to pay compensation where the damage was not caused by any wrongful act on his part or on the part of other persons acting on his behalf.[30]

This approach was not acceptable to the coastal States and the representatives of the environmental movement. Their view was that, whatever the reasons or justification there might have been for the "traditional" rule that liability should be based on fault, they did not apply to damage caused in connection with the transport of a substance such as oil, which had the potential to cause massive damage to persons and property totally unconnected with the contract of carriage. A régime of liability based on fault would impose an unreasonable burden on the claimant, who would have to prove that there had been fault on the part of the ship. The wrongful conduct in question would normally take place on board the ship or in connection with the management of the ship by the shipowner or his representatives. Information on any such matters would be in the knowledge of the shipowner, and it would be difficult for the victim to obtain the evidence to prove any misconduct in such circumstances. Hence the only acceptable solution was a system of "strict liability", under which the shipowner would be liable for all damage caused by his ship.[31] They were also unwilling to accept the "compromise" solution under which the liability of the shipowner would be based on fault, but with the "burden of proof reversed", so that it would be for the shipowner to prove that the incident that caused the

---

[29] French Delegation, *Official Records*, see note 11, 444-445.
[30] See note 20.
[31] "The passive nature of the part played by the victim compelled abandonment of the notion of liability based on fault in favour of the notion of strict liability". Canadian Delegation, *Official Records*, see note 11, 86.

damage did not result from the wrongful act of the shipowner or his employees or agents.[32]

C. The Types of Damage for Which Compensation is to be Payable

The issue here was whether compensation was to be limited to damage caused by "physical contamination" by oil, or whether it should extend also to other types of damage, such as damage from fire or explosion. Some delegations wished to restrict the scope of the compensation payable, while others insisted that compensation should be payable only for damage that resulted from actual pollution by oil. Other types of damage, such as damage by fire or explosion that resulted from an oil spill, would be dealt with by the applicable general law on liability. It was, however, agreed that compensation should be payable for personal injury as well as damage to property,[33] and that the damage to be compensated would include the costs incurred for clean-up activities, as well as expenses for reasonable measures taken to prevent or minimise damage.[34]

There was considerable disagreement whether any compensation was to be paid for what was referred to as "pure environmental damage", i.e., damage to the environment as a whole which could not be represented as damage to any particular person or persons. While some delegations wanted provision to be made for damage to the environment, others did not consider this to be necessary, or indeed feasible.[35]

D. The Geographical Scope of Application of the Convention

Three main choices were before the Conference with respect to the geographical scope of application of the new régime, namely:

i. The régime should apply to all damage caused by an incident, regardless of where the damage occurred. This would make the new

---

[32] "The United Kingdom favours liability based on fault with the burden of proof reversed". UK Delegation, *Official Records*, see note 11, 463.
[33] In proposing that pollution should be limited to "damage by contamination" the United Kingdom explained that this applies to all damage "whether to persons or property". *Official Records*, see note 11, 556. See also note 59 below.
[34] Article 1, para. 7, of the 1969 Civil Liability Convention, as amended by the Protocol of 1992.
[35] Article 2, para. 3(a), of the 1992 Protocol to the 1969 Civil Liability Convention.

régime apply to damage which was caused on the territory of a State, in its territorial waters and also in areas outside the jurisdiction of any State.

ii. the international régime should apply only to damage caused outside national jurisdiction. This would leave it to each State to determine the rules and procedures it considered appropriate to deal with damage caused within its jurisdiction.[36]

iii. the régime should apply to the same extent in all areas subject to the jurisdiction of States participating in the new régime. This would place all victims of damage on more or less the same footing. In fact, proposals were made to extend the application of the Convention to damage caused "on the high seas", but these did not attract sufficient support in 1969.[37]

E. The Extent to Which Compensation Should be Paid for Damage

The question was whether liability to pay compensation should be limited or unlimited and, if limited, what should be the basis and extent for such limitation. For those in favour of limitation of liability, the main argument was that it would be difficult, and perhaps impossible, for the owner of a ship to obtain insurance to cover unlimited liability. They noted that it would be pointless, from the point of view of the potential victim of damage, for compensation to be set at a level which made it impossible for the person liable to insure against the risk. Without insurance there would be no guarantee that any compensation would in fact be received by the victim if the person liable was unable to pay the total amount of compensation due. And there might actually be cases in which a victim might not get any compensation at all, for example, where the person liable had become bankrupt, possibly as the result of the incident.[38]

On the other hand a strong body of opinion considered it unfair and unacceptable that a victim of pollution damage should be denied compensation for part of the damage actually suffered. Many of those who held this view doubted that it would be so difficult for a shipowner to obtain insurance, as had been claimed. In any case, they maintained

---

[36] Proposals to this effect were made by the Delegations of Norway and Sweden, see *Official Records*, see note 11, 440 and 513.

[37] Proposal of the German Delegation in Document LEG/CONF/C.2/WP.4. *Official Records*, see note 11, 661.

[38] "The United Kingdom considers that, in order to enable the shipowner to obtain insurance cover, the principle of limitation of liability should be accepted, except in the case of an incident arising from the fault or privity of the owner". UK Delegation, *Official Records*, see note 11, 487.

that the non-availability of insurance was no justification for denying to an innocent victim the legitimate right to be compensated for damage that it has suffered.[39]

With regard to the level at which any limit of compensation was to be fixed, it was agreed that it should be sufficient to cater for all foreseeable cases of damage, whilst ensuring that it did not place an unnecessarily heavy burden on the shipowner. In this respect, the shipowners insisted that the most equitable arrangement would be to impose some of the burden on the shipper of the cargo.[40] Although it was not possible for the 1969 Conference to deal with this proposal within the framework of the 1969 Convention, the principle was accepted and finally incorporated in the system through the 1971 Fund Convention.[41]

F. Mechanisms to Ensure That Compensation Will be Available to Those Who Suffer Damage

It was generally agreed that a system of "compulsory insurance" would provide the most effective mechanism to ensure that compensation would in fact be available to those who suffered damage, but there was controversy about how feasible such a system might be. Many delegations felt that this would pose difficult and even insurmountable administrative problems. Others were opposed to the proposal that persons who suffer damage should have the right to bring their claims directly to the insurer. They wanted claims to be brought instead to the person liable, in accordance with the usual practice. These delegations argued that the contract of insurance would be between the shipowner

---

[39] "A seemingly pertinent objection against (a system of absolute liability) has been that the capacity of the insurance market would not suffice to cover absolute liability up to the amounts necessary to afford adequate protection; ... It seems, however, dubious that the impact on the insurance capacity should be so different if the alternative system based on fault with a reversal of the burden of proof were to be chosen". Swedish Delegation, *Official Records*, see note 11, 514. "Any argument based on the cost of insurance and the capacity of the insurance market was specious and out of place". Canadian Delegation, *ibid.*, 86.

[40] See proposal of the Government of Canada. *Official Records*, see note 11, 521.

[41] *ILM* 11 (1972), 284. The Resolution of the 1969 Conference on the establishment of a compensation fund stated that the purpose of the new fund was to ensure, firstly, "that victims should be fully and adequately compensated under a system based upon the principle of strict liability" and secondly, that "the fund should in principle relieve the shipowner of the additional financial burden imposed by the present (1969) Convention". Resolution 2 of the 1969 Conference, *Official Records*, see note 11, 185.

and the provider of insurance, and the victim of damage would be a "third party" for the purposes of this contract. However, other delegations maintained that it would be in the interest of the claimant to be able to go directly to the insurer, since it might not always be easy to identify or reach the registered shipowner at any of the locations where claims might be brought against him.[42]

G. Courts with Jurisdiction to Decide on Questions of Liability and Compensation for Damage

There was general agreement that jurisdiction in respect of a claim for damage should lie with the competent court of a State in which the damage was caused, but some delegations wanted to extend competence to other courts. One proposal was to give competence also to the courts of any Contracting State in which the owner, or the person providing insurance or other financial security, has his habitual residence. It was even suggested that competence might be given to "the courts of a Contracting State in which arrest of the ship involved or any other ship belonging to the owner was effected, or even a State where such arrest could have been effected". The general view was that while such a system might permit a wider choice of courts, it might lead to an inconsistent or contradictory assessment of damage if claims arising from the same incident were presented to a multiplicity of courts. The consensus, therefore, was to limit jurisdiction to the courts of States where damage has been caused. The reason given was that these courts, "being geographically close to the site of the pollution, [were] the best place to assess the consequences". It was also noted that "it is also easy for the victims to have access to [such courts]".[43]

V. The 1969 Civil Liability Convention and Subsequent Revisions

After extensive discussions, the 1969 Diplomatic Conference managed to resolve these issues, and agreement was reached on the provisions for an international convention, the International Convention on Civil Liability for Oil Pollution Damage of 1969, adopted in Brussels on 29 November 1969.[44] In 1976 a Protocol was adopted to the 1969 Convention in order to replace the "unit of account" used for calculating

---

[42] "In the interest of the victims themselves, it is essential that they should be able to take direct action against the insurer in all cases...". Finnish Delegation, *Official Records*, see note 11, 467.
[43] *Official Records*, see note 11, 493.
[44] UNTS Vol. 973, 3; *ILM* 9 (1970), 45.

the amounts of compensation under the Convention.[45] The 1969 Convention was subsequently revised by a Protocol adopted in 1992.[46] Together with the 1971 and 1992 Conventions establishing the International Oil Pollution Compensation Fund (IOPC),[47] the 1969/1992 Liability Conventions establish a comprehensive, and so far largely successful, international régime for dealing with pollution damage caused by the maritime carriage of oil as cargo.

The main elements of the liability and compensation régime under the 1969/1992 Conventions may be summarised as follows.

1. Liability of the Shipowner

The Convention channels all liability for pollution damage to the owner of the ship in which the oil was being transported at the time of the incident which caused the damage. Article III of the Convention states that "the owner of a ship at the time of an incident, or where the incident consists of a series of occurrences, at the time of the first occurrence, shall be liable for any pollution damage caused by oil which has escaped or been discharged from the ship as a result of an incident".[48] The owner of the ship is defined as "the person or persons registered as the owner of the ship or, in the absence of registration, the person or persons owning the ship".[49]

Article IV of the Convention states that if more than one ship is involved in an incident which causes pollution damage, the owners of all of them will be jointly and severally liable for any damage which is not reasonably separable.

The decision to "channel" the liability to the owner of the ship means that no other person is to be held liable for any pollution damage to which the Convention is applicable. In particular, the Convention expressly states that "no claim for pollution damage *under this*

---

[45] UNTS Vol. 1225, 356. The "gold Franc" unit of account was replaced by the Special Drawing Rights (SDRs) of the International Monetary Fund.
[46] IMO Document LEG/CONF.9/15 of 2 December 1992.
[47] *ILM* 11 (1972), 284 and IMO Document LEG/CONF.9/16 of 2 December 1992.
[48] In the 1969 Convention, "incident" was defined as any occurrence which "causes pollution damage". The 1992 Protocol revised this to "causes pollution damage *or creates a grave and imminent threat of causing such damage*". (emphasis added)
[49] Article 1, para. 3. When a ship is "owned by a State and operated by a company which in that State is registered as the ship's operator", this company shall be the "owner" of the ship for the purposes of the Convention. This provision was introduced to cover the case of ships owned or operated by state-owned enterprises or public corporations.

Convention or otherwise may be made against the servants, or agents of the owner".[50] However, the Convention makes it clear that purpose of the "channelling" mechanism is only intended to benefit potential victims of damage and does not give any indemnity to the servants or agents of the owner or other persons who might otherwise have responsibility for the occurrence which resulted in the damage. To make that point clear, the Convention provides that "nothing ... shall prejudice any right of the owner against third parties".[51]

Another important consequence of the "channelling" mechanism is that, where the Convention applies, the liability of an owner for pollution damage is determined only by reference to the provisions of the Convention. Paragraph 4 of Article IV of the Convention states that "no claim for compensation for pollution damage shall be made against the owner otherwise than in accordance with this Convention". Thus a person who suffers pollution damage covered by the Convention in a State which is party to the Convention can only claim compensation against the owner of the ship on the grounds provided in the Convention and in accordance with the procedures set out for making claims. In addition, the compensation available to the claimant will be as specified in the Convention, regardless of the damage actually suffered or the fact that a different provision of national or international law would have permitted a much higher level of compensation.

2. Basis of the Liability of the Shipowner

The liability of the shipowner is "strict", that is to say, it does not depend on any proven negligence or other fault on his part or on the part of his servants, employees or agents. All that a claimant needs to prove is that the damage was caused as a result of the escape or discharge of oil from a ship owned by the particular shipowner as a result of an incident.[52]

---

[50] Article III, para. 4, emphasis added. The 1992 Protocol amended this provision to extend the exoneration under the Convention to the members of the crew as well as to pilots, charterers, persons performing salvage operations or taking preventive measures, and the servants and agents of all such persons (Article 4, para. 2, of the 1992 Protocol). However, the revised provision states that the persons mentioned will lose the benefit of the exoneration if the "damage resulted from their personal act or omission, committed with the intent to cause such damage, or recklessly and with knowledge that such damage would probably result".

[51] Article III, para. 5.

[52] Article III, para. 1, simply provides that the owner "shall be liable for any pollution damage *caused by oil which has escaped or been discharged from the ship as a result of an incident*". (emphasis added)

However, to take account of the concerns of the shipowners and the insurance industry, certain clearly defined exceptions to the liability of the owner were provided. These exceptions apply if the damage:

1. resulted from "an act of war, hostilities, civil war, insurrection or a natural phenomenon of an exceptional, inevitable and irresistible character"; or
2. "was wholly caused by an act or omission done with ... the intention to cause damage by a third party"; or
3. "was wholly caused by the negligence or other wrongful act of any Government or other authority responsible for the maintenance of lights or other navigational aids in the exercise of that function".[53]

The owner might also be exonerated wholly or partially from liability to a person suffering damage if the pollution damage "resulted wholly or partially either from an act or omission done with the intention to cause damage by the person who suffered the damage or from the negligence of that person".[54]

In all cases, however, the burden of proving that the conditions exist for exoneration from liability rests on the owner. This is made clear in the *châpeau* of Article III, paragraph 2, of the Convention, which states that "no liability for pollution damage shall attach to the owner *if he proves* that" any of the facts relieving him of liability exist. (emphasis added) Under paragraph 3 of Article III, the exoneration of the owner may apply "if the owner proves" that damage resulted wholly or partially from the person who suffered the damage. The exoneration under paragraph 3 does not apply automatically. As the wording indicates, the owner "may be exonerated wholly or partially" from liability. This means that there may be some liability on the owner even if he is able to prove a wrongful act on the part of the victim. This would be the case, for example, where the wrongful act of the victim was only partially responsible for the damage (i.e., where there is contributory negligence of the person suffering the damage). In that case, the owner may only be entitled to partial exoneration from liability to that person.

3. The Types of Damage to be Compensated

A. Pollution Damage (i.e., "Damage by Contamination")

The liability and compensation régime of the Convention covers pollution damage. This damage is defined as "loss or damage caused

---

[53] Article III, para. 2.
[54] Article III, para. 3.

outside the ship carrying oil by contamination resulting from the escape or discharge of oil from the ship, wherever the escape or discharge may occur, and includes the costs of preventive measures ... and further loss or damage caused by preventive measures".[55] "Preventive measures" are defined as "any reasonable measures taken by any person after an incident has occurred to prevent or minimize pollution damage"[56]; "incident" is defined as "any occurrence, or series of occurrences having the same origin which causes pollution damage or creates a grave and imminent threat of causing such damage".[57]

Under this definition, damage for which compensation is available under the Convention is limited to damage resulting from "contamination" by oil. In other words, damage covered by the Convention must be related to actual pollution and does not, therefore, include damage resulting, for example, from "fire and explosion". Thus where damage is caused as a result of fire or explosion on board a tanker or even outside the tanker, liability for any resulting damage will have to be determined by reference to other applicable law, national or international.[58]

B. Personal Injury

Damage from oil pollution can also be in the form of "personal injury" and damage to property. For such damage compensation would be due to the person or entity that suffered the damage.[59] However, it was not considered necessary or possible to spell out in the Convention the precise types of damage that would be covered in particular situations. It was expected that the concept would be elucidated in the course of the practical application of the Convention. In the event, the International Oil Pollution Compensation Fund, which was established by the 1971 Fund Convention to supplement the compensation available under the

---

[55] Article 1, para. 6.
[56] Article 1, para. 7.
[57] Article 1, para. 8, as amended by the Protocol of 1992 (Article 2, para. 4).
[58] Such damage will, in most cases, be covered by the 1996 HNS Convention, when it enters into force.
[59] See note 31. In October 1995, the Executive Committee of the IOPC Fund in considering claims in respect of the *Braer* incident (United Kingdom, 1993) noted certain claims for "personal injury such as respiratory conditions resulting from the inhalation of oil vapour and skin complaints resulting from contact with oil". The Committee took the view that, "in the light of the discussions at the 1969 International Conference which adopted the 1969 Civil Liability Convention, the Convention in principle covered personal injury caused by contamination, whereas personal injury resulting from other causes was not admissible". *Annual Report of the IOPC Fund* (1996), 64.

Civil Liability for Pollution of the Marine Environment 1411

1969 Convention, has provided much of the practical guidance needed for this purpose.[60]

C. Economic Loss

As indicated in the IOPC Claims Manual and in the practice of the IOPC Fund in respect of claims dealt with by the Fund, pollution damage covers in general any quantifiable damage that results from pollution from the escape of discharge of oil to which the Civil Liability Convention applies. This includes contamination and consequential damage to property such as fishing boats and fishing gear, fisheries resources (such as oyster beds, fish farms and aquaculture facilities), recreational boats (yachts), piers and harbour works, and property on land, including beach huts and dwelling houses. Damage also includes "economic losses resulting from restriction of fishing activity or from closure of coastal industrial or processing installations, as well as loss of income by resort operators (hoteliers and restaurateurs)". Such economic loss should be "quantifiable".[61]

D. Pure Environmental Damage

There has been considerable controversy regarding the question whether, and if so to what extent, the definition of "pollution damage" in the 1969 Convention covered damage to the environment where such damage did not also result in "quantifiable loss" to any particular person ("pure environmental damage"). On the basis of the discussions at the 1969 diplomatic conference, the IOPC Fund, the first body to address the issue in a concrete manner, concluded that the term "pollution damage" as defined in the 1969 Civil Liability Convention was to be understood as referring to pollution damage that results in "quantifiable loss". In a resolution adopted in 1980, the Assembly of the IOPC Fund expressed the view that compensation under the 1969 Convention (and, in

---

[60] See IOPC Fund Claims Manual in E. Gold, *Gard Handbook on Marine Pollution*, 2nd ed., 1996, Appendix IX. See also CMI Guidelines on Oil Pollution Damage, adopted at the 35th International Conference of the Comité Maritime International (CMI) in Sydney, October 1994. Text reproduced in Gold, *ibid.*, Appendix I.

[61] See IOPC Fund Claims Manual, in Gold, see note 60, para. 7.4. Examples of the different types of damage for which compensation has been paid over the years are given in the various Annual Reports of the IOPC Fund. For example, "houses contaminated by smoke from burning oil". *Annual Report* (1996), 51.

consequence, the 1971 Fund Convention) was payable only if a claimant had suffered "quantifiable economic loss". Specifically, the Assembly "rejected the assessment of compensation for damage to the marine environment on the basis of an abstract quantification of damage calculated in accordance with theoretical models".[62]

This position of the IOPC Fund has not been accepted by some States, which have maintained that damage to the marine environment should be entitled to compensation. For example, the United States proposed a method for assessing the value of pure environmental damage (the "contingent valuation method") which would make it possible to quantify damage to the environment for the purposes of compensation, even in cases where no specific individual or group can prove any quantifiable economic loss as a result of the damage to the environment.[63] A position similar to this has been taken by the Italian Court of Appeal in the claim relating to the *Haven* incident. In response to the contention of the Fund challenging claims for environmental damage from the Government of Italy and some municipal authorities, the Court ruled that the 1969 and 1971 Conventions "did not exclude environmental damage". The Court held that the definition of "pollution damage" as laid down in Article 1, paragraph 6, of the 1969 Civil Liability Convention is wide enough to include damage to the environment (which is not quantifiable). The Court stated that "the environment must be considered as a unitary asset, separate from those of which the environment is composed (territory, beaches, fish etc.), and it includes natural resources, health and landscape. The right to the environment belongs to the State, in its capacity as representative of the

---

[62] Resolution No. 3 of the Assembly of the International Oil Pollution Compensation Fund (IOPC Fund), adopted in October 1980. The CMI Guidelines state that "compensation for impairment of the environment (other than loss of profit) shall be limited to the costs of reasonable measures of reinstatement actually undertaken or to be undertaken". CMI Guidelines, see note 60, Part III, para. 11.

[63] Gold, see note 60, 226. "Oil spill damages are rising in the United States following the court decision largely upholding the NOM natural resource damage assessment (NRDA) regulations. The owners of the *American Trader* were recently assessed natural resource damages in excess of $18 million for the 1990 spill of 400,000 gallons of crude oil off the Southern California coast. The major portion of the damages was based on the lack of public access to the beach for approximately five weeks. It was estimated that over 27,000 persons might have visited the 14-mile long beach each day (but for the spill) during that period and the value of the lost access was estimated by the jury to be $13.19 per person per day. The remainder of the natural resource damages focussed on small animals and plants and miscellaneous micro-organisms presumably killed by the oil spill". Haight Gardner Holland & Knight, *Maritime Round-Up* (April 1998), 2.

collectivities". It asserted that "the damage to the environment prejudices immaterial values, which cannot be assessed in monetary terms ... and consists of the reduced possibility of using the environment. This damage can be compensated on an equitable basis which may be established by the Court on the ground of an opinion of experts". On that basis the Court decided that the State of Italy (though not the municipalities) could claim compensation for damage caused to the environment which could not be quantified as economic loss. With regard to the amount of compensation to be paid, the Court held that such damage "could not be quantified according to commercial or economic valuation". Rather, it assessed the damage as a proportion of the total costs of the clean-up operations related to the accident. This amount would represent the "damage which was not repaired by these (clean-up) operations".[64]

This issue, which could have led to extensive litigation, was finally put to rest in the 1992 Protocol to the 1969 Convention. That Protocol adopted a revised definition of "pollution damage" which states that "compensation for impairment of the environment other than losses of profit from such impairment *shall be limited to costs of reasonable measures of reinstatement actually undertaken or to be undertaken*".[65]

This means that the damage to be compensated will include any economic losses suffered as a result of damage to the environment (such as loss of business or profits) and also expenses incurred in restoring the damaged environment. However, any other damage to the environment, for instance damage resulting in the loss or diminution of amenities to persons or the community as a whole, will not be entitled to compensation.[66]

E. Preventive Measures

As noted earlier, the definition of damage in the Convention was worded to include not only damage resulting from the initial escape or discharge

---

[64] *IOPC Annual Report* (1991), 30; see also *IOPC Fund Annual Report* (1996), 42-43. The environmental damage for which compensation was claimed by the Government of Italy related to "consequences of beach erosion caused by damage to phanerogams and irreparable damage to the sea and the atmosphere". *Ibid.*, 41.

[65] Article 2, para. 3, of the 1992 Protocol to the 1969 Civil Liability Convention, emphasis added.

[66] This approach has been adopted in a number of international conventions. Examples include the HNS Convention (see note 98, Article 1, para. 6(c)), and the Geneva Convention on Civil Liability for Damage Caused During Carriage of Dangerous Goods by Road, Rail and Inland Water Vessels, 1989, ECE/TRANS/79 of 10 October 1989, Article 1(10)(c).

of oil, but also damage in terms of costs of preventive measures and damage resulting from such preventive measures.[67] This approach was adopted to emphasise the "environmental objective" of the Convention. Although the main purpose of the Civil Liability Convention is to ensure that adequate compensation is available to persons who suffer damage caused by pollution resulting from the escape or discharge of oil from ships, the general wish was that every incentive should be given to all concerned to take measures necessary to prevent or minimise damage, wherever possible. For this reason it was unanimously agreed that costs incurred as a result of measures taken to prevent or minimise damage should be included in the concept of "damage" within the meaning of the Convention. For this purpose, the shipowner himself will be assimilated to "a victim" of pollution damage if he incurs costs as a result of measures taken by him or on his behalf to prevent damage, after his vessel is involved in an incident. The same applies also to any state authority or any private entity which takes preventive measures. Where such measures have been taken, the costs involved are to be treated as "damage" for the purposes of the Convention, and the person who incurred the costs will be entitled to compensation on the same basis as other persons who suffered personal or property damage as a result of the incident.[68]

The 1969 Convention set out specific conditions under which preventive measures qualify for compensation.[69] First, the measures should be "reasonable", having regard to the circumstances of the situation. Secondly, the measures should have been taken "after an incident has occurred". This means that the person taking preventive measures would be entitled to claim compensation for the costs involved only if the measures were taken after the incident. By to the definition of "incident" as given in the original 1969 Convention, an "incident" occurs only if an occurrence "causes pollution damage". On that definition, "preventive measures" under the Convention would not include measures taken before any oil has escaped or been discharged from a ship involved in an accident. On that basis, measures taken, for example, to prevent oil escaping from a stranded ship would not qualify for compensation, since they would not be measures taken "after an incident had occurred which causes pollution damage". This was clearly unreasonable and incompatible with the objective of encouraging action to prevent or minimise pollution damage. Accordingly, the provision

---

[67] Article 1, para. 6.
[68] The costs of preventive measures taken by the owner are also eligible for compensation. Article V, para. 8, of the Convention.
[69] "Preventive measures" are defined as "any reasonable measures taken by any person after an incident has occurred to prevent or minimise pollution damage". Article 1, para. 7, of the Convention.

was revised in the 1992 Protocol to the 1969 Convention. The definition of "incident" was amended to read: "any occurrence, or series of occurrences having the same origin, which causes pollution damage *or creates a grave and imminent threat of causing such damage*".[70] As a result of this change, entitlement for compensation now does not depend on whether there has been an actual discharge or escape of oil but rather whether the measures taken were reasonably intended to prevent or minimise pollution, either because pollution has actually occurred or because there was a "grave or imminent threat" of pollution at the time the measures were taken.

F. Damage from Oil in the Bunkers of Ships

At the 1969 Conference it was decided to limit the application of the new régime to what was regarded at the time to be the main culprit for pollution damage, i.e., crude oil carried in bulk in tankers. Although some delegations wanted the scope of coverage to be extended to other potential pollutants, this did not attract sufficient support. Many delegations felt that an attempt to extend coverage beyond hydrocarbon oils would raise complex issues that could not be dealt with satisfactorily in the time available.[71] Coverage of the 1969 Convention was, therefore, restricted to damage caused by the escape or discharge of oil "from a sea-going vessel actually carrying oil in bulk as cargo at the time of the incident, i.e., normally a laden tanker". Oil was defined as "any persistent oil". This included crude oil, heavy diesel oil and lubricating oil, but excluded non-persistent oils such as gasoline, light diesel oil and kerosene.[72] By implication, this excluded pollution damage caused by oil from the bunkers of a ship involved in an accident, since such oil would not be "carried in bulk as cargo". It also excluded damage caused by residues of oil in the tank of an unladen tanker, i.e., a tanker after it has discharged its cargo of oil.

---

[70] Article 3, para. 4, of the 1992 Protocol, emphasis added.

[71] "Although the Netherlands Government appreciates the reasons of a pragmatic nature that have led to the proposal to limit the scope of the Convention to pollution by oil, it would stress the need to lay down international law concerning liability for damage caused by noxious or hazardous cargo other than oil". Comments of the Netherlands, *Official Records*, see note 11, 439. "It is apparent ... that it will not at present be possible to arrive at a world-wide agreement (for financial protection against oil pollution, whatever its source) among the interested nations...". Sewdish Delegation, *Official Records*, see note 11, 512.

[72] The 1969 definition of oil included whale oil. This was done at the strong urging of the Japanese Delegation. However, the 1992 Protocol excluded whale oil from the coverage of the Convention.

It was subsequently agreed that these definitions unduly restricted the scope of coverage of the régime and could deny compensation for what could be serious damage from maritime casualties. The definitions were, therefore, revised in the 1992 Protocol to the 1969 Convention. The definition of oil was amended to "any persistent hydrocarbon mineral oil such as crude oil, heavy diesel oil and lubricating oil, whether carried on board a ship as cargo or in the bunkers of such a ship".[73] In addition, the definition of "ship" was amended so as to include a ship even after it has transported oil as cargo, provided it has "residues" of oil aboard. As a result of these revisions coverage of the régime is now extended to all damage caused by oil, whether it is carried on board a ship as cargo, whether it is from the bunkers of a ship or whether it is the residue of oil previously carried on the ship as cargo. However, the restriction of the régime to pollution from "persistent hydrocarbon oils" has been maintained.[74]

4. Territorial Application of the Convention

The 1969 Civil Liability Convention as amended by 1992 Protocol applies to damage caused in State Parties to those instruments. For the purposes of the Convention, pollution is caused in a State if it is "caused in the territory, including the territorial sea" of the State or in its "exclusive economic zone" established in accordance with international law.[75] The 1969 Convention applied only to damage caused in the "territory, including the territorial sea, of a Contracting State". However, when the 1992 Protocol was adopted, it was decided to extend the scope of application to take account of the provision in 1982 United Nations Convention on the Law of the Sea that the coastal State has jurisdiction to regulate activities within the new exclusive economic zone for a number of specific purposes, including jurisdiction with regard to the "protection and preservation of the marine environment".[76] Consequently, the 1969 provision was amended to include damage caused in the exclusive economic zone where such a zone has been declared by the State concerned. Where no such zone has been formally

---

[73] 1992 Protocol, Article 2, para. 2.
[74] A new Convention on liability and compensation for oil carried in bunkers was finally adopted by a diplomatic conference convened by IMO in April 2001.
[75] Article 3, para. (a)(ii), of the 1992 Protocol.
[76] The 1982 United Nations Convention on the Law of the Sea gives to the coastal State jurisdiction "as provided in the relevant provisions of this Convention with regard to ... the protection and preservation of the marine environment". Article 56, para. 1(b).

established, the coverage of the Convention is to extend to an area no more extensive than what is permitted under international law for such a zone.[77]

5. The Extent of Compensation to be Paid

On the basis of the compromise consensus reached at the conference, the 1969 Convention limits the liability of the owner of the ship for pollution. But this is subject to certain conditions. The shipowner is entitled to limit his liability for pollution damage "in respect of any one incident" to "an aggregate amount" specified in the Convention. This represents the maximum ceiling of compensation payable by the shipowner to all persons who suffer damage as a result of the incident. This maximum amount is obtained by multiplying the "limitation tonnage" of the ship by the "limitation amount" per ton as specified in the Convention. However, where the amount so obtained is more than the overall limit of liability in the Convention (the maximum limit of liability), compensation will only be payable up to that maximum limit.

6. Liability Limits

Under the 1969 Convention, the limit of liability was given as 2,000 gold francs for each ton of the ship's tonnage, up to an overall maximum limit of 210 million gold francs, regardless of the tonnage of the ship or the amount of damage actually caused by the incident. A Protocol adopted in 1976 replaced the "gold franc" unit of account with the Special Drawing Rights (SDRs) of the International Monetary Fund.[78] This brought the limitation figures to 133 SDRs for each ton of the ship's tonnage, and the maximum liability ceiling to 14 million SDRs.

The limits of liability were revised significantly by the 1992 Protocol. In addition to increasing the per tonnage limitation figure, the Protocol also introduced a system under which the amount of compensation due from a small ship was increased above what would normally have applied by multiplying the tonnage with the per ton limitation figure. For a ship not exceeding 5,000 tons there was a single

---

[77] Article 3, para. (a)(ii), of the 1992 Protocol.
[78] 1976 Protocol to the 1969 International Convention on Civil Liability for Oil Pollution Damage; UNTS Vol. 1225, 356. The Protocol provides that for States Parties that are not Members of the IMF, the limits would continue to be designated in gold units, but with the conversion to be made "according to the law of the State concerned". Article II of the 1992 Protocol, amending Article V of the 1969 Convention.

limitation amount of 3 million SDRs. For a ship over 5,000 tons, the limit was, as before, calculated multiplying its tonnage by the new per ton figure of 420 SDRs, up to a maximum ceiling of 59.7 million SDRs.

A. Conditions for Limitation of Liability

As indicated earlier, the right of the shipowner to limit his liability was made subject to a number of well-defined conditions. The first of these conditions was that the shipowner cannot limit his liability if the incident occurred as a result of his "actual fault" or "privity".[79] This expression, which had been used in previous maritime law conventions, was resisted by many delegations at the conference as being unclear and liable to lead to unnecessary litigation. However, there was a strong body of opinion that it would be inadvisable to change such a time-honoured concept. In the end the 1969 Conference decided to retain it. However, other views prevailed in the period between the adoption of the 1969 Convention and the 1984 Conference, and there was general agreement to use a more precise expression. Accordingly, the wording was changed in the 1992 Protocol to read: "The owner shall not be entitled to limit his liability ... if it is proved that the pollution damage resulted from his personal act or omission, committed with the intent to cause such damage, or recklessly and with knowledge that such damage would probably occur".[80]

B. Establishment of a Limitation Fund

The other condition for the limitation of liability is that the shipowner must establish a "limitation fund". Under Article V, paragraph 3, of the 1969 Convention, as amended by the 1992 Protocol, the shipowner wishing to limit his liability must "constitute a fund for the total sum representing the limit of his liability with the court or other competent authority" of anyone of the States Parties to the Convention in which an action for compensation has been brought. Where no such action has yet been brought, the fund may be constituted with the court or other competent authority of any of the States Parties in which such action may be brought pursuant to the Convention. The limitation fund can be

---

[79] Article V, para. 2, of the 1969 Civil Liability Convention.
[80] Article 6, para. 2, of the 1992 Protocol. The burden of proving that these conditions exist is on the claimant. The provisions state that the owner shall not be entitled to limit his liability *"if it is proved that pollution damage resulted from his personal act or omission"*.

constituted either by depositing the sum involved or a bank guarantee or other guarantee that is acceptable under the laws of the State in which the fund is to be constituted. The guarantee must also be accepted as "adequate by the Court or other competent authority" with which it is to be deposited.

7. Mechanisms to Ensure the Availability of Compensation

A. Compulsory Insurance

One of the main considerations which led the 1969 conference to agree to place liability for damage on the shipowner was the need to ensure that the compensation due under the Convention would in fact be available to those who suffered damage. The ship and cargo-owning interests, as well as the insurers, had maintained that a system based on unlimited liability would not benefit claimants, because it might not be possible to obtain insurance cover for such liability, and this could lead to situations in which the shipowner would be unable to pay the compensation due from his resources.[81] It was partly for this reason that the 1969 conference agreed to limit the liability of the shipowner, first by reference to the tonnage of the ship and then by reference to an overall ceiling, regardless of the size of the ship. But for this reason also the conference decided to include a system of "compulsory insurance" in the régime of civil liability and compensation. For this purpose the 1969 Convention provided that the owner of a ship to which the Convention applied would be "required to maintain insurance or other financial security ... in the sums fixed by applying the limits of liability prescribed in article V, paragraph 1 to cover his liability for pollution damage under this Convention".[82] As evidence that this requirement has been met, the owner will be given a certificate by the appropriate authority of the State of registry. This certificate is to be carried at all times on the ship, and every State Party is required to ensure that relevant ships which enter or leave its ports or offshore terminals have such certificates.[83]

---

[81] "It should be borne in mind that the capacity of the insurance market is restricted, thus calling for a lower limitation amount under a system of strict liability than would be adequate if liability is based on fault". Norwegian Delegation, *Official Records*, see note 11, 440.

[82] Article VII. The obligation to carry insurance or other financial security applies only to a ship which is carrying more than 2,000 tons of oil as cargo.

[83] Article VII, para. 2 and para. 11. A certificate is not required for a ship owned by a State Party, but the State Party must certify that the ship's liability is covered (para. 12).

B. Direct Recourse Against Insurers

A major innovation of the 1969 Convention is that it enables claims for compensation for damage caused by a ship to be brought directly against the person who has provided the insurance or other financial security required by the Convention. In such a situation, the insurer will be entitled to all the defences that would have been available to the owner if the claim had been brought against him. In particular, the insurer will be able to avail himself of the right to limit his liability to the same extent that the shipowner is able to do so.[84]

8. Courts with Jurisdiction to Determine Questions of Liability and Compensation

A. Jurisdiction to Decide on Claims

To ensure that the régime of the Convention is truly international it was necessary to have a uniform arrangement for determining questions of liability and compensation. This is particularly important in cases where the same incident has resulted in pollution damage in more than one country. The only solution generally acceptable was to give jurisdiction to the courts of the States in which pollution damage has been caused. Where pollution damage has been caused in more than one State, the courts of any of the States concerned will have jurisdiction.[85] A person who has suffered damage is free to choose the court before which to bring a claim for compensation. Once a claim has been brought before a particular court, that court has sole competence to determine both the question whether the shipowner is liable for the damage in terms of the Convention and also the issue of how much, if any, compensation is to be paid to the claimants in the case. Except in cases specifically provided for in the Convention, the decision of a competent court or authority in one State Party is to be recognised and enforced in all other States Parties.[86] Hence, where claims for compensation in respect of a

---

[84] Article VII, para. 8.
[85] Article IX of the 1969 Convention states that "where an incident has caused pollution damage in ... one or more Contracting States ... actions for compensation may only be brought in the Courts of any of such Contracting States", (para. 1). All Contracting States are required to ensure that their courts "possess the necessary jurisdiction to entertain such actions for compensation", (para. 2).
[86] The only exceptions are when "the judgment was obtained by fraud" or when "the defendant was not given reasonable notice and a fair opportunity to present his case". Article X, para. 1, of the 1969 Convention.

particular incident have been brought before the courts or competent authorities of two or more States Parties, the total compensation due in respect of the incident will be the aggregate of the amounts awarded by all the courts or authorities concerned. If this total exceeds the maximum limit of the owner's liability as prescribed by the Convention, the compensation payable to each claimant will be reduced *pro rata* by reference to the applicable limitation amounts.

B. Distribution of Limitation Amounts

An important issue that had to be decided was the location and procedure for the establishment of the "limitation fund", which is a condition precedent for the right of the shipowner to limit his liability in respect of any particular incident. According to the Convention, the limitation fund is to be deposited with the court or other competent authority of the State Party in which an action for compensation has been brought. Where no action has actually been brought, the limitation fund may be established in the court of any State Party where an action can be brought under the Convention.[87]

This means that, where claims for compensation have been brought, or may be brought, in two or more State Parties, the shipowner has the option to decide the jurisdiction in which to deposit the limitation fund. Once the limitation fund in respect of a particular incident has been established by the shipowner with a court or authority that is competent to receive such a fund, the shipowner as well as his assets become exempt from action by any person in connection with any damage caused as a result of that incident. The court or other competent authority with which a limitation fund has been deposited has the exclusive jurisdiction to "determine all matters relating to the appointment and distribution of the fund".[88] A consequence of this provision is that the share of the limitation fund available to claimants who have had their claims adjudicated in the courts of other States will be decided exclusively by the court or authority with which the limitation fund has been established. However, the court distributing the fund is obliged to accept both the decisions of the other courts regarding the liability of the shipowner as well as the amount of damage for which the claimants are entitled to be compensated.

---

[87] Article V, para. 3, of the 1969 Civil Liability Convention, as amended by the Protocol of 1992 (Article 6, para. 3).
[88] Article IX, para. 3, of the 1969 Convention.

9. Means of Enforcing the Convention

The enforcement mechanism of the Convention is based mainly on the provision relating to "compulsory insurance" or other financial security. As stated above, the Convention requires every ship carrying more than 2,000 tons of oil to carry insurance to cover the shipowner's liability up to the maximum amount to which he can limit his liability under the Convention.[89] A certificate attesting that the insurance required is in force for a ship, in the form and subject to the conditions prescribed by the government of the ship's registry, has to be carried at all times on board the ship. States Parties to the Convention are under an obligation to ensure that all ships covered by the Convention are duly insured as required. For this purpose, the Convention obliges each State Party not to permit a ship under its flag to which the compulsory insurance provisions apply, "to trade unless a certificate has been issued" to the ship in the manner prescribed in the Convention. And port States are required to ensure, under their national legislation, that the appropriate insurance or other financial security is in force in respect to any ship, regardless of where it is registered, that seeks to enter or leave ports in their territories or offshore terminals in their territorial seas.[90] Thus a ship for which the required insurance is not in force will not be able to operate. In this connection it is to be noted that the obligation of States Parties to prevent vessels entering or leaving their ports or terminals applies to all ships which are carrying the quantity of oil specified in the Convention. This is because the obligation on the States is in respect of ships, *wherever registered*. This means that even a ship registered in a State which is not a party to the Convention will be required to carry the necessary insurance or financial security if it wishes to enter or leave the port or terminal of a State Party to the Convention.

VI. Unresolved Questions

Although the 1969 diplomatic conference reached consensus on most of the main issues addressed in the 1969 Civil Liability Convention, a number of major problems which were raised at the conference could not be dealt with conclusively in the time available. Among these perhaps the most difficult was the limitation of the liability of the shipowner. The compromise reached at the conference did not really satisfy any of the different groups. Those who had argued for unlimited liability accepted the principle of liability limitation with great reluctance, and were dissatisfied with the actual limits adopted for the Convention. They

---

[89] Article VII, para. 1, of the 1969 Convention.
[90] Article VII, para. 11, of the 1969 Convention.

felt that the limits were too low and could lead to situations in which persons who suffered damage in very serious incidents would be compensated for only part of the damage suffered. On the other side, the shipowning interests were very unhappy at the fact that liability for compensation had been placed entirely on the shipowner. Apart from the fact that they considered this to be inequitable, they also believed that the obligations placed on them by the Convention, including the obligations of the compulsory insurance system, constituted an unduly heavy burden for which they needed some relief.[91]

These concerns were considered at the conference.[92] However, it was not possible to resolve all of them, having regard to the complexities of the issues involved and the limited time available at the conference. In a resolution, the conference noted that the convention adopted "does not afford full protection for victims in all cases". It also recalled the view that had emerged in the discussions that "some form of supplementary scheme in the nature of an international fund is necessary to ensure that adequate compensation will be available for victims of large-scale pollution incidents". Noting that the time available "had not made it possible to give full consideration to all aspect of such a compensation scheme", the Conference requested the IMO to elaborate a draft of such a scheme for submission to an international conference to be convened "not later than the year 1971". The scheme was to be based on two principles, namely, victims should be fully and adequately compensated under a system based upon the principle of strict liability, and the fund should in principle relieve the shipowner of the additional financial burden imposed by the present (1969 Civil Liability) Convention.[93]

1. Establishment of an International Fund for Compensation for Oil Pollution Damage

Pursuant to the request in the conference resolution, the Legal Committee of the IMO prepared a draft convention to establish an

---

[91] See paragraph 2 of Resolution 2 of the Conference. *Official Records*, see note 11, 185.

[92] A Working Group of the Committee of the Whole II of the Conference submitted a Report on the Subject (Document LEG/CONF/C.2/WP.45), *Official Records*, see note 11, 604. However, this report could not be discussed for final adoption in the time available. In a resolution, the Conference asked the IMO to consider the contents of the Report in connection with its further work with regard to the (International Compensation) "Fund". *Official Records*, see note 11, 186.

[93] Resolution II of the Conference. *Official Records*, see note 11, 185.

international compensation fund along the lines indicated in the resolution. This draft was considered by a diplomatic conference which was convened by the IMO in Brussels in December 1971. The conference adopted the 1971 Convention on the Establishment of an International Fund for Compensation for Oil Pollution Damage, which was described as "supplementary to the 1969 Liability Convention".[94] The 1971 Convention was revised by a Protocol adopted in 1992.[95] Under the régime constituted by the Conventions of 1969, 1971 and 1992, the Civil Liability instruments govern the liability of shipowners for oil pollution damage, whilst the Fund Conventions "establish a regime for compensation to victims when the compensation under the Civil Liability Conventions is inadequate".[96]

2. Liability and Compensation for Damage from Substances Other Than Oil

The other question which could not be resolved at the 1969 Conference related to the scope of coverage of the 1969 Civil Liability Convention. A number of States had suggested, both before and during the Conference, that the régime of liability and compensation to be established should not be restricted to pollution damage by oil, but should cover all damage caused by any ship-borne substances that are capable of causing serious environmental and other damage. Although all delegations agreed that there was a need for an international agreement on liability and compensation for damage caused by these "noxious and hazardous substances", many of them recognised that such an agreement raised complicated problems which could not be resolved in the time available.[97] It was agreed, therefore, to postpone action on this aspect of the matter to a suitable time in the future.

---

[94] *ILM* 11 (1972), 284. For the history and operation of this Convention, see R. Ganten, "HNS and Oil Pollution, Developments in the Field of Compensation for Damage to the Marine Environment", *Environmental Policy and Law* 27 (1997), 310 et seq. Also M. Göransson, "Liability for Damage to the Marine Environment", in A. Boyle and D. Freestone (eds.), *International Law and Sustainable Development*, 1999, 345-358.

[95] Text in IMO Doc. LEG/CONF.9/16 of 2 December 1992.

[96] Article 4, para. 1, of the 1971 Fund Convention.

[97] In its comments the Swedish Delegation stated: "Disastrous consequences may occur from pollution by other substances than oil. In principle there is no reason to limit the scope of the Convention to pollution by oil. Practical difficulties may, however, make it impossible for the time being to broaden the scope". Swedish Delegation, *Official Records*, see note 11, 513. See also note 71.

In the event, the issue turned out to be much more difficult than had been imagined, and almost two decades were to pass before the international community could finally agree on a régime to deal with damage from the maritime carriage of noxious and hazardous substances other than oil.

VII. A Convention on Liability and Compensation for Damage from Hazardous and Noxious Substances

In developing an international convention to deal with liability and compensation for damage caused in the course of the maritime carriage of "noxious and hazardous substances other than oil", the international community had to contend with other problems besides those dealt with in connection with oil pollution damage. And some of these problems were more complicated and difficult than those raised in the earlier exercise.[98]

Following the approach adopted in the 1969 and 1992 liability and compensation régime for oil pollution damage, it was agreed that primary liability for compensation for damage from hazardous and noxious cargoes should be placed on the shipowner. It was also decided that the shipowner's liability would be "strict", but that he should be entitled to limit this liability in certain circumstances. And, as in the case of oil pollution damage, the régime included an additional tier of compensation to be provided by the "cargo-owning interests". However, these decisions raised other problems, especially regarding the levels of compensation that would be available to victims of damage and the proportions of the total compensation package that should be the responsibility of the shipowners and cargo owners, respectively.

In the case of oil pollution damage, it had been relatively easy to discuss these issues and find solutions for them. There were several reasons for this. Ship-borne cargo oil is uniquely homogeneous and it was, therefore, fairly easy to define. It is also much easier to assess the nature and extent of the damage it is likely to cause when it escapes or is accidentally discharged from the ship transporting it. Another reason was that it is not too difficult to ascertain the parties who own or have interests in any cargo of oil. Moreover, it was possible to pinpoint a "focal point" that could speak with reasonable assurance for the industry and with which it was possible to discuss proposals which affected the

---

[98] On the problems relating to the adoption of the HNS Convention, see Gold, see note 60, 242-245. "The whole area (of pollution from hazardous and noxious substances) is much more complex as so many different interests: manufacturers, packers, forwarders, carriers, shippers, receivers etc. are involved". *Ibid.*, 82. See also Ganten, see note 94, and Göransson, *ibid.*

"cargo interests". Finally, because of the homogeneous nature of oil as cargo, it was relatively easy to reach agreement on criteria for determining the proportionate burden of liability to be imposed on the owners of different cargoes that might be involved in an incident.

In all these respects, the situation was different and more difficult when one came to deal with ship-borne hazardous substances other than oil. In the first place it was not easy to determine which substances were to be included in the scope of coverage of the new régime as well as the basis for inclusion or exclusion of the various substances. Secondly, the substances to be covered, regardless of the basis of the choice, could not be homogeneous. For the most part they would have little in common with each other apart from the capacity to cause damage to persons, property and the environment. Thirdly, there was no single body with capacity or mandate to speak on behalf of the "cargo interests" on the proposals which were of concern to those interests in the discussions. Above all, there was no dependable basis for determining the nature and extent of the damage that could be caused by the same quantity of the different substances. This makes it difficult to determine the level of liability to be imposed on the owner of any particular substance or the basis on which to calculate the limits of compensation to be expected from such substances.

The absence of an easy formula for apportioning responsibility for different substances presented a particularly acute complication in the discussions on a supplementary "compensation fund" to the shipowner's liability, since it made it difficult to agree on the basis on which to determine which parties should contribute to such a fund and how much should be contributed by the various parties. And it had been agreed from the outset that, without such a second and additional tier of compensation, it would not be possible to have a régime that would meet the needs of potential victims of damage while keeping the burden on shipowners within the limits that they could accept.[99]

Partly because of these difficulties, and partly because of the absence of a genuine consensus among governments that there was a real and pressing need for an international régime on damage from non-oil cargoes, the process of developing a convention on damage from noxious and hazardous cargoes took much longer than it took to develop the civil liability and compensation régime for oil pollution damage. It was only after many false starts, including a major and embarrassing failure in 1984, that the exercise was finally completed with the adoption in May 1996 of an international convention to deal with questions of

---

[99] "The economic consequences of damage caused by the carriage by sea of hazardous and noxious substances should be shared by the shipowning industry and the cargo interests involved". Preamble of the 1996 HNS Convention.

liability and compensation for damage resulting from the maritime carriage of substances other than oil.

1. The 1996 Convention on Liability and Compensation for Damage in Connection with Carriage of Hazardous and Noxious Substances by Sea (HNS Convention)

This convention is a follow-up to the 1969 Convention on Civil Liability for Oil Pollution Damage, and is intended to complete the international régime for liability and compensation for damage resulting from accidents involving ships carrying substances that have the potential to cause damage to persons, property or the environment.

The first draft of such a convention was considered by a diplomatic conference convened by the IMO in 1984. The discussions at the conference revealed sharp differences of views – among state delegations and also between the representatives of the industrial and commercial interest involved in the production and transport of the various substances – regarding some of the key questions that needed to be resolved in the proposed international agreement. After extensive discussions and negotiations, it became apparent that it would not be possible to reach a consensus on these questions and adopt an agreement that would receive sufficient support from States to come into force in the foreseeable future. Accordingly, the conference decided to postpone action on the draft convention. Instead it requested the IMO to give further consideration to the subject, taking into account the results of the discussions at the conference. The IMO was asked to prepare another draft and arrange for it to be considered at a conference to be convened at a later date.[100]

After long and difficult discussions, the Legal Committee of the IMO finally managed to agree on a new draft convention, which was submitted to a diplomatic conference convened by the Organization in May 1996. This draft was adopted by the conference in the form of the International Convention on Civil Liability and Compensation for Damage in Connection with the Carriage of Hazardous and Noxious Substances by Sea, 1996 (HNS Convention), which was opened for signature on 1 October 1996.[101]

As stated earlier, the 1996 HNS Convention follows the approach to liability and compensation adopted for oil pollution damage in the 1969

---

[100] Resolution of the 1984 International Legal Conference on Liability and Compensation for Pollution Damage. Reproduced in United Nations, *Annual Review of Ocean Affairs, Law and Policy, Main Documents 1985-1987*, Vols. I and II, 1989, 679-680.

[101] *ILM* 35 (1996), 1415.

Civil Liability Convention and its 1976 and 1992 Protocols. But unlike the 1969 conference, the 1996 conference was able to agree on a single treaty instrument dealing with all aspects of liability and compensation. This means that the corresponding provisions of the 1971/1992 Fund Convention (dealing with supplementary compensation from the cargo) were incorporated in the 1996 Convention as well. Like the régime for oil pollution damage, the 1996 HNS Convention adopted the two-tier system of compensation, with the shipowner (as the party bearing primary liability for the damage) responsible for the first tier of compensation and a second (supplementary) layer of compensation provided by a compensation fund that is to be funded by levies from the cargo interests.

The main provisions of the HNS Convention relating to liability for damage are as follows.

A. Party Liable for Damage and Basis of Liability

The Convention places liability for damage on the registered shipowner.[102] This liability is strict, but the shipowner is relieved of liability in certain well-defined circumstances. The circumstances in which the shipowner is exonerated from liability include all the grounds of exoneration that apply in the case of oil pollution damage, as provided for in the 1969 Civil Liability Convention and its 1992 Protocol.[103] In addition the owner is relieved of liability if the damage was caused by the failure of the shipper or any other person to furnish information concerning the hazardous and noxious nature of the substances, or if such failure has led the shipowner to obtain insurance as required by the Convention.[104]

B. Damage to be Compensated under the HNS Convention

Article 1, paragraph 6 of the Convention defines damage as:

a. loss of life or personal injury on board or outside the ship carrying the hazardous and noxious substances caused by those substances;
b. loss or damage to property outside the ship carrying the hazardous and noxious substances caused by those substances;

---

[102] Article 7 of the Convention.
[103] Article 7, para. 2(a)-(c).
[104] Article 7, para. 2(d). However, this exoneration does not apply if the owner, or his servants or agents "knew or ought reasonably to have known of the hazardous and noxious nature of the substance shipped".

c. loss or damage by contamination of the environment caused by the hazardous and noxious substances, provided that compensation for impairment of the environment other than loss of profit from such impairment shall be limited to costs of reasonable measures of reinstatement actually undertaken or to be undertaken; and
d. the costs of preventive measures and further loss or damage caused by preventive measures.

The same provision states that "where it is not possible to separate damage caused by the hazardous and noxious substances from that caused by other factors, all such damage shall be deemed to be caused by the hazardous and noxious substances, except if ... the damage caused by other factors is damage of the type referred to in Article 4, paragraph 3", i.e., pollution damage covered by the 1969 Civil Liability Convention or damage caused by radioactive material.

Hazardous and noxious substances are defined (in Article 1, paragraph 5, of the Convention) as including:

a. oils carried in bulk as listed in Appendix 1 of Annex I to the 1973/78 MARPOL Convention;
b. noxious liquid substances carried in bulk as listed in Appendix II of Annex II to MARPOL;
c. dangerous liquid substances carried in bulk as listed in Chapter 17 of the International Code for the Construction and Equipment of Ships Carrying Dangerous Chemicals in Bulk, 1983;
d. dangerous, hazardous and harmful substances, materials and articles in packaged form covered by the International Maritime Dangerous Goods Code;
e. liquefied gases as listed in Chapter 19 of the International Code for the Construction and Equipment of Ships Carrying Liquefied Gases in Bulk of 1983;
f. liquefied substances carried in bulk with a flashpoint not exceeding 60 degrees Centigrade;
g. solid bulk material possessing chemical hazards covered by Appendix B of the Code for Safe Practice for Solid Bulk Cargoes; and
h. residues of any of the above-listed substances.

The Convention expressly states that it does not cover "pollution damage" as defined in the 1969 Civil Liability Convention and its 1992 Protocol.[105] This means that if damage is caused "by contamination" by oil as defined in the 1969/1992 Civil Liability Conventions,

---

[105] Article 4, para. 3(a).

compensation will be payable under the 1969 Liability and, as necessary, also under the 1971/1992 Fund treaties. However, if the oil in question is not covered by the civil liability régime, the question of liability and compensation for any damage caused will be dealt with under the general law applicable to liability and compensation in the State in which the damage occurred.

It is also to be noted that damage caused by fire or explosion from a cargo of oil, which is not entitled to compensation under the 1969/1992 oil pollution régime, is covered by the 1996 HNS Convention.

The 1996 HNS Convention does not cover damage caused by radioactive material falling either within Class 7 of the International Dangerous Goods Code or Appendix B of the Code of Safe Practice for Solid Bulk Cargoes.[106]

C. Extent of Compensation Payable by the Shipowner

As in the instruments dealing with pollution damage, the shipowner is entitled to limit his liability for damage in all cases, except where the damage "resulted from his personal act or omission, committed with the intent to cause such damage, or recklessly and with knowledge that such damage would probably occur".[107]

Again, following the approach in the oil pollution instruments, the limits of the shipowner's liability are fixed by reference to the tonnage of the ship involved in the damage. There is basic minimum level of compensation of 10 million SDRs (US$14.3 million) for any ship not exceeding 2,000 gross tons. For a ship above 2,000 tons, the limit of liability is increased with the addition of an extra 1,500 SDRs (US$2,149) for every additional ton, up to 50,000 gross tons. Ships with a gross tonnage beyond 50,000 tons will attract a further 360 SDRs (US$515) for each ton above 50,000, but with an overall maximum ceiling of compensation of 100 million SDRs (US$143 million) for any one incident.[108]

The shipowner is required to maintain insurance or other financial security to cover his liability under the Convention up to the maximum limit of liability applicable to his ship, as provided for in the Convention. A certificate attesting that such insurance is in force is to be issued by the competent authorities of the State of the ship's registry.[109] By way of enforcement, States Parties to the Convention are under an

---

[106] Article 4, para. 3(b).
[107] Article 9, para. 2.
[108] Article 9, para. 1.
[109] Article 12.

obligation not to permit ships flying their flag to trade unless they have satisfied the insurance requirements of the Convention, and port States are required not to permit ships without valid certificates of insurance or other appropriate financial security to enter or leave their ports or offshore terminals.[110]

D. Geographical Scope of Application of the HNS Convention

The 1996 HNS Convention applies to any damage caused in the territory, including the territorial sea, of a State Party to the Convention. For pollution damage (i.e., damage by contamination), the scope of coverage is extended to the exclusive economic zone of States Parties.[111]

Because of the special nature of the substances covered by this Convention, it was found necessary to make provision for cases where damage is suffered in areas which lie outside the jurisdiction of a State Party by a State or a person who would otherwise be entitled to compensation under the Convention. This could happen, for example, where an incident involving a ship covered by the Convention causes damage to persons or property on the high seas. While it was not considered likely that pollution damage on the high seas would result in damage of the kind that would require compensation under the 1969 Convention or even under the 1996 Convention, it was quite realistic to envisage a situation in which hazardous and noxious substances could cause damage on board a ship – or even outside the ship – when the ship, or the person suffering damage, is outside the exclusive economic zone of any State. To cater for this situation the coverage of the 1996 Convention is extended, in respect of "damage, other than damage by contamination", to "damage caused outside the territory, including the territorial sea, of any State". However, in such a situation the Convention will only apply if the "damage has been caused by a substance carried on board a ship registered in a State Party or, in the case of unregistered ships, on board a ship entitled to fly the flag of a State Party".[112]

Like the 1969 Civil Liability Convention, the 1996 HNS Convention also covers preventive measures taken anywhere to prevent or minimise damage that would otherwise be covered by the Convention.[113]

---

[110] Article 12, para. 11.
[111] Article 3, para. (b).
[112] Article 3, para. (c).
[113] Article 3, para. (d).

E. Other Provisions

The other provisions of the 1996 HNS Convention relating to liability are in substance the same as the corresponding provisions of the 1969 Civil Liability Convention, as amended by the 1992 Protocol. These include the provisions on the channelling of liability to the shipowner,[114] the conditions under which the shipowner is entitled to limit his liability,[115] jurisdiction in respect of actions against the shipowner, recognition and enforcement of judgments by competent courts[116] and the procedure for the revision of limitation amounts.[117]

F. Entry into Force of the HNS Convention

The 1996 HNS Convention was opened for signature on 1 October 1996. By its terms it will enter into force eighteen months after

(i) no fewer than *twelve States*, including *four States* with not less than two million units of gross tonnage, have consented to be bound by the Convention;
(ii) the Secretary General of the IMO (as the depository of the Convention) has received information that States which receive 20,000 tons or more per year of the substances covered by the Convention had, in the preceding calendar year, received a total quantity of at least *forty million* tons of those substances.[118]

2. Additional Compensation for Damage

As happened with the 1969 Civil Liability Convention, the level of compensation that could be imposed on the shipowner was not considered adequate to compensate all the damage which could be foreseen from the substances covered by the 1996 HNS Convention. It was therefore agreed to establish a supplementary scheme, in the form of an international compensation fund, to provide additional compensation where the compensation due from the shipowner was either not sufficient to cover the actual damage caused or was not available for any reason. This fund, to be known as the International Hazardous and Noxious Substances Fund (HNS Fund), is to be made up of contributions

---

[114] Article 7, paras. 5 and 6.
[115] Article 9, para. 3.
[116] Article 38.
[117] Article 48. On the HNS Convention in general, see Ganten, see note 94.
[118] Article 46.

from persons and entities in the States Parties to the Convention for whom hazardous and noxious substances are transported by sea (the "receivers" of HNS substances).[119] The HNS Fund is established within the framework of the 1996 HNS Convention, and will operate along the same lines as the IOPC Fund operates in respect of oil pollution damage.[120]

VIII. Concluding Remarks

The development of the international instruments on liability and compensation for damage to the marine environment have served to highlight one of the important assumptions made both by the United Nations Convention on the Law of the Sea and by the Rio Declaration on Environment and Development. As stated above, the Law of the Sea Convention expects that States will cooperate in the development of appropriate régimes on liability and compensation for damage resulting from pollution of the marine environment. The same expectation was expressed in Principle 22 of the Stockholm Declaration on the Human Environment, and has been reiterated in the Rio Declaration and also in Agenda 21.

The civil liability approach is particularly appropriate in dealing with damage resulting from pollution of the marine environment. This is because a considerable part of the pollution of the seas results from activities of non-state entities, and it is overall more convenient to victims of such pollution to seek compensation in national courts rather than through international procedures. By giving the right of recourse to national courts, civil liability conventions provide realistic possibilities for claimants and an incentive to those whose acts or omissions may leave them open to such claims

The increasing use of the civil liability approach in many areas of international environmental law is evidence of its efficacy and usefulness as one of the tools for ensuring effective application of international regulations and standards for environmental protection. Civil liability and compensation régimes are not an alternative to state responsibility: they are complementary to the responsibility of the State under international law. The relative success of the civil liability conventions on damage of the marine environment from vessel-source pollution would seem to suggest that States have come to accept them as legitimate and useful additions to the developing armoury in the fight to protect and preserve the marine environment from pollution and other

---

[119] Articles 13 to 20.
[120] See Ganten, see note 94.

forms of degradation.[121] It is to be hoped that similar régimes can be developed in respect of damage from other sources of marine pollution.

---

[121] The 1969 Civil Liability Convention was accepted by more than 80 States before it was replaced by the 1992 Protocol. The 1992 Protocol has so far been ratified by more than 60 States.

XII

HUMAN RIGHTS

The Follow-up Procedure of the Human Rights Committee's Views

Nisuke Ando

I. Introduction

On 27 March 1976, the Optional Protocol to the International Covenant on Civil and Political Rights entered into force, and, in August 1977, the Human Rights Committee, the monitoring body established by the Covenant, began consideration of individual communications which it had received under the Optional Protocol. By the end of its 69th session in July 2000, the Committee had adopted 346 views (decisions on the merits of communications), in 268 of which it found violations of the Covenant.[1]

When the Committee finds such a violation, it recommends as a rule that the State Party concerned should provide the victim of the violation with some kind of appropriate remedy, compensation or release, for example. Being a recommendation, the Committee's views as such do not possess a legally binding character, but the Committee has been endeavoring to persuade States Parties to comply with its views through what it calls the "follow-up" procedure.

The purpose of the present article is to examine how this follow-up procedure has been developed by the Committee, and to ascertain the extent to which the procedure has been effective in securing States Parties' compliance with the Committee's views. In a sense, the article is a case study of compliance with non-binding decisions for the international protection of human rights. It is also a case study of the settlement of disputes between a State and individuals through the intermediacy of an international organ. For the sake of convenience, the development of the follow-up procedure will be examined in four

---

[1] Report of the Human Rights Committee, GAOR, 55th Sess., Supp. No. 40 (A/55/40), Vol. I, 90, §§ 596 and 598. Hereinafter, the form used for the session and supplement number, taking this footnote as an example, will be: "Report-HRC (A/55/40)".

different periods: first, from 1977 to 1982; second, from 1982 to 1990; third, from 1990 to 1995; and fourth, from 1995 to date.

II. Initial Period: 1977-1982 (2nd to 16th Session)

The first period covers the years 1977-1982, from the 2nd to the 16th session of the Human Rights Committee. This period may be termed the "nascent stage" for the consideration of communications. At its first session in March-April 1977, the Committee adopted its Rules of Procedure, including those dealing with the consideration of communications. It was only at its second session in August of the same year that the Committee started the consideration of communications.[2] However, during the succeeding four sessions, the Committee's consideration of communications was limited to the issue of their admissibility, leaving the consideration of the merits to later sessions.[3] Accordingly, the question of the follow-up did not arise at all.

It was at its seventh session in July-August 1979 that the Committee considered the merits of communications for the first time. In that session the Committee adopted one final view on the merits of a case, in which it declared that the relevant facts disclosed violations of several Covenant provisions by the State Party, Uruguay. Then, the Committee stated that the State Party was "under an obligation to take immediate steps to ensure strict observance of the provisions of the Covenant and to provide effective remedies to the victims".[4] From the 8th to the 16th session the Committee adopted 30 views altogether, in 26 of which it found a violation of various Covenant provisions. Of those 26 cases of violations, 21 were attributed to Uruguay, three to Colombia, two to Canada, and one to Mauritius.[5]

All the Uruguayan cases were related to the persecution of its citizens by the military dictatorship, and, in many of those cases, Uruguay was declared to be under an obligation to provide the victims and their families with effective remedies, including compensation,[6] and

---

[2] Report-HRC (A/32/44), 37, §§ 146-147.

[3] Report-HRC (A/33/40), 98, §§ 575-576. According to Rule 87 of the Rules of Procedure, "[t]he Committee shall decide as soon as possible ... whether or not the communication is admissible under the Protocol". See Report-HRC (A/32/44), 63.

[4] Report-HRC (A/34/40), 129.

[5] No violation was found in two Finnish cases, one Swedish and one Canadian case.

[6] Report-HRC (A/35/40), 110, 119, 126, 131 and 137; Report-HRC (A/36/40), 119, 124, 129, 146, 159, 183 and 188; Report-HRC (A/37/40), 129, 136, 178 and 192.

to take steps to ensure that similar violations would not occur in the future.[7] In some cases, the obligation was extended to include release of the victim,[8] provision of medical care,[9] assurance of political participation,[10] granting of a passport,[11] permission to leave the country,[12] opening of a new trial,[13] and investigation of alleged torture.[14] In two of the Colombian cases, the Committee decided that the State Party should adjust its domestic laws to give effect to Article 14, paragraph 5, and Article 9, paragraph 4, of the Covenant, respectively.[15]

Similarly, the Committee stated that Mauritius should adjust its Immigration (Amendment) Act and Deportation (Amendment) Act to exclude discriminatory treatment based on sex.[16]

Despite all these findings of violations and indications of specific remedies, the Human Rights Committee did not raise the question of the follow-up of its views during the initial period.

III. Case-by-Case Disposal of the Follow-Up: 1982-1990 (17th to 36th Session)

At the 17th session, however, the question of the follow-up of the Committee's views was raised by the Chairman of the Committee, and heated discussion thereupon ensued among the Committee's members.[17] Some were of the opinion that there was no clear legal mandate for the Human Rights Committee to monitor the implementation of its views, that the question of implementation was left to the goodwill of the States Parties concerned, and that no useful progress could be made in trying to press States Parties to do what they were not obliged to do.[18] On the other hand, the majority of members emphasized that the Committee could not let its work under the Optional Protocol degenerate into an exercise in futility, that the Optional Protocol allowed considerable latitude for interpretations whereby the Committee could take

---

[7] Report-HRC (A/35/40), 110, 119, 126 and 131; Report-HRC (A/36/40), 119, 124, 129, 146, 159, 183 and 188; Report-HRC (A/37/40), 121 and 178.
[8] Report-HRC (A/36/40), 119, 183 and 188.
[9] Report-HRC (A/37/40), 129 and 186.
[10] Report-HRC (A/36/40), 133.
[11] Report-HRC (A/37/40), 160.
[12] Report-HRC (A/36/40), 183 and 188.
[13] Report-HRC (A/37/40), 121.
[14] *Ibid.*, 192.
[15] *Ibid.*, 173 and 205.
[16] HRC-Report (A/36/40), 142.
[17] See *Yearbook of the Human Rights Committee* I (1983-1984), 19-21 and 39-44.
[18] *Ibid.*, 20, § 16; 21, § 24; 40, §§ 10-14 and 44, § 42.

appropriate action that was reasonably open to it and not expressly prohibited, that the preamble of the Protocol, together with Article 2, paragraph 3, of the Covenant, evidenced the States Parties' intention to implement Covenant provisions, and that it was the Committee's inescapable duty to monitor the implementation of its decisions.[19]

In any event, the Committee had decided earlier in its 15th session that it should not include in its views a final sentence requesting States Parties to inform it of the measures taken in response to its views.[20] Instead, at the same session, the Committee decided that the same sentence should be included in all letters of transmittal of its final views to the States Parties concerned.[21] It must be noted, in this context, that several States Parties did respond to the final views of the Committee during the period under consideration.

Thus, for example, in a letter dated 6 June 1983, Canada responded to the Committee's wish to receive information about measures taken in respect of the views of the Committee on Communication No. R 6/24.[22] The essence of the response was that, while the Human Rights Committee had left the issue of sexual equality unanswered in its views, Canada was anxious to amend the Indian Act, that a section of the Canadian Charter of Rights and Freedoms stipulating the equality of the sexes would come into force in April 1985, and that domestic remedies would be subsequently available for persons discriminated against on the basis of sex. Likewise, in a letter dated 15 June 1983, Mauritius responded in respect of the Committee's views on Communication No. R 9/35. According to the response, the two impugned Acts, the Immigration (Amendment) Act and the Deportation (Amendment) Act, had now been amended so as to remove the discriminatory effects on the grounds of sex.[23] Furthermore, in a letter dated 20 June 1983, Finland responded to the Human Rights Committee with respect to the

---

[19] Ibid., 40, § 6; 41, § 15; 42, §§ 22, 26 and 28; and 43, § 35.
[20] Ibid., 44 § 43.
[21] Ibid. See also at 44, § 46.
[22] Report-HRC (A/38/40), 249 ff. The communication concerned an American Indian woman of Canadian nationality who had lost her Indian status by marrying a non-Indian, but wanted to return to the reservation after the dissolution of the marriage. Under the Indian Act of Canada then in force, one could not live in the reservation without Indian status. Her claim was that the Act violated the Covenant provisions on sexual equality as well as minority rights, because Indian men, unlike Indian women, did not lose their Indian status by marrying non-Indians and could continue to live in the reservation. Her claim was upheld by the Human Rights Committee on the basis of minority rights. For the Committee's views on this case, see Report-HRC (A/36/40), 166 ff.
[23] Report-HRC (A/38/40), 254. For the Committee's views on case No. R 9/35, see Report-HRC (A/36/40), 134 ff.

Committee's views on Communication No. 40/1978. The response indicated that, while the Committee's views found the Finnish domestic law in compliance with the Covenant, Finland had taken further administrative measures to address the complaint at issue.[24] In a similar vein, responding to the Committee's views on Communication Nos. 10/1977 and 28/1978, Uruguayan notes dated 31 May and 10 July 1984 reported on the release from imprisonment of the persons in question,[25] and notes to the same effect followed in October 1984, and February and March 1985.[26] Madagascar also reported on a case of release in response to the Committee's views on Communication No. 49/1979.[27] Subsequently, by a note dated 19 July 1985, Madagascar responded to the Committee's views on Communication No. 132/1982, reaffirming its disagreement with the views but giving detailed information to alleviate the concern of the Committee.[28] A Canadian note dated 5 July 1985 informed the Human Rights Committee that the discriminatory provision of the Indian Act had been abrogated,[29] and a Finish note dated 27 July 1989 reported on the proposed Bill of the Law on Military Disciplinary Procedure which would comply with the Committee's views on Communication No. 265/1987.[30]

Certainly, these responses provided useful information for the Human Rights Committee to ascertain what measures the States Parties concerned had taken with respect to its final views. However, the responses came only from a small number of States Parties. Besides, their information was not always satisfactory and often fragmentary. In order to make up for these shortcomings, the Committee attempted to use the opportunity to consider state reports under Article 40 of the Covenant by requesting information from States Parties about the measures taken by them with respect to specific views of the Committee.[31] Again, this attempt was not always successful. Thus, this period can best be characterized as that of case-by-case disposal of the follow-up.

---

[24] Report-HRC (A/38/40), 255. For the Committee's views on case No. 40/1978, see Report-HRC (A/36/40), 147 ff.
[25] Report-HRC (A/39/40), 126 § 623.
[26] Report-HRC (A/40/40), 145 § 703.
[27] Report-HRC (A/39/40), 126 § 624.
[28] Report-HRC (A/40/40), 145 § 705.
[29] Ibid., 145 § 704.
[30] Report-HRC (A/44/40), 311. In addition, for the response of the Dominican Republic concerning Communication No. 188/1984, the Ecuadorian response concerning Communication No. 238/1987, and the Finish response concerning Communication No. 291/1988, see Report-HRC (A/45/40), Vol. I, 145, §§ 636-638.
[31] Report-HRC (A/45/40), Vol. I, 144, § 633; Report-HRC (A/46/40), 173, § 702.

IV. Special Rapporteur for the Follow-Up of Views: 1990-1995 (39th to 51st Session)

One result of the case-by-case disposal as described above was that a number of victims sent letters of complaint to the Human Rights Committee and stated that, despite the adoption by the Committee of its final views finding violations of the Covenant, their situation remained unchanged, or that no appropriate remedies had been provided by the States Parties.[32] This lead the Committee to conclude that it should seek information on the follow-up of its views in a more effective and regular manner. After examining the problem in the 37th and 38th sessions, the Committee decided as follows at the 39th session in July 1990:

(1) When the Committee finds a violation of the Covenant in its views, the State Party concerned will be asked in the views themselves to inform the Committee, within 6 months, what action it has taken in relation to the case.

(2) If no reply is received within the time-limit or if the reply shows that no remedy has been provided, that fact will be noted in the Committee's Annual Report submitted to the General Assembly of the United Nations. Positive responses and cooperation from the State Party are also included in the Report.

(3) In addition, a State Party to both the Optional Protocol and the Covenant will be asked to inform the Committee in its report (under Article 40 of the Covenant) of the measures which it has adopted to give effect to the Committee's views in favor of the victim. If no such information is given in the State Party's report, the Committee will include relevant questions in the "List of Issues" for the State Party when it considers the report.[33]

Furthermore, the Committee decided to appoint a Special Rapporteur for the Follow-Up of Views, whose duties were: (a) to communicate with States Parties and, if appropriate, with the individuals concerned in respect of all letters of complaint, for the purpose of seeking information on any action taken by the States Parties in relation to views adopted by the Committee; (b) to assist the Rapporteur of the Human Rights

---

[32] Report-HRC (A/45/40), 145, § 633.
[33] Report-HRC (A/45/40), Vol. II, 205. Before considering a State Party's report, the Committee customarily adopts a list of written questions called the "List of Issues", and sends the list to the State Party so that it can prepare answers. If the answers are not satisfactory, Committee members may add oral questions during the consideration of the report.

Committee in preparing the Annual Report so that the report contains detailed information on the follow-up of the views; and (c) to recommend to the Committee possible ways for rendering the follow-up procedure more effective.[34]

On 24 July 1990, one of the Committee members, Janos Fodor, was appointed as the first Special Rapporteur for the Follow-Up of Views.[35] At the 40th session, Fodor proposed and the Committee agreed that *notes verbales* be sent to those States Parties which had not informed the Committee of measures taken in pursuance of the Committee's views. During the 41st and the 42nd sessions, he presented a progress report at closed meetings of the Committee.[36] Subsequently, the Special Rapporteur informed the Committee that he was in the process of analyzing the replies received and intended to report to the Committee at the earliest possible opportunity.[37] However, due to Fodor's ill health, the Committee decided at its 47th session in March 1993 to appoint Andreas Mavrommatis to succeed Fodor in the post of Special Rapporteur for the Follow-Up of Views.[38]

According to the Annual Report of the Committee of 1994, by the beginning of the 51st session in July of that year, follow-up information had been received in respect of 65 views, while no information had been received regarding 55 views. The Secretariat also received information from many victims to the effect that the Committee's views had not been implemented.[39] Of the 65 replies received, about one fourth showed the willingness of the States Parties concerned to implement the views or to offer a remedy. A little more than one third did not respond to the views at all, responded to them only partly, or displayed the unwillingness of the States Parties to grant the remedy requested. Indeed, some of them explicitly challenged the Committee's findings either in fact or in law.[40] Believing that publicity regarding follow-up activities would make the procedure more effective, the Committee adopted a new rule of procedure at the same session to include follow-up information in its Annual Reports.[41]

It must not be overlooked that, during the corresponding period between 1990 and 1995, a few States Parties continued to respond

---

[34] Report-HRC (A/45/40), Vol. II, 205-206.
[35] *Official Records of the Human Rights Committee 1989-1990*, Vol. I, 312, §§ 46-49.
[36] Report-HRC (A/46/40), 173, § 704. See also *Official Records of the Human Rights Committee 1992-1993*, Vol. I, 234-236.
[37] Report-HRC (A/47/40), 164, § 686.
[38] Report-HRC (A/48/40), Vol. I, 176, § 830.
[39] Report-HRC (A/49/40), Vol. I, 84, § 461.
[40] Report-HRC (A/49/40), Vol. I, 84-85, § 462.
[41] Rule 95, para. 4. See Report-HRC (A/49/40), Vol. I, 111.

favorably to the Committee's views. For example, in response to the Committee's views on Communication No. 291/1988, a Finnish note dated 9 April 1990 reported that a Government bill amending the Aliens Act had been submitted to Parliament, which would guarantee alien asylum seekers the right to judicial review of their custody order by the administrative authorities,[42] and subsequently Parliament approved the amendment.[43] Similarly, in response to the Committee's views on Communication No. 305/1988, a Dutch note of 15 May 1995 stated that, while the State Party was unable to share the views of the Committee, it would make an *ex gratia* payment "out of respect for the Committee".[44] Moreover, in a letter dated 23 May 1990, the Dominican Republic reported to the Human Rights Committee that the Government had guaranteed the freedom of action of one of the national non-governmental human rights organizations in pursuance of the Committee's views on Communication No. 188/1984.[45] By the same token, an Ecuadorian note dated 13 February 1990 reported to the Committee that, following the Committee's views on Communication No. 238/1987, the victim had been found innocent and released after being granted a fair trial.[46]

V. Development of Follow-Up Activities: 1995 to Date (52nd Session and Onward)

On the initiative of the new Special Rapporteur for the Follow-Up of Views, the first comprehensive report on follow-up activities was published in the Annual Report of the Human Rights Committee in July 1995. The report was comprehensive in that it showed "a country-by-country breakdown" of follow-up replies received or requested and outstanding as of 28 July 1995. For example, in the case of Canada, the report indicated "6 views finding violations, 3 fully satisfactory follow-up replies, 2 (incomplete) follow-up replies, no follow-up reply in 1 case".[47] In total, the names of 30 States Parties were listed in the country breakdown,[48] and, of 208 views adopted by the Committee since its

---

[42] Report-HRC (A/45/40), Vol. II, 209 ff.
[43] Report-HRC (A/46/40), 174, §§ 705-706.
[44] *Ibid.*, §§ 707-708.
[45] Report-HRC (A/45/40), Vol. II, 207-208.
[46] Report-HRC (A/45/40), Vol. II, 209.
[47] Report-HRC (A/50/40), Vol. I, 92, § 549.
[48] They are Argentina, Austria, Australia, Bolivia, Cameroon, Canada, Central African Republic, Colombia, Dominican Republic, Ecuador, Equatorial Guinea, Finland, France, Hungary, Jamaica, Libyan Arab Jamahiriya, Madagascar, Mauritius, Netherlands, Nicaragua, Panama, Peru, Senegal,

inception, the Committee had found a violation of the Covenant in 154 cases.[49] In respect of all the views with a finding of a violation, the Special Rapporteur had systematically requested follow-up information from each of the States Parties concerned. He received the information in respect of 81 views (and no information in respect of 62 views). In five cases, the deadline for receipt of follow-up information had not yet expired.[50]

In this connection, it should be noted that, starting from the 52nd session in October-November 1994, the Special Rapporteur and another member of the Committee had contacted the government authorities of several States Parties with regard to the follow-up of the Committee's views.[51] Also, in June 1995, the Special Rapporteur visited Jamaica with a view to contacting the Government, judicial authorities and non-governmental organizations. During the visit, he had a thorough discussion with the authorities about the status of the implementation of the Committee's views adopted in respect of Jamaica. While the Special Rapporteur was informed by the authorities of various difficulties in implementing the views, the Minister for Foreign Affairs pledged cooperation with him and the Committee on matters relating to the follow-up procedure.[52]

The subsequent Annual Reports of the Human Rights Committee confirm that the Committee's follow-up activities have been making gradual progress. The Annual Report of 1996 indicated that, out of 168 views with a finding of a violation, the Committee had received follow-up information in respect of 90 cases.[53] The number of States Parties listed in the country breakdown had increased to 32.[54] The Special Rapporteur held follow-up consultations with the representatives of seven States Parties, and a summary of these consultations was published in the report.[55] In 1997, the Annual Report started to include a "Communication Number" for each and every Committee view in the country breakdown list, so that it would be possible to identify which views on what case had or had not been followed up by the State Party in question.[56] The country breakdown list was extended to include 33 States Parties,[57] and the Special Rapporteur met with the representatives

---

Spain, Surinam, Trinidad and Tobago, Uruguay, Venezuela, Zaire (later Democratic Republic of the Congo), and Zambia. See *ibid.*, 92-93.

[49] *Ibid.*, 91, § 544.
[50] Report-HRC (A/50/40), Vol. I, 91, § 546.
[51] *Ibid.*, 95, § 555 (f).
[52] *Ibid.*, 96, § 559.
[53] Report-HRC (A/51/40), Vol. I, 65, §§ 424 and 426.
[54] The Czech Republic and the Republic of Korea were added.
[55] Report-HRC (A/52/40), Vol. I, 88, § 524 ff.
[56] *Ibid.*, 88, §§ 518 and 521.
[57] Togo was added.

of 10 States Parties for follow-up consultations.[58] It should be added that, at the 59th session in March-April 1997, Prafullachandra N. Bhagwati succeeded Mavrommatis as the Special Rapporteur for the Follow-Up of Views.[59]

The Annual Report of 1998 indicated that, of 223 views with a finding of a violation, the Human Rights Committee had received follow-up information in 133 cases.[60] The report also published a summary of newly received follow-up information from States Parties.[61] In 1999, the Annual Report indicated that, of 253 views in which the Committee had found violations of the Covenant, the Committee had received follow-up information with respect to 152 cases.[62] The number of States Parties listed in the country breakdown increased to 35,[63] and the report included a summary of newly received follow-up information.[64] According to the latest Annual Report of 2000, the Human Rights Committee has adopted 346 final views during the entire period from the 7th session in 1979 to the 69th session in July 2000. In those 346 views, the Committee found violations in 268 cases,[65] and, by the beginning of the 69th session, follow-up information had been received from States Parties in respect of 180 views.[66] The number of States Parties in the country breakdown list increased again to 37,[67] and a summary of newly received follow-up information was included in the report.[68] It should be noted that Fausto Pocar succeeded Bhagwati as Special Rapporteur for the Follow-Up of Views at the 65th session of the Committee in March-April 1999,[69] and that Christine Chanet succeeded Pocar at the 68th session in March 2000.[70]

VI. Some Concluding Remarks

The current status of the follow-up procedure under the Optional Protocol as developed by the Human Rights Committee may be summarized as follows. With respect to any final views in which the

---

[58] Report-HRC (A/52/40), Vol. I, 93, § 526 ff.
[59] Ibid., 88, § 519.
[60] Report-HRC (A/53/40), Vol. I, 72, §§ 480 and 483.
[61] Ibid., 76, § 490 ff.
[62] Report-HRC (A/54/40), Vol. I, 90, §§ 456 and 458.
[63] Georgia and Guyana were added.
[64] Report-HRC (A/54/40), Vol. I, 96-98, § 464 ff.
[65] See note 1 above.
[66] Report-HRC (A/55/40), Vol. I, 90, § 598.
[67] Italy and Norway were added.
[68] Report-HRC (A/55/40), Vol. I, 97, § 604 ff.
[69] Report-HRC (A/54/40), Vol. I, 90, § 457.
[70] Report-HRC (A/55/40), Vol. I, 90, § 597.

Committee has found a violation of the International Covenant on Civil and Political Rights, it is now possible to see whether or not the State Party concerned has taken measures in relation to the views in question. If the State Party has taken measures to implement the Committee's views by providing the victim with the remedy required by the Committee, as exemplified by Finland,[71] the procedure is regarded as having been effective, and the latest Annual Report states that 30 percent of the replies received are categorized as satisfactory.[72] Sometimes, a State Party challenges the Committee's views in fact or law.[73] Because the Human Rights Committee does not possess the capacity to gather information on its own or make on-site investigations, and because the Committee "shall consider communications ... in the light of all written information made available to it" by the parties,[74] the Committee is not free from errors. In such cases, a State Party's *bona fide* challenges may help to clarify the situation. Such responses, even if embarrassing to the Committee, may result in constructive exchanges between the State Party and the Committee and promote the significance of the follow-up procedure.

Of course, the main problem is a comparatively large number of non-replies. The Committee has endeavored to deal with this difficult issue by sending reminders to non-replying States Parties or by arranging consultations with their representatives. In this connection, it is worth noting that the ratio of non-replies to replies has been gradually decreasing, and the number of non-replying States Parties is not so great. With limited resources available, the Human Rights Committee has no other choice, perhaps, than to continue to make every effort to obtain as many replies as possible.

As pointed out in the Introduction, this article is, in a sense, a case study of compliance with non-binding decisions for the international protection of human rights. The article is also a case study of the settlement of disputes between a State and individuals through the intermediacy of an international organ. As a case study of compliance with non-binding decisions, the follow-up procedure as developed by the Human Rights Committee seems to be producing a noticeable if not remarkable result. The same holds true with regard to the follow-up procedure as a case study of the settlement of disputes between a State and individuals through the intermediacy of an international organ.

---

[71] See notes 30, 42 and 43, above.
[72] Report-HRC (A/55/40), Vol. I, 90, § 599.
[73] See, for example, the challenges of Austria (*ibid.*, 97, § 606), Canada (*ibid.*, 98, § 608) and the Netherlands (*ibid.*, 98, § 612).
[74] Optional Protocol, Art. 5, para. 1.

Les droits de l'homme et la Cour internationale de Justice : une vision latino-américaine

Héctor Gros Espiell

I.

La présente communication, écrite en hommage du Juge S. Oda, évoque la précieuse contribution qu'il a faite à la jurisprudence de la Cour internationale de Justice pendant de longues années, depuis 1976, et qui est un exemple d'intelligence, de zèle, de dévouement et d'indépendance.[1]

En choisissant la question des droits de l'homme et la Cour internationale de Justice, j'ai remémoré les activités que nous avons conduites avec le Juge Oda dans le domaine des droits de l'homme pendant des années, au cours de plusieurs Conférences Armand Hammer (Droits de l'Homme = Paix ; Paix = Droits de l'Homme).

Si je me suis penché sur la question de la Cour internationale de Justice et les droits de l'homme, malgré l'existence de nombreuses études d'une très haute valeur dans ce domaine,[2] c'est que je pense

---

[1] E. McWhinney, *Judge Shigeru Oda and the Progressive Development of International Law, Opinions (Declarations, Separate Opinions, Dissents) on the International Court of Justice, 1975-1992*, 1993 ; S. Oda, « The International Court of Justice Viewed from the Bench (1976-1993) », *RdC* 244 (1993), VII.

[2] A. A. Cançado Trindade, « La Jurisprudence de la Cour internationale de Justice sur les Droits Intangibles », dans D. Premont (ed.), *Droits intangibles et états d'exception*, 1996 ; S. M. Schwebel, « The Treatment of Human Rights and of Aliens in the International Court of Justice », dans V. Lowe (ed.), *Fifty Years of the International Court of Justice, Essays in Honour of Sir Robert Jennings*, 1996, 327-350 ; R. Higgins, « The International Court of Justice and Human Rights », dans K. Wellens (ed.), *International Law : Theory and Practice, Essays in Honour of Erik Suy*,

qu'elle revêt un intérêt actuel et très probablement une projection future, qui mérite des recherches complémentaires et la présentation d'une vision latino-américaine.

L'intérêt de la doctrine sur la question des droits de l'homme et la Cour internationale de Justice – un sujet qui était pour ainsi dire ignoré avant les années 50 –, montre l'importance actuelle de cette question des droits de l'homme, sa portée internationale et son influence tant politique que juridique, dont tout porte à penser qu'elle ne fera qu'augmenter à l'avenir.

Je me suis proposé d'envisager cette question depuis une approche latino-américaine, et ce, non seulement pour souligner l'influence exercée par des juges latino-américains membres de la Cour internationale de Justice sur l'examen de cette matière,[3] mais également pour faire ressortir la poids de la question des droits de l'homme dans la réalité de l'Amérique latine actuelle, et sa signification pour des affaires dont la Cour internationale de Justice pourrait éventuellement être saisie à l'avenir. Mais aussi, parce qu'il existe en Amérique latine un système régional de protection des droits de l'homme, fondé sur l'application de la Convention américaine des droits de l'homme (Pacte de San José), qui prévoit l'intervention d'un organe juridictionnel – la Cour interaméricaine des droits de l'homme –, qui a pris et prend en considération dans sa jurisprudence les arrêts de la Cour internationale de Justice.

Pour situer l'analyse de la question que j'ai choisie, j'ai pensé qu'elle ne saurait rester limitée à la citation et au commentaire des références aux droits de l'homme qui sont présentes dans la jurisprudence de la Cour internationale de Justice.

Cette citation est certes nécessaire, mais il faut de plus rappeler les limites de la Cour lorsqu'elle examine des questions juridiques relatives aux droits de l'homme, en vertu de la nature de sa

---

1998, 691-705 ; S. M. Schwebel, « The International Court of Justice and the Human Rights Clauses of the Charter », *AJIL* 66 (1972) ; N. Singh, *Human Rights and the Future of Mankind*, 1981 ; E. Jiménez de Aréchaga, « El Derecho y la Justicia : Resguardos de la Libertad », *Revista del Instituto Interamericano de Derechos Humanos* 1 (1984), 25-38 ; H. Gros Espiell, « La Corte Internacional de Justicia y los Derechos Humanos », *La Nación*, San José, septembre 1987.

[3] Sh. Rosenne, « La Contribución de América Latina al Desarrollo de la Corte Internacional de Justicia », *Revista de la Facultad de Ciencias Jurídicas y Políticas* 102 (1997), 263-264 ; M. Bedjaoui, « Présences latino-américaines à la Cour internationale de Justice », dans C. A. Armas Barea (ed.), *Liber Amicorum « in memoriam » of Judge José María Ruda*, 2000, 367 et seq.

compétence, et tenir compte également des cas prévus dans les traités en vigueur en matière de droits de l'homme, ou qui comportent une ou plusieurs dispositions ayant trait à cette matière et qui font référence à la compétence de la Cour internationale de Justice.

II.

Conformément à l'article 34.1 du Statut de la Cour, qui fait partie intégrante de la Charte des Nations Unies (art. 92 de la Charte), l'organe judiciaire des Nations Unies (art. 92 de la Charte), n'est compétent que pour les cas des États parties.[4] Elle est ouverte aux États parties dans son Statut, sous réserve que les autres États puissent, dans certains cas et sous certaines conditions, saisir la Cour internationale de Justice (art. 35 du Statut).

Même si ces dispositions entraînent la conséquence que la compétence contentieuse de la Cour se limite aux controverses entre les États et que les individus, en tant que tels, ne peuvent être parties aux procédures engagées auprès de cette Cour, c'est-à-dire qu'ils ne peuvent entamer aucune action contre un État devant la Cour, ni suivre en qualité de parties des procédures précédemment intentées, il est évident que les contentieux entre les États peuvent avoir parmi leurs matières les questions juridiques relatives aux droits de l'homme. Ces controverses peuvent éventuellement être portées devant la Cour, qui devra statuer conformément au droit international (art. 38.1 du Statut).

Les parties, autrement dit, les États, peuvent convenir de saisir la Cour d'un litige relatif aux droits de l'homme, pourvu que les cas concernés soient déjà prévus dans la Charte des Nations Unies ou dans les traités et conventions en vigueur (art. 36.1 du Statut).

---

[4] L'article 34.1 des Statuts dit : « Seuls les États ont qualité pour se présenter devant la Cour ». Dans la version espagnole : « Sólo los Estados podrán ser partes en casos ante la Corte ». Le texte anglais dit : « Only states may be parties in cases before the Court ». Ces textes, ainsi que les versions russe et chinoise, font également foi (art. 11 de la Charte). L'article 34 du Statut de la Cour permanente de 1920 disait : « only states can », au lieu du « may » actuel. Compte tenu des écarts des différentes versions et des changements introduits aux Statuts précédents, certains ont préféré utiliser le texte français, plutôt que l'anglais, bien que les textes espagnol et russe soient plus proches de l'anglais et s'éloignent du français, comme l'a fait remarquer Rosenne, qui ne parle pas des « parties » et se borne à disposer que seuls les États ont qualité pour se présenter devant la Cour. Cf. Sh. Rosenne, « Reflexions on the Position of the Individual in International State Litigation in the International Court of Justice », dans P. Sanders (ed.), *Liber Amicorum for Martin Domke*, 1967, 240-251 ; Sh. Rosenne, *The Law and Practice of the International Court*, vol. I, 1965, Chapter VIII, « Parties in Cases », 267 et seq.

Or, la compétence de la Cour internationale de Justice peut également découler de l'application d'une clause facultative liant deux États (art. 36.2 du Statut), susceptible de porter sur une controverse qui, dans ses quatre cas de figure (art. 36.2, a, b, c et d), peut être une matière relative aux droits de l'homme.

Cependant, il faut remarquer qu'il s'agira toujours, nécessairement, d'une controverse entre des États, mettant en jeu une question relevant des droits et obligations des États selon le Droit international.[5]

III.

Outre cette compétence contentieuse, la Cour possède une compétence consultative (art. 96 de la Charte et art. 65 du Statut de la Cour).

Les avis consultatifs peuvent être demandés par l'Assemblée générale et par le Conseil de sécurité. Les autres organes des Nations Unies ainsi que les organismes spécialisés (art. 57 de la Charte) qui seraient à tout moment autorisés à le faire par l'Assemblée générale, peuvent également demander des avis consultatifs à la Cour sur des questions juridiques relevant de la sphère de leurs activités (art. 96.2 de la Charte).

Il va sans dire que la matière de ces avis peut porter sur une question juridique relative aux droits de l'homme, dont l'élucidation aurait été demandée par l'une quelconque des instances en droit de le faire, conformément aux dispositions de la Charte des Nations Unies et du Statut de la Cour.

Par la voie de ces avis consultatifs, tel que nous le verrons par la suite, la Cour internationale de Justice a fait des contributions conceptuelles très importantes à la matière relative aux droits de l'homme.

IV.

Plusieurs conventions internationales en matière de droits de l'homme considèrent la Cour internationale de Justice comme étant l'organe

---

[5] Déclaration du Juge S. Oda, dans l'affaire *Le Grand (Allemagne c. États-Unis)*, CIJ Recueil 1999, 18-20, alinéas 2-6 ; Déclaration du Juge S. Oda, dans l'affaire *Breard (Paraguay c. États-Unis)*, CIJ Recueil 1998, 260-262, alinéas 2-7. Ces deux déclarations ont été citées et commentées dans le vote favorable du Juge Antonio Cançado Trindade à l'avis consultatif OC 16/99 du 1er octobre 1999 (paragraphe 29, note en bas de page) n° 28, 144.

chargé de régler les controverses relatives à l'interprétation et à l'application de ces textes internationaux.

Qu'il nous soit permis de rappeler que la compétence de la Cour s'étend non seulement à tout litige soumis par les parties, notamment, à toutes les affaires spécialement prévues par la Charte des Nations Unies, mais également toutes les affaires prévues conformément « aux traités et conventions en vigueur » (art. 36.1 du Statut de la Cour). Parmi ces « traités et conventions en vigueur », figurent, naturellement, ceux relatifs aux droits de l'homme, notamment lorsqu'ils prévoient expressément l'intervention de la Cour concernant leur interprétation ou leur application.

Parmi ces conventions en matière de droits de l'homme qui envisagent et acceptent la compétence de la Cour internationale de Justice en cas de différend ou de controverse entre deux ou plusieurs États parties, portant sur l'interprétation ou l'application de la convention pertinente – bien que souvent à condition que soient au préalable utilisés d'autres moyens – il convient de citer les instruments suivants :

a) Convention pour la prévention et la répression du crime de génocide du 9 décembre 1984, en vigueur depuis le 12 janvier 1951, article IX.
b) Convention internationale sur l'élimination de toutes les formes de discrimination raciale, du 21 décembre 1965, en vigueur depuis le 4 janvier 1969, article 22.
c) Convention internationale sur l'élimination et la répression de l'Apartheid du 30 novembre 1973, en vigueur depuis le 18 juillet 1976, article XII.
d) Convention internationale contre l'Apartheid dans les sports du 10 décembre 1985, article 19.
e) Convention contre la discrimination dans le domaine de l'enseignement, UNESCO, du 14 décembre 1960, en vigueur depuis le 22 mai 1962, article 8. Le Protocole portant création d'une Commission de conciliation et de bons offices relatif à la Convention ci-dessus du 10 décembre 1962, en vigueur depuis le 24 octobre 1968, fait également allusion à l'article 12.3, à la Cour permanente d'arbitrage de La Haye.
f) Convention sur l'élimination de toutes les formes de discrimination à l'égard des femmes du 18 décembre 1979, en vigueur depuis le 3 septembre 1981, article 29.
g) Convention sur les droits politiques de la femme du 20 décembre 1952, en vigueur depuis le 7 juillet 1974, article IX.

h) Convention relative à l'esclavage du 25 septembre 1926, en vigueur depuis le 9 mars 1927, article 8, qui fait référence à la Cour permanente de Justice internationale.
i) Convention complémentaire relative à l'abolition de l'esclavage, de la traite des esclaves et des institutions et pratiques analogues à l'esclavage du 7 septembre 1956, en vigueur depuis le 30 avril 1957, article 10.
j) Convention pour la répression de la traite des êtres humains et de la prostitution d'autrui du 2 décembre 1946, en vigueur depuis le 25 de juillet 1951, article 27.
k) Convention relative à la torture du 10 décembre 1984, en vigueur depuis le 26 juin 1987, article 30.
l) Convention relative au statut des réfugiés du 25 juillet 1951, en vigueur depuis le 22 avril 1954, article 38.
m) Protocole relatif au statut des réfugies du 16 décembre 1966, en vigueur depuis le 4 octobre 1967, article IV.
n) Convention sur la nationalité de la femme mariée du 29 janvier 1957, en vigueur depuis le 11 août 1957, article 10.
o) Convention sur la réduction des cas d'apatridie du 4 septembre 1954, entrée en vigueur le 13 décembre 1975, article 14.
p) Convention relative au statut des apatrides du 26 avril 1954, entrée en vigueur le 6 juin 1960, article 34.

La question de l'intervention de la Cour internationale de Justice pour régler les différends et controverses susceptibles d'être posés par l'interprétation ou l'application de ces instruments reste encore posée, notamment en ce qui concerne la Convention relative au génocide (Bosnie Herzégovine contre la Serbie – Monténégro et la Croatie contre la Yougoslavie) et risque d'acquérir une importance croissante.[6] Pour les pays latino-américains, cette éventualité peut s'avérer fort significative, du fait, entre autres, de la compétence consultative attribuée à la Cour interaméricaine des droits de l'homme sur l'interprétation de ces mêmes instruments.[7]

---

[6] A. S. Osuna, « La Contribución de las Naciones Unidas a la Humanización del Derecho Internacional », *La ONU, 50 Años Después*, 1995.

[7] La Cour interaméricaine des droits de l'homme jouit d'une compétence consultative « concernant l'interprétation de cette Convention » (le Pacte de San José), ainsi que de tous « autres traités concernant la protection des droits de l'homme dans les États américains » (art. 64). La Cour, dans deux de ses avis consultatifs (OC 1/82 et OC 16/99), a correctement interprété cette norme de la Convention américaine, soulignant que l'expression « autres traités » vise les traités internationaux relatifs aux

Il faut toutefois éviter qu'une application trop large de la compétence de la Cour, et en conséquence, de son intervention dans ces cas (art. 36.1 du Statut de la Cour internationale de Justice), et dans les résultats de l'application de la clause facultative (art. 36.2), ne nuisent à l'essence et à l'efficacité de l'action de la Cour.[8]

Cette précision est applicable aux réflexions du paragraphe 10 ci-dessus et aux situations visées aux paragraphes 12, 13, 14 et 15 suivants.

Il existe des conventions internationales qui ne visent pas dans l'ensemble la matière relative aux droits de l'homme, mais contiennent une ou plusieurs normes, un ou plusieurs articles, relatifs à cette matière, et qui prévoient la compétence de la Cour internationale de Justice pour régler les différends ou les controverses suscitées par leur interprétation ou leur application.

Tel est le cas, par exemple, de l'article 36 de la Convention de Vienne sur les relations consulaires du 24 avril 1963, dont le « Protocole de souscription facultatif portant sur la juridiction obligatoire pour le règlement des controverses », de la même date, prescrit à l'article premier que les controverses posées par l'interprétation ou l'application de la Convention seront soumises obligatoirement à la Cour internationale de Justice, qui pourra à ce titre statuer à la requête de l'une quelconque des parties à la controverse, qui serait partie au présent Protocole.

À la demande du Mexique, la Cour interaméricaine des droits de l'homme a interprété cet article 36 de la Convention de Vienne sur les relations consulaires sur la base de l'article 64 de la Convention interaméricaine des droits de l'homme, aux termes de l'avis consultatif OC 16/99 du 1er octobre 1999.

La Convention de Vienne sur les relations diplomatiques, datée du 18 avril 1961, contient également des dispositions associées à des questions relatives aux droits de l'homme. Le Protocole facultatif relatif à la juridiction obligatoire pour le règlement des controverses, de la même date, dispose à l'article I que les controverses éveillées par l'interprétation ou l'application de la Convention seront obligatoirement soumises à la Cour internationale de Justice qui pourra

---

    droits de l'homme ou contenant des dispositions relatives à ces droits, et ce, même lorsqu'il s'agit de traités ne relevant pas du système interaméricain, pourvu qu'un ou plusieurs États américains en soient parties et qu'ils soient donc applicables « à la protection des droits de l'homme dans les États américains ».

[8] S. Oda, « The Compulsory Jurisdiction of the International Court of Justice : A Myth », *ICLQ* 49 (2000), 265 et seq.

à ce titre statuer à la requête de l'une quelconque des parties à la controverse, qui serait partie au présent Protocole.

Le Statut du Tribunal pénal international, adoptés à Rome le 17 juillet 1998, qui n'ont pas encore pris effet à ce jour, prévoient à l'article 119 la possibilité qu'une dispute entre deux ou plusieurs États parties sur l'interprétation ou l'application du Statut puisse être soumise à la Cour, conformément à son Statut.

Cette disposition ne doit pas être négligée, car l'application ou l'interprétation des Statuts de Rome peuvent concerner des questions relatives aux droits de l'homme.

V.

Passons maintenant en revue quelques-unes des affirmations conceptuelles les plus importantes de la Cour internationale de Justice, statuant en matière contentieuse et consultative sur les droits de l'homme et le Droit international humanitaire.[9]

La Cour permanente de Justice internationale avait déjà évoqué le « principe de licéité, l'État de droit et les droits fondamentaux de l'individu ».[10]

En 1948, le juge latino-américain Philadelpho de Azevedo, avait souligné la question « de la protection des droits de l'homme »[11] dans l'avis consultatif « Conditions de l'admission d'un État comme membre des Nations Unies ». Et dans un certain nombre d'avis dissidents ou séparés rendus à différents moments, on trouve également des affirmations d'un grand intérêt à cet égard.[12]

Dans l'affaire du *Détroit de Corfou*, la Cour internationale de Justice a fait référence en 1949 à certaines obligations des autorités

---

[9] J. F. Flauss, « La protection des droits de l'homme et les sources du droit international, La protection des droits de l'homme et l'évolution du droit international », Societé Française pour le Droit International, Colloque de Strasbourg, 1998, 53-56. Cf. une étude de la contribution de la jurisprudence de la Cour internationale de Justice en matière de droits de l'homme, entre 1970 et 1980, par H. Gros Espiell, « Las Naciones Unidas y los Derechos del Hombre », dans *Estudios sobre Derechos Humanos*, Tomo II, Instituto Interamericano de Derechos Humanos, 1988, 52-54, § 24.

[10] Avis consultatif sur les *Décrets législatifs de Dantzig*, 1935, CPJI, Série A/B, n° 65, 54-56.

[11] Avis consultatif sur les *Conditions de l'admission d'un État comme membre des Nations Unies*, CIJ Recueil 1948, 78.

[12] Read (*CIJ Recueil* 1950, 231) ; Guggenheim (*CIJ Recueil* 1955, 63-64) ; Jessup, Bustamante et Tanaka (*CIJ Recueil*, 1962, 355 et 425 ; 1966, 310) ; Riphagen et Morelli, (*CIJ Recueil* 1970, 234 et 338).

albanaises en temps de paix, qui n'étaient pas fondées sur la Convention de la Haye de 1907, applicable uniquement en temps de guerre – « mais sur certains principes généraux et bien reconnus, tels que des considérations élémentaires d'humanité, plus absolues encore en temps de paix qu'en temps de guerre . . . ».[13]

Cette affirmation de la Cour, réaffirmée et développée par des arrêts ultérieurs,[14] a eu, garde et gardera une projection importante et positive sur le droit international humanitaire, essentiellement lié aux droits de l'homme,[15] et sur les concepts relatifs aux principes généraux, au *jus cogens*, aux obligations *erga omnes* et aux droits intangibles, qui sont tous nécessairement liés à la question des droits de l'homme.[16]

Dans l'avis consultatif intitulé : *Interprétation des traités de paix conclus avec la Bulgarie, la Hongrie et la Roumanie*, du 30 mars 1950, la Cour a justifié la demande d'avis introduite par l'Assemblée générale faisant valoir que les Nations Unies sont tenues, en application de l'article 55 de la Charte, de promouvoir le respect universel et effectif des droits de l'homme et les libertés fondamentales de tous.

La Cour a rejeté l'opposition de ces trois gouvernements à la demande d'avis consultatif, qui avançait l'argument selon lequel l'Assemblée générale, en examinant la question du respect des droits de l'homme et des libertés fondamentales dans les trois États visés se serait « immiscée » ou serait « intervenue » dans des affaires que relèvent essentiellement de la compétence nationale des États. La Cour, sans se prononcer sur les

---

[13] *CIJ Recueil* 1949, 22-23. Voir P.-M. Dupuy, « Les considérations élémentaires d'humanité dans la jurisprudence de la Cour internationale de Justice », dans R.-J. Dupuy (ed.), *Mélanges en l'honneur de Nicolas Valticos, Droit et Justice*, 1999 ; A. Verdross, « Jus Dispositivum and Jus Cogens in International Law », dans ASIL, *International Law in the Twentieth Century*, 1969, 221 et seq.

[14] A. A. Cançado Trindade, voir note 2 ; Voir notamment l'arrêt du 27 juin 1986, *CIJ Recueil* 1986, 112 et 114, paragraphes 215-218, *(Nicaragua c. États-Unis)* ; Avis consultatif du 8 juillet 1996, *Licéité de la menace ou de l'emploi d'armes nucléaires*, *CIJ Recueil* 1996, paragraphe 79.

[15] H. Gros Espiell, « Derechos Humanos y Derecho Internacional Humanitario », dans C. Swinarski (ed.), *Etudes et Essais sur le Droit International Humanitaire et sur les Principes de la Croix Rouge en l'Honneur de Jean Pictet*, 1984, 699-711 ; « Droits de l'Homme et Droit international humanitaire », *Nations Unies, Bulletin des Droits de l'Homme* 91/1 (1992).

[16] A. A. Cançado Trindade, *La Jurisprudence de la Cour internationale de Justice sur les Droits intangibles, Droits intangibles et états d'exception*, 1996.

accusations faites lors de l'Assemblée générale sur les violations des droits de l'homme dans ces trois pays, a dit :

> « Aux fins du présent avis, il suffit de constater que l'Assemblée générale a justifié l'adoption de sa résolution en considérant qu'en vertu de l'article 55 de la Charte, les Nations Unies sont tenues de favoriser le respect universel et effectif des droits de l'homme et des libertés fondamentales pour tous, sans distinction de race, de sexe, de langue ou de religion ».[17]

Le Juge Zoricic a précisé, dans son avis dissident, que « les questions visant le respect des droits de l'homme ne rentrent aucunement dans le cadre des questions de la demande d'avis »[18] et le Juge Krilov, également dissident, ayant soutenu que la question des droits de l'homme était comprise dans la demande d'avis consultatif, a affirmé, en solitaire, l'interprétation plus limitative de l'article 55 de la Charte et dit que la question des droits de l'homme relevait uniquement de la compétence nationale.[19]

Commentant cet avis consultatif, Eduardo Jiménez de Aréchaga conclut que pour « la Cour, la question du respect et de l'efficacité des droits « de l'homme sans discriminations ne relève pas de la juridiction interne des États ».[20]

Dans l'avis consultatif sur les *Réserves à la Convention pour la prévention et la répression du crime de génocide* du 18 mai 1951,[21] la Cour estima que le génocide est « un crime de droit de gens », impliquant le refus du droit à l'existence de groupes humains entiers, refus qui bouleverse la conscience humaine, inflige de grandes pertes à l'humanité et qui est contraire à la fois à la loi morale et à l'esprit et aux fins des Nations Unies ». « Cette conception entraîne une première conséquence : les principes qui sont à la base des Conventions sont des principes reconnus par les nations civilisées comme obligeant les États, même en dehors de tout lien conventionnel ».[22]

---

[17] *CIJ Recueil* 1950, 70.
[18] *Ibid.*, 78.
[19] *Ibid.*, 112-113.
[20] E. Jiménez de Aréchaga, « Balance Sobre la Actuación de la Corte Internacional de Justicia en los Cuarenta años de su Funcionamiento », *Revista de la Facultad de Derecho de la Universidad Complutense de Madrid* 13 (1958).
[21] *CIJ Recueil* 1951.
[22] *Ibid.*, 23 et 24.

La protection de la vie humaine, et en particulier, le droit à l'existence des groupes humains, figure parmi ces principes dont la négation offense la conscience humaine, est contraire à la loi morale et viole l'esprit et les fins des Nations Unies.[23]

L'arrêt rendu par la Cour le 5 février 1970, statuant dans l'affaire de la *Barcelona Traction* (1970), affirmait :

> « Une distinction essentielle doit en particulier être établie entre les obligations des États envers la communauté internationale dans son ensemble et celles qui naissent vis-à-vis d'un autre État dans le cadre de la protection diplomatique. Par leur nature même, les premières concernent tous les États. Vu l'importance des droits en cause, tous les États peuvent être considérés comme ayant un intérêt juridique à ce que ces droits soient protégés ; les obligations dont il s'agit sont des obligations erga omnes.
>
> Ces obligations découlent par exemple, dans le droit international contemporain, de la mise hors la loi des actes d'agression et du génocide mais aussi des principes et des règles concernant les droits fondamentaux de la personne humaine, y compris la protection contre la pratique de l'esclavage et la discrimination raciale. Certains droits de protection correspondants se sont intégrés au droit international général (Réserves à la convention pour la prévention et la répression du crime de génocide, avis consultatif, Recueil CIJ 1951, p. 23) ; d'autres sont conférés par des instruments internationaux de caractère universel ou quasi universel ».[24]

---

[23] E. Jiménez de Aréchaga, voir note 2 ; M. Diez de Velazco, « El Sexto Dictámen del Tribunal Internacional de Justicia sobre las Reservas a la Convención sobre Genocidio », *Revista Española de Derecho Internacional* 4 (1951), 1029-1089 ; P. Akhavan, « Enforcement of the Genocide Convention Through the Advisory Jurisdiction of the International Court of Justice », *Human Rights Law Journal* 12 (1991), 285-299.

[24] CIJ Recueil 1970, *Affaire de la Barcelona Traction, Light and Power Company Limited*, Arrêt du 5 février 1970 ; A. Miaja de la Muela, « Aportación de la Sentencia del Tribunal de la Haya en el Caso Barcelona Traction a la Jurisprudencia Internacional », dans *Cuadernos de la Cátedra J. Brown Scott*, 1970 ; J. Justo Ruiz, « Las Obligaciones Erga Omnes », *Estudios de derecho internacional : homenaje al profesor Miaja de la Muela*, 1979, 219-233.

En 1971, la Cour s'est prononcée catégoriquement dans l'affaire de la *Namibie*,[25] faisant valoir que « la Charte des Nations Unies impose des obligations juridiques exigibles dans le domaine des droits de l'homme ».[26] La Cour souligna que conformément à la Charte des Nations Unies, l'Afrique du Sud s'était engagée à observer et à respecter les droits de l'homme et qu'en implantant l'apartheid, elle s'était rendue coupable d'« une violation flagrante » des principes et buts de la Charte. Cette affirmation figure au paragraphe 131 de l'arrêt, ci-après transcrit textuellement :

> « En vertu de la Charte des Nations Unies, l'ancien mandataire s'était engagé à observer et à respecter, dans un territoire ayant un statut international, les droits de l'homme et les libertés fondamentales pour tous sans distinction de race. Le fait d'établir et d'imposer, au contraire, des distinctions, exclusions, restrictions et limitations qui sont uniquement fondées sur la race, la couleur, l'ascendance ou l'origine nationale ou ethnique et qui constituent un déni des droits fondamentaux de la personne humaine, est une violation flagrante des buts et principes de la Charte ».[27]

Au sujet de l'affaire relative au *personnel diplomatique et consulaire des États-Unis à Téhéran*, arrêt rendu le 24 mai 1980, paragraphe 91, la Cour a dit :

> « Le fait de priver abusivement de leur liberté des êtres humains et de les soumettre dans des conditions pénibles à une contrainte physique est manifestement incompatible avec les principes de la Charte des Nations Unies et avec le droits fondamentaux énoncés dans la Déclaration universelle des droits de l'homme ».[28]

---

[25] CIJ Recueil 1971, *Conséquences juridiques pour les Etats de la présence continuel de l'Afrique du Sud en Namibie (Sud-Oeust Africain) nonobstant la résolution 276 (1970) du Conseil de Securité, Avis Consultatif 21/VI/71* ; B. Bollecker, « Avis Consultatif du 21 juin 1971 (Namibie) », *AFDI*, XVII (1971), 281-333 ; A. W. Rovine, « The World Court Opinion on Namibia », *Colum. J. Transnat'l L.* 11 (1972), 203-239 ; J. P. Jacqué, « Avis consultatif du 21 juin 1971 », *RGDIP* (1972), 1046-1097.

[26] E. Jiménez de Aréchaga, voir note 2, 190-191.

[27] CIJ Recueil 1971, 57, paragraph 131.

[28] CIJ Recueil 1980, 42, paragraph 91.

Au sujet de l'affaire relative aux *activités militaires et paramilitaires au Nicaragua*, (Nicaragua vs. États-Unis d'Amérique), arrêt rendu le 27 juin 1986, paragraphes 267 et 268, la Cour a dit :

> « La Cour relève par ailleurs que le Nicaragua est accusé de violer les droits de l'homme, selon la conclusion tirée par le Congrès des États-Unis en 1985. Ce point particulier doit être approfondi, indépendamment de l'existence d'un 'engagement juridique' pris par le Nicaragua envers l'Organisation des États Américains de respecter ces droits. L'inexistence d'un tel engagement ne signifierait pas que le Nicaragua puisse violer impunément les droits de l'homme. Toutefois, quand les droits de l'homme sont protégés par des conventions internationales, cette protection se traduit par des dispositions prévues dans le texte des conventions elles-mêmes et qui sont destinées à vérifier ou à assurer le respect de ces droits. La promesse politique avait été faite par le Nicaragua dans le cadre de l'Organisation des États Américains, de sorte que les organes de cette organisation se trouvent compétents pour en vérifier le respect. La Cour a relevé (paragraphe 168), que, depuis 1979, le Gouvernement du Nicaragua a ratifié plusieurs instruments internationaux relatifs aux droits de l'homme, dont la Convention américaine portant sur ce sujet (Pacte de San José, Costa Rica). Ces mécanismes ont fonctionné. Ainsi, la Commission interaméricaine des droits de l'homme a pris des mesures et élaboré deux rapports (OEA/Ser. L/V/II.53 et 62) après s'être rendue au Nicaragua à l'invitation de son gouvernement. L'Organisation des États Américains était donc à même, si elle le souhaitait, de statuer sur la base de ces constatations.
>
> De toute manière, si les États-Unis peuvent certes porter leur propre appréciation sur la situation des droits de l'homme au Nicaragua, l'emploi de la force ne saurait être la méthode appropriée pour vérifier et assurer le respect de ces droits. Quant aux mesures qui ont été prises en fait, la protection des droits de l'homme, vu son caractère strictement humanitaire, n'est en aucune façon compatible avec le minage de ports, la destruction d'installations pétrolières, ou encore l'armement et l'équipement des contras. La Cour conclut que le motif tiré de la préservation des droits de l'homme au Nicaragua ne peut justifier juridiquement la conduite des États-Unis et ne s'harmonise pas, en tout état de cause, avec la stratégie

judiciaire de l'État défendeur fondé sur le droit de légitime défense collective ».[29]

Dans l'avis consultatif du 8 juillet 1996 sur *la licéité de la menace ou de l'emploi d'armes nucléaires*,[30] la Cour a par ailleurs examiné des questions relatives aux droits de l'homme, notamment en ce qui concerne le droit à la vie et le droit international humanitaire.[31]

Parmi les nombreuses affirmations émises par la Cour sur cet sujet dans cet avis consultatif, il convient de rappeler les suivantes :

> « 25. La Cour observe que la protection offerte par le pacte international relatif aux droits civils et politiques ne cesse pas en temps de guerre, si ce n'est par l'effet de l'article 4 du pacte, qui prévoit qu'il peut être dérogé, en cas de danger public, à certaines des obligations qu'impose cet instrument. Le respect du droit à la vie ne constitue cependant pas une prescription à laquelle il peut être dérogé. En principe, le droit de ne pas être arbitrairement privé de la vie vaut aussi pendant des hostilités. C'est toutefois, en pareil cas, à la spécialité applicable, à savoir le droit applicable dans les conflits armés, conçu pour régir la conduite des hostilités, qu'il appartient de déterminer ce qui constitue une privation arbitraire de la vie. Ainsi, c'est uniquement au regard du droit applicable dans les conflits armés, et non au regard des dispositions du pacte lui-même, que l'on pourra dire si tel cas de décès provoqué par l'emploi d'un certain type d'armes au cours d'un conflit armé doit être considéré comme une privation de la vie contraire à l'article 6 du pacte.
>
> Certains États ont aussi avancé l'argument selon lequel l'interdiction du génocide, formulée dans la convention du 9 décembre 1948 pour la prévention et la répression du crime de

---

[29] *CIJ Recueil* 1986, 113, 114, 134, 217, 218, 220, 254, 267 et 268 ; G. Abi Saab, « Les Principes généraux du doit humanitaire selon la Cour internationale de Justice », *Rev. ICR* 766 (1987), 381-389.

[30] *CIJ Recueil* 1996. Voir P. Weil, « Avis consultatif sur la licéité de la menace ou de l'emploi d'armes nucléaires : deux lectures possibles », dans Weil, *Écrits de Droit International*, 2000, 57-65 ; J. M. Gómez Robledo, « Introduction », dans *Alegato de México en la Corte Internacional de Justicia, Sergio González Galvez, Opinión Consultiva sobre la Ilegalidad de la Amenaza o el Uso de las Armas Nucleares, Secretaría de Relaciones Exteriores, México*, 1999, 17-18.

[31] *CIJ Recueil* 1996, paragraphes 24, 25, 26, 74, 75, 76, 77, 78, 79, 86, 89.

génocide, serait une règle pertinente du droit international coutumier que la Cour devrait appliquer en l'espèce. La Cour rappellera que le génocide est défini à l'article II de la convention comme :

'l'un quelconque des actes ci-après, commis dans l'intention de détruire, en tout ou en partie, un groupe national, ethnique, racial ou religieux, comme tel :

a) meurtre de membres du groupe ;
b) atteinte grave à l'intégrité physique ou mentale de membres du groupe ;
c) soumission intentionnelle du groupe à des conditions d'existence devant entraîner sa destruction physique totale ou partielle ;
d) mesures visant à entraver les naissances au sein du groupe ;
e) transfert forcé d'enfants du groupe à un autre groupe.'

Il a été soutenu devant la Cour que le nombre de morts que causerait l'emploi d'armes nucléaires serait énorme ; que l'on pourrait, dans certains cas, compter parmi les victimes des membres d'un groupe national, ethnique, racial ou religieux particulier ; et que l'intention de détruire de tels groupes pourrait être inférée du fait que l'utilisateur de l'arme nucléaire aurait omis de tenir compte des effets bien connus de l'emploi de ces armes.

La Cour relèvera à cet égard que l'interdiction du génocide serait une règle pertinente en l'occurrence s'il était établi que le recours aux armes nucléaires comporte effectivement l'élément d'intentionnalité, dirigé contre un groupe comme tel, que requiert la disposition sous-citée. Or, de l'avis de la Cour, il ne serait possible de parvenir à une telle conclusion qu'après avoir pris dûment en considération les circonstances propres à chaque cas d'espèce.

. . .

79. C'est sans doute parce qu'un grand nombre de règles du droit humanitaire applicable dans les conflits armés sont si fondamentales pour le respect de la personne humaine et pour des 'considérations élémentaires d'humanité', selon l'expression utilisée par la Cour dans son arrêt du 9 avril 1949

rendu en l'affaire du Détroit de Corfou (Recueil CIJ 1949, 22), que la Convention IV de La Haye et les conventions de Genève ont bénéficié d'une large adhésion des États. Ces règles fondamentales s'imposent d'ailleurs à tous les États, qu'ils aient ou non ratifié les instruments conventionnels qui les expriment, parce qu'elles constituent des principes intransgressibles du droit international coutumier ».

Ni Guerrero, ni Alvarez, ni Fabela, ni Acevedo, dans l'affaire du *Détroit de Corfou* (1949), ni Guerrero, ni Alvarez, ni Acevedo dans l'affaire de la *Convention sur le génocide* (1951), ni Guerrero, ni Alvarez, ni Acevedo dans l'avis consultatif sur les *traités de paix* (1950), ni Bustamante y Rivero, Padilla Nervo et Armand Ugón dans l'Arrêt relatif à la *Barcelona Traction* (1970), ni Padilla Nervo, ni Jiménez de Aréchaga dans l'avis consultatif sur la *Namibie* (1971), ni Ruda et Sette-Cammara dans l'Arrêt concernant l'affaire relative *au personnel diplomatique des États-Unis à Téhéran* (1980), ni Ruda et Sette Cammara, dans l'Arrêt relatif à l'affaire des *activités militaires et paramilitaires au Nicaragua et contre celui-ci* (1986), autrement dit, tous les juges latino-américains ayant participé à l'étude de ces affaires se sont opposés aux critères ci-dessus énoncés.[32]

On peut donc affirmer que les juges latino-américains ont partagé à l'unanimité les points de vue de la Cour internationale de Justice, depuis 1949, sur le droit international humanitaire, les principes fondamentaux d'humanité, les droits de l'homme, les obligations qui émanent de la Charte des Nations Unies à cet égard et la portée de la Déclaration universelle des droits de l'homme.

Cette unanimité au sein de la Cour en ce qui concerne les droits de l'homme est renforcée, approfondie et généralisée par la doctrine latino-américaine qui dans ses nombreux exposés, a commenté et approfondi les affirmations conceptuelles du Tribunal de La Haye sur ces questions.

VI.

Il est intéressant de rappeler que la jurisprudence de la Cour internationale de justice a été à plusieurs reprises citée par la Cour

---

[32] Sur les juges latino-américains, en général, à la Cour internationale de Justice, notamment Bustamente y Rivero, Jiménez de Aréchaga et Ruda, cf. M. Bedjaoui, « Présences latino-americaines à la Cour internationale de Justice », dans *Liber Amicorum Ruda*, voir note 3, 367-392.

interaméricaine des droits de l'homme dans sa jurisprudence propre, à l'appui de certains des critères par elle retenue.

Cela prouve non seulement le prestige de la jurisprudence de la Cour internationale de Justice et sa force internationale, mais aussi l'influence grandissante de la Cour de la Haye, qui a eu des incidences sur la jurisprudence d'un organe juridictionnel comme la Cour interaméricaine, qui a une compétence spécifique en matière de droits de l'homme, conformément à la Convention américaine des droits de l'homme (Pacte de San José).

Voyons maintenant quelques exemples :

- Dans son avis consultatif n° 1 (OC 1/82), la Cour interaméricaine des droits de l'homme citait la Cour internationale de Justice pour préciser la nature de la compétence consultative.
- Dans son avis consultatif n° 2, pour distinguer les traités modernes sur les droits de l'homme et les traités multilatéraux du type traditionnel, la Cour cita « les idées similaires au sujet des traités humanitaires soutenues par la Cour internationale de Justice dans son avis consultatif sur les Réserves à la Convention pour la prévention et la répression du crime de génocide ».
- Dans son avis consultatif 3/83, elle cita l'avis consultatif sur le *Sahara Occidental (Western Sahara)*, notamment les questions relatives aux exceptions préliminaires, la distinction entre les compétences contentieuse et consultative et certains critères d'interprétation exposés par la Cour internationale de Justice dans son avis consultatif de 1980 relatif à l'Accord de 1951 entre l'OMS et l'Égypte.
- Dans son avis consultatif 4/84, elle cita le cas *Nottebohm* relatif à la nationalité.
- Dans son avis consultatif 6/86, elle invoqua la Cour permanente de Justice internationale concernant le « Principe de licéité ».
- Dans son avis consultatif 10/89 du 14 juillet 1989, elle cita l'avis consultatif de la *Namibie* (1970) portant sur l'exigence d'interpréter un instrument international « dans le cadre de l'ensemble du système juridique en vigueur au moment où l'interprétation a lieu » (paragraphe 37) et ce même avis consultatif de la CIJ et les cas de la Barcelona Traction et du personnel diplomatique et consulaire des États-Unis à Téhéran, concernant le devoir de respecter certains droits essentiels, considéré comme une obligation erga omnes (paragraphe 38).
- Dans son avis consultatif 14/94 du 9 décembre 1994, elle a évoqué un grand nombre de cas décidés par la Cour permanente de Justice

internationale et par la Cour internationale de Justice quant à l'obligation d'arbitrer (1988), étant entendu que les obligations internationales doivent être exécutées de bonne foi et que les pays ne peuvent invoquer le droit intérieur pour justifier leur inexécution (paragraphe 35).
- Dans son avis consultatif 15 du 14 décembre 1997, la Cour interaméricaine rappela sa position en ce sens qu'elle n'est pas forcée de s'abstenir de l'exercice de sa compétence consultative dans les situations où il existe une controverse sur ce point, et cité l'avis de la CIJ sur les cas des *Traités de paix* (1950), des *réserves à la Convention sur le génocide* (1951), de la *Namibie* (1970), du *Sahara Occidental* (1975) et la Convention sur les privilèges et immunités des Nations Unies (1989).
- En fin, dans son avis consultatif 16 du 1er octobre 1999, la Cour interaméricaine a pris en considération les cas *Breard* et *Le Grand* intentés auprès de la Cour internationale de Justice (paragraphes 54, 55 et 56).

Dans ce même avis consultatif, le Juge Cançado Trindade, motivant son vote en ce même sens, a longuement fait allusion, à plusieurs reprises, aux critères soutenus par la CIJ (paragraphes 8, 12, 26 et 27).

VIII.

Il est impossible de passer en revue dans ce bref article la totalité des cas contentieux où les arrêts de la Cour interaméricaine des droits de l'homme ont cité les critères jurisprudentiels de la Cour internationale de Justice.

Mais on ne peut manquer de faire ressortir que depuis ses premiers arrêts, en 1986, et jusqu'aux plus récents, de 1999,[33] la Cour interaméricaine a souvent cité les jugements rendus par la Cour de La Haye.

En particulier, mais non exclusivement, en ce qui se rapporte à la responsabilité de l'État, aux formalités de procédure et à différentes questions relatives à l'indemnisation, son influence s'est avérée essentielle.

Les citations de la jurisprudence de la Cour de La Haye ont toujours contribué à fonder ses propres critères, qui partageaient les arguments avancés par la Cour internationale de Justice.

---

[33] Par exemple : affaires *Ivcher Bronstein*, Compétence, Arrêt du 24 septembre 1999, paragraphes 44, 52 et 53, et affaire du *Tribunal constitutionnel*, Compétence, Arrêt du 24 septembre 1999, paragraphes 43 et 51.

Les droits de l'homme et la CIJ : une vision latino-américaine 1467

Cependant, dans deux affaires récentes, la Cour interaméricaine, ayant soutenu l'inadmissibilité du retrait de la reconnaissance de la compétence contentieuse par un État partie au Pacte de San José, s'est écartée de la jurisprudence de la Cour internationale de Justice concernant le retrait ou la modification de la clause facultative (art. 36.2 des Statuts). Ces deux arrêts de la Cour de San José, dont je ne partage pas le principe,[34] omettent curieusement la citation détaillée de la jurisprudence abondante et excellente de la Cour internationale de Justice en cette matière.

Sans préjudice de l'intérêt que revêt la conduite d'une étude détaillée d'une telle influence, il faut en souligner dès maintenant la signification et la portée, et faire remarquer que cette influence n'est nullement épuisée, qu'elle reste au contraire active, voire grandissante, et que tout porte à croire qu'elle persistera.

On pourrait penser que si la Cour internationale de Justice approfondit l'examen de cette matière, suite au traitement dans des affaires futurs, des questions relatives aux droits de l'homme, son influence sur la jurisprudence de la Cour interaméricaine ne fera qu'augmenter – l'élément essentiel n'étant plus les formalités de procédure[35] et de caractère juridique en général – mais plutôt les questions spécifiques, relatives au contenu, à la nature et aux éléments propres des droits de l'homme en tant que tels.[36]

---

[34] Affaire *Ivcher Bronstein*, voir note 34, § 47 ; affaire du *Tribunal constitutionnel*, voir note 34, § 46 ; H. Gros Espiell, « El Retiro del Reconocimiento de la Competencia Contenciosa de la Corte Interamericana de Derechos Humanos », dans *Estudios en Homenaje al Profesor Enrique Véscovi*, 2000, 5-15.

[35] Cf. Notamment, concernant la nature des mesures provisoires dans la jurisprudence de la Cour internationale de Justice et leur considération par la Cour interaméricaine des droits de l'homme – sous réserve des différences existant entre ces deux institutions dans l'une et l'autre juridictions – comme a été, souligné par A. A. Cançado Trindade dans son Prologue, Série E : Mesures provisoires, n° 2, Recueil = juillet 1996-juin 2000, Cour interaméricaine des droits de l'homme, San José, 2000, XII-XIII.

[36] Les références relatives à la jurisprudence de la Cour internationale de Justice ont été et restent importantes dans les avis individuels émis par les juges de la Cour interaméricaine des droits de l'homme, notamment celles du Juge Antonio Cançado Trindade. Il convient de souligner celle rédigée par ce juge conjointement avec son confrère A. Abreu Burelli, dans l'affaire *Villagrán Morales et autres*, arrêt rendu le 19 novembre 1999, 105-106, note 2, qui reprend en citation le vote dissident du Juge Tanaka, dans *South West Africa* (deuxième phase) *(Ethiopie et Liberia c. Afrique du Sud)*, *CIJ Recueil* 1966, 298, selon laquelle : « le droit fondamental à la

IX.

En dépit des limitations que comporte la nature de la compétence de la Cour internationale de Justice, elle peut et doit jouer un rôle important dans le domaine des droits de l'homme.

La tâche qu'elle a accomplie jusqu'à présent, bien qu'importante, n'est sans doute que le début d'un processus, nécessaire pour la communauté internationale, sur une question qui, comme c'est le cas droits de l'homme, est au cœur du contenu du droit international actuel et se trouve directement et nécessairement associé à la Paix et à la Sécurité internationales.

Conformément à sa tradition, l'Amérique latine qui a toujours appuyé ces tentatives en faveur de la garantie et de la protection juridique internationale des droits de l'homme, selon une approche juste et non discriminatoire, tant à l'échelle universelle que régionale, voit et verra toujours avec sympathie et compréhension les efforts accomplis dans ce sens par la Cour internationale de Justice, dans les limites de sa compétence.

---

vie relève du domaine du jus cogens ». Voir Y. Saito, « La non discrimination, serait-elle un principe du droit naturel ? L'opinion dissidente de M. K. Tanaka, juge à la CIJ, dans l'arrêt concernant le *Sud-Ouest Africain* », dans *Unterwegs zum Frieden, Beiträge zur Idee und Wirklichkeit*, 1973, 493-510 ; K. Tanaka, « Some Observations on Peace, Law and Human Rights », dans W. Friedmann, L. Henkin, O. Lisstzyn (eds.), *Transnational Law in a Changing Society, Essays in Honor of Philip C. Jessup*, 1972, 242-256.

The Impact of Judgments and Advisory Opinions of the PCIJ-ICJ on Regional Courts of Human Rights

Alexandre Kiss

International tribunals proliferated during the second half of the 20th century. The process accelerated during the last decade, with the creation of *ad hoc* international criminal tribunals, the Law of the Sea Tribunal, and the dispute-settlement mechanisms of the World Trade Organisation, the North American Free Trade Association, and other regional organisations. This multiplicity of international fora, to which may be added the prospect of a permanent International Criminal Court and a Human Rights Court for Africa, raises the question of the consistency of international norms applied by them. In each national legal order, a hierarchy among judicial bodies ensures the homogeneity of rules which such fora apply. Perhaps the oldest global international tribunal, the present International Court of Justice (ICJ), can perform a comparable function, not by imposing its precedents on other tribunals, which would be legally impossible in the present state of the world, but by the free acceptance of its jurisprudence by other courts.

A summary review of the judgments and advisory opinions issued by the two existing international courts in the field of the international protection of human rights and freedoms shows that the holdings and the *dicta* of the World Court are not ignored. The European Court of Human Rights (ECHR), created by the European Convention for the Protection of Human Rights and Fundamental Freedoms (Rome, 4 November 1950,[1] in force since 1953) and the Inter-American Court of Human Rights (IACHR), created by the American Convention on Human Rights (San José, 22 November 1969,[2] in force since 1978), have both utilised decisions and opinions of the ICJ or its predecessor, the Permanent Court of International Justice, in a fair number of their judgments and opinions. In other cases, such precedents have been cited by judges in separate opinions or have been invoked by governments parties to the

---

[1] UNTS Vol. 213, 221.
[2] OAS Treaty Ser., No. 36.

case, without being discussed in the judgment or opinion of the court itself. Altogether approximately 43 cases, more than two-thirds of which were decided by the IACHR, contain references to World Court texts. Some of those references concern jurisdictional matters, others apply substantive principles of international law, and quite a few are related to matters of procedure and evidence.

I. Jurisdiction

1. General Principles

A factor limiting the influence of the jurisprudence of the World Court is the substantial difference between the scope and nature of its jurisdiction and that of the tribunals whose competence is based on regional human rights conventions. The ECHR described this difference in a judgment of 23 March 1995[3]:

> "In the first place, the context within which the International Court of Justice operates is quite distinct from that of the Convention institutions. The International Court is called on *inter alia* to examine any legal dispute between states that might occur in any part of the globe with reference to principles of international law. The subject matter of a dispute may relate to any area of international law. In the second place, unlike the Convention institutions, the role of the International Court is not exclusively limited to direct supervisory functions in respect of a law-making treaty such as the Convention".

The European Court considered that such a fundamental difference in the role and the purpose of the respective tribunals, coupled with the existence of a practice of unconditional acceptance of the European Court's jurisdiction, provides a compelling basis for distinguishing Convention practice from that of the ICJ.[4] In particular, the long-established ICJ practice of accepting restrictions on its optional jurisdiction under Article 36 of its Statute flows from its character as a free-standing international tribunal which has no links to a standard-setting treaty such as the European Convention on Human Rights.[5]

In other cases, the European Court has stressed the differences between its powers and those of other universal or regional tribunals. In one case, the ECHR declared that the European Convention must be

---

[3] ECHR, *Loizidou* v. *Turkey*, 23 March 1995, para. 84.
[4] *Ibid.*, para. 85.
[5] *Ibid.*, paras. 68 and 83.

interpreted in the light of its special character. The Court specifically noted that the European Convention does not contain a provision authorising the Court to issue interim measures to preserve the rights of parties in pending proceedings.[6] This omission is in contrast to the powers granted by Article 41 of the Statute of the International Court of Justice and Article 63 of the American Convention on Human Rights. In another case, *Ahmet Sadik* v. *Greece*,[7] the ECHR described its place in the international legal order. During the Court's examination of the application, which alleged a violation of Article 10 of the European Convention of Human Rights, the European Commission of Human Rights observed that the applicant had challenged the State's actions in the domestic courts. The delegate of the Commission argued that such act afforded the concerned State the opportunity to put right the alleged violation and was thus sufficient for the purpose of exhaustion of local remedies. Referring to the case law of the International Court of Justice and the generally recognised rules of international law as expressed in Article 26 of the European Convention of Human Rights,[8] the delegate maintained that it was necessary for the domestic remedy to be based on the same ground as the international remedy. The ECHR, however, rejected the argument. It reiterated, quoting its own practice, that the supervisory machinery set up by the Convention is subsidiary to the national human rights protection systems. Accordingly, the principle of exhaustion of local remedies dispenses States from answering in an international body for their acts before they have had an opportunity to put matters right through their own legal system.[9]

2. Advisory Opinions

In a more specific field, concerning advisory opinions, the IACHR has followed the practice of the International Court of Justice. In its Advisory Opinion OC-1/82, it recalled that Article 64 of the American Convention confers on it an advisory jurisdiction that is more extensive than that enjoyed by any other existing international tribunal.[10] It also

---

[6] ECHR, *Santa Cruz Varas and Others*, 20 March 1991, para. 94.
[7] ECHR, *Ahmet Sadik* v. *Greece*, 25 October 1996, para. 29.
[8] "The Commission may only deal with the matter after all domestic remedies have been exhausted, according to generally recognised rules of international law, and within a period of six months from the date on which the final decision was taken".
[9] ECHR, *Ahmet Sadik* v. *Greece*, see note 7, para. 29.
[10] IACHR, *"Other Treaties" Subject to the Consultative Jurisdiction of the Court (Art. 64 American Convention on Human Rights)*, Advisory Opinion, OC-1/82, 24 September 1982, para. 14.

noted, however, the concerns and even the opposition of some States to the Court's exercise of consultative jurisdiction respecting specific questions submitted to it. In these instances the States viewed the process as a "disguised contentious proceeding", a method to evade the requirement that all States Parties to a legal dispute give consent before judicial proceedings may be instituted. The IACHR stressed, however, that when those objections were raised to requests for advisory opinions under the UN Charter, the International Court of Justice decided, for a variety of reasons, to render the opinions notwithstanding such objections.[11] Consistent with the jurisprudence of the International Court of Justice, the IACHR decided that its advisory jurisdiction is permissive in character, and thus that it empowers the Court to decide whether the circumstances of a request for an advisory opinion justify a decision granting or rejecting the request.[12]

The IACHR reiterated this approach a year later in its Advisory Opinion OC-3/83.[13] It recalled that the International Court of Justice has consistently rejected objections to the exercise of its advisory jurisdiction by States alleging that because the issue involved was in dispute, the Court was being asked to decide a disguised contentious case. In doing so, the Hague Court acknowledged that the advisory opinion might affect the interests of States which have not consented to its contentious jurisdiction and which are not willing to litigate the matter.[14]

The question of the preliminary examination of jurisdictional objections in the advisory proceeding was raised in the Inter-American Court by Guatemala. The Inter-American Court refused to admit such preliminary objections, because the purpose and utility of the advisory power would be seriously impaired. In doing so, it cited the practice of the International Court of Justice that adopted an amendment to its rules of Court in order to accelerate the consideration of requests for advisory

---

[11] *Ibid.*, para. 23. Reference is made by the Inter-American Court to the following cases: *Interpretation of Peace Treaties with Bulgaria, Hungary and Romania*, ICJ Reports 1950, 65 et seq.; *International Status of South West Africa*, ICJ Reports 1950, 128; *Certain Expenses of the United Nations*, ICJ Reports 1962, 151 et seq.; *Legal Consequences for States of the Continued Presence of South Africa in Namibia (South West Africa) notwithstanding Security Council Resolution 276 (1970)*, ICJ Reports 1971, 16 et seq.

[12] *Interpretation of Peace Treaties*, see note 11.

[13] IACHR, *Restrictions to the Death Penalty (Arts. 4(2) and 4(4) American Convention on Human Rights)*, Advisory Opinion, OC-3/83, 8 September 1983.

[14] *Ibid.*, para. 40. It makes reference to the following cases: *South West Africa*, see note 11; *Western Sahara*, ICJ Reports 1975, 12 et seq.

opinions.[15] The Inter-American Court's advisory opinion also noted that an amendment of 1972 to the Rules of the ICJ requires the Court in contentious cases to consider objections to its jurisdiction prior to dealing with the merits, but this rule has not been applied to advisory opinions.[16]

An important rule governing the substance of advisory opinions is that in such proceedings the Court does not exercise fact-finding functions; instead, it is called upon to render opinions interpreting legal norms, meaning that opinions lack the same binding force that attaches to judgments in contentious cases.[17]

The Inter-American Court rendered its most significant opinion to date on the scope of its advisory jurisdiction in OC-16/99 of 1 October 1999, concerning *The Right to Information on Consular Assistance in the Framework of the Guarantees of the Due Process of Law*. The matter posed particular difficulties, because at the time the request was submitted a contentious case involving the same treaty, the Vienna Convention on Consular Relations, had been submitted to the ICJ. The ICJ case was brought by Paraguay against the United States, both members of the OAS and thus potentially affected by any decision of the Inter-American Court.[18] Mexico sought the advisory opinion because of the situation of its nationals who had been condemned to death in ten States in the United States. Mexico specifically invoked the Vienna Convention, as well as the International Covenant on Civil and Political Rights, to which both States are party. Indicating the importance of the issue, eight States, the Inter-American Commission, and 22 organisations and individuals filed written pleadings in the matter. The United States argued that as the Vienna Convention on Consular Relations is a global treaty, there can be no differing interpretations of the States' obligations on a regional basis. The submission of a case to the ICJ on the same issue thus should lead the Inter-American court on the basis of "prudence, if not considerations of comity" to defer its consideration of the request until the ICJ rendered its decision.[19] To do

---

[15] *Ibid.*, para. 25. The Inter-American Court alluded to Article 103 of the ICJ Rules of Court.

[16] *Ibid.*, para. 26. Article 79 of the ICJ Rules of Court. Also *Western Sahara*, see note 14.

[17] IACHR, *Restrictions to the Death Penalty*, see note 13, para. 32; *Interpretation of Peace Treaties*, see note 11.

[18] Although the United States is not a party to the American Convention and thus subject to the Court's contentious jurisdiction, the Court's advisory jurisdiction is open to all Member States of the OAS. Further, its opinions affect OAS institutions such as the Inter-American Commission on Human Rights which have jurisdiction over all Member States.

[19] IACHR, *The Right to Information on Consular Assistance in the Framework of the Guarantees of the Due Process of Law*, OC-16/99, 1 October 1999,

otherwise would risk inconsistency between the findings of the IACHR and the ICJ and potentially create problems for the "vast number of States outside the hemisphere" that are party to the Convention. Moreover, the United States asserted that the Vienna Convention could not be considered a treaty concerning human rights within the advisory jurisdiction of the IACHR and that, in any case, the matter submitted was in fact a contentious case between itself and Mexico.

The Court found that the instruments invoked all concerned the protection of human rights in the American States. It also rejected the assertion that the request to give an opinion should be declined on the ground that it was a disguised contentious case. As in earlier opinions, the IACHR referred explicitly to the jurisprudence of the ICJ[20] on this issue. It also cited that jurisprudence to support its position that it would not settle questions of fact concerning the States in question, but would only advise about the meaning, object and purpose of international human rights norms.[21]

The Court paid particular attention to the issue of whether or not it should defer to the ICJ because of a pending contentious case involving similar issues.[22] The IACHR began by noting that it must interpret all provisions of the American Convention in such a way that the system for the protection of human rights has all its appropriate effects *(effet utile)*. Accordingly, the Court "cannot be restrained from exercising its advisory jurisdiction because of contentious cases filed with the International Court of Justice".[23] As an autonomous judicial institution, the Court must go forward. It recalled that

"... the possibility of conflicting interpretations is a phenomenon common to all those legal systems that have certain courts which are not hierarchically integrated. Such

---

Ser. A, No. 16, 178. The United States later informed the IACHR that Paraguay had discontinued the case it brought against the United States at the ICJ.

[20] Citing *Interpretation of Peace Treaties*, see note 11; *Reservations to the Convention on the Prevention and Punishment of the Crime of Genocide*, ICJ Reports 1951, 15 et seq.; *South West Africa*, see note 11; *Western Sahara*, see note 14; *Applicability of Article I, Section 22, of the Convention on the Privileges and Immunities of the United Nations*, ICJ Reports 1989, 177 et seq.

[21] Citing *Interpretation of Peace Treaties*, see note 11.

[22] Although Paraguay discontinued its case, Germany filed a case against the United States at the ICJ on 2 March 1999 on the same legal issue. The filing came more than one year after Mexico submitted its request for an advisory opinion to the IACHR and eight months after the Court concluded the oral phase of the proceedings. IACHR, *Consular Assistance*, see note 19, 230.

[23] *Ibid.*, 231.

courts have jurisdiction to apply and, consequently, interpret the same body of law. Here it is, therefore, not unusual to find that on certain occasions courts reach conflicting or at the very least different conclusions in interpreting the same rule of law. On the international law plane, for example, because the advisory jurisdiction of the International Court of Justice extends to any legal question, the UN Security Council or the General Assembly might ask the International Court to render an advisory opinion concerning a treaty which, without any doubt, could also be interpreted by this Court under Article 64 of the Convention. Even a restrictive interpretation of Article 64 would not avoid the possibility that this type of conflict might arise".[24]

Judge Cançado Trindade, in his concurring opinion, asserted that the Inter-American Court is particularly entitled to pronounce on the human rights issues, ones that are distinct from the two contentious cases submitted to the ICJ. In this context, he specifically referred to the declaration of Judge Oda in the *LaGrand* case *(Germany* v. *United States)* and the *Breard* case *(Paraguay* v. *United States)* to the effect that the ICJ contentious function is limited to settling international disputes pertaining to the rights and duties of States.[25]

In addressing whether or not the Vienna Consular Convention is a treaty concerning human rights, the IACHR discusses not only the language and *travaux préparatoires* of the Convention, but the jurisprudence of the ICJ, in particular the *Case Concerning United States Diplomatic and Consular Staff in Tehran*.[26] The IACHR cites the US pleadings in the *Tehran* case as well as the judgment of the Court in concluding that the treaty does concern the rights of nationals of the sending State.[27] According to the IACHR, the treaty endows a detained foreign national with individual rights that are the counterpart of the host State's correlative duties.

Again using ICJ decisions, the IACHR finds that the treaty's interpretation is governed by "application of a general principle of interpretation that international jurisprudence has repeatedly affirmed ...

---

[24] *Ibid.*, 231-232, quoting *"Other Treaties" Subject to the Advisory Jurisdiction of the Court (Art. 64 American Convention on Human Rights)*, Advisory Opinion, OC-1/82, 24 September 1982, Ser. A, No. 1, para. 50.
[25] IACHR, *Consular Assistance*, see note 19, concurring opinion of Judge Cançado Trindade, fn. 28.
[26] ICJ Reports 1980, 3 and 42.
[27] See the discussion in the concurring opinion of Judge Cançado Trindade, see note 25, paras. 18-20 and 24-26. He also refers to the ICJ's decision in the *North Sea Continental Shelf* cases (1969), para. 27.

so that appropriate effects *(effet utile)* are obtained".[28] Similarly, in examining the International Covenant on Civil and Political Rights and the OAS Charter, the IACHR refers to the doctrine of evolving obligations, quoting from the ICJ advisory opinion on *Namibia* to state that "an international instrument has to be interpreted and applied within the framework of the entire legal system prevailing at the time of interpretation".[29]

3. Structure of the Different Courts

A situation that might occur with some frequency led the IACHR to consider the problems raised by a change in the Court's composition during the pendency of a case. The Spanish, English and French versions of the relevant provisions of the American Convention on Human Rights are not exactly the same; they do not give a clear answer to the question of whether the Court deciding upon provisional measures has to remain to hear the preliminary objections and the merits of the case.[30] The Court found that the Statute of the International Court of Justice contains a provision similar to that contained in the English text of the American Convention. It reads as follows:

> "The Members of the Court shall continue to discharge their duties until their places have been filled. Though replaced, they shall finish any cases which they may have begun (Article 13(1))".

The IACHR stated that the International Court of Justice has given very liberal interpretation of its statutory provision, allowing the composition to change at one stage or another during a case. Thus, the ICJ judges dealing with provisional measures are not necessarily the ones to hear preliminary objections or the merits, as a long list of judgments of the World Court demonstrates.[31] The cases heard in that Court were seen,

---

[28] Citing *Free Zones of Upper Savoy and the District of Gex*, Order of 19 August 1929, PAG, PCIJ Ser. A, No. 22, 1929, 13.

[29] IACHR, *Consular Assistance*, see note 19, 255. *South West Africa*, see note 11, 16 and 31. Judge Cançado Trindade also quotes this statement in his concurring opinion, see note 25, para. 9.

[30] IACHR, *Neira Alegria* case, Resolution of 29 June 1992, Annual Report of the Court, 1992, 55.

[31] *Barcelona Traction, Light and Power Company Limited* (Second Phase), ICJ Reports 1970, 3 et seq.; *Fisheries Jurisdiction (United Kingdom v. Iceland)*, Jurisdiction, ICJ Reports 1973, 3 et seq.; *Fisheries Jurisdiction (United Kingdom v. Iceland)*, Merits, ICJ Reports 1974, 3 et seq.; *Fisheries Jurisdiction (Federal Republic of Germany v. Iceland)*, Jurisdiction, ICJ

however, to be different in nature from those handled by the IACHR. In the former, the sources applied had to take into account the equilibrium of relationships between States. The IACHR viewed the protection of human rights as very different, holding that modern treaties with this aim are not multilateral treaties of the traditional type concluded to accomplish the reciprocal exchange of rights for the mutual benefit of the Contracting State. In concluding these human rights treaties, the IACHR said, States can be deemed to submit themselves to a legal order within which they, for the common good, assume various obligations, not in relation to other States, but towards all individuals within their jurisdiction. Hence, the case law of the International Court of Justice cannot be applied in blanket fashion by the IACHR: to do so would not take into account the need to guarantee the victims the most efficient proceeding possible. In concluding, the IACHR held that the best judge of a case is the court that began to hear it, composed of those who already may have begun to address the merits, even when oral proceedings have not yet been initiated.[32]

III. Substantive International Law

1. Scope of Applicable Principles

The IACHR used a general statement of the International Court of Justice to indicate the scope of international principles that are to be applied by the different international tribunals:

> "[the] rule of international law, whether customary or conventional, does not operate in a vacuum; it operates in relation to facts and in the context of a wider framework of legal rules of which it forms only a part. Accordingly, if a question put in the hypothetical way in which it is posed in the request is to receive a pertinent and effectual reply, the Court must first ascertain the meaning and full implications of the actual framework of fact and law in which it falls for consideration". [33]

Reports 1973, 49; *Fisheries Jurisdiction (Federal Republic of Germany v. Iceland)*, Merits, ICJ Reports 1974, 175 et seq.; *Military and Paramilitary Activities in and against Nicaragua (Nicaragua v. United States of America)*, Jurisdiction and Admissibility, ICJ Reports 1984, 392 et seq.; *Military and Paramilitary Activities in and against Nicaragua (Nicaragua v. United States of America)*, Merits, ICJ Reports 1986, 14 et seq.

[32] IACHR, *Neira Alegria*, see note 30, paras. 26-32.
[33] IACHR, *Restrictions to the Death Penalty*, see note 13, para. 44.

Thus, in addressing a case or question that is submitted to it, the competent tribunal should begin by setting out the pertinent elements of fact and of law which, in its view, constitute the context in which the meaning and implications of the issue posed in the request have to be ascertained.[34] Answering a question submitted to it in a request for an advisory opinion, the Inter-American Court held that the power granted to it to render advisory opinions interpreting the American Convention on Human Rights or other treaties concerning the protection of human rights in the American States of necessity also encompasses jurisdiction to interpret the reservations attached to those instruments.[35]

2. Hierarchy of Norms

In its advisory opinion on *International Responsibility for the Promulgation and Enforcement of Laws in violation of the Convention (Articles 1 and 2 of the American Convention on Human Rights)*,[36] the IACHR had to answer the question: what are the legal effects of a domestic law that manifestly violates the obligations the State assumed upon ratifying the Convention? In responding to the question, the Court affirmed that:

> "[p]ursuant to international law, all obligations imposed by it must be fulfilled in good faith; domestic law may not be invoked to justify nonfulfillment. These rules may be deemed to be general principles of law and have been applied by the Permanent Court of International Justice and the International Court of Justice even in cases involving constitutional provisions".[37]

---

[34] *Interpretation of the Agreement of 25 March 1951 Between WHO and Egypt*, ICJ Reports 1980, 73 et seq., 76.
[35] IACHR, *Restrictions to the Death Penalty*, see note 13, para. 45.
[36] IACHR, *International Responsibility for the Promulgation and Enforcement of Laws in Violation of the Convention (Articles 1 and 2 of the American Convention on Human Rights)*, Advisory Opinion, OC-14/94, 9 December 1994.
[37] References in the *International Responsibility* case, para. 35: *Greco-Bulgarian "Communities"*, Advisory Opinion, PCIJ, Ser. B, No. 17, 1930, 32; *Treatment of Polish Nationals and Other Persons of Polish Origin or Speech in the Danzig Territory*, Advisory Opinion, PCIJ, Ser. A/B, No 44, 1932, 24; *Free Zones of Upper Savoy and the District of Gex*, Judgment, PCIJ, Ser. A/B, No. 46, 1932, 167 and ICJ Pleadings; *Applicability of the Obligation to Arbitrate under Section 21 of the UN Headquarters Agreement of 26 June 1947 (PLO Mission* case), ICJ Reports 1988, 12, para. 47.

In an earlier judgment, *Aloeboetoe et al. v. Suriname*,[38] the American Court stated that the remedial provisions of Article 63(1) of the American Convention on Human Rights – prescribing that when there has been a violation of a right or freedom protected by the Convention, the consequences of the violation must be remedied – codifies a "rule of customary law which, moreover, is one of the fundamental principles of current international law". As a consequence

> "[t]he obligation contained in Article 63(1) of the Convention is governed by international law in all of its aspects, such as, for example, its scope, characteristics, beneficiaries, etc. Consequently, this judgment must be understood to impose international legal obligations, compliance with which shall not be subject to modification or suspension by the respondent State through invocation of provisions of its own domestic law".[39]

In the more recent case of *Garrido and Baigorria*,[40] the Inter-American Court recalled the customary law duty of a State that has concluded an international agreement to introduce into its domestic laws whatever changes are needed to ensure execution of the obligations it has undertaken. Calling this principle one that is universally valid and "an evident principle", the IACHR cites the PCIJ opinion *Exchange of Greek and Turkish Populations*[41] in support of its statement.

3. Erga Omnes Obligations

On 14 July 1989, the Inter-American Court of Human Rights delivered an important advisory opinion interpreting Article 64 of the American Convention.[42] The Court expressed the view that "inter-American law" mirrors on the regional level the developments in contemporary

---

[38] IACHR, *Aloeboetoe et al. v. Suriname*, Judgment, 10 September 1993.
[39] References in the *Aloeboetoe* case, para. 43, to the jurisprudence of the PCIJ: *Jurisdiction of the Courts of Danzig*, Advisory Opinion, PCIJ, Ser. B, No. 15, 1928, 26 and 27; *Greco-Bulgarian "Communities"*, see note 37, 32 and 35; *Free Zones of Upper Savoy and the District of Gex*, Order of 6 December 1930, PCIJ, Ser. A, No. 24, 1930, 12; and Judgment, see note 37; *Treatment of Polish Nationals*, see note 37.
[40] IACHR, *Garrido and Baigorria* (Reparations), Judgment, 27 August 1998, Ser. C, No. 39, para. 68.
[41] *Exchange of Greek and Turkish Populations*, Advisory Opinion, PCIJ, Ser. B, No. 10, 1925, 20.
[42] IACHR, *Interpretation of the American Declaration of the Rights and Duties of Man Within the Framework of Article 64 of the American Convention on Human Rights*, Advisory Opinion, OC-10/89, 14 July 1989, para. 38.

international law, especially in human rights law, which distinguishes that law from classical international law. According to the Court, the duty to respect certain essential human rights is considered to be an *erga omnes* obligation, following the judgment of the ICJ in the *Barcelona Traction* case.[43] The same idea was developed in the separate opinion of Judge Cançado Trindade in the *Nicolas Chapman Blake* case,[44] involving a disappearance in Guatemala. Judge Cançado Trindade goes further in the *Villagran Morales et al.* case (the *Street Children* case),[45] asserting in his joint concurring opinion with Judge Abreu-Burelli that "[t]here can no longer be any doubt that the fundamental right to life belongs to the domain of *jus cogens*", citing, *inter alia*, the dissenting opinion of Judge Tanaka in the *South West Africa* cases at the ICJ.[46]

Related to this, the IACHR has quoted from the ICJ advisory opinion on *Reservations to the Genocide Convention* to emphasise the unique quality of human rights treaties, those where "the contracting states do not have any individual advantages or disadvantages nor interests of their own, but merely a common interest; hence the Convention's *raison d'être* is to accomplish its purposes".[47]

4. "Principle of Legality"

The "principle of legality", a term used by the Permanent Court of International Justice in its advisory opinion concerning the *Consistency of Certain Danzig Legislative Decrees with the Constitution of the Free City*,[48] triggered developments in Advisory Opinion OC-6/86 of the IACHR.[49] Referring to the principle, the Court stressed that in order to

---

[43] *Barcelona Traction*, see note 31. The Inter-American Court also mentions the ICJ's advisory opinion in *South West Africa*, see note 11, 57, and its judgment in *United States Diplomatic and Consular Staff in Tehran*, see note 26, 42.

[44] IACHR, Judgment, 24 January 1998.

[45] *Vellagran Morales* case, Judgment, 19 November 1999, Ser. C, No. 63, para. 2.

[46] *South West Africa* cases (Second Phase) *(Ethiopia and Liberia* v. *South Africa)*, dissenting opinion of Judge Tanaka, ICJ Reports 1966, 298 ("... surely the law concerning the protection of human rights may be considered to belong to the jus cogens".).

[47] IACHR, *Constitutional Court* case, Competence, Judgment, 24 September 1999, Ser. C, No. 55, quoting in para. 43 the ICJ's advisory opinion in the *Reservations to the Genocide Convention* case, see note 20. The same quotation appears in IACHR, *Ivcher Bronstein*, Competence, Judgment, 24 September 1999, Ser. C, No. 54, para. 44.

[48] PCIJ, Ser. A/B, No. 65, 1935, 56.

[49] *The Word "Laws" in Article 30 of the American Convention on Human Rights*, 9 May 1986, para. 22.

guarantee human rights, it is essential that state actions affecting basic rights not be left to the discretion of the government, but, rather, that these actions should be surrounded by a set of guarantees designed to ensure the non-impairment of the inviolable attributes of the individual. The most important of these guarantees is that basic rights can be restricted only by a law passed by the legislature in accordance with the Constitution. This procedure may not always prevent a law from being in violation of human rights, but it is an important obstacle to the arbitrary exercise of power.

5. Reparation for a Violation of International Law

The IACHR has faced repeatedly the problem of reparation for human rights violations. The resulting judgments have recalled the basic rules of international law in this field such as they were formulated by the World Court. On 21 July 1989, the Inter-American Court delivered its judgment in the *Velasquez Rodriguez* case stating that

> "[i]t is a principle of international law, which jurisprudence has considered 'even a general concept of law', that every violation of an international obligation which results in harm creates a duty to make adequate reparation. Compensation, on the other hand, is the most usual way to do so".[50]

In order to support its statement, the Court invoked two judgments and one advisory opinion of the World Court.[51] The statement was repeated with the same references in the judgment of the Inter-American Court handed down on the same day in the *Godinez Cruz* case.[52]

Subsequent judgments[53] have enlarged the basis of the principle of reparation thus stated, by invoking Article 63(1) of the American Convention, which reads as follows:

---

[50] IACHR, *Velasquez Rodriguez* case, Preliminary Objections, Advisory Opinion, Judgment, OC-6/86, 26 June 1987, para. 25.
[51] *Case Concerning the Factory at Chorzów*, Jurisdiction, Judgment No. 8, PCIJ, Ser. A, No. 9, 1927, 21; *Case Concerning the Factory at Chorzów*, Merits, Judgment No. 13, PCIJ, Ser. A, No. 17, 1928, 29; *Reparation for Injuries Suffered in the Service of the United Nations*, Advisory Opinion, ICJ Reports, 1949, 184.
[52] IACHR, 21 July 1989, para. 23.
[53] See also IACHR, *Suarez Rosero* case (Reparations), Judgment, 20 January 1999, Ser. C, No. 44, para. 40; *Blake* case (Reparations), Judgment, 22 January 1999, Ser. C, No. 48, para. 33.

"If the Court finds that there has been a violation of a right or freedom protected by this Convention, the Court shall rule that the injured party be ensured the enjoyment of his right or freedom that was violated. It shall also rule, if appropriate, that the consequences of the measure or situation that constituted the breach of such right or freedom be remedied and that fair compensation be paid to the injured party".

The Court added that this article codifies a rule of customary international law that, moreover, is one of the fundamental principles of current international law. In addition to its own jurisprudence, it invoked the two judgments of the Permanent Court of International Justice in the *Factory at Chorzów* case, as well as an advisory opinion of the International Court of Justice.[54] In the *Garrido and Baigorria* case,[55] the Court adds that the obligation to ensure reparations is "a necessary corollary of the right" that has been violated, citing, *inter alia*, the *Barcelona Traction* case.[56]

In a judgment of 27 November 1998, the Inter-American Court discussed the different forms of reparation,[57] quoting several judgments of the World Court concerning the rule of *restitutio in integrum*[58] and compensation for material as well as moral damage.[59]

6. Nationality

An advisory opinion of 19 January 1984 of the Inter-American Court, requested by the Government of Costa-Rica,[60] concerned certain aspects of the right of States to grant their nationality. The Court stated that it is

---

[54] *Reparation for Injuries Suffered in the Service of the UN*, see note 51. The IACHR's judgment in the *Aloeboetoe et al. v. Suriname*, see note 38, however, refers to a different advisory opinion of the ICJ, related to the *Interpretation of Peace Treaties with Bulgaria, Hungary and Romania* (Second Phase), ICJ Reports 1950, 228.
[55] IACHR, *Garrido and Baigorria*, see note 40.
[56] *Ibid.*, para. 40, citing *Barcelona Traction*, see note 31, 33.
[57] IACHR, *Castillo Caez* case, Judgment, 27 November 1998.
[58] *Ibid.*, para. 69, quoting *Chorzów Factory*, Jurisdiction, see note 51, 21; *Chorzów Factory*, Merits, see note 51, 29; *Reparation for Injuries Suffered in the Service of the UN*, see note 51.
[59] IACHR, *Castillo Caez*, see note 57, paras. 69 and 86, quoting *The Treaty of Neuilly, Article 179, Annex, Paragraph 4* (Interpretation), Judgment No. 3, PCIJ, Ser. A, No. 3, 1924, 9.
[60] IACHR, *Proposed Amendments to the Naturalisation Provisions of the Constitution of Costa Rica*, Advisory Opinion, OC-4/84, 19 January 1984, paras. 36 and 37.

natural that primarily the domestic law of the State should govern the conditions and procedures for acquisition of nationality. As long as such rules do not conflict with superior norms, it is the State conferring nationality that is best able to judge what conditions to impose to ensure that an effective link exists between the applicant for naturalisation and the system of values and interests of the society with which he seeks to fully associate himself. That State is also best able to decide whether the applicant has complied with these conditions. The IACHR stressed that the International Court of Justice voiced certain ideas in the *Nottebohm* case[61] which are consistent with its views.

7. Reservations

In his separate opinion joined to the judgment of the Inter-American Court of Human Rights in the *Nicolas Chapman Blake* case,[62] Judge Cançado Trindade recalled the system of reservations enshrined in the two Vienna Conventions on the Law of Treaties. This system was inspired by the criteria established by the International Court of Justice in its advisory opinion of 1951 on *Reservations to the Convention against Genocide*.[63] According to Judge Cançado Trindade, in that advisory opinion the International Court of Justice endorsed the so-called Pan-American practice relating to reservation in treaties, due to its flexibility and in search of a certain integrity of the text of the treaty and the universality of participation in it. This resulted in the criterion of the compatibility of the reservation with the object and purpose of the treaties.

8. Interpretation of Treaties

In its advisory opinion on the *Interpretation of the American Declaration of the Rights and Duties of Man within the Framework of Article 64 of the American Convention on Human Rights*,[64] the Inter-American Court recalled a fundamental rule stated by the International Court of Justice: "an international instrument must be interpreted and applied within the overall framework of the juridical system in force at

---

[61] *Nottebohm* case (Second Phase), Judgment, 6 April 1955, ICJ Reports 1955, 24.
[62] IACHR, see note 44. See further the discussion in the judgment in *Blake*, see note 53, paras. 9-33.
[63] ICJ Reports 1951, 15-30.
[64] IACHR, *Interpretation*, see note 42.

the time of the interpretation".[65] Hence, it concluded that to determine the legal status of the American Declaration it is appropriate to look to the present inter-American system in the light of the evolution it has undergone since the adoption of the Declaration, rather than to examine the normative value and significance which that instrument was believed to have had in 1948.

Judge Verdross pointed to a limitation on the interpretative powers of international courts in his separate opinion joined to the judgment of the ECHR in the *Ringeisen* case.[66] He affirmed that, like the International Court of Justice, it is the duty of the European Court "to interpret the treaties, not to revise them".[67]

In the case *Relating to Certain Aspects of the Law on the Use of Languages in Education in Belgium*,[68] the ECHR responded to a contention of the Belgian Government that the Court had no jurisdiction to pronounce on the merits of the case, because first it had to give a decision on the applicability of the European Convention on Human Rights and its First Protocol. The Court considered that it should, if need be, adopt the method followed by the Permanent Court of International Justice in its judgment in the *Mavrommatis* case,[69] when, before ruling on the merits of the case, it verified that the dispute fell within the treaty invoked. The use of this method was justified by the principle of economy of proceedings, by the logical sequence in which the various questions arise, and by the fact that the European Court, like the World Court, has only an attributed jurisdiction, derived purely from the consent of States.[70] The European Court also referred to the World Court's jurisprudence concerning questions of interpretation and application of treaties that are inseparable from the merits.[71]

In some of its most important decisions interpreting the American Convention, the IACHR considered whether a State Party to the Convention may withdraw from the jurisdiction of the Court without denouncing the treaty as a whole.[72] The IACHR relied on the

---

[65] *South West Africa* case, see note 11, 31, cited in IACHR, *Interpretation*, see note 42, para. 37.
[66] ECHR, Judgment, 16 July 1971.
[67] *Interpretation of Peace Treaties*, see note 11, 228.
[68] Case *Relating to Certain Aspects of the Laws on the Use of Languages in Education in Belgium*, Preliminary objection, 9 February 1967, ECHR, Series A, 5, 16.
[69] *Mavrommatis Palestine Concessions* case, Judgment No. 2, PCIJ, Ser. A, No. 2, 1924, 34.
[70] ECHR, *Languages in Education in Belgium*, see note 68, para. 1.
[71] *Electricity Company of Sofia and Bulgaria*, Judgment, 4 April 1939, PCIJ, Ser. A/B, No. 77, 83.
[72] IACHR, *Constitutional Court* case (Competence), Judgment, 24 September 1999, Ser. C No. 55, 50; IACHR, *Ivcher Bronstein*, see note 47, paras. 52-53.

jurisprudence of the ICJ in holding that the declaration of recognition of the contentious jurisdiction of an international tribunal, once made, does not give the State the authority to change its content and scope at will at some later date. It referred specifically to the *Nuclear Tests* cases[73] and the *Nicaragua* case[74] at the ICJ.

9. Procedural Rules

It might be considered surprising that more than one-third of the judgments adopted by the two regional courts of human rights that refer to decisions by the World Court concern procedural issues. One possible explanation is that both tribunals remain in the framework of the conventions on human rights which created them, while there are quite a few unsolved procedural questions where the practice of the World Court as the first international tribunal could be helpful.

In several cases, the Inter-American Court held that the failure to observe certain formalities required by domestic law is not necessarily relevant to cases heard on the international plane. It has stated that "what is essential is that the conditions necessary for the preservation of the procedural rights of the parties be not diminished or unbalanced, and that the objectives of the different procedures be met".[75] It quoted three judgments of the Hague Court which stated, in two instances, that the Court, whose jurisdiction is international, is not bound to attach to matters of form the same degree of importance that they might possess in municipal law.[76] The American Court then concluded that its task was to determine whether the essential points implicit in the procedural norms contained in the Convention have been observed.

The powers of the judges were discussed in the judgment of the Inter-American Court in the *Honduran* cases.[77] The American Commission had asked the Court to find that Honduras had violated the

---

[73] *Nuclear Tests* case *(Australia v. France)*, Judgment, 20 December 1974, ICJ Reports 1974, 268, para. 46; *Nuclear Tests* case *(New Zealand v. France)*, Judgment, 20 December 1974, ICJ Reports 1974, 473 and 267, paras. 49 and 43, respectively.
[74] *Nicaragua* case, Jurisdiction and Admissibility, see note 31, para. 63; cf. paras. 59 and 60.
[75] IACHR, *Velasquez Rodriguez*, see note 50, paras. 33-34; *Castillo Petruzzi v. Peru*, 4 September 1998, para. 77; *Garrido and Baigorria*, see note 40, para. 55.
[76] *Mavrommatis Palestine Concessions*, see note 69, 34; *Legal Status of Eastern Greenland*, Judgment, PCIJ, Ser. A/B. No. 71, 1933; *Aegean Sea Continental Shelf*, Judgment, ICJ Reports 1978, para. 42.
[77] IACHR, *Honduran* cases, 20 January 1989, para. 172.

specific rights to life and personal security guaranteed by Articles 4, 5 and 7 of the Convention. The Commission, however, did not specifically allege the violation of Article 1(1) of the Convention prescribing the general obligation of States to respect rights. The Court held that this omission did not preclude it from applying the provision, because the precept contained therein constitutes the generic basis of the protection guaranteed by the Convention. The provision would also be applicable, in any case, in virtue of a general principle of law, *iura novit curia*, on which international jurisprudence has repeatedly relied and under which a court has the power and the duty to apply the juridical provisions relevant to a proceeding, even when parties do not expressly invoke them. In support of its statement, the IACHR quoted the judgment of the Permanent Court of International Justice in the *Lotus* case.[78]

Two other judgments of the Inter-American Court concerned the validity of acts performed by those involved in judicial proceedings. First, in the *Loayza Tamayo* v. *Peru* case,[79] the Peruvian Government contested the powers of representation granted by the victim to certain individuals, because they did not conform to domestic procedural requirements. The Court declared that the State's argument was not acceptable in an international court, adding that one distinctive feature of international law is that no special formalities are required to lend validity to an act. Even oral statements are valid under the law of nations.[80] In the other case, the Court held that a note sent by the Ministry of Foreign Affairs of a State to the American Commission of Human Rights, which was transmitted by its Alternative Ambassador to the Organization of American States, is binding on the concerned State.[81]

The problem of weighing evidence also led the Inter-American Court to use the jurisprudence of the International Court of Justice. In the judgments concerning the *Fairen Gabri and Solis Corales* case[82] and the *Castillo Pàez* case,[83] the Court had to determine what the standard of proof should be. Neither the Convention, the Statute of the Court, nor its Rules of Procedure speak to this matter. The IACHR recalled that international jurisprudence has recognised the power of the courts to

---

[78] Reference was made (*ibid.*, para. 172) to the *Lotus* case, Judgment No. 9, PCIJ, Ser. A, No. 10, 1927, 31.
[79] IACHR, *Loayza Tamayo* case, Judgment of Reparations (art. 67 American Convention on Human Rights), Judgment of 27 November 1998.
[80] Reference was made (*ibid.*, para. 97) to the judgment in *Legal Status of Eastern Greenland*, see note 76.
[81] *Neira Alegria et al.* v. *Peru*, Judgment, 19 January 1995, No. 67, para. 67, quoting the judgment in *Legal Status of Eastern Greenland*, see note 76.
[82] IACHR, *Fairen Gabri and Solis Corales*, 15 March 1989, para. 130.
[83] IACHR, *Castillo Pàez* (Reparations), see note 57, para. 38.

weigh the evidence freely and it has always avoided a rigid rule regarding the amount of proof necessary to support the judgment.[84] In the first of the two cases, the Inter-American Court had to decide whether press clippings could be considered as documentary evidence. It stated that because many of the proffered clippings contained public and well-known facts, which, as such, do not require proof, while others were of evidentiary value,[85] they had to be accepted insofar as they textually reproduced public statements, especially those of high-ranking officials.

In one of the first judgments it handed down, the ECHR considered the conditions for the participation of individuals during the exercise of its jurisdiction. The European Commission, at that time the first instance for receiving complaints of human rights violations, invoked in the *Lawless* case various precedents drawn from the advisory opinion procedure of the Permanent Court of International Justice and, subsequently, of the International Court of Justice.[86] Those precedents allowed observations by individuals to be submitted through the international organisations that applied for the advisory opinion and for those observations to be taken into consideration by the Court, although the Statutes of both these bodies provided that States alone may be represented. The ECHR recognised the force of such precedents, albeit none of the examples involved an individual appealing against the action of his own government. The Court stressed that the solution to this question must be sought in the special nature of the procedure laid down in the European Convention on Human Rights.[87]

The European Court of Human Rights had to face the problem of *locus standi* in a case submitted by shareholders in a limited company.[88] After stressing the risks and difficulties involved in according shareholders *locus standi* to complain of a violation of their company's

---

[84] In *Castillo Pàez* the IACHR referred to the following judgments: *Corfu Channel* case, Merits, ICJ Reports 1949; *Nicaragua* case, Merits, see note 31, paras. 29, 30, 59, 60.

[85] IACHR, *Fairen Gabri and Solis Corales*, see note 82, para. 145. Here, the Inter-American Court referred to the judgment of the ICJ in *Nicaragua*, see note 31, paras. 62-64.

[86] "The present Commission, as an international tribunal, is not bound to treat questions of form with the same degree of strictness as might be the case in municipal law (*Mavrommatis Palestine Concessions* case, PCIJ, 1924, Ser. A, No. 2, 34)". Report of the Commission, 19 December 1959, 43.

[87] ECHR, *Lawless* case, Preliminary Objections and Questions of Procedure), Judgment, 14 November 1960.

[88] ECHR, *Agrotexim and Others* v. *Greece*, Judgment, 24 October 1995, Series A, 330, para. 66.

rights, the European Court considered that the "piercing of the corporate veil" or disregarding a company's legal personality is justified only in exceptional circumstances, in particular where it is clearly established that it is impossible for the company to apply to the Convention institutions through the organs set up under its articles of incorporation or – in the event of liquidation – through its liquidators. This judgment recalls that the principle has also been confirmed with regard to the protection of companies by the International Court of Justice in the case of the *Barcelona Traction, Light and Power Company Limited*.[89]

The European Court of Human Rights also recognised in its judgment *H. v. Belgium*[90] the requirement of fair hearing. In particular, judgments of courts and tribunals should adequately state the reasons on which they are based, following in that a statement of the International Court of Justice in its advisory opinion of 12 July 1973. According to the World Court, this statement must indicate in a general way the reasoning upon which the judgment is based, but it need not enter meticulously into every claim and contention on either side.

IV. Conclusions

This short review of the practice of the two regional tribunals of human rights reveals that both the ECHR and the IACHR found inspiration in the practice of the World Court in various situations. Indeed, although both courts are legally bound by the conventions that establish them, they also are aware that international human rights law is a part of general international law – something that practitioners sometimes ignore or forget. Thus, when the regional human rights conventions do not provide the solution to a question, the best way to resolve the problem is to make reference to general rules of international law. Interestingly, such rules often concern not only substantive issues, but also procedural ones.

It is clear, however, that no formal hierarchy can be established between the World Court and the regional human rights tribunals. The superiority of the regional international tribunals is only established and must be maintained in their relations to domestic fora that have to comply with their decisions.

---

[89] See note 88, ICJ Reports 1970, see note 31, 41.
[90] ECHR, *H. v. Belgium*, 28 October 1987, joint concurring opinion of Judges Lagergren, Pettiti and MacDonald, Series A, No. 127, B, 43, citing *Application for Review of Judgment No. 158 of the United Nations Administrative Tribunal*, Advisory Opinion, ICJ Reports 1973, 166.

The relations between the International Court of Justice, on the one hand, and the European and American Courts, on the other hand, are thus mainly based on the wisdom of solutions adopted by the World Court and to some extent by comity. Such a conclusion is particularly appropriate when international lawyers of the whole world honour Judge Oda, since he largely contributed to the adoption of wise solutions by the ICJ.

Cultural and Ideological Pluralism in Public International Law

C. G. Weeramantry

The retirement from the Court of a distinguished Asian judge who also has been the longest serving member of the Court is an occasion for reflection on the extent to which the world's cultural traditions have made an input into the ideological base and governing principles of public international law. This article contains a few observations on this topic which, in the author's view, needs to be the subject of deep and ongoing research if international law is to be in the 21st century a truly universal body of learning such as the world community of this century would increasingly expect.

I. General Introduction

International law is an area of law par excellence where law and morality overlap. It has always been aspirational. It has always sought to steer a moral and teleological course through the tangled thicket of *realpolitik*. Hence the vast question of global order cannot be addressed from a purely legalistic point of view but needs an overarching moral framework. There have to be principles of general applicability which stand above the individual sovereignty of States. Lacking sheriffs or constables or soldiers to enforce it, this set of general principles needs to rely upon its own moral strength and upon its own universality and must at the same time be strong enough to prevail over the physical might of States. It must, therefore, be a set of concepts as universally supported and as deeply rooted as research can reveal. Shut out a view of its universality and you cut off the tap roots from which it derives its nourishment. Little wonder then if it is weak in the allegiance it commands.

It is a truism that international law is a discipline that was born of the idealism which saw with dismay the law of the jungle that tended to prevail among the newly emergent nation States released in that era from an overarching international legal order. Against such a background the trend one sometimes discerns towards a strict legalism in international law is to be deprecated. It constricts within a straitjacket a discipline which by its very nature needs the fresh air of liberalism. It can be truly said that the dependence of international law on philosophy is fundamental to its health, its development and its acceptability. Since the system it administers is universal, the philosophy from which it draws must be universal, garnering together in a common storehouse the wisdom to be drawn from the entire human experience.

This is a factor considerably reinforcing the need to search globally for the traditions embodying the wisdom of the whole of humanity rather than to confine oneself within the limits of any one cultural tradition, however rich. "One of the great tasks, perhaps the greatest task, weighing on modern international lawyers is to craft a universal law and legal process capable of ordering relations among diverse people with differing religions, histories, cultures, law and languages".[1]

When Article 9 of the Statute of the International Court of Justice required that the body of judges as a whole should assure the representation of "the main forms of civilization and the principal legal systems of the world", there was a clear indication of an expectation that the insights from all these systems would be fed into the developing body of international law. That has not happened thus far in any significant measure, and such input, if any, has been minuscule. Hence the need for the further studies envisioned in this article.

II. Phases in the Interrelationship between International Law and Philosophy

For the purpose of this discussion, and at the risk of oversimplification, it is useful to view our topic against five phases in the evolution and development of international law.

The first phase was mainly philosophical. In particular the natural law philosophy prevalent at the time of its modern formulation in the 16th and 17th centuries was a primary stimulus that triggered the emergence of the new discipline. Grotius and the other founding fathers were philosophers attempting to translate their philosophical ideas into norms of morally acceptable conduct for States. They did what they could within the limited framework of the knowledge available to them to search for their sources of inspiration from knowledge ancient and modern, religious and secular. It was that philosophical inspiration that

---

[1] See M. W. Janis and C. Evans (eds.), *Religion and International Law*, 1999, xi.

provided the general contours of international law and the framework for its emergence as a new discipline.

A second phase through which international law passed was the phase in which it came to be structured as a framework for the enterprise of territorial expansion and colonialism. This was sought to be accommodated within the basic moral framework within which it was conceived, but a number of self-serving rules suitable to a club of expansionist nations came to be incorporated within it. International law as thus adapted from its original natural law contours served well the needs of world empires. It allowed for the acquisition under law of territories that had for centuries been under indigenous occupation and rule. It was able to cast a cloak of legality over the subjugation of conquered peoples, the appropriation of their wealth and even such practices as the slave trade. These were concepts which would have been anathema to the moral framework within which the discipline was conceived.

International law in this second phase continued for some centuries until the age of traditional empires came to an end.

There then commenced a third phase in which a large number of new members of the family of nations emerged from the state of non-recognition to which international law had consigned them for centuries and began to claim equality of rights with all the prior members of the international club that had dominated the world in the imperial era.

It became necessary in this context to rethink many of the self-serving rules to which international law had given its sanction in the past, and to construct an ostensibly universalist body of norms. Theoretically at any rate, it was necessary to jettison concepts of privilege, which the traditional members of the club had claimed for themselves in the past.

This trend commenced after World War I but really moved into full swing after World War II, for the spirit of empire was still alive and well even after World War I. International law thus entered what may be described as a third phase.

However, there was a pronounced limiting factor retarding the move towards universalism. The momentum towards full equality of the new members of the world community of States was impeded and overshadowed by the ideological confrontation that now emerged between capitalism and socialism. These two ideologies tended to block out a vision of the numerous other traditions and ideologies which should in this new phase have made their impact upon the world of international law. In the world of realpolitik the influence wielded by these two ideologies was out of all proportion to the share of these ideologies in the totality of the world's ideological and cultural inheritance. That share was minuscule, but these ideologies bestrode the world stage like competing colossi. The obvious historical opportunity for other world cultures to make their impact on the content of international law was thus lost by the dominance of the two rival philosophies which overshadowed all else.

We have now entered a fourth phase, with the cessation of the Cold War. We are not in a stage now of confrontation between two powerful ideologies nor are we dominated by their ideological conflict. One would think therefore that the time has at last emerged for all the rich cultural traditions of the world to make their impact upon the body of international law. However, the voice of these other cultural traditions is muted and the stage has been taken over by the capitalist ideology, which tends to dominate all departments of state activity in both theory and practice It tends to assess all institutions, customs and values by the yardstick of profits and profitability, without which their worth tends to be denied and they face oblivion. Its mercantile ethic is fast tending to become the measure of all things and has been freed of the powerful challenge it faced during the Cold War. In a needy world it tends to drown out other perceptions, standards and values.

The rich cultural inheritances which can contribute so many perspectives, values and norms to international law continue to remain largely untapped as they have been for the past four centuries. It is time the scholars of the world started delving into the rich veins of untapped experience and wisdom contained in these mines of traditions.

The fifth phase, yet before us, will be the phase beyond globalisation. The inadequacies of globalisation, the paucity of its philosophical base, the inability of earth resources to sustain its requisite of continuous expansion, its monolithic nature and the tendency it breeds of accentuating economic divisions both domestically and globally will all combine to force upon the scholarly community a consideration of alternatives, and out of this will emerge a new realisation of the importance of making international law a truly multicultural system drawing on the richness of the universal cultural inheritance. That phase may be closer than we tend to think, and with it will come a blossoming of international law to reflect its multicultural background. The present author sees signs of this around him, especially in the reawakened vigour of the scholarship of the developing world.

This phase is especially interesting and challenging because the revolutionary force that will bring about this fundamental change will come not from armed might or economic force but from the world of scholarship. It is scholars alone who will be able to illuminate the principles which lie at the foundation of international law and show how universal they are. It is scholars alone who can stimulate a wider popular perception of these truths.

The responsibility lying upon the world of scholarship is therefore enormous. The challenge has not been taken up in many quarters but there are hopeful signs of that stirring of scholarly interest which will be

1. The First Phase

Natural law has performed a number of functions in the course of its voyage down the centuries. It ranged from building the Roman Civil Law into a cosmopolitan system to enabling the American judges to interpret the constitution. Not the least of its many achievements has been the role it played in the evolution of international law by giving it a transcendental perspective reaching beyond the particular rules and self-interests of individual States.

In the expressive words of Julius Stone, natural law was "the vessel in which various criteria of justice sailed on to western horizons".[2] Modern international law was one of the principal destinations to which the vessel sailed and to which it delivered its goods.

The cargo came from many sources. There was from classical Greece the Aristotelian idea that "if the written law tells against our case, clearly we must appeal to the universal law and insist on its greater equity and justice".[3] That Aristotelian cargo, incidentally, was not picked up directly from Greece but from Arab ports of call whither it had found its way through the work of Islamic philosophers such as Avicenna (Ibn Sina, b. 980 AD) and Averroes (Ibn Rushid, 1126-1198 AD).

From ancient Rome came Cicero's classic description which no doubt exercised a profound influence on the minds of the founding fathers: "There is in fact a true law, namely right reason, which is in accordance with nature, applies to all men and is unchangeable and eternal... It will not lay down one rule at Rome and another at Athens, nor will it be one rule today and another tomorrow. But there will be one law eternal and unchangeable binding at all times and upon all peoples".[4]

The horizontal spread of this universalism was complemented by the vertical penetration from the highest levels down to all subject classes through the Judaeo-Christian idea[5] of a law coming directly down to every layer of society from God himself.

This idea was further elaborated by St. Augustine (354-430 AD) who arranged law in three tiers, *lex temporalis*, *lex naturalis*, and *lex aeterna*. Natural law thus obtained a position well above enacted or state

---

[2] See J. Stone, *Human Law and Human Justice*, 1965, 36.
[3] *Ibid.*, 39, citing *Rhetorics* I, 15.
[4] *Ibid.*, 41-42, citing *De Republica* III, 22.
[5] Stone, see note 2, 43.

law and somewhere close to the eternal law, which is the reason or will of God. The element of reason received further emphasis from Thomas Aquinas. Eternal law was the incorporation of divine wisdom and natural law that part of it which revealed itself in natural reason. All positive law enacted by human authority must keep within the divine limits.[6]

Nor must we lose sight of the effect of Islam's philosophical thought upon the work of Aquinas and the Christian scholars of his age. Scholars such as Averroes and Avicenna in their work on the doctrine of double truth grappled with the problem of the place of human reason in the context of divine revelation – a problem faced by Islam and Christianity alike. They worked out a rapprochement between the two categories of knowledge and gave a liberal place to human reason, which was thus freed of the shackles which had confined it in the past. The freedom of inquiry thus achieved passed on to the Christian world in one of the most interesting intellectual movements of history and stimulated the work of scholars such as Aquinas, thus giving freer rein to human inquiry than it had enjoyed in the past.

These were some of the ideas that, in the hands of Grotius and others, provided the driving philosophical force behind the evolution of international law. The idea of an overarching law based upon higher principles standing above the specific laws of specific States dominated this first phase in the history of international law. This system was conceived as standing above the mundane economic, political and military considerations which shape the conduct of States.

There were thus many builders of the edifice of the natural law, which Grotius took over as the foundation of his new system.

Grotius selected from all these ideas those which could best build a bridge between the medieval world and the modern secular world.[7] From the philosophies of the past a structure was erected under which States were to conduct their affairs with due regard to universally accepted standards of morality and reason. The whole edifice rested upon the bedrock of philosophy and idealism.

2. The Second Phase

The second phase based itself on the proposition that it was not possible to work out a universal system in the sense that it represented a universal set of norms and aspirations. Rather in the manner in which Chinese and Roman civilisations saw themselves as the cultured section of the world population and all the others as being in an outside realm "without the law" or as "barbarians", so also the international law of this era gave up

---

[6] *Ibid.*, 44.
[7] *Ibid.*, 68.

any attempt to find a system based upon universal sentiments. An often quoted passage from Henry Wheaton reads as follows: "Is there a uniform law of nations? There certainly is not the same one for all the nations and States of the world. The public law, with slight exceptions, has always been, and still is, limited to the civilized and Christian people of Europe or to those of European origin".[8]

There could not therefore be a universal natural law of overarching authority but an imposed law, which governed "civilised" and "uncivilised" people alike, not because it was accepted by the latter but because those who thought they were in the former group considered it essential for all, especially as in their uninformed view the latter had none.

Statements such as Wheaton's tend in one form or another to be re-echoed even in modern writings. This is a strongly alienating factor taking away from the discipline of international law that unqualified universality of allegiance which it should command if it is to be an effective system in the 21st century. Thus, for example, it has been claimed even of international environmental law that it is rooted in basic Judeo-Christian values, as is the concept of the common heritage of mankind. Such statements ignore the particular richness of traditional legal systems in regard to principles which today fall within the rubric of international environmental law. Modern international law needs to acknowledge the wide range of world cultural traditions which, for example, in the field of environmental law have shown much more understanding of the environment than modern Western law.

The international legal thinking of this era exhibited many shortcomings. Among these were:

1. positivism
2. unawareness of other cultures
3. an emphasis on sovereignty
4. a failure to appreciate the large number of basic international law principles on which there would be a common agreement across the cultures
5. a total insensitivity to the rights of other peoples
6. the imposition upon other cultures of forms and rituals – and indeed of terminology – of Western law.

Much could be written on each of these, but considerations of space prevent this.

---

[8] See H. Wheaton, *Elements of International Law*, G. Grafton Wilson (ed.), The Carnegie Institute of Washington, rpt. 1964, 15.

3. The Third Phase

As already observed, during the Cold War two ideologies commanded an influence out of all proportion to the share of these ideologies in the totality of the world's ideological and cultural inheritance. The gentler ideologies – the pacifist love-thy-neighbour ideology of Christianity, the Buddhist ideologies of compassion for all beings, the Islamic ideology that all humans are as equal as the teeth of a comb irrespective of colour, race, caste or creed, the warm and sharing fellowship of the Pacific, the African ideology that the human family consists not only of those existing in the here and now but also those who went before us and those who are yet to come – all these and many more were submerged in the political reality that power rested with the two ideologies first mentioned. In the result the potentially rich contribution of these other ideologies to international law tended to be overshadowed and inhibited. This writer believes that the dampening effect of this ideological confrontation on the contribution of the traditional ideologies to international law was significant.

The dominance of those two ideologies meant also a preoccupation with military considerations owing to the resultant power struggle. This obstructed the progress of public international law in the directions most needed for the betterment of humanity. The advancement of areas such as peace, the common heritage of mankind, the environment and social, economic and cultural rights was retarded. This was especially so in view of the negative impact of arms expenditure upon the expenditure of resources in pursuit of higher values such as education, culture, human rights and development. Development and all the human rights connected with it depend heavily on the international financial system. With that system heavily geared towards armaments, the material resources and human effort needed for the provision of basic human rights were diverted (as indeed they still are) to the tune of over a billion dollars a day. This wasteful channelling of resources tended to be taken as a fixed and immutable feature of the international landscape, thus diverting attention from the fundamental need to mould the principles of the international rule of law into an affirmative instrument for global peace.

The ideologies of free enterprise placed the manufacture of weapons within the hands of private enterprise and indeed gave them a stake in international unrest and tension. The armaments industry was thus a factor constantly militating against global peace and exercising a strong inhibiting influence over the processes of tension resolution. The power of the military-industrial-bureaucratic complex over politicians and policies kept growing through the era of the Cold War, with a vastly negative impact on the issues that were really important for the betterment of the human condition. A reflection of the effect of these attitudes on the world of international law is the fact that, to this day, a blind spot exists in international law and human rights in regard to the armaments trade, which flourishes despite its violation of every known canon of human rights. An even more extreme example is the global trade in instruments of torture by 150 firms which, according to the most recent report of Amnesty International released in March 2001, are openly marketed by firms throughout the world and have become a global business.

In the socialist system likewise, increasing militarisation meant that so much the less resources were available for development. While it was true that private profit from the manufacture of weapons was not possible, military needs often took precedence over urgent requirements in such fields as health, education and overseas aid. Those areas of international law which relate to development thus received less attention than they would otherwise have attracted and it was inevitable that the universalist norms that should have been building up in this field of international law were compelled to yield, in case of conflict, to the exigencies of the power confrontation.

Apart from the military confrontation there was a struggle between both systems to win the allegiance of different countries. Much of the world came under the protective umbrella of one or other of the two ideologies and those who received these benefits had naturally to lend their support to these ideologies, even at the cost of their traditional value systems and cultural traditions. With the newly emergent countries thus drawn into one camp or the other, there was little opportunity for them to make a real cross-cultural input into the content of international law. Just as powerful pressures often impaired their independence, so also these pressures muted the forces of cultural independence which might otherwise have been more productive.

Economic development and respect for human rights are combined together in Article 55 of the UN Charter as necessary for peaceful and friendly relations among nations. Economic development and human rights are thus interlinked and interdependent, and they are both firmly wedded to the notion of peace. They were both special victims of the superpower confrontation.

This conflict of two powerful ideologies was not without its affirmative legal effects, however, for the interaction of each ideological tradition upon the other produced a composite result which took us further than either ideology on its own could have achieved. The ideology of the welfare state was probably one such result. Each competing world view achieved certain modifications in its doctrines and applications as far as they could be accommodated without compromising its basic philosophical tenets. This mutual leavening influence was largely lost upon termination of the Cold War.

Also, despite the restraining factors of two preponderant ideological systems, the new membership of the family of nations was able to introduce into international law certain basic new concepts which, but for their influence, may not have received such recognition,. This is an important indication of the full contribution they could make if international law were freed of the inhibitions referred to.

One of these areas in which the ideologies of the newly emergent nations made an important impact was in regard to the elevation of social, economic and cultural rights. Ideologies based upon the right of the individual against the State tended to concentrate upon the civil and political rights of the individual. This was indeed the almost exclusive concentration of Western oriented systems until after World War II. The view that human rights reach into the social, economic and cultural areas at least as intensely as into the civil and political areas was widely shared by the ideologies of many of the newly emergent nations. It is true that socialist thinking played a very significant role in achieving a realisation of the essentiality of these rights, yet it needed the multicultural input of many newly independent States to achieve this breakthrough. The Covenants on Economic, Social and Cultural Rights and on Civil and Political Rights, both passed in 1966, embodied perhaps for the first time an international acceptance of their co-equal nature. Indeed the interval of 18 years between the Universal Declaration and these twin Covenants was an indication of the difficulties encountered in bringing about a recognition of these rights.

4. The Fourth Phase

With the end of the Cold War the power confrontation eased, to leave only one superpower in a situation of dominance in world affairs. The monopolistic power situation was matched by a monopolistic ideological situation for, with the demise of the Soviet Union, the philosophy of the open market place did not any longer need to compete with the philosophy of socialism for world dominance. It enjoyed a free run and with this came a trend towards globalisation in all areas of activity. The free trade principle began to take over in areas into which it might not otherwise have had entry. In association with global financial institutions it began to enter even such areas as education, health care and public utility services. The World Bank, the IMF, the ADB and the very powerful World Trade Organization began to lay down principles by which various countries were required to direct their governmental and economic activity, whatever their traditional philosophies might have indicated, in such fields as education and health care. Globalisation, driven by the economic imperative, submerged other inputs into international law.

To take one example, if principles of environmental protection required one form of action and principles of free trade required another, the balance tended to tilt in favour of the latter and this trend manifested itself even in decisions of WTO tribunals. The dice came to be heavily weighted against all principles and philosophies other than free trade.[9]

However this is only a passing phase, for globalisation has served the important purpose of forcing people all over the world to look inwards to their own cultures to see what they contain; globalisation and the commercial ethic lying behind it lack this element. People also see globalisation as a form of neo-imperialism which, for that very reason, they tend to counter by resort toother philosophical systems than the commercial.

In the field of international law scholars throughout the world, especially in the developing countries, are now searching afresh for the real bases of international law which lie beyond the confines of any one culture or any one approach to global problems – and especially beyond the mercenary ethic which often smothers higher ethical codes on which international law must rest. We are today witnessing the emergence of this multicultural approach.

The potential of other cultures to reinforce the basic principles of public international law is enormous, as will be noted in the ensuing discussion.

5. The Fifth Phase

This is the phase that lies ahead, the commencement of which is long overdue and the flowering of which will give new life and vigour to international law. It is an area awaiting an input from cross-cultural and interdisciplinary scholarship which, hopefully, will delve deep into these sources and uncover much international law-related material which has not hitherto been brought to the notice of the global international law community. In addition to specific principles, philosophies and attitudes of entire cultures towards some of the fundamental postulates of the international legal system will also be revealed and thus an indication of possible directions for development in the future.

I shall here address the potential of various cultures to contribute to the development of public international law, drawing attention to past contributions and to elements which can be of service in the future.

The different philosophies comprising the cultural inheritance of mankind have, in various ways which are not sufficiently recognized,

---

[9] See J. H. Jackson, "World Trade Rules and Environmental Policies: Congruence or Conflict?", *The Jurisprudence of GATT and the WTO*, 2000, 414-448.

already made contributions to the current content and development of international law. The reason for this non-recognition is a belief still current among many jurists that the substance of public international law is almost in its entirety a creation of the Western Christian tradition.

Eight hundred years before Grotius, treatises on public international law appeared in the Islamic world as we would now describe it.[10] These contained discussions on the rights of captives, humanitarian rules of warfare, diplomatic protection, contracts with enemies, the honouring of treaties, the recognition of sovereign States and numerous other matters pre-eminently the province of public international law. Islam looks upon all humans as being "as equal as the teeth of a comb". In the farewell sermon of the Prophet of Islam, one of the great human rights documents in world history, he stressed, "the Arab is not superior to the non-Arab, and the non-Arab is not superior to the Arab. Nor is the fair-skinned superior to the dark-skinned, nor the dark-skinned superior to the fair-skinned".[11] We also know that the Prophet as head of the Islamic State himself received foreign embassies with great honour in his personal mosque, wearing special robes for the occasion.[12] All of these – Qu'ranic passages as well as the traditions of the Prophet – were the subject of elaborate commentaries by the jurists and were assembled together in treatises such as that of Al-Shaybani. We also know that some of this material found its way into Western jurisprudence. For example, the monumental *Siete Partidas* of King Alfonso X of Castile, which has sections on international law, contains some such material which thus became more easily accessible to later European writers.

The extent to which Islamic thinking fertilised the Western European tradition in international law has yet to be researched, though we do know that the Crusades acted as major vehicles for the movement of ideas between East and West.

It is important to the global observance of public international law norms that there be a better understanding of the cross-cultural traffic on which this discipline is based.

Returning, however, to more modern times, the development of notions underlying the human rights concept is another fascinating field of cultural interaction. The philosophers of individual liberty such as Locke did not produce their theories full blown and unaided. They drew heavily upon the past. Locke, for example, was part of a tradition which

---

[10] See generally M. Khadduri, *Islam and the Modern Law of Nations*, AJIL 50 (1956), 358-372; *idem*, "The Islamic System: Its Competition and Co-existence with Western Systems", *Proceedings of the ASIL*, 1959, 49-52.

[11] See this writer's *Invitation to the Law*, 1982, 273, citing M. U. Akbar, *The Orations of Mohamed*, revised ed., 1966.

[12] For these and other reference, see H. R. Hussan, *The Reconstruction of Legal Thought in Islam*, 1974, 172.

went back, as recounted earlier, to the elevation of the status of human reasoning at the hands of Aquinas. From Aquinas the thread can be traced further back to the Jewish philosopher Maimonides and the Islamic philosopher Averroes and through them further back to Aristotle. The philosophy of individual liberty and the dignity of the individual thus contain within themselves the confluence of at least four major cultural streams – the Greek, the Islamic, the Jewish and the Christian. Modern human rights and the principle of the rule of law are a cross-cultural product and not the exclusive invention of any one cultural group.

Moreover, with the development of the concept of human dignity, other cultural forces from the rich mosaic of world cultural traditions are joining together in reinforcing and entrenching human rights doctrine. The cultural traditions of other civilisations – Hindu, Buddhist, Chinese, African, Melanesian, Polynesian and others – can contribute new facets to public international law. Outstanding among these is the realization that the current generation of humanity is not the exclusive repository of all legal and human rights. Generations yet unborn also have rights which must be recognised in the present if those rights are not to be destroyed for the future. Such unborn generations have little, if any, legal status in the Western tradition and any legal system, municipal or international, that is built upon that supposition, is a legal system which is necessarily incomplete. We are seeing manifestations of this in the necessities for preservation of the environment and the conservation of the earth's resources, both of which are being harmed by the current generation in a manner which negates the rights of posterity. Belatedly, public international law is taking note of this danger and is still evolving the concepts to deal with it. A rich body of cultural tradition ranging from Africa to Polynesia reminds modern-day jurists how much the formalistic concentration of rights in individuals who are alive today is a departure from global and futuristic perspectives which must be the lodestar of public international law.

These cultural traditions also remind us that just as individuals have rights, so also have groups and that while the concept of human rights for the individual is admirable, one must not lose sight of the rights of groups. All too often the exclusive concentration on the individual tends to fragment and scatter the groupings which in the past cushioned and protected the individual. In that portion of public international law which deals with human rights, we need a continual reminder of the dangers of depriving the individual of these group protections and of thereby dissolving the cement which held societies together in the past. The task of balancing individual and group rights is thus an area of public international law where the world's cultural traditions can make a significant contribution.[13]

---

[13] On group rights in Third World society, see C. G. Weeramantry, *Equality of Freedom: Some Third World Perspectives*, 1976, 20-23.

Given the strength in the modern world of religious traditions such as the Buddhist, Christian, Hindu and Islamic, and given that they command the allegiance of over three billion members of the world's population, there cannot be any doubt that future thinking on international law can benefit deeply from the teachings contained in these traditions. The cynic may say that in a world of harsh practical realities such religious teachings can find no place, but it is a reality of the geo-political scene that these religions command deep allegiance on a massive scale. Their moral authority is immense. Consequently their impact upon the world of practical politics, whether domestic or international, cannot be gainsaid. It would indeed be unrealistic to look at the practical world scene without making due allowance for the power of these religions.

This becomes especially important in public international law, where the interaction between the moral principle and the legal precept becomes particularly compelling in such contexts as nuclear war, genocide, slavery, torture and fair contractual dealing in international trade. Such topics form an important part of the corpus of public international law.

The fact that the world has many cultures and that each culture is many faceted does not preclude certain underlying norms receiving universal recognition. Locke, who wrote in his "Essay Concerning Human Understanding" that scarcely any principle of morality or rule or virtue accepted as necessary by one society was not condemned at some time by other societies,[14] himself formulated certain principles of human rights intended to be of universal applicability. The framers of the Universal Declaration of Human Rights, while conscious of cultural diversities, were fortunately not put off by them in their quest for universal norms. We must bear these in mind in the sphere of international law, which is feeling the impact of cultural traditions that had not directly affected it during the period between Grotius and the emergence of the United Nations. The magnitude of the change between that period and ours can be appreciated when we consider that it was with some difficulty that Turkey came to be admitted into the "club" of European powers that was laying down general principles of international law.

The world scene today is so markedly different from that which prevailed at the end of the 19th century that accommodation must necessarily be made for cultural and ideological pluralism. International law needs to mount a special effort in this direction, especially as the inheritance from the 19th century public international law modes of thinking still pervades some areas of international law. Unless there is a

---

[14] See J. Locke, *An Essay Concerning Human Understanding*, I.ii.10.

special awareness of the importance of maintaining the concept of cultural and ideological pluralism, there is an ever-present danger of its suffocation in the midst of inherited 19th century thought patterns. The cause of pluralism is not served by despairing reflections on the apparent irreconcilability of divergent traditions in different countries.

III. Concepts of Traditional Law Which Would Enrich Modern International Law

Numerous developing areas of international law could profit greatly from some of the perspectives ingrained in traditional legal systems which have been obscured by having superimposed upon them the individually oriented and contemporaneously focussed legal systems of the West.

With no attempt at being comprehensive, the following are offered as a specimen sampling of concepts and areas of study where a knowledge of such legal systems could be especially rewarding:

Intergenerational rights
Obligations *erga omnes*
The common heritage of humanity
Sustainable development
Trusteeship of land
Sanctity of land
Environmental protection
Group rights
The notion of group duties
The subjection of state sovereignty to a higher law
The notion of the oneness of humanity
Friendly relations between States
The inherent dignity of every human being
The notion of peace
Non-violence
Reconciliation
World order
Active cooperation as opposed to passive coexistence
Cross-cultural understanding
Industrial relations
Humanitarian rules of warfare
The sanctity of treaties
Diplomatic representation
The problem of violence
Non-aggression
Peaceful coexistence

The psychology of conflict
The causes of conflict
The resolution of conflict

I will now proceed to make some observations about major global traditions which have not yet been sufficiently used to enrich and invigorate public international law.

1. Buddhism

Buddhism is a profound source of support for many of the fundamental principles of international law. Unfortunately this body of learning has not yet been sufficiently researched from the standpoint of international law, with a few notable exceptions.[15]

The deeply philosophical approach of Buddhism not unsurprisingly gives us many psychological insights into the causes of war. The mental attitudes that lead to conflict are minutely analysed, ranging from righteous indignation to profound hatred. Righteous indignation, which is one of the principal causes of conflict, can very easily be debased and during the course of conflict does in fact become so debased. The importance of peaceful resolutions of disputes and the various ways in which peaceful resolution can be achieved are also discussed.

The causes of world disorder are examined. The question posed to the Buddha, adopted by Buddhaghosha as the opening stanza of his *Path of Purification (Visuddhi Magga)* is as follows

> "The inner tangle and the outer tangle
> This generation is entangled in a tangle
> And so I ask of Gotama[16] this question
> Who succeeds in disentangling this tangle".[17]

One way of disentangling the tangle is to know its causes, and Buddhism is unequivocal in pointing to what is probably the principal cause – the greed and craving from which not only individuals but also nations suffer. This blinds their perception of the fact that in seeking more possessions they are depriving others of their legitimate share. An

---

[15] Such as that of K. N. Jayatilleke, "The Principles of International Law in Buddhist Doctrine", *RdC* 120 (1967-I), 445-556.
[16] Gotama or Gautama was the name of The Buddha.
[17] See C. G. Weeramantry, "Some Buddhist Perspectives on International Law", *Boutros Boutros-Ghali Amicorum Discipulorumque Liber, Peace, Development, Democracy*, 1998, 782.

answer to the problem of national greed is education of the citizenry at an individual level.

According to Buddhist teaching one of the key concepts towards international peace is an equitable distribution of the world's resources. An equitable world order cannot, in other words, be achieved without an equitable economic order.

Buddhism was among the first major belief systems to elaborate the theme of the equality of all humans in considerable detail. The unity of the human species and the lack of any material distinction between its members thus go back over 2,500 years in world tradition. Nor was this a teaching of equality in the limited Greek sense of equality among free citizens with the slave population being left rightless. Buddhism drew no distinction of this sort between members of the human family. The concept of equality was expounded and explained on ethical,[18] legal,[19] moral,[20] religious,[21] sociological,[22] anthropological[23] and biological[24] grounds.[25]

Every individual member of the human family owed duties to every other.

The advice of the Buddha was in fact sought by many rulers of his time. He was therefore called upon to apply his teachings to actual political facts and circumstances. The duties of kings, the welfare of subjects, the resolution of hostilities, equitable adjudication of the matter in contention – all of these received his attention.

The absence of dualism in Buddhism is of great importance for international law. No law of any individual sovereign had validity if it offended the basic principles as taught in the *Dhamma*. There was a profound universalism that swept through his teaching and this affords enormous reinforcement to the fundamentals of human rights and international law. International law has almost totally ignored the wealth of insights available from Buddhism to strengthen its fundamental principles and increase their universal acceptance.

One reason for this may be the enormous extent of the Buddhist scriptures ranging through multiple volumes of discourses: the *Digha*

---

[18] *Digha Nikāya*, Vol. IV, Part III, 173, 1995, translated by T. W. and C. A. F. Rhys Davids.

[19] *Majjihima Nikāya*, Vol. II, Pali Text Society Translation Series No. 30, 113-114, 1997, translated by I. B. Horner.

[20] *Ibid.*, 224 and 299.

[21] *Ibid.*, 128.

[22] *Digha Nikāya*, Vol. IV, Part III, 168-184, 1995, translated by T. W. and C. A. F. Rhys Davids.

[23] *Ibid.*, 93.

[24] *Ibid.*

[25] For these and other references, see L. P. N. Perera, *Buddhism and Human Rights*, 1991, 35-36.

Nikāya (34 long discourses), the *Majjhima Nikāya* (152 middle length discourses), the *Samyutta Nikāya* (2889 short discourses), the *Anguttara Nikāya* (2308 short sayings, often in the form of maxims or aphorisms) and the *Khuddaka Nikāya* (over a 1000 sayings in the form of stanzas or aphorisms, covering the whole range of Buddhist philosophy).[26] All of these discourses and sayings, though available in English, need to be researched by international lawyers as many principles pertinent to international law, though not directly stated, lie concealed within them.

Buddhism would lend powerful support to natural law thinking and would discourage legalism. It would extol social and economic rights and offer us pointers to ways in which the emotional causes of war can be modified if not overcome.

Buddhist analyses of violence are also relevant to our studies because they deal with institutional violence, ideological violence, defensive violence, political violence and others. Ideological violence, the cause of much conflict, and terrorist violence also come in for scrutiny.

Peaceful coexistence and non-aggression, non-alignment, the tracking down of the roots of violence to their sources, the sanctity of promises, the arms trade, the futility of collective punishment, tolerance, the right of asylum, environmental protection, individual responsibility, social duties, industrial relations, integrity in trade – all of these are of deep relevance to the development of future international law and are intensely analysed in Buddhism. It is surely not possible to describe modern international law as a universal system if it neglects this huge source of support and inspiration.

2. Confucianism

As pointed out in a recent study, books on international law and order do not include any meaningful discussion of Confucius or Confucianism. This is despite the fact that Confucian teaching did have much to say on such questions as world order,[27] the world community, the principle of harmony, the importance of the moral order to legal order, the goals of the State, the goals of the world community and the principles of harmony.

The notion of *Ping* denotes peace, harmony, evenness, equality, fairness and the like. The ultimate goal of the world order is *Ping*.

Every individual is urged to attempt eight virtuous accomplishments of which the last is the *Ping* of the world. In other words it is the obligation of each individual to do what he or she can to achieve global

---

[26] Weeramantry, see note 17, 787.
[27] Janis and Evans, see note 1, 27.

harmony. The duty of every individual to attempt this eighth requirement is an integral part of Confucian teaching. The eight virtuous accomplishments, which every individual should attempt, are laid done in the *Book of Great Learning*.

Confucian teaching envisaged the world community as all under heaven, and all under heaven are treated as one family. The term *tien-hsia* (literally, all under heaven), although translated by some as referring to the kingdom of China, was extensive enough to comprehend the entire world. It was essential for a peaceful and happy China that the rest of the world under heaven also enjoyed peace and happiness.

A cardinal principle that emerges from Confucian teaching, which in the author's view is most relevant to international law, is the emphasis it places on the conduct of each individual. Modern international law tends to suffer because it confines the bulk of its attention to the conduct of States, ignoring the fact that the conduct of States is largely the impact upon those States of the individuals comprising them. There is hence a lack of emphasis on what each individual can do to advance the goal of world harmony. That tends to be looked upon as the preserve of States and not the business of the individual. This personalised approach of Confucianism is fundamental to the future of international law, for unless education systems sensitise the thinking of each individual to his or her role and importance in the world order, an improvement in the world order will not be achieved. Each State would continue to play the selfish and self-interested role that has resulted in the world problems of today

Confucian teaching is thus of cardinal value in emphasising that it is the individual human being who is the ultimate actor in all community processes. It is possibly an idea of this sort that lay behind H.G. Wells' observation that the future of humanity is "a race between education and catastrophe".

Western law has made the mistake – probably largely under the influence of positivism – of considering that Confucianism had no legal tradition as such. A well-known text on jurisprudence[28] exhibits such a misapprehension, under which generations of lawyers have been trained, when it says: "the Chinese had never had a legal tradition at least as that term is understood in the West. Legality has no roots in Chinese civilisation, law being regarded as a sign of an imperfect society". It was the total law that Confucianism contemplated and not that part of it which falls under the rubric of positive law.

It is surely the failure to distinguish the part from the whole that causes a large segment of the world's wisdom to be discarded as irrelevant to international law. There is a great contrast between

---

[28] See J. Lloyd, Introduction of Jurisprudence, 4th ed., 1979, 760-761.

positivist law and a total legal system based upon *Ping*. The error of elevating positive law to being the only law is an error into which international law, above all else, must not fall.

The rounded concept of *Li* in Chinese law contrasts strongly in this respect with enacted or positivist law as understood in Western systems. The Chinese concept covered not merely ethical or moral rules of conduct but also juridical norms and laws.[29] Confucian teaching has insights to offer on such matters as humanitarian intervention, punitive expeditions, the "right" to revolution, economic, social and cultural rights, freedom of opinion, and expression, and the honouring by rulers of what would today be described as human rights. The stress throughout is on duties but, as Leon Duguit taught in the West, one's rights fall into place when one's duties are properly performed.

It would strengthen the acceptability of the Universal Declaration of Human Rights if, for example, it were more widely realised that there is a substantial concurrence of view between the Confucians and the Universal Declaration, just as it would strengthen international law if there was a better understanding of the importance of the harmony of "all under Heaven as one family"

I conclude with a passage from the *Analects*:

> "When the Master went to the state of Wei, Zan Yu acted as driver to his carriage. The Master observed, 'How numerous are the people!' Zan Yu asked, 'Since they are thus numerous, what more shall be done for them?' 'Attend to their economic well-being', was the reply. 'And when their economic well-being has been attended to, what more shall be done?' The Master said 'Teach them'".[30]

Education of the citizenry towards global consciousness and duty towards all under heaven are perhaps the essence of the Confucian message for world order and international law.

3. Hinduism

The ancient law books of Hinduism had a finely nuanced concept of justice. Righteous or fair or just conduct was the subject of elaborate commentaries.

Dharma or righteousness stood supreme above the ruler and any law he enacted. This was so not only domestically but also on the international plane. It was theoretically possible for the whole world to

---

[29] Janis and Evans, see note 1, 31-33.
[30] *Ibid.*, 49, n. 120.

be brought under the just rule of a universal monarch. But even such a person would not be the ultimate ruler of the world. The ultimate ruler of the world would be not a physical person but "the kingless authority of the law". There could scarcely be a better description of international law and of the international rule of law than this eloquent description handed down to us more than 3000 year ago.

If we were researching the humanitarian laws of war, we would find a wealth of information in the ancient Hindu law which imposed fine limitations upon the extent to which combatants could go. There were limitations on the weapons they could use, the circumstances in which they could attack, and the person on whom an attack could be launched. The ancient books tell us that in warfare peasants working in the fields could not be attacked. Civilians had to be spared. I personally derived much inspiration when writing my Dissenting Opinion in the International Court on the illegality of nuclear weapons[31] from the ancient Hindu texts, which categorically stated that hyper-destructive weapons which went beyond the purposes of warfare could not be used lawfully. In giving modern law an overall view and a sense of direction such wisdom can be a continuing source of inspiration.

This is just one illustration of ways in which the deeply philosophical background of Hinduism can offer insights in our day and age, but many more may be cited.

Nor has Hinduism remained static in its legal development. Its volume of recorded legal literature rivals the Roman, and its modern proponents such as Sri Aurobindo and Mahatma Gandhi have gone far in developing concepts integrally related to law such as non-violence. Indeed the power of this concept in practical affairs was demonstrated by Mahatma Gandhi and played no little part in the eventual dismantling of the most widespread empire the world has seen.

Hinduism, like many other Eastern philosophies, stresses not merely external conduct but turns the searchlight of scrutiny on the internal mental state that leads to that conduct. In the words of Sri Aurobindo: "It says strongly enough, almost too strongly, 'Thou shouldst not kill', but insists more firmly on the injunction, 'thou shalt not hate, thou shalt not yield to greed, anger or malice', for these are the roots of killing".[32]

The Law of *Dharma* exempted no one and, in anticipation of the social contract theory, rulers were required to take an oath of obedience to the Laws of *Dharma*. Their right to rule was dependent on their observance of this promise.[33] Consequently the theory of sovereignty did

---

[31] *Legality of the Threat or Use of Nuclear Weapons*, Opinion Requested by the General Assembly, ICJ Reports 1996, 429-555.

[32] See S. Aurobindo, *The Foundation of Indian Culture*, Sri Aurobindo Ashram Pondicherry, 1988, 91.

[33] *Ibid.*, 18.

not give absolute power to the ruler, who was subject to this overriding rule of conduct. Unlike Austinian theory, which required an enforcing arm to give authority to a law, the Hindu concept of international law based itself on the moral authority of that law and thus placed international law well above the level of state sovereignty. The Hobbesian idea that "covenants without the sword are vain" had no place in the Hindu legal system.[34]

Not only the Code of Manu but also the text of the Mahabharata and the Ramayana and also the work of other jurists such as Naradh and Sukra exhibit an outstanding humanitarian approach in dealing with the laws of warfare. Acts prohibited include night attacks, strategies of deceit, booby traps, treachery, the molestation of non-combatants, the destruction of crops, the ill-treatment of prisoners of war and the destruction of an enemy's land by fire. Indeed the Mahabharata states that: "Enemies captured in war are not to be killed but to be treated as one's own children".[35] There was also a principle that the prisoner had to be released after a certain period and sent back to his home with a proper escort. There is evidence that this was in fact followed, for Chandragupta Maurya set free prisoners captured in war. The wounded were to be well looked after and there is evidence that regular camps were established with medicine and surgical instruments to look after the wounded on the battlefield.

At the hands of a galaxy of 20th century intellectuals the humanism and the universalism of ancient Indian philosophy has been expounded, and their relevance to the modern world demonstrated. Among these figures are Rabindranath Tagore, Mahatma Gandhi, S. Radhakishnan and Sri Aurobindho. Their philosophies and writings have received scant attention in the world of formal international law.

4. Islam

Reference has already been made to the Islamic contribution to the development of international law. This section deals in somewhat greater detail with the contents of Islamic international law.

Verses xvi: 91,92 of the Qu'ran embody the rule of *pacta sunt servanda*, beautifully elaborated in terms to the effect that fine interpretations of a treaty, like unravelling a strong thread into fine filaments, can depart from the spirit of treaties by making weak what was strong before, and should in no circumstances be used as a means of defeating its intent and becoming a basis of deceit between nations.

---

[34] *Ibid.*, 107.
[35] See *Mahabharatha*, Shanthi Parva 102.32.

Islam's philosophy of all power issuing from God makes rulers only trustees of power and not absolute wielders of power. Any dualism of national and international law is thereby rejected.

Al-Shaybani (749-805) wrote an extensive treatise on international law and Shafi'i's *Kitab-al-umm* contains an exposition of numerous principles of international law. Islamic law saw all human beings as "alike as the teeth of a comb", and there is a *hadith* of the Prophet that the "whole universe is the family of Allah...".

Humanitarian law was expounded in detail. Among acts expressly forbidden were the use of poisonous weapons, the killing of envoys, the killing of captive women, the cutting of forests and the killing of non-combatants and the destroying of harvests. Indeed concern for prisoners went so far as to ordain that their dignity was to be respected, near relatives were not to be separated and they were to be permitted to draw up their last wills, which were to be transmitted to the enemy through some appropriate means.[36] Even modern conventions on humanitarian warfare could be improved through concepts and inspiration derived from the study of such ancient writings.

On the battlefield itself honourable conduct was required. Practical examples of this are to be found in Islamic history, such as Calif Abu Bakr's instruction to his commander: "I enjoin upon you ten commandments. Remember them. Do not embezzle, do not cheat, and do not break trust...".[37]

European literature at the time of the Crusades records its surprise at the translation of some of these principles into actual practice on the field of battle. One writer, Oliverus Scholasticus, relates how the Sultan al-Malik-al-Kamil supplied a defeated Frankish army with food:

> "Who could doubt that such goodness, friendship and charity come from God? Men whose parents, sons, and daughters, brothers and sisters had died in agony at our hands, whose lands we took, whom we drove naked from their homes, revived us with their own food when we were dying of hunger and showered us with kindness even when we were in their power".[38]

With regard to the law of international trade it must not be forgotten that there was an enormous expansion of trade in the Islamic world and a growth of fresh principles of commercial law against the background of the fundamental principle of good faith. Among matters involved in such

---

[36] See S. R. Hussan *The Reconstruction of Legal Thought in Islam*, 1974, 177.
[37] See Ahmed Ibrahim, "Religious Belief and Humanitarian Law with Special Reference to Islam", *Journal of Malaysian and Comparative Law* 11 (1984), 133.
[38] See F. Heer, *The Medieval World: Europe 1100-1350*, 1968, 144.

trade were international trading treaties between States, international trading agreements between individuals, and safe conduct of traders in foreign territory.[39] The prevalence of a state of war did not necessarily mean the suspension of trade. In this context, there is the tradition that the Prophet once sent dates from Medina to Mecca and received hides in return at a time when hostilities prevailed between Medina and Mecca.

Apart from the strict legal material, there is a vast amount of information regarding treaty practice in Islamic States. The treaties entered into between the Ottoman Empire and Christian powers (such as that between Suleiman the Magnificent and King Francis I of France in 1535), those between Islamic States *inter se*, trading agreements between Moghul emperors and non-Moghul Indian rulers are some examples. A well-known early example of a treaty in relation to the large-scale exchange and ransoming of prisoners was the treaty of 804 between the Calif Harun-al-Rashid of Baghdad and the Emperor Nicephorus.[40]

Other matters covered were the law of diplomatic protection and the law of asylum. In the human rights field some of the aspects covered were the notions of fair industrial relations, human dignity, trusteeship of property, sharing, assistance to those in need or distress, abuse of rights and privacy. There is no space here to elaborate on these, but an interesting reference to the last of these is a tradition of the Prophet, which would have enormous relevance to modern times. In relation to prying into the correspondence of others, he condemned it in terms that "he who reads a letter of his brother without his permission will read it again in hell".[41] This is as eloquent a description of the privacy of correspondence as one can find anywhere in world tradition.

These few glimpses of Islamic law will, it is hoped, convey some idea of the wealth that can be gained from this system if modern international law looked into its vast literature with a view to reaping from it insights and concepts that can assist considerably in its development.

5. Judaism

The *Torah*, or *Pentateuch*, the *Prophets* and the *Writings* together constitute an enormous corpus of sacred and juridical literature, the full depth of which has yet to be researched from the standpoint of their relevance to the principles of international law. The *Talmuds* (the

---

[39] See C. G. Weeramantry, *Islamic Jurisprudence: An International Perspective*, 1988, 139.
[40] See also A. Nussbaum, *A Concise History of the Law of Nations*, 1954, 47.
[41] See M. C. Bassiouni (ed. and trans.), *The Islamic Criminal Justice System*, 1982, 69.

Babylonian and Palestinian) were immense compilations of juristic expositions notable for their thoroughness and minuteness of detail. All this literature embodies not only a deep philosophical approach but also a very imaginative and sometimes futuristic view of the problem under discussion – as where, centuries ago it discussed whether a robot with all the appearances of a human could be counted as a human for the purpose of a prayer quorum. There can be little doubt that there are many nuggets of wisdom here which are yet to be quarried.

There is no doubt regarding the substantial input of Judaic learning into international law in its formative age. Indeed if one looks at the sources cited by Grotius himself, he makes 65 references to the Old Testament as compared with 50 references to the New.

Yet there is room for concern that all the rich literature of Judaic jurisprudence has not yet been sufficiently tapped by scholarly investigation. References are mainly to Old Testament sources, and the later literature is scantily referred to. For example, one looks in vain among many of these primary writers for references to the work of the master jurist Moses Maimonides (1135-1205). Even when such material is referred to, it is often second hand – as writers like Shabtai Rosenne have pointed out.[42]

This extensive body of philosophical and analytical literature must surely contain many insights that can strengthen modern international law as it steers its way through the many complexities confronting it.

Grotius does indeed refer to post-Biblical writers such as Flavius Josephus and Philo Judaeus and so also does Gentili. Some later writers, like John Selden (1584-1654), have shown a closer acquaintance with Jewish writings but much research has yet to be done on the influence of Jewish thought upon the modern founders of international law. In the words of Rosenne: "There is here a vast field for research".[43]

As with the other religions discussed, the source material needs to be carefully perused by scholars with an eye to their international law significance. A rich harvest of concepts and discussions can be expected. Materials such as the *Jewish Antiquities* of Josephus are thought to contain a "wealth of source material awaiting critical analysis from this point of view".[44]

Other material to be researched would include, importantly, Maimonides' *Law Concerning Kings and their Wars*, not to speak of his *Guide for the Perplexed*. In the former we find references to the essential features of international law and in the latter illuminating discussions of

---

[42] See Sh. Rosenne, "The Influence of Judaism on the Development of International Law: An Assessment", in Janis and Evans, see note 1, 71.
[43] Ibid., 68.
[44] Ibid., 74.

such matters as equity and industrial relations, which have an intimate bearing on human rights.

There is also much to be learned from discussions of such topics as the pursuit of peace and the love of mankind from teachers such as Hillel (70 BC - 10 AD). His philosophy of gentleness occupies an important place in Judaic tradition, and would be an important aid to the interpretation and application of principles of international law.

It was characteristic of Jewish law that it made no sharp distinction between law and morality, viewing the two as a common system, as did other religious systems we have referred to. Out of this fusion come important perspectives which can only do lasting good to international law.

I refer finally to the teachings which placed all rulers under the authority of law and thus diminished the self-centredness of the sovereignty concept. This again would have lasting value for international law. As Maimonides taught, the king's duty was to fill the world with righteousness and whatever he did must be done by him for the sake of Heaven – a remark which reminds one of the universalism of Confucian teaching.

The words: "and they shall beat their swords into plowshares and their spears into pruning hooks; Nation shall not lift up sword against nation, neither shall they learn war any more"[45] are given a prominent place at the United Nations Headquarters in New York. They may truly be said to embody the base line of international law.

There are more than 20 centuries of juristic commentary upon passages such as these, for they go back to sacred scripture. They are well worthy of further examination from the standpoint of the basic principles which will bring this Biblical vision closer to attainment.

6. Christianity

I have left Christianity to the last because both at the inception of the era of modern international law and until at least the close of the 19th century, it was the dominant influence in international law. It is not possible to touch on its many facets of interaction with international law in the course of such a short discussion.

We have already referred to some of the thoughts of Augustine and Aquinas and to the role they played in developing concepts of natural law. In the course of the development of international law through the later centuries it played many roles in fashioning the attitudes of scholars to international law and the area of its applicability. The role

---

[45] Janis and Evans, see note 1, 83.

played by writers such as the Dominican Francisco de Victoria (1480-1546) and the Jesuit Francisco Suarez (1548-1617) constitutes another important chapter in merging international law with theology.

When international law had broken free of theology another trend became apparent which damaged the universality of international law. This continued until well into the 19th century. It pertained to the circle of nations to whom international law applied. At the hands of the most respected scholars up to the end of the 19th century, Christianity was seen as virtually providing the admission ticket to the exclusive club of independent nations. For James Lorimer (1818-1890), for example, membership of this select group required a "reciprocating will". There needed to be a mutual recognition of the existence of other States and a judgment that their right to existence corresponded to our own. This judgment became easier because governments influenced by Christianity were thought by the holders of this view to be superior to other religious or secular modes of government. "If Christianity be not the test of the presence of reciprocating will in religious creeds, what have we?"[46] For Robert Phillimore (1810-1885) as well, Christian nations deserved a privileged place in international law. "Unquestionably, however, the obligations of International Law attach with greater precision, distinctness, and accuracy to Christian States in their commerce with each other".[47] Even Oppenheim[48] could write that "the civilized states are, with only a few exceptions, Christian states".

All this did not bode well for the universalism which is essential to the acceptance of international law. It constructed an attitudinal barrier which needed to be broken through, regarding the competence of other cultures to make a contribution to international law. What is more, all these interpretations misread the simplicity and universality of the basic Christian message. They made a rule of particular exclusion out of a message of universal inclusion. That message has tremendous relevance to modern international law and to its future development. The teaching "love thy neighbour as thyself" and the elevation of this principle to the level of a pre-eminent commandment gives us a prime index of the tenderness and the sweep of the Christian message. Who is my neighbour? The parable of the Good Samaritan places in position the basic answer to this question. The concept of neighbourhood transcends all boundaries of race and religion and privilege. What is my duty to my neighbour? The parable answers this in terms of affirmative duty rather than passive toleration. One must go that extra mile to help a neighbour in distress.

---

[46] See J. Lorimer, *The Institutes of Law: A Treatise of the Principles of Jurisprudence as Determined by Nature*, 1872.
[47] See R. Phillimore, *Commentaries upon International Law*, 1855, 23-24.
[48] See Oppenheim, *International Law: A Treatise*, 1905, 10.

A careful study of the Christian scriptures and juristic writings with a view to their international law and human rights content would certainly reveal vast funds of further inspiration, which are not often brought into modern legal discussions. Among the concepts that would receive support from such a systematic study are the following:

The sanctity of treaties
Limitations on the absoluteness of state sovereignty
The brotherhood and sisterhood of all humanity
Peaceful and friendly relations between States
The concept of universalism
The concept of peace
The duty of environmental protection
The higher or universal law standing over the law of States
The principle of intergenerational fairness
Humanitarian conduct in peace and war
The criminality of genocide in any shape or form or for whatever reason
The duty of active cooperation of States
The peaceful resolution of disputes
The observance of good faith in all international dealings
The rejection of colonialism
The need for the elimination of the hatreds which are the cause of disputes
The rejection of legalism
A duty-related rather than rights-related view of human rights
Non-discrimination among all human beings
Equality of all human beings as members of a common family, irrespective of differences in race, nation, caste, creed, gender or age
The duty of affirmative action to assist those in distress
The duty to observe non-violence in the settlement of disputes
The prohibition of slavery
The condemnation of torture in every shape or form
Forced labour

With regard to the higher law this has been expounded at great length by commentators down the ages. Reference has already been made to some aspects of this in the discussion of natural law.

Another aspect of the Christian scriptures that needs to be emphasised in a study of international law is the condemnation of legalism. Christianity is particularly strong on the rejection of legalism in all its forms, and this would apply to international law as well. An important message of the Christian scriptures is that the letter of the law should yield to the spirit. Harsh and literal interpretation is to yield to softer interpretations more consonant with the object and purposes of the text. This was the basis for the severe castigation of lawyers by Jesus in at least three places in the New Testament. Jesus was protesting against formal interpretations that belied the spirit or purpose of a law and it is for these reasons that commentators have observed "lawyers generally get a bad press in the New Testament".[49] While adhering to the letter and the ritual they "have neglected the more important matters of the law – justice, mercy and faithfulness".[50]

Another aspect which needs attention is the concept of active assistance to others, as indicated in the Good Samaritan example. This duty of assistance must be taken to refer not merely to individuals but should apply to nations as well.

It would be an enormous task beyond the scope of this article to examine the vast literature that needs to be surveyed for the purposes of such studies. The writer will content himself by referring to the numerous insights that can be gained from a law-related study of even one small piece of scripture such as the Lord's Prayer. The author has attempted to do this[51] and would refer the reader to this small study as an example of the wealth of material yet awaiting research conducted with an eye to its relevance to international law.

IV. Article 38(1)(c) of the Statute of the International Court of Justice

This article has designedly been written as an overview of major world cultures. The value of a more intimate acquaintance with their traditions and literature is nothowever confined to a macro vision of the cultures concerned. It also has value at the micro level, and this assumes special importance in view of Article 38(1)(c) of the Statute of the International Court of Justice. Since international law specifically includes "the general principles of law recognized by civilized nations", a heavy burden is thrown upon international lawyers and judges to ascertain what these are and not to form their conclusions on a survey of only part of the field. The greater familiarity with other systems that is advocated in this article as a *sine qua non* of the work of the international lawyer of the future will facilitate a search for those general principles – a search which is often only a partial search under prevailing conditions.

There is not the space here to illustrate the many principles that have entered the corpus of public international law under this heading, but they range all the way from principles such as estoppel to overarching principles such as *pacta sunt servanda* and *abus de droit*, not to speak of the seminally important rules relating to interpretation of treaties.

---

[49] See K. Mason, *Constancy & Change*, 1990, 29.
[50] See Matthew 23:24.
[51] See C. G. Weeramantry, *The Lord's Prayer: Bridge to a Better World*, 1998.

The approach advocated in this article will enrich the development of international law through a more fruitful use of the comparatively neglected provisions of Article 38(1)(c).

Indeed this is a fruitful area for enlisting the discipline of comparative law in the service of public international law. In the words of a leading treatise on comparative law: "To do this would avoid reducing the valuable notion of 'general principles of law' to a mere minimum standard". We have too long been satisfied with minimum standards of research into this vital repository of concepts and principles wherewith to enrich the developing body of international law. The broader perspectives pleaded for in this article can revitalise this search.

V. Conclusion

I have attempted here to depict a few areas of possible cross-cultural and interdisciplinary research. This can be universally valuable in strengthening the integrity of the fabric of international law which, while being so extensive as to form a canopy around the entire world, must at the same time be so strong as to resist wear and tear in areas where it comes under particular resistance and tension. The task is immense because of the extensive nature of the literature involved. This literature must be subjected to searching study and examination by minds attuned to the legal nuances embedded in these teachings and traditions. It is a task for lawyers and gives them an unrivalled opportunity to use their skills for the construction of a more peaceful world order.

The task is overdue. The work is intense. The reward will be rich. A world wracked by conflict cannot afford to neglect it much longer.

XIII

DEFENCE, THE USE OF FORCE AND THE LAW OF ARMED CONFLICT

Contemporary Law Regulating Armed Conflict at Sea

L. C. Vohrah
Kelly Dawn Askin
Daryl A. Mundis

I. Introduction

Technological advances in the 20th century have had a significant impact on the means and methods of conducting warfare. Yet, the international law regulating armed conflict at sea has not been updated for nearly a century and thus the applicable law is largely customary international law. Moreover, it appears highly unlikely that an international conference will be convened in the near future to address this situation although academics periodically call for the negotiation of a comprehensive treaty governing naval warfare.[1]

The primary instruments regulating armed conflict at sea are the Hague Conventions of 1907.[2] However, as several prominent experts on armed conflict at sea have noted, the 1907 treaties, with a few important exceptions, have very little continuing significance in modern naval warfare.[3] Other treaties, including the 1949 Geneva Conventions and the

---

[1] See for example, N. Ronzitti, "The Crisis of the Traditional Law Regulating International Armed Conflicts at Sea and the Need for Its Revision", in N. Ronzitti (ed.), *The Law of Naval Warfare*, 1988, 1 et seq.

[2] Hague Convention [VI] Relating to Status of Enemy Merchant Ships at the Outbreak of Hostilities, 205 CTS 305; Hague Convention [VII] Relating to the Conversion of Merchant-ships into Warships, 205 CTS 319; Hague Convention [VIII] Relating to the Laying of Automatic Submarine Contact Mines, 205 CTS 331; Hague Convention [IX] Respecting Bombardment by Naval Forces in Time of War, 205 CTS 345; Hague Convention [XI] Relating to Certain Restrictions on the Right of Capture in Maritime War, 205 CTS 367; and Hague Convention [XIII] Respecting the Rights and Duties of Neutral Powers in Maritime War, 205 CTS 395, 19 October 1907.

[3] See, for example, J. A. Roach, "The Law of Naval Warfare at the Turn of Two Centuries", *AJIL* 94 (2000), 64 et seq., 65:

> "Of the seven treaties relating to naval operations adopted by the second Hague Peace Conference in 1907, only Convention (No. VIII) Relative to the Laying of Automatic Submarine Contact Mines, and portions of Nos.

1977 Additional Protocols to the 1949 Geneva Conventions, supplement the 1907 agreements.[4] Although the Second Geneva Convention does not regulate the means and methods of sea warfare, it does provide important protection to wounded, sick and shipwrecked members of armed forces at sea.[5] Finally, although the 1958 and 1982 Conventions on the Law of the Sea primarily regulate the uses of the seas during times of peace, these treaties impact upon the use of the seas by non-belligerent States, and establish the maritime zones in which naval forces operate.[6]

XI and XIII, respectively concerning certain restrictions regarding the exercise of the right of capture in naval war and the rights and duties of neutral powers in naval war, have continuing significance". (footnotes excluded)

See also C. J. Greenwood, *International Humanitarian Law (Laws of War): Revised Report for the Centennial Commemoration of the First Hague Peace Conference 1899*, May 1999, 59, para. 4.43 (on file with the authors):

"While [the 1907] treaties lay down a detailed code of rules, they have proved to be far from satisfactory. Parts were already anachronistic when they were drafted and they were largely disregarded in both World Wars, when the doctrine of reprisals was invoked to justify widespread departures from their provisions".

[4] See especially Geneva Convention (II) for the Amelioration of the Condition of Wounded, Sick and Shipwrecked Members of Armed Forces at Sea, 12 August 1949, 75 UNTS 85, 6 UST 3217; Geneva Convention (IV) Relative to the Protection of Civilian Persons in Time of War, 12 August 1949, 75 UNTS 287, 6 UST 3516; Protocol (I) Additional to the Geneva Conventions of 12 August 1949 and Relating to the Protection of Victims of International Armed Conflicts, 8 June 1977, 1125 UNTS 3; Protocol (II) Additional to the Geneva Conventions of 12 August 1949 and Relating to the Protection of Victims of Non-International Armed Conflicts, 8 June 1977, 1125 UNTS 609. Certain protection is provided to wounded, sick, or shipwrecked belligerents at sea, regardless of the nature of the armed conflict as international or internal. See, e.g., *Prosecutor* v. *Duško Tadić*, Opinion and Judgment, Case No. IT-94-1-T, 7 May 1997, 230. (Common Article 3 of the Geneva Conventions extends protection to "at the least, all of those protected persons covered by the grave breaches regime applicable to conflicts of an international character: civilians, prisoners of war, wounded and sick members of the armed forces in the field and wounded sick and shipwrecked members of the armed forces at sea".)

[5] See discussion in J. F. Rezek, "Protection of the Victims of Armed Conflicts: Wounded, Sick and Shipwrecked Persons", in *International Dimensions of Humanitarian Law*, 1988, 153-166. Protection to civilians or others who have been rendered *hors de combat* is provided in Article 3 to the Second Geneva Convention, and civilian protection is also afforded in the Fourth Geneva Convention and in the 1977 Additional Protocols.

[6] For analyses of the impact of the 1958 and 1982 Law of the Sea Conventions on naval warfare, see R. R. Churchill and A. V. Lowe, *The Law of the Sea*,

This paper will revisit the law of naval warfare, with an emphasis on the *San Remo Manual*[7] and the customary law of armed conflict at sea. Following a brief discussion of the *Manual*, the topics to be addressed include zones of naval operations, the methods and means of conducting naval warfare, neutral vessels, measures short of attack and protected persons and vessels.

II. The San Remo Manual on International Law Applicable to Armed Conflicts at Sea

Because the 1907 Hague Conventions are inadequate with respect to contemporary naval warfare, international lawyers and naval experts undertook a review of the application of the laws regulating armed conflict at sea in the late 1980s and early 1990s.[8] This review ultimately culminated in the 1994 *San Remo Manual on International Law Applicable to Armed Conflicts at Sea*.

The *San Remo Manual* attempts to identify and clarify the current status of conventional and customary international law as it relates to armed conflict at sea. It was deemed important to undertake this process for several reasons:

> "First, new technology and modern methods and means of warfare have put into question the continued viability of the whole of the traditional legal regime, which was based on nineteenth-century conditions... Second, the law applicable to armed conflict on land was updated by the 1977 Protocols I and II Additional to the Geneva Conventions of 1949. Although

---

3rd ed., 1999, especially Chapter 17, "Military Uses of the Sea", 421-432; S. Oda, *The Law of the Sea in Our Time*, 1977; H. B. Robertson, Jr., "The 'New' Law of the Sea and the Law of Armed Conflict at Sea", Newport Paper No. 3, Naval War College (Newport, RI), October 1992; V. A. Lowe, "Some Legal Problems Arising from the Use of the Seas for Military Purposes", 10 *Marine Policy* 3 (July 1986), 171 et seq.; V. A. Lowe, "The Laws of War at Sea and the 1958 and 1982 Conventions", 12 *Marine Policy* 3 (July 1988), 286 et seq.; B. H. Oxman, "The Regime of Warships under the United Nations Convention on the Law of the Sea", *Va. J. Int'l L.* 24 (1984), 809 et seq.

[7] L. Doswald-Beck (ed.), *San Remo Manual on International Law Applicable to Armed Conflicts at Sea, 12 June 1994*, 1995 (hereinafter, *San Remo Manual* or *Manual*).

[8] For an analysis of the contribution the Hague instruments have made and continue to make in regard to naval warfare, see D. G. Stephens and M. D. Fitzpatrick, "Legal Aspects of Naval Mine Warfare", *Loyola L.A. Int'l & Comp. L.J.* 21 (1999), 553 et seq.

some provisions of Protocol I affect naval operations[9] ... Part IV of the Protocol, which protects civilians against the effects of hostilities, is applicable only to naval operations that affect civilians and civilian objects on land. Nevertheless, some of the basic concepts in this part have affected thinking in naval operations, in particular the principle of distinction and the concept of limiting attacks to military objectives. Third, important developments have taken place since the beginning of this century in other branches of international law, namely, the law of the sea, the law of the United Nations Charter, environmental law and air law, which inevitably have had some effect on the law applicable to armed conflicts at sea".[10]

The *San Remo Manual* comprises 183 articles arranged into six themes.[11] In drafting the *Manual*, an attempt was made to take into account certain factors which are specific to naval warfare, such as measures of economic warfare related to the taking of prize and the effect belligerent naval operations has on neutrals.[12] The substance and history of the

---

[9] For example, Article 49 of Protocol I states:
  "1. 'Attacks' means acts of violence against the adversary, whether in offence or defence.
  2. The provisions of this Protocol with respect to attacks apply to all attacks in whatever territory conducted, including the national territory belonging to a Party to the conflict but under the control of an adverse Party.
  3. The provisions of this Section apply to any land, air or *sea warfare* which may affect the civilian population, individual civilians or civilian objects on land. They further apply to all attacks *from the sea* or from the air against objectives on land but do not otherwise affect the rules of international law applicable in *armed conflict at sea* or in the air.
  4. The provisions of this Section are additional to the rules concerning humanitarian protection contained in the Fourth Convention, particularly in Part II thereof, and in other international agreements binding upon the High Contracting Parties, as well as to other rules of international law relating to the protection of civilians and civilian objects on land, *at sea* or in the air against the effects of hostilities".
  (emphasis added)
[10] L. Doswald-Beck, "The San Remo Manual on International Law Applicable to Armed Conflict at Sea", *AJIL* 89 (1995), 192 et seq., 193-194.
[11] Part I: General provisions; Part II: Regions of operation; Part III: Basic rules and target discrimination; Part IV: Methods and means of warfare at sea; Part V: Measures short of attack: interception, visit, search, diversion and capture; and Part VI: Protected persons, medical transports and medical aircraft.
[12] The *Manual* also incorporates provisions concerning actions of the United Nations Security Council and norms of self-defence. See *San Remo Manual*,

Contemporary Law Regulating Armed Conflict at Sea 1527

articles contained in the *San Remo Manual* are set forth in a commentary[13] that highlights which provisions were considered controversial, progressive, customary in nature or reflective of conventional law.[14] In examining regulations governing the conduct of sea warfare, this chapter places primary emphasis on the contemporary restatement of the law and customs of war as reflected in the *Manual*.

III. Zones of Naval Operations

Naval warfare may generally be conducted in, on, or over the high seas[15] and the exclusive economic zone and continental shelf of neutral and belligerent States,[16] as well as in, on, or over the territorial sea,

---

paras. 3-6 (self-defence) and paras. 7-9 (action by the Security Council) and in the *San Remo Manual*, Explanation, 75-84. See also discussion in Doswald-Beck, see note 10, 196-197; and N. Ronzitti, "The Right of Self-Defense and the Law of Naval Warfare", 14 *Syracuse Journal of Int'l Law & Commerce* 14 (1988), 571 et seq.

[13] The commentary is attached to the *San Remo Manual* and is referred to as the "Explanation".

[14] Doswald-Beck states that the primary features of the *Manual* that differ significantly from traditional instruments regulating armed conflict "are the section on the effect of the UN Charter, the use of the concept of the military objective, the reference to military activities in sea areas that did not exist in law at the beginning of the century, the references to the environment and the provisions relating to aircraft". Doswald-Beck, see note 10, 196.

[15] Marine warfare is not limited to battleships or other such vessels, as submarine warfare, and use of aircraft that supplement the maritime vessels, form part of contemporary armed conflict at sea. See, e.g., J. Gilliland, "Submarines and Targets: Suggestions for New Codified Rules of Submarine Warfare", *Geo. L.J.* 73 (1985), 975 et seq.

[16] As regards the exclusive economic zone and continental shelf of neutral States, paragraphs 34-35 of the *San Remo Manual* elaborate upon the duties of belligerent States more fully, and specify:

"34. If hostile actions are conducted within the exclusive economic zone or on the continental shelf of a neutral state, belligerent States shall, in addition to observing the other applicable rules of the law of armed conflict at sea, have due regard for the rights and duties of the coastal State, inter alia, for the exploration and exploitation of the economic resources of the exclusive economic zone and the continental shelf and the protection and preservation of the marine environment. They shall, in particular, have due regard for artificial islands, installations, structures and safety zones established by neutral States in the exclusive economic zone and on the continental shelf.

35. If a belligerent considers it necessary to lay mines in the exclusive economic zone or the continental shelf of a neutral State, the

internal waters, land territories and archipelagic waters of belligerent States.[17]

Hostile actions by belligerent forces in contravention of the laws of war are forbidden within and over neutral waters,[18] which consist of the internal waters, territorial sea and any archipelagic waters of neutral States.[19] Such acts include attacking or capturing persons or objects located in, on, or over neutral waters or territory, using neutral waters as a base of operations, laying mines in neutral waters or conducting visit, search, diversion or capture operations in neutral waters.[20]

On a non-discriminatory basis, a neutral State may impose conditions on or prohibit belligerent warships and auxiliary vessels from entering or passing through its neutral waters unless the maritime vessels have a right of transit or innocent passage through such waters in armed conflict situations.[21] When exercising their right of transit through, under, or over neutral waters, belligerents may take defensive measures consistent with maintaining their security but may not engage in offensive operations.[22]

Neutral waters cannot be used as a sanctuary by belligerent forces.[23] A neutral State may, however, subject to the duty of impartiality and to general rules concerning international straits and archipelagic sea lanes,[24] permit the following acts within its waters without losing its neutral status:

---

belligerent shall notify that State, and shall ensure, inter alia, that the size of the minefield and the type of mines used do not endanger artificial islands, installations and structures, nor interfere with access thereto, and shall avoid so far as practicable interference with the exploration or exploitation of the zone by the neutral State. Due regard shall also be given to the protection and preservation of the marine environment".

[17] *San Remo Manual*, para. 10. As to the high seas and the sea-bed beyond national jurisdictions, paragraphs 36 and 37 stipulate:
"36. Hostile actions on the high seas shall be conducted with due regard for the exercise by neutral States of rights of exploration and exploitation of the natural resources of the sea-bed, and ocean floor, and the subsoil thereof, beyond national jurisdiction.
37. Belligerents shall take care to avoid damage to cables and pipelines laid on the sea-bed which do not exclusively serve the belligerents".

[18] *San Remo Manual*, paras. 14-15.
[19] *San Remo Manual*, para. 14.
[20] *San Remo Manual*, para. 16.
[21] *San Remo Manual*, paras. 19, 29, and 33.
[22] *San Remo Manual*, paras. 27-30.
[23] *San Remo Manual*, para. 17.
[24] Regarding international straits and archipelagic sea lanes, see *San Remo Manual*, paras. 23-26 (General rules); paras. 27-30 (Transit passage and archipelagic sea lanes passage); and paras. 31-33 (Innocent passage).

"(a) passage through its territorial sea, and where applicable its archipelagic waters, by warships, auxiliary vessels and prizes of belligerent States; warships, auxiliary vessels and prizes may employ pilots of the neutral State during passage;
(b) replenishment by a belligerent warship or auxiliary vessel of its food, water and fuel sufficient to reach a port in its own territory; and
(c) repairs of belligerent warships or auxiliary vessels found necessary by the neutral State to make them seaworthy; such repairs may not restore or increase their fighting strength".[25]

Similarly, belligerent military and auxiliary aircraft are not allowed to enter neutral airspace, and in the event they do so, the neutral State is expected to take measures to neutralise and intern the aircraft and its crew.[26] A neutral State has an obligation to terminate any violations relating to neutral waters.[27]

One of the most controversial issues in the *San Remo Manual* concerns the inclusion of provisions relating to naval exclusion zones.[28] As one prominent commentator has written,

---

[25] *San Remo Manual*, para. 20. Belligerent warships of auxiliary vessels must not remain in neutral waters longer than 24 hours, unless the vessel is unable to leave due to damage or stress of weather. *San Remo Manual*, para. 21. This 24 hour rule does not apply in international straits or waters in which the right of archipelagic sea lanes passage exists.

[26] *San Remo Manual*, para. 18.

[27] *San Remo Manual*, paras. 21-22. See also discussion on both the rights and the duties of neutral States in M. Bothe, "The Law of Neutrality", in D. Fleck (ed.), *The Handbook of Humanitarian Law in Armed Conflict*, 1995, 485-515. "The neutral State is bound to repel any violation of its neutrality, if necessary by force. This obligation, however, is limited by the international legal prohibition on the use of force. The use of military force to defend neutrality is permissible only if it is legitimate self-defence against an armed attack". *Ibid.*, at 495.

[28] *San Remo Manual*, paras. 105-108. For a discussion of use of maritime zones during armed conflict, see W. J. Fenrick, "The Exclusion Zone Device in the Law of Naval Warfare", *Canadian YIL* 24 (1986), 91 et seq.; and J. Astley and M. N. Schmitt, "The Law of the Sea and Naval Operations", *Air Force Law Review* 42 (1997), 119 et seq. Bothe explains that what "became known as the 'exclusion zone' during the Falklands/Malvinas conflict resembles quite closely the notion of 'war zones' instituted by Germany around the British Isles during World War II". M. Bothe, "Neutrality in Naval Warfare:

"An exclusion zone, also referred to as a military area, barred area, war zone or operational zone, is an area of water and superjacent air space in which a party to an armed conflict purports to exercise control and to which it denies access to ships and aircraft without permission".[29]

Although such zones are controversial among commentators,[30] they have become fairly common during periods of armed conflict at sea, and consequently, regulations for these zones were included in the *Manual* to provide guidance and limitations to their use. However, the drafters of the *Manual* considered the designation of such zones to be an exceptional measure,[31] and noted that "[a] belligerent cannot be absolved of its duties under international humanitarian law by establishing zones which might adversely affect the legitimate uses of defined areas of the sea".[32]

IV. Methods and Means of Conducting Naval Hostilities

As with armed conflict on land or in the air, the parties to an armed conflict at sea are under an obligation to respect international humanitarian law from the moment armed force is used.[33] Moreover, the

---

What is Left of Traditional International Law", in A. J. M. Delissen and G. J. Tanja (eds.), *Humanitarian Law of Armed Conflict: Challenges Ahead*, 1991, 387 et seq., 400.

[29] Fenrick, see note 28, 92.

[30] For example, in the preliminary remarks to the "Explanation" of paragraphs 105-108 of the *San Remo Manual*, the drafters acknowledged that:
"Some of the participants [in the San Remo process] were of the view that zones were simply unlawful and that the topic should not be addressed. The majority were of the view, however, that the existence of such zones was a reality and that it was desirable to develop guidelines for them". *San Remo Manual*, Explanation, 181.
See also G. P. Politakis, "Waging War at Sea: The Legality of War Zones", *NILR* 38 (1991), 125 et seq., 170:
"[T]he concept of war zones should remain well outside the bounds of legal approbation. Unerasable memories of the World War practices and fresh experience of the latest decade should suffice to convince us that, if anything, the benefit of doubt is what the concept of war zones should imperatively never deserve to enjoy".

[31] *San Remo Manual*, para. 106.

[32] *San Remo Manual*, para. 105.

[33] *San Remo Manual*, para. 1. International humanitarian law can be defined as "international rules, established by treaties or custom, which limit the right of parties to a conflict to use the methods or means of their choice, or which

right to choose methods or means of warfare is limited by the laws of war and customary norms.[34] Consequently, the same fundamental rules and principles of warfare that apply to armed conflicts on land and in the air are also applicable to sea warfare, including the principles of distinction, necessity, proportionality and humaneness.[35] The conduct of hostilities must not be excessive and the degree and kind of force used must be limited to that required for self-defence or conducting legitimate military operations.[36]

Only military objectives may be targeted for attack. Such objectives are limited strictly to "objects which by their nature, location, purpose or use make an effective contribution to military action and whose total or partial destruction, capture or neutralisation, in the circumstances ruling at the time, offers a definite military advantage".[37] Civilians and civilian objects – such as merchant vessels and civil aircraft – may not be targeted for attack unless they satisfy the criteria for a military objective. Even if an object is deemed a legitimate target of attack because it amounts to a military objective, it is forbidden to use methods and means of warfare which tend to cause superfluous injury or unnecessary suffering or which are indiscriminate in nature or effect or which could be expected to cause disproportionate injury in relation to the actual gains.[38] Precautions must be taken, *inter alia*, to avoid or minimise "collateral casualties or damage that would be excessive in relation to the concrete and direct military advantage anticipated from the attack".[39]

---

protect states not party to the conflict or persons or objects that are, or may be, affected by the conflict". *San Remo Manual*, para. 13(a).

[34] *San Remo Manual*, para. 38. That the means and methods of warfare are not unlimited is the most fundamental principle of international humanitarian law.

[35] On these principles, see generally W. Fenrick, "The Role of Proportionality and Protocol I in Conventional Warfare", *Military Law Review* 98 (1982), 91 et seq.; R. K. Goldman, "International Humanitarian Law: Americas Watch's Experience in Monitoring Internal Armed Conflicts", *Am. U.J. Int'l L. & Pol'y*, 9 (1993), 49 et seq., 64-79; M. N. Schmitt, "The Principle of Discrimination in 21st Century Warfare", *Yale Human Rights & Development L.J.* 2 (1999), 143 et seq.

[36] See generally *San Remo Manual*, paras. 3-6.

[37] *San Remo Manual*, para. 40. This language is taken from Additional Protocol I, Article 52. For a discussion of targeting, see Fenrick, see note 28; and Astley and Schmitt, see note 28.

[38] *San Remo Manual*, paras. 39-42.

[39] *San Remo Manual*, para. 46. This language is taken from Additional Protocol I, Article 57. The *San Remo Manual* defines collateral damage or collateral casualties as "loss of life of, or injury to civilians or other protected persons, and damage to or destruction of the natural environment or objects that are not in themselves military objectives". *San Remo Manual*, para. 13(c). As to

As has long been recognised, protected persons and objects often sustain incidental damage during the course of attack against legitimate military objectives and the law requires that the incidental effects not be excessive in relation to the real advantage anticipated or the attack must be suspended.

As to the means of sea warfare, the *San Remo Manual* provides more guidance concerning the conduct of hostilities than did the 1907 Conventions, and sets forth specific requirements as to prohibitions or use of particular weapons.[40] Affirmative duties are placed on belligerents as to the use of certain weapons, e.g., torpedoes that do not sink or become harmless after completing their run are prohibited,[41] and particular emphasis is placed on the use of sea mines.[42] For example, only mines that become effectively neutralised when they are not in or under the control of the belligerent are allowed. In addition, the State laying the mines must record the location where the mines are laid and the mines must be removed or rendered harmless after the cessation of hostilities.[43] Sea mines have not received the same degree of attention that land mines have received, so their extensive treatment in the *Manual* is an important inclusion:

> "Military planners herald the naval mine as an extremely effective, yet unglamorous option in the national weapons arsenal. A silent, passive, and pernicious weapon, the naval mine,

---

environmental protection, Article 55 of the 1977 Additional Protocol I stipulates:
"Care shall be taken in warfare to protect the natural environment against widespread, long-term and severe damage. This protection includes a prohibition of the use of methods or means of warfare which are intended or may be expected to cause such damage to the natural environment and thereby to prejudice the health or survival of the population".
Attacks against the natural environment by way of reprisals are prohibited.

[40] *San Remo Manual*, paras. 78-92. For information on conventional provisions concerning methods of mine warfare, torpedoes, missiles, and submarine warfare, see H. von Heinegg, "The Law of Armed Conflict at Sea", in D. Fleck (ed.), *The Handbook of Humanitarian Law in Armed Conflict*, 1995, 405 et seq., 442-459; Astley and Schmitt, see note 28, 119; Roach, see note 3, 71-73; and J. Gilliland, "Submarines and Targets: Suggestions for New Codified Rules of Submarine Warfare", *Geo. L.J.* 73 (1985), 975 et seq.

[41] *San Remo Manual*, para. 79.

[42] *San Remo Manual*, paras. 80-92. For an excellent discussion on naval mines, see J. J. Busuttil, *Naval Weapons Systems and the Contemporary Law of War*, 1998, Chapter 2, "Naval Mines", 12 et seq.; and G. P. Politakis, *Modern Aspects of the Laws of Naval Warfare and Maritime Neutrality*, 1998, Chapter 2, "Legal Aspects of Mine Warfare", 166 et seq. See also Stephens and Fitzpatrick, see note 8.

[43] *San Remo Manual*, paras. 80-92.

with its multiplier effect, has influenced the strategic outcomes of conflict this century. It is described as both an 'immoral weapon' and one that is an 'inseparable element of naval power'. ... [I]t is timely to consider the rules regulating the deployment of naval mines because these weapons remain a very important aspect of the catalogue of weapons available to most States".[44]

Blockades, which are one of the oldest methods of naval economic warfare,[45] may legally be used as a method of warfare provided the blockade is declared and the declaration specifies all relevant details.[46] Notification must be given to all belligerents and neutral States. Efforts must be made to ensure that the blockade is effective and that it is applied impartially to vessels of all States. The blockade must be enforced and maintained consistent with international law. Certain proscriptions prohibit, *inter alia*, using the blockade for the purpose of starving the civilian population or from barring access to the ports and coasts of neutral States. If reasonable grounds exist for believing a merchant vessel is breaching a blockage, it may be captured; resisting merchant vessels may be attacked after a warning is given.[47]

In conformity with long established norms, perfidy is prohibited.[48] Ruses of war are permitted, however, provided the deception does not involve simulating the status of protected persons, objects or symbols.[49] Thus, a belligerent may attempt to trick or lure the enemy, but it may not use protected persons, objects or symbols as part of the deception.[50]

Protection of the natural environment is strongly encouraged and the parties are urged to refrain from engaging in hostilities in marine areas

---

[44] Stephens and Fitzpatrick, see note 8, 554.
[45] von Heinegg, see note 40, 470.
[46] While Bothe notes that a "specific means of interdicting intercourse between the enemy and the outside world, which affects both enemy and neutral shipping, is the blockade", he also questions whether the concept of blockade is still valid with the development of new technological options. Bothe, see note 28, 388-389.
[47] *San Remo Manual*, paras. 93-104. See also von Heinegg, see note 40.
[48] Perfidy is defined as an act "inviting the confidence of an adversary to lead it to believe that it is entitled to, or is obliged to accord, protection under the rules of international law applicable in armed conflict, with intent to betray that confidence". *San Remo Manual*, para. 111. See also Politakis, see note 42, Chapter 3, "Ruses and Perfidy at Sea", 268-341; and S. Oeter, "Methods and Means of Combat", in Fleck, see note 27, 199-203.
[49] Such as hospital ships or vessels transporting cultural property under special protection or the use of other such objects or persons having exempt, civilian, or neutral status, such as the UN or Red Cross symbols.
[50] *San Remo Manual*, paras. 109-111. For a discussion of the use of deception during armed conflict at sea, see Astley and Schmitt, see note 28.

containing rare or fragile ecosystems and endangered species.[51] the rights of neutral States are not to be impinged during the course of hostilities, so that any attack – whether offensive or defensive – must be conducted without infringing the rights of other States, including neutral parties.[52]

V. Neutral Vessels

There are certain instances or situations in which neutral merchant vessels and civil aircraft may be attacked, but merely being armed does not cause a neutral vessel or aircraft to lose its protected status of being exempt from attack.[53] Neutral merchant vessels and aircraft lose their exempt status and are considered military objectives which may legitimately be attacked, subject to rules of warfare, when they:

> "(a) are believed on reasonable grounds to be carrying contraband or breaching a blockade, and after prior warning they intentionally and clearly refuse to stop, or intentionally and clearly resist visit, search or capture;
> (b) engage in belligerent acts on behalf of the enemy;
> (c) act as auxiliaries to the enemy's armed forces;
> (d) are incorporated into or assist the enemy's intelligence system;
> (e) sail under convoy of enemy warships or military aircraft; *or*
> (f) otherwise make an effective contribution to the enemy's military action, e.g., by carrying military materials, and it is not feasible for the attacking forces to first place passengers

---

[51] *San Remo Manual*, para. 11. While certain treaties impose a duty on States to protect marine environment, the language here is hortatory and not mandatory and is reflective of international environmental law during peacetime. *San Remo Manual*, Explanation, 82-83. Roach notes that it "is not unlawful to cause collateral damage to the natural environment during an attack upon a legitimate military objective. However, the commander has an affirmative obligation to avoid unnecessary damage to the environment to the extent that it is practicable to do so consistent with mission accomplishment". Roach, see note 3, 69. For an excellent analysis of international environmental law and naval warfare, see S. A. J. Boelaert-Suominen, *International Environmental Law and Naval War: The Effect of Marine Safety and Pollution Conventions During International Armed Conflict*, Doctoral dissertation, University of London, The London School of Economics and Political Science, Law Department, April 1998.

[52] *San Remo Manual*, para. 12. See also Bothe, see note 28.

[53] *San Remo Manual*, para. 69. On maritime neutrality in general, see Politakis, see note 42, Chapter 4, "The Notion of Maritime Neutrality", 345 et seq.

and crew in a place of safety. Unless circumstances do not permit, they are to be given a warning, so that they can re-route, off-load, or take other precautions".[54]

VI. Measures Short of Attack

The *San Remo Manual* also contains detailed provisions for using measures short of attack, including invoking rights of interception, visit, search, diversion and capture. According to Doswald-Beck:

"The law relating to capture, in particular to prize, was retained in the *Manual* and is largely based on the traditional law. The possibility of diversion, in lieu of visit and search, was added by way of progressive development, in recognition of the difficulty of carrying out visit and search at sea, especially if a container vessel is concerned".[55]

The *Manual* also provides rules for determining the enemy character of vessels and aircraft and assessing certain rights to attack or implement measures short of attack depending upon the character of the object as enemy or neutral. For example, the fact that a merchant vessel is flying the flag of an enemy State is considered conclusive evidence of its enemy character. By contrast, the fact that a merchant vessel is flying the flag of a neutral State is considered *prima facie* evidence of its neutral character. A merchant vessel that is flying a neutral flag but which is suspected of having enemy character may be legitimately visited, searched and diverted.[56] After a merchant vessel flying a neutral flag has

---

[54] *San Remo Manual*, para. 67.
[55] Doswald-Beck, see note 10, 202. For a discussion of traditional laws and customs related to prize, see Roach, see note 3, 65-66.
[56] *San Remo Manual*, paras. 112-115; see also G. K. Walker, "The Interface of Criminal Jurisdiction and Actions under the United Nations Charter with Admiralty Law", *Maritime Lawyer* 20 (1996), 217 et seq. As to diversion, *San Remo Manual*, para. 121, provides: "If search and visit at sea is impossible or unsafe, a belligerent warship or military aircraft may divert a merchant vessel to an appropriate area or port in order to exercise the right of visit and search". See also W. J. Fenrick, "The Merchant Vessel as Legitimate Target in the Law of Naval Warfare", in A. J. M. Delissen and G. J. Tanja (eds.), *Humanitarian Law of Armed Conflict: Challenges Ahead*, 1991, 425 et seq. See also R. S. McClain, "The Coastal Fishing Vessel Exemption from Capture and Targeting: An Example and Analysis of the Origin and Evolution of Customary International Law", *Naval Law Review* 45 (1998) 77 et seq. Note that the right to search and visit certain vessels does not apply to warships. Indeed, "[n]ot all vessels are subject to visit and

been visited and searched, if reasonable grounds exist for suspecting that the merchant vessel has enemy character, it may be captured as prize subject to adjudication.[57] Similar provisions apply to civil aircraft bearing the markings of a neutral or enemy State[58] and the *Manual* sets forth detailed provisions governing the interception, visit and search of civil aircraft.[59]

As a basic rule, when reasonable grounds exist for suspecting that merchant vessels outside neutral waters are subject to capture, belligerent warships and military aircraft may visit and search these vessels.[60] With its consent, a neutral aircraft may be diverted from its declared destination in lieu of visit and search.[61] A neutral merchant vessel is exempt from visit and search if it meets the following conditions:

"(a) it is bound for a neutral port;
(b) it is under the convoy of an accompanying neutral warship of the same nationality or a neutral warship of a State with which the flag State of the merchant vessel has concluded an agreement providing for such convoy;
(c) the flag State of the neutral warship warrants that the neutral merchant vessel is not carrying contraband or otherwise engaged in activities inconsistent with its neutral status; *and*
(d) the commander of the neutral warship provides, if requested by the commander of an intercepting belligerent warship or military aircraft, all information as to the character of the merchant vessel and its cargo as could otherwise be obtained by visit and search".[62]

In an effort to prevent unnecessary search and visit of neutral vessels, rules also allow belligerent States to establish reasonable supervisory measures for inspecting cargo of neutral merchant vessels and obtaining certification that a vessel is not carrying contraband.[63]

---

search. Under international law, a warship is a sovereign platform, thereby exempt from visit and search". Astley and Schmitt, see note 28, 144.

[57] *San Remo Manual*, para. 116. Registration, ownership, charter or other such criteria can determine enemy character.
[58] *San Remo Manual*, para. 116.
[59] *San Remo Manual*, paras. 125-134.
[60] *San Remo Manual*, paras. 118-124.
[61] *San Remo Manual*, paras. 118-119.
[62] *San Remo Manual*, para. 120.
[63] See *San Remo Manual*, paras. 122-124. As a practical matter, it is extremely difficult to conduct a search of modern cargo vessels, since the vast majority of such ships carry sealed containers, which are stacked on top of each other.

Subject to certain vessels that are exempt from capture, enemy vessels and goods on board such vessels may be captured outside neutral waters without exercise of a visit and search measure.[64] Similar provisions apply to enemy civil aircraft and goods.[65] In general, the vessels and aircraft that are exempt from attack are also exempt from capture.[66]

Neutral merchant vessels and goods are subject to capture outside neutral waters if they engage in any activity which renders them subject to attack or if a visit and search or other such efforts lead to a determination that they:

"(a) are carrying contraband;
(b) are on a voyage especially undertaken with a view to the transport of individual passengers who are embodied in the armed forces of the enemy;
(c) are operating directly under enemy control, orders, charter, employment or direction;
(d) present irregular or fraudulent documents, lack necessary documents, or destroy, deface or conceal documents;
(e) are violating regulations established by a belligerent within the immediate area of naval operations; or
(f) are breaching or attempting to breach a blockade".[67]

Exempt enemy vessels are subject to capture if they are not innocently employed in their normal role, if they commit acts harmful to the enemy, if they fail to immediately submit to identification and inspection when required, and if they intentionally hinder the movement of combatants and fail to obey orders to stop or move when required.[68] Goods of neutral vessels are only subject to capture – which is exercised by the

---

It is virtually impossible to shift such loaded containers at sea while the vessel is underway.

[64] *San Remo Manual*, para. 135.
[65] *San Remo Manual*, paras. 141-145. Roach notes that the "Hague Conventions did not address which vessels were exempt from attack. A century later, the *San Remo Manual* lists ten classes of enemy vessels as exempt from attack, including a new category, vessels designed or adapted exclusively to respond to pollution incidents in the marine environment. Recognising that today many scientific missions have a military purpose, the *Manual* exempts only those vessels engaged in non-military scientific activities. The *Manual* specifies both the conditions of exemption and the circumstances under which it may be lost. These provisions reflect developments in customary law during this century". Roach, see note 3, 69.
[66] See *San Remo Manual*, paras. 136-137. See also para. 47.
[67] *San Remo Manual*, para. 146.
[68] *San Remo Manual*, paras. 136-137.

taking of the goods as prize[69] for adjudication – if they are contraband.[70] To constitute contraband, the goods on such vessels must be "ultimately destined for territory under the control of the enemy and which may be susceptible for use in armed conflict".[71] Before the right of such capture may be exercised, the belligerent must compile a list, with reasonable specificity, of goods considered to be contraband. The belligerent must publish this list and any goods not on the belligerent's contraband list are not subject to capture.[72] "Free goods" are those that are not listed on the belligerent's contraband list and are thus not subject to capture. Such items also include, as a minimum:

"(a) religious objects;
(b) articles intended exclusively for the treatment of the wounded and sick and for the prevention of disease;
(c) clothing, bedding, essential foodstuffs, and means of shelter for the civilian population in general, and women and children in particular, provided there is not serious reason to believe that such goods will be diverted to other purpose, or that a definite military advantage would accrue to the enemy by their substitution for enemy goods that would thereby become available for military purposes;
(d) items destined for prisoners of war, including individual parcels and collective relief shipments containing food, clothing, educational, cultural, and recreational articles;
(e) goods otherwise specifically exempted from capture by international treaty or by special arrangement between belligerents; *and*
(f) other goods not susceptible for use in armed conflict".[73]

The final item is a "catch-all" provision, included to curtail abuse, but clearly some discretion is available and the question which goods might be "susceptible for use in armed conflict" will tend to be very controversial.[74]

---

[69] For an analysis of modern prize law, see Politakis, see note 42, Chapter 6, "Prize Law Revisited", 526 et seq.
[70] *San Remo Manual*, para. 147. Regarding contraband, see Politakis, see note 42, Chapter 5, "The Theory of Absolute Contraband", 409 et seq.
[71] *San Remo Manual*, paras. 146-148.
[72] *San Remo Manual*, paras. 149-150. Goods on the list vary depending on the particular circumstances of the armed conflict.
[73] *San Remo Manual*, para. 150.
[74] *San Remo Manual*, Explanation, paras. 148.1-150.2.

VII. Protected Persons and Vessels

The final section of the *San Remo Manual* contains limited provisions concerning protected persons. With respect to medical transports and aircraft, these provisions are generally consistent with the Second Geneva Convention and Additional Protocol I. The *Manual* departs from these provisions primarily in one respect, stipulating that:

> "In order to fulfil most effectively their humanitarian mission, hospital ships should be permitted to use cryptographic equipment. The equipment shall not be used in any circumstances to transmit intelligence data nor in any other way to acquire any military advantage".[75]

By contrast, Article 34(2) of the Second Geneva Convention provides that "hospital ships may not possess or use a secret code for their wireless or other means of communication". However, in contemporary society, this provision has been considered to jeopardise the ability of hospital ships to operate effectively.[76] The *Manual* focuses instead on precluding the ship from engaging in acts of military intelligence and thus allows hospital ships to use cryptographic equipment exclusively in fulfilling their humanitarian purpose.

Protected persons include those on board vessels and aircraft that have fallen into the hands of a belligerent or neutral State; members of the crew of hospital ships and rescue craft; persons on board other vessels or aircraft that are exempt from capture; religious and medical personnel attending to wounded, sick and shipwrecked persons; certain nationals of an enemy State; nationals of a neutral State; and civilians.[77] Each of these groups of persons is given protections consistent with the provisions of the Geneva Conventions and Additional Protocols. For example, religious and medical personnel may not be considered prisoners of war. Additionally, nationals of an enemy State, except those who are crew members of hospital ships or rescue vessels, religious and medical personnel, or those who are otherwise exempt from capture, may be made prisoners of war if they fall into any of the enumerated

---

[75] *San Remo Manual*, para. 171. See also discussion in Roach, see note 3, 76.
[76] *San Remo Manual*, Explanation, paras. 171.1-171.6, explaining that cryptographic equipment (i.e., a secret code) is an "integral part" of modern communication systems, and prohibiting its use by hospital ships prevents the ship from fulfilling effectively its humanitarian mission.
[77] *San Remo Manual*, paras. 161-167. For a more detailed discussion of protected persons and medical transports, see Roach, see note 3, 74-76.

categories of listed persons.[78] It has long been an established norm that members of an enemy's armed forces and certain persons accompanying them are generally entitled to prisoner-of-war status upon capture.

Certain enemy vessels and aircraft, such as hospital ships, vessels transporting cultural property and vessels carrying civilian passengers, are deemed to be exempt from attack.[79] All protected vessels and aircraft may lose their protected status under certain conditions.[80] Enemy merchant vessels may only be attacked if they satisfy the definition of a military objective. In particular, it was determined that the following activities may turn enemy merchant vessels into military objectives subject to attack in conformity with the basic principles of warfare:

"(a) engaging in belligerent acts on behalf of the enemy, e.g., laying mines, minesweeping, cutting undersea cables and pipelines, engaging in visit and search of neutral merchant vessels or attacking other merchant vessels;
(b) acting as an auxiliary to an enemy's armed forces, e.g., carrying troops or replenishing warships;
(c) being incorporated into or assisting the enemy's intelligence gathering system, e.g., engaging in reconnaissance, early warning, surveillance, or command, control and communications missions;
(d) sailing under convoy of enemy warships or military aircraft;
(e) refusing an order to stop or actively resisting visit, search or capture;
(f) being armed to an extent that they could inflict damage to a warship; this excludes light individual weapons for the defence of personnel, e.g., against pirates, and purely deflective systems such as chaff; or
(g) otherwise making an effective contribution to military action, e.g., carrying military materials".[81]

Traditionally, only enemy merchant vessels refusing an order to stop or actively resisting visit, search or capture, would have justified attack.

---

[78] *San Remo Manual*, para. 165. These generally include being a member of or accompanying the enemy's armed forces, being a crew member of auxiliary vessels or aircraft, enemy merchant vessels, or civil aircraft not exempt from capture, or of neutral merchant vessels or aircraft that have taken a direct part in the hostilities on the side of the enemy or served as an auxiliary for the enemy.

[79] *San Remo Manual*, paras. 47-58. See also paras. 135-137, containing similar language in relation to capture of enemy vessels and goods.

[80] See *San Remo Manual*, paras. 49-52 (regarding vessels), and *San Remo Manual*, paras. 57-58 (regarding aircraft).

[81] *San Remo Manual*, para. 60.

Thus, the *San Remo Manual* is progressive in incorporating Additional Protocol I standards for determining military objectives.[82]

The *Manual* also contains specific provisions establishing exemption from attack for certain categories of aircraft. For example, medical aircraft, aircraft granted safe conduct and civil aircraft of a belligerent State are exempt from attack only if certain conditions are met, and this exemption may be lost if they breach any of the applicable conditions of their exemption. However, even when this exemption is lost, the aircraft may be attacked only if all of the following four conditions are satisfied:

"(a) diversion for landing, visit and search, and possible capture, is not feasible;
(b) no other method is available for exercising military control;
(c) the circumstances of non-compliance are sufficiently grave that the aircraft has become, or may be reasonably assumed to be, a military objective; and
(d) the collateral casualties or damage will not be disproportionate to the military advantage gained or anticipated".[83]

In case of doubt, the presumption is that the enemy vessel or aircraft has not lost its protected status and is thus exempt from lawful attack.[84]

VIII. Conclusion

The 20th century was the most violent in history, with both World Wars and scores of other armed conflicts raging in every region of the world.[85] So long as war continues to exist, it must be regulated in an effort to lessen its inevitable horrors on people, property and the environment.
Sea warfare, like land and air warfare, has an enormous impact upon belligerents, neutrals, civilians and other protected persons and property, and has significant implications for the natural environment and economic welfare.[86] As a restatement of the law of naval warfare, the

---

[82] See discussion in Doswald-Beck, see note 10, 200.
[83] *San Remo Manual*, para. 57.
[84] *San Remo Manual*, para. 58.
[85] See K. D. Askin, "The ICTY: An Introduction to its Origins, Rules, and Jurisprudence", in May, Tolbert, Hocking, Roberts, Jia, Mundis and Oosthuizen (eds.), *Essays on ICTY Procedure and Evidence In Honour of Gabrielle Kirk McDonald*, 2001, 13 et seq., 15.
[86] For instance, for a description of the impact naval mines can have on the environment and commercial maritime activity, and the danger they pose to

provisions contained in the *San Remo Manual* are not legally binding *per se* but are largely reflective of customary international law.[87] The *San Remo Manual* could serve as the basis for an updated convention on the laws of armed conflict at sea. In the event this proves to be politically impossible, the approach suggested by Professor Christopher Greenwood, a noted authority on the law of armed conflict (and participant in the San Remo process), should be followed: "international efforts should be directed towards the further refinement of the customary law, using the San Remo statement as a starting point, and attempts to improve compliance with that law".[88]

---

people, vessels, and the environment even years after an armed conflict has ended, see Stephens and Fitzpatrick, see note 8, 557-559.

[87] See *San Remo Manual*, Explanation, 61-62.
[88] Greenwood, see note 3, 61, para. 4.48.

The Relationship Between the UN Charter and General International Law Regarding Non-Use of Force: The Case of NATO's Air Campaign in the Kosovo Crisis of 1999

Shinya Murase[1]

I. Introduction

Much has been written on the legality of NATO's air campaign in the Kosovo crisis of 1999.[2] I have chosen this topic not to either justify or condemn the NATO action, but rather because it offers a good example of *unilateral measures* whose legal effects are characterized in terms of *opposability* rather than legality.[3] Unilateral measures are prevalent in various fields of international life today, such as trade, the environment, human rights and security. As pathological as this phenomenon may be considered to be, however, I believe that it should be considered within the framework of general international law.

Nonetheless, by taking up the case of the NATO action for discussion, I must admit that I am a bit dissatisfied with some

---

[1] This paper is an abbreviated version of my earlier article written in Japanese for *Sophia Law Review* 43(3) (December 1999), 1 et seq., and is based on my "Friday lecture" given at the Lauterpacht Research Centre for International Law, Cambridge University, on 24 November 2000.

[2] B. Simma, "NATO, the UN and the Use of Force: Legal Aspects", *EJIL* 10 (1999), 1 et seq.; A. Cassese, "Ex injuria ius oritur: Are We Moving Towards International Legitimation of Forcible Humanitarian Countermeasures in the World Community?", *ibid.*, 23 et seq.; L. Henkin et al., "Editorial Comments: NATO's Kosovo Intervention", *AJIL* 93 (1999), 824 et seq.; I. Brownlie et al., "Kosovo: House of Commons Foreign Affairs Committee 4th Report, June 2000", *ICLQ* 49 (2000), 876 et seq.

[3] S. Murase, "Unilateral Measures and the Concept of Opposability in International Law", in *Might and Right in International Relations, Thesaurus Acroasium*, Vol. 28, Institute of International Public Law and International Relations of Thessaloniki, 1999, 397 et seq.

distinguished international lawyers in America and Europe who maintain that NATO's actions were somehow "illegal, but justified", or "unlawful, but necessary". Those who believe that the actions were plainly illegal have nothing to be concerned about. Those who feel obliged to respond to the extreme humanitarian situations, by contrast, are faced with the most difficult dilemma of conflicting dictates of law and morality. I agree that this latter position is a more responsible one than simply saying that the actions were illegal without demonstrating effective alternative solutions for the then on-going atrocities. I believe, however, that the necessity and justification for the measures taken by NATO should be given *in legal terms* and not simply on extra-legal or moral grounds when one is faced with the question as a professional international lawyer.

I understand, and strongly support, the position that legal evaluations should be distinguished from appraisals of political processes; mixing law and politics is a dangerous business, even a professionally fatal mistake, for any international lawyer. However, I also believe that it is very important to have a methodological framework which facilitates the proper legal analysis of political realities in international relations. If one limits the role of international law to the narrowly defined province of "legality", the result will be the *marginalization* of the function of international law as a tool to regulate the conduct of States. Frankly, I feel that such an attitude would be irresponsible for those involved professionally in international law, and I believe that one should try to establish a certain normative relevance to the actual relations of States. It is from this perspective that I propose to incorporate the concept of opposability into the usual criteria for the determination of legality or illegality in the debate on NATO.

I share the view held by many that NATO's air strikes were in violation of the provisions of the UN Charter, particularly Article 2(4). As is well known, there has been much debate regarding the interpretation of Article 2(4), and whether the use of force is prohibited altogether, or whether certain exceptions are permitted. In particular, regarding "humanitarian intervention", some take the position that it is absolutely unlawful, while others feel that it may be permissible depending on the purpose, form and scale of the action.[4] Some of those who try to justify the NATO actions also rely on a theory of the Security Council's "implied authorization".[5] However, as long as one considers

---

[4] For an excellent account on humanitarian intervention, see V. Lowe, "The Principle of Non-Intervention: Use of Force", in V. Lowe and C. Warbrick, *The United Nations and the Principles of International Law*, 1994, 66 et seq.

[5] For a critical review of the theory, see J. Lobel and M. Ratner, "Bypassing the Security Council: Ambiguous Authorization to Use Force, Cease-fires and the Iraqi Inspection Regime", *AJIL* 93 (1999), 124 et seq.

the UN Charter to be the applicable law, no matter how broadly or strictly one tries to interpret the relevant provisions, large-scale uses of force carried out without the explicit authorization of the Security Council, such as the NATO air strikes, cannot be considered acceptable *under any interpretation of the UN Charter* at the very least.

What is the point of emphasizing this kind of violation of the UN Charter in cases such as the NATO air strikes? My position is that the UN Charter is not properly maintained as the exclusive applicable law in these cases, and that, therefore, the application of general international law should be considered. I believe that when the Security Council is stalled in the resolution of a conflict, and one is faced with a situation where the measures ordered by the Security Council have not been fulfilled, then Chapter VII must be deemed "inoperative". As a result, *lex specialis* ceases to be used and *lex generalis* comes back into effect. In my view, a shift in the applicable law from the UN Charter to general international law takes place in such a situation. It is appropriate to look first into some precedents which bear out this argument.

II. A Shift in the Applicable Law from the UN Charter to General International Law

1. Precedents

A shift in the applicable law can be recognized in several precedents. For example, the Korean War began in 1950 as a result of the application of an enforcement action of the United Nations under the UN Charter, although there was some irregularity with regard to the absence of the Soviet Union from the Security Council. Then, with the return of the Soviet Union to the Security Council in August 1950, the Council's activities were paralyzed, and, beginning in January 1951, the issue of Korea was removed from the Security Council's agenda. The conflict ended with an armistice agreement concluded under general international law, following well-known processes. The General Assembly at that time merely welcomed the truce, and recommended that the Korean question should be settled by a political conference of "the Member States contributing the forces", rather than the United Nations itself. This action can only be interpreted as an expression by the United Nations of its intention to withdraw from involvement in the Korean situation.[6] Thus, the Korean conflict, which began as an

---

[6] O. Schachter, "Authorized Uses of Force by the United Nations and Regional Organizations", in L. Damrosch and D. Scheffer (eds.), *Law and Force in the New International Order*, 1991, 65 et seq., 73.

application of the UN Charter, was concluded under the system of general international law.

The hostage crisis in Teheran can also be considered from this viewpoint. Immediately after the crisis began in November 1979, the Security Council passed a unanimous resolution calling for the release of the hostages. At the end of December, it again demanded the immediate release of the hostages. The Security Council decided to reconvene and take effective measures based on Chapter VII of the UN Charter if its most recent demand was not met within a week. However, the US proposal for economic sanctions against Iran was vetoed by the Soviet Union, subsequent mediation attempts by the United Nations and other organizations failed, and the hostages remained captive.

Seeing no other feasible alternative, the United States attempted its own hostage rescue operation in April 1980. The International Court of Justice criticized this use of force by the United States as having undermined respect for judicial process.[7] However, the Court never considered this action from the viewpoint of illegality and responsibility, but rather evaluated it as a matter of opposability,[8] and, in effect, the Court appeared to consider the problem on the basis of general international law rather than the UN Charter.[9]

Another case that is relevant to this approach is the Falklands/Malvinas dispute of 1980. The Security Council recognized Argentina's military invasion of the Falklands as a breach of the peace and demanded the immediate withdrawal of the Argentine army. However, Argentina refused to comply, and subsequent mediation attempts by the UN Secretary-General, the United States and others all proved unsuccessful. As a result of this turn of events, the UN Charter took a back seat to the applicable law, and a war was fought in the form of a military skirmish under general international law. In the end, the matter was resolved when the United Kingdom forcibly took back the islands.[10] It was the position of the British Government that this was the exercise of the right of self-defense on the part of the United Kingdom, but, even so, it seems that the right of self-defense in this case was based on general international law rather than Article 51 of the UN Charter due to the Security Council's inaction.

---

[7] ICJ Reports 1980, 3 et seq., 43.

[8] *Ibid.* See also Murase, see note 3, 411.

[9] J. R. D'Angelo, "Resort to Force by States to Protect Nationals: The US Rescue Mission to Iran and Its Legality under International Law", *Va. J. Int'l L.* 21 (1981), 485 et seq., 493, 510-511.

[10] A. Parsons, "The Falklands Crisis in the United Nations, 31 March - 14 June 1982", *International Affairs* 59 (1983), 169 et seq.; W. M. Reisman, "The Struggle for the Falklands", *Yale L.J.* 93 (1983), 287 et seq.

2. The Relationship Between Article 2(4) and Chapter VII of the Charter

This shift in the applicable law from the UN Charter to general international law should not be linked unconditionally to the inability of the Security Council to function due to the use of veto power. Rather, this shift in the applicable law should be recognized under the following sets of conditions. First, the Security Council determines that a certain incident falls under Chapter VII by referring to a breach of the peace, etc., as specified in the Chapter. Secondly, the Council orders the aggressor State to take specific measures, for example, the cessation of hostilities and withdrawal of army units. And thirdly, the Security Council is unable to secure compliance with the conditions it has demanded. In these circumstances, Chapter VII may be deemed unenforceable or inoperative, and the legal evaluation of the actions taken by the injured State or other related States to restore the original *status quo* should be performed under general international law rather than under the UN Charter.

It appears that the UN Charter and, in particular, the principle of the non-use of force have been rather excessively praised in Japan, while it is my impression that Europeans and Americans seem to have a more realistic view of the Charter. In any event, it is natural to think that the reason that a State accepts the prohibition of the use of force under the Charter to begin with is because collective security functions effectively, and that, if worst comes to worst, the State's own security is guaranteed through this collective security system. In this sense, Article 2(4) should be considered a "function" of Chapter VII; the principle of non-use of force relies on the actual functioning of Chapter VII.[11]

3. Paralysis of Lex Specialis and the Return to Lex Generalis

It is said that the UN Charter has two sets of norms: constitutional and institutional. These norms are integrally linked and cannot be separated. Moreover, I submit that the UN Charter is, despite its extreme importance, only one of a number of ordinary multilateral treaties. The idea that it is the constitution, quasi-constitution or higher law of the

---

[11] W. M. Reisman, "Coercion and Self-Determination: Construing Charter Article 2(4)", *AJIL* 78 (1984), 642, et seq.; J. Combacau, "The Exception of Self-Defence in UN Practice", in A. Cassese (ed.), *The Current Legal Regulation of the Use of Force*, 1986, 9 et seq., 32; J. Stone, *Aggression and World Order*, 1958, 96, 98-101.; *contra* O. Schachter, *International Law in Theory and Practice*, 1991, 126-131.

international community is difficult to support in light of the realities of international relations, much less that it be defined as *jus cogens*.[12]

The foregoing discussion shows that there are two ways to justify the NATO actions. One is to rely on the "margin of flexibility" in the interpretation of Article 2(4), with a view to expanding the scope of *exceptions* to the non-use of force principle.[13] The other approach, which I employ, shifts the applicable law from the UN Charter to general international law. One may think in terms of "who killed Article 2(4)?"[14] for the first type of approach, while for the latter, the question "who killed the UN Charter?" might ring true. Of course, neither proposition is valid. I have never had even the slightest intention of undermining the United Nations or the UN Charter. However, one should not uphold it inappropriately, nor entertain disproportionate expectations or illusions about it. If one gives unduly high status to the UN Charter, thereby loading an excessive burden on the United Nations, it may invite the paradoxical result of threatening the very life of the Charter – a *"Titanic paradox"*. In any event, I believe that there is a need to decide upon the positioning of the UN Charter appropriately within the context of general international law. Doing so would lead the way to the Charter's sustainable development over the long term.

Next is the question how to assess forcible humanitarian measures, as in the case of the NATO air attacks, under general international law. I cannot agree with the view that the principle of non-use of force has been established as a principle of international customary law with normative content identical to, but independent of, Article 2(4). It is true that Article 2(4) was also recognized as a principle of customary law in

---

[12] Due to limitations of space, I cannot elaborate on these points here. Nonetheless, I would like to point out that Article 103 of the UN Charter, the so-called supremacy clause, seems to have been unduly overestimated in order to give the Charter an incorrect characterization of a "higher law". Note, first, that the article is placed only as one of the "Miscellaneous Provisions" rather than as one of the basic principles of the Charter (for example, Article 2(2)). Secondly, while the provision refers to a conflict of obligations, it should be construed, in my view, that the existence not only of the "positive" conflict, but also of an "active" conflict, is necessary for the application of Article 103, which means that the conflict arises as a result of a binding decision of a UN organ, most notably the Security Council. In other words, the scope of application of Article 103 of the Charter is quite limited. Needless to say, the provision has nothing to do with the relationship between the UN Charter and general international law.

[13] T. M. Franck, "The Law Pertaining to the Use of Force Within and Without the UN", Hersch Lauterpacht Memorial Lectures, Cambridge University, 21-23 November 2000.

[14] T. M. Franck, "Who Killed Article 2(4)? Or: Changing Norms Governing the Use of Force by States", *AJIL* 64 (1970), 809 et seq.

the ICJ's majority opinion in the *Nicaragua* case.[15] However, no basis whatsoever for this was indicated in the judgment, which was one of the points criticized, among others, by Judge Jennings in his dissenting opinion.[16] While one may refer in this regard to the 1970 General Assembly declaration on Friendly Relations of States, I would take a rather restrictive view of its legal effect, namely, that the declaration is no more than the authoritative interpretation of the Charter, and is not of itself an instrument declaratory of customary international law.

As I discussed earlier, I find it difficult to support the method of interpretation of the existing international law on the issue of prohibition of force that equates the UN Charter, which embodies collective security mechanisms, with general international law, which lacks such a system, in the same argument. The distinction between the two is very important, because, as known, the method of legal evaluation, particularly the burden of proof, becomes quite different depending on which of the prohibitive or permissive norms are set as preconditions.

When I refer to general international law, I am speaking, of course, of the general international law of 1999, and not that of the 18th or 19th century. Cases of humanitarian intervention in those days were primarily based on the egoistic and selfish motives of the intervening States, which were nonetheless considered permissible then. The measures taken by NATO countries in more recent times can well be said to have been based on unselfish intentions to safeguard the general interests of the international community rather than each country's individual interests, however unwise or improper the actions might have been considered from the political point of view.

III. Unilateral Humanitarian Measures under General International Law

1. Distinction Between Unilateral Acts and Unilateral Measures

At this point, I would like to offer a provisional opinion on the NATO bombings under general international law. This action by NATO can be seen, in my view, as a *unilateral measure* taken by a group of States, and must, as such, be regarded as having *opposability vis-à-vis* the Federal Republic of Yugoslavia.

With regard to the concept of opposability, it may be recalled that the ICJ first referred to it in the *Fisheries* case in 1951,[17] and again in

---

[15] ICJ Reports 1986, 14 et seq., 98-101, 187-190.
[16] *Ibid.*, 528 et seq., 530-531.; see also R. Jennings, "Treaties as 'Legislation'", *Collected Writings of Sir Robert Jennings*, 1998, 727-729.
[17] ICJ Reports 1951, 116 et seq., 131.

more detail in the *Fisheries Jurisdiction* case in 1974.[18] Professor J. G. Starke[19] shed light on this concept in the late 1960s, and, in Japan, Professor Soji Yamamoto,[20] now a judge at the Law of the Sea Tribunal, made significant contributions in clarifying the implications of the concept of opposability in international law.[21]

First of all, it is very important to distinguish between a *unilateral act* and a *unilateral measure*. A unilateral *act* is a juridical act whose conditions and legal effects are clearly defined in the established rules of international law, whether by customary law or treaty law, and is thus presumably based on the prior consent of the States concerned. In contrast, a unilateral *measure* is an action taken by a State, a group of States or, in some cases, an international organization with external or extraterritorial effects, and is based on urgency and equity.[22] With regard to such unilateral measures, the ICJ has rendered judgments on the basis of opposability rather than legality in several cases.[23]

2. Opposability of Unilateral Measures

The most striking aspect of a unilateral measure is the fact that it is taken in a situation where there is a conspicuous *lacuna* in international law, with the law emerging or undergoing change. Thus, for example, in *Fisheries Jurisdiction*, the Court concluded that the Icelandic Regulations extending Iceland's fisheries zone to 50 nautical miles

---

[18] ICJ Reports 1974, 3 et seq.

[19] J. G. Starke, "The Concept of Opposability in International Law", *Australian YIL* 5 (1968-1969), 1 et seq.

[20] S. Yamamoto, "The Function of Unilateral Domestic Measures by a State in the International Law-Making Process", *Sophia Law Review* 33 (1991), 47 et seq. (in Japanese).

[21] See also S. Murase, see note 3, 397 et seq.; S. Murase, "Perspectives from International Economic Law on Transnational Environmental Issues", *RdC* 253 (1995), 283 et seq., 354-372.

[22] Unilateral measures in this sense may be close to what the French jurists have called *les actes unilatéraux hétéronormateur* (heteronormative unilateral acts) imposing obligations or burdens on other States without their consent, distinct from *les actes unilatéraux autonormateurs* (auto-normative or self-imposing unilateral acts). P. Daillier and A. Pellet, *Nguyen Quoc Dinh: Droit international public*, 6ème éd., 1999, 361-363.

[23] In addition to the *Fisheries* and *Fisheries Jurisdiction* cases, the idea of opposability (or non-opposability) and similar concepts have been adopted by the Court in its decisions in the *Rights of Nationals of the United States in Morocco* (1952), *Nottebohm* (1955), *Temple of Preah Vihear* (1962), *Barcelona Traction, Light and Power Company, Limited* (1970), *United Diplomatic and Consular Staff in Teheran* (1980), *Frontier Dispute* (1986), and *East Timor* (1955) cases.

"[were] not opposable to the United Kingdom".[24] In other words, the Court reached no conclusion regarding the United Kingdom's initial position that the extension was "*ipso jure* illegal and therefore invalid *erga omnes*".[25] Rather, the Court held only that the measures in question were not opposable. Such a judgment was inevitable in view of the fact that, while the case was pending before the Court, some countries were already beginning to assert a 200-mile exclusive economic zone.

Thus, a special problem with regard to unilateral measures occurs when the content of the applicable law is not yet clearly established. The response to large-scale human rights violations within a country's borders, as in the Kosovo crisis, is a perfect example of a situation where the international law has not yet been clearly established. Although primary substantive rules of international law stating that human rights violations are actions that violate international law are quite well developed, this is not a case with secondary international law rules regarding procedures and mechanisms that should be used to handle such grave violations. The development of such secondary rules is not yet complete. In this kind of ambiguous normative situation, I believe that the actions against Yugoslavia should at least be recognized as being opposable "humanitarian measures".

3. Legal Assessment of the NATO Action on the Opposability Criteria

At this point, it is necessary to review the component elements of opposability. The objective elements necessary to comprise opposability are *effectiveness* and *legitimacy*. Effectiveness refers to the factor of power needed to guarantee realization of the measure in question. If a measure is not implemented effectively, it is non-opposable. The measure in question must also be supported by legitimacy, and must conform to the general interest of the international community in a manner that outweighs the special interests of a particular State or group of States. The NATO actions appear to have been fully in accordance with the effectiveness requirement due to NATO's overwhelming military power. As to legitimacy, NATO had the strong support of the international community, including the G-8 Summit, the OSCE and the so-called Rambouillet Process. Security Council Resolution 1160 and subsequent resolutions also support the legitimacy of the actions, which can be assessed here from the viewpoint of general international law.

---

[24] ICJ Reports 1974, 3 et seq., 29.
[25] *Ibid.*, 7. See also Judge Waldock's separate opinion, *ibid.*, 106 et seq., 118-119.

Whereas effectiveness and legitimacy are objective elements of opposability, the principle of "good faith" is very important as the subjective standard in evaluating whether the measures in question can be considered opposable under an imminent situation in which there are no available alternatives. I believe that the efforts made by the NATO countries in this regard should also be assessed positively.

Therefore, the NATO actions can be considered "opposable" as measures undertaken to prevent further deterioration of the situation, while no effective measures were forthcoming from the Security Council. I do not think that one can rely on simple legal/illegal, black or white criteria in assessing a situation like the Kosovo crisis. On the contrary, one should look at the gray area of normativity, and apply a "relative" normative scale. In other words, instead of all-or-nothing, or zero or ten argument, an assessment should be balanced, say, in the three to seven range, recognizing the legal effect of a relative degree of opposability.

If this analysis is accepted, then the NATO bombings can be regarded as an "opposable" use of force taken as an unavoidable measure for the purpose of preventing a worsening of conditions in an emergency situation in which both large-scale human rights violations were being committed and UN Security Council measures had no effect whatsoever. I hope that, if the current dispute between Yugoslavia and some NATO countries pending before the ICJ proceeds to the merits phase, the Court will examine the case on the basis of opposability criteria. It will certainly encourage the development of customary international law under which certain humanitarian measures are legally justified in exceptional circumstances and within the limits clearly prescribed by specified criteria.[26]

In this connection, the idea of "the emerging customary principle of forcible countermeasures" elaborated by Professor Antonio Cassese is quite attractive.[27] However, the legal basis for countermeasures *by third parties* is not yet clearly established. In addition, when one considers that the Draft Articles on State Responsibility proposed by the International Law Commission prohibits the use of force in countermeasures,[28] the invocation of this idea seems somewhat difficult.

---

[26] V. Lowe, "International Legal Issues Arising in the Kosovo Crisis", *ICLQ* 49 (2000), 934 et seq., 941); S. Sur, "Le recours à la force dans l'affaire du Kosovo et le droit international", les notes de l'Ifri No. 22, 2000, 17-19.

[27] A. Cassese, see note 2, 23-27.

[28] See Article 51 of the Draft Articles on State Responsibility provisionally adopted by the International Law Commission on second reading (11 August 2000), UN Doc. A/CN.4/L.600, 14; see also the section on "forcible countermeasures" in the Third Report on State Responsibility by James Crawford, Special Rapporteur, UN Doc. A/CN.4/507/Add.3, 23.

IV. Conclusion

Measures based on opposability are provisional and transitional by their very nature, and their legal effect is limited, in principle, to particularized relations between the States concerned. Therefore, the clear and early authorization of these measures by a competent organization must occur. The opposability of the NATO air attacks was confirmed and legalized *ex post facto* by Security Council Resolution 1244. It may be argued that this Resolution, if read in context, recognized in effect the factual situation created by the NATO actions and laid the foundation for an international peace presence in Kosovo comprised primarily of NATO troops.

It is interesting to note that the first paragraph of Resolution 1244 quoted the Chairman's statement from the G-8 Foreign Ministers' Meeting of May 1999, noting that a political settlement of Kosovo crisis should be based on that statement. Especially because the G-8 includes Germany and Japan, the citation of the G-8 summit in the resolution speaks to the inability of the current Security Council system to adequately deal with the security of the present-day international community. In this sense, I believe that the reform of the Security Council is the most pressing task of all.[29]

Clearly, unilateral measures are, as a general rule, undesirable. However, it is largely because of structural defects in the international system, including the UN system, that unilateral measures are prevalent in the present-day world. The ineffectiveness of institutions in responding to crises creates a situation ripe for unilateral actions. Because of their status, the world's major powers are most capable of taking such actions. Precisely because of their preponderant power, however, abuses of unilateral action are difficult to control. Some unilateral measures are opposable, while others are not. Opposability is assessed according to the normative criteria of effectiveness and legitimacy. To be opposable, the action taken should meet international standards for disciplined behavior, and all unilateral actions must ultimately meet tests of international lawfulness. Opposability is thus a useful concept in differentiating between cases of permissible and non-permissible unilateral measures.

Needless to say, no use of force is desirable, and the NATO actions can be condemned from many points of view. However, I believe that there are circumstances in which the *unilateral* use of force is not only

---

[29] A. Peacht, "Kosovo as a Precedent: Towards a Reform of the Security Council? International Law and Humanitarian Intervention", Draft Special Report of the Civilian Affairs Committee of the NATO Parliamentary Assembly, 16 September 1999, para. 49. Text available at http://www.nqq.be/publications/comrep/1999/as244cc-e.html.

permitted but also required in order to prevent the worst conceivable situations from taking place. Under such circumstances, we, as international lawyers, cannot and should not merely say that the actions were illegal, but necessary and legitimate. I believe that one should try to make every effort to accommodate the ethical considerations of necessity and legitimacy, as well as the normative elements reflecting the actual power relations, in the province of international law.

A Review of the Debate in the United States Senate over the Deployment of Ground Troops to Kosovo

*Myron H. Nordquist**

I. Introduction

The most critical decisions made at the domestic level by the leaders of sovereign States concern issues of peace and national security. Likewise, these decisions are equally important in the international realm. We see, for example, that the primary purpose of the United Nations, as cited in Article 1, paragraph 1, of the UN Charter, is to maintain international peace and security. Indeed, the driving force for convening the multilateral conference to form the United Nations was to outlaw aggressive war. Thus, we can expect that decisions by sovereign leaders to use force against or in other States will be carefully considered and carried out in strict conformity to existing international agreements and the rule of law.

Prior to the establishment of the United Nations, efforts to outlaw war had failed mainly because there was no effective means of enforcement. What was fundamentally unique about the United Nations was that, for the first time, a Security Council was empowered to authorize enforcement actions against nations that endangered international peace and security. The power to back up the new legal enforcement rights granted to the Security Council was largely guaranteed by requiring a consensus among the five declared nuclear powers who were given permanent voting status and a veto.

In a democratic society, characterized by free speech and free press, one would expect an informed public to demand that its government leaders fully justify the use of deadly military force against foreign States on the basis of existing international law. This is especially true when the use of that force amounts to an act of war against another sovereign nation. Particularly after the US experience in Vietnam, one would expect, at minimum, widespread demands from the general public

---

* The views expressed by Dr. Nordquist are entirely his own.

that acts of war initiated by the elected leaders of the United States be conducted in strict conformity with the UN Charter, the American Constitution and the rule of law in general.

Yet, these expectations and demands were not met in the United States when NATO bombed the capital of Serbia and began deploying ground troops in Kosovo. These events, in fact, generated very limited public discussion or legal scrutiny in the United States. Few questioned the wisdom, morality or legality of NATO's actions or the leadership role played by the United States in taking these drastic actions. The relative silence among the media and intellectuals in the United States was puzzling. With the onset of the Kosovo crisis, the United States became the principal proponent of a breath-taking new mission for NATO: offensive intervention in an internal conflict outside of the territory NATO was legally authorized to defend under the North Atlantic Treaty. Thus, NATO shed its identity as a self-defense organization and adopted a self-defined regional peacekeeping role. The role was self-defined in that, even if NATO were a Chapter VIII regional peacekeeping organization, it would be required by the UN Charter to use force only pursuant to a mandate from the Security Council. In addition, under US domestic law, a fundamental change in NATO's mission clearly requires an amendment to the North Atlantic Treaty, which, in turn, requires the advice and consent of the United States Senate. Neither the Security Council nor the Senate approved NATO's deployment to Kosovo as required by the UN Charter and US Constitution. Despite the failure of the United States to obtain authorization for its actions under either of these constituting documents, the legal community and the public remained silent.

Can it be that no one noticed or cared that the UN Charter and the US Constitution were ignored? Why did the US Congress passively accept a new role for NATO without the benefit of serious public debate or careful compliance with the US Constitution and the UN Charter? What is the meaning of adherence to the rule of law if fundamental provisions of the UN Charter and the US Constitution are not seriously discussed when fundamental issues of peace and war are at issue?

II. Kosovo Crisis

In the post World War II era, it is difficult to conceive of a more classic act of war than to begin bombing a foreign capital without a mandate from the UN Security Council or without articulating a viable self-defense rationale under customary international law. NATO's air campaign against Belgrade was initiated by the President of the United States and his fellow heads of government in NATO with no meaningful attempt to lay out a legal justification for their actions. Even today, the

Office of Legal Adviser in the US Department of State does not justify the US military action under any recognized theory of international law.

In the governance of international peace and security and the conduct of States, the rule of law has enduring meaning only if nations respect it as a matter of principle. The UN Charter is based in part on a solemn agreement between sovereign States to refrain from interfering in internal matters of sovereign States unless there is a threat to international peace and security. The rule of law requires nations to observe each provision of the UN Charter; it does not permit Member States to pick and choose articles that suit mercurial political whims. Moreover, States cannot justify deviation from international legal norms simply for reasons of political expediency or convenience.

Was NATO's bombing Belgrade in early 1995 a "crime against the peace" as the concept is defined through customary international law?[1] To be sure, former President Milosevic of Serbia is the quintessential example of an international criminal. The decision of the Milosevic régime to eliminate the autonomy of its Kosovo province in 1989 caused widespread internal violence in 1998. The human rights abuses that followed, on the part of both the Serbs and the ethnic Albanians in Kosovo, were inexcusable, by any standard. But the rule of law cannot be based on media-enhanced "bad guy" personalities. Milosevic's actions may have merited humanitarian intervention, but did they justify US use of military force at the point at which NATO began bombing the country's capital under undisputed rules of international law? What happened to crimes against the peace under international law? As a matter of principle, the rule of law must be honored even if experts agree that the sanctions and other forms of pressure on Milosevic were ineffective and that Kosovars today are better off than they would have been if NATO had not acted.

As a practical matter, Kosovo's future status remains unresolved, and there is no guarantee that Kosovo will see no more bloodshed. Indeed, most indications are, unfortunately, to the contrary. The prospects for the future in Kosovo was very much in the minds of US Senators in mid-2000 when they briefly debated whether the Kosovo campaign was in the US national interest and, if so, to what extent continued US contribution to the Kosovo effort was warranted.

III. Background of the Senate Debate

Before, during and after NATO's Kosovo operation against Serbia in early 1999, several members of Congress challenged President Clinton's

---

[1] Yugoslavia certainly believes the bombing was illegal. See International Court of Justice, Press Communiqué 99/33 (2 June 1999).

authority under the US Constitution to engage US military forces in the Balkans without congressional approval. At the same time, many of these members had reluctantly voted to support the President's requests for funding military operations in the Balkans because the troops were already there. Historically, Congress lends practical and ideological support to American troops in a hostile environment even when members disagree with the actions of the Commander in Chief who orders US forces into harm's way. As the debate below shows, many Senators, despite the support of Congress for the troops' efforts, harbor longstanding reservations about the deployment of US ground troops to Kosovo.

The Clinton administration was well aware of the paper-thin political support for its Kosovo policies. That is one of the principal reasons for the administration's "frantic" reactions to debates in Congress about disallowing further appropriations for Kosovo. To combat the build-up of public resistance to its decisions, the administration successfully kept the focus on emotionally charged issues such as ethnic cleansing to justify the bombing action. As part of its media strategy, the White House called almost daily attention to refugees fleeing from Kosovo. Understandably, the Kosovo refugees related tragic stories about the civil war, ethnic strife and family losses. As a result of this focus by the press, the American people adopted a largely uncritical view of "Mr. Clinton's War" in Kosovo. Support for NATO's actions came from both ends of the political spectrum: those who believe that any action taken in the name of human rights is laudable and those who believe any US military action is *de facto* justified.

Although, in the end, politics, not principle, governed the actions of Congress, a number of thoughtful Senators felt that the Clinton administration had played fast and loose with the "truth" about why the bombing was necessary and about the expected duration of the US military commitment in Kosovo. Certain Senators, most notably Robert Byrd, a Democrat from West Virginia, and John Warner, a Republican from Virginia, clearly understood what was happening. They, and almost half of their colleagues in the Senate, challenged the President's use of force against Serbia and the consequent occupation of Kosovo by NATO. The debate that follows traces many of the key arguments made on the floor of the Senate by Senators on various sides of the issues.

This review is consciously limited in time and place to the debate on the floor of the Senate, on 17 and 18 May 2000, over an amendment dealing with the deployment of US ground troops to Kosovo. The focus on this debate is sufficient to spotlight most of the war and peace issues of principle that remain of critical importance to the United States and to the world as a whole. One must bear in mind, however, that, on 10 June 1999, Operation Allied Force was brought to a close. Whatever merit there was in questioning the wisdom, morality or legality of NATO's bombing of Belgrade greatly diminished after the bombing stopped. The significant "aggressive use of force" or "crime against the peace" issues were largely ignored in the Senate debate because they were largely moot. Thus, the May 2000 debate in Congress concentrated mainly on the proper role for the United States in the reconstruction of Kosovo and the US role in the peacekeeping force deployed there.

The Senate debate also took place prior to the time when Slobodan Milosevic lost the direct presidential elections in late September 2000 to nationalistic challenger Vojislav Kostunica. This change in government did make Kosovo policy a prominent campaign issue in the 2000 presidential elections in the United States. In that regard, the Bush campaign indicated shortly before the election that it aimed to withdraw US armed forces from the Balkans and hand over peacekeeping responsibilities to the Europeans.[2] Nearly a year after the Senate debates covered in this paper, there are still 5,500 American ground troops in Kosovo, constituting 14 percent of KFOR's strength.[3] The future outlook for the deployment of these troops remains uncertain and ultimate resolution of the issues may depend upon a thoroughgoing review of the national security strategy of the United States now being undertaken by the Bush administration. One prediction seems safe, however: Congress will not object if the President decides to bring the US troops home.

IV. Kosovo Funding and Military Reconstruction Bill

On 18 May 2000, the Senate approved S. 2521, the military construction appropriation bill for fiscal year 2001, by an overwhelming vote of 96-4. The bill provided over $1.8 billion in fiscal year 2000 for supplemental funding of Department of Defense peacekeeping costs in Kosovo. By a narrow vote of 53-47, the Senate voted to delete Section 2410 of the bill, which would have required the President to obtain confirmation of commitments from European countries engaged in Kosovo prior to the United States' providing additional financial support to maintain troops in Kosovo. This paper traces the debate on that vote.

From time to time, throughout 1999 and 2000, the 106th Congress debated various scenarios pertaining to US military involvement in Kosovo. After the peace accords at Rambouillet, Congress considered legislation authorizing, shaping, or prohibiting US participation in enforcement measures that might occur in the region. As such measures

---

[2] J. Kim, *Kosovo and the 106th Congress*, 2001, Congressional Research Service, Library of Congress, 23.

[3] Letter from General Wesley Clark, Commander-in-Chief, United States European Command, to US Senators (February 9, 2001), see note 5 *infra*.

became a reality, Congress occasionally debated the propriety of NATO's actions, with the Senate (but not the House) voting in favor of air strikes by a narrow margin. After 3 June 1999, when Milosevic finally signed on to a peace plan, Congress considered in greater detail the proper role for the United States in a European-led, non-military reconstruction effort, as well as subsequent scenarios for the withdrawal of US troops, and for US foreign policy towards Serbia. Permeating all the debates was the fundamental issue of the extent to which Congress, and especially the Senate, owed a duty to endorse, influence or even direct the policy of the United States in using and committing military force overseas.

President Clinton had unilaterally deployed the armed forces of the United States to the Balkans and subsequently engaged them in active combat without congressional endorsement. The Members of Congress that sought to impose congressional controls upon the exercise of Presidential prerogatives in this realm tried a variety of measures. These included amendments to mandatory spending or authorizing legislation that was defense related or involved foreign assistance. While the form of efforts varied, they included congressional hearings, informal consultations with administration officials, media discussions and field trips by individual Members to the Balkans.

After the air strikes ended, Senator John Warner, a Republican and Chairman of the Senate Armed Services Committee, and Senator Robert Byrd, a Democrat and Ranking Member of the Senate Appropriations Committee, addressed the issue of whether America's NATO allies were carrying their fair share of the burden of the Kosovo intervention. Senator Warner and Senator Byrd were concerned that the United Nations, the European Union, and the Organization for Security and Cooperation in Europe were not fulfilling their promised commitments in Kosovo. This, in turn, they argued, was delaying the return of US ground troops to the United States. They drafted Section 2410 to solve this problem. Consequently, in response to President Clinton's request for emergency supplemental funding in early 2000, the Senate Appropriations Committee voted 23-3 to add the Byrd-Warner Amendment to the fiscal year military construction bill set to be heard on the Senate floor. Normally, the military construction bill is the first appropriation bill completed each year because it is seldom controversial. The Byrd-Warner Amendment made the 2000 bill very different.

The Byrd-Warner Amendment contained three main provisions. The first would have cut off funding for the deployment of US ground troops in Kosovo after 1 July 2001, unless the President requested, and Congress approved, a joint resolution specifically authorizing the deployment. The President was obligated to submit a report requesting specific authorization for continued deployment, along with a justification. The President was granted a waiver power for 180 days to cover emergencies, with further exemptions for military personnel supporting intelligence and surveillance efforts.

The second provision required the President, by July 2001, to develop a plan for changing the force in Kosovo from one comprised of US troops to one that did not include US troops. The President was to report regularly on the remaining number of US troops in Kosovo and the costs of the Kosovo operation.

The third provision of the Amendment provided that not more than 75 percent of the funds provided by the bill for fiscal year 2000 could be used until the President certified that the European allies had obligated at least 33 percent of the amounts they pledged for reconstruction assistance in Kosovo, 75 percent of pledges for humanitarian assistance, 75 percent of pledges for the Kosovo budget, and 75 percent of pledges of police personnel for the UN international force. This certification had a 15 July 2000 deadline. If the President did not submit the report by 15 July 2000, the remaining 25 percent of funds would be available only for the purpose of withdrawing US military personnel from Kosovo, unless Congress enacted a joint resolution that authorized that amount to be used for purposes other than withdrawal.

In commenting upon the amendment adopted by the Appropriations Committee, Senator Byrd said that the intent of Section 2410 was not to force a pull-out of US armed forces from Kosovo. Rather, it was to restore congressional oversight and to return to Congress its constitutional authority in such matters. Senator Warner noted that the Amendment had been circulating in draft form since March and had already served as a "wake-up call" to the European allies to expedite the process of fulfilling their commitments on rebuilding Kosovo. While acknowledging that the Allies had improved the pace of contributions, he felt that the amendment was still needed as a means for Congress to "exercise its constitutional duty".[4] The Amendment would allow the next President (after the 2000 presidential elections) to seek and receive, in mid-2001, congressional authorization to continue the deployment of US armed forces in Kosovo.

After considerable procedural maneuvering that is only marginally germane to this review, Senator Levin (D-Michigan), on behalf of himself and 13 co-sponsors, offered an amendment on the floor of the Senate to strike Section 2410 of the military construction bill that embodied the Byrd-Warner Amendment.[5] Levin outlined the principal arguments for striking the amendment, including the creation of a year

---

[4] 146 Cong. Rec. S3887-93 (daily ed. May 11, 2000) (statement of Sen. Warner).

[5] 146 Cong. Rec. S4071-72 (daily ed. May 17, 2000) (statement of Sen. Levin, including Letter from General Clark, see note 3).

and a half of uncertainty in the Balkans. Levin introduced into the record a letter from General Wesley Clark who had just stepped down as the head of European Command. In his letter, Clark argued that the Amendment would be seen as a *de facto* pull-out decision that would undercut the NATO allies as well as the US troops.

Senator Warner announced on the Senate floor that the House had just voted 264-153 to adopt "one-half of the provision we are debating...".[6] Senator Biden (D-Delaware) responded that he opposed the House provision, which he characterized as providing that if Congress refused to authorize the deployment by July 2001, the troops would return. Biden argued that stability in the Balkans was in the fundamental national interest of the United States. He believed that chaos in the Balkans would endanger stability in Europe.

Senator Byrd responded that the content of the amendment was being misrepresented, and he said it would be "irresponsible to vote now to take the troops out".[7] The Byrd-Warner Amendment contained an orderly procedure for more than a year, after which the President had to justify the continued deployment to Congress. Byrd argued that authorities of Congress had been usurped because Congress had slept on its rights. He stressed that the Amendment did not mandate the withdrawal of US ground combat troops from Kosovo. Byrd said it was "preposterous for General Clark or the administration to suggest that the Byrd-Warner Amendment could undermine" the bond between the United States and its European allies. He assumed that the administration could come up with a supportable case for continued US involvement in Kosovo and that Congress would endorse it.[8]

The leader of the majority party, Senator Trent Lott (R-Mississippi) also rose to correct misinformation about the Amendment. He said that the language would do two things: "require the President to certify by 15 July 2000 that our allies were fulfilling their commitments and require Congress to authorize the continuation of ground combat troops".[9] Lott felt it was time for Congress to step up to its responsibilities. He observed that the American people were "not really aware of the commitment ... and when people find out what we are doing there – the commitment we have there in terms of the facilities and the troops involved, and how much it will be costing; and the fact that we have never voted to authorize; we do not know where we are headed, how long it is going to take, how much it is going to cost, what the plan is – they are horrified".[10]

---

[6] 146 Cong. Rec. S4073 (daily ed. May 17, 2000) (statement of Sen. Warner).
[7] 146 Cong. Rec. S4080 (daily ed. May 17, 2000) (statement of Sen. Byrd).
[8] *Ibid.*
[9] 146 Cong. Rec. S4081 (daily ed. May 17, 2000) (statement of Sen. Lott).
[10] 146 Cong. Rec. S4082 (daily ed. May 17, 2000) (statement of Sen. Lott).

Senator Hollings (D-South Carolina) called the Kosovo deployment a mistake, not unlike Vietnam. He asked what kind of Kosovo policy and what kind of a military policy the US Government had. Hollings favored a European defense force and hoped it would take over. He noted: "We are not trying to send a message to Milosevic. We are trying to send a message to ourselves...".[11]

Senator Reed (D-Rhode Island), a member of the Armed Services Committee, saw destabilizing consequences from enactment of the Amendment. He expressed misgivings about the "open-ended commitment" in Kosovo but did not believe the Byrd-Warner Amendment was the proper means to deal with the problems.[12]

Senator Levin continued to assert that the issue was whether the Senate would set a deadline for the withdrawal of US ground forces from Kosovo by the middle of 2001. He referred to mass torture, rape, and looting as the norm in Kosovo just a year prior to the current debate. He cited a weekly murder rate of 50 now having been brought down to five. He emphasized that Congress had the power to cut off funds, but that this debate was about the wisdom of doing so. He saw the Amendment as creating dangerous uncertainty and instability. Levin referred to a letter from the Secretary General of NATO, expressing deep concern at the prospect of any NATO ally unilaterally deciding not to take part in a NATO operation.[13]

Levin also referred to an amendment a year earlier that would have prohibited the use of funds for the deployment of US ground forces in Yugoslavia, except for peacekeeping personnel, unless authorized by a joint resolution authorizing the use of military force. That amendment lost in the Senate by a vote of 52-48 and Levin felt that the Byrd-Warner Amendment likewise went "just a step too far".[14]

Senator Sessions (R-Alabama) considered the Kosovo intervention as a "colossal failure of diplomacy and a colossal failure of foreign policy".[15] He asked rhetorically why people were afraid to have a debate and vote on what he saw as a duty of a Member of Congress. He accused the President of consistently misleading the people of the United States and Congress. Sessions stated:

> "[W]e have forgotten the true facts of the situation, but we were told we were commencing and carrying out this war to stop ethnic cleansing. There had not been ethnic cleansing until the

---

[11] 146 Cong. Rec. S4084 (daily ed. May 17, 2000) (statement of Sen. Hollings).
[12] 146 Cong. Rec. S4085 (daily ed. May 17, 2000) (statement of Sen. Reed).
[13] 146 Cong. Rec. S4089 (daily ed. May 17, 2000) (statement of Sen. Levin).
[14] 146 Cong. Rec. S4090-91 (daily ed. May 17, 2000) (statement of Sen. Levin).
[15] 146 Cong. Rec. S4092 (daily ed. May 17, 2000) (statement of Sen. Sessions).

bombing started. It was 3 days after the bombing started that Milosevic sent his troops south into one of the most vicious displays of violence ... against a basically defenseless people".[16]

Sessions pointed out that relations deteriorated with Russia which "didn't like the idea of NATO attacking an independent sovereign nation".[17] He noted that the United States had troops committed all over the world and, while American taxpayers were paying the bill, the Senate "had not discussed the issue seriously".[18] He observed that 39 out of 100 Senators had voted against the bombing and that the House had never voted for it.

Senator Roberts (R-Kansas) welcomed an end to the open-ended peacekeeping operations in Kosovo and therefore supported the Byrd-Warner Amendment. He observed, tongue in cheek, that the Amendment has caused "so much of a fuss that the Senate of the United States is actually in the midst of a foreign policy debate, some $15 billion and 6 or 7 years into intervention in the Balkans".[19] Roberts felt it was arrogant to suggest that Europe could not do the job without the United States. He asked: how long "will [we] keep draining the limited US resources when we still cannot define what our long-term objectives in Kosovo are...?"[20]

Senator Lautenberg (D-New Jersey) reminded the Senate that he had been one of only three senators who had opposed the Amendment when offered in the Appropriations Committee. He saw the Serb's actions in Kosovo as genocide and cited with favor Senate Concurrent Resolution 21 authorizing US offensive military air strikes. The issue presented in the Byrd-Warner Amendment was over the continued deployment of US troops in a peacekeeping mission with NATO allies. Lautenberg supported this deployment to sustain US leadership around the world, to burden-share on peacekeeping and reconstruction, to promote stability in the Balkans and Europe, and to maintain credibility in NATO.[21]

Senator Inhofe (R-Oklahoma), the Chairman of the Senate Armed Services Subcommittee on Readiness, thought that he had made more field trips, both to Serbia and Kosovo, than any other Member. Inhofe began by quoting the UK Foreign Secretary, Robin Cook, who had to answer claims that government leaders misled the public on the scale of civilian deaths in Kosovo. Mr. Cook said:

---

[16] 146 Cong. Rec. S4093 (daily ed. May 17, 2000) (statement of Sen. Sessions).
[17] 146 Cong. Rec. S4094 (daily ed. May 17, 2000) (statement of Sen. Sessions).
[18] *Ibid.*
[19] 146 Cong. Rec. S4122 (daily ed. May 18, 2000) (statement of Sen. Roberts).
[20] 146 Cong. Rec. S4123 (daily ed. May 18, 2000) (statement of Sen. Roberts).
[21] 146 Cong. Rec. S4123-24 (daily ed. May 18, 2000) (statement of Sen. Lautenberg).

"At the height of the war, western officials spoke of a death toll as high as 100,000. President Bill Clinton said the NATO campaign had prevented 'deliberate, systematic efforts at ethnic cleansing and genocide'".[22]

Inhofe then quoted a Spanish pathologist responsible for accurate body counts: "I calculate that the final figure of dead in Kosovo will be 2,500 at the most".[23]

Inhofe did not see Kosovo involvement as being vital to American national security interests. "There is no clear mission objective or schedule to accomplish it ... no exit strategy".[24] He reminded the Senate that a resolution of disapproval to stop the President from sending troops lost by three votes. "We lost it because the President said all the troops they would send here, in December of 1995, would be home for Christmas, 1996".[25] Kosovo "... was a propaganda effort deliberately to make the American people believe things were going on there that were not going on there. ... So I know a lot of lies got us into this thing".[26] Inhofe quoted questions taken from an article by Henry Kissinger about future deployments: "What consequences are we seeking to prevent? What goals are we seeking to achieve? In what way do they serve the national interest?"[27] Kissinger concluded: "I cannot bring myself to endorse American ground forces in Kosovo. ... Each incremental deployment into the Balkans is bound to weaken our ability to deal with Saddam Hussein and North Korea".[28]

Senator DeWine (R-Ohio) expressed strong support for the Levin Amendment to strike the Byrd-Warner Amendment. He opposed setting an arbitrary deadline for the withdrawal of US forces from Kosovo. DeWine felt the United States needed to demonstrate its resolve and to maintain its credibility.

Senator Torricelli (D-New Jersey) stated that Congress had not been a "jealous guardian of its own constitutional prerogatives".[29] The President had to deploy troops in emergencies, but a near-permanent

---

[22] 146 Cong. Rec. S4127 (daily ed. May 18, 2000) (statement of Sen. Inhofe). Even in March, 2001, Pentagon officials refuse to answer the question of the number of confirmed civilian deaths from ethnic cleansing. Telephone Interview with Department of Defense, Office of Congressional Affairs (March 3, 2001).
[23] 146 Cong. Rec. S4127 (daily ed. May 18, 2000) (statement of Sen. Inhofe).
[24] *Ibid.*
[25] 146 Cong. Rec. S4128 (daily ed. May 18, 2000) (statement of Sen. Inhofe).
[26] *Ibid.*
[27] 146 Cong. Rec. S4129 (daily ed. May 18, 2000) (statement of Sen. Inhofe).
[28] *Ibid.*
[29] 146 Cong. Rec. S4131 (daily ed. May 18, 2000) (statement of Sen. Torricelli).

presence in Kosovo was no longer a crisis. Defending interests that Europe should cover was causing the United States to forgo defending more vital interests elsewhere.

Senator Kerry (D-Massachusetts) commented: "We are being asked to vote today as to whether or not we think the investment we made in the war itself is worthwhile".[30] His reply to himself was that the investment was clearly worthwhile.

Senator Conrad Burns (R-Montana) was the Chairman of the Subcommittee on Military Construction to which the Byrd-Warner Amendment was attached. Burns stated: "Congress has a constitutional responsibility to vote on long-term military commitments, especially when they are offensive and not defensive in nature".[31] He highlighted the congressional role in the case at hand, since "the administration has already spent $21.2 billion since 1992 in the Bosnia/Kosovo area".[32] The administration was taking congressional appropriations as a tacit approval by Congress for American involvement in Kosovo. Burns observed that Congress had a duty to scrutinize and authorize defense spending before, and not after, the fact.

Senator Feingold (D-Wisconsin) opposed the Levin Amendment and supported the Byrd-Warner Amendment because congressional approval was essential to the commitment of US troops in dangerous situations abroad. He favored regional leadership in regional conflicts and wanted Europeans to take over.[33]

Senator Thomas (R-Wyoming) saw two issues: the role of Congress and an exit strategy for Kosovo. He viewed "Kosovo foreign policy as a sort of oxymoron".[34] He recalled that the Senate Foreign Relations Committee had been told by the administration that "we would not be in Bosnia more than 18 months". "We were told we were not going to be in Kosovo".[35] Thomas flatly stated that no one was going to vote against support for troops who are already committed. Byrd-Warner mandated a plan to get the United States back on track and he supported the Amendment.

Senator Cleland (D-Georgia) supported the Byrd-Warner Amendment because it helped the Government articulate an exit strategy to bring US military forces out of Kosovo. He felt that US "allies are quite willing for us to stay there forever and ever and ever".[36] He was

---

[30] 146 Cong. Rec. S4132 (daily ed. May 18, 2000) (statement of Sen. Kerry).
[31] 146 Cong. Rec. S4133 (daily ed. May 18, 2000) (statement of Sen. Burns).
[32] *Ibid.*
[33] 146 Cong. Rec. S4134 (daily ed. May 18, 2000) (statement of Sen. Feingold).
[34] *Ibid.*
[35] *Ibid.*
[36] 146 Cong. Rec. S4135 (daily ed. May 18, 2000) (statement of Sen. Cleland).

against overextending the US military, and wanted to "Europeanize" the peace in Bosnia and Kosovo. He reminded the Senate that it had adopted, by voice vote on 30 June 1999, his amendment to immediately convene an international conference on the Balkans to develop a final political settlement.

Senator Smith (R-Oregon) wanted to maintain American leadership in Europe and was worried that the Kosovars would rearm if the United States withdrew. He did not want to tie the hands of the President and he could not stand by in the face of "mass murder" by Milosevic. Smith also noted that George W. Bush was opposed to the Byrd-Warner Amendment.[37]

Senator Robb (D-Virginia) argued against deadlines which would undermine the confidence of America's allies and shake the world's confidence in US leadership. He thought passage of the Byrd-Warner Amendment would place the United States in an untenable position in the Balkans.

Senator Kay Bailey Hutchison (R-Texas) addressed herself to two major points. The first was that the Byrd-Warner Amendment did not set a deadline for withdrawing troops. The second was that the Amendment was responsible action by the Senate as the President had been spending money without authorization and then coming to Congress asking for emergency funds for the Department of Defense.

Senator Hegel (R-Nebraska) supported the McCain-Levin Amendment to strike the Byrd-Warner Amendment, noting that 85 percent of the ground troops were provided by Europe. He argued that withdrawal was not a policy and that the responsible act was to leave the President a little latitude and flexibility.

Senator Voinovich (D-Ohio) thought that the Byrd-Warner Amendment was the wrong approach. He did not agree with threatening a unilateral troop pull-out from Kosovo but believed that the United States should provide leadership and "a little bit of financial commitment".[38] This was not a time to abandon US allies.

Senator Warner, the floor manager for the Byrd-Warner Amendment, pointed out that the approaching vote was an historic one. To strike the Amendment was to go back to business as usual of writing blank checks, which would total $8 billion for Kosovo alone.

Senator Levin, who acted as floor manager for the Levin-McCain Amendment to strike, responded that the Byrd-Warner Amendment was inconsistent in that it required the Europeans to do certain things by certain dates, but then said the United States was pulling out its troops anyway. He again emphasized that the debate was not about the power

---

[37] 146 Cong. Rec. S4139 (daily ed. May 18, 2000) (statement of Sen. Smith).
[38] 146 Cong. Rec. S4142 (daily ed. May 18, 2000) (statement of Sen. Voinovich).

of Congress to control the purse strings; it was about the wisdom of exercising that prerogative in this case.

Senator Lieberman (D-Connecticut) supported the motion to strike the Byrd-Warner Amendment. He argued that US military and economic ties to Europe were deep and that the ethnic cleansing and mass forced exile were unacceptable. He did not believe in a cut-and-run approach and observed that both presidential candidates Gore and Bush opposed the Byrd-Warner Amendment.[39]

Senator Byrd commented in support of his amendment that the debate was about power, the "arrogance of power". This referred to a White House that insisted on putting US men and women in harm's way and spending taxpayer's dollars without the consent of their elected representatives. He noted this was the first debate in the Senate since US ground forces had entered Kosovo 11 months earlier. He argued that the Amendment gave the President a year to come up with an exit strategy. As to NATO allies, they were being encouraged to take full responsibility for the ground combat troops. "The NATO alliance will not collapse if the United States does not have ground combat troops in Kosovo".[40]

Senator Byrd condemned the arrogant practice of spending the money first and then asking Congress after the fact to pay the bills. He saw this as an affront to the separation of powers in the US Constitution. Congress had a solemn duty to decide whether to ask young Americans to put their lives at risk, and to ignore Congress was to circumvent the Constitution. When he turned to war powers, Byrd held up a copy of the Constitution and declared:

> "These are the powers of Congress. Congress shall have the power 'To declare War'. Congress shall have the power to 'grant Letters of Marque and Reprisal'. Congress shall have the power to 'make Rules concerning Captures on Land and Water'. ... Congress also has the general power 'To raise and support Armies'. Congress shall have the power 'To provide and maintain a Navy'. Congress has the power 'To make Rules for the Government and Regulation of the land and naval Forces'. Congress shall have the power 'To provide for calling forth the Militia to execute the Laws of the Union, suppress Insurrections and repeal [sic] Invasions'. Congress shall have the power 'To provide for organizing, arming, and disciplining the Militias, and for governing such Part of them as may be employed in the Service of the United States'. Add to these powers contained in

---

[39] 146 Cong. Rec. S4145 (daily ed. May 18, 2000) (statement of Sen. Lieberman).

[40] 146 Cong. Rec. S4147 (daily ed. May 18, 2000) (statement of Sen. Byrd).

this Constitution the power 'to exercise exclusive legislation ... over all places ... for the erection of forts, magazines, arsenals, dock-yards, and other needful buildings...'. ... Congress has the power 'To lay and collect Taxes" to defend this country. Congress shall have the power to 'provide for the common Defense'. ... Congress shall have the power 'To borrow money on the credit of the United States'. ... Congress shall have the power 'To make all Laws which shall be necessary and proper for carrying into Execution the foregoing Powers'.

And finally, this Constitution says, Congress has the greatest power of all. Congress is given the power in section 9, article I: 'No money shall be drawn from the Treasury, but in Consequence of Appropriations made by law'. Thus, the scope of the warpower [sic] granted to Congress is, indeed, remarkable. The intent of the framers is clear.

Now let us examine the war powers that flow from the Constitution to the President of the United States. In section 2, article II, the Constitution states: 'The President shall be Commander in Chief of the Army and Navy of the United States, and of the Militia of the several States, when called into the actual Service of the United States'.

That is it. That is it, lock, stock, and barrel, except the Constitution says that the President 'shall Commission all the officers of the United States'. But that is it. So compare what the Constitution says with respect to the powers of the Congress when it comes to warmaking [sic], when it comes to the military, with the powers the Constitution gives to the President".[41]

Byrd then reviewed pertinent views on war powers expressed by founding fathers who were directly involved in writing the US Constitution, including Pinckney, Wilson, Hamilton, Madison, Gerry, Sherman, Ellsworth, Mason and Jefferson. Byrd also made reference to other constitutional scholars.[42] Byrd highlighted as a great principle in free government the separation of the sword from the purse.[43] He asked why the administration was "hysterical" and "panic stricken". Was its case so weak that it could not withstand the scrutiny of Congress? Byrd

---

[41] 146 Cong. Rec. S4149 (daily ed. May 18, 2000) (statement of Sen. Byrd).
[42] 146 Cong. Rec. S4149-50 (daily ed. May 18, 2000) (statement of Sen. Byrd).
[43] 146 Cong. Rec. S4150 (daily ed. May 18, 2000) (statement of Sen. Byrd).

warned that "Congress can sleep on its rights until it can no longer claim those rights".[44]

Senator John McCain (R-Arizona) began by criticizing the executive branch and Congress for ignoring the War Powers Act. McCain complained that the Byrd-Warner Amendment was the wrong vehicle for so grave a debate. He faulted the committee chairmen in the Senate for having "abrogated their responsibilities" in failing to have the debate in proper Senate channels.[45] He felt Congress should be voting to fund or not fund the Kosovo operation. McCain quoted Governor Bush as labeling the Byrd-Warner Amendment as an overreach of congressional authority.[46] McCain chided the administration for getting into Kosovo and for carrying out "one of the more immoral military actions in the history of this country".[47] He referred specifically to the flying of US planes at high altitudes to avoid being shot down, but in so doing inflicting needless civilian casualties. McCain wanted the vote to be on whether or not it was in the US national security interest to have a military presence in Kosovo. He commented on the "propensity of the Administration to deploy American military forces with seemingly wanton abandon on ill-defined missions of indeterminate duration [which] is repeatedly met with efforts by Members of Congress to legislate the terms of those deployments".[48] He opposed the Byrd-Warner Amendment because it would set a negative precedent for future Presidents.

Senator Wellstone (D-Minnesota) distinguished between real hostilities and the peacekeeping operations in Kosovo. He argued that the "peacekeepers deserve a chance to stay".[49]

Senator Chris Dodd (D-Connecticut) stated that it would be incredibly foolish to announce the exact date for withdrawal to the enemy and he opposed the Byrd-Warner Amendment. He felt that it would be irresponsible for the Senate to approve legislation mandating an end to US participation in Kosovo.

Senator Roth (R-Delaware) supported the Levin Amendment. He believed that the Byrd-Warner Amendment would undercut the Europeans since the bill would be seen as a decision to withdraw forces at a given date.

Senator Diane Feinstein (D-California) had serious concerns about the potential impacts of the Byrd-Warner Amendment. She worried that

---

[44] Ibid.
[45] 146 Cong. Rec. S4151 (daily ed. May 18, 2000) (statement of Sen. McCain).
[46] 146 Cong. Rec. S4152 (daily ed. May 18, 2000) (statement of Sen. McCain).
[47] Ibid.
[48] Ibid.
[49] 146 Cong. Rec. S4154 (daily ed. May 18, 2000) (statement of Sen. Wellstone).

technical reasons might preclude securing the presidential certifications required by the Amendment, thereby forcing a withdrawal of US forces from Kosovo.

Senator Slade Gorton (R-Washington) commented that he had voted against the initial Senate resolution to authorize air attacks against Yugoslavia. This was a civil war and NATO's actions had "fostered the creation of an entirely new class of refugees; the US military has been required to police the region for an unspecified amount of time...".[50] It was time to leave the Balkan conflict to the Europeans.

Senator Olympia Snowe (R-Maine) supported the Byrd-Warner Amendment, as it would force consideration of the open-ended commitment to Kosovo. She chided the Europeans for lack of support for the Kosovo Police Force. She concluded that "we just cannot afford to unilaterally deploy troops and provide monetary support to each global hot spot for an indefinite period of time, with tepid and inconsistent support from the UN, NATO, and out other allies".[51]

Senator Robert Smith (R-New Hampshire) began by stating that the air war against Serbia was "inaccurately portrayed".[52] The administration had grossly exaggerated "the results of the air campaign in an attempt to buy public support for the war".[53] The main result of the bombing was to trigger a refugee crisis. The real threat to NATO is that it had abandoned its defensive posture and "blundered and contorted into a post-cold war crisis management agency with a lost sense of mission".[54] NATO's bombing killed innocent civilians and raised regional tensions. By demonizing Milosevic, the United States had become a tacit ally of the Kosovo Liberation Army, a terrorist group. Smith supported the Byrd-Warner Amendment and wanted to stop using the US military as a police force. Kosovo was a colossal mistake and the United States was blundering towards a crisis in Montenegro.

Senator Warner summed up the argument for the Byrd-Warner Amendment by pointing out the twofold purpose of the section. First, it required Congress to fulfill its co-equal constitutional responsibility with the President to make decisions on the deployment of armed forces to Kosovo. Secondly, the amendment sent the message that other nations and organizations must follow through with their commitments.[55]

The Minority Leader, Senator Daschel (D-South Dakota), was first concerned that passing the Amendment would increase the risk to US forces. Secondly, passage would reward Slobodan Milosevic for his

---

[50] 146 Cong. Rec. S4157 (daily ed. May 18, 2000) (statement of Sen. Gorton).
[51] 146 Cong. Rec. S4158 (daily ed. May 18, 2000) (statement of Sen. Snowe).
[52] 146 Cong. Rec. S4159 (daily ed. May 18, 2000) (statement of Sen. Smith).
[53] Ibid.
[54] Ibid.
[55] 146 Cong. Rec. S4160 (daily ed. May 18, 2000) (statement of Sen. Warner).

ethnic cleansing campaign. Thirdly, passage would "rupture NATO, and, finally, would undermine the US position as a global leader".[56]

Senator Lott, the Majority Leader, complimented the Senate on a constructive and healthy debate. Lott noted that, under the Byrd-Warner Amendment, the United States would still be able to provide logistics, intelligence and other support to the Kosovo deployment. The Amendment would deal with the issues of "... no long-term plan for Kosovo, ... how long we are going to be there ... how much it is going to cost ... allies not ... meeting their commitments...".[57] Moreover, the Amendment would deal with Congress "abdicating" its prerogatives.[58]

The Clinton administration was sufficiently worried about the outcome of the vote on the Byrd-Warner Amendment that it took the highly unusual action of asking Vice President Gore to preside over the Senate in the event of the necessity to break a tie. With Gore in the chair, the Byrd-Warner Amendment was defeated when the Senate passed the amendment to strike, yeas 53, nays 47.[59]

---

[56] 146 Cong. Rec. S4162 (daily ed. May 18, 2000) (statement of Sen. Daschle).
[57] 146 Cong. Rec. S4163 (daily ed. May 18, 2000) (statement of Sen. Lott).
[58] *Ibid.*
[59] *Yeas: 53*

Abraham	DeWine	Kennedy	Murray
Akaka	Dodd	Kerrey	Reed
Baucus	Dorgan	Kerry	Robb
Bayh	Durbin	Landrieu	Rockefeller
Biden	Edwards	Leahy	Roth
Bingaman	Feinstein	Leiberman	Sarbanes
Boxer	Frist	Levin	Schumer
Breaux	Graham	Lincoln	Smith (OR)
Bryan	Hagel	Lugar	Thompson
Chafee, L.	Harkin	Mack	Voinovichy
Cochran	Hatch	McCain	Willstone
Conrad	Jeffords	Mikulski	Wyden
Daschle	Johnson	Moynihan	

Nays: 47

Allard	Craig	Hollings	Santorum
Ashcroft	Crapo	Hutchinson	Sessions
Bennett	Domenici	Hutchison	Shelby
Bond	Enzi	Inhofe	Smith (NH)
Brownback	Feingold	Inouye	Snowe
Bunning	Fitzgerald	Kohl	Specter
Burns	Gorton	Kyl	Stevens
Byrd	Gramm	Lott	Thomas
Campbell	Grams	McConnell	Thurmond
Cleland	Grassley	Murkowski	Torricelli
Collins	Gregg	Nickles	Warner
Coverdell	Helms	Roberts	

V. Concluding Thoughts

In spite of the importance of the issues involved, the debate in the Senate received little media coverage in the United States. The Clinton White House and members of the Senate, however, fully understood the fundamental principles at stake. The vote in the Senate Appropriations Committee (yeas 23, nays 3) in favor of the Byrd-Warner Amendment probably was a more accurate reflection of the true sentiment in the Senate than was the narrow vote to strike on the full Senate floor. The administration clearly had to come from behind to win on the floor after the overwhelming support for the amendment evidenced in the Senate Appropriations Committee vote. The Clinton administration prevailed largely because of Democratic Party discipline and the support of two Republicans concerned about Presidential prerogatives: George W. Bush and John McCain.[60]

It would be an error to think that the debate in the Senate on Kosovo is over. The differences remain and the results could well be different if the vote is not taken at the height of a US presidential campaign.[61]

---

[60] The Clinton administration was so concerned about the outcome of the vote that it flew in former Senators who continue to enjoy floor privileges to assist with administration lobbying efforts to defeat the Byrd-Warner Amendment. They included Ambassador George McGovern, who was brought in from his ambassadorial post to the FAO in Rome.

[61] In the past eight years, only twice has the Vice President cast tie-breaking Senate votes on budget issues. See G. Kessler, *Cheney Breaks Tie on Tax Plan*, Washington Post, April 4, 2001, A4.

International Organizations and Use of Force

Sreenivasa Rao Pemmaraju[*]

In a number of instances, States or groups of States, including NATO, have employed military action and engaged in large-scale armed attacks on other States by way of countermeasures or sanctions in response to alleged wrongful acts or in defense of human rights where egregious abuses are involved.[1] In almost all these cases, it is claimed that such unilateral use of force is unavoidable and necessary, as no action by the United Nations is possible. The lawfulness of such unilateral use of force without authorization from the United Nations is then a question. This issue will be the subject of this article. We shall deal with the principle of use of force under the UN Charter and the requirements and limitations governing the right of self-defense; the problems concerning the collective security system of the United Nations; the powers of the Security Council and the General Assembly; justifications offered for humanitarian intervention, including the scheme of the ILC dealing with state responsibility in respect of violation of *erga omnes* obligations; and the attempts at legal regulation of unilateral acts or countermeasures,

---

[*] Views expressed in this article are exclusively those of the author in his personal capacity and do not represent the position of the government and the other organizations with which he is officially associated.

[1] Some examples are the Israeli attack on Egypt in early 1950 in response to Fadeyen attacks on Israel; the Turkish invasion of Iraq in pursuit of insurgent Turkish Kurds (1995); Turkish actions in Cyprus in 1964 and in 1974 to defend the Turkish Cypriot population; Israel's raid on Entebbe airport in 1976; US bombing of a terrorist training camp in Afghanistan and a factory in Sudan (1998); humanitarian intervention; the Tanzania-Uganda incident (1978); Vietnam-Kampuchea (1978-1979); France, United Kingdom and the United States-Iraq (the Kurdish question, 1991); ECOMOG-Liberia, Sierra Leone (1988 to 1999); and NATO-Yugoslavia (Kosovo, 1999). For a discussion and other instances, see T. M. Franck, "The Use of Force by States Without Prior Security Council Authorization", The Lauterpacht Lectures, 21-23 November 2000, lectures 2 and 3.

including some academic suggestions in this regard and difficulties arising therefrom. Finally, we conclude with a statement on current law.

I. The Principle of Non-Use of Force under Article 2(4) and the Right of Self-Defense under Article 51 of the UN Charter

The authority for international organizations other than the United Nations to use force is well defined and very limited. Under Article 53 of the UN Charter, they can only engage in an enforcement action under the authority of the Security Council.[2] However, only in one respect – that is, measures to be taken against an enemy State provided for pursuant to Article 107 – can such regional arrangements or agencies engage in enforcement action without authorization from the United Nations.[3] This provision is irrelevant to our present consideration and has never been invoked.

Article 53 is part of a wider scheme in the UN Charter under which use of force is prohibited against the territorial integrity or political independence of any State or in any other manner inconsistent with the purposes of the United Nations. Secondly, under Article 24, the Security Council has been given the primary responsibility for the maintenance of international peace and security. The UN Member States conferred this primary responsibility on the Security Council in order to ensure prompt and effective action by the United Nations and further agreed that, in carrying out its duties under this responsibility, the Security Council acts on their behalf. It is also interesting to note that Article 24(2) specifically directs the Security Council that, in discharging its duties to maintain international peace and security, it is to act in accordance with the purposes and principles of the United Nations.

The only exception to the prohibition against the use of force is the inherent right of individual or collective self-defense recognized under Article 51: if an armed attack occurs against a UN Member State. However, measures of self-defense taken should immediately be reported to the Security Council and may only be engaged in until the Security Council has taken measures necessary to maintain international peace and security. Article 51 makes it clear that the inherent right of

---

[2] G. Gaja, "Use of Force Made or Authorized by the United Nations", in C. Tomuschat (ed.), *The United Nations at Age Fifty: A Legal Perspective*, 1995, 39 et seq., 46. One incidental question is whether the United Nations, invoking Article 53 of the UN Charter, can authorize a regional organization to take action against a State which is not a party to the regional arrangement. According to Gaja, such an authorization is not possible, given the context of Chapter VIII.

[3] See H. Kelsen, *The Law of the United Nations*, 1951, 327-328.

individual or collective self-defense recognized under that article does not in any way affect the authority and responsibility of the Security Council, under the Charter, to take such action as it deems necessary in order to maintain or restore international peace and security at any time.

It is generally agreed that international organizations may use force without the authorization of the Security Council as a measure based on the inherent right of collective self-defense under Article 51 when one or more of its members are under an armed attack.[4]

The occurrence of an armed attack is an objective fact. In the *Nicaragua* case, the Court specified elements constituting an armed attack. These included not merely action by regular armed forces across an international border, but also the sending by or on behalf of a State of armed bands, groups, irregulars or mercenaries that carry out acts of armed force of such gravity against another State as to amount to an actual armed attack conducted by regular forces, or its substantial involvement in it.[5] However, the Court distinguished an armed attack from assistance to rebels in the form of the provision of weapons or logistical or other support which amounted to unlawful intervention.[6] Similarly, isolated acts of armed incursion on a limited scale for a limited duration do not constitute an armed attack. It is also suggested that the intention of the attacking country in launching or stationing a

---

[4] See Y. Dinstein, *War, Aggression and Self-defense*, 1988, 230-253. At a minimum, it is understood that the right to self-defense is conceived as an inherently individual right of a State specifically concerned. Only through the conclusion of the specific defense alliances could States broaden the right to self-defense to all members of that alliance. See the dissenting opinion of Sir Robert Jennings in the *Nicaragua* case, Merits, ICJ Reports 1986, 545. It is also clear that States may help either with use of force or extend any other help, such as imposing economic sanctions or a total boycott of the aggressor State. See J. A. Frowein, "Reactions by Not Directly Affected States to Breaches of Public International Law", *RdC* 248 (1994-IV), 370.

[5] ICJ Reports 1986, 103, para. 195. According to one commentary, it appears beyond doubt that an armed attack has occurred "when armed forces of one State, regular, irregular, armed bands composed of private individuals controlled by and in fact remaining under the orders of, the State, start using violence in or against the territory of another State, or against its forces on or over the open sea or its forces that stay in foreign territory either by agreement of the sovereign or by virtue of a lawful military occupation". K. Skubiszweski, "Use of Force by States. Collective Security. Law of War and Neutrality", in Max Sorensen (ed.), *Manual of Public International Law*, 1968, 777.

[6] ICJ Reports 1986, 104, para. 195. Furthermore, frontier incidents consisting of shooting across the border or incursions of a short duration into the territory of another State by small detachments of frontier guards or other units controlled by the State are not to be regarded as armed attack. Skubiszweski, see note 5, 777.

missile is relevant when considering whether such an act constituted an armed attack.[7] Armed attacks are also distinguished and treated differently from the threat or use of force or intervention in the internal affairs of another State. In the latter cases, while the right of self-defense is said to be unavailable, other proportional measures short of armed attack to redress the wrongful act committed are permitted.[8]

While Article 51 makes an *armed attack* a precondition for the exercise of the inherent right of self-defense, the question arises whether armed attack can be anticipated and military action in self-defense can be mounted even before the occurrence of an armed attack. A number of influential writers have taken the view that Article 51 does not abrogate the right of self-defense under customary international law, which itself does not restrict the right only to the case of armed attack.[9] Several other equally prominent authorities on the subject oppose this view. According to them, the UN Charter provides a total prohibition on the use of force in international relations and places a monopoly on enforcement action in the hands of the Security Council. Following this scheme, Article 51 is a deliberate and specific exception and which does not admit any residuary right of self-defense.[10]

---

[7] A missile launched with the intent to attack, even if it did not hit the target, could be regarded as an armed attack. Similarly, the positioning of missiles in the absence of any intention to attack may not amount to an armed attack. Skubiszweski, see note 5, 778.

[8] The Court also ruled that under contemporary international law, there existed no general right of intervention in support of an opposition within another State. It also concluded that acts constituting a breach of the customary principle of non-intervention would also, if they directly or indirectly involve the use of force, constitute a breach of the principle of non-use of force in international relations. *Nicaragua* case, see note 4, para. 209, 110. In other words, where there existed no armed attack, no military action is possible by way of collective self-defense.

[9] See Bowett, *Self-defense in International Law*, 1958, 185-186; J. Stone, *Aggression and World Order*, 1958, 43 and 95-96, H. Waldock, "General Course on Public International Law", *RdC* 106 (1962), 231-237, J. Brierly, "The Law of Nations", 6th ed., 1963, 417-418; and D. P. O'Connell, *International Law*, 2nd ed., Vol. 1, 1970, 317.

[10] Brownlie, rejecting the argument of the "partisans of the broader doctrine of self-help", who contend that Article 51 formally reserved the right of self-defense but does not define its content, stated that: "Such an interpretation is perverse and would render the exception to the principle stated in Article 2, paragraph 4, substantially subversive of the principle. At least, it can be accepted that the right of self-defense exists in customary international law and this was recognized by the International Court in the judgment on the Merits in the *Nicaragua case*. The Court, not very surprisingly, adopted the view that the customary law right depended on the existence of an armed attack". See I. Brownlie, "International Law at the Fiftieth Anniversary of

In the *Nicaragua* case (1986),[11] the ICJ endorsed the latter view and held that not only should a State regard itself as having been subject to armed attack, but it should also specifically seek third-party assistance for the right of collective self-defense to operate. Following this decision, Brownlie observed that Article 51 "falls to be considered as part of the general evidence of the content of the right of self-defense in customary law".[12]

The right of self-defense can further be justified only in those circumstances in which "the necessity of the self-defense is instant, overwhelming and leaving no choice of means and no moment for deliberation".[13] This principle generally rules out anticipatory self-defense, except in the case of an attack that is well prepared and about to be mounted. However, in modern times, given the sea change that took place in the development of nuclear weapons and missile technology, and the limited response time that is allowed to States to exercise the right of self-defense, the application of this principle cannot be too rigid.[14] In spite of this, confining the right of self-defense to armed attacks and allowing armed attacks to be interpreted in a flexible manner is considered the best alternative compared to attempts to define armed attack too broadly.[15] The State claiming the right of self-defense

---

the United Nations, General Course on Public International Law", *RdC* 255 (1995), 203-204. For a similar view, see also J. de Aréchaga, "International Law in the Past Third of the Century", *RdC* 159 (1978), 87-98; and Kelsen, see note 3, 914. Given the special scheme of the UN Charter dealing with collective security and the new approach it adopted to self-defense, Kelsen felt that the right of self-defense under the Charter scheme offered protection only against the illegal use of force and not against other violations of law. H. Kelsen, "Collective Security and Collective Self-defense under the Charter of the United Nations", *AJIL* 42 (1948), 783 and 874. As also noted by Skubiszewski, the decisive factor becomes not the content of the right in question, and the measure or extent of its violation, but the form in which such violation takes place: that form must be an armed attack. Consequently, it is noted that "any preventive, anticipatory or preemptive use of force to the occurrence of an armed attack cannot be regarded as action in self-defense". It is also noted that in several instances the United Nations gave its backing to the restrictive interpretation and refused to regard offensive use of armed force as self-defense, although it did not deny that important interests of States were involved. Skubiszewski, see note 5, 767. For a description of traditional means of compulsion other than war: retorsion, reprisals, pacific blockade and armed intervention, see *ibid.*, 753-760.

[11] *Nicaragua* case, see note 3, 14 et seq.
[12] Brownlie, see note 10, 204.
[13] *The Caroline* case reported in Moore, *Digest of International Law*, Vol. 2, 409. This is known as the Webster formula endorsed in *The Caroline* case between the States and the United Kingdom.
[14] I. Brownlie, *International Law and Use of Force by States*, 1963, 368.
[15] M. N. Shaw, *International Law*, 4th ed., 1997, 790.

inevitably has the right to judge when the armed attack occurred. However, the Security Council and the General Assembly and, where appropriate, an international tribunal should ultimately be the final judge to determine the correctness of such a claim.

It has been noted that the Security Council authorized the use of force in a situation of civil war, as in Somalia and Rwanda, or in order to restore democratic rule, as in Haiti, even when there was no instance of armed attack in the sense of Article 51 of the UN Charter. In these instances and others, it was asserted that such authorization was compatible with the Charter system. One substantive requirement, however, is the determination of a threat to the peace by the Security Council under Article 39.[16] In this connection, it is further noted that the term "threat to the peace" was given a wide meaning, even when there is no imminent or actual armed attack, that includes gross violations of fundamental norms of international law and those involving especially egregious civil wars. In this connection, a denial of freedom and self-rule to a majority of the population by a minority government, as in the case of Rhodesia; massive violence against and killing of the population by a racist régime, as in the case of South Africa; past conduct, military policy and repression of the civilian population in many parts of the country, as in the case of Iraq; and a failure to surrender to foreign States nationals that were indicted for the bombing of an aircraft, as in the case of Libya in the Lockerbie affair, were all cited as situations which the Security Council considered a threat to the peace within the meaning of Article 39 of the UN Charter.[17]

Finally, the Security Council considered the civil war-based crises which enveloped the former territories of Yugoslavia as a threat to international peace and security. This case further extended the meaning given to the words "threat to the peace" for invoking sanctions under Chapter VII, even in the absence of actual or anticipated armed attack.[18] Accordingly, it was possible for Combacau to define a threat to the peace in the sense of Article 39 as "a situation which the organ, competent to impose sanctions, declares to be an actual threat to peace".[19] This complete discretion of the Security Council, which is said

---

[16] See Gaja, supra note 1, 39 et seq., 45.

[17] See T. M. Franck, "The Security Council and 'Threats to the Peace': Some Remarks on Remarkable Recent Developments", in R.-J. Dupuy (ed.), *Peace Keeping and Peace Building: The Development of the Role of the Security Council*, The Workshop of the Hague Academy of International Law, The Hague, 21-22 July 1992, 1993, 83 et seq.

[18] For a review of the above situations, see *ibid.*

[19] Author's translation of J. Combacau, *Le pouvoir de sanction de l'ONU*, 1974, 100, quoted by P. H. Kooijmans, "The Enlargement of the Concept 'Threat to the Peace'", in Dupuy (ed.), see note 17, 111 et seq., 111, n.1.

to be beyond any challenge,[20] is in fact questionable because of the unrepresentative character of the Security Council and the fact that most of the decisions related to cases involving developing countries. In addition, such a determination made under Article 39, not with a view to triggering the exercise of the Council's power to take enforcement action – namely to enforce the peace – but with a view to affect the rights and obligations of States and thereby to determine, declare or enforce international rights or obligations, could also be challenged as outside the scope of the Security Council's competence. Holding such a view, Arangio-Ruiz felt that "an absolute monarchy of Security Council (or its permanent members) would not mark an improvement over the anarchy of the interstate system".[21]

II. Use of Force for Enforcement of International Human Rights in the Absence of Action by the UN Security Council

However, in most of the cases where the Security Council decided on action under Chapter VII after determining the existence of threat to the peace, even in the absence of an armed attack, such action did not involve enforcement action under Article 42, but only military interventions that were limited in scope and time in order to provide temporary humanitarian relief.[22] Accordingly, the question arises whether these instances could be cited as a legitimate basis for taking full-fledged military action to enforce human rights and prevent their serious abuse.

Furthermore, it is even more questionable whether States can unilaterally determine a threat to the peace on their own and take unilateral military action to deal with egregious abuses of human rights. Even if these actions are open to review by the Security Council later, it does not appear to meet the proper requirements of the UN collective security system and may almost amount to imposing a *fait accompli* on the international community. In addition, it is against "the spirit of the Charter to impose sanctions if the threat is not actual and efforts to resolve the dispute peacefully have not yet been completely exhausted".[23]

It is for these reasons that many unilateral enforcement actions to defend human rights and so-called humanitarian interventions, when reviewed by the United Nations and particularly the General Assembly,

---

[20] Kooijmans, see note 19, 111.
[21] See G. Arangio-Ruiz on the Security Council's "Law-Making", *Riv. Dir. Int.* LXXXIII, Fasc. 3 - 2000, 609 et seq., 631, 725.
[22] Gaja, see note 2, 45.
[23] Kooijmans, see note 19, 117.

actually met with disapproval or mild approbation and could not get approval or endorsement.[24] As Schachter has pointed out "no united resolution has supported the right of a State to intervene on humanitarian grounds with armed troops in a State that has not consented to such intervention. Nor is there evidence of state practice and related *opinio juris* on a scale sufficient to support a humanitarian exception to the general prohibition against non-defensive use of force.[25] Brownlie also concurs and does not believe that there is any support in customary international law for the proposition that a modern customary law of humanitarian intervention, which may condone action to protect lives, providing it is short and results in fewer casualties than would have resulted from non-intervention, is beginning to take shape.[26] "It is accepted", he notes, "that the occasions on which States have invoked humanitarian considerations to justify the use of force within and against another State do not inspire confidence in the new doctrine". He adds: "Such interventions are commonly based on a collateral political agenda and involve considerable loss of life, the existence of which is obscured by the manipulation of news media".[27] Kamalesh Sharma, Permanent Representative of India to the United Nations, also notes that "deep concern at humanitarian crisis should not obscure the reality that action is prone to being viewed through a political prism".[28]

---

[24] Franck, see note 1, Lecture 3, 1-21, 22. "States do not accept that any unilateral, unauthorized use of force in the national interest is *per se* acceptable".

[25] O. Schachter, *International Law in Theory and Practice*, 1991, 124. It is also observed that "although State governments have often violated human rights, genuine instances of 'humanitarian intervention' have been rare, if they have occurred at all". It is further concluded that, "in the light of state practice since 1945", the norm prohibiting humanitarian intervention "is both authoritative and controlling". A. C. Arend and R. J. Beck, *International Law and Use of Force*, 1993, 137.

[26] Brownlie, see note 10, 206. For the view of Franck to which Brownlie referred to in this connection, see T. Franck, "Fairness in the International Legal and Institutional System", *RdC* 240 (1993-III), 256-257.

[27] Brownlie, *ibid.*

[28] Statement of Kamalesh Sharma, UN General Assembly, 6 October 1999. Sharma also noted:
"It is clear that the emergence of a principle of armed intervention to redress humanitarian issues would set us on a perilous slope because, in principle at least, there would be no limits to it; because its underlying premise would be based on a dubious presumption that external forces can resolve all problems in every part of the world; and because the UN and the international community have neither the resources nor the capability to undertake it... Another danger is that theories of intervention seeking to justify interference and use of force to fight alleged repression may end up strengthening the hands of covert interventionist".

A Danish report on the legality of humanitarian intervention states that humanitarian intervention in the absence of authorization from Security Council or justification based on customary international law may only be conceived of as admissible in extreme cases and as an emergency exit. The Advisory Council on International Affairs (AIV) and Advisory Committee on Issues of Public International Law (CAVV) of the Netherlands came to a similar conclusion. In their opinion, neither the Charter nor the customary law doctrine of state responsibility, particularly a state of necessity or distress could provide justification for unauthorized humanitarian intervention.[29] It is also their view that there is no basis to conclude that there is any new legal basis for unauthorized humanitarian intervention as new customary international law, in spite of a limited number of cases of unauthorized humanitarian intervention in Iraq (the Kurdish question), Somalia, Bosnia Herzegovina, Liberia, Rwanda, Haiti, Albania, Sierra Leone, the Central African Republic, Kosovo and East Timor. In this connection, they point out that even though States used such terms as "legitimate", "justified", or "in accordance with the provisions and/or purposes and principles of the UN Charter" and "in accordance with general international law" in approving one or more of the interventions in question, no *opinio juris* existed to give rise to a new customary international law, particularly because of the differing views of major countries such as Russia, China and India.[30] It is also noteworthy that this report pointed out that there was no agreement even among the members of the NATO on the legal basis for the action they took in Kosovo. Although there was a consensus on the need for intervention, each Member State had its own views regarding the justification for it.

III. Collective Security System of the United Nations: Relative Powers of the Security Council and the General Assembly

Against the preceding background it is difficult to accept, as argued by Reisman, that international law now sets as an imperative objective a peremptory standard by which the behavior of governments is to be tested and, where necessary, restrained and sanctioned, with the ultimate mandate to use force.[31] Furthermore, equally against present world

---

[29] Copy on file with the author, 20.
[30] *Ibid.*, 22.
[31] See W. M. Reisman, "Kosovo's Antinomies", *AJIL* 93 (1999), 862. Cassese, after reviewing the views of several States expressed during and since the Kosovo crisis, noted that even though *opinio necessitatis* has been forcefully and loudly proclaimed by the States engaged in military actions to avoid genocide, there were also protests and criticisms. Countries like Germany

constitutive process as reflected in the UN Charter is his argument that when human rights enforcement by military means is required, while it is the responsibility of the Security Council to take action, in the face of its inaction, "the legal requirement continues to be to save lives however one can and as quickly as one can".[32]

However, it should be admitted that the collective security system of the United Nations leaves a lot to be desired. The idea that the Security Council is a trustee on behalf of all the UN Member States to maintain international peace and security or to restore them when they are threatened, breached or otherwise broken has hardly taken any root in the practice of the United Nations. It is equally clear that whereas prompt and effective action should be taken by the Security Council as a matter of duty to maintain or restore international peace and security, keeping in view the purposes and principles of the United Nations, the Security Council's decisions are based on the political interests of the members of the Security Council, particularly the permanent members of the Security Council.

While the inaction of the Security Council is attributable to the right of veto of a permanent member, it is rarely, if ever, examined whether such a member is entitled to exercise its right of veto contrary to the purposes and principles of the United Nations.[33] It may also be noted

---

and Belgium explicitly stated that the episode should not set a precedent. Opposition or reservation of States might also be due to the conduct of hostilities by the armed forces of NATO countries and the heavy causalities they caused. In view of this, he came to the conclusion that "although the psychological element of customary law has come into being, it does not yet possess, however, the requisite elements of generality and non-opposition". It goes without saying, according to him, "that enforcement action taken by the Security Council, or authorized, remains the lynch pin of the present system for the maintenance of international peace and security". A. Cassese, "A Follow up: Forcible Humanitarian Countermeasures and Opinio Necessitatis", *EJIL* 10 (1999), 791-99, 798 and 799.

[32] *Ibid.* This point was further elaborated by W. M. Reisman, "Unilateral Action and the Transformations of the World Constituting Process: The Special Problem of Humanitarian Intervention", *EJIL* 11 (2000), 3 et seq. For the view that the UN Charter represents a world constitution, see B. Simma, *The Charter of the United Nations: A Commentary*, 1995. See also H. Mosler, "The International Society as a Legal Community", *RdC* 140 (1974); C. Tomuschat, "Obligations Arising for States Without or Against their Will", *RdC* 241 (1993), 216-240, particularly references cited at note 24, 218; and B. Simma, "From Bilateralism to Community Interest in International Law", *RdC* 250 (1994), 221 et seq.

[33] In at least one incident it was noted that one permanent member, after having first vetoed, adjusted its policy in response to criticism in the General Assembly to allow approval of a resolution in the Security Council to send an observer group to Guatemala as part of implementation of the peace

that under Article 27(3) a permanent member of the Security Council is obliged not to participate in decisions under Chapter VI and under Article 52(3) concerning a dispute to which it is a party. This provision does not sufficiently inhibit the freedom of a permanent member from participating in the decisions of the Security Council under Chapter VI for various reasons. First, the Security Council generally does not specify any chapter of the Charter as a basis for its action, except in some cases when it has taken action under Chapter VII. Secondly, a permanent member of the Security Council is rarely directly involved in a dispute under consideration, even though it might have substantial political interest in the matter. Thirdly, it is difficult often to distinguish disputes arising under Chapter VII and Chapter VI once a determination is made under Article 39, or when dealing with disputes which eventually lead to such a decision.[34]

As the Security Council's actions are essentially based on political considerations, the question has often arisen as to whether its decisions (or inability to make decisions because of the use of the veto) could be judicially reviewed by the ICJ. The system of judicial review is inherent in the constitution of several countries, but not directly envisaged in the UN Charter.[35]

However, it is clear that the Court is not prevented from passing judgment on the decisions, the lack of decision or the failure to make a decision by the Security Council if these are matters of contention in a dispute in which it has established its jurisdiction. These matters have on a few occasions come up for consideration before the Court, but it stopped short of pronouncing upon its right of judicial review. Judge Fitzmaurice, in his separate opinion in the advisory opinion of the Court in the *Namibia (South West Africa)* case (1970), noted that States are not bound to give effect to decisions of the Security Council that are not in accordance with its powers and functions under the Charter.[36] The UN Security Council, being a creature of the Charter, is bound by its

---

process. See W. M. Reisman, "Toward a Normative Theory of Differentiated Responsibility for International Security Functions: Responsibilities of Major Powers", in N. Ando (ed.), *Japan and International Law: Past, Present, and Future*, 1999, 43 et seq., 51-52. In this connection, Reisman favored Member States of the United Nations insisting that permanent members of the Security Council use the right of veto conferred upon them not in pursuit of their special interests, but in the maintenance of minimum world public order.

[34] See B. Simma, *The Charter of the United Nations: A Commentary*, 1995, 455-56.

[35] On judicial review, see B. Martenczuk, "The Security Council, the International Court and the Judicial Review: What Lessons from Lockerbie?", *EJIL* 10 (1999), 517 et seq.

[36] See ICJ Reports 1971, 16, para. 293.

provisions. A different view was also expressed to the effect that the claimed right of Member States to question a decision of the Security Council on substance using its own value judgments would be contrary to the spirit of Article 25, as it would otherwise frustrate the proper functioning of the Security Council or even destroy it altogether, besides weakening the peacekeeping system of the United Nations. According to this view, any decision taken by the Security Council in accordance with the procedure provided for under the Charter is a decision taken in accordance with the Charter.[37] Any disagreement a State may have with the decisions of the Security Council can only be resolved by political means in the ultimate analysis, in the absence of any willingness of the Security Council to submit its decisions for judicial review.

If the Security Council is prevented from discharging its primary responsibility, it is open to the UN General Assembly to take up matters concerning international peace and security. Article 10 of the Charter empowers the General Assembly to discuss any question or matter within the scope of the Charter or relating to the powers and functions of any organs provided for in the Charter. This includes its power to deal with any failure of the Security Council in a given situation with a view to making appropriate recommendations to the members of the United Nations, the Security Council, or both, on any such questions or matters.

More specifically, the General Assembly may discuss any questions relating to the maintenance of international peace and security. Apart from its powers under Article 10 of the Charter, any question on which action is necessary should be referred to the Security Council either before or after discussions. "Action" in the sense of this article is action to be taken by the United Nations, not action recommended for consideration by the UN Member States to be taken on their own responsibility.

It is not clear whether the UN Member States are under an obligation to report to the Security Council any action taken or to be taken which involves the use of force on the basis of, or in accordance with, the recommendations of the General Assembly. However, it is reasonable to assume the existence of such a duty to report. Any action taken in this regard is still without prejudice to the primary responsibility of the Security Council to take any action which it may deem necessary in order to maintain or restore international peace and security.[38]

---

[37] See Simma, supra note 31, 414.
[38] However, it is argued that a regional organization could take non-military or economic sanctions in response to use of force by a third State against a Member State of the regional organization, provided the force involved is in violation of Article 2(4) of the UN Charter but does not amount to an "armed attack". In such a case, it is suggested that there is no need to report to or

Once action is taken by Member States in accordance with a recommendation of the General Assembly and the same is reported to the Security Council, several possibilities exist. The Security Council could take note of it and not do anything else. In such a case, the States are entitled to pursue their action until further action by the Security Council or achievement of the desired objective. Secondly, the Security Council may fail to decide to take action of its own or suggest modification of measures in progress. Equally, a resolution denouncing the action taken may also not be passed. In all these cases of inaction, the better view is that Member States are free to pursue the action taken in accordance with the recommendations of the General Assembly.[39]

---

seek subsequent sanction from the UN Security Council. According to Frowein,

"No general prohibition of economic sanctions or other reprisals exists in public international law including the United Nations Charter. This is the crucial difference between the application of economic sanctions on the one hand and the use of armed force on the other. While the use of armed force has to be justified under the United Nations Charter to overcome the prohibition in Article 2, paragraph (4), no similar problem arises with regard to economic sanctions".

In such cases, it is further suggested that the application of economic sanctions to third States be governed by general public international law. Frowein, see note 4, 388. In any case this argument does not apply in respect of economic sanctions or boycotts approved or authorized by the Security Council as "measures" decided under Article 41 of the UN Charter. In such a case, the Security Council remains supreme in deciding whether measures not involving the use of armed force employed by States is in conformity with its directions. Furthermore, the Security Council is also competent to revoke the sanctions it ordered at its own discretion. States including members of regional collective security system are bound by such decisions, when taken under Chapter VII, in accordance with Article 25 of the UN Charter.

[39] There is some disagreement on the effect of a resolution of the Security Council dealing with any action taken by States in collective self-defense if the resolution authorizing such an action is vetoed. It was argued that if a proposed resolution authorizing force such as Resolution 678 was vetoed, any collective self-defense action would have been barred and if continued would have been illegal. M. E. O'Connell, "Enforcing the Prohibition on the Use of Force: The UN's Response to Iraq's Invasion of Kuwait", *S. Ill. U.L.J.* 15 (1991), 453 and 478. Schachter however, disagreed with this view. According to him, a positive decision of the Council is necessary to debar or stop an action already underway by way of right of self-defense. Thus, if the resolution is vetoed and no decision was possible, the action undertaken in pursuance of the right of self-defense under Article 51 would continue as is otherwise permissible by the Charter and international law. O. Schachter, "United Nations Law in the Gulf Conflict", *AJIL* 85 (1991), 452 et seq., 459. Frowein further suggests that the measures a State voluntarily agrees to

In practice, these provisions of the UN Charter have come up for consideration and application a number of times. On the basis of an analysis of this practice, some further comments concerning the interpretation of the relevant articles is in order.

The Security Council, after determining that a threat to the peace, a breach of the peace or an act of aggression exists, can decide on non-forcible measures which are binding upon Member States under Article 41 read together with Article 25 of the Charter. A decision in this regard authorizes a willing and participating group of States to take enforcement action or continue any action in which they are already engaged to maintain international peace and security.

Any decision taken under Article 42 automatically implies a decision by the Security Council on the inadequacy of measures taken under Article 41. Thus, the Security Council need not first take an express decision under Article 41 to institute non-forcible measures, only to find them inadequate afterwards, and then proceed to take action under Article 42 to authorize use of force by a consenting group of States.[40]

---

apply in pursuance of the recommendations of the General Assembly may create a presumption that the measures are lawful under international law. However, he also noted that for a State to react against a breach of the peace or an act of aggression merely on the basis of recommendation by the Security Council would generate an important procedural presumption concerning the lawfulness of the measures, but they could become unlawful if these measures were in violation of international law and there is no other rule of public international law justifying the actions concerned. Such rules could be either Article 51 or the right to peaceful reprisals. Frowein, see note 4, 382-383.

[40] Normally, measures under Article 42 should be considered by the Security Council only if the measures decided under Article 41 are not adequate. It is not clear whether, for taking measures under Article 42, the UN Security Council should first take a decision concerning measures to be taken under Article 41 and allow some time for those measures to be seen as adequate or not. It is reasonable to assume, however, that taking measures under Article 41 and allowing them some time to succeed may be regarded as a precondition in the case of situations determined to be a threat to peace or breach of the peace, if not in the case of situations determined to be aggression under Article 39. The question arose whether the Council should have declared inadequacy of the provisional measures ordered under Article 40 in the case of Iraqi aggression against Kuwait demanding withdrawal and negotiation before authorizing the cooperating States to use all necessary means (including use of force) under Resolution 678 of the Security Council. It was submitted that Resolution 678, in authorizing cooperating States to use force, impliedly recognized that sanctions under Article 41 would be inadequate. In any case, according to Schachter, the defiant position taken by the Iraqi régime even after six months of sanctions added support to the

If arrangements envisaged under Article 43 did not come through, it is open to the Security Council to authorize States to employ all necessary means at their disposal, including the use of force, to maintain or restore international peace and security. Even though this is not an express mode of maintaining peace and security, it is implied in the powers of the Security Council.[41]

While the Security Council can authorize action employing the use of force by a designated group of States under Articles 42 and 48, Article 42 can also be used to authorize action not involving the use of force. However, such decisions do not oblige States that are not willing to participate in the enforcement action to do so, although they are still under an obligation to cooperate with the States taking action and certainly not to oppose it.

The General Assembly can recommend peacekeeping arrangements. The power of the General Assembly in regard to the maintenance of international peace and security is recognized as secondary to the primary power given to Security Council under Article 24.[42] In the *Certain Expenses* case,[43] the ICJ approvingly interpreted the power of the General Assembly in this regard by referring to "action" in Article 11(2) of the Charter as "coercive or enforcement action". Furthermore, it held that the General Assembly may in the implementation of its recommendations, either under Article 11 or under Article 14 of the Charter, set up commissions or other bodies in connection with the maintenance of international peace and security. Such implementation is a normal feature of the functioning of the United Nations. It can establish subsidiary organs under Article 22 of the Charter and these may include organs for investigation, observation and supervision subject to the consent of the State or States concerned. Furthermore, recommendations of the General Assembly may include coercive measures: as the Court itself pointed out, the functions and powers conferred by the Charter "are not confined to discussion, consideration, the initiation of the studies, and the making of recommendations; they are not merely hortatory".[44] It is clear from this that the general power of the General Assembly to recommend enforcement measures (including military measures) is not limited by Article 11(2), and such a power can be exercised, according to the Uniting for Peace Resolution, only in

---

belief that military action was needed to bring about its compliance. Furthermore, in his view, there is no doubt that the Council had the legal right to decide on the need for military action to remove Iraqi occupation of Kuwait. Schachter, see note 39, 463.

[41] Frowein, see note 4, 379.
[42] Kelsen, see note 3, 975-979.
[43] ICJ Reports 1962, 151 et seq.
[44] *Ibid.*, 163.

cases when the Security Council is unable to function. On that occasion, the powers of the General Assembly are not limited by Article 106 of the Charter. In this regard, the General Assembly is not required to make any determination that there exists a threat to the peace or breach of the peace and an act of aggression – a power that is only given to the Security Council under Article 39 of the Charter.[45]

The preceding discussion sets out certain possible options open to States to utilize the system of the United Nations even in its imperfect state to deal with the humanitarian crises. However, given the difficulties involved in the system and the separate political interests which make the achievement of international consensus difficult, States continue to take unilateral, unauthorized action involving the use of force in such situations.

IV. Armed Intervention in Case of Crimes Against Humanity: Could the Concept of Erga Omnes Obligations Be a Legal Basis?

The Secretary-General of the United Nations considered the matter of use of force without authorization from the United Nations and noted that such intervention embraced a wide continuum of responses from diplomacy to armed action. Referring to the controversy[46] concerning armed intervention, he noted:

> "We confront a real dilemma. Few would disagree that both the defense of humanity and the defense of sovereignty are principles that must be supported. Alas, that does not tell us which principle should prevail when they are in conflict. Humanitarian intervention is a sensitive issue, fraught with political difficulty and not susceptible to easy answers. But

---

[45] B. Simma (1995), see note 32, 234-235. Kelsen, see note 3, 979.
[46] He summarized the criticism expressed against humanitarian intervention thus:
> "Some critics were concerned that the concept of humanitarian intervention could become a cover for gratuitous interference in internal affairs of sovereign States. Others felt that it might encourage secessionist movements deliberately to provoke government into committing gross violations of human rights in order to trigger external interventions that would aid their cause. Still others noted that there is little consistency in the practice of intervention, owing to its inherent difficulties and costs as well as perceived national interests – except that weak States are far more likely to be subjected to it than strong ones".

See the Millennium Report of the Secretary-General of the United Nations, *"We the Peoples": The Role of the United Nations in the Twenty-First Century*, 2000, paras. 218-219.

surely no legal principle – not even sovereignty – can ever shield crimes against humanity. Where such crimes occur and peaceful attempts to halt them have been exhausted, the Security Council has a moral duty to act on behalf of the international community. The fact that we cannot protect people everywhere is no reason for doing nothing when we can. Armed intervention must always remain the option of last resort, but in the face of mass murder it is an option that cannot be relinquished".[47]

Attempts are usually made to justify the use of force against a State accused of committing acts of genocide and engaged in systematic and widespread violations of human rights on the grounds that such acts constitute a threat to the peace or breach of the peace. In such cases, the violations involved are also considered as violations of *erga omnes* obligations owed to the entire international community. Hence, such violations give rise to a legal interest in favor of every State and a right to take action, including the use of military force, to stop the abuse and defend the rule of law in the international community. Use of force in such a context is also defended as being in conformity with Article 2(4) of the UN Charter as (a) no territorial dismemberment is conceived, and (b) the human rights defended are fully consistent with the purposes of the United Nations. These arguments raise a number of issues and are based on certain policy preferences.

To argue that the use of force is legitimate where no territorial dismemberment is sought is to ignore the Charter's scheme of collective security. In any case where use of force is employed for intervention within the territory of a State without its consent, it invariably amounts to use of force against the territorial integrity and political independence of that State.[48] The justification offered for the unilateral use of force as a defense of *erga omnes* obligations, particularly when massive violations of human rights obligations are involved, namely, that they are in accordance with the purposes of the United Nations, is equally unconvincing.

An *erga omnes* obligation is an evolving concept. These obligations are owed towards the international community as a whole.[49] *Erga omnes*

---

[47] Ibid.
[48] Schachter, see note 25, 118.
[49] M. Ragazzi, *A Concept of International Obligations Erga Omnes*, 1997; see also C. Chinkin, *Third Parties in International Law*, 1993. The Vienna Convention on Law of Treaties of 1969 does not deal with *erga omnes* rights and obligations. However, the International Law Commission considered that the rule in Article 36(1) of the Convention, by which a right can be accorded to "all States", and the process recognized by Article 38 of the Convention,

obligations are to be distinguished from the *jus cogens* obligations defined in Articles 53 and 64 of the Vienna Convention on the Law of Treaties. Principles of *jus cogens* are norms "accepted and recognized by the international community of States as a whole as a norm from which no derogation is permitted and which can be modified only by a subsequent norm of general international law having the same character". There is no agreement, however, on the method or manner of formation of these obligations. According to one view,[50] this is essentially a legislative function or the result of the voluntary acceptance by States, like any other international obligation, even if these obligations can be given a higher status at any given point in time. Another view is that these obligations are a result of the crystallization of higher community values.[51] The process by which such crystallization

---

that is, a rule set forth in a treaty becoming binding upon a third State as a customary rule of international law, "furnished a sufficient legal basis for the establishment of treaty rights and obligations valid *erga omnes*". See A. Aust, *Modern Treaty Law and Practice*, 2000, 209.

[50] See R. P. Dhokalia, "Problems Relating to Jus Cogens in the Law of Treaties", in S. K. Agrawala, *Essays on Law of Treaties (with Special Reference to India)*, 1972, 149-177, where the author notes:

"The introduction of the article on *jus cogens* into the codification of the law of the Treaties, providing the effect of preemptory norm of international law from which no derogation is permitted, is an example of the development of law *de lege ferenda*. It may be desirable on grounds of law and morality but from the practical point of view it has not been feasible to lay down precise and clear substantive rules relating to identification of the *jus cogens*" (175).

He also noted that even though the Vienna Convention provided for an effective institutional machinery for authoritative settlement of disputes relating to denunciation of treaties, the problem of identifying international *jus cogens* rules remains unsolved since arbitral and judicial tribunals cannot exercise legislative functions such as establishing these norms (176).

[51] This approach is discussed and favored by Simma. First, relying upon an assertion of Yasseen, Chairman of the Drafting Committee of the Vienna Conference on Law of Treaties, he noted that an isolated State or a small number of States could not prevent a general rule of international law from acquiring the preemptory character. Secondly, he suggested that the formation of a *jus cogens* rule can be derived less from state practice and more from "obligations based on certain general and well recognized principles", among them "elementary consideration of humanity" (*Corfu Channel* case, ICJ Reports 1949, 2); "Principles which are recognized by civilized nations as binding on States even without any conventional obligations" (*Reservations to the Genocide Convention*, ICJ Reports, 1951, 23); "the principles and rules concerning the basic rights of human person" (*Barcelona Traction*, Judgment, ICJ Reports 1970, 32) and "fundamental principles" of human rights and humanitarian law (*Tehran Hostages* case, ICJ Reports 1980, 42 and the *Nicaragua* case, ICJ Reports 1986, 113).

takes place is independent of any practice of States and unrelated to normal sources of international obligations. The ICJ itself did not refer to any process by which these obligations are formed or can be identified.[52] Despite this lack of agreement concerning the formation of *erga omnes* and *jus cogens* obligations, these concepts are increasingly relied upon for the progressive development of international law. Many of the examples cited illustrating these obligations appear to be the same. As Simma points out, they may be regarded as obligations on the two sides of the same coin, with different implications.[53] Some of the

---

Simma (1994), see note 32, 289-293. In the commentary to Article 53, the ILC stated: "The Commission considered the right course to be to provide in general terms that a treaty is void if it conflicts with the rule of *jus cogens* and to leave the full content of the rule to be worked out in state practice and in the jurisprudence of international tribunals". See V. Nageswar Rao, "*Jus Cogens* and the Vienna Convention on Law of Treaties", *Indian JIL* 14 (1974), 363 et seq., 379. Also, see Rao, *ibid.*, 377-379, for a review of the views of the members of the ILC on the criteria for identification of rules of *jus cogens*: Yasseen: "accepted by a large number of States but also must be found necessary to the international life and deeply rooted in the international conscience"; Sh. Rosenne: "it was more society and less the law itself which defined the content of the *jus cogens*"; M. Bartos: "Minimum rules of conduct necessary to make orderly international relations possible"; M. Lachs: "interest of the international community as a whole"; Tunkin: "Rules of interest to all". Ago and Gross took a view similar to the commentary of the ILC on Article 53. According to Murty, a search for the norms of *jus cogens* ought to be directed not merely to concrete rules and relatively generalized principles, but as well to the ultimate goals sought in the establishment of these rules and principles. In addition, in his view the fact that the customary prescription which are *jus cogens* are not capable of precise formulation cannot be a ground for the denial of their existence. No one suggests, he pointed out, their non-existence in municipal law by virtue of the fact that they do not admit of precise statement. *Ibid.*, 377-378. See B. S. Murty, "Jus Cogens in International Law", in *The Concept of Jus Cogens in International Law*, Papers and Proceedings of the Conference on International Law, Lagonissi (Greece), 3-8 April 1966, 1967, 79-83.

[52] *Ibid.*

[53] Bruno Simma explained in a succinct manner the nature and implications of the two sets of obligations. The notion of obligation *erga omnes* is not identical to that of obligations arising from general international law. Only obligations owed to the collectivity of States can be regarded as obligations *erga omnes*; on the other hand, not all obligations owed to several or even many States represent obligations *erga omnes*. Only if the community of States is entitled to demand fulfillment of an obligation are we in the presence of a true rule with effect *erga omnes*. Only in that case the *"omnes"* are deemed to have a legal interest in the fulfillment of obligation.

On the other hand, *jus cogens* obligations are preemptory norms and these are defined precisely as norms, which have the legal power to invalidate

examples given are prohibition of use of force under Article 2(4) of the UN Charter (the only principle cited by the ILC when it first adopted the article on *jus cogens*), the prohibition of genocide, slavery, piracy, colonialism and apartheid, and other fundamental human rights.[54] In spite of these examples, as pointed out by one authoritative treatise, "such a category of rules of *jus cogens* is comparatively a recent development and there is no general agreement as to which rules have this character".[55]

V. Erga Omnes Obligations: The Treatment of the ILC

According to the draft articles of the ILC being finalized at second reading (August 2000), which is the latest attempt to codify and progressively develop the law of state responsibility, the legal right to invoke the responsibility of a State in defense of an *erga omnes* obligation could arise only if the obligation involved is essential for the protection of the fundamental interests of the international community as a whole.[56] In addition, the wrongful act involved should be a serious breach of an obligation in the sense of a gross or systematic failure by the responsible State to fulfill the obligation, risking substantial harm to the fundamental interests protected thereby.[57]

---

conflicting treaties. Therefore, it is quite conceivable in theory that some obligations *erga omnes* do not have this effect. Thus, obligations *erga omnes* flow from a certain class of norms the performance of which is owed to the international community as a whole, whereas the concept of *jus cogens* invests some or all these norms with particularly far-reaching legal effects. "Therefore, *jus cogens* and obligations *erga omnes* are but two sides of one and the same coin, namely, that of the existence of certain rules of international law which are, in the words of the Court, "the concern of all States". Simma (1994), see note 32, 300. For a fuller appreciation of the relationship between obligations *erga omnes* and *jus cogens* and the concept of international crimes of States, see G. Gaja, "Obligations Erga Omnes, International Crimes and Jus Cogens: A Tentative Analysis of Three Related Concepts", in J. H. H. Weiler, A. Cassese and M. Spinedi (eds.), *International Crimes of States*, 1988, 151-160. See also C. Dominice, "The International Responsibility of States for Breach of Multilateral Obligations", *EJIL* 10 (1999), 17 et seq.

[54] See also the *Barcelona Traction* case, see note 51, 32, para. 31, for citation of some examples of *erga omnes* obligations: aggression, genocide, basic rights of human person including protection from slavery and racial discrimination.. Simma (1994), see note 32, 292, also adds torture to this list.

[55] Jennings and Watts (eds.), *Oppenheim's International Law*, 9th ed., Vol. I: Peace, 1992.

[56] Article 41.

[57] Article 41(2).

An invocation of responsibility in the case of a serious breach of an *erga omnes* obligation by States not directly injured is subject to the same requirements of notice, admissibility of the claim and other conditions on loss of the right that are applicable to the invocation of responsibility by the injured State.[58] In addition, where countermeasures are undertaken by States other than the injured State, these can only be taken at the request and on behalf of any State injured by the breach and to the extent that that State may itself take countermeasures.[59] Furthermore, before taking countermeasures, States are required to fulfill several conditions: to call on the responsible State to fulfill its obligation, to notify any decision, to take countermeasures and offer to negotiate with the State, not to engage in countermeasures while negotiations are in progress and have not been unduly delayed, and not take the countermeasures if the wrongful act has ceased or the dispute is submitted to a court or tribunal. These measures are without prejudice to the right of the injured State to take such provisional and urgent countermeasures as may be necessary to preserve its rights.[60] In any case, countermeasures should not derogate from several obligations: to refrain from the threat or use of force, to protect fundamental human rights, to honor obligations of a humanitarian character, to respect peremptory norms of general international law and the inviolability of diplomatic or consular agents, premises, archives and documents. Apart from these obligations, the taking of countermeasures does not relieve the State from fulfilling its obligations under any applicable dispute settlement procedure in force between it and the responsible State.[61] There is also the requirement of proportionality, according to which countermeasures must be commensurate with the injury suffered, taking into account the gravity of the internationally wrongful act and the rights in question.[62]

It is also important to note that any right to take countermeasures under the general law of state responsibility does not operate in the case of arrangements made or sanctions prescribed in specific international agreements.[63] Furthermore, the legal consequences of an internationally wrongful act, including the right to take countermeasures, are without prejudice to the provisions of UN Charter.[64]

Some of the conditions governing the issue of countermeasures were also noted by the ICJ in the *Gabčíkovo-Nagymaros Project* case.

---

[58] Article 49(3).
[59] Article 54.
[60] Article 53.
[61] Article 51.
[62] Article 52.
[63] Article 56.
[64] Article 59.

According to the Court, in the first place, countermeasures must be taken in response to a previous international wrongful acts of another State and must be directed against that State. Secondly, the injured State must have called upon the State committing the wrongful act to discontinue its wrongful conduct or to make reparation for it. In addition, a most important consideration is that the effects of countermeasures must be commensurate with the injury suffered, taking account of the right in question. Furthermore, for a countermeasure to be lawful, its purpose must be to induce the wrongdoing State to comply with the obligations under international law and the measure must therefore be reversible.[65]

In addition, countermeasures are not just another means to enforce international law and are to be conceived, if at all, only as a measure of last resort, subject to the criterion of necessity.[66] It must be pointed out that the "state of necessity" is dealt with by the ILC as one of the circumstances precluding wrongfulness. Article 26 in the current draft before the ILC (2000) corresponds to Article 33 of the draft articles finalized in 1980. It is interesting that the "criterion of necessity" noted as a requirement of legitimate countermeasures is different from the "state of necessity" as a circumstance precluding the wrongfulness of the act.[67] In the latter case, an act that is inherently wrongful cannot be treated as such in the presence of a state of necessity, and hence cannot attract countermeasures, as no state responsibility can thereby be invoked. However, in the former case, even where such a responsibility could be invoked, a countermeasure can only be treated as lawful subject to the criterion of necessity. Viewed thus, even in the case of non-forcible measures, the criterion of necessity brings in a additional condition to the conditions prescribed in present Article 53. There is good reason to include the criterion of necessity in Article 53 as distinct from the state of necessity as a circumstance precluding wrongfulness.

It is necessary that a State should not be easily allowed to resort to countermeasures, which is nothing but self-help unless it is established as the sole means of safeguarding an essential interest. In addition, even as a measure of last resort, countermeasures should only be taken under strictly defined conditions that must be cumulatively satisfied, and the State concerned is not the sole judge of whether these conditions have been met. In this connection, in the *Gabčíkovo-Nagymaros Project* case,

---

[65] See *Gabčíkovo-Nagymaros Project* case *(Hungary v. Slovakia)*, Judgment, ICJ Reports 1997, 7, paras. 82-88.

[66] See statement of Brownlie on behalf of the United Kingdom in the Sixth Committee of the United Nations, 7 November 1966.

[67] For a discussion of the "state of necessity" as a condition precluding wrongfulness of an act, see the commentary to Article 33 of the Draft Articles on State Responsibility adopted by the International Law Commission on first reading, January 1997, 240-258.

the ICJ has emphasized while dealing with the state of necessity (under draft Article 33 of the ILC, 1980) that a state of necessity could not exist without a "peril", duly established at the relevant point in time, that is both grave and imminent. "Imminence" is synonymous with "immediacy and proximity" and goes beyond the concept of "possibility". That does not exclude, as the Court pointed out, that a "peril" appearing in the long term might be held to be "imminent" as soon as it is established, at the relevant point in time, that the realization of that peril, however far off it might be, is not thereby any less certain and inevitable.[68] Extending this logic to the criterion of necessity, it could be said that countermeasures cannot be taken against a wrongful act which took place at a different time and place as a measure of punishment or a sanction to induce the wrongdoing State to comply with the international obligation.

The draft articles of the ILC (2000) are still to be refined following the comments of the Member States expressed during the debate in the General Assembly concluded in 2000.[69] In this connection, questions have been raised regarding one or more aspects of the régime of state responsibility and the lawfulness of countermeasures in response to violations of *erga omnes* obligations. Some States questioned the very need to give legitimacy and legal recognition to the concept of countermeasures, which so far has not been well recognized in international law. Others have questioned the concept of community interest and the procedures prescribed as preconditions for taking countermeasures. Yet another group of States did not believe that the unilateral character and arbitrariness inherent in countermeasures were sufficiently regulated and subordinated to broader, more objective and neutral interests of the international community. In this connection, the need for the compulsory settlement of disputes, both for establishing the wrongfulness of the conduct and the responsibility of the State concerned and for settling the differences between the injured and the responsible State before embarking upon countermeasures, were also noted. The importance for preserving and promoting the United Nations

---

[68] The *Gabčikovo-Nagymaros Project* case dealt with the requirement of the state of necessity under Article 33, see note 65, 4, para. 54.

[69] See, for example, the view of Algeria (the issue of countermeasures was a delicate and controversial subject and it had been resistant to its conclusion in the text); the Russian Federation (draft articles on countermeasures needed to be concise, as they could be abused by powerful States an additional study is required); Indonesia (rules concerning countermeasures were not sufficiently developed); States (believed that the draft article suggested restrictions on use of countermeasures that did not reflect customary international law). See Press Release of the 18th meeting, Sixth Committee, Fifty-Fifth General Assembly, GA/L/3158, 27 October 2000.

as a universal forum and subordinating the right of self-help to the collective security system under the Charter was also stressed.

VI. Implementation of Erga Omnes Obligations and Countermeasures

In addition, the implementation of *erga omnes* obligations through unilateral action is also based on several assumptions that are open to debate and are short of any reasonable consensus among different sections of the international community, particularly among States. States still largely subscribe to consent-based régimes and bilateralism. There is no denying that there is greater nexus or interconnectivity or transnationality to individual actions in international relations. The world is also more interdependent than ever before. However, this does not account for a greater commonality or common interest and, more importantly, the commonly established strategies for implementation of such common interests.

This being a controlling conditioning factor, to base *erga omnes* obligations on a theory of consolidated community values, and to suggest that any and every State could assume a supervisory and enforcing function in respect of such higher community values or *erga omnes* obligations, is to overstate the case. At best or at worst, it may amount to imposing personal preferences or political conveniences on others against their will. We may agree with Simma, a strong votary of community interest, who notes distinct dangers if individual States were "to act as self-appointed guardians of what they see as a community interest".[70] With reference to the permanence of private justice in the international sphere, which can find lawful expression in "legitimate countermeasures" or "unacceptable unilateralism" as soon as it comes to the use of force without authorization from the Security Council or, in the last resort, the UN General Assembly, Alain Pellet pointed out that it "must be firmly rejected". According to him, there is an urgent need to promote effective institutional machinery to regulate and defend the conceptual advances and progress in community values if the latter are not used or misused as justifications for unilateral and self-serving actions.[71]

Moreover, any argument to promote unilateral action to use force to prevent and deter serious human rights abuses in effect amounts to supporting a discretionary right of a State to use force only when it serves its interests, even if incidentally serious human rights abuses are

---

[70] Simma, see note 53, 364.
[71] A. Pellet, "Brief Remarks on the Unilateral Use of Force", *EJIL* 11 (2000), 385-392, 391.

prevented or deterred. Unless the right to take unilateral action is cast as a duty and a mandatory obligation of all States to enforce international human rights, irrespective of their political considerations, the interests of the international community and rule of law cannot be promoted. In fact, actions so far taken by States in the name of humanitarian intervention have belied any such concern for upholding international law and order as an obligation. Accordingly, it is fair to say that

> "[i]t seems that humanitarian intervention is a restricted notion, operates in a location that is relatively cheap, is against a militarily weak nation, operates in a location that is accessible and strategically important, where public emotion is in favour and the intervention does not interfere with other political and economic objectives. States apparently wish to be able to justify unilateral intervention on humanitarian grounds, where the claim is contextually feasible, but not to undertake a duty to offer such assistance when objective criteria satisfies. This is a limited Samaritan, and way of policing the world".[72]

All unilateral acts are *per se* beyond the pale from the point of view of the law, because as a matter of policy they cannot be treated as an acceptable method of enforcement of international law. The failure of the UN collective security system or the functioning of the Security Council cannot be a ground to invent a new and controversial technique of enforcement where the political and special interest of the countries taking action and not the common interests of the international community generally control the operation of enforcement action and the objectives sought to be achieved thereby.

The failure of the Security Council to take action in a given case in accordance with the UN Charter is matter of concern for all the Member States of the United Nations. Inaction of the Security Council cannot always and automatically be treated as a failure. Use or likely use of the veto is a power given to a permanent member of the Security Council, and its use signifies the need for all the five permanent members of the Security Council to work together in concert and in accordance with the principles and purposes of the United Nations. Lack of agreement on one occasion or the other does not allow the permanent member to bypass or claim inaction on the part of the Security Council and proceed to take action which suits its own political interest. Any such unilateral action is a deliberate attempt to bypass the collective security system of the United Nations by first downgrading it when it does not suit the political

---

[72] C. Chinkin, "The State That Acts Alone: Bully, Good Samaritan or Iconoclast", *EJIL* 11 (2000), 30 et seq., 37.

interest of the State concerned. The imperative of having to continue to work with the UN system as long as it is not abandoned altogether cannot be overestimated. As Bowett points out, "to concede a general right of intervention without the authorization of a competent organ, would, in our view, be entirely inconsistent with the system of collective security envisaged by the Charter...". Referring to the impotency of the Security Council and the decided advantage of the potential aggressor arising out of a restrictive view of the freedom of individual action, he further noted: "These arguments are valid enough for the purpose of a possible revision of the Charter; they may be valid enough for a complete abandonment of the Charter. They are not, in our view, valid as reasons for distorting the meaning of the obligations contained in the Charter so long as those obligations are accepted as subsisting".[73]

Pierre-Marie Dupuy also made a similar point:

> "The danger is that if collective unilateralism takes the place of collective action under the UN Charter, we may see reappearing the reign of 'sphere of influence' or of 'backyards' in which some parties are barred from intervention... This means that is the dismantling of the whole system of collective security that we risk seeing. But the UN Charter does not just set up organs and procedures. It also asserts basic principles. By weakening or marginalizing the former we risk reducing the scope of the latter".[74]

Reacting cautiously to NATO's intervention in Kosovo, and observing that in his view "only a thin red line separates NATO's action on Kosovo from international legality", Bruno Simma shared a similar concern and warned against repetition of such interventions as a matter of deliberate future strategy:

> "But should the Alliance now set out to include breaches of the UN Charter as a regular part of its strategic programme for the future, this would have an immeasurably more destructive impact on the universal system of collective security embodied in the Charter. To resort to illegality as an explicit *ultima ratio* for reasons as convincing as those put forward in the Kosovo case is one thing. To turn such an exception into a general policy is quite another. If we agree that the NATO treaty does

---

[73] D. W. Bowett, *Self-Defense in International Law*, 1958, 273-274.
[74] P.-M. Dupuy, "The Place and Role of Unilateralism in Contemporary International Law", *EJIL* 11 (2000), 19 et seq., 29.

have a hard legal core ... it is NATO's subordination to the principles of United Nation's Charter".[75]

VII. Legitimate But Unlawful Countermeasures: Can the Unlawfulness Be Cured?

In recognition of the need to meet dangers posed by grave and widespread violations of human rights involving genocide, particularly within States where all structures of authority have either lost their force and effect, or completely disintegrated, some commentators have noted that, subject to certain conditions, unilateral use of force including use of force by international organizations should be permitted. Based on nascent trends in the world community, Cassese has argued that, under certain strict conditions, resort to armed force of the type that NATO was engaged in may become gradually justified. According to him these conditions are:

1. Crimes against humanity carried out in the territory of a sovereign State either by the authorities of the States themselves or by others "with their connivance or support, or because the total collapse of such authorities cannot impede those atrocities";
2. In the case of crimes against humanity, resulting "from anarchy in a sovereign state, proof is necessary that the central authorities are utterly unable to put an end to those crimes", refused to seek help from other States or international organizations and prevented them from entering into the territory to offer assistance, or otherwise "have consistently withheld their cooperation" from international organizations or the United Nations or "have systematically refused to comply with appeals, recommendations or decisions of such organizations";
3. The "Security Council is unable to take any coercive action" because of a veto or "refrains from any action or only confines itself

---

[75] B. Simma, "NATO, the UN and the Use of Force: Legal Aspects", *EJIL* 10 (1999), 1 et seq., 22. A contrary view, however, was expressed by Reisman, who came to the conclusion that the Kosovo intervention was like no other humanitarian interventions, which were tainted as instruments of policy of particular States. He stressed that if Security Council cannot save lives when human rights enforcement by military means is required, "the legal requirement continues to be to save lives however one can and as quickly as one can, for each passing day, each passing hour means more murders, rapes, mutilations and dismemberments – violations of human beings that no prosecution is expunged nor remedy repair". Reisman, see note 31, 860-862.

to deploring or condemning the massacres, plus possibly terming the situation a threat to the peace";
4. All available peaceful means have been exhausted without reaching any solution;
5. Action is taken by a group of States with the support of or at least without any opposition from the majority of UN Member States; and
6. Armed force "is exclusively used for the limited purpose of stopping the atrocities and restoring respect for human rights" and not for any other goal.[76]

Ruth Wedgewood supports this line of thinking. She notes that, in an era with an expanded account of human rights and human security, there is impetus to permit effective action in humanitarian emergencies through an expanded reading of Chapter VIII of the UN Charter with new latitude for regional action. In her view, unilateral or regional action will gain greater comfort when it is bracketed by a Council diagnosis of a crisis situation. For example, even though no *ex ante* or *ex post* authorization was available for the NATO military action in Kosovo, the Council rejected Russia's draft resolution that would have condemned the NATO bombing. In addition, after the NATO campaign, the Security Council also acted under Chapter VII to authorize the deployment of an "international military presence" to supervise Serb withdrawal from the province. According to her, this implied an indirect approval of the NATO campaign.[77] As conditions justifying a unilateral use of force by an international organization other than the United Nations, she also suggested the following[78]:

1. The "potential cost to the Council authority must be prominently" kept in mind and "[a]t a minimum, the underlying reasons for Council involvement should be part of the decision process";
2. Avoid escalation of the conflict and the intervention must have a limited aim;
3. Consult with other interested countries, where possible;
4. Maintain impartiality between warring factions; and
5. The action must be urgent enough to exclude any alternative course of action.

---

[76] See A. Cassese, "*Ex iniuria jus oritor*: Are We Moving towards International Legitimation of Forcible Humanitarian Countermeasures in the World Community?", *EJIL* 10 (1999), 23 et seq., 27.

[77] R. Wedgwood, "Unilateral Action in the UN System", *EJIL* 11 (2000), 349 et seq., 358.

[78] *Ibid.*, 359

Similarly, the Netherlands Report takes the view that there is a duty to develop a new basis for further development of customary law as a justification for humanitarian intervention without a mandate from the Security Council or the United Nations to meet grave emergencies and protect fundamental human rights, particularly the right to life. For this purpose, it submitted a framework for assessment of legitimate and hence lawful use of force to defend human rights when these are gravely threatened. According to this framework, the following steps are important:

1. A State or group of States should first obtain Security Council authorization for the use of force for humanitarian purposes by means of a draft resolution. Such a resolution should contain the clearest possible terms of reference (identity of the States authorized to intervene, objectives of the operation, scale and duration of the authorization and duty to report).
2. If the Security Council fails to give authorization because of a veto, the matter should be submitted to the General Assembly by a resolution of the Security Council, which is possible as the right of veto does not apply to a procedural decision to refer a matter to the General Assembly.
3. The General Assembly must then adopt a resolution recommending action by at least a two-thirds majority.
4. If the General Assembly is also not in a position to recommend action to deal with the humanitarian crisis, either because the debate is not sufficiently decisive or swift, unauthorized action may be taken, as an emergency solution, under conditions which are similar to the UN system of collective security, on the basis of the principle of "approximate treaty application".[79]
5. When such unauthorized humanitarian intervention is inevitable, the following factors are also considered important:
   a. The States involved, preferably States in the region for operational reasons, should be party to the regional and universal conventions for the protection of human rights. Furthermore, any such operation should be carried out by a group of States acting under the auspices

---

[79] This principle is described by Sir Hersch Lauterpacht:
"It is a sound principle of law that whenever a legal instrument of continuing validity cannot be applied literally owing to the conduct of one of the parties, it must, without allowing that party to take advantage of its own conduct, applied in a way approximating most closely to primary object. To do that is to interpret and give effect to the instrument – not change it".
Separate opinion, *South West Africa*, Admissibility, ICJ Reports 1956, 46, quoted in the *Gabčíkovo-Nagymaros Project* case, see note 64, para. 75.

of an international organization within an institutional context to avoid unilateral or arbitrary action or abuse.

b. Violation of human rights must be serious and widespread, thus both qualitative and quantitative criteria must be fulfilled. Grave or fundamental violations of human rights include not only extermination by means of summary executions and deliberate armed or police attacks on arbitrary civilian targets, but also torture, hostage taking and grave infringements of human dignity. Other conditions that must be fulfilled include urgency resulting from the systematic nature of the violations. Violations committed by non-state actors also constitute grounds for humanitarian intervention. Furthermore, the government should either be unable or unwilling to protect the victims and refuse third States or international organizations access to its territory to provide the necessary assistance or protection.[80]

c. Humanitarian intervention may take place during a national crisis or humanitarian emergency with international implications (massive flow of refugees across borders or regional destabilization). However, the mere threat of internal or international armed conflict is not in itself sufficient to satisfy the conditions for humanitarian intervention. In such situations, there is no need for a threat to international peace and security to be established as a separate condition for intervention.

d. The intervening State or States must make the humanitarian objectives of intervention clearly known in advance to the international community and to the State on whose territory the action will take place. Even though national security or other interests may play a part in the decision to intervene, these must be clearly subordinate to the humanitarian objective of the intervention.

e. The intervening State or States must exhaust in good faith all the appropriate non-military means of action against the offending State, but without allowing a rapid deterioration of humanitarian emergency to occur, which may necessitate immediate military action.

f. Humanitarian intervention must be proportional to the gravity of the situation. It should observe all the requirements of international humanitarian law as provided in the four 1949 Geneva Conventions and in the 1977 Additional Protocols. The humanitarian intervention

---

[80] The UN Secretary-General had indicated this as a ground to uphold rule of law and to prevent large scale violation of human rights (UN Doc. S/1999/957). See also Resolution 1265 (1999), 17 September 1999, Secretary-General's Report to the Security Council on the Protection of Civilians in Armed Conflict.

should not constitute a graver threat to international peace and security while attempting to remedy a great threat to human rights.

g. The humanitarian intervention must be limited, but may include designs to alter the structure and forms of authority of a State or direct attacks on the régime in power if this forms an essential part of the humanitarian objective.

h. The States engaged in the humanitarian intervention must report the matter to the Security Council immediately.

i. The intervening States must undertake in advance to suspend the intervention as soon as the State concerned is willing and able to end the violation of human rights by itself, or if the Security Council, or a regional organization acting with Security Council authorization, undertakes enforcement measures involving the use of force for the same humanitarian purpose.

j. The intervening States must end their intervention when its objective, namely, the cessation of violations of human rights, has been attained.[81]

Yet another perspective[82] is to develop a culture of prevention through early crisis warning, better ways to act both politically and economically and better legal and institutional instruments for enhancing the observance of our common values. According to this scheme, such conflict prevention as a strategy should address development of measures which can be implemented before a difference or a dispute escalates into violation. The measures should be designed to counteract the spreading of the conflict into other geographical areas and to prevent violence from flaring up again after the signing of a peace agreement or a cease-fire. This is not entirely a new perspective even though it appears to acquire fresh significance in view of the more recent violent conflicts in Bosnia, Kosovo, Rwanda and Sierra Leone.

One of the major problems with any approach on early intervention in a situation where a difference or a dispute between States exists is to identify the optimum point at which such international intervention would be beneficial. Early third-party interventions without the consent of the Parties involved would only complicate the matter further and internationalize the dispute sooner. Informal approaches may not always succeed in having an impact. The principle of national sovereignty and non-intervention in the internal affairs of the State are important considerations over and above the need for third parties to maintain neutrality.

---

[81] See the Netherlands Report, copy on file with the author, 28-32.
[82] See the Swedish Action Plan, *Preventing Violent Conflict*, 1999, 10, 35 and 36.

VIII. The Need for Strengthening the UN System: An Approach to Universal and Uniform World Order

While the suggestions made to improve accountability of unilateral armed action in defense of humanitarian objectives is laudable, it is highly doubtful that the framework of assessment and the other conditions suggested would achieve the desired purpose. A uniform and universal system of response to such humanitarian crises as a matter of duty and not as a matter of discretion is what is essentially required. In addition, any criteria suggested will only have their desired effect if they are acceptable to States in general. While some of the suggestions might come before the United Nations for consideration in different forms, there appears for the present to be no general support or consensus for them among States.

In the meantime, to doubt the validity of the suggestions made and to take a strict construction of the Charter provisions, in the face of certain inability of the United Nations to deal with crisis situations, is not to elevate law over ethics or to ignore altogether urgent humanitarian considerations. As Vera Gowlland-Debbas notes, a certain autonomy of operation of the law from the social and external environment within which it operates is not only desirable but also essential for preserving the sanctity of the legal process. The legal process itself determines mechanisms for lawful change. Furthermore, unbridled unilateralism, general disdain for multilateral institutions, hijacking of institutions for unilateralist and regional claims and the exclusion of the large majority of Third World States from the arena of decision making concerning such unilateral or regional actions are equally important considerations for opposing or expressing unease concerning the justifications offered in favor of "legitimate but unlawful" military action.[83] In this connection she also notes that: (i) where the Security Council authorized the use of force it "serve[d] to delegate the Council's powers and competences" to States, and use of force by such States is no longer wrongful (this is a natural consequence of a lack of means at the disposal of the Security Council); (ii) lack of authorization from the Security Council for the resort to unilateral action remains wrongful and prohibited under international law; (iii) the choice is not between protection of human rights on the one hand and preservation of despotic power in the name of State sovereignty on the other; but (iv) a "debate over the means not the ends, for remedial action can encompass a number of reactions to human rights violations".[84] As the ICJ noted in the *Nicaragua* case, a strictly

---

[83] V. Gowlland-Debbas, "The Limits of Unilateral Enforcement of Community Objectives in the Framework of UN Peace Maintenance", *EJIL* 11 (2000), 361 et seq.

[84] *Ibid.*, 378-379.

humanitarian objective cannot be compatible with the mining of ports, the destruction of oil installations or with the training, arming and equipping of the contras.[85]

In any case, grave breakdowns of international order resulting in mass killings or failed statehood cannot be tackled only though military sanctions or the use of force. It is rightly pointed out that sanctions are a blunt instrument and that they must be used sparingly and after careful consideration, and that they must include obligatory, immediate and enforceable humanitarian exemptions. It is also valid to note that the international community must focus on the concerns of the vast majority of its Member States, which continue to grapple, without much international community help or involvement, with the challenges of development and eradication of poverty, which impede durable and just peace.[86]

IX. Conclusions: Current State of International Law

In conclusion, the following points appear to be relevant to be highlighted:

1. International organizations other than the United Nations can engage in the use of force only in the event of an armed attack or upon authorization from the UN Security Council as provided under Article 53.
2. Where one or more States, or an international organization, engage in the use of force without authorization from the UN Security Council, it can only be justified on the basis of Article 51 of the UN Charter. While the States involved are entitled to assess in the first instance the need and legitimacy of acting in individual or collective self-defense, this is subject to the authority of the UN Security Council to review and decide upon necessary action. Furthermore, any individual or collective self-defense action by States or international organizations should immediately be reported to the Security Council.
3. Where the Security Council cannot take any action, the matter should be brought before the General Assembly, and any action to be taken thereafter by States should be in accordance with the recommendations of the General Assembly.

---

[85] *Ibid.*, 379.
[86] See the statement of Mr. Kamalesh Sharma, Permanent Representative of India to the United Nations, in the UN General Assembly on 6 October 1999, while speaking on the Report of the Secretary General on the Work of the Organization.

4. Inaction of the Security Council should not lightly be presumed. An effort to negotiate in good faith commonly acceptable responses to threats to the peace and breaches of the peace or acts of aggression is an obligation for all Member States, particularly the permanent members of the Security Council.
5. The primary responsibility for maintaining or restoring international peace and security is entrusted to the Security Council, as a matter of trust, by the larger membership of the United Nations. The permanent members of the Security Council are therefore required to cooperate with each other. The exercise of their power in the decision making of the Security Council should be done not merely with the aim of serving their personal political interests, but also in order to enable the Security Council to discharge its duties in accordance with the purposes and principles of the United Nations.
6. There is no acceptance in customary international law of a general rule in favor of humanitarian intervention as an exception to the basic rule against the use or threat of force.
7. The argument that the wider meaning accorded to the term "threat to the peace" in Article 39 of the UN Charter in the practice of the Security Council can validate unilateral measures by States that interfere in matters which are essentially within the domestic jurisdiction of a State, including – where necessary – the use of force, runs counter to Articles 2(7) and 24 of the UN Charter. The Security Council, being a political body authorized under the UN system, is at liberty to characterize any situation as a threat to the peace. However, the same liberty is not given to the States individually. Furthermore, where countermeasures are involved, except in the case of a direct armed attack, use of force is prohibited.
8. Arguments about the lawfulness of unilateral countermeasures against the State responsible for a wrongful act involving *erga omnes* obligations are at best arguments based on a law that is still being developed and is not crystallized as *lex lata*. At worst, they are arguments to support naked power plays in which only the powerful can engage against the weak.
9. While mass killings, genocide and other egregious violations of elementary principles of humanity should not be tolerated, the answer does not lie in formulating legal principles or guidelines on a general level to legitimize humanitarian intervention. What is needed is an approach to strengthen the UN system and to develop a uniform and universal world order wherein all States as members of the international community have a right to participate in the development and implementation of such an order in the common interest.

Réflexions sur l'intégration européenne dans les domaines des relations extérieures et de la défense

Daniel Vignes

Sommaire

Introduction : Jusqu'au traité de Maastricht (1992), l'Europe sera celle des marchands, pas celle de l'Union politique, ni celle de la défense.

I

Le traité de Maastricht répudie certaines des idées force classiques de l'intégration et de la construction supranationale, pour créer la Pesc (politique étrangère et de sécurité commune) :

1. la Pesc abandonne la règle de « l'effet direct »,
2. la Pesc ignorerait la majorité qualifiée,
3. la Pesc dédaignerait la commission européenne et le parlement européen.

II

Après 1992, le bon sens des États membres les amène à mettre progressivement en place une forme pragmatique d'intégration pour les activités propres de l'Union européenne et notamment pour la Pesc :

1. des échelons centraux institutionnels pour la Pesc : le SG/HP (secrétaire général haut représentant) et la troïka,
2. après une tentation intergouvernementale, un financement communautaire pour la Pesc, avec gestion par la commission.

III

Le développement récent et fortement intégré de la sécurité et de la défense européennes sous le nom de Pecsd (politique européenne commune de sécurité et de défense) :

1. mise en place de comités intergouvernementaux nécessaires et d'un état-major intégré,
2. des forces armées européennes multinationales intégrées, avec en plus des avions gros porteurs,
3. un contenu matériel original pour la Pecsd, avec des actions civiles de gestion des crises et des missions de maintien et de rétablissement de la paix, relations avec les Nations Unies, l'OTAN et les États-Unis et problème de l'UEO.

Conclusion : les développements de la politique européenne commune de sécurité et de défense sont satisfaisants sous réserve d'une question concernant l'UEO.

***

Introduction

C'est bien en vain – contrairement à tout ce que pourrait faire croire le terme fréquemment utilisé d'intégration européenne – qu'on chercherait dans les traités européens des années cinquante un appel formel à l'intégration, le terme est absent du traité CECA (1951) comme des traités de Rome (1957). Il ne l'est en revanche pas du traité « mort-né » entre ces deux dates, celui établissant la communauté européenne de défense, dit traité de la CED, signé le 27 mai 1952 et ultérieurement complété de divers instruments, mais rapidement rejeté sans débats de fond par l'assemblée nationale française le 31 août 1954. Il faut reconnaître que l'idée de forces armées intégrées occupait une place majeure dans le vocabulaire de ce traité. On le verra dans un instant.

Mais au fait, qu'est-ce que l'intégration ?

Le Littré, édition de 1962, laisse le juriste sur sa soif : l'intégration est un terme de mathématique, l'action d'intégrer consiste à « trouver l'intégrale d'une quantité différentielle . . . » ce qui permet au prix d'une certaine agilité d'esprit et d'un peu de trapèze volant d'imaginer qu'est de nature intégrée un corps unique composé de plusieurs éléments homogènes différents

C'est d'ailleurs ce qui résulte du texte du traité CED – notamment ses articles 1, 15 et 64-66 – qui est assez clair en disant que « les Forces européennes de défense sont des forces intégrées . . . que (leurs) corps d'armée sont formés d'unités de base, . . . (que celles-ci) sont (elles-mêmes) formées d'éléments de la même nationalité d'origine, (qu'elles) sont aussi légères que le permet le principe d'efficacité », que ces unités de base sont appuyées par des « unités de soutien tactique et de support logistique, qui (sont elles-mêmes) de nationalité homogène », mais qu'une « répartition entre nationalités se fait » entre ces unités de base et ces éléments d'appui, compte-tenu des « nécessités tactiques et d'organisation ». Pour parler plus simplement, les forces armées européennes intégrées devaient être constituées d'un assemblage d'unités de base de nationalités différentes, chacune de ces unités étant elle-même homogène de nationalité.

L'idée était très nouvelle, trop nouvelle, trop proche aussi de la seconde guerre mondiale. De là son rejet brutal de 1954. De là également la défaveur du terme d'intégration dans les trente premières années de la construction européenne ainsi que son remplacement par une autre idée-force, semblant plus technique et juridique, et donc moins dangereuse, celle d'une construction *supranationale*, celle-ci bâtie autour de deux prémices : les dirigeants européens – les membres de la haute autorité ou ceux de la commission – doivent être indépendants des États membres, ne pas avoir à quotidiennement dépendre d'eux, ne pas pouvoir être sanctionnés par eux, même par celui dont ils ont la nationalité, d'une part, le droit communautaire doit être immédiatement applicable, de l'Europe aux particuliers, de la communauté à l'opérateur économique sans le relais obligatoire des États membres, lesquels sont néanmoins soumis juridiquement à son système, d'autre part.

En même temps que la sécurité et la défense se constituaient – suite à l'échec de 1954 – en tabous dans la société européenne, l'Europe politique aussi pâtissait un peu de cette même exclusion. Les trois communautés de 1951 et 1957 étaient sectorielles, reposant l'une sur un marché commun charbon-acier, la deuxième sur un marché commun généralisé, la dernière sur un marché commun des produits nucléaires. Triplement l'Europe était celle des marchands.

Finalement le terme d'intégration ne sortira du réfrigérateur que dans les années 90, et pas spécialement en Europe-communautaire, mais sur la scène mondiale quand dans les cinq continents les États se regrouperont au sein de mouvements « d'intégration régionale », alena, asean, cedeao, mercosur... au nombre d'une demi douzaine.

De 1952-58 à 1992, l'Europe communautaire se construira, elle, en dépit du tabou politico-militaire pesant sur les six États membres et sur leur communauté (leurs communautés), celle-ci s'élargira, passant en 1973 à neuf membres, en 1980 à dix, à douze en 1986, à quinze en 1995 ; treize États de l'est et du sud-est de l'Europe actuellement négocient leur adhésion.

Parallèlement elle approfondit ses objectifs, en préemptant aux États membres les nouvelles activités sectorielles qui apparaissent, telles la protection de l'environnement et la politique commune des pêches maritimes, se substituant à eux pour d'autres de leurs tâches classiques et régaliennes, telle la monnaie. Enfin elle affirmera ses mécanismes institutionnels : le déficit démocratique qu'elle connaissait indéniablement à sa création aura totalement disparu en l'an 2000 après le traité de Nice, les pouvoirs du Parlement européen sont redoutables.

La conquête de son identité politique et sécuritaire a pour la communauté et l'union peut-être été plus difficile à apparaître : en 1973 à un de ses premiers sommets de chefs d'État et/ou de gouvernement, à Copenhague, dérisoirement, une déclaration sur l'identité européenne s'en remet aux États-Unis et au pacte atlantique du soin d'assurer la sécurité militaire du continent européen. Un timide pas est franchi en 1983, à Stuttgart, avec une intensification des « consultations dans les domaines de la politique étrangère y compris la coordination des positions des États membres sur les aspects politiques et économiques de la sécurité intérieure et extérieure ». Le saut de l'Acte unique, en 1987, est très limité : « davantage de coordination ».

Un grand pas sera en revanche franchi vers une mise en commun de la compétence de sécurité avec la signature à Maastricht en 1992 du traité du même nom, encore appelé de l'Union européenne, en ce qu'il crée celle-ci. Ce traité répudie toutefois en partie le mode supranational au profit de celui de la coopération *intergouvernementale*. Ce n'est donc plus au sein d'une super-communauté que va s'exercer cette nouvelle compétence, ou plutôt ces nouvelles compétences, car il y a la sécurité extérieure, exercée dans une politique étrangère et de sécurité commune et la sécurité intérieure, exercée au sein d'une coopération en matière de justice et d'affaires intérieures.

Peut-être, avec ces nouveautés, y aura-t-il dans le domaine des idées-force et des concepts un grand remue-ménage, des principes auront été écornés, des larmes ruisselleront de la crainte d'une disparition des mécanismes supranationaux jugés par certains essentiels. Mais, avec un certain bon sens, les États membres mettront en place, dès l'entrée en vigueur du traité de Maastricht, progressivement une forme pragmatique d'intégration, spécialement dans le domaine de la sécurité. *A priori*, coopération intergouvernementale et intégration sembleraient antithétiques, pourtant de Maastricht à Amsterdam et surtout à Nice, la Pesc et *maintenant* la Pecsd deviennent des instruments d'intégration.

Successivement, on examinera la répudiation, probablement plus verbale que réelle, de la plupart des idées-force et des concepts de l'intégration, pour la Pesc et la JAI, par substitution partielle en 1992 au mode supranational de celui de la coopération intergouvernementale (I), puis l'apparition pragmatique, du fait d'une volonté diffuse des États membres, d'une sorte d'intégration rénovée, avec plus d'institutions centralisées et plus de financement commun (II) ; cela aboutira, en 1999 à une politique européenne commune de sécurité et de défense, la Pecsd (III).

I. La répudiation des idées forces de l'intégration par le traité de 1992

Calmons d'emblée le jeu, la répudiation n'a peut-être existée que dans les joies et craintes des deux catégories d'adversaires du traité de Maastricht, ceux qui se réjouissaient comme manifestation d'un retour aux règles classiques de la société internationale et ceux qui inversement, étant les thuriféraires de la société supranationale et de l'intégration, se lamentaient du délaissement de celle-ci.

Que le délit ait été intentionnel ou non, peu importe : il semble en revanche constant que les États membres aient abandonné pour la Pesc (et pour la JAI) la sacro-sainte idée de l'effet direct (1), qu'ils aient ignoré celle non moins fondamentale de la majorité qualifiée (2), qu'ils aient dédaigné la commission et le parlement européen, pour sur-doter le conseil européen (3).

Sans doute dans l'application pratique des choses et plus encore dans les deux traités (Amsterdam et Nice), qui ont suivi (et révisé) celui de 1992, ces erreurs fort attaquées ont-elles été atténuées, puisque la majorité qualifiée, par exemple, a regagné du terrain, de même la commission a-t-elle également retrouvé un rôle à sa mesure.

Il n'en reste pas moins que dans trois de ses règles essentielles concernant l'intégration, le système de 1957 a en 1992, subi une véritable « *capitis deminutio* ».

1. La Pesc abandonne l'effet direct

La Pesc abandonne l'effet direct du droit communautaire à l'égard des particuliers (ainsi que ses instruments que sont le règlement et la directive) au profit d'un nouvel arsenal d'actes, adressés aux États membres et non soumis pour eux à une obligation de transposition, telles les positions communes, les actions communes et (depuis Amsterdam) les stratégies communes.

Cela est toutefois compréhensible car s'agissant, avec la Pesc, d'une politique qui concerne au premier chef les États et leurs compétences de souveraineté, ce n'était pas vraiment les individus et autres opérateurs économiques (comme le prescrivent de nombreuses dispositions du TCE) qui sont les premiers visés par elle ; l'ensemble du TUE ainsi – avec des exceptions toutefois dans le secteur de la justice et des affaires intérieures – ne contient pas de règles applicables/opposables aux individus, mais seulement des règles destinées aux États membres.

Et d'ailleurs pour rendre la Pesc plus opérationnelle dans les secteurs économiques, eux restés purement communautaires, diverses passerelles ont été lancées en 1992 et depuis lors :

Le système Pesc comporte par exemple des dispositions destinées à assurer que si cela était nécessaire le droit issu de la Pesc pourra devenir applicable aux personnes et opérateurs intéressés par lui, ainsi s'explique l'article 301 TCE (ex 228A) permettant de transposer dans la politique commerciale commune des décisions politiques Pesc (telles la rupture des relations politiques et économiques avec un pays tiers, appelée à se traduire par des contre-mesures commerciales)

On pensera aussi à ces accords de coopération, à contenu à la fois économique et politique, passés avec des pays tiers et prévoyant, outre des relations commerciales, un dialogue politique, il est normal que celui-ci s'effectue au sein de la Pesc (art. 26 TUE) et que le SG/HP (secrétaire général haut représentant, cf. *infra*) conduise celui-ci.

On pensera encore à l'existence d'accords internationaux conclus dans le cadre de la Pesc, article 24 TUE ; innovation d'Amsterdam qui, si elle ne prévoyait pas dans le texte alors rédigé que les États membres soient liés par ces accords, a été modifiée à Nice par l'ajout à cet article que les institutions communautaires également étaient liées par ces accords (§5).

Par ces trois dispositions, le TUE ensemble le TCE, ont donné à la Pesc une structure complète d'ordre juridique autonome qui ne le met pas en retard par rapport à l'ordre juridique communautaire, même si la Pesc n'est pas intégrée au niveau de l'individu, mais presque exclusivement à celui des États.

2. La Pesc ignorerait la majorité qualifiée

La Pesc *ignorerait* – a-t-il déjà été indiqué – *la majorité qualifiée* et ne fonctionnerait qu'à l'unanimité. On sait combien la majorité qualifiée et la proscription à terme de l'unanimité sont des règles de base du système communautaire.

Il est de fait que la majorité qualifiée n'occupe qu'une place très restreinte dans le traité signé à Maastricht. De là à crier au scandale de l'abandon des règles fondamentales, il y a une certaine mesure à savoir garder : rappelons que la majorité qualifiée n'était, dans les premières années des communautés, légalement pas applicable (ou presque) avant le début de la troisième étape de la période transitoire, c'est-à-dire du 1er janvier 1958 au 1er janvier 1966 (on se souviendra d'ailleurs que la crainte de cette échéance fût la cause de la première des grandes crises internes – la crise de « la chaise vide » – ayant secoué les communautés) ; bien plus, dans les communautés, et malgré un demi-siècle de progression, l'unanimité n'a pas encore totalement disparu comme mode de prise des décisions : même après Amsterdam en effet (mais avant Nice) il reste, dans le seul TCE, cinquante-neuf articles où perdure l'emploi de l'unanimité ; alors pourquoi tirer le signal d'alarme parce que le TUE fait, pour la Pesc, de l'unanimité la règle générale, la majorité qualifiée n'ayant qu'une place marginale : pour les décisions de procédure et pour celles du second degré, c'est-à-dire pour l'adoption des mesures d'application d'une action déjà décidée elle à l'unanimité (arts. J.3 et J.8 de Maastricht combinés) ?

En 1997 (Amsterdam) et en 2000 (Nice), les choses ont beaucoup changé : sans doute la majorité qualifiée n'est pas devenue le droit commun, celui-ci se trouve indiqué dans l'article 23.1 : unanimité, mais celle-ci est assortie de substantielles exceptions.

D'abord celle née d'un assouplissement imaginé en France et qui s'est vu consacré à Amsterdam, il s'agit de ce qu'on dénomme « l'abstention constructive » ; sans doute est classique dans les traités communautaires le fait que l'abstention n'empêche pas la prise d'une décision requérant l'unanimité, mais l'abstention constructive est beaucoup plus sophistiquée : l'abstenant fait la déclaration qu'il accepte que la décision lie l'Union et les autres États membres, au surplus il indique qu'il s'abstiendra de toute mesure entravant les effets de celle-ci, tout cela sans pour autant être lui-même lié par la décision.

Ensuite, la majorité qualifiée va être d'application si antérieurement a été prise un première décision, elle-même ayant requis l'unanimité et qu'il s'agit d'appliquer, c'est le cas des actions ou positions prises dans le cadre d'une stratégie commune (elle-même adoptée par le conseil européen, donc d'un commun accord), c'est aussi le cas - et c'est là l'apport du traité d'Amsterdam - des décisions d'application d'une action ou position commune déjà adoptée (art. 23.2) ; toutefois, une exception fondée sur « des raisons de politique nationale importantes » va limiter le jeu de ce recours à la majorité qualifiée (art. 23.2.3). Le traité de Nice ajoute par ailleurs plusieurs exceptions à l'unanimité dans le domaine de l'adoption des accords internationaux - notamment si la matière concernée par l'accord relève dans l'ordre interne de la majorité (art. 24.2) - aussi en prévoyant quelque chose qui ressemble à l'abstention constructive (art. 24.5).

Enfin le traité de Nice, en ce qu'il élargit le recours à la coopération renforcée - c'est-à-dire à une coopération non plénière entre États membres - permet à la Pesc de se développer hors le concert de tous les États membres.

Précisons inversement - et si cette exception à l'exception est de taille, elle se justifie aisément -, le traité d'Amsterdam, ayant à faire face à l'émergence d'une politique européenne de défense, a exprimé dans ce domaine un maintien général de l'unanimité, contenu dans l'article 23.2.4.

Pour conclure, disons que si le reproche d'ignorer la majorité a pu être fait contre les dispositions Pesc en 1992, il semble bien que les deux révisions ultérieures aient corrigé la situation ; la même évolution va se constater au sujet du grief formé parfois contre le traité de Maastricht d'avoir dédaigné les Institutions supranationales créées en 1957 pour la Communauté économique européenne.

3. La Pesc dédaignerait la Commission

La Pesc *dédaignerait la Commission* (et le *Parlement européen*) au profit du Conseil européen (composé des chefs d'État et de gouvernement des États membres) et du Conseil UE (composé des ministres des affaires étrangères) ; pour la créer, on aurait au surplus fait éclater le système communautaire en instituant ce qu'on a appelé les « *trois piliers* », plus exactement en bardant la Communauté des deux mécanismes que sont la Pesc pour les relations extérieures et la JAI pour la justice et les affaires de sécurité intérieure, lesquels sont plus internationaux que supranationaux ; pour couronner le tout, on refuserait à l'Union européenne la *personnalité juridique internationale* ; toutefois – ultime concession de nature intégrationniste –, la règle dite du « *cadre institutionnel unique* » de l'Union sauvegarderait le pouvoir de décision intégré de l'Union.

Tout cela était dans la logique des choses dès l'instant où pour des raisons de souveraineté on ne voulait pas que les compétences étatiques en matière de défense et de sécurité (externe comme interne), bien que devant s'exercer en commun et non comme avant en ordre quasi dispersé, aient le même statut, spécialement à l'égard du principe de la supranationalité, que les règles de fonctionnement du « marché commun ». Ainsi était-il dans la logique de la création de la Pesc que la commission perde ses deux plus beaux joyaux : celui du monopole de l'initiative décisionnelle et celui résultant de l'article 250 TCE (ex 189A, ex 149) lui permettait librement de s'opposer à un vote majoritaire du conseil UE contredisant une de ses propositions, c'était là pourtant deux armes majeures dans la maïeutique décisionnelle communautaire. Il était dans la même logique des choses que l'on s'abstint de lui donner dans les domaines Pesc et Jai l'exercice du pouvoir d'agir devant la Cour de Justice, en manquement, en carence ou en contrôle de la légalité. C'est en vain que l'on chercherait dans les articles J de Maastricht un quelconque pouvoir réel à son profit, certes elle est « associée » à tout et « pleinement », voir article J.9 ; certes elle peut faire des propositions, mais non au sens de l'article 250 TCE ; finalement elle n'a qu'un rôle d'assistant de la Présidence et des fonctions de relai informatif. Une vraie *capitis deminutio*. Fini son rôle majeur – et très (peut-être trop) largement claironné – dans l'intégration européenne ?

À l'inverse de l'adage sur l'éloignement du Capitole et de la roche Tarpéienne, la pratique de la mise en œuvre de la Pesc et complémentairement les révisions d'Amsterdam et de Nice ont permis à la Commission d'opérer sinon un rebondissement *in integrum*, du moins de jouer à la Pesc dans la cour des États membres. Elle fait partie de la nouvelle troïka ; ses propositions sont fréquemment sollicitées et ses avis demandés ; elle gère le budget de la Pesc, ou peu s'en faut.

Une semblable chute et un semblable rétablissement d'audience sinon de pouvoirs se seront produits à l'égard du *Parlement européen*. D'aucuns avaient prétendu que si la CIG 1991 s'était montrée si favorable à l'attribution de nouveaux pouvoirs décisionnels au Parlement européen dans le domaine communautaire, tels celui de co-décision, c'était pour être plus économe dans le domaine de la Pesc : un maigre débat annuel, le droit de poser des questions écrites et orales seraient ses seuls droits réels. Substantiellement les révisions d'Amsterdam et de Nice n'ont pas fait grand chose pour le Parlement, mais avec le budget, sans fracas, il a sérieusement amélioré ses capacités (cf. *infra*).

La *Cour de justice* serait-elle alors la principale grande perdante ? Certes tous les articles la concernant sont absents d'être cités dans l'article 28.1 TUE qui énumère les articles institutionnels du TCE applicables à la Pesc ; s'il en était de même dans le traité de Maastricht à l'égard de la JAI, le traité d'Amsterdam a-t-il corrigé cet oubli en rétablissant en faveur de la cour de substantielles fonctions dans ce domaine ; mais inversement, pour ce qui concerne la Pesc, même dans un État de droit rigoureux comme le pratique l'Union, on voit mal que le non exercice de la compétence de guerre puisse faire l'objet d'un recours en carence ! Le juriste de droit administratif ne connaît-il pas en droit interne la théorie de l'acte de gouvernement, même si le domaine de cette théorie est de plus en plus exigu ?

Dans un autre ordre d'idées, si la découpe de la Communauté/Union en trois piliers, a été et reste déplorée – quand ce ne serait que pour des raisons de clarté et de simplicité de l'édifice total –, des passerelles entre piliers existent, dans le TCE, l'article 301 et dans le TUE, l'article 42, en outre des délimitations ont pu être revues, telle pour corriger le transfert lors de Maastricht des problèmes de circulation des personnes au TUE et les ramener par le traité d'Amsterdam au TCE.

Un des plus sensibles « oublis » du traité de Maastricht a longtemps été considéré quant à la Pesc dans l'absence pour celle-ci de tout *treaty-making power* tant à l'égard de pays tiers que des organisations internationales. Si une puissante logique théorique interdisait en effet dans un système reposant strictement et exclusivement sur de la coopération intergouvernementale, que l'Union se substitue aux États membres pour exercer des compétences extérieures, une évolution vers une reconnaissance pragmatique au bénéfice de l'Union de pouvoirs internationaux était toutefois possible – et logique – dès lors où celle-ci se trouvait mêlée sur le terrain à des activités internationales. Par une curieuse circonstance, c'est un universitaire yougoslave qui le premier a vu dans des accords, plutôt des arrangements ou des mémorandums, trilatéraux, de trêves ou de cessez le feu, conclus entre deux adversaires ex-yougoslaves sous le chaperonnage (ou en présence) de la présidence de l'Union européenne intervenue à titre de médiateur ou comme on dit en langue anglaise de *go-between*, des *accords internationaux* de l'Union, c'est la théorie de Lopandic, du nom d'un chercheur de Belgrade (Revue du marché commun et de l'Union européenne, 1995, n° 392, 557). Quelques mois plus tard, le traité d'Amsterdam reconnaissait la juridicité de ces accords Pesc, article 24 TUE, sans toutefois dire clairement qui, de l'Union, de ses États membres ou de ses Institutions, ils lieraient, probablement ni celle-là, ni celles-ci, mais ceux-là (puisqu'il est prévu que par une déclaration spéciale un État membre peut ne pas se lier à l'égard d'un tel accord). Depuis lors le traité de Nice a bouclé la boucle, puisque son article 24.6 indique que de tels accords « lient les institutions ».

Quoiqu'il en soit, on comprend le refus de la CIG 1991 – en raison du caractère intergouvernemental patent des activités de l'Union et notamment de celles afférentes à la Pesc – de reconnaître à l'Union, la personnalité juridique ; on comprend aussi en 1997 le maintien de cette non-reconnaissance, étant donné le caractère trop cursif – au point de vue des engagements à l'égard de l'Union de ces accords – du texte d'Amsterdam ; et on comprend enfin très bien que le texte de Nice n'évoque plus la question puisqu'elle est tranchée de manière pragmatique par la règle de droit international qu'une organisation internationale n'a pas besoin d'un texte pour se voir reconnaître la personnalité juridique dès l'instant où elle est habilitée expressément par son statut à accomplir des actes internationaux. Un demi siècle après San Francisco, la Charte des Nations Unies reste muette sur la personnalité juridique de l'Organisation des Nations Unies ! Pourquoi l'Union devrait-elle faire plus ?

Qui va piano va sano, telle semble une règle générale de l'intégration européenne avec ce quinquennalisme qui semble régir beaucoup de ses activités, ses CIG de révision, ses plans budgétaires (les paquets), ses engagements financiers extérieurs, ses périodes de transition au début des élargissements. À partir de compromis obtenus à l'arrachée, elle met progressivement en place de solides institutions, mécanismes et marchés intégrés

Il semble en être ainsi aussi pour la Pesc, outre ce qu'on vient de dire pour ses acquis institutionnels, avec d'autres règles et notamment ses institutions centrales, son financement et ce dossier délicat qu'est la nouvelle *politique européenne commune de sécurité et de défense*, la Pecsd. Chacun d'eux montre en effet des efforts d'intégration réels.

II. L'apparition pragmatique d'une forme d'intégration rénovée ?

Le bon sens des États membres les amène à mettre progressivement en place une forme pragmatique d'intégration pour les activités propres de l'Union européenne, notamment pour la Pesc. Cela se produira aux échelons centraux institutionnels (1) et dans les règles budgétaires concernant la Pesc (2) et dans la mise en place de la Pecsd.

1. Les échelons centraux institutionnels

Par échelon institutionnel centralisé responsable de la Pesc, nous pensons aux fonctions de celui que pendant les négociations de 1997 on désignait sous le vocable de Monsieur ou de Madame Pesc (cette féminisation évoquée à la fois pour faire preuve de modernisme et par métempsychose de Madeleine Albright) mais le vrai problème était de savoir où celui-ci serait recruté ? du côté de la Commission ? le président ou le commissaire relations extérieures politiques, mais ceci aurait eu l'inconvénient que M. Pesc soit très « indépendant » des États membres – et ceci cadrait mal avec le caractère intergouvernemental qu'on désirait donner à la Pesc. Serait-il alors élu ? pour x années ? x=3 ? directement, au suffrage universel par les citoyens communautaires ? Autre indépendance donnée à M. Pesc et ne plaisant pas forcément aux États membres. N'était-il pas alors plus conforme au concept intergouvernemental qu'il soit désigné par le Conseil européen, voire le Conseil UE, soit pour x années, soit révocable à tout moment, ce qui, compte tenu de l'importance du poste ne pouvant être fait qu'à l'unanimité, assurant ainsi à son titulaire une certaine stabilité de poste. Sur cette dernière lancée qui fut retenue, on transforma les fonctions du Secrétaire général du Conseil de Ministres qui, de plus haute autorité gestionnaire de celui-ci – et ayant récemment acquis, selon la personnalité de son titulaire des fonctions aux confins d'un haut conseiller politique –, devint son représentant politique sous l'appellation de « Haut représentant pour la politique étrangère et de sécurité commune » et pour bien marquer la promotion de ces fonctions, on institua un Secrétaire général adjoint auquel fut confié l'ensemble des fonctions de gestion du précédent secrétaire général (art. 207.2 TCE). Il est en général connu par l'acronyme des initiales de son titre : le SG/HP. C'est, à l'égal de la présidence de la Commission, un poste qui convient à un ancien premier ministre ou à un ancien ministre des affaires étrangères, des finances ou de la défense, ayant longtemps tenu son poste...

Quant à ses fonctions, on dira que le SG/HP « assiste » le Conseil pour toutes les questions de Pesc, « en contribuant notamment à la formulation, à l'élaboration et à la mise en œuvre des décisions de politique et, le cas échéant, en agissant au nom du Conseil et à la demande de la présidence, en conduisant le dialogue politique avec des tiers » (art. 26 TUE). De telles fonctions évoquent peut-être, plus que toutes autres, celles du Secrétaire d'État américain, *primus inter pares* auprès du Président des États-Unis, chargé des affaires étrangères, et en cas de disparition simultanée du président et du vice-président le troisième dans l'ordre successoral à pouvoir devenir président. En fait les développements récents de la Pecsd donnent – on le verra – au SG/HP des fonctions simultanées de Secrétaire aux affaires étrangères et de Secrétaire à la défense de l'Union.

Il fait aussi partie d'un organisme au nom insolite, existant tour à tour officiellement ou officieusement, officiellement sous le traité de Maastricht (art. J.5.3), officieusement depuis lors, la *troïka*, composée du président du conseil des ministres des affaires étrangères, éventuellement – en surnombre – du successeur dans le temps de celui-ci à la présidence, du SG/HP et du commissaire relations politiques extérieures. Cet organe qui est plus envoyé en missions au loin que confiné à de la représentation à Bruxelles, donne par sa composition une haute idée de ce qu'une collègue appela un jour « l'Europe Puissance ».

Avant même que l'on ait pensé au poste de SG/HP, en étudiant les réformes à apporter au système de la Pesc, était conçu un organe dont le premier surnom fut la « cellule » et qui finalement vit le jour sous le nom moins poétique de « *unité de planification et d'alerte rapide* », la déclaration jointe au traité d'Amsterdam qui la crée est encore plus explicative – s'il en était besoin – que les mots figurant dans sa dénomination.

On hésitera à parler d'organes centraux pour les « *représentants spéciaux* de l'Union » que le conseil UE peut désigner en leur conférant « un mandat en liaison avec des questions politiques particulières ». Sont-ce des ambassadeurs *at large* ? Ne sont-ils pas plutôt des médiateurs (enquêteurs, observateurs ou autres) désignés par l'Union et en mission extraordinaire dans un point chaud du globe ? Leur appellation ne les apparente-t-elle pas alors avec les envoyés du même nom assistant le Secrétaire général dans le cadre des Nations Unies ? Le fait qu'il en a existé longtemps pour l'Union quatre, un pour les affaires ex yougoslaves, un pour le Moyen-Orient ; un pour l'Afrique des grands lacs et un pour la Méditerranée, c'est-à-dire quatre zones de conflits où l'Union s'est particulièrement investie, y pousserait. Ces représentants spéciaux sont dès lors pour l'Union des agents importants sur l'échiquier de sa politique de prévention des conflits dont l'Union est en cours de doter la Pesc (cf. *infra*).

Ne parlons en revanche pas maintenant du Comité politique, créé au temps de l'Acte unique européen (1986) comme l'organe de la « Coopération politique européenne», maintenu en 1992 dans le cadre de la Pesc, mais peut-être un peu dépassé actuellement où le suivi de la Pesc et de la Pecsd est une tâche quotidienne, on en parlera plus loin (cf. *infra* III.1). Ne parlons pas non plus – parce que ce sont des mécanismes non intégrés – des coordinations entre États membres soit de leurs postes diplomatiques (en compagnie des délégations de la commission) dans les pays tiers, soit aux réunions d'organisation internationales ou de conférences diplomatiques..

La mise en place progressive de ces organes intégrés centraux est en tous cas un signe de l'intérêt de l'Union pour l'efficacité de sa politique étrangère et de sécurité commune. Une telle intention se manifeste également dans une très rapide évolution des mécanismes budgétaires concernant le financement des dépenses opérationnelles de la Pesc pour intégrer ceux-ci dans le budget communautaire.

2. Budgetisation du financement de la Pesc

Les dispositions du traité de Maastricht, article J.11.2, et celles du traité d'Amsterdam, article 28 TUE, relatives au *financement de la Pesc* présentent de profondes différences les unes des autres. Peut-être doit-on en trouver la cause dans le caractère presque trop théorique et le peu de cohérence financière du système établi en 1992 – ce qu'on verra plus bas. Quoiqu'il en soit, s'est produit, dans la pratique budgétaire de la communauté, au cours des cinq années qui séparent ces deux traités fondamentaux pour la Pesc, et cela pour des raisons aussi diverses que pragmatiques, un recul très net de l'idée initiale que la Pesc étant essentiellement de caractère intergouvernemental, il ne devait pas – sauf exception votée à l'unanimité – y avoir de prise en charge de dépenses par le budget de la communauté. On est dès lors arrivé en 1997 à l'inverse, ou presque, de ce qu'on avait fait en 1992 et la prise en charge par le budget communautaire de la majeure partie des dépenses Pesc est un net signe d'intégration ; seules les « dépenses afférentes à des opérations ayant des implications militaires ou dans le domaine de la défense » restent à la charge de l'État membre qui engage ses troupes ou exerce des prérogatives de défense. Ceci se comprend étant donné le lien étroit entre l'utilisation des forces armées d'un État et sa mouvance souveraine. En revanche les dépenses administratives de la Pesc ainsi que la généralité des dépenses opérationnelles de la Pesc sont, sauf décision casuelle de dérogation, inscrites au budget communautaire (art. 28 TUE).

On n'insistera pas sur la distinction qui est de base en droit budgétaire étatique, entre dépenses administratives et dépenses opérationnelles. L'idée existe évidemment aussi dans le système financier communautaire et aboutit à ce que les premières sont inscrites normalement dans le budget général alors que les dépenses opérationnelles peuvent connaître des règles spéciales de répartition – une clé de répartition spéciale entre États membres – de leur financement. S'y ajoute l'absence d'une définition précise de ce qui est dépenses administratives et ce qui est dépenses opérationnelles. Par exemple, les frais d'administration d'une mission d'observation ou de médiation, exposés à Bruxelles et ceux exposés sur le terrain sont-ils administratifs par leur nature intrinsèque ou opérationnels en raison de l'objectif de la dépense ? On peut longtemps en discuter ; en revanche, le fait que selon l'article 28 TUE les unes et les autres de ces dépenses soient (maintenant) à la charge du budget communautaire rend un peu vain le débat.

Trois raisons, totalement extérieures à la Pesc, ont poussé les États membres à abandonner subrepticement le texte de 1992. Ceci s'est passé dans le cadre proprement budgétaire, riche en joutes communautaires et cela peu après l'entrée en vigueur du traité de Maastricht :

- Étant donné qu'à défaut d'accord unanime au sein du conseil pour inscrire une dépense au budget communautaire – et c'est à cela que le système de 1992 aboutissait –, la dépense allait rester à la charge des États membres, cela obligeait chaque ministre à un retour devant son autorité budgétaire nationale, d'où quinze débats nationaux avec risque d'absence de cohérence dans les financements ; n'était-il pas dès lors hautement souhaitable de prévoir en toute analyse un financement inscrit au budget des Communautés ? sans doute y avait-il alors un autre problème lui de nature budgétaire à résoudre : était-ce au budget du conseil ou à celui de la commission que les crédits seraient inscrits ? Normalement pour des dépenses opérationnelles, c'est la commission qui les gèrent et les crédits sont à son budget, mais cela était-il conforme à la nature intergouvernementale de la Pesc où le pouvoir de décision appartient au conseil ? La solution pragmatique de compromis adoptée dès 1995 – outre une subtile répartition entre budget de Conseil et budget de la Commission – aboutit à la fois à augmenter le contrôle du Parlement sur la Pesc et à étendre le pouvoir de gestion de la commission sur les dépenses ; il présente également une plus grande cohérence budgétaire. C'est par ailleurs de ce compromis budgétaire que résulte l'importance du binôme formé par le HP/SG et le Commissaire chargé des relations extérieures dans nombre de contacts relatifs à la gestion.

- Autre motif de cette budgétisation communautaire, elle plus fortuite : en 1993-1994, années où les ministres des finances des États membres de l'Union se préoccupaient, en vue du passage à l'Euro, de limiter – dans l'optique du sacro-saint respect des « critères de convergence », fondement de la monnaie unique – leurs dépenses nationales, le transfert d'une dépense des budgets nationaux au budget communautaire n'allait-il pas être un intéressant tour de passe-passe pour alléger les dépenses nationales ?
- Troisièmement les États membres étaient, à l'égard des dépenses Pesc, facilement dépensiers, comme l'était également le Parlement européen, lui pour des raisons corporatives de classifications budgétaires – ces raisons, connues des initiés sous le nom de dépenses obligatoires/dépenses non obligatoires – poussaient aussi à mettre les dépenses Pesc à la charge du budget communautaire.

Devant une telle convergence de raisons les changements introduits par le traité d'Amsterdam dans l'article devenu 28 TUE n'ont fait qu'entériner une pratique des enceintes budgétaires inscrite dans les budgets 1994-1997. C'est donc fortuitement – et on doit s'excuser de la technicité hors-sujet du raisonnement nécessaire pour l'expliquer – que le caractère intégré du financement de la Pesc a été décidé, à contre-courant du caractère intergouvernemental de la Pesc qui aurait du tendre au renvoi absolu aux budgets nationaux.

Un troisième domaine où la Pesc de l'Union est en cours de développement fortement intégré est de celui de la sécurité et de la défense, avec ce que le conseil européen de Cologne a dénommé la *politique européenne commune de sécurité et de défense* (la Pecsd, on trouve encore employé le terme de politique européenne de sécurité et de défense).

III. Le développement récent et fortement intégré de la sécurité et de la défense européennes sous le nom de Pecsd

Trois aspects de la *Pecsd* (politique européenne commune de sécurité et de défense) nous retiendront ici : d'abord, ci-après, sous (1) l'établissement progressif dans le cadre de celle-ci d'instances délibérantes intergouvernementales composées des représentants des États membres ainsi que d'un service multinational d'état-major intégré, les uns et l'autre à vocation spécialisée dans le sécuritaire, voire le militaire ; cela est intéressant comme structure institutionnelle d'intégration européenne ; ensuite, ci-après sous (2) l'émergence de ce que les militaires appellent des capacités militaires ou en hommes et que plus simplement nous appellerons les forces armées européennes multilnationales intégrées, une force européenne d'intervention rapide notamment voit le jour ; enfin, ci-après sous (3) la mise en place (actuellement à ses débuts seulement) d'une politique de sécurité et de défense, se caractérisant, d'abord par des mesures dites de prévention des conflits ensuite par un fort recours à ce qu'on a appelé les aspects civils de la gestion des crises, enfin par le recours à des missions confiées aux forces militaires intégrées européennes, qui sont des missions de nature humanitaire (au sens large) mais aussi de maintien, voire de rétablissement de la paix, ces vieilles connaissances des Nations Unies.

Sans doute s'agit-il là de fonctions nouvelles pour l'Union et posant pour elle de délicats problèmes d'engagements de ses États membres et de rapports entre eux ainsi que de relations avec l'Alliance atlantique (dont cinq États membres sur quinze, rappelons-le, ne sont pas membres, mais par ailleurs que deux ou trois des dix ne sont pas loin de préférer à la Pecsd) et au delà de celle-ci de relations avec le Conseil de sécurité des Nations Unies (aux yeux duquel l'Union devrait être considéré comme un organisme régional au sens de l'article 52 et suivants de la Charte des Nations Unies), le sujet est si vaste, si nouveau et mouvant que dans certains de ses prolongements il ne pourra qu'être effleuré.

L'Union s'est lancée dans cette voie dans le cadre de la Pesc, quand le traité de Maastricht instituant cette politique (art. J.4.1) a inclus dans celle-ci « l'ensemble des questions relatives à la sécurité de l'Union européenne, y compris la définition à terme d'une politique de défense commune, qui pourrait conduire, le moment venu, à une défense commune ». Etaient en outre prévu que l'Union de l'Europe occidentale agirait à la demande de l'Union européenne pour l'assister dans la mise en place de cette politique (art. J.4.2 et une déclaration relative à l'UEO jointe au traité), ainsi que diverses dispositions pouvant concerner les États membres de l'Union en même temps membres de l'Alliance atlantique (J.4.4 et 5).

Par la suite, le traité d'Amsterdam (1997) en révisant celui de Maastricht, apporta deux sortes de précisions.

Les premières semblent plus formelles que de fond, il est toujours question comme objectif d'une politique de défense commune et au delà que celle-ci doive conduire à une défense commune, mais alors qu'en 1992, dans la logique de Maastricht, cette défense commune était seulement une éventualité (« le moment venu ») et requérait en outre des modifications des traités, elle se rapproche dans la logique d'Amsterdam puisque le conseil européen reçoit mandat de recommander aux États membres une décision à cette fin – bien sûr sous réserve qu'ils satisfassent à leurs exigences constitutionnelles. Ce passage à la défense commune, le traité de Nice et la CIG de Nice le préciseront : d'une part l'article 17.1 TUE sera allégé de tout recours à l'UEO (laquelle, sous réserve de son article 5 (cf. *infra*) sera ainsi quasi-vidée de toute fonction), mais d'autre part, une déclaration de la CIG, reprise en annexe au traité, tendra à ce que la Pecsd « soit rapidement opérationnelle » et définie par le conseil européen de fin 2001 au plus tard.

Par ailleurs est ajoutée l'indication que relèvent de la politique européenne de sécurité et de défense certaines interventions humanitaires ou de maintien ou de rétablissement de la paix, connues sous l'appellation de « missions de Petersberg » et qui avaient été élaborées par l'UEO dans le cadre de son assistance à l'Union pour établir sa politique de défense (voir article 17.3 TUE). Que représentent ces missions décidées à la réunion de l'UEO à Petersberg le 19 juin 1992 ? Disons qu'à une époque où le pacifisme, le neutralisme voire le non-engagement, montent en importance dans la société internationale, celle-ci, pour légaliser le droit des conflits armés met l'accent plus sur le maintien de la paix voire son rétablissement que sur les sanctions et l'usage de la force. Qu'il s'agisse des Nations Unies, de l'OTAN ou d'autres enceintes, le vocabulaire s'adoucit.

L'emploi de telles formules était peut-être dans le cas de l'Union le prix à payer en raison de l'accroissement de la présence en son sein d'États se déclarant plus ou moins fermement neutres, ou simplement n'étant pas membres de l'OTAN, ou encore faisant montre de tendances pacifistes... Rappelons d'ailleurs qu'entre la négociation de Maastricht et celle d'Amsterdam, l'Union avait acquis trois nouveaux États membres tous de cette mouvance. Il est certain qu'en introduisant cette formule dans le cadre de son assistance à l'Union pour mettre en place le volet défense de la Pesc 1992, l'UEO a avec succès contourné une difficulté majeure de celle-ci. On examinera plus loin comment de telles missions vont pouvoir relever de la Pecsd.

À une époque encore plus récente, depuis 1999, on doit souligner les importants progrès dans la mise en œuvre de la politique de sécurité et de défense résultant spécialement des fonctions « d'orientation » exercées par la plupart des sessions du Conseil européen. Chacune de ces sessions (ou presque) depuis celle de Cologne, les 3 et 4 juin 1999, sous présidence allemande, a été fructueuse ; on ne saurait énumérer les apports de chacune d'elles ; au surplus les idées inventées à un session, souvent n'ont pris forme qu'à la suivante ou à une autre ; on évoquera dès lors dans le cours de l'examen matériel du contenu de la Pecsd l'apport de telle ou telle session du Conseil européen.

1. Trois comités intergouvernementaux et un état-major integré

Il a ci-dessus déjà été question des organes centraux de la Pesc, le SG/HP notamment : même si beaucoup du temps de celui-ci est voué à l'exercice du dialogue avec des pays tiers ainsi qu'à des problèmes de médiation et de prévention des crises, il n'est toutefois pas une autorité spécialisée exclusivement dans la sécurité et la défense, comme le sont en revanche *trois comités et un organe* dont on voudra parler ici : le comité politique et de sécurité, encore appelé le COPS, le comité militaire de l'Union, le CMUE et enfin le comité chargé de la gestion civile des crises ; quant à l'organe, il s'agit de l'État-major de l'Union. Pour chacun, on examinera ses fonctions et sa place dans l'échiquier de la Pecsd.

Les trois comités sont des organes dépendant du conseil de l'Union européenne, donc comme tous les organes de ce type composés de délégués gouvernementaux, librement choisis par ces gouvernements. Ils travaillent sous la houlette du comité des représentants permanents, le coreper, auquel ils font rapport.

Certaines particularités de la composition et de la structure de ces comités doivent toutefois être signalées. D'abord pour le COPS, en ce qu'outre sa dépendance du coreper, il en a en outre une – presque d'une manière maintenant historique – à l'égard du comité politique, composé des directeurs des affaires politiques des ministères et dont il est quelque peu l'échelon permanent et bruxellois. Un cas un peu différent se présente pour le comité militaire de l'Union lequel « est composé des chefs d'état-major des États membres, représentés par leurs délégués » étant entendu que le comité « se réunit au niveau des chefs d'état-major lorsque ce sera nécessaire » ; ici l'emploi du mot délégué et non de celui de représentant semble signifier que le délégué est un suppléant-permanent, probablement résident à Bruxelles, que c'est à son niveau qu'ont lieu les réunions habituellement, ceci n'excluant pas la présence à l'une d'elles des (ou d'un) chefs état-major, selon le sujet étudié.

Par ailleurs une autre exception à la règle habituelle au conseil de l'Union que les réunions des organes délégués sont présidées par un représentant de l'État qui assure la présidence (semestrielle) du Conseil de ministres est prévue dans le cas de deux comités : le COPS « peut être présidé ... notamment en cas de crise » par le SG/HP, ce qui n'est pas sans rappeler que la présidence du conseil atlantique est toujours – depuis Paul Henry Spaak – assuré, non par un ministre ou par un ambassadeur représentant permanent d'un État partie (désigné par rotation annuelle), mais par le secrétaire général de l'OTAN. La seconde exception est celle du président du comité militaire lequel est désigné pour trois ans par le conseil de l'Union parmi les membres du comité sur recommandation de ses pairs qui présentent éventuellement en outre un vice-président. Ces deux exceptions donnent à la présidence de ces comités un indéniable lustre et concourent ainsi à donner à l'organe, même si celui-ci est intergouvernemental dans sa composition, un label de corps intégré.

Quant à leurs fonctions, les remarques suivantes peuvent être faites :
Le COPS (comité politique et de sécurité) est né (cf. *infra*) le jour où la Pecsd prenant de plus en plus d'importance, il a fallu que l'examen de la conjoncture politique internationale ne soit pas faite bimestriellement – comme le faisait le comité politique (composé des directeurs des affaires politiques des États membres) – mais quasi quotidiennement (ou du moins trois fois par semaine, mais pourquoi pas d'heure à heure ?) ; pour cela il fallait un comité composé de hauts fonctionnaires, occupant leur fonction à temps complet et donc résidant à Bruxelles. De là les conclusions des conseils européens d'Helsinki et de Nice (décembre 1999 et 2000) tendant à l'institution (d'abord à titre intérimaire puis définitivement) du COPS. Celui-ci, n'est pas un organe différent de comité politique prévu au TUE (depuis l'Acte unique et ayant subsisté dans les révisions de Maastricht et d'Amsterdam), il est l'appellation de ce comité politique quand celui-ci ne siège pas au niveau de ses titulaires. Le comité politique subsiste donc ; tous les deux mois, avec les directeurs, il se livre à un examen (appelons-le général, global ou horizontal) de la conjoncture ; bi hebdomadairement, sous le nom de COPS, il examine minutieusement, au peigne fin, cette conjoncture.

Il est difficile de résumer en dix lignes les fonctions d'un organe quand le Journal officiel les décrit en trois pages ; disons, en employant des mots qui reflètent les idées en soulignant quelques mots employés au texte, « qu'il fait » la Pesc (y compris la Pecsd) *au jour le jour*, qu'il *suit la situation, émet des avis* à l'intention des instances supérieures (conseil UE et conseil européen), établit une relation particulière régulière avec le conseil de l'OTAN, *donne des orientations* aux autres comités et leur *adresse des directives*, il exerce le « *contrôle politique et la direction stratégique* » de la réponse militaire de l'UE à la crise. Rappelons que le SG/HP peut le présider chaque fois que nécessaire. Resterait à régler ses rapports avec une instance plus chevronnée que lui et surtout qui possède une prérogative politique, voire juridique – exclusive, le coreper, en ce que ce dernier est l'organe qui « prépare » le Conseil.

À côté du COPS, le CMUE (comité militaire de l'Union européenne) fait presque figure d'organe spécialisé, il l'est d'ailleurs en sa composition, des militaires de haut rang, les chefs d'État-major des États membres ou leurs délégués.

Rappelons en outre cette particularité que son président est désigné pour trois ans par le conseil UE sur recommandation du CMUE siégeant au niveau des chefs d'état-major et que ce président permanent doit être un « officier quatre étoiles », soit le chef d'état-major d'un État membre, soit un ancien chef d'état-major d'un État membre ; cette désignation fait de lui l'interlocuteur sur un pied d'égalité des personnalités à la tête des autres organes de l'Union, président du Conseil européen et président du conseil UE.

Le CMUE est en effet « l'organe militaire le plus élevé mis en place auprès du Conseil UE », il « donne des avis militaires et fournit des recommandations » destinées au COPS et au delà de celui-ci au conseil de l'Union, il « assume la direction militaire de toutes les activités militaires dans le cadre de l'Union . . . en fournissant des directives militaires à l'État-major de l'UE ».

Il est donc l'organe militaire de préparation pour le conseil UE, d'avis et d'évaluation pour le COPS, de formulation et de transmission des directives à l'État-major de l'Union et de contrôle opérationnel de celui-ci.

Ses fonctions ont donc à la fois un caractère préparatoire, au stade de la gestion des crises, mais s'exercent aussi au niveau du suivi ; par le contrôle des opérations militaires.

L'institution du CMUE résulte d'une décision du conseil UE du 22 février 2001 (n° 2001/79/Pesc au JOCE L 27/4 du 30), il a remplacé l'organe militaire intérimaire en fonction depuis mai 2001.

Avec le *comité chargé des aspects civils de la gestion des crises*, nous sommes devant un organe d'un niveau moins important, un organe subsidiaire du COPS et du coreper, plus spécialisé, dans la gestion d'un secteur de la Pecsd, ce secteur qu'est la gestion (non militaire) des crises (cf. *infra* II.3.C). Cette gestion avait fait l'objet dans le cadre de l'élaboration de la Pecsd d'un examen par le Conseil européen, notamment à ses sessions de Cologne (juin 1999) et Lisbonne et Santa Maria de Feira (mars et juin 2000) ; cet examen avait été préparé par des rapports du conseil de l'Union élaborés eux-mêmes par le coreper et des groupes *ad hoc* ; sans doute ces rapports ont-ils été approuvés par le Conseil européen, mais à ce jour ils n'ont pas fait l'objet de textes exécutoires, sauf la constitution du comité par décision du conseil UE du 22 mai 2000 (n° 2000/354/Pesc au JOCE 127/1 du 27). Ce comité est un groupe de travail du conseil, il fait donc rapport au coreper, conseille le COPS et autres instances appropriées . . . sur les aspects civils de la gestion des crises (cf. *infra*).

Le dernier organe, l'État-major de l'Union européenne, n'est pas un organe de nature intergouvernementale, mais un organe spécialisé de nature intégrée et de composition multinationale et composé – truisme en vérité – de militaires des États membres Institué par décision du Conseil du 22 janvier 2001 (n° 2001/80/Pesc au JOCE 27/7 du 30), il est devenu opérationnel le 1er juillet 2001. Il dépend des organes supérieurs et notamment reçoit des directives du comité militaire de l'Union et a sous ses ordres les Forces européennes.